Insect Viruses and Pest Management

Insect Viruses and Pest Management

FRANCES R. HUNTER-FUJITA
School of Animal and Microbial Sciences, University of Reading, UK

PHILIP F. ENTWISTLE
NERC Institute of Virology and Environmental Microbiology, Oxford, UK

HUGH F. EVANS
Forestry Commission Research Agency, Farnham, Surrey, UK

and

NORMAN E. CROOK
Formerly of Horticultural Research International, Wellsbourne, UK

JOHN WILEY & SONS
Chichester • New York • Weinheim • Brisbane • Singapore • Toronto

Baffins Lane, Chichester,
West Sussex PO19 1UD, England

National 01243 779777
International (+44) 1243 779777
e-mail (for orders and customer service enquiries):
cs-books@wiley.co.uk
Visit our Home Page on http://www.wiley.co.uk or
httw://www.wiley.com

Other Wiley Editorial Offices

John Wiley & Sons, Inc., 605 Third Avenue,
New York, NY 10158-0012, USA

Weinheim • Brisbane • Singapore • Toronto

Library of Congress Cataloging-in-Publication Data

Insect viruses and pest management/Frances Hunter-Fujita . . . [et al.].
p.cm.
Includes bibliographical references and index.
ISBN 0–471–96878–1 (hc. alk. paper)
1. Viral insecticides. 2. Baculoviruses. I. Hunter-Fujita, Frances.
SB933.37.1571998
632′.9517-dc21 97–17448

British Library Cataloguing in Publication Data
A catalogue record for this book is available from the British Library

ISBN 0 471 96878 1

Typeset in 10/12pt Times by Florencetype Ltd, Stoodleigh, Devon.
Printed and bound in Great Britain by Bookcraft (Bath) Ltd, Midsomer Norton.
This book is printed on acid-free paper responsibly manufactured from sustainable forestry, in which at least two trees are planted for each one used for paper production.

This book is dedicated to Norman E. Crook
1946–96

Contents

Contributors

Margaret Brown
Natural Resources Institute, University of Greenwich, Chatham Maritime, Chatham, Kent ME4 4TB, UK

Andrew Cherry
Natural Resources Institute, University of Greenwich, Chatham Maritime, Chatham, Kent ME4 4TB, UK

Norman E. Crook
Formerly of Horticultural Research International, Wellsbourne, UK

John T. Cunningham
R.R. Number 1, Hilton Beach, Ontario, Canada P0R 1G0

S. Easwaramoorthy
Sugarcane Breeding Institute, Coimbatore 641007, Tamil Nadu, India

Philip F. Entwistle
NERC Institute of Virology and Environmental Microbiology, Oxford, UK

Hugh F. Evans
Forestry Commission Research Agency, Farnham, Surrey, UK

David Grzywacz
Natural Resources Institute, University of Greenwich, Chatham Maritime, Chatham, Kent ME4 4TB, UK

Jurg Huber
Institute for Biological Control, Biologische Bundesanstalt, Heinrichstr 243, D-64287 Darmstadt, Germany

Frances R. Hunter-Fujita
School of Animal and Microbial Sciences, Whiteknights, PO Box 228, University of Reading, Reading, Berkshire, RG6 2AJ, UK

Keith A. Jones
Natural Resources Institute, University of Greenwich, Chatham Maritime, Chatham, Kent ME4 4TB, UK

Uthai Ketunuti
Biological Control Section, Entomology and Zoology Division, Department of Agriculture, Bangkok 10900, Thailand

Yasuhisa Kunimi
Faculty of Agriculture, Tokyo University of Agriculture and Technology, Saiwai-cho 3-5-8 Fuchu, Tokyo 183-8509, Japan

Edna Kunjeku
Department of Biological Sciences, University of Zimbabwe, PM 196, Mount Pleasant, Harare, Zimbabwe

Jerzy J. Lipa
Department of Pest and Disease Control, Institute of Plant Protection, Miczurina 20, 60–138 Poznan 8, Poland

Jun Mitsuhashi
Department of Biosciences, Tokyo University of Agriculture, Sakuragaoka 1-1-1, Setagaya-ku, Tokyo 156-0054, Japan

Galal M. Moawad
Plant Protection Institute, Dokki, Cairo, Egypt

M. Regina V. de Oliveira
Centro Nationale de Recursos Geneticos, Cavia Postale 10-2372, 70-770 Brasilia DF, Brazil

Robert E. Teakle
Entomology Branch, Department of Primary Industries, Meiers Road, Indooroopilly, Brisbane, Queensland 4068, Australia

Bernhard Zelazny
FAO, Plant Protection Service (AGPP), Via delle Terme di Caracalla, 00100 Rome, Italy

Acknowledgements

We are indebted to all those who contributed so generously of their time, patience and knowledge. In particular we should like to thank the following: Andy Cherry, John Cunningham, S. Easwaramoorthy, David Grzywacz, Jurg Huber, Keith Jones, Uthai Ketunuti, Edna Kunjeku, Yasuhisa Kunimi, Jerzy Lipa, Galal Moawad, M. Regina V. de Oliveira, Bob Teakle and Bernard Zelazny, for their contributions to Part Two: World Survey. When it became necessary to update this area because of the passing of time, they uncomplainingly provided fresh information. We would also like to thank Jun Mitsuhashi for his contributions on cell culture (some additions were made on the virological side to this section for which we take full responsibility); Margaret Brown for the section on PCR techniques; and David Grzywacz for his contribution on microbiologial testing. We wish to mention, also, Keith Jones for his valuable comments on Chapter 7; D. Grainger, Department of Atmospheric Physics, University of Oxford, for his very kind assistance with Chapter 31; Tim Carty, Institute of Virology and Environmental Microbiology, for discussions on insect rearing; Ronalde Estrada, Agricola el Sol, Guatemala, for generously supplying information on the situation in Central America; Luisa B. de Lugo and all the staff of the Laboratoria de Control Biologico, Facultad de Ciencias, UNAN, Leon, Nicaragua for the supply of information and for happy times working together; and Fanji Zeng, Department of Virology, University of Wuhan, People's Republic of China, for his courtesy, help and valuable discussions. Chuan fan Quian (Department of Applied Chemistry, Beijing Agricultural University, People's Republic of China) also supplied information on some of the work being done in China and Li Ying Wang (Department of Plant Protection, Beijing Agricultural University, Beijing) supplied cell culture information. Doreen Winstanley (Horticulture Research International, Wellesbourne) made many helpful suggestions concerning the sections on cell culture; Stephen Pountney (School of Animal and Microbial Sciences, University of Reading) was a fund of information on technical aspects of general management; and Alan Jenkins and Liz Wheeler (also of the School of Animal and Microbial Sciences, University of Reading) were most helpful with notes on electron microscopy. Kazimierz Dyk, a student on vacation from Warsaw to improve his command of English, is to be congratulated on having learned how to use CorelDraw! in a foreign language and for having succeeded in instructing one of us

(FRH-F) in its use. He also produced some of the diagrams presented here. Finally we wish to thank all those who kindly gave permission for the use of copyright material.

Electron photomicrographs of viruses were kindly provided by Dr Brian Federici, Department of Entomology, University of California, Riverside, CA 92521, USA and Drs Linda King and Susan Marlow, Oxford Brookes University, Gipsy Lane Campus, Headington, Oxford, OX3 0BP, UK.

PART ONE

BASIC PRINCIPLES

1 Rationale for the use of microbial pesticides

Losses from the activities of insect pests remain one of the major limitations to full productivity in crop production. This applies regardless of the ultimate purpose of the crop. Management of insect pests is, therefore, a balance between the costs of that management, both in money and environmental terms, and the benefits that could be gained in terms of crop yield. The massive increases in yield that have been achieved during the twentieth century have been the result of improved plant varieties, better soil and nutrient management and, perhaps most importantly, the development of a suite of chemical pesticides to combat pests, diseases and weeds. This is illustrated clearly in the scale of the use of agrochemicals. Thus, sales of agrochemicals worldwide were US$27 billion in 1991, of which 30% were insecticides. These data are summarized in Figure 1.1.

The estimated market for bioinsecticides in 1995 was US$380 million including $3–4 million for viruses. By 2000 the market for bioinsecticides is predicted to have increased by $116–141 million, with viruses making up $5–6 million of the total (Georgis, 1997).

Among the crops treated, corn, rice, cereals, soybeans, fruit, vegetables and cotton remain dominant. By contrast, the use of biopesticides in forestry remains relatively small, being dominated by the use of *Bacillus thuringiensis* (*B.t.*) against lepidopterous pests in north America and, increasingly, elsewhere (Navon, 1993).

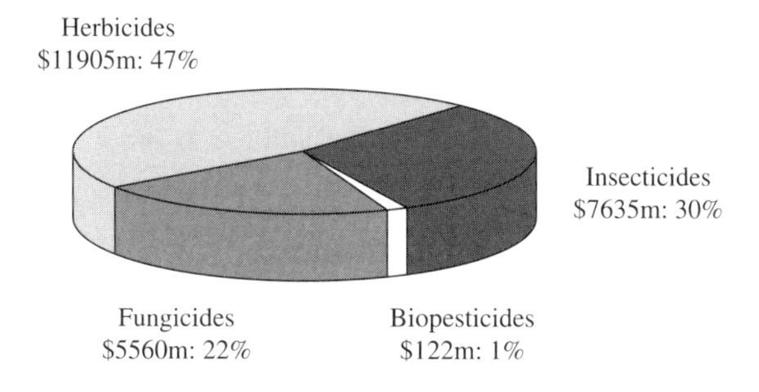

Figure 1.1 World pesticide usage in US dollars (millions) and by percentage usage during 1991. (Redrawn and adapted from Powell and Rhodes, 1994.)

However, as is illustrated clearly in Figure 1.1, the total use of biopesticides, including bacteria, fungi, viruses, protozoa and nematodes, is still less than 1% of the total usage of pesticides.

Although the impetus to develop new agents with increased activity and ease of use remains, there are increasing difficulties in bringing new products to market. Expense and time are two of the main constraints in further development of chemical pesticides. For example, it is now typically necessary to screen over 10 000 chemicals per year in order to identify up to three for further development. Programmes of this scale could take up to 10 years at a cost of around $200 million before a product is registered for commercial use. In addition, the period for exclusive marketing that might be granted under initial patents is only 20 years. When the 10 years of development is taken into account, followed by a period of marketing to cover the development costs, there may be as little as 5 years when the product is performing profitably. These statistics are not likely to change in the future and it is not surprising, therefore, that new ways of pest control are being sought that avoid the enormous costs and short commercial exploitability of chemical pesticides.

Environmental constraints add a further dimension to the difficulty of developing or, indeed, continuing to use chemical pesticides. Increasing levels of resistance to many of the main chemical groups are being noted, which are problems in their own right, but it is also apparent that cross-resistance between chemical groups is increasing. These attributes add to the difficulty of developing novel active ingredients. Resistance to chemical pesticides results from the very large quantities of active ingredients used to maintain freedom from pests in crops with low economic thresholds. Greater quantities of pesticide are then used to try to maintain the same levels of control that were achieved prior to the development of resistance. This results in the 'insecticide treadmill' where increasing quantities or frequencies of application are employed but with decreasing efficacy over time. This can lead to almost complete resistance to a given chemical group, making it virtually impossible to achieve economic crop yields. One of the clearest examples of this phenomenon is the use of pesticides to control the diamond-back moth, *Plutella xylostella*, which is one of the most important pests of crucifers in both temperate and, particularly, tropical and subtropical climates. Virtually all the recognized groups of pesticides, including organochlorines, organophosphates, pyrethroids, insect growth regulators and *B.t.* have been used against this moth. The scale of usage has given rise to significant resistance to all these agents, including *B.t.* Enormously increased dosages are now required in some areas where many years of pesticide usage has resulted in selection of races of *P. xylostella* with greatly enhanced resistance. Most worrying is the very rapid development of resistance to the pyrethroids, an extreme example being >11 000-fold resistance to fenvalerate (Hung and Sun, 1989). Insect growth regulators, which have previously been regarded as relatively selective, have also given rise to major resistance in this pest, with, for example, >1000-fold increase in resistance to teflubenzuron in 6 months (Cheng, Kao and Chiu, 1990). Cross-resistance between different pyrethroids and between pyrethroids, organophosphates and carbamates has also been demonstrated (Zoebelein, 1990). These examples are typical of the situation that can arise under intensive management based entirely around chemical pesticides.

MICROBIAL CONTROL AS AN ENVIRONMENTALLY BENIGN ALTERNATIVE TO CHEMICAL PESTICIDES

The economic and environmental constraints within the chemical pesticide industry have provided fresh incentives to develop alternatives to chemicals. This is not always possible and it is not envisaged that chemicals can be replaced completely by more sustainable management techniques.

Use of natural enemies has many attractions, not least of which is the low level of active management required if they become established and exert long-term regulation of pest population densities. Predators, parasites and microbial agents can all be regarded as regulatory agents for extended pest control, but the practicality of their use is determined by the economic threshold of the crop being protected. When economic thresholds are high and a certain amount of damage can be tolerated then any delay in exerting control does not present a major problem. Such situations pertain in forestry, where there are a number of examples of successful use of natural enemies in their own right or within an integrated pest management (IPM) programme. When economic thresholds are low, often driven by customer preferences for blemish-free produce, the time period between appearance of the pest and use of control measures is very short, requiring a knock-down approach that, previously, has only been achieved by the use of chemical agents.

The drive to find alternatives or supplementaries to chemical pesticides must, therefore, take into account the mode of action of the alternatives as well as the increased environmental benefits that might accrue from their use. As in all pest management decision making, a balance has to be struck between the costs and benefits of the proposed approach. Notwithstanding these provisos, it is clear that a great deal of promise is provided by the use of microbial control agents for pest management. Like more conventional natural enemies, microbial agents have the capacity to maintain themselves in the environment but, more significantly from the pest manager's point of view, they can be used in a manner similar to the familiar chemical pesticides. The ability to use microbial agents in conventional spray machinery and to integrate their use within decision-making processes based on pest monitoring etc. provides an interface that can be adapted easily to existing management systems. Speed of action and questions of specificity have to be taken into account, but, once these are solved, the major environmental advantages of microbial agents can far outweigh the technical constraints. The predicted annual growth rate for the biopesticides industry over the next 10 years is 10–15% as compared with 1–2% for synthetic chemical pesticides (Menn, 1996).

The purpose of this book is to provide a practical guide to the use of one group of microbial agents, the Baculoviridae, within pest management regimes. As a necessary prelude to the more practical aspects, we present a synopsis of the characteristics of the viruses, their ecology and epizootiology and a general overview of their use as practical pest control agents. This is placed in a world context with contributions from leading experts in the field (Part 2). The third section of the book gives detailed guidance on the procedures involved in working with baculoviruses (BVs), from laboratory to field. *Oryctes rhinoceros* virus, although not now classified as a BV, is included for completeness.

REFERENCES

Cheng, E.Y. Kao, C.H. and Chiu, C.S. (1990) Insecticide resistance study in *Plutella xylostella* L. X. The IGR-resistance and the possible management strategy. *Journal of Agricultural Research of China* **39**, 208–220.

Georgis, R. (1997) Commercial prospects of microbial insecticides in agriculture. In *BCPC Symposium Proceedings No. 68: Microbial Insecticides: Novelty or Necessity?* Farnham, UK: British Crop Protection Council, pp. 243–252.

Hung, C.F. and Sun, C.N. (1989) Microsomal monooxygenases in diamondback moth larvae resistant to fenvalerate and piperonyl butoxide. *Pesticide Biochemistry and Physiology* **33**, 168–175.

Menn, J.J. (1996) Biopesticides, has their time come? *Journal of Environmental Science and Health Part B, Pesticides, Food Contaminants and Agricultural Wastes* **31**, 383–389.

Navon, A. (1993) Control of lepidopteran pests with *Bacillus thuringiensis*. In Bacillus thuringiensis, *an Environmental Biopesticide: Theory and Practice*, P.F. Entwistle, J.S. Cory, M.J. Bailey and S. Higgs (eds). Chichester: John Wiley, pp. 125–146.

Powell, K.A. and Rhodes, D.J. (1994) Strategies for the progression of biological fungicides into field evaluation. In *1994 BCPC Monograph No 59: Comparing Glasshouse and Field Pesticide Performance II*, H.G. Hewitt, J. Caseley, L.G. Copping, B.T. Grayson, and D. Tyson, (eds). Farnham, UK: British Crop Protection Council, pp. 307–315.

Zoebelein, G. (1990) Twenty-three year surveillance of development of insecticide resistance in diamondback moth from Thailand (*Plutella xylostella* L., Lepidoptera, Plutellidae). *Mededelingen van de Faculteit Landbouwwetenschappen, Rijksuniversiteit Gent* **55**, 313–322.

2 Characteristics of insect pathogenic viruses

INTRODUCTION

The use of pathogens as pest control agents is much more likely to be successful if the pathogen has been carefully selected and characterized. In many instances, several isolates of one or more viruses capable of infecting the target species may be available. Field isolates frequently consist of mixtures of different viruses or of different strains of one virus. These viruses may differ in their pathogenicity for the target insect and also in their ability to infect other species. They may differ also in other characteristics relevant to their use as biological insecticides, for instance the yield of virus obtained from larvae or their ability to replicate in cell culture. To examine these characteristics, it is necessary to obtain pure strains of each virus and to be able to identify each of these strains unambiguously.

Although this book concentrates predominantly on BVs as these have by far the greatest potential for insecticidal use, other viruses may be encountered in field-collected insects and it is important to be able to identify them. The insecticidal use of viruses other than BVs has so far been very limited but their possibilities should not be completely overlooked.

IDENTIFICATION OF VIRUS TYPES

If new isolates are being made from natural occurrences of diseased larvae, it is preferable to collect individual moribund or freshly dead larvae. Care should be taken to identify the larvae accurately; in some instances this may involve collecting healthy larvae of the same species and rearing them through to adults. The pathology of the disease provides the first indication of the type of virus involved as viruses vary in the tissues they infect, in their ability to cause acute or chronic infections and in the appearance of moribund or dead larvae (Figure 2.1). Examination of infected tissues by light and/or electron microscopy will usually provide a good indication to which family, and sometimes to which genus, the virus belongs. Precise identification will usually require biochemical or molecular techniques. Some of the general properties of insect viruses are, therefore,

Figure 2.1 Larva of the pine beauty moth, *Panolis flammea*, characteristically infected with NPV.

described below. Information on their taxonomy and characteristics can be obtained by referring to the *Sixth Report of the International Committee on Taxonomy of Viruses* edited by Murphy *et al.* (1995). Updated information is available via the world wide web at the following address: http://life.anu.edu.au/viruses/Ictv/index.html.

GENERAL FEATURES OF INSECT VIRUSES

Insect viruses belong to many different virus families, some of which occur exclusively in arthropods and some of which include representatives that occur in vertebrates and/or plants (Table 2.1). A feature of many insect viruses, which does not occur in viruses infecting plants or vertebrates, is that they are occluded, i.e. the virions are embedded within a proteinaceous body. Occlusion bodies (OBs) vary in size from about 0.5 to over 20 μm across but are all visible under the light

Table 2.1 Classification of virus families containing insect pathogenic viruses[a]

Family	Genera occurring in insects	OBs[b]	Examples of virus	Examples of genera occurring in vertebrates	Examples of genera occurring in plants
DNA viruses					
Ascoviridae	Ascovirus	–	*Trichoplusia ni* AV	None	None
Baculoviridae	Nuclear polyhedrosis virus (NPV)	+	*Autographia californica* MNPV	None	None
	Granulosis virus (GV)	+	*Cydia pomonella* GV	None	None
Iridoviridae	Iridovirus	–	*Chilo* iridescent virus	Ranavirus	None
	Chloriridovirus	–	Mosquito iridescent virus	Lymphocystivirus	
Parvoviridae	Densovirus	–	*Galleria mellonella* densovirus	Parvovirus Dependovirus	None
Polydnaviridae	Ichnovirus	–	*Campoletis sonorensis* PV	None	None
	Bracovirus	–	*Cotesia melananoscela PV*		
Poxviridae	Entomopoxvirus	+	*Amsacta moorei* EPV	Orthopoxvirus Capripoxvirus Avipoxvirus Leporipoxvirus Parapoxvirus Suipoxvius + others	None
Unclassified			*Oryctes* virus Hz-1 virus Bee filamentous virus Tsetse virus Narcissus bulb fly virus		
RNA viruses					
Birnaviridae	Birnavirus	–	*Drosophila* X virus	Birnavirus (infectious pancreatic necrosis virus)	
Caliciviridae		–	Chronic stunt virus	Calicivirus	None
Nodaviridae	Nodavirus	–	Black beetle virus	None[c]	None
Picornaviridae		–	Cricket paralysis virus	Enterovirus	None

continued overleaf

Table 2.1 (*continued*)

Family	Genera occurring in insects	OBs[b]	Examples of virus	Examples of genera occurring in vertebrates	Examples of genera occurring in plants
RNA viruses continued					
Reoviridae	Cytoplasmic polyhedrosis virus	+	*Bombyx mori* CPV	Reovirus Orbivirus Rotavirus	Phytoreovirus Fujivirus
Rhabdoviridae		–	Sigma virus	Vessiculovirus Lyssavirus	Lettuce necrotic yellows Potato yellow dwarf
Tetraviridae	Tetravirus	–	*Nudarelia* β virus	None	None
Unclassified		–	Various bee paralysis viruses Various *Drosophila* viruses		

[a] A large number of other vertebrate and plant viruses also occur in insects and may replicate in certain insect tissues. In general, the insect is considered to be simply a vector for transmission of these viruses although it may be that some of these viruses evolved first in the insect host. These viruses do not have significant pathogenicity for insects and therefore are not included here.
[b] Absent –, present +.
[c] Nodamura virus infects suckling mice in the laboratory.

microscope. This is such a distinctive feature, particularly for diagnosis, that viruses have been grouped below according to whether or not they produce OBs. A general review of the various groups of viruses occurring in insects can be found in Evans and Entwistle (1987). Listings of viruses infecting insects are given in *Virology, Directory & Dictionary of Animal, Bacterial and Plant Viruses* (Hull, Brown and Payne, 1984).

OCCLUDED VIRUSES

Occluded viruses belong to three virus families. Each of these families appears to have independently evolved the ability to produce OBs consisting largely of a single protein. Within each family the OB proteins from different viruses are related but there appears to be no homology between proteins from different families. The OB allows these viruses to persist for long periods in the environment and it is presumably no coincidence that nearly all of these viruses replicate only in the larval stages of insects that undergo complete metamorphosis. Since in many species larvae occur for only a limited period each year, the virus must be able to complete its replication cycle within the larval life span and then survive for many months in the absence of susceptible host insects. In some instances occluded virus is known to have survived for decades in the absence of any host insects and still be infectious. Almost all occluded viruses are infectious *per os*, though *Tipula paludosa* (Diptera) nuclear polyhedrosis virus (NPV) seems to be an exception.

Baculoviruses (BVs, family Baculoviridae)

The BVs make up a large family of viruses, with more isolates recorded than for any other group of insect viruses. They have a large circular double-stranded (ds) DNA genome and structurally complex rod-shaped enveloped virus particles. BVs have a narrow host range and generally cause an acute infection resulting in high mortality. Lepidopteran viruses often cause a generalized infection though some, especially among the granulosis viruses (GVs), infect only the fat body and a few other tissues. Hymenopteran viruses cause infection mainly in the gut epithelium. The fact that BVs occur exclusively in arthropods has provided a strong positive argument in assessing the safety of using these pathogens for pest control since the chance of any BV, even a mutated BV, being able to replicate in a plant or any vertebrate animal is effectively zero.

Because of the predominance of BVs over all other viruses for pest control, a more detailed account of BVs is given later in this chapter.

Cytoplasmic polyhedrosis viruses (CPVs, family Reoviridae)

The cytoplasmic polyhedrosis viruses (CPVs) are isolated only from arthropods. CPVs have been recorded from about 250 species of insects, predominantly Lepidoptera but also Diptera and a few Hymenoptera. Viruses replicate in the cytoplasm of midgut epithelial cells, producing large numbers of OBs which cause

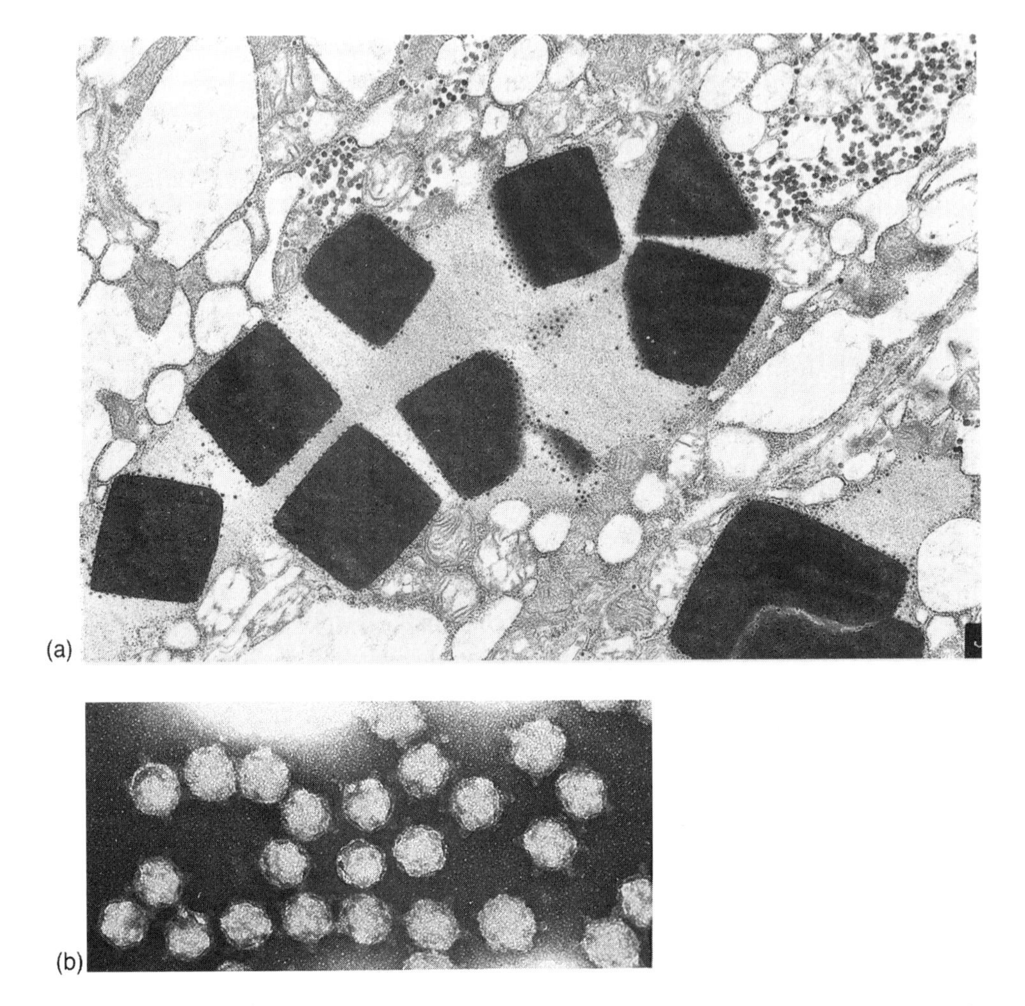

Figure 2.2 (a) Section of CPV OBs in the gut of the silkworm, *Bombyx mori* (× 8 400). (b) Purified virus particles of the same virus (× 13 200).

the gut to turn white (Figure 2.2). Many infected cells are sloughed off and frass from infected larvae may contain large numbers of polyhedra. CPVs cause chronic infection; infected larvae may pupate and develop into adults, though they are often malformed and have reduced fecundity. Viruses frequently have a broad host range, which may include several families of Lepidoptera. The host species is, therefore, not a very useful guide to virus identity.

The CPV genome consists of 10 segments of dsRNA. The profile generated by electrophoresis of these fragments has been used as a basis to group isolates into different virus 'types'. The icosahedral virus particles are 50–65 nm in diameter, which is smaller than other reovirus particles. Also, unlike other reoviruses, which have a double capsid structure, CPVs have a single capsid shell. A distinguishing feature is the presence of 12 spikes on the icosahedral particles.

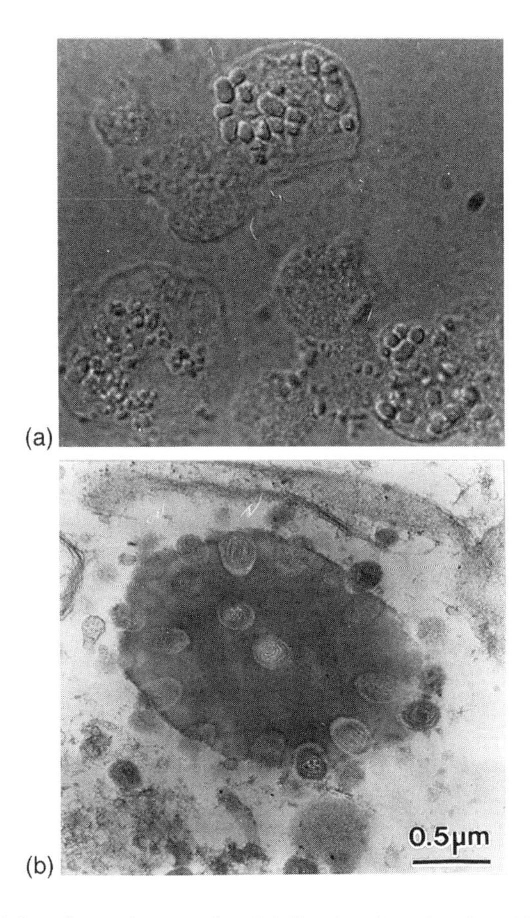

Figure 2.3 (a) Light photomicrograph of fully developed spheroids of *Amsacta moorei* EPV in cultured cells of the gypsy moth, *Lymantria dispar*, 96 h after infection. (b) Electron photomicrograph of spheroidin-occluded mature virions of the same virus. (Photographs kindly supplied by Linda King and Susan Marlow.)

Entomopoxviruses (EPVs, subfamily Entomopoxvirinae)

Entomopoxviruses (EPVs) form a distinct subfamily within the Poxviridae family and have been recorded from about 60 insect species within Lepidoptera, Coleoptera, Diptera, Orthoptera and Hymenoptera. The virus particles are brick-shaped or ovoid and, with the exception of hymenopteran EPVs, are contained within spheroidal OBs called spherules or spheroids (Figure 2.3). These mainly comprise a single matrix protein known as spherulin or spheroidin with a M_r of 109 000–115 000. A second type of paracrystalline proteinaceous body called spindle occurs in some lepidopteran and coleopteran EPVs. Spindles comprise predominantly a single spindle protein (fusolin) with a M_r of approximately 40 000–50 000 and do not occlude any virus particles. Like other poxviruses, EPVs replicate in the cytoplasm of infected cells. The fat body is the main site of virus replication but some infections are systemic.

The genomes of EPVs are large linear molecules of dsDNA with a low GC content. Very little work has been done on the molecular biology of EPVs but recently more interest has been shown in these viruses and they are being developed as an expression system for foreign genes.

NON-OCCLUDED VIRUSES

Ascoviruses (family Ascoviridae)

Ascoviruses have been recorded, so far, only from the lepidopteran family Noctuidae. They cause a chronic, fatal disease of larvae with an unusual cytopathology in which the infected cell nucleus enlarges and then disintegrates into fragments. The cell also grows much larger than uninfected cells and, later in infection, large numbers of virion-containing vesicles form. Ascoviruses are very poorly infectious *per os* and the main route of infection appears to be by injection when parasitic wasps that have previously become contaminated from an infected larva lay their eggs in a host larva.

Ascoviruses form large enveloped virions (130 × 400 nm) with complex symmetry (reniform to bacilliform) (Figure 2.4) containing a linear dsDNA genome of about 170 kb.

Birnaviruses (family Birnaviridae)

The only known insect birnavirus is *Drosophila* X virus, which was isolated from a laboratory culture of *Drosophila melanogasta*. Virus-infected flies died in 5–15 days and were unusually sensitive to anoxia for 3–4 days before death. The virions are approximately 60 nm in diameter, single-shelled, non-enveloped icosahedrons and contain a bipartite dsRNA genome.

Caliciviruses (family Caliciviridae)

The Caliciviridae family of small, single stranded (ss) RNA viruses occurs predominately in vertebrates. Chronic stunt virus isolated from the navel orangeworm, *Amyelois transitella* (Lepidoptera), is the only calicivirus so far identified from

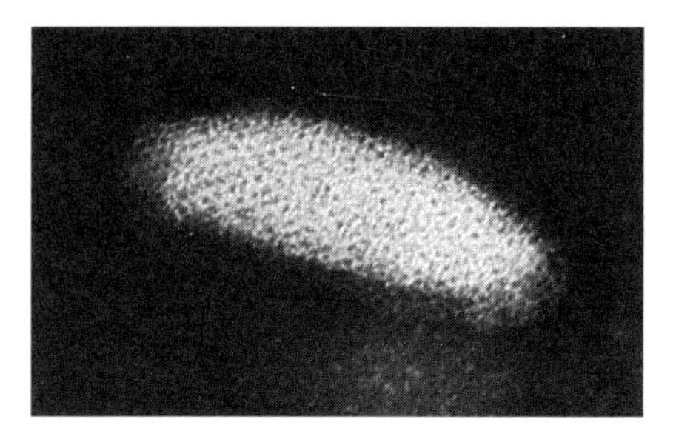

Figure 2.4 Virion (approximate length, 400 nm) *Trichoplusia ni* ascovirus. (Photograph kindly supplied by Brian Federici.)

insects. The virions are 30–38 nm in diameter with 32 cup-shaped depressions arranged in T = 3 icosahedral symmetry and contain a genome of ssRNA. The virus is infectious *per os* and is pathogenic to early instar larvae. Later instar larvae are more likely to develop a chronic infection, resulting in stunted adults with reduced fecundity.

Iridoviruses (family Iridoviridae)

The Iridoviridae occur in many insect families and also in other invertebrates. Invertebrate iridoviruses are divided into two genera, iridovirus and chloriridovirus, based on the size of the icosahedral virions. The iridovirus or small iridescent viruses are 125–140 nm across whilst the chloriridovirus or large iridescent viruses are 180–200 nm across. The normal route of infection is unknown since it is very difficult to infect larvae *per os* and it is not known to cause epizootics. If larvae become infected during the early instars, the disease is fatal. During viral replication, the nucleus remains intact, and virions form paracrystalline arrays in the cytoplasm (Figure. 2.5) giving rise to an iridescent hue in the infected host.

Iridoviruses have large (about 200–250 kb) linear dsDNA genomes, though because the DNA is cyclically permuted, iridovirus maps are often presented in circular form.

Nodaviruses (family Nodaviridae)

The Nodaviridae family is named after nodamura virus, which was isolated from mosquitoes in Japan and originally was thought to be an arbovirus as it can be transmitted to mice. However, the nodamura virus genome, unlike any other arbovirus, consists of two ssRNA molecules encapsidated within a single non-enveloped icosahedral virion about 30 nm in diameter. Five other viruses, all isolated in Australasia, including black beetle virus and Flock House virus, share these properties and are classified within the Nodaviridae. They are all insect pathogenic viruses able to infect a range of species, particularly within the Lepidoptera and Coleoptera, though some species are susceptible only by injection. None of these other nodaviruses is able to replicate in vertebrates.

Because of the simplicity of these viruses, their molecular biology is relatively well characterized and black beetle virus was the first insect virus for which the complete genome sequence was determined. However, little is known about their ecology and distribution and there is no evidence that they are of any economic significance.

Parvoviruses (family Parvoviridae)

Densonucleosis viruses, or densoviruses, are insect-pathogenic viruses belonging to the Parvoviridae. They have small non-enveloped virions (ranging in size from 19 to 24 nm) with icosahedral symmetry and a genome consisting of both positive- and negative-strand DNA, which are complementary but separately encapsidated. They have been isolated primarily from Lepidoptera but also occur in Diptera, Odonata, Orthoptera and Dictyoptera. There is considerable variation in the

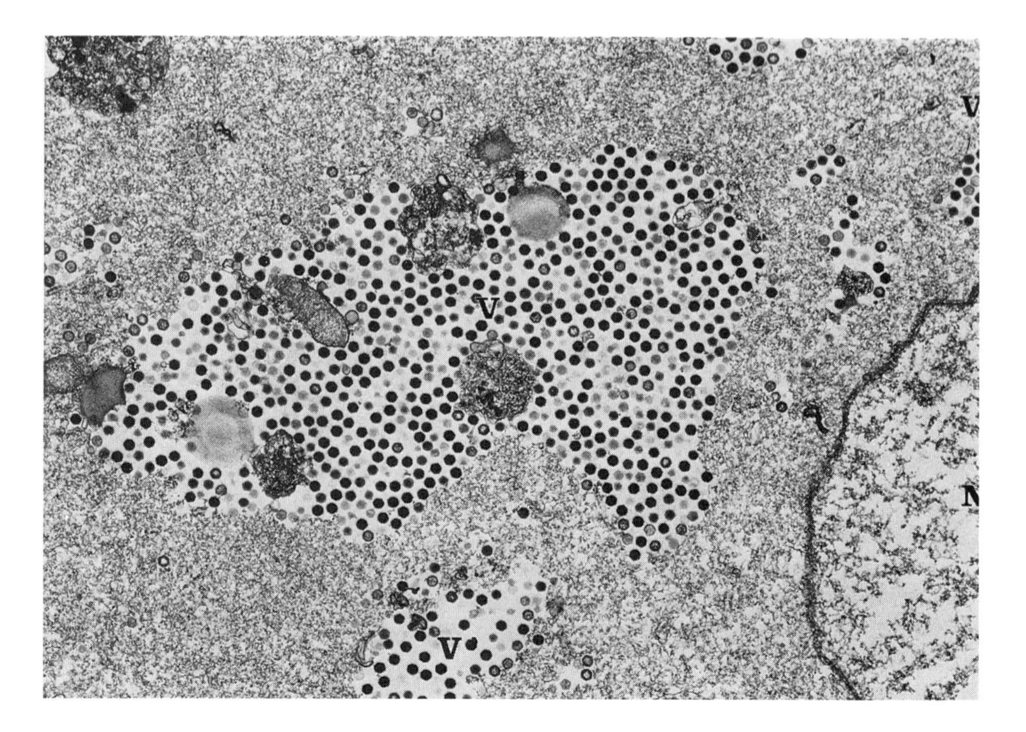

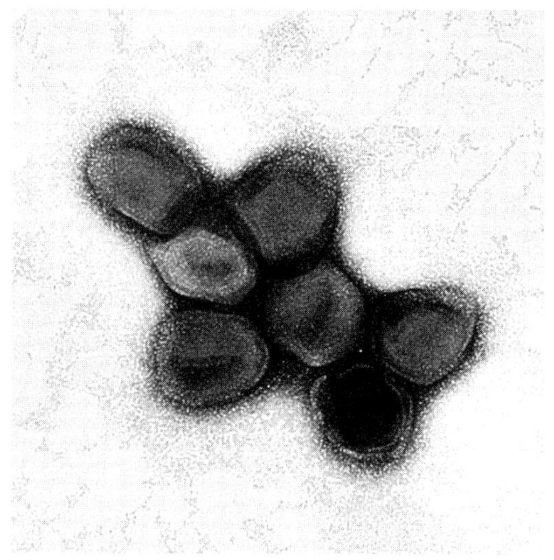

Figure 2.5 (a) Section of *Simulium* infected with an iridovirus. Paracrystalline arrays of virus particles (V) are visible in the cytoplasm while part of the intact nucleus (N), bounded by the nuclear membrane, can be seen on the right hand side (× 10 080). (b) Purified virions (× 100 000) shown in greater detail.

biological characteristics of these viruses. Some, such as *Galleria mellonella* densonucleosis virus, have a very limited host range while others are able to infect species across several families. Pathogenesis can vary, as demonstrated by the disease in *Bombyx mori* from which different types of densonucleosis virus have been isolated. Type 1 virus causes an acute infection leading to death in about 1 week whereas type 2 causes a more chronic infection taking up to 3 weeks to kill larvae. Detailed studies have been conducted on resistance to infection in *B. mori*, but different genes are responsible for resistance to types 1 and 2 and insects do not normally develop resistance to both virus types.

The small genome size (4–5 kb) has made it straightforward to obtain a complete nucleotide sequence and, therefore, some of these viruses are well characterized in terms of their molecular biology and genetics.

Picornaviruses (family Picornaviridae)

The Picornaviridae include small, ssRNA viruses that are a common cause of disease in mammals, including the common cold and polio in humans. The virus particles (diameter 27–32 nm) are non-enveloped, exhibit icosahedral symmetry and contain four major capsid polypeptides. In addition, there is a small protein covalently bound to the 5′ terminus of the RNA genome. Over 30 viruses isolated from insects have been considered as possible picornaviruses but very few of these have been sufficiently well characterized to allow their formal classification in this family. The two best studied insect picornaviruses are cricket paralysis virus and *Drosophila* C virus. In the laboratory, these viruses have a wide and overlapping host range but little is known about their natural distribution and ecology. They are not known to cause natural epizootics but may cause high mortality in laboratory insects particularly when infection is by injection. An insect virus that is probably a picornavirus and does cause epizootics is *Gonometa* virus, isolated in Uganda from *Gonometa podocarpi*, a lepidopterous defoliator of pine trees. The virus caused high mortality when applied to larval populations and was shown to be an effective control agent.

Polydnaviruses (family Polydnaviridae)

The polydnaviruses are unusual and complex viruses with a genome consisting of multiple supercoiled DNAs of variable size from 2 to more than 28 kb; they replicate only in the calyx (part of the ovaries) of parasitic Hymenoptera. In the species in which they occur, they are present in 100% of females and appear to have a symbiotic relationship with their host. When the parasite lays its egg in a host larva, virus is injected and may aid the successful development of the parasite by suppressing the host's immune response. These viruses are, therefore, an advantage to their host and have no role in insect control.

Rhabdoviruses (family Rhabdoviridae)

Many rhabdoviruses occur in insects but nearly all are transmitted by arthropods. The virions are bullet-shaped (200 nm long and 75 nm diameter) and contain ssRNA. Only one, the sigma virus of *Drosophila*, is restricted to insects. Unusually, this virus is transmitted only vertically in nature and occurs in about 10% of the natural populations of *Drosophila*. Under normal circumstances it is not pathogenic but it confers sensitivity to CO_2 and infected adults suffer fatal paralysis when exposed to the gas.

Tetraviruses (family Tetraviridae)

The Tetraviridae is a small family of insect viruses all isolated from Lepidoptera. Virus particles are non-enveloped with diameters from 35 to 40 nm. This group of viruses was formerly named after the type species *Nudaurelia* β virus (genus '*Nudaurelia capensis* β-like viruses'), one of a complex of viruses (α–ε) isolated from the pine tree emperor moth, *Nudaurelia cytherea*. They contain a major capsid protein of M_r 60 000–70 000 and a small protein of M_r approximately 8000 and have a genome comprising a single 5 kb molecule of ssRNA. Two other viruses, *Nudaurelia* ω virus and *Helicoverpa armigera* stunt virus, are included in this family (genus '*Nudaurelia capensis* ω-like viruses') but have bipartite ssRNA genomes. The complete sequence of the *H. armigera* stunt virus genome has been determined and consists of RNA 1 of 5311 nucleotides and RNA 2 of 2478 nucleotides.

UNCLASSIFIED VIRUSES

There remain many insect viruses that do not appear to belong to any of the above families. In particular, a large number of small RNA viruses have been isolated from bees and from *Drosophila melanogasta*. No doubt if other insect species were studied to the same extent as these two species, many more viruses would be found and there may be large families of viruses of which we are unaware. There is a smaller number of DNA viruses that do not appear to belong to any of the above families but from a pest control point of view, at least one of these is of considerable importance.

Oryctes virus, which was isolated from *Oryctes rhinoceros*, has short rod-shaped virions that contain a large circular dsDNA genome. It was for a long time classified in a separate subgroup within the Baculoviridae, but differences in virion morphology, lack of an OB and no evidence of genetic similarity caused it to be removed from the Baculoviridae along with other viruses formerly in this subgroup.

Two other insect viruses which occur as long flexuous rods and which may have considerable potential for host control should be mentioned. The linear dsDNA virus of the tsetse fly, *Glossina palidipes* (Odindo, Payne and Crook, 1984), causes hypertrophy of the adult male salivary glands and associated arrest of spermatogenesis, with consequent sterility. Gonadal lesions in females are associated with severe ovariole necrosis (Jura *et al.*, 1988). The widespread occurrence of salivary gland hypertrophy (e.g. Gouteux, 1987) suggests several other tsetse fly species may be affected by a similar virus. A morphologically similar but uncharacterized DNA virus of the large narcissus bulbfly, *Merodon equestris*, is also associated with adult salivary gland hypertrophy and gonadal atrophy (Armagier *et al.*, 1979).

CLASSIFICATION OF BACULOVIRUSES

The taxonomy of BVs, as with viruses in general, has undergone various changes since the late 1960s and will no doubt continue to be modified as our knowledge of the relationships between viruses increases. Until recently, the family Baculoviridae was divided into three subgroups: subgroup A, nuclear polyhedrosis viruses (NPVs); subgroup B, granulosis viruses (GVs); and subgroup C, non-occluded viruses. Previously, the only division within subgroups A, B and C was that NPVs (group A) were split into multiply enveloped (M) NPVs and singly enveloped (S) NPVs. Currently, Baculoviridae comprises only two genera, polyhedrovirus (NPVs) and granulovirus (GVs) and the terms nuclear polyhedrosis virus and granulosis virus have been replaced by **nucleopolyhedrovirus** and **granulovirus**, respectively (Murphy *et al.*, 1995). The OBs of GVs are smaller (0.3–0.5 μm in length) than those of NPVs (0.15–15 μm in diameter) and usually contain a single enveloped nucleocapsid (the virus particle) only (Figures 2.6 and 2.7a). The OBs of NPVs contain several hundred virus particles each of which may contain one (SNPV) or many (MNPV) nucleocapsids (Figures 2.6 and 2.7b). There is no longer a subdivision relating to the morphology of NPVs as either multiply or singly enveloped as this does not appear to be very significant taxonomically. For example, *Bombyx mori* SNPV (BmNPV) and *Autographa californica* MNPV (AcMNPV) are far more closely related to each other than they are to most other NPVs whether SNPVs or MNPVs. Also, the non-occluded viruses, formerly in subgroup C, are no longer included within the Baculoviridae. Although they have a large dsDNA genome and infect only arthropods, they show little other similarity to baculoviruses (Figure 2.7c). The abbreviations commonly used for baculoviruses are based on the first letter of the insect species and the viral subgroup, e.g. *Bombyx mori* SNPV as BmSNPV and *Pieris brassicae* GV as PbGV.

IDENTIFICATION OF BACULOVIRUSES

Although BVs are usually named after the host species from which they were isolated, this does not provide a satisfactory form of identification. In several instances, quite different BVs have been isolated from the same insect species while a single virus, or a group of closely related strains, can occur in more than one insect species, e.g. *Autographa californica* MNPV (Payne, 1986). Occasionally this difficulty has been resolved by foregoing the usual convention and naming a new virus after another virus to which it is closely related even though it was isolated from a different species. Although there is general agreement that a better system of nomenclature is needed that reflects more accurately the identity of the virus, there has so far been no consensus on how this could be achieved. Identification of a BV must, therefore, depend upon other characteristics in addition to host range.

For the purpose of this book, the terms granulosis virus and nuclear polyhedrosis virus are used synonomously with granulovirus and nucleopolyhedrovirus, respectively (Murphy *et al.*, 1995).

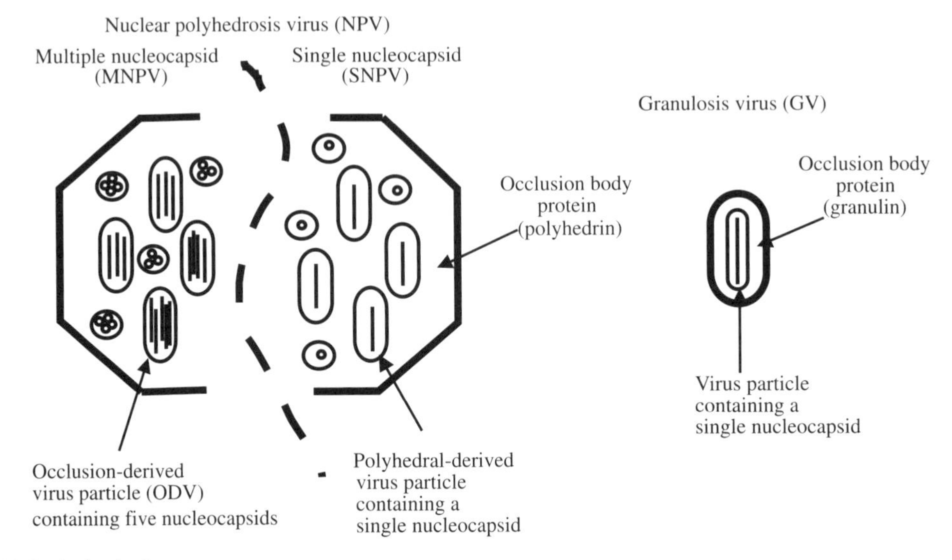

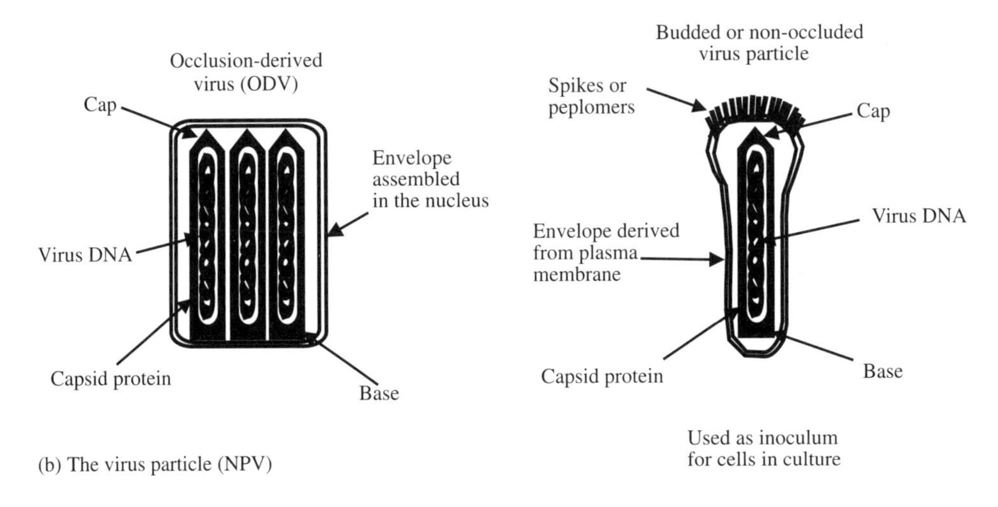

Figure 2.6 Diagrammatic representation of the morphology of members of the Baculoviridae family of insect pathogenic viruses.

CHARACTERIZATION OF BACULOVIRUSES

BVs have been studied in far more detail than any other family of insect viruses. This is largely because of their much greater potential as pest control agents, but the use of AcMNPV as an expression system for foreign proteins has also encouraged detailed studies on BV promoters, gene transcription, large-scale cell culture and other basic aspects of BV biology. The complete genomes of three BVs

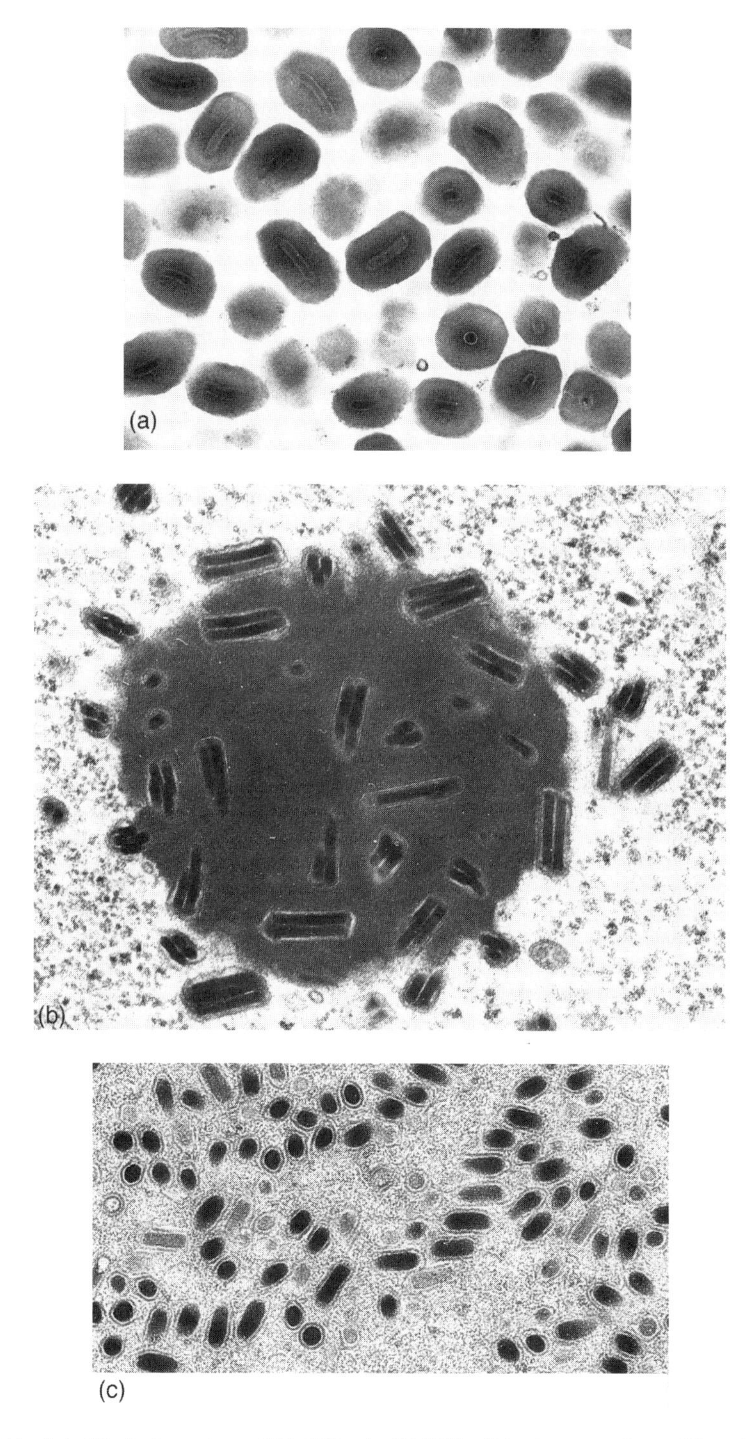

Figure 2.7 (a) *Pieris brassicae* GV OBs (×50 000). (b) *Autographa californica* MNPV (alfalfa looper) OB (approximately 2.5 μm diameter). (c) Non-occluded virus particles (approximately 200 nm in length) of the phantom midge, *Chaoborus astictopus*. (Photographs for (b) and (c) kindly supplied by Brian Federici.)

(AcMNPV, BmNPV and *Orgyia pseudotsugata* NPV (OpNPV)) have been sequenced but these have turned out to be very closely related. Varying lengths of DNA sequence information have been obtained from another dozen or so viruses. However, for the vast majority of BVs, few or no molecular or biochemical data have been obtained and it is not yet clear to what extent characteristics found in those few well-studied viruses will be typical of the majority of BVs. In particular, the vast majority of studies on BV genetics and replication have been carried out with NPVs, and much more information is required about GVs before the relationship between the two genera is more fully understood.

Although a detailed knowledge of BV molecular biology is not essential for using these viruses for pest control, some background information on this subject is relevant to both the replication and ecology of these viruses as well as to understand work that is underway to produce genetically engineered viruses with improved properties for insecticidal use. The following section, therefore, summarizes what is currently know about BV genomes.

There are probably between 100 and 200 genes distributed on both strands with little overlap and relatively short intergenic regions. Gene expression is temporally regulated by a cascade of early, late and very late gene transcription. Early gene expression precedes viral DNA replication and appears to be dependent on the host RNA polymerase II, which is sensitive to α-amanitin. Late gene expression commences at the onset of DNA replication and also coincides with the inhibition of host gene expression. BV late and very late promoters normally contain a TAAG sequence motif, which is recognized by an α-amanitin-resistant RNA polymerase.

Genes can be divided into several categories according to their function. Since there are both budded and occluded forms of the virus, both with complex structures, there are a large number of genes encoding structural proteins, including polyhedrin/granulin, polyhedral envelope protein, capsid proteins, DNA-binding (basic) protein, gp64 (major budded virus envelope protein), etc. A large number of genes have been identified that regulate the expression of other viral genes. These include the *trans*-activating factor genes *ie-1* and *ie-2*, which increase the transcription of other early genes, and a large number of late expression factor (*lef*) genes. It has been shown that a total of 18 genes are necessary and sufficient to support expression from a late viral promoter in a transient expression assay. These include genes with sequence motifs shared by DNA polymerases, DNA helicases and RNA polymerases. Two genes, *p35* and *iap*, have been identified that block apoptosis, one of the natural defence mechanisms of the host cell. This mechanism may be involved in host-range determination since AcMNPV lacking *p35* is able to replicate quite normally in *Trichoplusia ni* cells and insects but not in *Spodoptera frugiperda* cells and insects. Another group of genes are those that are homologues of genes which are already present in the host, such as superoxide dismutase and ubiquitin, although why the virus needs to carry its own copies of these genes is not known. The viral ecdysteroid UDP-glucosyl transferase (*egt*) gene also has homology to an insect gene. Expression of this enzyme alters levels of ecdysone within the larva and, thus, interferes with moulting and pupation. For reasons that are not clear, deletion of the *egt* gene in AcMNPV reduces the time it takes the virus to kill larvae. One of the best studied of baculoviral genes is *p10*. The p10 protein is highly expressed very late in the infection cycle yet is non-essential for viral replication and OB formation. Although p10 is a small protein, three functional domains have been identified: for aggregation, nuclear disintegration and fibrillar structure formation.

An interesting feature found on the genomes of several BVs is a series of homologous repeat (hr) regions. The best studied of these are on the AcMNPV genome, which has a total of eight hr regions containing variable numbers of repeats of an imperfect palindromic sequence about 70 nucleotides in length. In transient expression assays, these regions greatly enhance expression from early viral promoters. They are also able to act as origins of viral replication in transient replication assays. It has been more difficult to determine the effect of these regions during normal viral replication since deletion of a single region has no apparent effect and deletion of many or all of these regions has not yet been attempted.

BIOLOGY OF BACULOVIRUSES

Virus particles of BVs exist in two antigenically and morphologically distinct forms, depending on their environment (Figure 2.6). Those within the OB are known as occlusion- or polyhedral-derived virus (ODV or PDV) particles and are responsible for initiating infection in the epithelial cells of the midgut. The other form, known as budded virus, or non-occluded virus (NOV), is responsible for spread and secondary infection within the insect body. The composition of the envelopes of the two forms differs. PDV envelopes are synthesized within the nucleus during OB morphogenesis. Those of budded virus are acquired when newly synthesized nucleocapsids bud through the host cell cytoplasmic membrane and are released into the haemolymph. Spikes of a virus-coded glycoprotein (budded virus envelope fusion protein, gp64) protrude through one end.

The main features of the biology of BVs are illustrated in Figure 2.8. Briefly, OBs are ingested and dissolved in the alkaline conditions of the midgut. PDV enters the epithelial cells by fusion, travels to the nucleus and is uncoated either before (GV) or after (NPV) passing through the nuclear pores. One round of replication follows, the newly formed nucleocapsids bud through the nucleus gaining an envelope of nuclear membrane. This is shed in the cytoplasm and an envelope comprising cytoplasmic membrane and virus-coded glycoprotein spikes is acquired by budding through the midgut basal membrane. This budded virus is released into the haemolymph and undergoes further rounds of multiplication in the cells of susceptible tissues. Entry in this instance is by cell-mediated endocytosis. Late in the replication cycle, PDV is formed around which OB protein crystallizes to form OBs. These are released into the environment when the insect dies and disintegrates. A diagrammatic representation of a dissected lepidopterous larva is given in Figure 2.9.

HOST RANGE

The host specificity or limited host range of BVs may be both an advantage and a disadvantage. From an environmental and safety point of view, it is an advantage

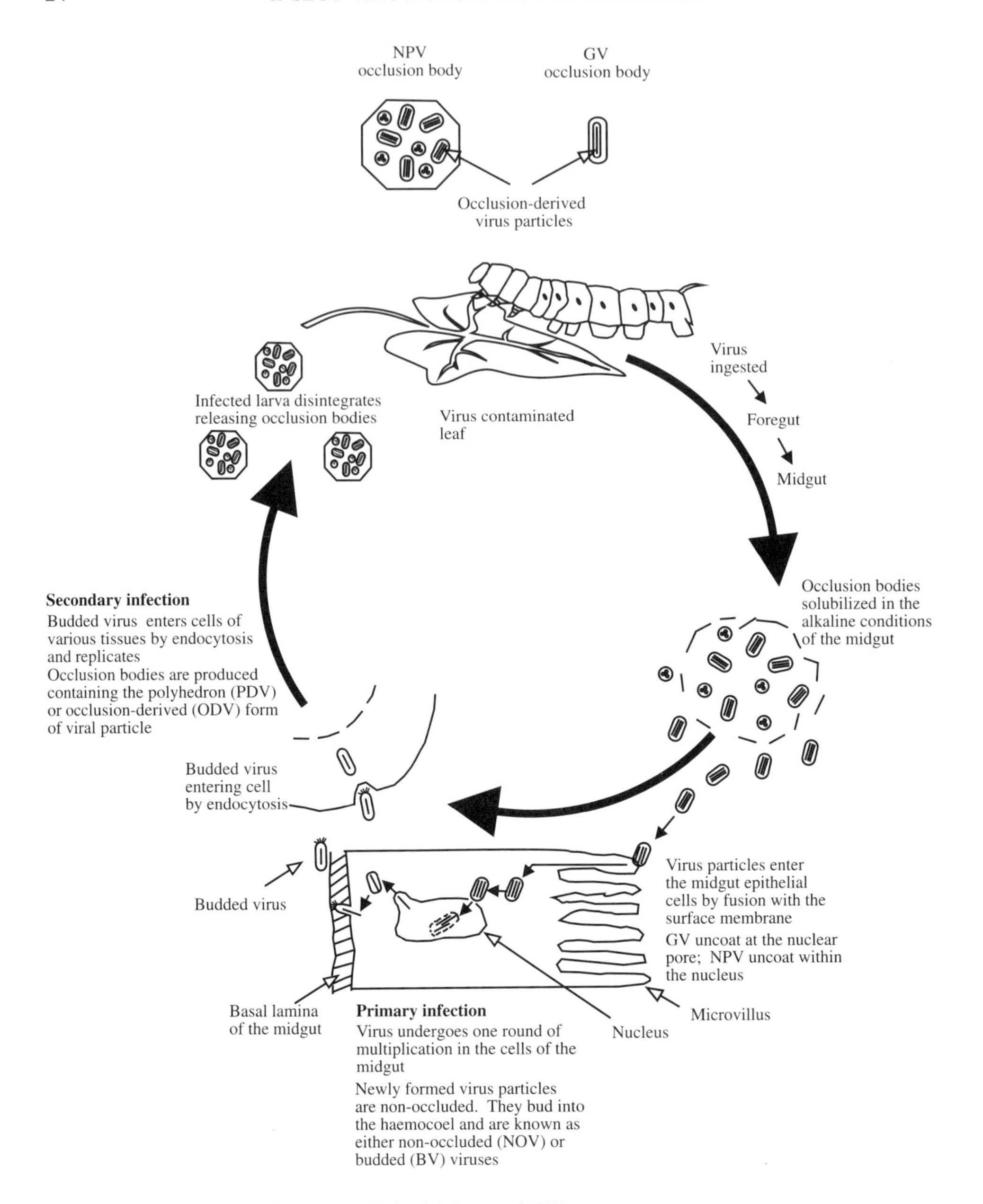

Figure 2.8 The main features of the biology of BVs.

to have a control agent that is highly specific for a pest and has no detrimental effects on any other organisms. However, there may be commercial disadvantages to highly specific pesticides. Firstly, if several pests occur together on a crop, it would be more expensive to have to use different control agents for each pest rather than one broad-spectrum pesticide. Secondly, producing a different product for every different pest would require a large number of different products to be

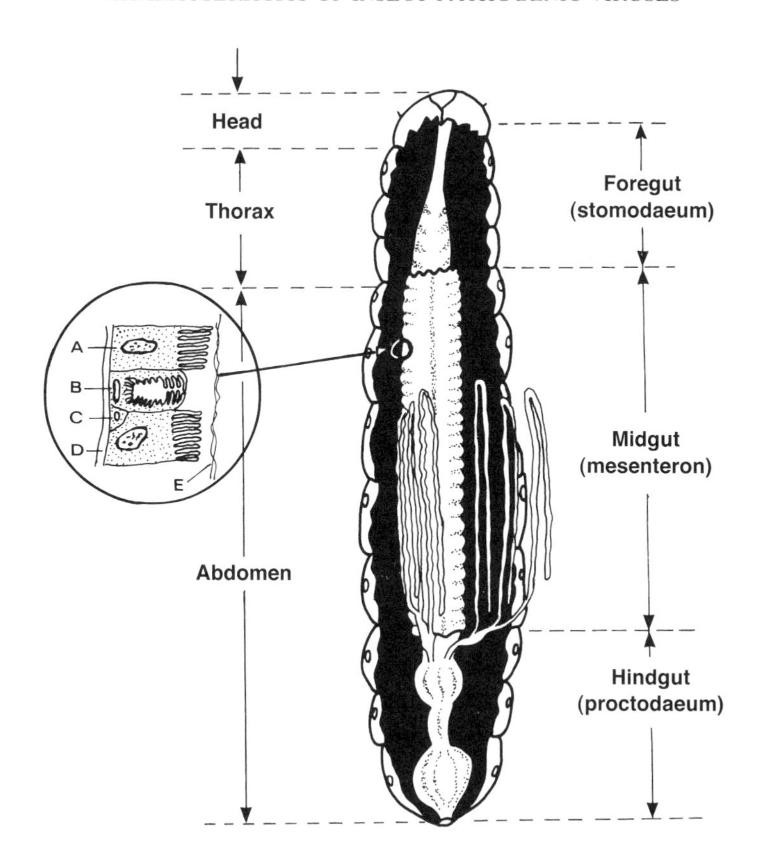

Figure 2.9 Stylized lepidopterous larva dissected from the dorsal side to show the alimentary canal. For clarity, the salivary glands, fat body, tracheal system and heart have been omitted. The malpighian tubes on the right-hand side have been 'exploded' but are normally closely apposed to the midgut. The inset shows the basic histology of midgut: A, columnar cell; B, goblet cell; C, interstitial cell; D, basement lamina; E, peritrophic membrane. The brush border is especially pronounced on the columnar cells.

available. An ideal virus would, therefore, have a host range that included several pest species, particularly species which occurred together but which did not infect any non-target insects. Most viruses have not been extensively tested to determine their host range, particularly on non-pest species, so the true host range of most viruses is not known.

REFERENCES

Amargier, A., Lyon J.-P., Vago, C., Meynardier, G. and Veyrunes, J.-C. (1979) Mise en évidence et purification d'un virus dans la prolifération monstreuses glandulaire d'Insectes. Etude sur *Merodon equestris* F. (Diptère, Syrphidae). *Compte Rendu Academiè Scientifique* (Paris) **289D**, 481–484.

Evans, H.F. and Entwistle, P.F. (1987) Viral diseases. In *Epizootiology of Insect Diseases*, J.R. Fuxa and Y. Tanada (eds). New York: Wiley, pp. 257–322.

Gouteux, J.-P. (1987) Prevalence of enlarged salivary glands in *Glossina palpalis*, *G. pallicera* and *G. nigrofusca* (Diptera: Glossinidae). From the Vavoua area, Ivory Coast. *Journal of Medical Entomology* **24**, 268.

Hull, R., Brown, F. and Payne, C. (1989) *Virology, Directory & Dictionary of Animal, Bacterial and Plant Viruses*. London: Macmillan Press, p. 325.

Jura, W.G.Z.O., Odhiambo, T.R., Otieno, L.H. and Tabu, N.O. (1988) Gonadal lesions in virus-infected male and female tsetse, *Glossina pallidipes* (Diptera: Glossinidae). *Journal of Invertebrate Pathology* **52**, 1–8.

Murphy, F.A., Fauquet, C.M., Bishop, D.H.L., Ghabrial, S.A., Jarvis, A.W., Martelli, G.P., Mayo, M.A. and Summers, M.D. (eds) (1995) *Virus Taxonomy, Classification and Nomenclature of Viruses. Sixth Report of the International Committee on Taxonomy of Viruses*. Vienna: Springer-Verlag, p. 586.

Odindo, M., Payne, C.C. and Crook, N. (1984) A novel virus isolated from the tsetse fly *Glossina pallidipes*. *Abstracts of the 6th International Congress of Virology*, Japan: Sendai, p. 336.

Payne, C.C. (1986) Insect pathogenic viruses as pest control agents. In *Biological Plant and Health Protection*, J.M. Franz (ed.). Stuttgart: Fischer Verlag, pp. 183–200.

3 Assessment of biological activity

One of the fundamental requirements in studying and using BVs is the need to assess the infectivity of the preparations against both target and non-target organisms. At the most basic level, this is simply the application of a known dosage to a given population of the test organisms and scoring of response. However, there are many variables that have to be taken into account in designing and, particularly, in interpreting the responses of populations of organisms to challenge by the pathogen.

It is a fortunate characteristic of most BVs that the virions are occluded within inclusion bodies, thus enabling them to be counted accurately and making it a relatively straightforward task to apportion dosages appropriate to the test organisms. Bioassays of occluded BVs are, therefore, all based on delivery of known concentrations of OBs to representative numbers of the test organisms (Hughes and Wood, 1987).

Consideration also needs to be given to the purpose of the assay, which, in turn, will influence the methods used to analyse the dosage-response results. There are fundamental differences between the modes of action of pathogens, including BVs, and chemical pesticides, but most of the methods of analysis have been developed to use data from the latter category. Probit analysis (Finney, 1971), the most commonly used statistical method for dosage-response work, applies to situations where the effects of increasing dosage are cumulative, implying a threshold over which the test organism is certain to die. For viruses, however, it is assumed that each individual virus particle is capable of inducing infection, independently of any others that might be present. This theory of independent action has been discussed by Ridout, Fenlon and Hughes (1993) who point out that the increases in responses to increasing dosage do not result from additive effects within the organism but from increased probability of one or more virus particles successfully passing through the various natural barriers to infection (the peritrophic membrane, alkaline midgut milieu, enzyme action, etc.) and also the distribution of dosages between droplets. On this basis it is, theoretically, not possible to have dosage–mortality responses with slopes greater than 2.0, even when there is very tight control over insect variability (usually assessed by weight) and delivery of dosage (such as the droplet feeding method of Hughes and co-workers (Hughes and Wood, 1981; Hughes *et al.* 1986)). There are a number of instances in the literature of dosage–mortality relationships with slopes over 2.0, but it is possible that these reflect intrinsic variability.

QUANTAL RESPONSES

Quantal responses can be measured in a number of ways depending on the purpose of the assay. Death of the organism, with appropriate confirmation that it has died from the virus, is the most commonly used response. However, it may not always be easy to determine precisely when the insects have died, in which case other symptoms such as general appearance, responses to external stimuli (prodding etc.) or rate of food consumption can be used.

Where precise dosage ingestion is determined, the endpoint is expressed as the median lethal, infective or effective dose (LD_{50}, ID_{50} or ED_{50}, respectively) and where it is not possible to determine the precise dosage ingested the endpoint is the median lethal, infective or effective concentration (LC_{50}, IC_{50} EC_{50}, respectively).

Another property of importance in determining the suitability of a candidate virus for development as a biopesticide is the time it takes to kill its host. The method used for investigating this property is essentially the same as that for estimating the amount of virus required to kill the insect. Once virus has been administered, larvae must be monitored daily until the first larva has died, and then at 8-hourly intervals, or less, until all larvae are dead. Strictly speaking, median survival times (ST_{50}, also known as median lethal times, LT_{50}) should be calculated from life tables, but many of the published data have been derived by probit analysis.

TRENDS IN POPULATION RESPONSES TO BACULOVIRUSES

The literature abounds with examples of bioassays of insect responses to challenge by BVs. Regardless of the methods used to deliver the dosages, there is a general trend in expected response of test insects (usually the larval stages) at various stages in their development (Hughes and Wood, 1987). In particular, susceptibility to virus decreases with increasing larval age, usually measured as weight. In lepidopterous larvae, this trend can be quite dramatic (Evans, 1986), with interesting differences between the responses of larvae to NPV relative to GV. In general, older (larger) larvae are significantly less susceptible than younger larvae, this difference often spanning many orders of magnitude. For example, the LD_{50} for cabbage moth, *Mamestra brassicae*, larvae ranges from 7 OBs in the first instar to around 240 000 OBs in the fifth instar (Evans, 1981), while sixth instar larvae are virtually resistant to infection (Evans, 1983). In all cases, the LD_{50} was related to larval weight, although this did not fully explain the variability with age. Methods of bioassay must take account of these age-related changes in responses, particularly if the purpose is to attempt to define the dosage requirements for field application of BVs (Evans, 1994). It is also difficult to determine the precise stage of development of test insects, especially the length of each instar and the time at which ecdysis takes place. However, since the main changes in insect size take place through the process of moulting, with the head capsule width remaining constant within each instar, it is possible to measure head capsule size of selected individuals on a daily basis and to correlate these measurements with the known distribution of head capsule widths for the population of insects being tested. There is a reasonably precise linear relationship between larval instar and head capsule width when the measurements are plotted on a logarithmic scale. This relationship, known as Dyar's rule, must be determined on a sufficiently large sample of the population to differentiate clearly the peaks characteristic of each instar. Selection of insects for assay, therefore, involves determination of both the instar and the weight within that instar, weight being the main determinant of response.

VARIABILITY BETWEEN ASSAYS

Variability in carrying out bioassays is an inherent feature of both the test organisms and of the methodology employed to carry out the tests. The methods themselves can be designed to reduce this variability and may, technically, be very precise in maintaining reproducible conditions. However, there is also an element of 'learning' in carrying out bioassays and it may, and indeed should, be necessary to carry out a number of assays to ensure that operator experience is optimized. This has been investigated by J. S. Fenlon (personal communication), who assessed the reproducibility of bioassays of the GV of the diamondback moth, *Plutella xylostella,* carried out over a period of 2 years. Twenty assays were carried out over the assessment period and Fenlon examined the way in which variability changed from the first year, when assays were carried out infrequently and at irregular intervals, to the second year when a series of 11 assays was carried out over a short time period. The assays in the first year gave very variable results, with logarithmic LD_{50} values ranging from 4.07 to 7.5. The second year results were much less variable (mean LD_{50} was 6.1, with much greater control over sample standard deviation and probit standard error). The main conclusion from the study was that repetition of assay procedures over short time periods results in improved consistency in operator performance, particularly in reducing probit standard errors. If such repetition can be carried out the range of dosages can be tailored to span the LD_{50} without loss of information at the 0% and 100% quantal responses. There were also important lessons in determining the range of dosages to use within a given assay.

THE USE OF BIOASSAY DATA

Information on dosage–mortality relationships is essential to the understanding of the processes of infection during the study of BVs. They have value in assessing susceptibility *per se* and are also essential in quality control and assessment of infectivity against standards. However, translation of the information to field performance is difficult because so many biotic and abiotic variables have to be taken into account. This has been discussed by Evans (1994), who concluded that laboratory studies have value in development of microbial control programmes, particularly in initial screening and in development of effective *in vivo* mass production systems. However, he stressed the need to gather effective field information on performance of BVs within an integrated approach to pest management, which he termed the **control window**. Within this concept, laboratory assay information can prove very valuable in assessing field performance against known criteria.

REFERENCES

Evans, H.F. (1981) Quantitative assessment of the relationships between dosage and response of the nuclear polyhedrosis virus of *Mamestra brassicae*. *Journal of Invertebrate Pathology* **37**, 101–109.

Evans, H.F. (1983) The influence of larval maturation on responses of *Mamestra brassicae* L. (Lepidoptera: Noctuidae) to nuclear polyhedrosis virus infection. *Archives of Virology* **75**, 163–170.

Evans, H.F. (1986) Ecology and epizootiology of baculoviruses. In *Biology of Baculoviruses* Vol. 2. *Practical Application for Insect Pest Control*, R.R. Granados and B.A. Federici (eds). Boca Raton, FL: CRC Press, pp. 89–132.

Evans, H.F. (1994) Laboratory and field results with viruses for the control of insects. In *BCPC Monograph No 59: Comparing Glasshouse and Field Pesticide Performance II*, H.G. Hewitt, J. Caseley, L.G. Copping, B.T. Grayson and D. Tyson (eds). Farnham, UK: British Crop Protection Council, pp. 285–296.

Finney, D.J. (1971) *Probit Analysis*. London: Cambridge University Press.

Hughes, P.R. and Wood, H.A. (1981) A synchronous peroral technique for the bioassay of insect viruses. *Journal of Invertebrate Pathology* **37**, 154–159.

Hughes, P.R. and Wood, H.A. (1987) *In vivo* and *in vitro* bioassay methods for baculoviruses. In *The Biology of Baculoviruses:* Vol. II. *Practical Application for Insect Control*, R.R. Granados and B.A. Federici (eds). Boca Raton, FL: CRC Press, pp. 1–30.

Hughes, P.R., Beek, N.A.M., Wood, H.A. and van Beek, N.A.M. (1986). A modified droplet feeding method for rapid assay of *Bacillus thuringiensis* and baculoviruses in noctuid larvae. *Journal of Invertebrate Pathology* **48**, 187–192.

Ridout, M.S., Fenlon, J.S. and Hughes, P.R. (1993) A generalized one-hit model for bioassays of insect viruses. *Biometrics* **49**, 1136–1141.

4 The ecology of baculoviruses in insect hosts

The purpose of this chapter is to provide an account of BV ecology and epidemiology as a basis for understanding the use of viruses for insect pest control. As a discrete subject, it cannot be said that BV ecology has yet been adequately reviewed though the following are some useful contributions: Evans and Harrap (1982), Jaques (1985), Entwistle and Evans (1985), Entwistle (1986), Fuxa and Tanada (1987) and Fuxa (1995). Although not considered to be a BV, the virus infecting *Oryctes rhinoceros* (OrV) is included in this section.

BACULOVIRUS STRUCTURE AND ECOLOGICAL ASPECTS OF FUNCTION

In ecological and, particularly, pest-control terms, the primary ecological characteristics of occluded viruses are the production of very large quantities of OBs and the numerical relationships between these OBs and their potential to induce infection in susceptible life stages. It is the combination of inherent safety and the ability to visualize and count OBs for use in application programmes that has made the BVs the most successful viral agents in pest management. In addition, survival of OBs between generations is a prime evolutionary characteristic of the BVs, thus enabling them to endure periods when the primary hosts are not available and to induce and maintain infection in target host populations.

When an infected insect dies, virus in all stages of development is gradually released and enters the environment. This may also occur before death in the case of insects where the BV infection is localized to the midgut, e.g. in NPVs in sawflies (Hymenoptera, Symphyta) and in GV of *Harrisina metallica* (*brillians*) (Lepidoptera, Zygaenidae) (Federici and Stern, 1990). These stages may include naked DNA, nucleocapsids, virions, virions within OB protein masses and, eventually, the full OB complete with its envelope. However, the environmental fate of DNA, nucleocapsids and virions has only been investigated superficially. Virions probably have only short environmental persistence, as has been indicated in risk assessment field studies on polyhedrin gene-deletion NPVs (Bishop *et al.*, 1988; Wood and Hughes, 1995). Virions liberated from OBs can initiate infections following ingestion (van Beek *et al.*, 1987) though the infectivity of these may be very low. DNA *per se* is known to have the potential for prolonged environmental survival. For instance, bacterial DNA persisted well in synthetic simplified 'soil' systems from where it could be acquired by bacteria (Lorenz and Wackernagel, 1988). However, the primary structure of DNA may be unstable (Lindahl, 1993) and in its free form it is unlikely to have any environmental significance.

ENVELOPED AND NON-ENVELOPED OCCLUSION BODIES

Populations of mature NPVs are often a mixture of enveloped and non-enveloped OBs. The latter can be distinguished easily by scanning electron microscopy, which reveals surface pits in the polyhedrin matrix formed to accommodate superficially placed virions. Separation of these two groups can be effected by zonal electrophoresis because the 'pitted' variety carry a greater charge (Small, 1986).

Entry to the gut of potential host species is, in the great majority of cases, the first step in the infection process and it is, therefore, important that the effects of enzyme and gut pH activities are understood to help to determine the likelihood of infection. The nature of the OB envelope and its properties requires fuller investigation, although a few basic details are known. On the basis of spectrophotometric and histochemical staining, the envelope of *Heliothis virescens* NPV has been shown to be mainly carbohydrate (Minion, Coons and Broome, 1979) and for *Trichoplusia ni* NPV it is suggested that it is more heavily glycosylated than is the polyhedral matrix protein (Small, 1985). Nucleocapsid envelopes are also glycosylated (Dobos and Cochran, 1980; Kelly and Lescott, 1983; Stiles and Wood, 1983; Russell and Consigli, 1985).

The possibility of a differential response of envelope and matrix proteins to proteases was first suggested by Harrap (1969), who showed that pronase (pH not stated) digested the matrix but not the envelope in thin sections of *Aglais urticae* and *Lymantria dispar* NPVs in infected tissues embedded in methacrylate. This response was confirmed in studies of *Chrysodeixis* (*Pseudoplusia*) *includens* NPV (embedded in eponaraldyte following glutaraldehyde and OsO_4 fixation) (Gipson and Scott, 1975) and *T. ni* NPV (Small, 1985). However, it appears that gut proteases can have an effect on the polyhedral envelope, at least of *Autographa californica* MNPV leading to its rupture and hence to the release of virions. When the gut fluid was heated to 60°C, only the matrix protein dissolved, presumably a non-enzymatic effect mediated by high pH, suggesting that enzymes cannot be more than marginally involved in envelope disruption (Prichett, Young and Yerian, 1982). Proteases that attacked the matrix protein but not the envelope also seem to be inactive against virion envelopes (Harrap, 1969; Gipson and Scott, 1975). The envelopes are required for attachment to midgut cell microvilli in the first step of the cellular phase of the infection process.

Therefore, in general, the characteristics of the OB envelope probably adequately explain why BVs are largely unaffected by passage through the gut of both predatory invertebrates and vertebrates (reviewed by Entwistle and Evans (1985) and see below). In such circumstances, however, the fate of immature OBs lacking the envelope has not been investigated. The pH of predator gut varies from acid to neutral, which is conducive to maintenance of OB integrity. The tolerance of occluded BVs to microbial action has been formally demonstrated (Jaques and Houston, 1969) and helps to explain why these viruses can persist for so long in the soil despite the presence of microbial proteases. Again the fate of OBs lacking the envelope has not been resolved.

NPV preparations from living insect hosts seem less potent than those from dead hosts but no explanations have been offered (Ignoffo and Shapiro, 1978). The difference, however, may be linked to the poorer stability of non-enveloped NPVs, which may be in greater proportion in the first type of preparation. In addition, the numbers of virions in 'unfinished' OBs may be smaller.

BACULOVIRUS REPLICATION STRATEGIES: INFLUENCE ON INFECTION

Pathways and patterns of infection growth

The replication strategy of a virus can have a strong bearing on the disease transmission process and on the pattern of infection growth within each host generation. However, regardless of the tissues ultimately infected within the host, the initial routes of infection are similar, being influenced mainly by the requirement for susceptible individuals to ingest the virus. The significance of latent infections and their possible expression remains a rather indefinite area, the relevance of which to ecology or to control strategy has not yet become clear. Latent infections of insect viruses were reviewed by Podgwaite and Mazzone (1986).

Oral infection

Ingestion of virus is the normal route of entry for the great majority of BVs. The question of what constitutes an infective dose and the manner in which this is made available in the insect gut has been reviewed by Hughes and Wood (1981) and is discussed more fully in Chapter 3. However, once virus is ingested, the process of infection begins, with some or all of the replication taking place in the midgut secretory cells. It is, therefore, essential to know where an insect acquires inoculum in lethal quantities and how long it takes to achieve this dosage. In many cases, this will be a matter of assessing the site and rate of feeding for each developmental stage of the insect, usually the younger stages, and also determining any characteristics that may alter the apparent observed rate of feeding. This is not always immediately clear from observation of feeding itself. There is a growing body of information indicating very substantial acquisition of virus from sites that cannot be appreciably involved in feeding. For instance, during the first two larval instars, pine beauty moth, *Panolis flammea* (Lepidoptera: Noctuidae), eats small holes in new needles of pines. This activity occupies about 20% of each day, the remainder being passed in resting on, or wandering over, older and inedible foliage (Ballard, 1987). However, in experiments when only old foliage was treated with NPV, about 50% of larvae became infected compared with about 75% when all foliage was treated (Cory and Entwistle, 1990). *L. dispar* (Lepidoptera: Lymantriidae) larvae are tree foliage feeders but can acquire NPV infections by exposure to contaminated bark (Woods, Elkinson and Podgwaite, 1989) or even soil (Weseloh and Andrews, 1986). Acquisition from the soil is probably common in the larvae of cutworms such as *Agrotis segetum* (Lepidoptera: Noctuidae) (de Oliveira 1988; Bourner 1993). Where the egg of codling moth, *Cydia pomonella* (Lepidoptera: Tortricidae), is not laid directly on a fruit, the neonate larva takes small sporadic bites in adjacent leaves and so can acquire sprayed GV before entering a fruit. Probably such 'environmental quality testing' is common in insects, representing an aspect of food location. The extent of the behaviour, especially during the earlier more susceptible larval stages, clearly has practical importance in determining dosage rate and its ideal distribution.

Virus may also be acquired directly during emergence from the egg (Martignoni, 1962). The source of the virus on the egg surface may be the adult insect, as in sawflies, or may be general environmental contamination especially by water movement, e.g. the egg chorion of various moths may accumulate lethal dosages of BV: *Malacosoma fragile* (Clark, 1955, 1956, 1958), *L. dispar* (Doane, 1969) and *Lymantria ninayi* (Entwistle, 1986).

Infection by parasites

Insects that are parasitic, mainly Hymenoptera and Diptera (in the family Tachinidae), are involved in the dispersal of BVs but probably only the former group inoculates virus directly into the host haemocoel. This is because tachinid oviposition does not involve piercing the host, whilst in the endoparasitic Hymenoptera insertion of an ovipositor through the cuticle is the normal method of oviposition.

It is assumed that, because occluded BVs are not infective when introduced directly into the haemocoel of the host, parasitoids must carry other forms of the virus, notably virions and possibly nucleocapsids. After oviposition in an infected host larva, a parasitoid may be able to transfer infection to two or three subsequent healthy hosts before its load of virus is exhausted (Thompson and Steinhaus, 1950). However, it seems possible that parasitoids 'vector' BVs by external contamination, as well as by the 'flying needle' effect (Young and Yearian, 1990).

The interrelationships between hymenopterous parasitoids and BV-infected hosts are complex. For instance, there is evidence that some species may avoid infected hosts and that the LD_{50} for a parasitized host may be higher than normal. These relationships have been reviewed by Entwistle (1982) and Kaya (1982). The impact of polydnaviruses (Chapter 2), which are carried by some parasitic Hymenoptera (notably species in the genera *Apanteles* and the closely related *Cotesia*), on the host is interesting. Evidence exists for two effects. Firstly, polydnavirus can affect host susceptibility to BV infection and, secondly, host growth can be retarded. A possible consequence of such growth effects is the potential for the host to remain susceptible for longer.

Transovarial infection

Consideration of the potential for transovarial transmission (i.e. within the egg) of BVs is basically a question of whether BVs can exist as latent or transmitted sublethal infections. Attempts to demonstrate the latent passage of virus infections, especially where, like BVs, the virus has considerable extra-host stability will always be criticized on the grounds that the results could have resulted from contamination. Despite the difficulty of proving the absence of contamination, some good quality evidence of latency exists. For instance, Kelly *et al.* (1981) showed latent NPV in *Heliothis* cell cultures and Skuratovskaya *et al.* (1984) of *Bombyx mori* and *Galleria mellonella* NPVs in cell systems. *In vivo* demonstration of latency has usually been by challenging feeding caterpillars with heterologous BVs in order to 'stress out' latent infections. In such studies, for instance, serological methods were used to show that homologous NPV infections could be induced in *Spodoptera* spp. (McKinley *et al.*, 1981). Restriction endonuclease (REN) techniques were used to show clearly that cross-feeding of NPVs between *Mamestra brassicae* and *Adoxophyes orana* resulted solely in infection by homologous NPVs (Jurkovicova, 1979). Many earlier studies stressed the value of chemical and even inert (e.g. quartz) stressors capable of breaking BV latency (Krieg, 1956; Nuorteva, 1972). This approach has been unfashionable for some years and its relevance either to fundamental epizootiology or to pest control is not clear. Indeed, the evolutionary significance of latency may be low because, without overt expression of the virus at some stage in the life cycle, there is a high risk of the 'carrier' host dying from other mortality factors and, thence, preventing further replication of the virus.

SITES OF VIRUS REPLICATION AND THE DYNAMICS OF VIRUS TRANSMISSION

Biologically, BVs can be divided into those which multiply exclusively in midgut (endodermal) cells and those which multiply mainly or only in cells of mesodermal and ectodermal embryonic origin (Table 4.1). NPVs in Lepidoptera fall into the latter category but in sawflies (Hymenoptera, Symphyta) replication of NPVs is associated only with midgut cells. The biology of sawfly NPVs differs further in that adults carry and disperse virus. The virus of the rhinoceros beetle, *O. rhinoceros*, also replicates in both larval and adult stages from which transmission can similarly occur. GVs are known only from Lepidoptera where they have a pattern of replication similar to NPVs with the exception of one species, *Harrisina metallica*, where replication seems to occur purely in larval and adult midgut (Federici and Stern, 1990).

Gut infections

Where replication is gut-associated, virus can be excreted before host death so that horizontal disease spread can be rapid. Virus is released both during defecation and in vomiting, whether as a defence or as a result of the infection itself.

Table 4.1 Replication strategies of BVs and OrV in various life stages of their hosts

Virus group	Host taxon	Host life stage			
		Larva	Prepupa	Pupa	Adult
NPV	Lepidoptera	Yes[a]	Yes	Yes	No
	Hymenoptera (sawflies)	Yes[b]	No	Yes[c]	Yes[c]
	Diptera (mosquitoes)	Yes[c]	No	No	?
GV	Lepidoptera (most species)	Yes[a]	?	?	No
	Lepidoptera (*Harrisina metallica*)	Yes[b]	?	?	Yes[b]
OrV	Coleoptera (*Oryctes rhinoceros*)	Yes[c]	No	no	Yes[b]

[a] Minor replicative phase in midgut; significance in horizontal transmission unknown.
[b] Replication exclusively in midgut.
[c] Replication in all/most tissues.

Yields of virus tend to be relatively low compared with 'whole body' infections but this is offset by the more rapid release of secondary inoculum and the generally lower LD_{50} values in the sawflies. The form of infection curve associated with NPV disease in sawfly populations is typical of one where the release of secondary inoculum is rapid. This is illustrated in Figure 4.1 which shows the infection growth-pattern associated with the NPV disease of the European spruce sawfly, *Gilpinia hercyniae*. The rapid rise of instantaneous infection reflects the release of viable inoculum soon after replication commences, which is, thus, available to the remaining population while it is still young and highly susceptible. In addition, the range of LD_{50} is only of the order of around 500 times from the most to the least susceptible larval stages.

Another aspect of replication strategy important for transmission is where it occurs in adult gut so that it may be transmitted sexually (e.g. *Oryctes* spp.) or by surface contamination of eggs where it may be acquired by larvae during hatching. In *Oryctes*, it is possible that the activity of some virus may persist long enough to infect hatching larvae when acquired either from the egg surface or from the surrounding breeding site material. Transmission of this type is described as vertical and, because the virus is on the egg surface, as transovum.

'Whole body' infection

Where replication is essentially associated with tissues other than the gut, horizontal transmission is only possible once virus has been released from the dead host (except by intervention by parasites and predators). This is typical of both NPV and most GV infections in Lepidoptera. The slow release of secondary inoculum may be partially compensated for by the very large yields of OBs typical of even the smallest larval stages (p. 39). Infection curves, therefore, are typically bimodal, reflecting the considerable time required for a full replication cycle and release of secondary inoculum on host death (Figure 4.1). A further factor is the generally greatly decreased susceptibility of late-stage larvae such that infectivity factors of up to 250 000 have been recorded between least and most susceptible life stages (Evans, 1981).

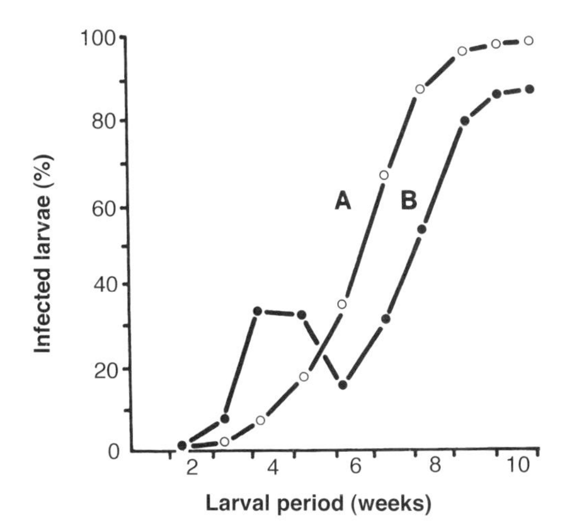

Figure 4.1 The natural growth of BV infections in a sawfly (based on *Gilpinia hercyniae*) (A) and a lepidopteran (based on *Panolis flammea*) (B). A. Infection is monophasic because of continuous release of virus inoculum by infected larvae. B. Infection growth is biphasic because inoculum is not released into the environment until larvae die and break down. The data employed are for natural epizootics. They represent 'instant' infection (see text) and the values are not cumulative but depict a progressive change in the status of infection within a single host larval generation.

BACULOVIRUS REPLICATION STRATEGIES

The period from initial infection to host death, the incubation period, is dependent principally on the combination of dosage and temperature, but it also varies with the virus type and the host in which the virus is replicating. Recent work on the molecular basis of infection has shown that BVs carry the enzyme UDP-glucosyl transferase (*egt*), which prevents normal hormone control of moulting and, effectively, prolongs the life of the infected host (O'Reilly and Miller, 1989). Such a strategy increases the yield of virus and is, presumably, of evolutionary advantage to the virus. Indeed, deletion of *egt* has been used as a strategy to speed up the rate of kill of BVs for use in microbial control programmes (O'Reilly and Miller, 1991). Most NPVs of Lepidoptera have relatively short incubation periods so that death may occur in the larval instar following infection. With some GVs and low doses of NPV, however, the incubation period is prolonged and the host may even be abnormally large at death; during this period, feeding may continue. Prolonged incubation in a host that is large at death results in a large amount of inoculum per host individual and, by late release of virus, may mean that the inoculum has a shorter period during which it would have to withstand extra-host attritional factors.

Infected larvae of Lepidoptera and Hymenoptera commonly die at the feeding site, where inoculum may be later released to contaminate food that is readily accessible to other susceptible larvae. However, where infection is acquired in the final feeding instar, death in Lepidoptera may be delayed until the prepupal or pupal stages. As pupation is not necessarily at the feeding site, inoculum may be eventually released elsewhere, e.g. in the soil or on the trunks of trees. In sawflies, late infection may not be lethal. At metamorphosis to the prepupal (eonymphal) stage, the secretory midgut epithelium is sloughed off and replaced by a stratified epithelium of small undifferentiated and apparently non-susceptible cells. Sawflies may enter an overwinter resting period, or even diapause for several years in this condition. During the subsequent, relatively brief, pupal period the secretory midgut epithelium regenerates and may become infected, perhaps as a result of inoculum persisting in the gut lumen during the eonymphal period (Bird, 1953). The resultant adults will also be infected but as they are not necessarily affected in any other way (weight, longevity, fecundity) they may effectively disperse inoculum to oviposition sites, as described above. The source of adult infection in *Oryctes* is different as virus does not persist through the pupal stage from infected larvae but must be acquired by the adult either orally from contaminated feeding sites or sexually.

The existence of gut-infecting GV in a gregarious lepidopteran (*H. metallica)* is interesting and suggests that similar patterns of infection and, therefore, of horizontal transmission may be found in other gregarious species, e.g. NPVs in the nesting *A. urticae, Euproctis chrysorrhoea* and *Malacosoma* spp.

EFFECTS OF TEMPERATURE ON DEVELOPMENT OF INFECTION

As host development time decreases with rising temperature, so the incubation period of BV diseases may decrease. However, in the Pine sawfly, *Neodiprion sertifer*, as temperature rises (studied between 12 and 24°C) the NPV incubation period decreases relative to host development. At 12°C NPV LT_{50} was 42.2% of the LP_{50} (median larval period) while at 24°C it was only 20.8% of the LP_{50} (Tvermyr, 1969). Bird (1949, 1955) was the first to observe that high host developmental temperatures can inhibit BV disease expression. If larvae of the sawfly *G. hercyniae* are both infected and reared at 29.4°C there is no disease expression; however, the larvae die rapidly of NPV disease if transferred to 22.2°C or even if they are alternated between these two temperatures at 12 or 24 h intervals. It seems, however, that temperature inhibition in *G. hercyniae* is rather lower than is general. The NPVs of *T. ni* and *Helicoverpa zea* do not cease to be able to infect until 39°C while that of *Spodoptera* species can infect at temperatures as high as 46°C. When *T. ni* and *H. zea* were infected and kept for 2 days at 26.7°C and then incubated at 39°C, they died of NPV disease (Thompson, 1959). Also working with *H. zea*, Ignoffo (1966) provided some evidence of a decrease in infection with increasing incubation temperatures above 29°C but he felt that at normal cotton field temperatures (13–35°C) viral infections would not be inhibited though there might be an effect above 40°C. NPV development in third instar *Anticarsia gemmatalis* larvae began to be apparent at 32.2°C where it was 30–50% less than at lower temperatures (Boucias, Johnson and Allen, 1980). From these and similar experiments, Stairs (1978) concluded that the effect of high temperature was to render virus receptor sites ineffective, a similar conclusion to that of

Thompson (1959), who suggested that it is the mode of invasion which is affected since disease is not inhibited by high temperatures after infection is initiated. Work on the NPV of the silkworm, *B. mori*, resulted in similar findings. When inoculated at 35°C and transferred to 25°C, the longer they were kept at the higher temperature the longer they survived at the lower. Conversely, when inoculated at 25°C the longer they were kept there before transference to 35°C the sooner they died. The authors (Kobayashi, Inagaki and Kawase, 1981) concluded that 'the virus replication mechanism itself is more sensitive to high temperatures than that related to other events necessary for viral infection to be initiated'.

BACULOVIRUS YIELDS

In keeping with both the sites of replication and with the virus types themselves, there is a great deal of variation in the total yield of virus. Perhaps the most constructive generalization is to state that the yield is generally weight related so that much of the variability between larvae can be removed if yield is expressed as OBs per mg of body weight (Figure 6.1, p. 93). Entwistle and Evans (1985) have analysed published yields and found that quantities of NPVs in Lepidoptera were up to 4.3×10^7/mg but reached a maximum of 5×10^6/mg in sawfly larvae, reflecting the restriction of infection to the midgut. Yields of GV OBs were up to 2×10^7/mg in lepidopterous larvae. While such information is useful as a 'ready-reckoner' of virus yield when it is not known precisely what the true yield of a given larval stage is, the ecological significance of virus yield rests in both the total productivity and in the rate of increase of inoculum within the host. In virtually all studies of virus growth curves, the rate of increase of inoculum in the larvae is logistic, being described by a sigmoid response of $\log_{10}$ OBs against time (Figure 6.2, p. 94). The critical portion of this curve is clearly the central logarithmic growth phase where yield rises very rapidly to the peak value and then virtually stabilizes. This may be well before insect death and ecologically it is important to know when this inoculum is released and becomes available for further infection in susceptible individuals, as discussed above.

BACULOVIRUS PERSISTENCE

BVs must retain activity for considerable periods between susceptible host generations. Some account has already been given of the quantities of inoculum produced as a result of infection. A number of abiotic and biotic factors influence the persistence of this inoculum in the environment. In the main, these factors are attritional, with some factors acting more rapidly than others. The main attritional agents tending to reduce BV deposits on plants act in two different ways. Solar ultraviolet (UV) irradiation (Chapter 31) inactivates the virus (see below) leading to a true environmental loss. Rain and leaf fall remove virus to the soil, which almost certainly tends to make it less available to insects on trees. With insects on low-growing plants, however, rain splash from soil is a major source of inoculum return.

Solar irradiation

Because of the very great importance of this subject both in insect virus ecology and in the use of viruses as pest control agents (and to other taxa of entomopathic microorganisms such as protista, fungi and bacteria, including their crystalline toxins), a separate chapter (31) describes the nature of solar radiation and especially the UV component and, in general terms, its impact on nucleic acids and proteins. Chapter 25 discusses the laboratory generation of UV radiation and the means by which UV can be measured.

The literature concerning the responses of insect viruses to UV irradiation largely concerns the BVs and is especially concentrated on the NPVs. In analysis of BV reactions to irradiation, the viral response as a whole is invariably considered and the possibility that viral nucleic acids and virus-associated structural proteins may each be affected has not been approached. This is surprising since it is known that exposure to sunlight can detoxify the *B.t.* δ-endotoxin and that Raman spectroscopic analysis has demonstrated post-exposure differences in crystal protein structure (Pozsgay *et al.*, 1987). A considerable variety of approaches has been taken in the study of the extent to which UV/sunlight can adversely effect the activity of BVs. For instance, earlier studies in the laboratory (e.g. Gudauskas and Canerday, 1968; Jaques, 1968; David, 1969; Witt and Stairs, 1975; Witt and Hink, 1979) tended to employ germicidal lamps generating UV at around 254 nm, a component of the UV spectrum absent in solar radiation at the earth's surface. The use of monochromators (e.g. David, Gardiner and Woolner, 1968; Griego, Martignoni and Claycomb, 1985) permitted analysis of the relative importance of particular parts of the solar spectrum, while the advent of sunlight lamps and solar simulators introduced a means of generation of controlled 'natural' irradiation (e.g. Mubuta, 1985; Carruthers *et al.*, 1988a; Ignoffo and Garcia, 1992). In addition, other studies have been made in which BVs have been exposed to natural sunlight (e.g. Cantwell, 1967; Morris, 1971) but at various latitudes, times of year and atmospheric conditions. This wide variety of experimental conditions has resulted in many sets of data that are difficult to compare with one another. However, not withstanding this caveat, some concept of the types of inactivational response that can be expected from these various methods can be provided. In Figure 4.2a, curves of inactivation against exposure duration are shown for two NPVs to natural sunlight and for one NPV to simulated sunlight, while three NPVs exposed at 'far' UV at 254 nm and to 'near' UV at circa 280–300 nm are shown in Figure 4.2b. While differences in the actual energy levels (watts/unit area per unit time) are not taken into account in these depictions, the very great difference (>100-fold) in the speed of inactivation between exposure to the 'unrealistic' wavelength of 254 nm and exposure to real/simulated sunlight is evident and is entirely characteristic of such studies.

Various parameters for viral inactivation have been employed, for instance the proportion of OBs surviving (e.g. Witt and Hink, 1979), extrapolation from previously constructed log dose–probit mortality curves of the actual OB dose corresponding to a particular observed mortality level (Richards and Payne, 1982) or, most commonly, in terms of mortality relative to the original activity (OAR) of the unexposed virus isolate. However, largely because the dose–mortality

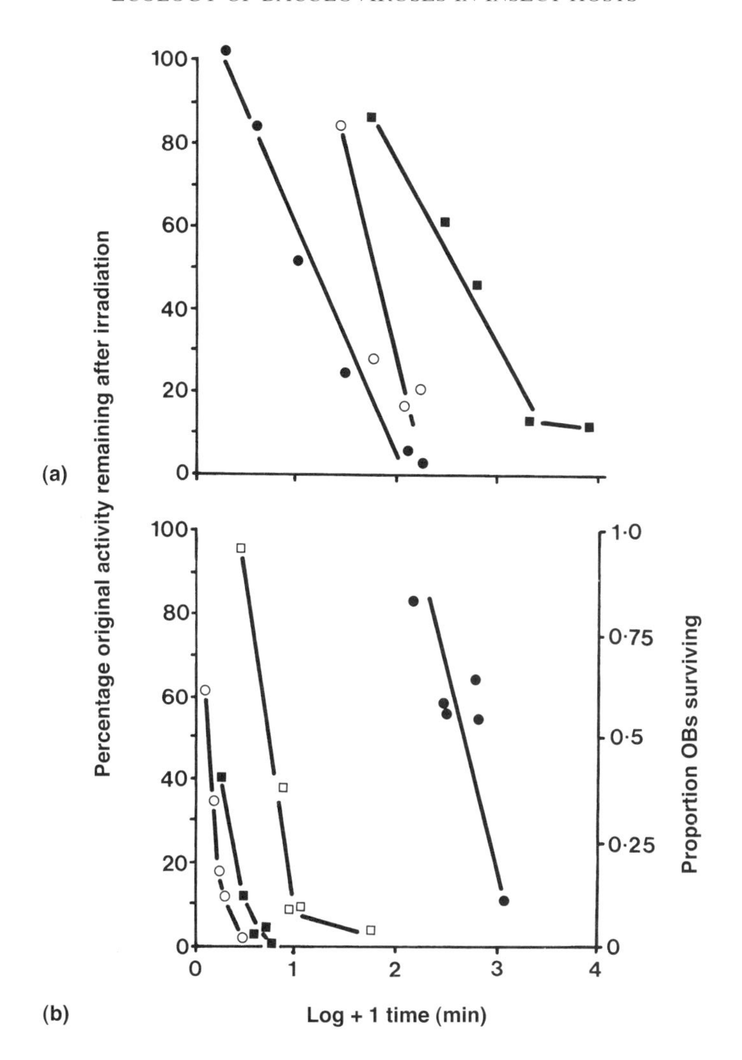

Figure 4.2 Responses of BVs to a variety of irradiational sources. (a) Responses to natural sunlight: *Lambdina fiscellaria* NPV (■) (Morris, 1971); *Trichoplusia ni* (○) (Cantwell, 1967); response to simulated sunlight: *Panolis flammea* NPV (●) (Mubuta, 1985). (b) Responses to restricted wavelengths. Response to 254 nm: AcMNPV (○); *Trichoplusia ni* MNPV(■); *Helicoverpa zea* NPV (□) (the last two, Gudauskas and Canerday, 1968). Response to near-UV 280–300 nm; AcMNPV (●). The AcMNPV response data are in terms of the proportion of OBs with surviving activity. (Reproduced from Witt and Hink, 1979. *Journal of Invertebrate Pathology*, **33**. © Academic Press.)

response relationship is sigmoid, rather than linear, the latter convention provides a misleading guide to the actual extent of virus inactivation. The proper understanding of the UV inactivational response, such as shown in Figure 4.2a, requires the use of a conventional log dose–probit analysis. This aspect of the calibration of the impact of irradiation is discussed very clearly by Richards and Payne (1982).

Unsurprisingly, the rate of sunlight inactivation varies with season, this being especially pronounced with increasing distance from the equator. For instance, the

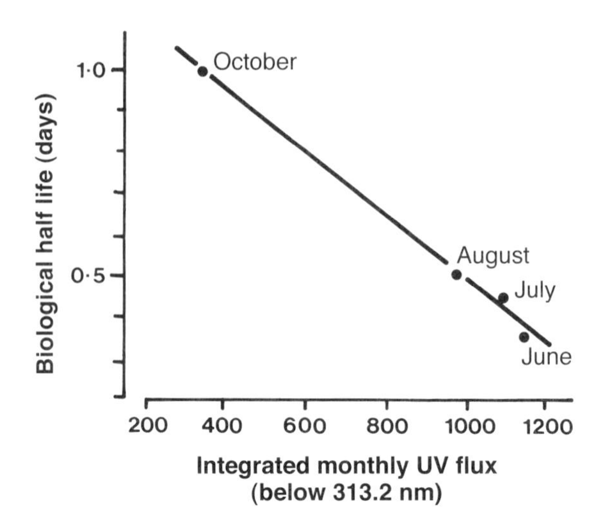

Figure 4.3 Probable relationship between biological decay of purified *Pieris brassicae* granulosis OBs on cabbage leaves and solar UV light. Half-life data from Richards and Payne (1982) and UV data for Washington extrapolated from Figure 6 in Barker (1968) (Reproduced with permission). Because of latitude and locally different O_3 levels, actual UV flux in the UK would differ from Washington, though there would be a common seasonal pattern.

biological half life of *Pieris brassicae* GV (PbGV) deposits on cabbage leaves in England differed by a factor of more than twofold between June and October and correlated well with integrated monthly UV flux data (Figure 4.3) (Richards and Payne, 1982).

At the plant level, the impact of solar UV is greatly influenced by crop architecture and aspectual position of virus on the crop. Working with the NPV of pine beauty moth, *P. flammea*, in lodgepole pine plantations in Scotland, Killick and Warden (1991) quantified UV penetration at different heights and compass aspects of the canopy and recorded large differences that they were able to correlate with the biological survival of the virus and, *ipso facto*, to demonstrate the very considerable degree of internal shading in the forest. It had already been shown (Richards, 1984) that *C. pomonella* GV (CpGV) survived better on larger than on smaller apple trees (Figure 4.4), a result that is also probably consequent on differences in internal shading. A valuable analysis of the inactivation of conidia of the entomopathic fungus *Entomophaga grylli* at different sites in a plant canopy provides corroborative evidence (Carruthers *et al*., 1988a).

However, in field studies on the impact of solar irradiation there has been an almost uniform tendency to neglect the estimation of physical loss of OB deposits over time from the test plant surfaces. Proper interpretation of observations on the decline of pathogen deposit effectiveness will depend on: the nature of plant surfaces, the influence of potentially attritional factors such as rainfall and dust excoriation and, possibly, some influence of plant surface chemistry on OB tenacity. While studying the impact of solar irradiation on PbGV on cabbages in the field over 7 days, Richards and Payne (1982) were also able to show that

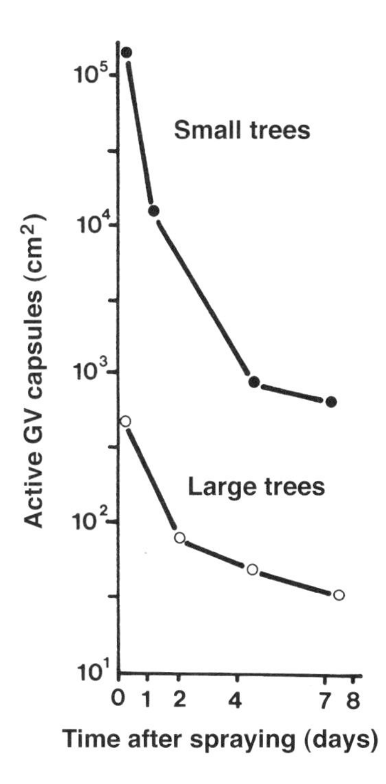

Figure 4.4 Decay in the activity of *Cydia pomonella* GV on small and large apple trees. Small trees (1.2 m) sprayed on 13 August 1981 with 10^{11} OBs/tree, maximum solar elevation 48°. Large trees (4.0 m) sprayed on 20 June 1981 with 10^{10} OBs/tree, maximum elevation 61°. (From Richards, 1984; reproduced with permission of the University of London.)

populations of GV capsules (^{32}P-labelled for the purpose) remained constant. However, over a similar period, deposits of a purified preparation of *Neodiprion sertifer* NPV were in rapid decline and had been largely lost by about 75 days: impure virus persisted better (Figure 4.5).

The subject of UV inactivation in the natural environment cannot be left without considering the possibility of interactions with other factors. Using a sunlight simulator, the interaction of UV-B with water, pH (at 3, 6 and 9) and temperature (at 10, 22, 35 and 50°C) was studied by Ignoffo and Garcia (1992). There were no interactions with either pH or temperature but when the test *H. zea* SNPV was exposed to water for 24 h it was about three times more sensitive than dry occluded virus. The possibility of reactions with plant surface materials adverse to virus activity are noted below (pp. 49–50); this could be a confusing factor in the interpretation of some UV inactivational studies.

Temperature

Interpretation of BV ecology and the use of BVs in pest control requires information on the effects of temperature in relation to the survival of activity in particular environmental zones, e.g. plant surfaces (especially the leaves), the soil and the usually very dry conditions of stored products such as grains. It is also

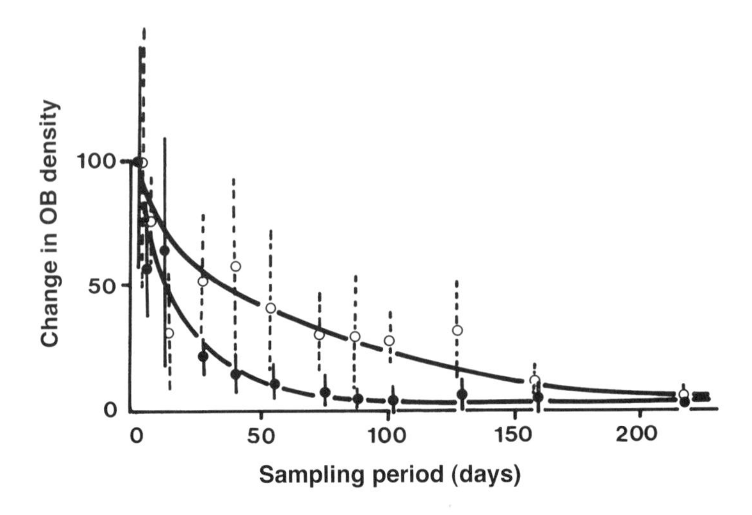

Figure 4.5 Decay curves for deposits of purified (●) and impure (○) preparations of OBs of *Neodiprion sertifer* NPV on needles of *Pinus contorta*. Note swifter decline in density of purified OBs and the greater deviation in counts of unpurified OBs. Range is SD.

desirable to understand the effect of temperature fluctuations, e.g. freezing and thawing cycles, and the interactions of temperature with other environmental variables such as humidity, pH and UV.

Temperature changes

In general, occluded BVs are little affected by cycles of freezing and thawing; for example, the infectivity of PbGV was unaltered by 10 such cycles (in the presence of 0.1% Teepol) in 12 days (David, Ellaby and Taylor, 1971c). Freezing and thawing of *Heliothis* NPV as a wettable powder SAN 240 I was 'not detrimental' (Bassard and Knutti, 1977) but *M. brassicae* NPV frozen at –20°C in soil lost activity in the first 5 days though it was then stable to day 25 of the test (Evans, Bishop and Page, 1980). However, a single freezing/thawing cycle may cause almost total activity loss of free BV virions in buffer.

Effects on isolated baculovirus virions

Fundamental as it may be to understanding the thermal relationships of BV infection there have been few investigations on the heat sensitivity of BV virions. The infectivity of an NPV of the sawfly *G. hercyniae* (GhNPV) was tested on diatomaceous filtrates that appeared free of polyhedra. Infectivity was lost after 5 min incubation at 65°C and there was an indication that the heat inactivation 50% point (HI_{50}) was at less than 60°C (Bird, 1949). Free, cell-budded NPV of *G. mellonella* in larval haemolymph was inactivated at a rate directly related to exposure above a 38°C threshold. The HI_{50} was measured following various times at 158°C (Stairs and Milligan, 1979). A method for the long-term storage of NPV virions in haemolymph was described by Vaughn (1972).

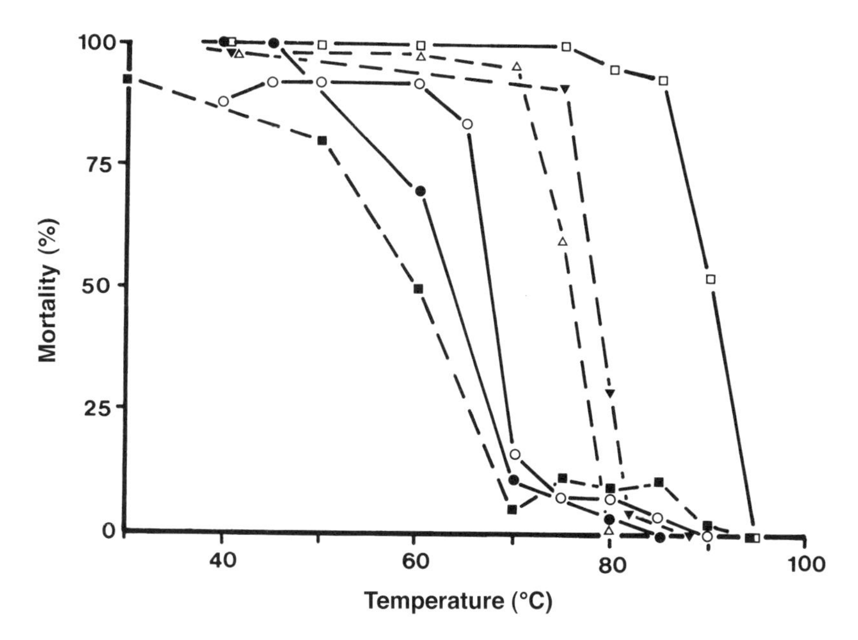

Figure 4.6 Heat inactivation curves for several NPVs in aqueous suspension for 10 min at each temperature step. △, *Helicoverpa zea* (Gudauskas and Canerday, 1968); ○, *Mythimna separata* (Manjurnath and Mathad, 1978); ●, *Spilosoma oblique* (Chaudhari and Ramakrishnam, 1988); ■, *Spodoptera littoralis* (Harpaz and Raccah, 1974); □, *Spodoptera litura* (Pawar and Ramakrishnam, 1971); ▼, *Trichoplusia ni* (Gudauskas and Canerday, 1968).

Effects on occluded baculoviruses in aqueous suspension

Most information relates to BVs in this condition. For the two NPVs of *Or. pseudotsugata* heat inactivation was a first-order reaction when the viruses were exposed to 55°C for variable times, i.e. inactivation fits a classical negative exponential equation (Martignoni and Iwai, 1977). The heat responses of the naturally non-occluded virus of citrus red mite held at 38°C did not change in 28 days but virus was destroyed in 1 h at 60°C (Reed (1974) and see Gilmore and Munger (1963) and Tashiro *et al.* (1970)). The median infectivity temperature HI_{50} appears to be very variable. For instance it was about 60°C for *Spodoptera littoralis* NPV while for *S. litura* it was 90°C. Most other BVs appear to fall between these extremes (Figure 4.6). An SNPV of *O. pseudotsugata* was inactivated at about 5°C less than an MNPV from the same host species (Martignoni and Iwai, 1977). In another experiment, the SNPV of *H. zea* was inactivated at a slightly lower temperature than the MNPV of *T. ni* (Gudauskas and Canerday, 1968). However, packing frequency of nucleocapsids within the virion envelope does not appear to contribute greatly to explaining the wide range in heat response recorded in the literature.

Information is scarce for GVs. The HI_{50} for a GV of *Mythimna* (*Pseudaletia*) *unipuncta* was just over 70°C (for 10 min), about 4°C higher than for an MNPV of the same host species (Tanada, 1959). The point of total inactivation of

PbGV was between about 66 and 70°C (David and Gardiner, 1967). Probably the lowest HI_{50} value recorded for a BV is that of the shrimp *Penaeus vannamei* with an inferable value of >50°C and <60°C (LeBlanc and Overstreet, 1991). Stairs and Milligan's (1979) data for *G. mellonella* free virions translates to an HI_{50} of 53.8°C. As this is appreciably lower than that recorded for any BV in the occluded state, it may be suggested that the OB contributes to virus heat stability.

Effects on dry occluded virus

Much of the academic work on the influence of heat on BVs concerns aqueous suspensions of OBs. Work on the persistence of dried virus tends to relate to commercial storage questions. Impure air-dried PbGV was stored at 0–3°C and at 15–28°C in the dark. During 1 year there was no alteration in activity but slight differences appeared at 2 years, while at 3 years, activity was low, though some remained to 4 years. The rate of activity loss was greatest at the higher temperature range (David and Gardiner, 1967). Such long-term studies are rare but Lewis and Rollinson (1978) reported on activity changes in air-dried *L. dispar* NPV powder over 24–25 months at three temperatures, as shown in Figure 4.7. Storage at 38°C led to an LC_{50} increase by two logs in 2 months but at 4°C there was a 0.5 log LC_{50} increase developing fairly steadily over the 24-month test period. At 4°C, lyophilization produced similar results to air-drying. Dry, dark storage of *H. zea* NPV was especially injurious at 45°C for 48 h and was equal to 92 and 85% losses over the same period at 15 and 30°C, respectively. The time–temperature

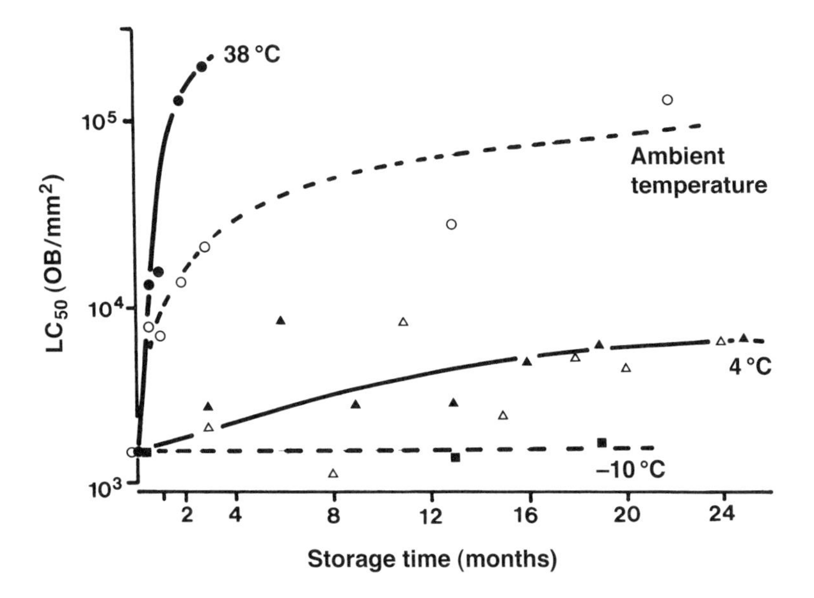

Figure 4.7 Infectivity changes in dry preparation of *Lymantria dispar*, gypsy moth, NPV during long-term storage at three temperatures: ▲—▲, lyophilized powder; all other symbols refer to an air-dried preparation. (From data in Lewis and Rollingson, 1978.)

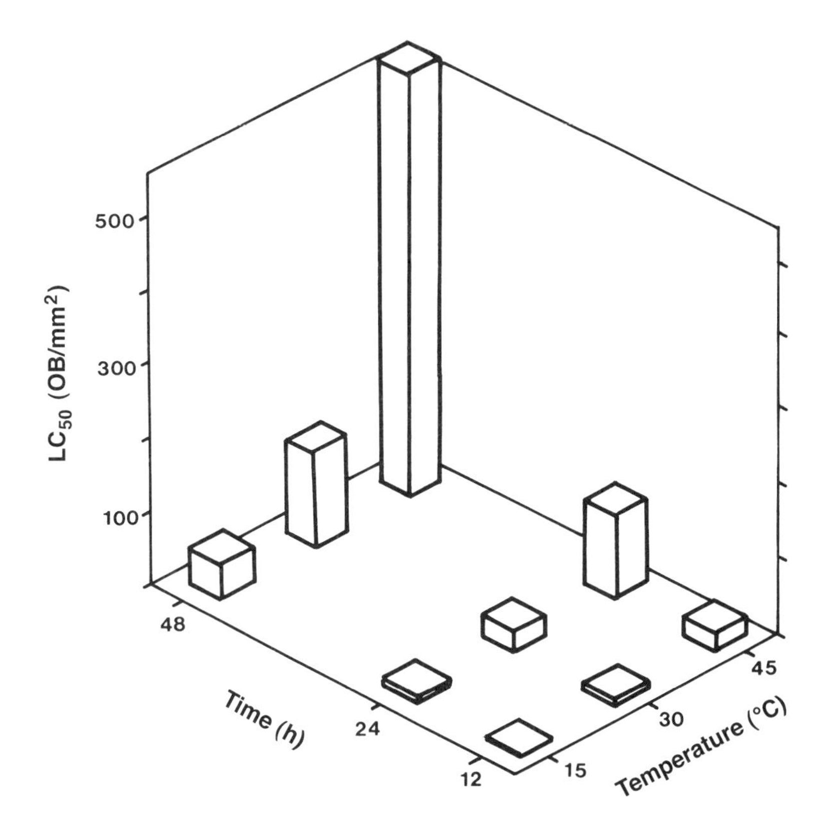

Figure 4.8 Changing activity of a dry preparation of *Helicoverpa zea* NPV exposed at three temperatures in the dark. (From data in McLeod *et al.*, 1977.)

interaction is illustrated in Figure 4.8 (McLeod, Yearian and Young, 1977). Commenting on *Heliothis* NPV formulated as a wettable powder (SAN 240 I), Bassard and Knutti (1977) advised that it be stored at 22°C or less but if 'held beyond the season, it should be refrigerated or frozen until the following season' and that its activity 'may be impaired . . . above 27°C'.

Exposing dried films of *M. separata* NPV in the dark at 50°C showed inactivation commencing between 49 and 72 h, with a very high loss at 96 h (Manjunath and Mathad, 1978).

The survival of BVs in dry stored products is of practical concern. The NPV of the almond moth, *C. cautella*, when incubated in wheat for 7, 14 and 28 days from 4 to 42°C was only seriously impaired after 14 and 28 days at about 32°C (Hunter, Hoffman and Collier, 1973) whilst the Indian meal moth, *Plodia interpunctella*, GV activity in wheat bins did not decline appreciably in the first year at –19°C to 48°C, though when held at a constant 42°C it was 90% inactivated in 42 weeks (Kinsinger and McGaughy, 1976).

A penaeid shrimp NPV was very susceptible to dry conditions, being totally inactivated by 48 h exposure to 22°C (LeBlanc and Overstreet, 1991).

Interactions of temperature with other environmental factors

The majority of accounts of environmental influence are of single factors, especially temperature and UV. However, the interaction of heat (15–28°C) and light (in a greenhouse and, therefore, likely to be deficient in UV wavelengths) was studied on crude dried PbGV. While virus stored in the dark retained considerable infectivity for 4 years, by 2 years almost all light-stored virus was inactive and inactivation was total by 3 years. Activity of GV stored in the light in a CT room at 20°C was also lost by 3 years (David and Gardiner, 1967). Using solar lamps, Manjunath and Mathad (1978) did not demonstrate UV-related losses in *M. separata* NPV exposed in water suspension or to dry conditions for 4 h. Irradiation of the dry deposit, which was studied longer, showed an adverse effect setting in at 72 h. When *Heliothis* NPV was studied under a regime of variable temperature and UV insulation, there was significantly greater inactivation at 45°C than at 30°C but exposure at 15°C or 30°C resulted in no significant inactivation (McLeod *et al.*, 1977) (Figure 4.8).

Retention of occlusion bodies on plant surfaces

The plant surface is important for the accumulation and persistence of BVs. With annual plants (e.g. most crop plants) the duration of inoculum persistence is governed by the length of life of the plant in cultivation. At plant death any surviving inoculum reaches the soil or it may be removed from fields with the plant crop or may be burned *in situ*. For perennial plants, particularly trees in forests, woodlands and orchards, there is some evidence that BV persistence increases with increasing complexity of plant architecture. Thus Richards (1984) observed a greater persistence of CpGV on older apple trees with apparently increased shading from UV light (Figure 4.4). Further evidence that persistence is to some extent dependent on shading was provided by Elgee (1975), who observed that deposits of *Orgyia leucostigma* NPV lasted longer on the north side of spruce trees. In measurements of sawfly NPVs on needles of *Picea* and *Pinus* spp. (unaccompanied by estimates of changes in biological activity) it was shown that OBs could physically persist for more than 250 days. The rate of loss of an NPV applied to pine foliage as an unpurified suspension was less than for purified OBs (Figure 4.5). The fate of OBs 'disappearing' from needles in the sample zones on these trees was not determined, but it seems possible they could have been trapped by lower foliage (P. F. Entwistle, H. F. Evans and S. R. Hoyle, unpublished data). It was also shown that if spruce foliage was surface sterilized *in situ* on trees infective *G. hercyniae* NPV gradually reappeared in following weeks, presumably as a result of rain drip from contaminated foliage above (these observations were made outside the period of larval activity on the trees and so concern movements in a pre-existing inoculum pool) (Evans and Entwistle, 1982).

As with annual plants, the persistence of BVs on the foliage of deciduous perennials is governed by seasonal events, and any inoculum remaining in the autumn will descend to the soil at leaf fall. However, BVs persist very effectively on bark and, for instance, bark was the source of NPV contamination of overwintering

Malacosoma eggs (Clark, 1955). Later instars of *L. dispar* dying of NPV disease largely do so under the lower branches and on the lower trunk of trees. These bark areas are the day time resting sites of older larvae and also of pupation and then oviposition. Egg surface contamination can occur leading to infection of first instar larvae (Doane, 1969).

Despite very high rainfall and intense solar insulation, BVs can persist in tropical forests and tree plantations. In Papua New Guinea, rain mobilization of the bark reservoir of *L. ninayi* NPV inoculum appears to be the main factor triggering epizootics (Entwistle, 1986). In oil palm plantations, virus diseases are major mortality factors of limacodid moths. Frond replacement on palms is slow and individual fronds may be present for over 30 months. Virus-infected (NPVs, GVs, picornaviruses, *Nudaurelia* β viruses) limacodid larvae tend to die on the underside of fronds. Here they adhere strongly and epizootic initiation is often associated with rain, which presumably spreads the virus and increases the area of leaf over which it is available (Entwistle, 1987).

In addition to bark itself, BVs may persist on trees in deposits of decaying vegetable material, which accumulate in the angle of branch and trunk. Polyhedral OBs were recovered from such sites at over 6 m height on *Pinus patula* in Papua New Guinea.

The capacity to endure precipitation is not restricted to BVs. A CPV persisted well on pine foliage (Burgerjon and Grison, 1965).

Mechanisms for retention of baculovirus occlusion bodies on leaf surfaces

The forces involved in attachment of the OBs of NPVs has been investigated by Small (1985). Attachment was considered on a micro-level, in terms of purely physicochemical criteria and on a macro-level in terms of specific biological determinants. Micro-forces were subdivided into groups: long-range (class 1) forces, operating over distances of greater than 10 nm, and short-range (class 2) forces operating at below 0.4 nm. Short-range forces were considered to be primarily involved in OB adhesion and included the various types of chemical bond. Macro-forces referred to specific biological modifications and aspects of the contacting surfaces. Aspects of this subject have been discussed by Small (1985), Small and Moore (1987), Small, Moore and Entwistle (1986) and Carruthers *et al.* (1988b). Natural adhesion of OBs to plants is, of course, not solely a function of plant and OB surfaces, because the process is often affected by host breakdown products.

Plant surface chemistry

Much remains to be elucidated concerning the impact of the extra-host chemical environment on the survival of BV biological activity. This is an important area not only in the ecology of insect viruses but also in relation to their use in control.

For a time it was considered that the alkaline nature of the surface of cotton plants was the cause of the loss of *Heliothis* NPV activity. More detailed investigations revealed the presence of small 'salt glands' that are probably specialized trichomes. The glands excrete deliquescent salt complexes that have been partially characterized (Elleman, 1983). The divalent cations Mg^{2+} and Ca^{2+} are well represented. Exposure of NPV OBs to Mg^{2+} alters their solubility characteristics and is associated with a reduction in infectivity *in vivo*. *In vitro*, this effect can be reversed by the chelating agent EDTA. However, when OBs are exposed to salts gathered from the leaf surface EDTA is not effective, presumably because other, unidentified substances are also involved in inactivation (Elleman and Entwistle, 1982, 1985a,b). The impact of salt gland secretions is not uniform across the cotton cultivation zones of the world, a situation which is not yet understood but that may be associated with the influence of soil nutrient status on the composition of salt gland exudates and the innate differences between species and varieties (Elleman, 1983). For instance, the salt gland secretions of *Gossypium barbadense* were found to have a less alkaline pH and to cause less inactivation of *S. littoralis* NPV than *G. hirsutum* (P. F. Entwistle, unpublished data).

Further insights have come from research into the survival of *B.t.* δ-endotoxins. For instance Luthy (1986) demonstrated that quite short periods of incubation with leaf extracts of cotton or red spruce resulted in significant inactivation of δ-endotoxins. The mode of action is uncertain but it has been suggested that there may be adverse interactions with plant polyphenols, especially tannins. Cotton and spruce are both rich in tannins.

The influence of tannins on the infectivity of NPVs has been studied in relation to the Gypsy moth, *L. dispar*. Infectivity falls with increasing hydrolysable tannins whether in plant food or added under controlled conditions to semi-synthetic diet (Keating, Yendol and Schultz, 1988; Keating, Hunter and Schultz, 1990). Plant catechols also influence NPV infectivity (Felton *et al.*, 1987). Tannins are known to bind to proteins and it seems likely that, in their presence, the solubility of OB matrix proteins could be affected and the numbers of virions available for infection reduced. The extraction of tannin from leaves and the selection of tannin assay methods are described by Haggerman (1988) and Haggerman and Butler (1989). For fall armyworm, *Spodoptera frugiperda*, there was a strong host plant–NPV interaction: the LC_{50} on signal grass was 142 OB/mm^2 but on soybean it was only 13 OB/mm^2 (Richter, Fuxa and Abdel-Fattah, 1987). Acidity in food also reduces BV infectivity, again possibly because of an adverse effect on OB dissolution and virion release.

BACULOVIRUS DISPERSAL

An understanding of disease spread is central in virus ecology and epizootiology and can make valuable contributions to the development of control strategies. The present section concerns the agencies involved in inoculum dispersal. A clear distinction is drawn between inoculum dispersal and disease spread. Disease spread is a reflection of inoculum dispersal. Spatial and temporal aspects of disease spread are discussed below (and are mentioned again in Chapter 5). Both abiotic and biotic modes of dispersal can be distinguished and may act together (Figure 4.9).

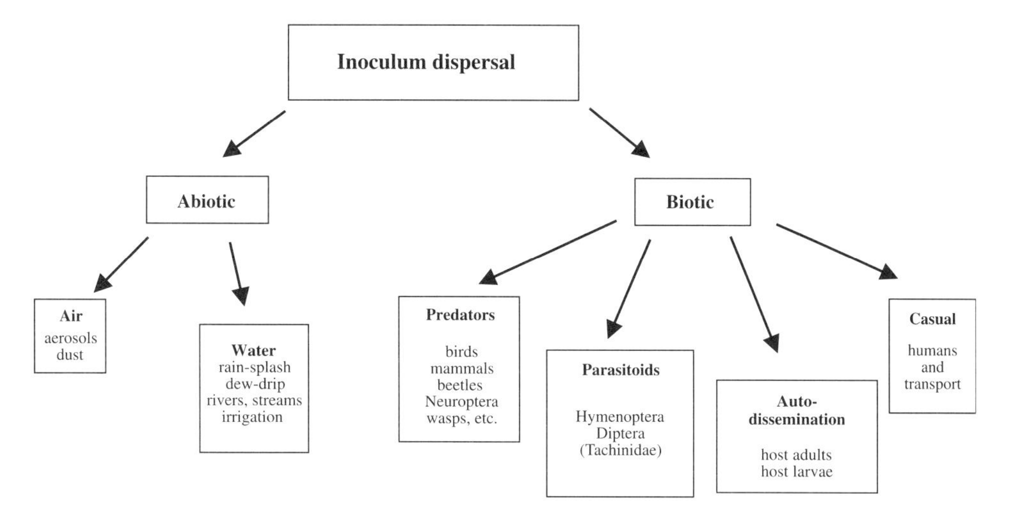

Figure 4.9 Principal factors involved in the environmental dispersal of BVs and other insect pathogenic viruses.

Abiotic agencies

Air

Air dispersal is an inadequately researched area. Dispersal in association with naturally generated aerosols, for instance when wet foliage is disturbed by wind or by rain splash, has not been demonstrated. Methods for concentrating aerosols to increase the likelihood of detecting virus include cyclone-type air scrubbers (devices that collect very small air-borne particles by sucking them into a chamber from which they can be collected by washing with a very fine spray of a suitable collection fluid). These could be used to process large volumes of air in the vicinity of epizootics, particularly during the period of active breakdown of infected larvae that occurs after peak mortality takes place.

In fact, the best evidence for aerial movement of inoculum is through the medium of airborne soil or debris. For example, the spread of *N. sertifer* NPV has been associated with wind-blown dust (Olofsson, 1988) and that of *O. pseuotsugata* with dust generated by cattle driven through the forests (Thompson, 1978).

Water

Rain plays a major part in the initiation of epizootics through the mobilization and spread of inoculum. For instance, epizootics of NPV disease in cabbage looper, *T. ni*, are rain correlated (Hoffmaster and Ditman, 1961). The subject has been reviewed by Entwistle (1986, 1987). Substantial run-off of *N. sertifer* NPV from pine trees was demonstrated by Kaupp (1981), who collected OBs in pots containing washed sand beneath the trees. The spread of an NPV of *Colias eurytheme* in irrigation water was noted as a major cause of epizootics in alfalfa fields in California (Steinhaus and Thompson, 1949). Spread between spruce trees of *G. hercyniae* NPV disease by rain splash was identified by Bird (1961) in Canada.

Biotic agencies

Predators

In 1916 Allen suggested that the introduced carabid beetle *Calosoma sycophanta* was involved in dispersal of gypsy moth, *L. dispar,* NPV in the USA. This subject has since been well explored in a qualitative sense and the literature refers to the capacity of animals in many vertebrate and invertebrate taxa to excrete, after feeding on infected hosts, biologically active BVs. The subject has been reviewed by Entwistle (1982), Kaya (1982) and Andreadis (1987). Among the vertebrates, the role of birds as effective dispersal agents is correlated with their high metabolic rates, such that food, and virus with it, may begin to pass through the gut in less than 30 min, but virus may continue to be present in faeces for several days. The contribution of birds to BV dispersal can, thus, be important both locally and, through migration, over very long distances. In Scotland in August, willow warblers, *Phylloscopus trochilis*, migrating south and east to winter quarters carried NPV of *P. flammea.* It is interesting to note that at that time of the year *P. flammea* was in the pupal stage in forest soils and that the birds, therefore, probably acquired the virus by feeding on larval cadavers from the trees (Entwistle *et al.*, 1993). A similar situation was recorded with birds in spruce forest, which were trapped in January carrying *G. hercyniae* NPV. It was shown that birds could disperse this NPV for about 10 months of the year (Entwistle, Adams and Evans, 1977b). During peak periods of NPV availability more than 70% of bird individuals in populations, which were composed of more than 10 species, were involved in dispersal (Entwistle, Adams and Evans, 1977a). The role of haematophagous midges (Diptera: Ceratopogonidae) requires further investigation (Wirth, 1956; 1972).

Surprisingly, the excretion of BVs by predatory arthropods may take longer than in birds. Invertebrate predators may also retain some virus for longperiods, e.g. *Nabis tasmanicus* up to 4 days, *Ephippiger bitterensis* up to 15 days and nymphs of *Oechalia schellenbergi* 10–15 days. Being poikilothermic, retention times in arthropods are likely to be connected closely with environmental temperatures.

The capacity of occluded BVs to tolerate predator gut conditions results from the almost uniformly acidic conditions and, as has been discussed above in considering structure–function in OB components, the strong protease resistance of the OB envelope.

Parasitoids

Parasitoids have already been referred to above in terms of their capacity to mediate infection of hosts with BVs. A model has been proposed for the interaction of parasitoids and BV diseases in an epizootic and control context

(Entwistle, 1982). Essentially this states that, to a considerable extent, parasitoids and BV diseases compete for hosts. Therefore, for example, because larvae of *Apanteles* spp. are easily killed and only survive minimal infection of hosts by BVs, they tend to attack the younger host individuals of which the proportion diseased in epizootics tends to be low. This is strong and effective competition. Such parasitoids may be termed 'isolationists' from their tendency to select young host individuals linked to a positive avoidance of diseased hosts or oviposition sites. Some tachinids, however, tolerate infection in hosts (e.g. *Voria ruralis*; Vail, 1981) or even host death from BV infection (e.g. *Drino bohemica*; Bird, 1961) and hence do not need to compete for hosts. These can be termed 'non-isolationists'. Therefore, 'isolationists' are more vulnerable in an evolutionary context. Epizootics may thus alter the proportions of species within a parasite complex. When BVs are used as biological insecticides, usually applied when host larvae are young, not surprisingly populations of parasitoids become depressed, as was seen to occur with the orchard pest *A. orana* (Shiga *et al.*, 1973) and *Pieris rapae* on cabbages (Kelsey, 1960).

Auto-dissemination

Spread of BVs may be brought about naturally by a host species. This may either be casual or can be an integral part of disease biology, as in *O. rhinoceros* and diprionid sawflies. Auto-dissemination has not been well documented, though the effectiveness of surface contamination of adult Lepidoptera with OBs (resulting in egg-surface contamination and some early infections) is referred to in Chapter 5. Auto-dissemination is not restricted to adult insects but can be effected over long distances by first instar larvae of those Lepidoptera that are wind-borne on strands of silk (ballooning). This is possibly a major dispersal mechanism for *L. dispar* NPV in North America where adult females are virtually flightless (discussed in Murray *et al.* (1989)), and for winter moth, *Operophtera brumata,* the adult females of which are wingless.

Auto-dissemination by larvae can also result from disease-directed modification of behaviour. Restlessness is a frequent result of infection: NPV-infected larvae of *M. brassicae* tend to wander from cabbage plant to cabbage plant (Evans and Allaway, 1983). The classic example of disease-modified behaviour is the photo-positive or photo-negative geotactic behaviour resulting in *wipfelkrankheit* ('tree top disease') where in nun moth, *Lymantria monacha*, NPV-infected larvae ascend to the tops of trees to die. This phenomenon can be observed with other insect diseases; for instance, fungus-infected larvae of dolerine sawflies ascend to the tops of grass stems, where they later die. Martignoni (1964) discussed this and other types of 'Abnormal Behaviour' in a review of insect pathophysiology.

Casual dissemination

Accidental dissemination by humans has not been quantified, though the spread of BV diseases through insect cultures can often be attributed to an absence of disciplined hygiene.

Alternative host species

The environmental persistence of BVs can be favoured by a wide host range especially when these hosts are not phenologically synchronized but form something of a temporal succession. In addition, infection in different hosts with widely and essentially allopatric distributions would also confer evolutionary and ecological flexibility. In such situations, the capacity of birds to carry BVs over considerable distances would ensure a degree of fluidity in the gene pool.

Synecological studies involving two or more hosts have yet to be initiated with BVs. Potentially interesting agro-ecosystems in which such studies are conducted are AcMNPV in *Plutella xylostella*, *Hellula phidilealis* and *T. ni* on brassicas and MbNPV in *H. armigera* and *Diparopsis watersi* on cotton.

Predators as a baculovirus reservoir

Predators provide a locus for BV OB accumulation even though the duration of passage through an individual animal may in total be not much greater than 5–15 days, a period applicable to both vertebrates and invertebrates. In temperate region forests, BVs may be present in the bird population through most of the year. This is because insect remains are gleaned from the trees during periods when live insects are unavailable. In UK pine forests, the coal tit, *Parus ater*, consumes about 1100 separate food items a day in winter, often each with a weight of only 2 mg (Gibb, 1960). The proportion of the total BV population in birds at any one time may be small but during a host extra-larval period of 9 or more months it may be large. The bird-borne inoculum body represents a major contribution to the total pool of virus dispersing.

THE ECOLOGY OF BACULOVIRUSES IN THE SOIL ENVIRONMENT

The discontinuity of host availability that is typical for the great majority of BVs means that mechanisms for survival over long periods outside the host are a necessity. An outstanding attribute of BVs, which has presumably evolved in response to the need for persistence outside the host, is the polyhedral OB. This provides a protective entity that has potential for persistence in a range of environments, including the soil. Much has been written concerning the general question of persistence of BVs outside the host insect (Evans and Harrap, 1982; Mohamed, Coppel and Podgwaite, 1982; Barbercheck, 1992). It is, however, almost inevitable that BVs will eventually reach the soil environment. This stems from breakdown of insect cadavers, removal of OBs from leaf surfaces by various physical attrition factors and the ultimate decay of all plants as they fall to the soil surface.

BV epizootiology in relation to this soil reservoir is determined largely by the probability, in space and time, of that virus being ingested by permissive insect hosts. Encounter between host and virus requires susceptible stages to be in direct contact with soil or for soil to be moved onto the normal feeding substrate of the host. Of equal importance is the quantitative retention of OBs in soil. Availability of OBs is determined more by the physical and chemical structure of the soil than by differences between the BVs themselves. Paramount in this are a range of physicochemical factors, such as cation exchange capacity, that reflects the ability of soil components, especially clays, to adsorb cations as a reaction to the generally negative charges present on the surfaces of soil particles (Tapp and Stotzky, 1995). Within this continuum of variability, the main characteristic is the presence of clay particles and organic matter, both of which act to increase the cation exchange capacity of the soil. A further complication arises from the pH of the soil, such that stability of OB–clay complexes may change from stable to unstable as the soil changes from acid to alkaline. Detailed studies of these forms of interaction were carried out for bacterial cells in soils by Hattori and Hattori (1993), the same principles applying to OBs. More recent studies by Tapp and co-workers (Tapp and Stotzky, 1995) provided detailed information on the fate of *B.t.* in various clay soils. These authors showed that the two clay minerals montmorillonite and kaolinite, either washed (clean soils) or in the presence of ferric compounds (dirty soils), had differing abilities to adsorb the crystal toxins of *B.t.* They also showed that adsorption was greater at pH 6–8 for clean clays compared with a wider range of pH 5–9 for dirty soils. Interestingly, adsorption of the crystal toxins to fixed quantities of clay increased with toxin concentration but eventually reached a plateau. Washing of the clay–toxin complexes gave only between 10 and 30% removal during initial washing, with no further loss in subsequent washings. This indicates extremely strong binding of the toxins to the clay structure, a feature that appears also to be the case for BVs which, as crystal proteins, have similar surface properties to *B.t.* toxin crystals. Indeed, Evans (1982) showed that adsorption of *M. brassicae* NPV polyhedra was directly related to the clay content of various soils, although this was biphasic, with a greater linear rate of adsorption for soils with 20% clay content.

Excellent persistence of BVs in the upper layers of soil was demonstrated by Fuxa and Richter (1996). They studied *A. gemmatalis* NPV (AgNPV) released into soybean plots after which the soil was sampled for virus for a period of 10 months. The purpose was to determine whether normal agricultural practice and precipitation affected vertical distribution of the OBs. AgNPV was found to soil depths of 37.5–50 cm if virus-contaminated plant material was present on the soil surface. Agricultural operations, such as disking, harrowing, mowing, planting or cultivating, did not alter the vertical distribution of AgNPV in the soil. However, after disking the fields in November there was a decrease in OBs at all soil depths, and Fuxa and Richter concluded that recontamination of soybeans planted in the plots should be possible from the soil reservoir.

Adsorption of OBs to clay and other particles can result in very long persistence in the upper layers of soil, dependent on the degree of disturbance of that soil. Evans (1982) carried out experiments to determine, quantitatively, how the presence of *M. brassicae* NPV changed over time in a brown earth soil with a neutral pH. He showed that, over a 12-month period, in excess of 98% of the NPV polyhedra were lost from the discrete area of soil on which they had been placed. Comparison of physical presence with biological activity of extracted OBs indicated that physical loss rather than inactivation accounted for the changes. Similar results were obtained by Ogaard *et al.* (1988) for PbGV using a similar experimental design. Other evidence for extremely long total persistence comes from the demonstration of viable OpNPV in soils below white fir trees that had not experienced high populations of the moth for over 41 years (Thompson, Scott and Wickman, 1981). The authors estimated that over 99% of activity of the virus was lost during the 41-year period compared with only 50–70% activity in the first 10 years.

Quantitative estimates of the presence of BVs in soils vary enormously, reflecting both the methods employed for enumeration and also the large differences in holding capacity of different soil types. Unfortunately, although basic information on soil type and pH is often provided, it is far less common to have data on the colloidal properties of the soil, notably the ratios of clay, sand and silt. In particular, measurement of clay content combined with the pH could provide much useful information on the holding capacity of the soil. Recent work on genetically modified AcMNPV, using the so-called co-occlusion strategy, has concentrated on the fate of the modified virus on both foliage and soil (Wood, Hughes and Shelton, 1994). These authors used an SDS extraction method that was very similar to that developed by Evans *et al.* (1980) and were able to detect polyhedral concentrations down to 7 OB/g soil.

Availability of virus in soil for ingestion by suitable hosts

The fact that BVs are present in soil and remain viable biologically implies that this medium represents a significant reservoir that might drive epizootics. Indeed, the mathematical models of Anderson and May (1981) recognize the need to quantify the mortality rate of free-living infective stages. Of more significance is the reciprocal of this, the survival rate of these stages, which ultimately determines the transmission rate of infective stages. This is particularly important when host populations are univoltine with well-synchronized larval recruitment. However, the role of soil in natural or induced epizootics is normally difficult to demonstrate with certainty since it is not always possible to provide a quantitative link between the presence of OBs in soil and any infections observed in larvae feeding on those plants. Intuitively, low-growing plants should have a higher probability of being contaminated with virus–soil mixtures than more established plants such as shrubs and, ultimately, trees. However, there is little quantitative information available to test this simple hypothesis. Jaques (1985) concluded that soil-borne BV was a significant source of inoculum for infection of *T. ni* by NPV and *P. rapae* by GV. He showed that the BVs were moved between plots in contaminated soil and produced epizootics even in fields that had not been treated with virus. Less convincing links between soil inoculum and infection on cabbage plants grown in the plots was provided by Evans and Allaway (1983). They estimated that only a very small percentage (expressed in LD_{50} units) of the very large quantities of virus produced on death of larvae in the plots in the previous year was recycled back to cabbages planted in the same plots in the following year. However, the recent work by Fuxa and Richter (1996) appears to indicate that, provided virus remains in the top layer of soil, there is likely to be significant contamination of low-growing plants in that soil. Stability of *N. sertifer* NPV in forest soils has also been shown to be high, even in the presence of simulated acid rain, thus confirming the role of soil as the most stable medium for long-term BV persistence (Sakkonen, 1995).

The possibility that soil-dwelling organisms and microorganisms may influence the fate of insect viruses has received very little attention. Among the invertebrates, earthworms (Annelida), especially in temperate zones, and termites (Isoptera) in warmer climatic regions are of great ecological importance in the processing of organic matter in soils. The role of termites is being actively investigated in relation to BV control of teak defoliator moth, *Hyblaea puera*, in India (H. F. Evans, personal communication). In all regions, Collembola are universal feeders on soil organic matter. In an unique study, it was shown that the earthworm *Lumbricus terrestris* ingested NPV OBs on filter paper disks (a neutral leaf substitute) and excreted them intact with no loss of infectivity (P. F. Entwistle, unpublished information). Although having the potential to recycle or breakdown BVs, the capacity of Protozoa to do so is unknown.

THE STRUCTURE OF EPIZOOTICS

The term epizootic concerns the development of disease in an animal population. In an idealized sense, epizootics commence with a single infected individual host from which disease-causing inoculum spreads, over time, to infect further individuals. Because of the natural spatially dispersed occurrence of hosts, inoculum and resultant disease also spreads spatially. In essence, then, an epizootic is a spatio-temporal event. (The term 'epizootic' is also frequently used to indicate a high level of disease in a population whilst 'enzootic' suggests that disease has returned to a low level of presence.) In nature it is generally difficult to discern the exact basic structure of an insect OB virus epizootic because when a disease increase commences it generally does so from a multitude of spatial loci, which usually represent the disease-producing impingement of environmentally persistent virus on permissive host individuals. Such disease centres, as they spread, may rapidly interact in a manner that may present the observer with a confusing pattern of disease incidence.

If disease can be deliberately initiated at a single spatial point, the epicentre, of an initially healthy host population, or if the observer is fortunate enough to encounter a young and relatively isolated epicentre, the progress of disease may be followed from an early limited to a later more general stage. The patterns involved in the temporal and spatial evolution of an epizootic can then be measured. Two notable examples in which this has been possible are NPV disease in the European spruce sawfly, *G. hercyniae*, in Canada (Bird and Burk, 1961) and in the UK (Entwistle *et al.*, 1983) and a virus disease (OrV) in coconut rhinoceros beetle, *O. rhinoceros*, as observed in the Tonga Islands (Young, 1974); both examples are analysed in Entwistle *et al.* (1983). Parasitoids were absent or of negligible importance, as were other pathogens, but the two host species would have been exposed to an unquantified impact of predators. Further supportive information exists for NPV in *N. sertifer* (P. F. Entwistle and H. F. Evans, unpublished data).

An idealized view of epizootic structural processes, based to a large extent on these examples, is presented in Figure 4.10. Here the process of infection growth

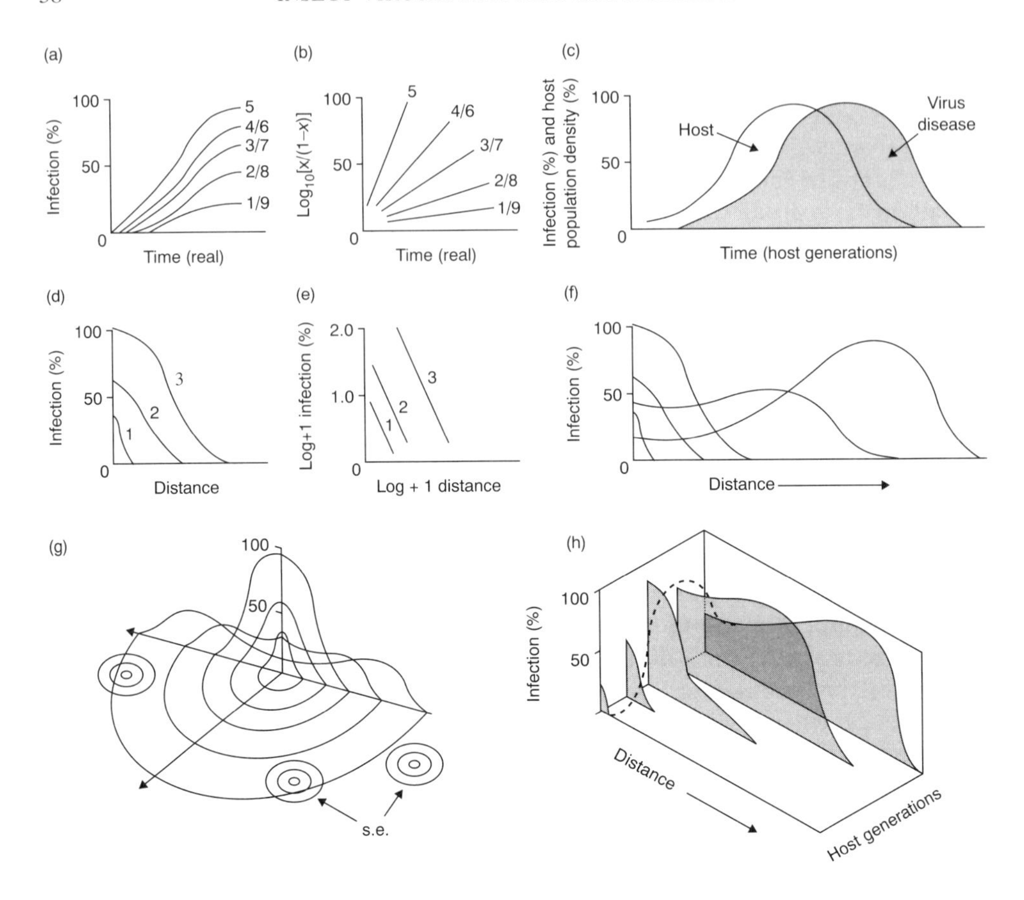

Figure 4.10 The structure of an idealized epizootic. (a–c) Temporal changes in infection level. (a) Development of infection (generations 1–5, as the epizootic grows and 6–9 as it declines) is illustrated; for simplicity, a sawfly infection curve has been employed (Figure 4.1). (b) Normalization of infection growth curves (x is the proportion of disease where 100% = 1.0); the gradient, b, in the simple linear regression is one measure of the infection growth. (c) The overall relationship between host population fluctuation under the influence of an infectious virus (shaded); as in classical host–parasitoid interactions, the rise and fall of the proportion of hosts infected is in lag phase with that of the host population itself (for strict comparison with the parasitoid model, infection rates should actually be translated into a numerical measure of the number (population size) of virus particles generated. (d–g) Spatial patterns in infection levels. (d) Development of the primary phase of disease dispersal from the epicentre, 0. (e) Primary dispersal curves can be normalized; the gradient of dispersal, b, is always negative. (f) The primary dispersal pattern evolves into a wave form; the peak of the wave was at 450 m from the epicentre for *G. hercyniae* but at 23 km for *O. rhinoceros*. (g) An epizootic is shown in elevation (as in f) and plan views; increasingly as disease spreads from the initial epicentre, secondary (s.e.) and tertiary (etc.) epicentres are generated through the action of those dispersal agents that can especially easily convey virus appreciable distances (e.g. insectivorous birds); the terms discontinuous and jump spread have been employed by others to describe such events. The clarity of the primary disease outbreak then tends to deteriorate as these secondary centres themselves grow. (h) The interrelationship of the generation of temporal and spatial waves of infection is depicted over five host generations.

through a rise and eventual decline over a sequence of host generations is depicted to illustrate the evolution of the classical, and well-known, temporal epizootic wave. The figure also illustrates the evolution of the classical spatial wave of infection spread from the early, simple indented wave of primary dispersal (known also to characterize the spread of disease in plant populations (Gregory, 1968)). The gradient of the normalized wave of primary dispersal varies with different host and virus taxa interactions, and no doubt also with varying environmental circumstances, but appears to tend to consistency for any one system. For instance, its value for four separate data sets collected in the UK and one in Canada for the spread of *G. hercyniae* NPV disease was -1.979 ± 0.159. There were very different values for gradients of primary dispersal of OrV disease in *O. rhinoceros* (–0.253) and of an NPV disease in the moth *Malacosoma disstria* (–0.368): the flatter gradients in these last two examples indicate more distant earlier spread, though not necessarily the dispersal of a greater amount of inoculum. However, within any particular system, the presence of more inoculum will result in further spread of disease, but the gradient of dispersal is unaltered (van der Plank, 1963; Gregory, 1968).

The initiation of temporal infection in each succeeding host generation is essentially dependent on inoculum persisting between generations: but infection growth is not immune from the influence of those agencies (e.g. predatory arthropods and vertebrates, and parasitoids) much involved in spatial dispersal of inoculum. Adults of the host species itself may also be concerned in spatial dispersal of inoculum giving rise to disease, this being especially true for NPV in sawflies and CPV in Lepidoptera (as midgut infections).

Clearly, the patterns of temporal rise and fall of infection and spatial spread are two aspects of an unitary epizootic process. To visualize this integral relationship it is easily possible to regard the temporal epizootic wave as a cross-section of the (essentially radial) pattern of spatial dispersal as the infection wave develops and progressively moves away from the epicentre (Figure 4.10h).

The relevance of epizootic concepts to control is discussed below and in Chapter 5.

THE USE OF MODELS IN UNDERSTANDING AND USING BACULOVIRUSES IN PEST MANAGEMENT

Separate measurements of host and pathogen populations and the subsequent infection arising from their interaction provide empirical data for inclusion in descriptive and mathematical models. These models can provide insights into the key factors determining the degree of infection in a host population. Modelling has been used to improve the use of bioinsecticides through quantitative descriptions of the biology of the target host, the distribution of pathogen applied and the roles of secondary inoculum in overall mortality. Brand and Pinnock (1981) were early advocates of this approach, emphasizing delivery of an effective dose of pathogen when allowing for rates of attrition and degree of host exposure to the pathogen. Such models are useful for quantifying the effects of primary inoculum but do not provide details on the long-term effects of the pathogen.

Further refinement was provided by the models of Anderson and May (1981), who included descriptions of the rates of recruitment of both susceptible and infected individuals and also took account of persistence and vertical transmission of pathogens, recognizing that both healthy and sublethally infected individuals may reproduce. In their model, vertical transmission lowers host-density thresholds for infection in the next generation. Interestingly, the model predicts that pathogens are unable to maintain themselves through vertical transmission alone. Brown (1987) later discussed a range of modelling approaches, concentrating on three components of the host population: susceptible individuals, infected individuals that are not yet infectious (i.e. are not yet releasing secondary inoculum) and infectious individuals delivering secondary inoculum. Density dependence and changing susceptibility with age were also incorporated into a generalized simulation model that, although a simplification, provides useful insights into the dynamics of a host-pathogen interaction. A more comprehensive evaluation of epizootiological models was provided by Onstad and Carruthers (1990), who made the point that validation of models has seldom been attempted. However, the concept of a threshold above which density-dependent mortality occurs was recognised to be a central feature of epizootiological models, although it is important to realise that the threshold can change depending on the amount of primary inoculum and initial infection levels, as demonstrated in their complex model of the *Ostrinia nubilalis*–*Nosema pyrausta* (Microsporidia) system (Onstad and Carruthers, 1990). In this case, inclusion of differential susceptibility with age, the spatial dynamics of both pathogen and host and the number of generations of the host improved the predictions of the model.

More recent models have incorporated realistic estimates of parameter values obtained by either laboratory or field observations or by a combination of the two. Feng *et al.* (1988) developed a model for *Beauvaria bassiana* infection of *O. nubilalis* that included the effects of primary inoculum density and host phenology on disease expression. The accuracy of the model depended on whether the pathogen was already on the plants when host larvae commenced feeding. Hajek *et al.* (1993) took account of disease dynamics at both local and wider geographic scales in modelling *Entomophaga maimaiga* fungal epizoosis in *L. dispar* populations in north-eastern North America. Data on rates of lethal infection at a range of temperatures as well as transmission rates in the field were included in a simulation model based on SERB (Simulation Environment for Research Biologists) (Larkin, Carruthers and Soper, 1988). The model was sensitive to the onset of primary infection, host density and relative humidity, suggesting that expression of *E. maimaiga* infection in the field is dependent on exceeding a critical host-density threshold and that changes in larval behaviour in late instars increases local host density and, hence, encounter rate with the pathogen.

DEVELOPMENT OF THE 'CONTROL WINDOW' CONCEPT

The use of parameter estimation has been included in a Control Window for the principal variables in a given pathogen-host interaction and to optimize application methods for pest management. In effect, the purpose here is to remove the density-

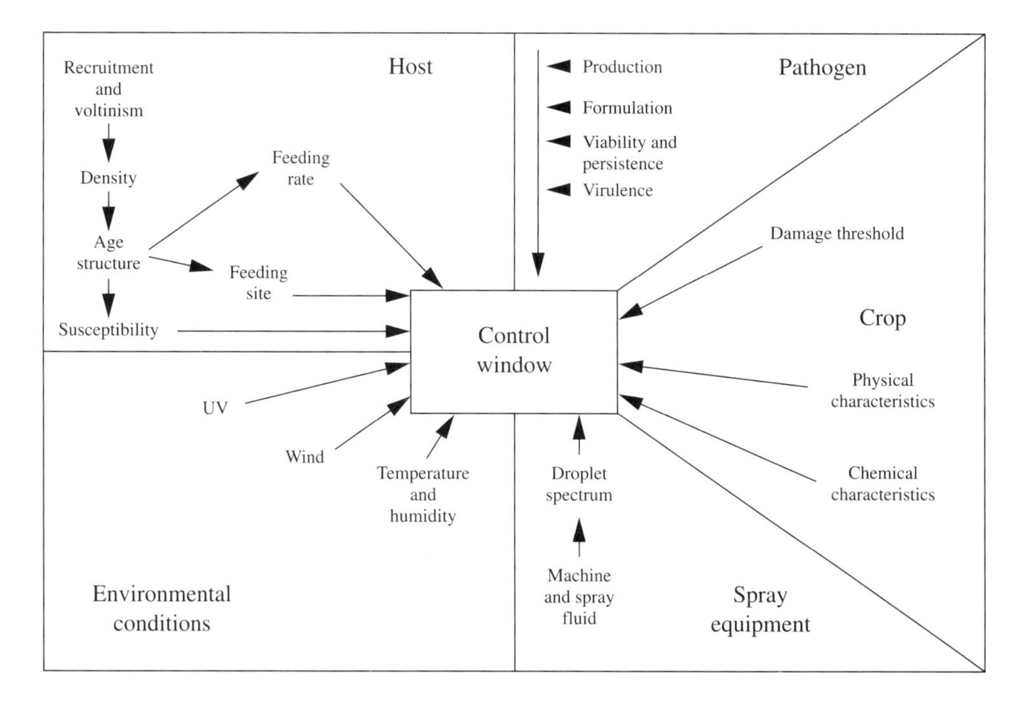

Figure 4.11 The interactions of host, pathogen, crop, environment and spray technology in defining a Control Window for use of pathogens in pest management.

dependent component of naturally occurring infections by saturating the host environment with pathogens. Developments in pest management employing pathogens have tended to be *ad hoc*, with consequent variability in success (Entwistle and Evans, 1985; Navon, 1993). Many of the processes driving epizootics must also be taken into account in field use of pathogens; Evans (1994) brought these together in a concept termed the Control Window (Fig. 4.11).

Using this approach, appropriate management strategies can be developed that combine the biological attributes of the host–pathogen interaction with the physical attributes of droplet generation and deposition in relation to the structure of the crop being protected (Chapter 5). Such an approach has been used to optimize the application of NPV for control of pine beauty moth, *P. flammea*, in Scotland (Entwistle *et al.*, 1990). Knowledge of host distribution and the droplet emission and deposition characteristics of the ultra-low-volume equipment employed enabled 95% of the droplets, and hence of the virus, to be deposited precisely where the target first instar larvae were feeding. Although improvements in field use of *B.t.* have been achieved, this has not always been accompanied by improved knowledge of the interactions between pathogen and host, so that inconsistency in performance remains a problem in large-scale use of *B.t.* (van Frankenhuyzen, 1993). A Control Window approach to use of fungal pathogens, based on use of ultra-low-volume oil-based formulations has dramatically increased efficacy in use of *Metarhizium flavoviride* for control of desert locusts, overcoming the problems of lack of humidity for spore germination (Bateman *et al.* 1993; Moore *et al.* 1993).

REFERENCES

Anderson, R.M. and May, R.M. (1981) The population dynamics of micro-parasites and their invertebrate hosts. *Philosophical Transactions of the Royal Society, Series B* **291**, 451–524.
Allen, H.W. (1916) Notes on the relation of insects to the spread of the wilt disease. *Journal of Economic Entomology* **9**, 233–235.
Andreadis, T.G. (1987) Transmission. In *Epizootiology of Insect Diseases*. J.R. Fuxa and Y. Tanada (eds). New York: Wiley pp. 159–176.
Ballard, J. (1987) Larval behaviour of the Pinebeauty moth. In *Population Biology and Control of the Pine Beauty Moth*, S.R. Leather, J.T. Stoakley and H.F. Evans (eds). London: Forestry Commission (UK) Bulletin 67, pp. 31–36.
Barbercheck, M.E. (1992) Effect of soil physical factors on biological control agents of soil insect pests. *Florida Entomologist* **75**, 539–548.
Barker, R.E. (1968) The availability of solar radiation below 290 nm and its importance in photomodification of polymers. *Photochemistry Photobiology* **7**, 275–295.
Bassard, O. and Knutti, H.T. (1977) SAN 240 I, the first commercial virus based insecticide. In *British Crop Protection Council Symposium Proceedings: Pests and Diseases*. Farnham, UK: British Crop Protection Conference, pp. 547–554.
Bateman, R.P., Carey, M., Moore, D. and Prior, C. (1993) The enhanced infectivity of *Metarhizium flavoviride* in oil formulations to desert locusts at low humidities. *Annals of Applied Biology* **122**, 145–152.
Bird, F.T. (1949) A virus (polyhedral) disease of the European spruce sawfly *Gilpinia hercyniae* (Hartig). PhD thesis, McGill University, Montreal, Canada.
Bird, F.T. (1953) The effect of metamorphosis on the multiplication of an insect virus. *Canadian Journal of Zoology* **31**, 300–303.
Bird, F.T. (1955) Virus diseases of sawflies. *Canadian Entomology* **87**, 124–127.
Bird, F.T. (1961) Transmission of some insect viruses with particular reference to ovarial transmission and its importance in the development of epizootics. *Journal of Insect Pathology* **3**, 352–380.
Bird, F.T. and Burk, J.M. (1961) Artificially disseminated virus as a factor controlling the European spruce sawfly, *Diprion hercyniae* (Htg.), in the absence of introduced parasites. *Canadian Entomologist* **93**, 228–238.
Bishop, D.H.L., Entwistle, P.F., Cameron, I.R., Allen, C.J. and Possee, R.D. (1988) Field trials of genetically engineered baculovirus insecticides. In *The Release of Genetically engineered Microorganisms*, M. Sussman, C.H. Collins, F.A. Skinner and O.E. Stewart-Tull (eds). London: Academic Press, pp. 143–179.
Boucias, D.G., Johnson, D.W. and Allen, G.E. (1980) Effects of host age, virus dosage, and temperature on the infectivity of a nucleopolyhedrosis virus against Velvetbean caterpillar, *Anticarsia gemmatalis*, larvae. *Environmental Entomology* **9**, 9–61.
Bourner. T.C. (1993) 'The control of the common cutworm, *Agrotis segetum* (Lepidoptera: Noctuidae)' using baculoviruses. PhD thesis, Imperial College, London.
Brand, R.J. and Pinnock, D.E. (1981) Application of biostatistical modelling to forecasting the results of microbial control trials. In *Microbial Control of Pests and Plant Diseases 1970–1980*, H.D. Burges (ed.). London: Academic Press, pp. 667–693.
Brown, G.C. (1987) Modeling. In *Epizootiology of Insect Diseases*. J.R. Fuxa and Y. Tanada (eds). New York: Wiley, pp. 43–68.
Burgerjon, A. and Grison, P. (1965) Adhesiveness of preparations of *Smithiavirus pityocampae* on pine foliage. *Journal of Invertebrate Pathology* **7**, 281–284.
Cantwell, G.E. (1967) Inactivation of biological insecticides by irradiation. *Journal of Invertebrate Pathology* **9**, 138–140.
Carruthers, R.I., Ziding, F., Ramos, M.E. and Soper, R.S. (1988a) the effect of solar radiation on the survival of *Entomophaga grylli* (Entomophthorales Entomophthoraceae) conidia. *Journal of Invertebrate Pathology* **52**, 154–162.
Carruthers, W.R., Cory, J.S. and Entwistle, P.F. (1988b) Recovery of pine beauty moth (*Panolis flammea)* nuclear polyhedrosis virus from pine foliage. *Journal of Invertebrate Pathology* **52**, 27–32.

Chaudhari, S. and Ramakrishnan, N. (1988) Effect of temperature sunlight and UV-rays on the infectivity of nuclear polyhedrosis virus of Bihar hairy caterpillar, *Spilosoma obliqua* (Walker). *Journal of Entomology Research* **12**, 109–112.

Clark, E.C. (1955) Observations on the ecology of a polyhedrosis of the Great Basin tent caterpillar, *Malacosoma fragile*. *Ecology* **36**, 373–376.

Clark, E. (1956) Survival and transmission of a virus causing polyhedrosis in *Malacosoma fragile*. *Ecology* **37**, 728–732.

Clark, E. (1958) Ecology of the polyhedrosis of tent caterpillars. *Ecology* **39**, 132–139.

Cory, J.S. and Entwistle, P.F. (1990) The effect of time of spray application on infection of the pine beauty moth, *Pannolis flammea,* (Den and Schift) (Lep., Noctuidae) with nuclear polyhedrosis virus. *Journal of Invertebrate Pathology* **110**, 235–214.

Cox, C.S. (1987) *The Aerobiological Pathway of Microorganisms*. Chichester, UK: Wiley.

David W.A.L. (1969) The effect of ultraviolet radiation of known wavelength on a granulosis virus of *Pieris brassicae*. *Journal of Invertebrate Pathology* **14**, 336–342.

David, W.A.L. and Gardiner, B.O.C. (1967) The effect of heat. cold and prolonged storage on a granulosis virus of *Pieris brassicae*. *Journal of Invertebrate Pathology* **9**, 555–562.

David, W.A.L., Gardiner, B.O.C. and Woolner, M. (1968) The effect of sunlight on a purified granulosis virus of *Pieris brassicae* applied to cabbage leaves. *Journal of Invertebrate Pathology* **11**, 496–501.

David, W.A.L., Clothier, S.E., Woolner, M. and Taylor, G. (1971a) Bioassaying an insect virus on leaves II. The influence of certain factors associated with the larvae and the leaves. *Journal of Invertebrate Pathology* **17**, 178–185.

David, W.A.L., Ellaby, S.J. and Gardiner, B.O.C. (1971b) Bioassaying an insect virus on leaves I. The influence of certain factors associated with the virus application. *Journal of Invertebrate Pathology* **17**, 158–163.

David, W.A.L., Ellaby, S.J. and Taylor, G. (1971c) The stability of a purified granulosis virus of the European cabbageworm, *Pieris brassicae*, in dry deposits of intact capsules. *Journal of Invertebrate Pathology* **17**, 228–233.

de Oliveira, M.R.V. (1988) The ecology of *Agrotis segetum*, Turnip moth (Lepidoptera, Noctuidae) nuclear polyhedrosis virus and its uses as a control agent. MSc thesis. University of Oxford, UK.

Doane, C.C. (1969) Trans-ovum transmission of a nuclear-polyhedrosis virus in the gypsy moth and the inducement of virus susceptibility. *Journal of Invertebrate Pathology* **14**, 199–210.

Dobos, P. and Cochran, M.A. (1980) Protein synthesis in cells infected by *Autographa californica* nuclear polyhedrosis virus (AcNPV): the effect of cytosine arabinoside. *Virology* **103**, 446–464.

Elgee, D.E. (1975) Persistence of a virus of the white marked tussock moth on balsam fir foliage. *Canadian Forestry Service. Bi-monthly Research Notes* **31**, 33–34.

Elleman, C.J. (1983). The interrelationship between a baculovirus of *Spodoptera littoralis* and the leaf surface of *Gossypium hirsutum*, with comparative observations on *Brassica oleracea*. DPhil thesis. University of Oxford.

Elleman, C.J. and Entwistle, P.F. (1982) A study of glands on cotton responsible for the high pH and cation concentration of the leaf surface. *Annals of Applied Biology* **100**, 553–558.

Elleman, C.J, and Entwistle, P.F. (1985a) Inactivation of a nuclear polyhedrosis virus on cotton by the substances produced by the cotton leaf surface glands. *Annals of Applied Biology* **106**, 83–92.

Elleman, C.J. and Entwistle, P.F. (1985b) The effects of magnesium ions on the solubility of polyhedral inclusion bodies and its possible role in the inactivation of the nuclear polyhedrosis virus of *Spodoptera littoralis* by the cotton leaf gland exudate. *Annals of Applied Biology* **106**, 93–100.

Entwistle, P.F. (1982) Passive carriage of baculoviruses in forests. *Proceedings of the 3rd International Colloquium on Invertebrate Pathology*, Brighton, pp. 344–351.

Entwistle, P.F. (1986) Epizootiology and strategies of microbial control. In *Biological Plant and Health Protection*. J.A.M. France (ed.). Stuttgart: Fischer Verlag, pp. 257–278.

Entwistle, P.F. (1987) Virus diseases of Limacodidae. In *Slug and Nettle Caterpillars*. M.J.W. Cock, H.C.J. Godfray and J.D. Holloway (eds). Wallingford, UK: CAB International, pp. 213–221.

Entwistle, P.F. and Evans, H.F. (1985) Viral control. In *Comprehensive Insect Physiology, Biochemistry, and Pharmacology*. I. Kerkut and L.I. Gilbert (eds). Oxford: Pergamon Press, pp. 347–412.

Entwistle, P.F., Adams, P.H.W. and Evans, H.F. (1977a) Epizootiology of a nuclear polyhedrosis virus in European spruce sawfly (*Gilpinia hercyniae*): the status of birds as dispersal agents during the larval season. *Journal of Invertebrate Pathology* **29**, 354–360.

Entwistle, P.F., Adams, P.H.W. and Evans, H.F. (1977b) Epizootiology of a nuclear polyhedrosis virus in European spruce sawfly *Gilpinia hercyniae*: birds as dispersal of the virus during winter. *Journal of Invertebrate Pathology* **30**, 15–19.

Entwistle, P.F., Adams, P.H.W., Evans, H.F. and Rivers, C.F. (1983) Epizootiology of a nuclear polyhedrosis virus (Baculoviridae) in European spruce sawfly (*Gilpinia hercyniae*): spread of disease from small epicentres in comparison with spread of baculovirus diseases in other hosts. *Journal of Applied Ecology* **20**, 473–487.

Entwistle, P.F., Evans, H.F., Cory, J.S. and Doyle, C.J. (1990) Questions on the aerial application of microbial pesticides to forests. *Proceedings of 5th International Colloquium on Invertebrate Pathology*, pp. 159–163.

Entwistle, P.F., Forkner, A.C., Green, B.M. and Cory, J.S. (1993) Avian dispersal of nuclear polyhedrosis viruses after induced epizootics in the Pine beauty moth, *Panolis flammea* (Lepidoptera: Noctuidae). *Applied Environmental Microbiology* **45**, 493–501.

Evans, H.F. (1982) The ecology of *Mamestra brassicae* NPV in soil. In *Proceedings of 5th International Colloquium on Invertebrate Pathology*, Brighton, Vol. 3, pp. 159–163.

Evans, H.F. (1994) Laboratory and field results with viruses for the control of insects. In *BCPC Monograph No. 59: Comparing Glasshouse and Field Pesticide Performance II*, H.G. Hewitt, J. Caseley, L.G. Copping, B.T. Grayson and D. Tyson (eds). Farnham: British Crop Protection Council, pp. 285–296.

Evans, H.F. and Allaway, G.P. (1983) Dynamics of baculovirus growth and dispersal in *Mamestra brassicae* L. (Lepidoptera: Noctuidae) larval populations introduced into small cabbage plots. *Applied and Environmental Microbiology* **45**, 493–501.

Evans, H.F. and Entwistle, P.F. (1982) Epizootiology of a nuclear polyhedrosis virus in European spruce sawfly with emphasis on persistence of virus outside the host. In *Microbial and Viral Pesticides*. E. Kurstak (ed.). New York: Marcel Dekker, pp. 449–461.

Evans, H.F. and Harrap, K.A. (1982) Persistence of insect viruses. In *Virus Persistence, SGM Symposium No. 33*. W.J. Mahy, A.C. Minson and G.K. Darby (eds). Cambridge: Cambridge University Press, pp. 57–96.

Evans, H. F., Bishop, J.M. and Page, E.A. (1980) Methods for the quantitative assessment of nuclear polyhedrosis virus in soil. *Journal of Invertebrate Pathology* **35**, 1–8.

Federici, B.A. and Stern, V.M. (1990) Replication and occlusion of a granulosis virus in larval and adult midgut epithelium of the Western grape-skeletonizer, *Harrisina brillians*. *Journal of Invertebrate Pathology* **56**, 401–414.

Felton, G.W., Duffey, S.S., Vail, P.V., Kaya, H.K. and Manning, J. (1987) Interactions of nuclear polyhedrosis virus with catechols: potential for host-plant resistance against noctuid larvae. *Journal of Chemical Ecology* **13**, 947–957.

Feng, Z., Carruthers, R.I., Roberts, D.W. and Robson, D.S. (1988) A phenology model and field evaluation of *Beauveria bassiana* mycosis of the European corn borer, *Ostrinia nubilalis*. *Canadian Entomologist* **120**, 133–144.

Fuxa, J.R. (1995) Ecological factors critical to the exploitation of entomopathogens in pest control. In *Biorational Pest Control Agents: Formulation and Delivery, American Chemical Society Symposium Series No. 595*. F.R. Hall and J.W. Barry (eds). Washington DC: American Chemical Society, pp. 42–66.

Fuxa, J.R. and Richter, A.R. (1996) Effect of agricultural operations and precipitation on vertical distribution of a nuclear polyhedrosis virus in soil. *Biological Control* **6**, 324–329.

Fuxa, J.R. and Tanada, Y. (1987) *Epizootiology of Insect Diseases*. New York: Wiley.

Gibb, J.A. (1960) Populations of tits and goldcrests and their food supply in pine plantations. *Ibis* **102**, 163–208.

Gilmore, J.E. and Munger, F. (1963) Stability and transmissibility of a viruslike pathogen of the Citrus red mite. *Journal of Invertebrate Pathology* **5**, 141–151.

Gipson, L. and Scott, H.A. (1975) An electron microscope study of effects of various fixatives and thin-section enzyme treatments on a nuclear polyhedrosis virus. *Journal of Invertebrate Pathology* **26**, 171–179.

Gregory, P.H. (1968) Interpreting plant disease dispersal gradients. *Annual Review of Phytopathology* **6**, 189–212.

Griego, V.M., Martignoni, M.E. and Claycomb, A.E. (1985) Inactivation of nuclear polyhedrosis virus (baculovirus subgroup A) by monochromatic UV radiation. *Applied and Environmental Microbiology* **49**, 709–710.

Gudauskas, R.T. and Canerday, D. (1968) The effect of heat, buffer salt and H-ion concentrations, and ultraviolet light on the infectivity of *Heliothis* and *Trichoplusia* nuclear polyhedrosis viruses. *Journal of Invertebrate Pathology* **12**, 405–411.

Haggerman, A.E. (1988) Extraction of tannin from fresh and preserved leaves. *Journal of Chemical Ecology* **14**, 453–461.

Haggerman, A.E. and Butler, L.G. (1989) Choosing appropriate methods and standards for assaying tannin. *Journal of Chemical Ecology* **15**, 1795–1810.

Hajek, A.E., Larkin, T.S., Carruthers, R.I. and Soper, R.S. (1993) Modeling the Dynamics of *Entomophaga maimaiga* (Zygomycetes, Entomophthorales) epizootics in gypsy moth (Lepidoptera, Lymantriidae) populations. *Environmental Entomology* **22**, 1172–1187.

Harpaz, I. and Raccah, B. (1974) Nucleopolyhedrosis virus (NPV) of the Egyptian cottonworm, *Spodoptera littoralis* (Lepidoptera: Noctuidae): temperature and pH relations, host range and synergism. *Journal of Invertebrate Pathology* **32**, 368–372.

Harrap, K.A. (1969) The structure and replication of some insect viruses. DPhil thesis. University of Oxford, UK.

Hattori, R. and Hattori, T. (1993) Soil aggregates as microcosms of bacteria–protozoa biota. *Geoderma* **56**, 493–501.

Hoffmaster, R.N. and Ditman, L.P. (1961) Utilization of a nuclear polyhedrosis to control the cabbage looper on cole crops in Virginia. *Journal of Economic Entomology* **54**, 921–923.

Hughes, P.R. and Wood, H.A. (1981) A synchronous per oral technique for the bioassay of insect viruses. *Journal of Invertebrate Pathology* **37**, 154–159.

Hunter, D.K., Hoffman., D.F. and Collier, S.J. (1973) Pathogenicity of nuclear polyhedrosis virus of the Almond moth, *Cadra cautella*. *Journal of Invertebrate Pathology* **21**, 282–286.

Ignoffo, C.M. (1966) Effects of temperature on mortality of *Heliothis zea* larvae exposed to sublethal doses of a nuclear polyhedrosis virus. *Journal of Invertebrate Pathology* **8**, 290–292.

Ignoffo, C.M. and Garcia, C. (1992) Combinations of environmental factors and simulated sunlight affecting activity of inclusion bodies of the *Heliothis* (Lepidoptera: Noctuidae) nuclear polyhedrosis virus. *Environmental Entomology* **21**, 210–213.

Ignoffo, C.M. and Shapiro, M. (1978) Characterisation of baculovirus preparations processed from living and dead larvae. *Journal of Economic Entomology* **71**, 186–188.

Jaques, R.P. (1968) The inactivation of the nuclear polyhedrosis virus of *Trichoplusia ni* by gamma and ultraviolet radiation. *Canadian Journal of Microbiology* **14**, 1161–1163.

Jaques, R.P. (1985) Stability of insect viruses in the environment. In *Viral Insecticides for Biological Control*. K. Maramorosch and K.E. Sherman (eds), New York: Academic Press, pp. 285–360.

Jaques, R.P. and Houston, F. (1969) Tests on microbial decomposition of polyhedra of the nuclear-polyhedrosis of the cabbage looper, *Trichoplusia ni*. *Journal of Invertebrate Pathology* **14**, 289–290.

Johnson, D.W., Boucias, D.B., Barfield, C.S. and Allen, G.E. (1982) A temperature-dependent developmental model for a nucleopolyhedrosis virus in Velverbean caterpillar, *Anticarsia gemmatalis*, (Lepidoptera: Noctuidae). *Journal of Invertebrate Pathology* **40**, 292–298.

Jurkovicova, M. (1979) Activation of latent virus infections in larvae of *Adoxphyes orana* (Lepidoptera: Tortricidae) and *Barathra brassicae* (Lepidoptera: Noctuidae) by foreign polyhedra. *Journal of Invertebrate Pathology* **34**, 213–223.

Kaupp, W.J. (1981) Studies of the ecology of the nuclear polyhedrosis virus of the European pine sawfly, *Neodiprion sertifer*. DPhil thesis. University of Oxford, UK.

Kaya, H.K. (1982) Parasites and predators as vectors of insect diseases. In *Proceedings of the 3rd International Colloquium of Invertebrate Pathology*, Brighton, pp. 39–44.

Keating, S.T., Yendol, W.Y. and Schultz, J.C. (1988) Relationship between susceptibility of gypsy moth larvae (Lepidoptera: Noctuidae) to a baculovirus and host plant foliage conditions. *Environmental Entomology* **17**, 952–958.

Keating, S.T., Hunter, M.D. and Schultz, J.C. (1990) Leaf phenolic inhibition of gypsy moth nuclear polyhedrosis virus. *Journal of Chemical Ecology* **16**, 1445–1457.

Kelly, D.C. and Lescott, T. (1983) Baculovirus replication: glycosylation of polypeptides synthesized in *Trichoplusia ni*, nuclear polyhedrosis virus infected cells and the effect of tunicamycin. *Journal of General Virology* **64**, 1915–1926.

Kelly, D.C., Lescott, T., Ayres, M.D., Carey, D., Coutts, A. and Harrap, K.A. (1981) Induction of a nonoccluded baculovirus persistently infecting *Heliothis zea* cells by *Heliothis armigera* and *Trichoplusia ni* nuclear polyhedrosis viruses. *Virology* **112**, 174–189.

Kelsey, J.M. (1960) Interaction of virus and parasites of *Pieris rapae* L. In *Proceedings of the 11th International Congress of Entomology*, Vol. 2, pp. 790–796.

Killick, H.J. and Warden, S.J. (1991) Ultraviolet penetration of pine trees and insect virus survival. *Entomophaga* **36**, 87–94.

Kinsinger, R.A. and McGaughey, W.H. (1976) Stability of *Bacillus thuringiensis* and a granulosis virus of *Plodia interpunctella* on stored wheat. *Journal of Economic Entomology* **69**, 149–154.

Kobayashi, M., Inagaki, S. and Kawase, S. (1981) Effect of high temperature on the development of nuclear polyhedrosis virus in the Silkworm, *Bombyx mori*. *Journal of Invertebrate Pathology* **38**, 386–394.

Krieg, A. (1956) 'Endogene Virusentstelung' und Latenzproblem bei Insektenviren. *Archiv Virus-Forschung* **6**, 472–481.

Larkin, T.S., Carruthers, R.I. and Soper, R.S. (1988) Simulation and object-oriented programming: the development of SERB. *Simulation* **51**, 93–100.

LeBlanc, B.D. and Overstreet, R.M. (1991) Effect of desiccation, pH, heat, and ultraviolet irradiation on viability of *Baculovirus penaei*. *Journal of Invertebrate Pathology* **57**, 277–286.

Lewis, F.B. and Rollinson, W.D. (1978) Effect of storage on the virulence of gypsy moth nucleopolyhedrosis inclusion bodies. *Journal of Economic Entomology* **71**, 719–722.

Lindahl, T. (1993) Instability and decay of the primary structure of DNA. *Nature* **362**, 709–715.

Lorenz, M.G. and Wackernagel (1988) Impact of mineral surfaces on gene transfer by transformation in natural bacterial environments. In *Risk Assessment for Deliberate Releases*. W. Klingmuller (ed.). Berlin: Springer-Verlag, pp. 110–119.

Lühl, von R. (1974) Versuche mit insektenpathogenen Polyederviren und Chemischen Stressoren zur Bekamfung forstschadlicher Raupen. *Zeitschrift für Angewandte Entomologie* **76**, 49–65.

Luthy, P. (1986) Insect pathogenic bacteria as pest control agents. In *Biological Plant and Health Protection*. J.M. Franz (ed.). Stuttgart: Paul Parey, Fischer Verlag, pp. 201–216.

Manjunath, D. and Mathad, S.B. (1978) Temperature tolerance, thermal inactivation and ultraviolet light-resistance of nuclear polyhedrosis virus of the armyworm *Mythimna separata* (Walk.) (Lepid.; Noctuidae) *Zeitschrift für Angewandte Entomologie* **87**, 82–90.

Martignoni, M.E. (1962) A 1ist of papers concerning the transovum transmission of insect viruses. *Mimeographed Series No. 5*, Berkeley, CA: Department of Insect Pathology, University of California.

Martignoni, M.E. (1964) Pathophysiology in the insect. *Annual Review of Entomology* **9**, 179–206.

Martignoni, M.E. and Iwai, P.J. (1977) Thermal inactivation characteristics of two strains of nucleopolyhedrosis virus (baculovirus subgroup A) pathogenic for *Orgyia pseudotsugata*. *Journal of Invertebrate Pathology* **30**, 255–262.
McKinley, D.J., Brown, D.A., Payne, C.C. and Harrap, K.A. (1981) Cross-infectivity and activation studies with four baculoviruses. *Entomophaga* **26**, 79–90.
McLeod, P.J., Yearian, C. and Young, S.Y. (1977) Inactivation of *Baculovirus heliothis* by ultraviolet irradiation, dew and temperature. *Journal of Invertebrate Pathology* **30**, 237–241.
Minion, F.C., Coons, L.S. and Broome, J.R. (1979) Characterization of the polyhedral envelope of the nuclear polyhedrosis virus of *Heliothis virescens*. *Journal of Invertebrate Pathology* **34**, 303–307.
Mohamed, M.A., Coppel, H.C. and Podgwaite, J.D. (1982) Persistence in soil and on foliage of nucleopolyhedrosis virus of the European pine sawfly, *Neodiprion sertifer* (Hymenoptera: Diprionidae). *Environmental Entomology* **11**, 1116–1118.
Moore, D., Bridge, P.D., Higgins, P.M., Bateman, R.P. and Prior, C. (1993) Ultra-violet radiation damage to *Metarhizium flavoviride* conidia and the protection given by vegetable and mineral oils and chemical sunscreens. *Annals of Applied Biology* **122**, 605–616.
Morris, O.N. (1971) The effect of sunlight, ultraviolet and gamma radiation and temperature on the infectivity of a nuclear polyhedrosis virus. *Journal of Invertebrate Pathology* **18**, 292–294.
Mubuta, D. (1985) Ultraviolet protectants of *Panolis flammea* nuclear polyhedrosis virus. MSc thesis. University of Newcastle-upon-Tyne, UK.
Murray, K.D., Elkinton, J.S., Woods, S.A. and Podgwaite, J.D. (1989) Epizootiology of Gypsy moth nucleopolyhedrosis virus. In *Lymantriidae: A Comparison of Features of New and Old World Tussock Moths, General Technical Report NE-123* W.W. Walner and M.A. McManus (eds). Washington, DC: USDA Forest Service, pp. 439–453.
Navon, A. (1993) Control of lepidopteran pests with *Bacillus thuringiensis*. In Bacillus thuringiensis, *An Environmental Biopesticide: Theory and Practice*. P.F. Entwistle, J.S. Cory, M.J. Bailey and S. Higgs (eds). Chichester, UK: Wiley, pp. 125–146.
Nuorteva, M. (1972) Use of the nuclear polyhedrosis virus in the control of the European pine sawfly, *Neodiprion sertifer* (Geoffr.). *Silva Fennica* **6**, 172–186.
Ogaard, L., Williams, C.F., Payne, C.C. and Zethner, O. (1988) Activity persistence of granulosis viruses (Baculoviridae) in soils in United Kingdom and Denmark. *Entomophaga* **33**, 73–80.
Olofsson, E. (1988) Dispersal of the nuclear polyhedrosis virus of *Neodiprion sertifer* from soil to pine foliage with dust. *Entomologia, Experimentalis et Applicata* **46**, 181–186.
Onstad, D.W. and Carruthers, R.I. (1990) Epizootiological models of insect diseases. *Annual Review of Entomology* **35**, 399–419.
Onstad, D.W., Maddox, J.V., Cox, D.J. and Kornkven, E.A. (1990) Spatial and temporal dynamics of animals and the host-density threshold in epizootiology. *Journal of Invertebrate Pathology* **55**, 76–84.
O'Reilly, D.R. and Miller, L.K. (1989) A baculovirus blocks molting by producing ecdysteroid UDP-glucosyl transferase. *Science* **245**, 1110–1112.
O'Reilly, D.R. and Miller, L.K. (1991) Improvement of a baculovirus pesticides by deletion of the *egt* gene. *Bio/Technology* **9**, 1086–1089.
Pawar, V.M. and Ramakrishnan, N. (1971) Investigations on the nuclear-polyhedrosis of *Prodenia litura* Fabricius. II Effect of surface disinfectants, temperature and alkalis on the virus. *Indian Journal of Entomology* **33**, 428–432.
Podgwaite, J.D. and Mazzone, H.M. (1986) Latency of insect viruses. *Advances in Virus Research* **31**, 293–320.
Pozsgay, M., Fast, P., Kaplan, H. and Carey, P.R. (1987) The effect of sunlight on the protein crystals from *Bacillus thuringiensis* var *kurstaki* HD1 and NRD12: a Raman spectroscopy study. *Journal of Invertebrate Pathology* **50**, 246–253.
Pritchett, D.W., Young, S.Y. and Yearian, W.C. (1982) Dissolution of *Autographa californica* nuclear polyhedrosis virus polyhedra by the digestive fluid of *Trichoplusia ni* (Lepidoptera: Noctuidae) larvae. *Journal of Invertebrate Pathology* **39**, 354–361.

Reed, D.K. (1974) Effects of temperature on virus–host relationships and on activity of the nonoccluded virus of citrus red mites, *Panonychus citri*. *Journal of Invertebrate Pathology* **24**, 218–223.
Richards, M.G. (1984) The use of a granulosis virus for control of codling moth, *Cydia pomonella*. PhD thesis. Imperial College, University of London.
Richards, M.G. and Payne, C.C. (1982) Persistence of baculoviruses on leaf surfaces. In *Proceedings of 3rd International Colloquium on Invertebrate Pathology*. Brighton, pp. 296–301.
Richter, A.R., Fuxa, J.R. and Abdel-Fattah, M. (1987) Effect of host plant on the susceptibility of *Spodoptera frugiperda* (Lepidoptera: Noctuidae) to a nuclear polyhedrosis virus. *Environmental Entomology* **16**, 1004–1006.
Russell, D.L. and Consigli, R.A. (1985) Glycosylation of purified enveloped nucleocapsids of the granulosis virus infecting *Plodia interpunctella* as determined by lectin binding. *Virus Research* **4**, 83–91.
Sakkonen, K. (1995) Nuclear polyhedrosis virus of the European pine sawfly, *Neodiprion sertifer* (Geoffr.) (Hym., Diprionidae), retains infectivity in soil treated with simulated acid rain. *Journal of Applied Entomology* **119**, 495–499.
Shiga, M., Yamada, H., Oho, N., Nakaguwa, H., and Ito, Y. (1973) A granulosis virus, possible biological agent for control of *Adoxophyes orana* (Lepidoptera: Tortricidae) in apple orchards II. Semipersistent effect of artificial dissemination into an apple orchard. *Journal of Invertebrate Pathology* **21**, 149–157.
Skuratovskaya, I., Strokovskaya, L. Alexeenico, L., Miriuta, N. and Kok, L. (1984) Replication of baculovirus genome at productive and latent infections: structural aspects. *Abstracts 6th International Congress in Virology, W38–10*, Sendai, Japan. p. 334.
Small, D. (1985) Aspects of the attachment of a nuclear polyhedrosis virus from the Cabbage looper *Trichoplusia ni* to the leaf surface of cabbage (*Brassica oleracea*). DPhil thesis. University of Oxford, UK.
Small, D.A. (1986) Identification of subpopulations within baculovirus polyhedra: possible relevance to attachment to leaf surfaces. *Microbios Letters* **32**, 91–96.
Small, D.A. and Moore, N.F. (1987) Measurement of surface charge of baculovirus polyhedra. *Environmental Microbiology* **53**, 598–602.
Small, D.A., Moore, N.F. and Entwistle, P.F. (1986) Hydrophobic interactions involved in attachment of a baculovirus to hydrophobic surfaces. *Environmental Microbiology* **52**, 220–223.
Stairs, G.R. (1978) Effects of temperature on the rate of development of *Galleria mellonella* and its specific baculovirus. *Environmental Entomology* **7**, 297–299.
Stairs, G.R. and Milligan, S.E. (1979) Effects of heat on non-occluded nuclear polyhedrosis virus (Baculovirus) from *Galleria mellonella* larvae. *Environmental Entomology* **8**, 756–759.
Steinhaus, E.A. and Thompson. C.G. (1949) Preliminary field trials using a polyhedrosis virus to control alfalfa caterpillar. *Journal of Economic Entomology* **42**, 301–305
Stiles, B. and Wood, H.A. (1983) A study of the glycoproteins of *Autographa californica* nuclear polyhedrosis virus (AcNPV). *Virology* **131**, 230–241.
Tanada, Y. (1959) Descriptions and characteristics of a nuclear polyhedrosis virus and a granulosis virus of the armyworm *Pseudaletia unipuncta* (Haworth) (Lepidoptera, Noctuidae). *Journal of Insect Pathology* **1**, 197–214.
Tapp, H. and Stotzky, G. (1995) Insecticidal activity of the toxins from *Bacillus thuringiensis* subspecies *kurstaki* and *tenebrionis* adsorbed and bound on pure and soil clays. *Applied and Environmental Microbiology* **61**, 1786–1790.
Tashiro, H., Beavers, J.B., Grosa, M. and Moffit, C. (1970) Persistence of a nonoccluded virus of the citrus red mite on lemons and in intact dead mites. *Journal of Invertebrate Pathology* **16**, 63–68.
Thompson, C.G. (1959) Thermal inhibition of certain polyhedrosis virus diseases. *Journal of Insect Pathology* **1**, 189–190.
Thompson, C.G. (1978) Nuclear polyhedrosis epizootiology. In *The Douglas-fir Tussock Moth: A Synthesis, Technical Bulletin* 1585. M.H. Brooks, R.W. Stark and R.W. Campbell (eds.). Washington, DC: USDA Forest Service, pp. 136–140.

Thompson, C.G. and Steinhaus, E.A. (1950) Further tests using a polyhedrosis virus to control the alfalfa caterpillar. *Hilgardia* **19**, 411–443.
Thompson, C.G., Scott, D.W. and Wickman, B.E. (1981) Long-term persistence of the nuclear polyhedrosis virus of the douglas-fir tussock moth, *Orgyia pseudotsugata* (Lepidoptera: Lymantriidae), in forest soil. *Environmental Entomology* **10**, 254–255.
Tvermyr, S. (1969) Effect of nuclear polyhedrosis virus in *Neodiprion sertifer* (Geoffr.) (Hymenoptera : Diprionidae) at different temperatures. *Entomophaga* **14**, 245–250.
Vail, P.V. (1981) Cabbage looper nuclear polyhedrosis virus–parasitoid interactions. *Environmental Entomology* **10**, 517–520.
van Beek, N.A.M., Derksen, A.C.G., Granados, R.R. and Hughes, P.R. (1987) Alkaline liberated baculovirus particles retain their infectivity *per os* for neonate Lepidopterous larvae. *Journal of Invertebrate Pathology* **50**, 339–340.
van der Plank, J.E. (1963) *Plant Diseases: Epidemics and Control*. New York: Academic Press.
van Frankenhuyzen, K. (1993) The challenge of *Bacillus thuringiensis*. In Bacillus thuringiensis, *An Environmental Biopesticide: Theory and Practice*. P.F. Entwistle, J.S. Cory, M.J. Bailey and S. Higgs (eds). Chichester: Wiley, pp. 1–35.
Vaughn, J.L. (1972) Long term storage of hemolymph from insects infected with nuclear polyhedrosis virus. *Journal of Invertebrate Pathology* **20**, 367–368.
Weseloh, R.M. and Andreadis, T.G. (1986) Laboratory assessment of forest microhabitat substrates as sources of the gypsy moth nuclear polyhedrosis virus. *Journal of Invertebrate Pathology* **48**, 27–33.
Whitt, M.A. and Manning, J.S. (1988) Stabilization of the *Autographa californica* nuclear polyhedrosis virus occlusion body matrix by zinc chloride. *Journal of Invertebrate Pathology* **51**, 278–280.
Wirth, W.W. (1956) New species and records of biting midges ectoparasitic on insects (Diptera, Heleidae). *Annals of the Entomological Society of America*, **49**, 356–364.
Wirth, W.W. (1972) Midges sucking blood of caterpillars (Diptera: Ceratopogonidae). *Journal of Lepidopterists Society* **26**, 65.
Witt, D.J. and Stairs, G.R. (1975) The effects of ultraviolet radiation on a baculovirus infecting *Galleria melonella*. *Journal of Invertebrate Pathology* **26**, 321–327.
Witt, D.J. and Hink, W.F. (1979) Selection of *Autographa californica* nuclear polyhedrosis virus for resistance to inactivation by near ultraviolet, far ultraviolet, and thermal radiation. *Journal of Invertebrate Pathology* **33**, 222–232.
Wood, H.A., Hughes, P.R. and Shelton, A. (1994) Field studies of the co-occlusion strategy with a genetically altered isolate of the *Autographa californica* nuclear polyhedrosis virus. *Environmental Entomology* **23**, 211–219.
Wood, H.A. and Hughes, P.R. (1995) Development of novel delivery strategies for use with genetically enhanced baculovirus pesticides. In *Biorational Pest Control Agents: Formulation and Delivery, American Chemical Society Symposium Series 595*. F.R. Hall and J.W. Barry (eds). Washington, DC: American Chemical Society, pp. 221–228.
Woods, S.A., Elkinton, J. and Podgwaite, J.D. (1989) Acquisition of a nuclear polyhedrosis virus from tree stems by newly emerged gypsy moth (Lepidoptera: Lymantriidae) larvae. *Environmental Entomology* **18**, 298–301.
Young, E.C. (1974) The epizootiology of two pathogens of the coconut palm rhinoceros beetle. *Journal of Invertebrate Pathology* **24**, 82–92.
Young, S.Y. and Yearian, W.C. (1990) Transmission of nuclear polyhedrosisvirus by the parasitoid *Microplitis croceipes* (Hymenoptera: Braconidae) to *Heliothis virescens* (Lepidoptera: Noctuidae) on soybean. *Environmental Entomology*, **19**, 251–256.

5 Control strategies

INTRODUCTION

There are wide opportunities to use BVs for the development of insect pest control systems relying completely on the use of BVs or for IPM systems that involve several other control methodologies combined with BV use.

Essentially BV control strategies can be divided into two broad categories: biological introductions and spray applications. The effectiveness of either of these can be enhanced by incorporation of approaches that may not in themselves provide control at an adequate level. As an example, while the mainstay for control of a lepidopterous pest on a field crop may be BV spraying, control may be reinforced by methods of soil cultivation conducive to maintaining a BV reservoir close to the surface from where it is available to contaminate crops by dust and by rain splash.

BIOLOGICAL CONTROL

Classical control

At present the potential strength of BVs as classical agents appears to rest on a single example. This is the appearance in the field in North America, as a presumed accidental introduction from Europe in 1938, of a NPV of European spruce sawfly, *Gilpinia (Diprion) hercyniae*. With very little deliberate assistance from humans this NPV spread effectively throughout a severely infested area of about 12 000 square miles and by 1940 *G. hercyniae* had ceased to be a pest (Balch and Bird, 1944; Bird, 1949). That there have been no further outbreaks of *G. hercyniae* in North America is now attributed to multifactor suppression involving insect parasitoids and NPV disease. Some occasional short-lived host resurgences have been observed in areas where chemical insecticides (e.g. DDT) have been applied against Spruce budworm, *Choristoneura fumiferana*, which are presumed to have had a greater suppressive effect on the parasitoids than on the sawfly. After a lag of three generations parasitoids reappeared and after seven generations (just over 3 years) virus also reappeared. Pest populations were subsequently reduced to their previous stable level (Neilson, Martineau and Rose, 1971). Another *G. hercyniae* outbreak in Wales during 1968–74 also collapsed when a naturally occurring

NPV spread throughout the infestation area from very localized disease epicentres (Entwistle *et al.*, 1983). Published work on this example of classical microbiological control has been summarized by Adams and Entwistle (1981). Attention is drawn to a discussion of the role of parasitoids and NPV disease in the context of the regulation of *G. hercyniae* as an r–K intermediate strategist (p.85; Southwood, 1977).

A detailed study of *G. hercyniae* in Canada and in Wales provided information on the patterns and rates of spread of NPV disease in insect populations (Entwistle *et al.*, 1983) and is the basis of lattice or point introduction BV control methodologies.

Semi-classical biological control

The term semi-classical control is used in the sense of introductions of 'natural enemies', the spread of which results in effective control of limited duration but which require reinforcement by further introductions at a later date if and when the pest shows signs of escaping from control. This situation has also been described as 'long-term control' (Payne, 1986).

The best documented example is the virus control of the coconut palm rhinoceros beetle, *Oryctes rhinoceros* (Coleoptera: Dynastidae). OrV causes a general infection in larvae of *O. rhinoceros* but in adults replicates only in the midgut. The time from infection to death in larvae is between 9 and 25 days but, though longevity is curtailed in adults, the adults may survive for 25–30 days (Figure 5.1) (Bedford, 1981). Adult males are sterilized by infection, while female fecundity is drastically reduced; these are considered to be key aspects of control. General adult behaviour (Zelazny and Alfiler, 1991) appears to be unaffected by disease and so during their movements in the environment they act as mobile virus reservoirs. Adults feed by boring into the apex of living palms, where disease is spread by excretion in feeding galleries, sexually and by excretion in breeding sites, thus compensating for the very limited extra-host persistence of this non-occluded virus. Larvae develop in separate breeding sites – compost heaps, rubbish pits, decaying logs, including palm trunks – so that the inoculum they produce can be available to quite high larval densities. Many *O. rhinoceros* populations are naturally affected by the virus but, in economic terms, only at an enzootic level. Epizootics can be initiated by increasing the amount of available inoculum. Initially this was attempted by introducing macerated infected larvae into breeding sites (e.g. Young, 1974; Bedford, 1981). A more recently adopted method is to collect or rear adults (methods described by Bedford, 1976; Schipper, 1976; Alfiler and Zelazny, 1985), infect them in the laboratory and then release them in the field. Because OrV tends to be unstable, a simple method has been developed whereby inoculum can be extracted from infected adult midguts and stored for several weeks at room temperature or for much longer in a domestic refrigerator (Zelazny, Alfiler and Crawford, 1987) (described in Chapter 25). In this way, effective virus can be distributed to farmers and shipped between countries. Prior to field release, individual adults are inoculated by placing them upside down on the laboratory bench, where they will remain quiescent for a while, and pipetting a small droplet (*c.* 10 μl) of an inoculum in 10% sucrose mixture onto the mouth parts.

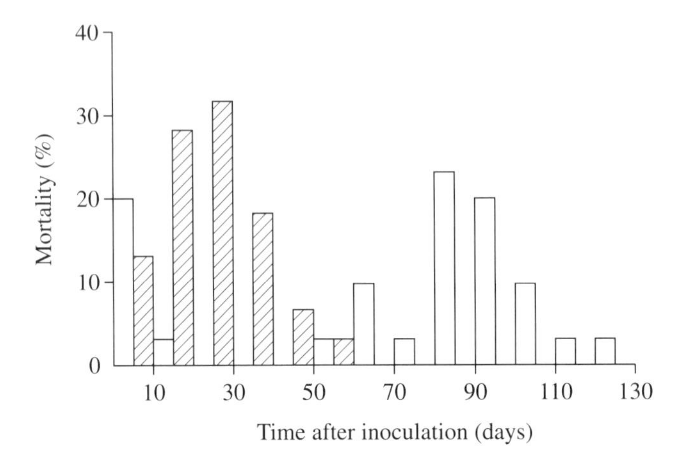

Figure 5.1 Longevity of laboratory-reared *Oryctes rhinoceros* adults injection with 2×10^{-4}/ml of virus-containing haemolymph (▨) or water (□). Combination of three replicate tests where each gave similar results. The percentage mortality is compared for 60 inoculated and 30 control beetles. In field-collected adults in Western Samoa, those found infection survived 38.5 ± 2.4 days and those found healthy survived 59.8 ± 7.8 days. (From Zelazny (1973) *Journal of Invertebrate Pathology*; reproduced by permission of B. Zelazny, Academic Press, Inc.)

This system has been modelled by Hochberg and Waage (1991). The impact of a virus release programme on an outbreak population of *O. rhinoceros* in the Tonga Islands is illustrated in Figure 5.2. Here artificial breeding sites were constructed at one end of the 30 km long island of Tongatapu and inoculated with virus. The incidence of diseased breeding sites throughout the island was then monitored over a subsequent period of 450 days. The pattern of spread graphically depicted in Figure 5.2 shows the development of a classical spatial wave of infection. By day 450, disease had spread over 25 km and was followed by a strong decrease in the incidence of damage to palm fronds (Young, 1974). There is some evidence to show that the rate of spread of healthy beetle infestations (*c.* 1.7 km/month) into areas being newly invaded by the pest and spread of disease are very similar: Tongatapu 1.7–1.9 km/month (Young, 1974), New Britain *c.* 1.0 km/month (Gorick, 1980) and Samoa 0.8–1.6 km/month (Bedford, 1981). This indicates that adult beetle dispersal was not disturbed by infection. From the early stage of dispersal (150 days), the rate of primary dispersal can be calculated for employment, for instance, in designing lattice introductions of the virus to ensure more rapid establishment of control than is possible from a single point or unique epicentre release.

Some idea of the sort of impact that can be achieved from the release of OrV into *O. rhinoceros* infestations can be obtained from Figure 5.3. This suggests a decline in frond injury occurring over the 24–30 months post-introduction, with a subsequent stable phase of low damage for at least the ensuing 24 months (data from Bedford (1981)).

Evidence from Samoa, the area to which this virus was first introduced from Malaysia, suggests commencement of a population upsurge within 2 to 3 years of complete cessation of virus release. Following re-release of virus, there appeared to be a new phase of decline in beetle numbers (Marschal and Iaone, 1982).

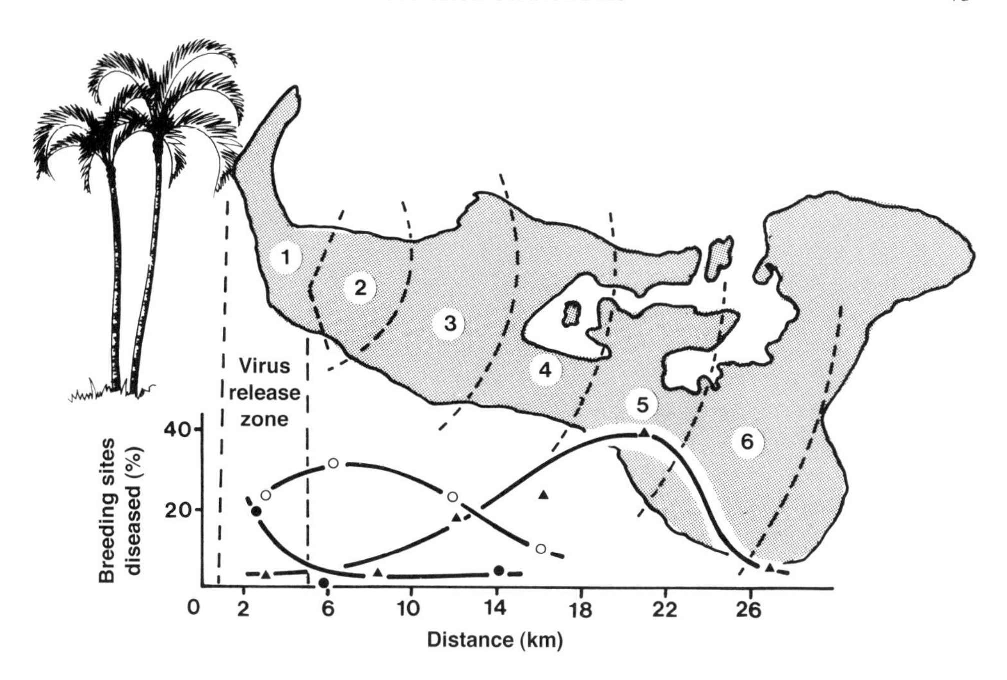

Figure 5.2 Development of an epizootic temporal wave during spread of OrV in *O. rhinoceros*, a coconut palm pest, on the island of Tongatapu, Tonga Islands. The virus was released into rhinoceros beetle breeding sites in zone 1 and the incidence of breeding sites with disease recorded in zones 1–6 at 150 days (●); 250 days (○); 350 days (for graphic clarity not depicted here but see Entwistle *et al.* (1983)) and 450 days (▲). (Drawn from original data in Young (1974) *Journal of Invertebrate Pathology*; reproduced by permission of E. C. Young, Academic Press, Inc.)

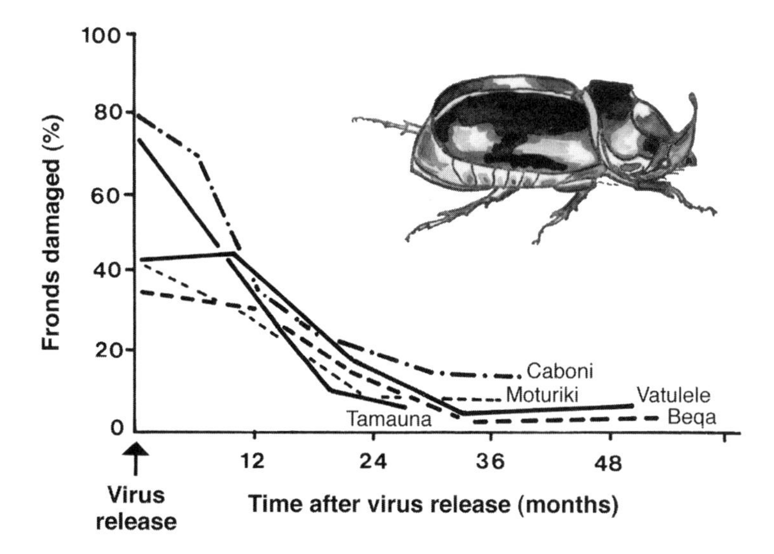

Figure 5.3 The influence of release of OrV into infestations of *O. rhinoceros* in coconut palms in five Fijian Islands, 1970–72. Lines corrected to common zero. (From Bedford (1981) In *Microbial Control of Pests and Plant Diseases*; reproduced by permission of G. O. Bedford, Academic Press, Inc.)

VIRUS RESOURCE MANAGEMENT

Many habitats contain large populations of persistent BV OBs. In some situations, practical use may be made of such reservoirs. The most notable examples relate to insect pests on low-growing crops. *Pieris rapae* GV (Jaques, 1974a) and *Trichoplusia ni* NPV (Jaques, 1974b) accumulate in soil when insecticides are not used and contribute especially to late season epizootics and hence to pest control. The capacity of virus from the soil reservoir to initiate infection cycles has also been strongly indicated for the NPVs of *Pseudoplusia includens* (Young and Yearian, 1979) and *Anticarsia gemmatalis* (Young and Yearian, 1986) on soybean. In a stepwise multiple regression analysis of factors affecting NPV infection in *Spodoptera frugiperda*, it was found that overwintering virus in the soil, and not host population density, was an important variable (Fuxa and Geaghan, 1983). Virus in soil is almost certainly the most important long-term link between sequential natural NPV disease epizootics in *Orgyia pseudotsugata* (Thompson, Scott and Wickman, 1981; Shepherd *et al.*, 1988). When larvae of porina (*Wiseana* spp; Lepidoptera: Hepialidae), which are serious pests of pasture in New Zealand, are affected by an NPV disease they tend to die on, or close to, the soil surface. Under permanent pasture ley conditions, the existence of this localized, concentrated stratum of NPV close to the larval feeding horizon aids the maintenance of populations at subeconomic levels. By comparison, periodic ploughing and reseeding, as a means of maintaining pasture productivity, causes dispersal of the NPV through the top soil and removes a restraint on *Wiseana* population growth. Development of simple management systems for existing pasture, or for developing native tussock grassland, involving reseeding accompanied by minimal soil surface disturbance permitted continued pasture productivity together with maintenance of *Wiseana* control without the need for any specific microbiological or chemical insecticidal treatments (Kalmakoff and Crawford, 1982).

The possibilities of management of soil-borne reservoirs of viruses of other pests have been demonstrated but have almost certainly been unexploited. An NPV epizootic in the South African butterfly *Colias electo* can be induced by overhead irrigation or even by dragging a branch, to knock down caterpillars to contact the soil, through lucerne fields (Polson and Tripconey, 1970). Identification of movement of NPV of *Colias (eurytheme) philodice* in irrigation water in California as a cause of epizootics suggested another method of virus resource management (Thompson and Steinhaus, 1950). It is suggested water sprays might be employed to mobilize viral and other disease inocula on the underside of oil palm fronds, and that this might be especially useful in the dry season when outbreaks of Limacodidae and other groups of Lepidoptera tend not to be suppressed by diseases (Entwistle, 1986). Appropriate tractor-mounted spray rigs exist in, for instance, Malaysian oil palm plantations.

Auto-dissemination

The dispersal of viruses by the host itself, especially by the adult, is common in insects. It is the major dissemination route of virus infection in coconut rhinoceros beetle, *O. rhinoceros*, of an NPV in such diprionid sawflies as *G. hercyniae* and with the GV of the zygaenid moth *Harrisina metallica (brillians)* which is known, *inter alia*, to infect adult midgut (Federici and Stern, 1990). Two other putative BVs are almost certainly adult dispersed. These are the non-occluded, rod-shaped viruses that replicate in the salivary glands of adult tsetse flies (*Glossina* spp.) (Odindo, 1988) and the large narcissus bulb fly (*Merodon equestris*) (Amargier *et al.*, 1979). In addition, the short rod-shaped non-occluded virus of citrus red mite, *Panonychus citri*, is also adult transmitted, probably most effectively when the adult is wind blown (Gilmore and Munger, 1963; Tashiro *et al.*, 1970). Though outside the scope of this book, it is worth recording that CPVs can have a pattern of tissue tropisms very similar to those of NPVs infecting diprionid sawflies. Thus adult Lepidoptera can disperse CPVs which can be transmitted to their progeny via egg surface contamination. In contrast to NPVs in sawflies, CPVs in Lepidoptera frequently affect pupal stages and can also result in the emergence of deformed adults or adults with reduced fecundity (see Katagiri 1981; Hukuhara, 1985). Iridoviruses in, for example, *Aedes taeniorrhynchus* (Linley and Nielsen, 1968a, 1968b), and probably also *Simulium* species, also disperse with the host adult stage.

In contrast to these last examples, Lepidoptera, which of course constitute a major taxon of pest insects, appear seldom to disperse BV diseases effectively. The possible exception to this statement is the carriage of latent, inapparent infections, the impact of which has yet to be assessed. The spread of infection by ballooning (aerial drift on threads of silk) neonate larvae appears to be especially associated with the lepidopterous family Lymantriidae (e.g. *L. dispar*, the gypsy moth) and represents a habit essentially absent from most of the major pest genera. It appears to be involved in the spread of NPV infection.

In discussing the possible employment of auto-dissemination as a contribution to pest control, it is convenient to consider two cases: one, where auto-dissemination is a natural major component of the host–virus biology and the other in those pests where it is absent or is a minor phenomenon but can be induced.

Natural auto-dissemination

Aspects essential to the effectiveness of this method of pathogen dispersal are that any infection in the adult should not cause an excessive decrease in dispersal capacity or that longevity should not be unduly curtailed. The longevity of *O. rhinoceros* adults is reduced by about half, but this is not sufficient to prevent a satisfactory degree of disease spread. Following demonstration of the susceptibility to OrV of *Papuana uninodis* (Zelazny *et al.*, 1988), a dynastid beetle pest of taro and other common commercial tubers in the Pacific region, it is possible that control may be implemented using a system similar to that successful with *Oryctes*.

As far as is known, the longevity and reproductive capacity of diprionid sawflies is unaffected by NPV infection (James, 1974; Smirnoff, 1962) but any influence of infection on their dispersal capacity has yet to be investigated. Smirnoff (1962) demonstrated that adults of *Neodiprion swainei* developed from NPV-infected larvae could propagate the disease. Between 33 and 95% of their progeny died within 15 days of hatching. Spraying a low concentration of 3×10^4 OB/ml (volume of spray fluid, and hence total virus dose, not stated) in the field on fourth to fifth instar larvae permitted about 70% of individuals to form cocoons from which infected adults emerged in the following year. Smirnoff suggested that an inexpensive approach to control would be spatially discontinuous low-dose application, which would, by a combination of transovum passage and local larval movement, lead to development of disease throughout the sawfly-infested area. The acceptability of such an approach depends on several factors, predominant among which will be the density of the pest population and the damage it can be expected to inflict before control is established in this relatively slow manner. However, the predictability of such an approach will be greatly aided by acquisition and application of information on the innate rate of primary dispersal of disease, which will allow selection of an appropriate spacing for NPV introduction centres within the forest.

Assisted auto-dissemination

The idea of using adults to spread occluded BVs through populations of pest Lepidoptera has been inspected several times. The approach usually adopted has been to attract wild moths to traps in which, before escape, they are surface contaminated by mildly adhesive OB preparations. For instance, adults of *Heliothis* and *Helicoverpa* spp. attracted to a network of ‘black light’ (UV) traps were contaminated by contact with *Helicoverpa zea* NPV formulated in an oily base of ground almond shells (Falcon, 1975; Gard, 1975). Adult males of *Spodoptera littoralis* were attracted to female sex pheromone-baited traps where they were exposed to a talc plus *S. littoralis* NPV dust (McKinley, 1985). Contamination of females, through mating, led to infection of some of their progeny. A pheromone-baited trap was also devised to contaminate *Plutella xylostella* adults with spores of *Zoophthora radicans* (Fungi: Entomophthorales) (Pell, Wilding and Macauley, 1991). Jackson *et al.* (1992) used a powder bait containing AcMNPV in pheromone traps to contaminate male *Heliothis virescens* moths that subsequently passed the NPV to females during mating. However, they found that only a maximum of around 8% of eggs laid by the females were NPV contaminated and, of these, only a maximum of 12% of larvae died from the virus. The authors concluded that although the method resulted in some dissemination of virus it was not economically viable. A similar approach was adopted by Vail, Hoffman and Tebbets (1993), who contaminated males of *Plodia interpunctella* with GV in pheromone traps. They demonstrated successful transfer of GV to females and subsequent larval infection of 60% and 50% in the first and second generations following release.

However, the value of such methods is almost undoubtedly to be seen as a contribution to control and not as leading to fully effective control itself, except in situations where a particular virus is able to become epizootic within an acceptably brief time. This view is supported by work in which virus was applied to the abdominal terminalia of Lepidoptera. Though the proportion of progeny initially infected (via egg surface contamination) may be quite high, decline with time can be rapid (Figure 5.4). To counter such limited persistence, it might be worth attempting to feed occluded BVs to adult Lepidoptera provided, as seems likely, the OBs would pass unaffected through the adult gut. In this way

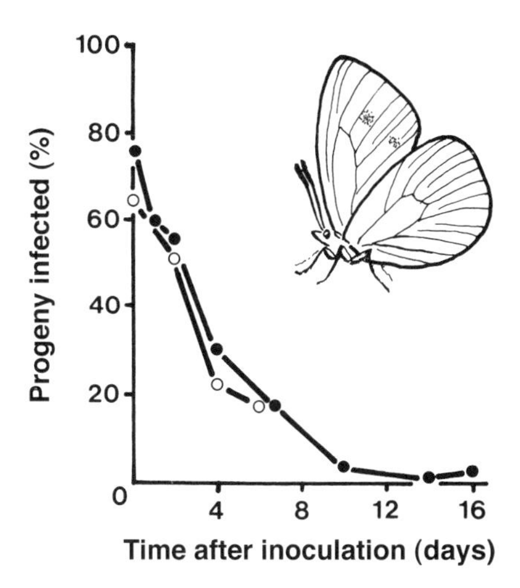

Figure 5.4 Transovum passage of BVs applied to the tips of the abdomen of adult female Lepidoptera to progeny: PbGV (●) (from Tatchell, 1981); NPV of *Colias philodice* (*eurytheme*) (○) (from Martignoni and Milstead, 1962).

NPV might be induced to act like CPVs in Lepidoptera and NPVs in diprionid sawflies.

Other methods have been employed to attempt auto-dissemination, including a hive-mounted device to use honey bees, *Apis mellifera*, to transfer NPV to *Heliothis* spp. in fields of *Trifolium incarnatum* (Gross, Hamm and Carpenter, 1994). The authors showed a significant increase in NPV infection in *Heliothis* and *Helicoverpa* larvae in bee-foraged fields compared with control plots.

Early introductions of virus

Early introductions of BVs are probably justified only where there is good survival of the virus up to the time of occurrence of susceptible life stages of the host. For instance, spray application of an NPV to the foliage of pine trees too soon before *Panolis flammea* egg hatch led to less control than spraying at about 95% hatch, possibly because of the loss of viral biological activity caused by environmental factors or a 'dilution' of the virus deposit resulting from expansion of the new needles on which the young larvae feed (Cory and Entwistle, 1990).

In addition to preventing early attack, the objective of early introductions in field crops is to avoid the necessity of a series of spray applications later on. Such introductions have employed two main methods: soil treatments and dipping transplants.

Soil treatments

It has been shown that virus may be rain-splashed from the soil onto plants and also that seedlings emerging from NPV-contaminated soil may carry an infecting load of viral inoculum (Bishop *et al.*, 1988). Some Lepidoptera, notably cutworms (*Agrotis* and *Euxoa* spp.), can acquire BVs directly from the soil, which they may regularly contact during natural daily movements. Thus de Oliveira (1988) showed substantial infection in *Agrotis segetum* exposed purely to inoculum applied to the soil.

When a single heavy application jointly of *T. ni* NPV (1.5×10^{13} OB/ha) and *P. rapae* GV (6×10^{13} OB/ha) was made to soil in Canada, economically acceptable pest suppression resulted on cole (brassica) crops at a similar cost to that incurred in application of a series of sprays throughout the crop season (Jaques, 1970). Attempts to duplicate this method in the USA (Baugher and Yendol, 1976) and in the UK (Payne, 1982) were unsuccessful, possibly because of differences in weather, pest biology and the rate of BV inactivation connected with differences in the soils or the rate of return of inoculum from soil to the crop plants. In contrast to persistence of greater than 95 weeks in the top 20 cm of soil for *T. ni* NPV in Canada (Jaques, 1964, 1975), the concentration of *Mamestra brassicae* NPV in the UK declined by 90% in 1 year (Evans, 1982) and in soil with a similar pH and mechanical analysis the half life of *P. rapae* GV was only 8–9 weeks (Payne, 1982).

Dipping transplants

Some field crops are grown from transplanted seedlings rather than by direct seed drilling, e.g. most brassica (cole) crops. Pre-planting dipping of brassicas in an NPV suspension gave 90% protection against *T. ni* during 84 days of sampling. Each plant received about 3×10^{8} OBs, the equivalent of about 7×10^{12} OB/ha (Ignoffo *et al.*, 1980), a figure which compares favourably with dose levels for spraying against Lepidoptera in general on field crops (Entwistle and Evans, 1985) though it is somewhat greater than the total quantity of OBs employed on cole crops in Canada (4–6 sprays at 7.5×10^{11}, equal to 3.0×10^{12} to 4.5×10^{12} OB/ha) (Jaques, 1976). The extra cost in inoculum, however, may be offset against that of spraying operations. Addition of a UV protectant to the BV dip formulation may be a further improvement involving very little extra cost.

Lattice introductions and strip treatments

It has long been realised that, with those BV diseases that have a strong capacity to spread, potential economies are possible by applying inoculum topically rather than overall. Natural spread from these epicentres can lead to widespread disease prevalence. As referred to above, this approach was suggested from the NPV control of Jack pine sawfly, *N. swainei*, employing a weak virus application late in the larval period to produce a population of diseased adults in the next generation (Smirnoff, 1962). The method has greatest value in permanent plantation and forest crops, where the environment is more stable (not subject to the severe annual disturbance imposed by cultural activities in the field crop environment) thus achieving continuity in epizootic growth.

The basis for discontinuous introduction of BVs can be provided by studies on the dispersal rates of disease (Chapter 4). Observations of this type suggest considerable consistency in two major aspects of disease spread. Firstly, the gradient of primary dispersal of disease seems very constant for any particular BV–host

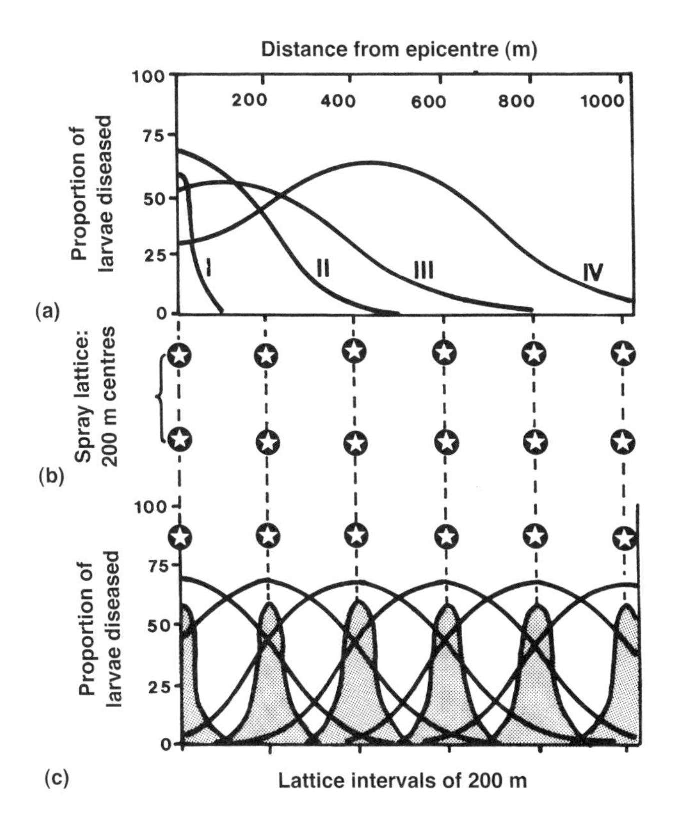

Figure 5.5 The relationship of the observed pattern and the rate of spread of an NPV disease in European spruce sawfly, *Gilpinia hercyniae*. (a) Patterns of primary dispersal of virus from a single epicentre at the *y* axis over four generations (I–IV) following virus introduction. (b) Based on the observed rate and pattern of spread in generations I and II in (a), a spray lattice at 200 m centres is predicted to give adequate virus control within two generations of application. (c) The anticipated pattern of disease over the two generations following NPV introduction at the lattice spacing in (b). It should be noted in (c) that, where the anticipated pattern of dispersal from adjacent points overlaps, no allowance has been made for the possibility of disease incidence being reinforced from other centres. Hence it is likely that (c) indicates a minimum overall view of the incidence of infection at this stage of spread. (From data in Bird and Burk (1961) and Entwistle *et al.* (1983).)

species association. Where it is proposed to introduce a BV at a series of points in a lattice of applications, the gradient of primary dispersal can apparently be reliably employed to select an appropriate lattice spacing. Where such a method is adopted, it is suggested that for maximum economy introduction should be on a square rather than on a triangular lattice (Figure 5.5). Secondly, it seems likely that further spread arising from the primary dispersal gradient will also be constant though, of course, over time more variation may have developed within the system. The phenomenon of the increasing generation of discontinuous, subsidiary epicentres has not been incorporated in Figure 5.5. The actual speed of epizootic growth depends on the development rate of the host insect species. In much of Canada, *G. hercyniae* has two generations a year while in Wales it has only one.

Thus the epizootic pattern will be expressed twice as rapidly in Canada. However, the degree of crop damage involved depends on biological and not real time and hence will be similar in both localities. The situation may also arise where the initiation of control measures must be rapid but where the inoculum supply is poor. In such a case, basing introductions on a wider lattice permits at least a prediction of the likely outcome. If more inoculum can be produced at a later stage, a second, spatially staggered, lattice can be superimposed.

The possibility of strip application has sometimes been considered but the spacing of treated strips is less clear than that for a system of points in a lattice of introduction. This is because, as Gregory wrote (1968), 'close to a strip or area source the gradient is often flattened and near a large area source the flattening may persist for a long distance'. Further studies are needed to explore the quantitative possibilities of the strip-treatment approach to inoculum economy.

The method of implementing lattice, or discontinuous introductions will depend in part on the nature of the crop and of the terrain. Where the crop is not too tall and the terrain accessible, introduction may be by, say, motor-powered knapsack or even tractor-mounted spray equipment. For taller crops (e.g. trees) and uneven terrain pulse spraying from an aircraft will be necessary. The reduced spray volume involved in spatially interrupted applications will be reflected by a considerable saving in aircraft turnaround frequency.

The lattice approach can, of course, be ideally used in the release of diseased adults of *Oryctes* in palm plantations, based on the quantitative data on dispersal available from the Tongatapu study (Young, 1974) and the strong likelihood that the rate of disease spread is a reflection of the innate dispersal rate of the adult beetle, whether healthy or diseased (see above). Pre-infected adults are placed in an open-topped box, containing a layer of vegetable detritus (e.g. decayed wood, bark, etc.), suspended from a tree to avoid rat predation. When conditions are appropriate, adults fly from the boxes.

Integration of beneficial insects

Beneficial insects (parasitoids and predators) are often involved in the effective dispersal of BVs. There are two aspects to the overall picture of their interaction.

Short-term factors

When a BV is sprayed onto a field crop or forest, existing beneficial insects are likely to be heavily involved in disease dispersal once the first phase of infected hosts become available.

Long-term factors

As the joint effects of introduced disease and natural enemies lead to host population decline, so the natural enemy populations will decline, albeit in lag phase. The result will be that as host populations begin to recover (which is not always a certainty since BV introduction can lead to long-term classical biological control, as in the above example of *G. hercyniae*) the incidence of beneficial insects will be low for a while and their contribution to disease dispersal will, therefore, be reduced. However, they will always be relatively higher than in equivalent areas that have been treated with chemical insecticides, as these indiscriminately destroy host, parasitoid and predator.

Beneficial insects may be used directly to introduce and disseminate virus into pest populations and the possibility (so far unexploited) exists to select parasitoids for vagility and/or for their innate frequency of contact with hosts. Ideally such selection needs to be married to a capacity to rear selected parasitoids on a large scale, as can be readily achieved, for instance, with trichogrammatid wasps. A single release into second instar larval populations of *N. sertifer* in North America of about 100 adult *Lophyroplectrus oblongopunctatus* (Hymenoptera: Ichneumonidae) externally contaminated with NPV resulted in disease establishment, albeit at a low level of incidence (Mohamed *et al.*, 1981). Experiments with the predator *Podisus maculatus* (Hemiptera: Pentatomidae) contaminated with NPV reduced *T. ni* populations to zero in 21 days compared with an average of 14 larvae per plant in the presence of *P. maculatus* alone (Biever, Andrews and Andrews, 1982). The presence of 'predators' improved suppression of *H. zea* on NPV-sprayed cotton, but it is not clear if the effect was synergistic or additive (Falcon, 1975).

Birds have been heavily implicated as mortality factors for forest insects, and attempts to increase bird populations by supplying nesting boxes have been made. Whether this would improve the virus control of forest defoliators by elevating the level of avian-disseminated inoculum has not been examined. The western silvereye, *Zosterops gouldi*, in Australia was implicated in the initiation of fresh and distant epicentres of GV disease in populations of the potato tuber moth, *Phthorimaea operculella*; this GV virus can be a very effective control agent (Matthiessen and Springett, 1973). The interaction of birds with infected lepidopterous larvae can be very favourable. NPV-infected larvae of *Spodoptera littura* were preferentially predated by sparrows, *Passer domesticus*, in India, possibly because they became pinkish and more conspicuous and also because they were less active (Chaudhari and Ramakrishnan, 1983); this type of avian predator response is probably quite usual.

Predator activity can be very intense and its encouragement should always be a component of control programmes. For instance, in dry-season-irrigated soybean in Nicaragua the conduct of NPV control experiments was made very difficult because predators swiftly destroyed most of the experimental larvae that had been deliberately introduced.

WORKING WITH PATHOGEN–PATHOGEN AND PATHOGEN–INSECTICIDE (CHEMICAL) INTERACTIONS

A considerable number of studies have been published on the impact, usually measured in terms of host mortality, of an insect pathogen in the presence of another factor such as a second pathogen or a chemical insecticide. Such studies have tended to have one or other of two main aims. Firstly, they may search for interactions in which the total mortality is greater than could be expected from the sum of the mortalities from each of the two mortality-causing entities acting separately, i.e. true synergism. Secondly, they may seek a means of reducing, perhaps for environmental reasons, the dosage of a chemical insecticide by identifying a compatible pathogen for simultaneous application. Here an effect that is the sum of two independently acting mortalities (an additive effect) might be acceptable whilst a synergistic effect would be a bonus. Interactions in which the total mortality is less than the sum of the mortalities to be expected of the individual entities acting separately are, of course, unacceptable; it is defined as antagonism (see below).

To detect and identify the types of interaction involved, it is essential that the test dose of each entity alone be calculated to give less than 50% mortality. Often an ELD or ELC (expected lethal dose or expected lethal concentration) of about 20% is selected as a starting point for studies. This would allow, for instance, a synergistic action to be measured from greater than 40% to about 99%. Interactions are generally measured over a range of concentrations of each entity in combination.

Most studies have concerned measurement of mortality when two substances are administered simultaneously. A good practical justification for this is that it is more economic to spray once than twice. However, it appears that sometimes the optimal type and level of interaction, synergism, is seen only in time-lag presentations, i.e. when a period of time separates exposure of the host to each of the entities under study (e.g. McVay, Gudauskas and Harper, 1977; Savanurmath and Mathad, 1981). Very often pathogens may already be naturally present in a pest population so that a single application of an appropriate second entity may essentially provide a lag phase combination, though with much less precise regulation than when each entity is separately and deliberately introduced. Lag-phase studies have been rather little explored and present a potentially fruitful field for further investigation. Another aspect of the study of interactions that requires clarification concerns pathogen–chemical insecticide combinations. Most, if not all, studies have employed the chemical as a commercial formulation without questioning its constitution. Therefore, the results obtained could not be definitely attributed to the active ingredient itself as they could also be caused by one or more formulating substances or even by a more complex interaction.

In summary, proper studies, which will most frequently seek synergistic interactions with the attendant possibilities of savings in the use of active ingredients (pathogens or pesticides), should incorporate the following considerations:

- the rate of concentration in combinations of the entities
- the possibility of lag-phase benefits obtained by the sequential applications of entity A then B, and entity B then A, covering a range of time intervals
- where one or both components is already formulated, separate testing of the formulation components should ideally be conducted to identify the true synergist/antagonist.

Terminology

Benz (1971) proposed a series of quantitatively based definitions for the classification of interactive mortality data involving two factors.

Independent synergism

Independent synergism is a system of two components acting independently and not interfering with each other (independent action with zero correlation). If P_A is the probability of death by microorganism A and P_B is the probability of death by the second microorganism or chemical, the probability of death by combined action is:

$$P_{A+B} = P_A + P_B\,(1 - P_A) \qquad 5.1$$

or in percentage terms:

$$M_{a+B} = M_A + M_B(1 - M/100) \qquad 5.2$$

Subadditive synergism

A system of two components that together produce an effect greater than independent synergism but less than the algebraic sum of the two single effects is known as a subadditive synergism. A weak potentiating effect is necessary to produce such a result.

Supplemental synergism

Supplemental synergism occurs in a system of two effective components that together produce an effect greater than the algebraic sum of the single effects ($M_A + M_B > M_A + M_B$).

Potentiating synergism

A system of a component A causing an effect M_A and a synergist S whose sole influence is to increase the effect of the actual component ($M_S = 0$) is referred to as potentiating synergism. This type of synergism may be found when non-lethal concentrations of an insecticide are combined with a microorganism.

Coalitive action

A system of two components of which each alone causes no measurable effect but which together produce a significant effect can be termed coalitive. Such a system may be found when an inapparent infectious disease becomes acute under the influence of a chemical stressor in sublethal concentration, or when infection with an externally applied microorganism is possible only in combination with non-lethal or sublethal doses of an insecticide.

Temporal synergism

To these definitions, Benz added the term temporal synergism, where two entities interact to enhance the speed of kill so as to achieve an economic synergism.

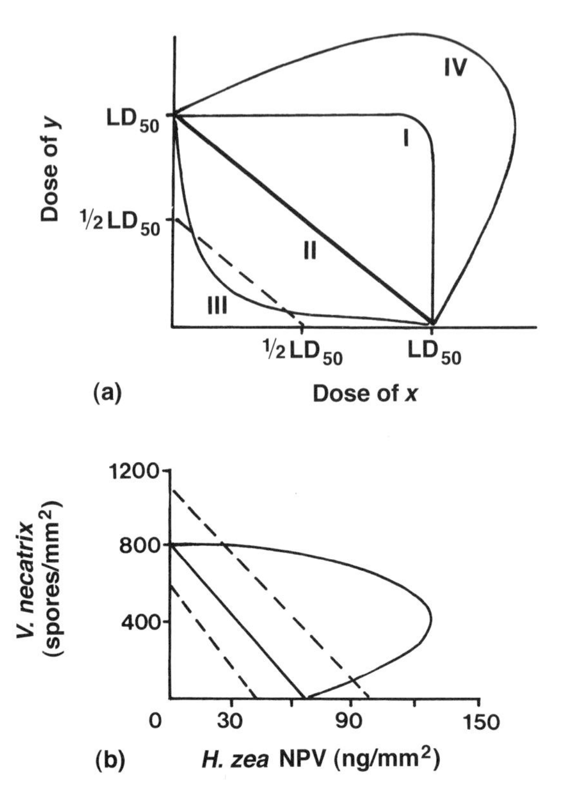

Figure 5.6 (a) The Tammes–Bakuniak method of graphic analysis of interactions of pesticides; I, independence; II, additivity; III, synergism; IV, antagonism. (b) Tammes–Bakuniak analysis indicating an antagonistic interaction of the microsporidian *Variomorpha necatrix* and an NPV in *Helicoverpa zea*. Dotted lines indicate upper and lower 95% confidence limits. (From Fuxa (1979) *Journal of Invertebrate Pathology*; reproduced by permission of J. Fuxa, Academic Press, Inc.)

Analysis of interactions

The Tammes-Bakuniak graphical method

The graphic method of LC_{50} 'isoboles' was first used by Tammes (1964) and was later modified by Bakuniak (1973). It is a simple method of graphical analysis permitting identification of the general types of interaction occurring. It depends on the plotting of 'isoboles' of LC_{50} or LD_{50} values against concentrations or doses of each pathogen/toxicant, as shown in Figure 5.6. It is generally accepted that synergism exists when the isobole crosses a line drawn between the half-LC_{50} or half-LD_{50} value for each agent. This method was employed by Fuxa (1979) to analyse interactions between a microsporidian and representatives of three other major taxa of pathogen in larvae of *H. zea*.

A binomial analysis

McVay *et al.* (1979) employed a simple method to compare expected and observed percentage mortalities. To dctermine expected mortality if the two pathogens act independently of each other:

$$E = O_A + O_B (1 - O_A) \quad 5.3$$

where E is the expected percentage mortality; O_A is the observed percentage mortality from entity A; and O_B that from entity B.

Equation 5.3 is from Finney (1971) and is essentially the same as Equations 5.1 and 5.2 from Benz (1971). Using the Chi-square test:

$$\chi^2 = (O_C - E)^2/E$$

where O_C is the observed mortality for the combination and E is the expected value.

The results can be compared with Chi-square tables for $1df$, $p = 0.05$. If the table value exceeds the calculated, it is concluded that observed mortality for the combination is within the range expected for additive effects. However, if the calculated value exceeds the table value, synergism or antagonism are indicated; these can be identified from the orientation of the test results.

The literature on pathogen–pathogen and pathogen–insecticide interactions has been reviewed a number of times (e.g. Benz, 1971; Krieg, 1971; Entwistle and Evans, 1985). A notable aspect presented by the literature is an absence of pattern in the results and hence an inability to predict the likely outcome of any associations which might be proposed in the future. This must largely be because few studies have been directed towards achieving a fundamental understanding of the nature of interactions so that principles cannot yet be established. A notable exception occurs in the long line of studies on the synergistic action of *Pseudaletia (Mythimna) unipuncta* NPV with one of its GV strains, work reviewed by Tanada (1985) and extended by Uchima, Egerter and Tanada (1989). The nature of the interactive mechanism has been further illuminated by Granados and Corsaro (1990).

A FRAMEWORK FOR CONTROL STRATEGIES

Reference has already been made to the position of *G. hercyniae* in the context of the r–K continuum (p. 71) and the way in which parasitoids and NPV regulate its populations. The r–K theory views the different life strategies of organisms as each representing selection to a part of a continuum. Essentially, r-strategist (r-selected) organisms are characterized by rapid development, a high reproductive rate, great vagility, a body size small for their taxon, polyphagy and a capacity to make use of evanescent habitats. They are essentially geared to an opportunist life style. Many field crop pests tend towards r-selection. At the other extreme of the continuum, K-strategists develop slowly, have a low reproductive rate, display limited dispersality, are large for their taxon and essentially exploit in a specialist manner particular types of resource within their special habitat (Southwood, 1981).

Table 5.1 Routes of intervention with insect viruses in relation to host r–K selection (Entwistle, 1986)

Route of intervention[a]	Pest life strategy		
	r	r–K intermediate	K
Epizootic prediction[b]	Unlikely	Valuable in plantation crops	Unlikely
Classical biological control	Unlikely	Occasionally valuable	Unlikely
Semi-classical biological control	Unlikely	Occasionally valuable	Valuable
Virus resource management	Valuable in pasture and forage crops	Valuable in pasture, forage and plantation crops	Unlikely
Parasite and predator support	Possibly valuable	Possibly valuable	Unlikely
Auto-dissemination	Additive value	Additive value	Valuable
Early introductions	Possibly valuable	Possibly valuable	Occasionally valuable
Lattice introductions	Unlikely	Possibly valuable	Occasionally valuable
Sprays			
Single	Unlikely	Valuable	Sometimes valuable
Multiple	Often essential	Occasionally necessary	Sometime essential

a Terms are discussed in the text and in Entwistle (1986).
b Epizootic prediction: it is sometimes possible to predict epizootics reliably, e.g. on the basis of known environmental virus load and meteorological conditions.

Many pests originating in such comparatively stable habitats as forests tend to be strongly K-selected. Between these poles lie the r-K intermediates with a gradation of life strategies.

The position of insect pests in relation to the r–K continuum and the related importance of several essentially non-microbial control strategies has been discussed by Southwood (1977) and Conway (1981). It is possible to examine pathogenic microorganisms to try to identify those characteristics that make them successful in hosts occupying different parts of the r-K continuum. Such an attempt has been made for entomopathic viruses (Entwistle, 1986). It appears, for instance, that BV diseases of r-strategist Lepidoptera tend themselves to have a very high reproductive rate and to have adapted to the spatially erratic movements of their hosts by the evolution of a capacity for extreme environmental persistence, through the evolution of the OB (which permits prolonged survival in the soil and elsewhere). Large 'resting' BV populations 'await' the eventual and possible return of the r-strategist hosts to their area. In K-strategist hosts, viruses have often adapted more intimate means of transmission between host individuals, which are 'always' present in a particular habitat, which is itself stable, so that prolonged extra-host environmental persistence is unnecessary. An example is OrV. In such situations, the evolution of an OB may not be essential to success. In so far as such a theory is acceptable, and its support or further development depends on enlargement of our knowledge of insect virus biology and ecology, it is possible to suggest which virus-control methodologies are likely to be most successful for particular insect pests given an understanding of the position of the host and its virus(es) in relation to their r–K characteristics. Some indication of the possibilities is indicated in Table 5.1. This conceptual approach has also been discussed by Fuxa (1995) in the context of a valuable examination of factors critical to the exploitation in general of entomopathogens for pest control.

REFERENCES

Adams, P.H.W. and Entwistle, P.F. (1981) An annotated bibliography of *Gilpinia hercyniae* (Hartig), European spruce sawfly. *Commonwealth Forestry Institute (C.F.I.). Occasional Papers No 11*. Oxford: Department of Forestry, University of Oxford.

Alfiler, A.R. and Zelazny, B. (1985) Mass rearing of *Oryctes rhinoceros* (L.) (Coleoptera: Scarabaeidae). *Philippine Entomologist* **6**, 231–234.

Amargier, A., Lyon, J.-P., Vago, C., Meynadier, G. and Veyrunes J.-C. (1979) Mise en évidence et purification d'un virus dans la proliferation monstreuse glandulaire d'Insectes. Etudes sur *Merodon equestris* F. (Diptère, Syrphidae). *Comptes Rendus de l'Academiè des Science Series D (Paris)* **289**, 481–484.

Bakuniak, E. (1973) The synergism evaluating method of two component insecticidal mixtures. *Polish Pismo Entomology* **43**, 395–414.

Balch, R.E. and Bird, F.T. (1944) A virus disease of the European spruce sawfly, *Gilpinia hercyniae* (Htg.), and its place in natural control. *Science and Agriculture* **25**, 64–80.

Baugher, D.G. and Yendol, W.G. (1976) Foliar and soil applications of nuclear polyhedrosis virus to control *Trichoplusia ni* larvae on cabbage. *Proceedings of the 1st International Colloquium on Invertebrate Pathology*, Kingston, Ontario, Canada, pp. 354–355,

Bedford, G.O. (1976) Mass rearing of the coconut palm rhinoceros beetle for release of virus. *Pesticide Abstracts and News Summaries* **22**, 5–10.

Bedford, G.O. (1981) Control of the rhinoceros beetle by baculovirus. In *Microbial Control of Pests and Plant Diseases 1970–1980*. H.D. Burges (ed.). London: Academic Press, pp. 409–426.

Benz, G. (1971) Synergism of microorganisms and chemical insecticides. In *Microbial Control of Insects and Mites*. H.D. Burges and N.W. Hussey (eds). London: Academic Press, pp. 327–355.

Biever, K.D., Andrews, P.L. and Andrews, P.A. (1982) Use of a predator, *Podisus maculiventris*, to distribute virus and initiate epizootics. *Journal of Economic Entomology* **75**, 150–152.

Bird, F.T. (1949) A virus (polyhedral) disease of the European spruce sawfly *Gilpinia hercyniae* (Hartig). PhD thesis. McGill University, Montreal, Canada.

Bird, F.T. and Burk, J.M. (1961) Artificially disseminated virus as a factor controlling the European spruce sawfly, *Diprion hercyniae* (Htg.) in the absence of introduced parasites. *Canadian Entomologist* **93**, 228–238.

Bishop, D.H.L., Entwistle, P.F., Cameron, I.R., Allen, C.J. and Possee, R.D. (1988) Field trials of genetically-engineered baculovirus insecticides. In *The Release of Genetically-engineered Micro-organisms*. M. Sussman, C.H. Collins, F.A. Skinner and D.E. Stewart-Tull (eds). London: Academic Press, pp. 143–179.

Chaudhari, S. and Ramakrishnan, N. (1983) A note on the recovery of a virulent nuclear polyhedrosis virus of *Spodoptera litura* (F.) from the bird droppings. *Indian Journal of Entomology* **45**, 491–492.

Conway, R.G. (1981) Man versus pest. In *Theoretical Ecology*, 2nd edn. R.M. May (ed.). Oxford: Blackwell Scientific, pp. 356–386.

Cory, J.S. and Entwistle, P.F. (1990) The effect of time of spray application on infection of the Pine beauty moth, *Panolis flammea* (Den. and Schiff.) (Lep., Noctuidae) with nuclear polyhedrosis virus. *Journal of Applied Entomology* **110**, 235–241.

de Oliveira, M.R. (1988) The ecology of *Agrotis segetum*, turnip moth (Lepidoptera, Noctuidae) nuclear polyhedrosis virus and its use as a control agent. MSc thesis, University of Oxford, UK.

Elmore, J.C. and Howland, A.F. (1964) Natural versus artificial dissemination of nuclear polyhedrosis virus by contaminated adult cabbage loopers. *Journal of Insect Pathology* **6**, 430–438.

Entwistle, P.F. (1986) Epizootiology and strategies of microbial control. In *Biological Plant and Health Protection*. J.M. Franz (ed.). Stuttgart: Fischer Verlag, pp. 257–278.

Entwistle, P.F. and Evans, H.F. (1985) Viral control. In *Comprehensive Insect Physiology Biochemistry and Pharmacology*, Vol. 12. G.A. Kerkut and L.I. Gilbert (eds). Oxford: Pergamon Press, pp. 347–412.

Entwistle, P.F., Adams, P.H.W., Evans, H.F. and Rivers, C.F. (1983) Epizootiology of a nuclear polyhedrosis virus (Baculoviridae) in European spruce sawfly (*Gilpinia hercyniae*): spread of disease from small epicentres in comparison with spread of baculovirus diseases in other hosts. *Journal of Applied Entomology* **20**, 573–587.

Evans, H.F. (1982) The ecology of *Mamestra brassicae* NPV in soils. In *Proceedings 3rd International Colloquim on Invertebrate Pathology* Brighton, UK. pp. 307–312.

Falcon, L.A. (1975) Patterns of use as they influence virus levels in the environment: chemical controls, biological controls, and application methods. In *Baculoviruses for Insect Pest Control, EPA-USDA Working Symposium*, Bethesda, MD. M. Summers, R. Engler, L.A. Falcon and P. Vail (eds). Washington, DC: American Society for Microbiology, pp. 134–137.

Federici, B.A. and Stern, V.M. (1990) Replication and occlusion of a granulosis virus in larval and adult midgut epithelium of the western grape skeletonizer *Harrisina brillians*. *Journal of Invertebrate Pathology* **56**, 401–414.

Finney, D.J. (1971) *Probit Analysis*. Cambridge: Cambridge University Press.

Fuxa, J.R. (1979) Interactions of the microsporidian *Vairimorpha necatrix* with a bacterium, virus and fungus in *Heliothis zea*. *Journal of Invertebrate Pathology* **33**, 316–323.

Fuxa, J.R. (1995) Ecological factors critical to the exploitation of entomopathogens in pest control. In *American Chemical Society Symposium Series 595*, *Biorational Pest Control Agents: Formulation and Delivery*. F.R. Hall and J.W. Barry (eds). Washington, DC: American Chemical Society, pp. 42–66.

Fuxa, J.R. and Geaghan, J.P. (1983) Multiple-regression analysis of factors affecting prevalence of nuclear polyhedrosis virus in *Spodoptera frugiperda* (Lepidoptera: Noctuidae) populations. *Environmental Entomology* **12**, 311–316.

Gard, I.E. (1975) Utilization of light traps to disseminate insect viruses for pest control. PhD thesis, University of California, Berkeley, CA.

Gilmore, J.E. and Munger, F. (1963) Stability and transmissibility of a viruslike pathogen of the Citrus red mite. *Journal of Insect Pathology* **5**, 141–151.

Gorick. B.D. (1980) Release and establishment of the baculovirus disease of *Oryctes rhinoceros* (L.) (Coleoptera : Scarabaeidae) in Papua New Guinea. *Bulletin of Entomological Research* **70**, 445–453.

Granados, R.R. and Corsaro, B.G. (1990) Baculovirus enhancing proteins and their implication for insect control. In *Proceedings of the 5th International Colloquium on Invertebrate Pathology and Microbial Control*, Adelaide. pp. 174–178.

Gregory, P.H. (1968) Interpreting plant disease dispersal gradients. *Annual Review of Phytopathology* **6**, 189–212.

Gross, H.R., Hamm, J.J. and Carpenter, J.E. (1994) Design and application of a hive-mounted device that uses honey bees (Hymenoptera: Apidae) to disseminate *Heliothis* nuclear polyhedrosis virus. *Environmental Entomology* **23**, 492–501.

Hochberg, M.E. and Waage, J.K. (1991) A model for the biological control of *Oryctes rhinoceros* (Coleoptera: Scarabaeidae) by means of pathogens. *Journal of Applied Ecology*, **28**, 512–532.

Hostetter, D.L. and Beiver, K.D. (1970) The recovery of a virulent polyhedrosis virus of the cabbage looper, *Trichoplusia ni*, from the faeces of birds. *Journal of Invertebrate Pathology* **15**, 173–176.

Hukuhara, T. (1985) Pathology associated with cytoplasmic polyhedrosis viruses. In *Viral Insecticides for Biological Control*. K. Maramorosch and K.E. Sherman (eds). Orlando, FL: Academic Press, pp. 121–162.

Ignoffo, C.M., Garcia, C., Hostetter, D.L. and Pinnell, R.E. (1980) Transplanting: a method of introducing an insect virus into an ecosystem. *Environmental Entomology* **9**, 153–154.

Jackson, D.M., Brown, G.C., Nordin, G.L. and Johnson, D.W. (1992) Autodissemination of a baculovirus for management of tobacco budworms (Lepidoptera: Noctuidae) on tobacco. *Journal of Economic Entomology* **85**, 710–719.

James, R.A. (1974) The effect of nuclear polyhedrosis virus and other factors on the reproductive biology of the European spruce sawfly, *Gilpinia hercyniae* (Htg.). Hymenoptera: Symphyta. Thesis in partial fulfilment of MSc. Imperial College of Science, London, UK.

Jaques, R.P. (1964) The persistence of nuclear-polyhedrosis virus in soil. *Journal of Insect Pathology* **6**, 251–254.

Jaques, R.P. (1970) Application of viruses to soil and foliage for control of the cabbage looper and imported cabbageworm. *Journal of Invertebrate Pathology* **15**, 328–340.

Jaques, R.P. (1974a) Occurrence and accumulation of the granulosis virus of *Pieris rapae* in treated field plots. *Journal of Invertebrate Pathology* **23**, 351–359.

Jaques, R.P. (1974b) Occurrence and accumulation of viruses of *Trichoplusia ni* in treated field plots. *Journal of Invertebrate Pathology* **23**, 140–152.

Jaques, R.P. (1975) Persistence, accumulation and denaturation of nuclear polyhedrosis and granulosis viruses. In *Baculoviruses for Insect Pest Control: Safety Considerations*. M. Summers, R. Engler, L.A. Falcon and P. Vail (eds). Washington, DC: American Society for Microbiology, pp. 90–101.

Jaques, R.P. (1976) Usage of microbial agents for insect control in Canada. In *Proceedings of the 1st International Colloquium on Invertebrate Pathology*, Kingston, Ontario, Canada, pp. 64–68.

Kalmakoff, J. and Crawford, A.M. (1982) Enzootic virus control of *Wiseana* spp. in a pasture environment. In *Microbial and Viral Pesticides*. E. Kurstak (ed.). New York: Marcel Dekker, pp. 435–448.

Katagiri, K. (1981) Pest control by cytoplasmic polyhedrosis viruses. In *Microbial Control of Pests and Plant Diseases 1970–1980*. H.D. Burges (ed.). London: Academic Press, pp. 433–440.

Krieg, A. (1971) Interactions between pathogens. In *Microbial Control of Insects and Mites*. H.D. Burges and N.W. Hussey (eds). London: Academic Press, pp. 459–468.

Linley, J.R. and Nielsen, H.T. (1968a) Transmission of a mosquito iridescent virus of *Aedes taeniorrhynchus*. I. Laboratory experiments. *Journal of Invertebrate Pathology* **12**, 7–16.

Linley, J.R. and Nielsen, H.T. (1968b) Transmission of a mosquito iridescent virus of *Aedes taeniorrhynchus*. II. Experiments related to transmission in nature. *Journal of Invertebrate Pathology* **12**, 17–24.

Marschal, K.J. and Ioane, I. (1982) The effect of re-release of *Oryctes rhinoceros* baculovirus in the biological control of Rhinoceros beetles in Western Samoa. *Journal of Invertebrate Pathology* **39**, 267–276.

Martignoni, M.E. and Milstead, J.E. (1962) Transovum transmission of the nuclear polyhedrosis virus of *Colias eurytheme* Boisduval through contamination of female genitalia. *Journal of Insect Pathology* **4**, 113–121.

Matthiessen, J.N. and Springett, B.P. (1973) The food of the silvereye, *Zosterops gouldi* (Aves: Zosteropidae), in relation to its role as a vector of the granulosis virus of the potato moth *Phthorimaea operculella* (Lepidoptera: Gelechiidae). *Australian Journal of Zoology* **21**, 533–540.

McKinley, D.J. (1985) Nuclear polyhedrosis virus of *Spodoptera littoralis* Boisd. (Lepidoptera, Noctuidae) as an infectious agent in its host and related insects. PhD thesis, University of London.

McVay, J.R., Gudauskas, R.T. and Harper, J.D. (1977) Effects of *Bacillus thuringiensis*-nuclear polyhedrosis virus mixtures on *Trichoplusia ni* larvae. *Journal of Invertebrate Pathology* **29**, 367–372.

Mohamed, M.A., Coppel, H.C., Hall, D.J. and Podgwaite, J.D. (1981) Field release of virus-sprayed adult parasitoids of the European pine sawfly (Hymenoptera : Diprionidae) in Wisconsin. *Great Lakes Entomology* **14**, 177–178.

Neilson, M.M., Martineau, R. and Rose, A.H. (1971) *Diprion hercyniae* (Hartig) European spruce sawfly (Hymenoptera: Diprionidae); biological control programmes in Canada 1959–1968. *Technical Bulletin No. 4*. Farnham, UK: Commonwealth Institute of Biological Control, pp. 136–143.

Odindo, M.O. (1988) *Glossina palidipes* virus: its potential for use in biological control of tsetse. *Insect Science and its Application* **9**, 399–403.

Payne, C.C. (1982) Insect viruses as control agents. *Parasitology* **84**, 35–77.

Payne, C.C. (1986) Insect pathogenic viruses as pest control agents. In *Biological Plant and Health Protection*. J.M. Franz (ed.). Stuttgart: Fischer Verlag, pp. 183–200.

Pell, J.K., Wilding, N. and Macauley, E.D.M. (1991) The use of pheromones to enhance control of *Plutella xylostella* with *Zoophthora radicans*. In the *Working Group Insect Pathogens and Insect Parasitic Nematodes of the 3rd European Meeting on Microbial Control of Pests*, Wageningen, The Netherlands, P.H. Smits (ed.). p. 19.

Polson, A. and Tripconey, D. (1970) A virus disease of the lucerne caterpillar, *Colias electo* Linn. *Phytophylactica* **2**, 17–20.

Savanurmath, C.J. and Mathad, S.B. (1981) Efficacy of fenitrothion and nuclear polyhedrosis virus combinations against the armyworm *Mythimna separata* (Lepidoptera: Noctuidae). *Zeitschrift für Angewandte Entomologie* **91**, 464–474.

Schipper, C.M. (1976) Mass rearing the coconut rhinoceros beetle, *Oryctes rhinoceros* L. (Scarab., Dynastinae). *Zeitschrift für Angewandte Entomologie* **81**, 21–25.

Shepherd, R.F., Bennett, D.D., Dale, J.W., Tunnock, S., Dolph, R.E. and Thier, R.W. (1988) Evidence of synchronized cycles in outbreak patterns of Douglas-fir tussock moth, *Orgyia pseudotsugata* (McDonnough) (Lepidoptera: Lymantriidae). *Memoirs of the Entomological Society of Canada* **146**, 107–121.

Smirnoff, W.A. (1962) Trans-ovum transmission of virus of *Neodiprion swainei* Middleton (Hymenoptera, Tenthredinidae). *Journal of Insect Pathology* **4**, 192–200.

Southwood, T.R.E. (1977) The relevance of population dynamic theory to pest status. In *The Origins of Pest, Parasite, Disease and Weed Problems*. J.M. Cherrett and G.R. Sagar (eds). Oxford: Blackwell, pp. 35–54.

Southwood, T.R.E. (1981) Bionomic strategies and population parameters. In *Theoretical Ecology*, 2nd edn. R.M. May (ed.). Oxford: Blackwell Scientific, pp. 30–52.

Tammes, P.M.L. (1964) Isoboles, a graphic representation of synergism in pesticides. *Netherland Journal of Plant Pathology* **70**, 73–80.

Tanada, Y. (1985) A synopsis of studies on the synergistic property of an insect baculovirus: a tribute to Edward A. Steinhaus. *Journal of Invertebrate Pathology* **45**, 125–138.

Tashiro, H., Beavers, J.B., Grosa, M. and Moffit, C. (1970) Persistence of a nonoccluded virus of the citrus red mite on lemons and in intact dead mites. *Journal of Invertebrate Pathology* **16**, 63–68.

Tatchell, G.M. (1981) The transmission of a granulosis virus following the contamination of *Pieris brassicae* adults. *Journal of Invertebrate Pathology* **37**, 210–213.

Thompson, C.G. and Steinhaus, E.A. (1950) Further tests using a polyhedrosis virus to control the alfalfa caterpillar. *Hilgardia* **19**, 411–445.

Thompson, C.G., Scott, D.W. and Wickman, B.E. (1981) Long-term persistence of the nuclear polyhedrosis virus of the Douglas-fir tussock moth, *Orgyia pseudotsugata* (Lepidoptera: Lymantridae), in forest soil. *Environmental Entomology* **10**, 254–255.

Uchima, K., Egerter, D.E. and Tanada, Y. (1989) Synergistic factor of a granulosis virus of the army worm *Pseudaletia unipuncta*: its uptake and enhancement of virus infection in vitro. *Journal of Invertebrate Pathology* **54**, 156–164.

Vail, P.V., Hoffmann, D.F. and Tebbets, J.S. (1993) Autodissemination of *Plodia interpunctella* (Hübner) (Lepidoptera, Pyralidae) granulosis virus by healthy adults. *Journal of Stored Product Research* **29**, 71–74.

Young, E.C. (1974) The epizootiology of two pathogens of the coconut palm rhinoceros beetle. *Journal of Invertebrate Pathology* **24**, 82–92.

Young, S.Y. and Yearian, W.C. (1979) Soil application of *Pseudoplusia* NPV: persistence and incidence of infection in soybean looper caged on soybean. *Environmental Entomology* **8**, 860–864.

Young, S.Y. and Yearian, W.C. (1986) Movement of a nuclear polyhedrosis virus from soil to soybean and transmission in *Anticarsia gemmatalis* (Hübner) (Lepidoptera: Noctuidae) populations on soybean. *Environmental Entomology* **15**, 573–580.

Zelazny, B. (1973) Studies on *Rhabdionvirus oryctes*. II. Effect on adults of *Oryctes rhinoceros*. *Journal of Invertebrate Pathology* **22**, 122–126.

Zelazny, B. and Alfiler, A.R. (1991) Ecology of baculovirus-infected and healthy adults of *Oryctes rhinoceros* (Coleoptera, Scarabaeidae) on coconut palms in the Philippines. *Ecological Entomology* **12**, 227–238.

Zelazny, B., Alfiler, A.R. and Crawford, A.M. (1987) Preparation of a baculovirus inoculum for use by coconut farmers to control rhinoceros beetle (*Oryctes rhinoceros*). *FAO Plant Protection Bulletin* **35**, 36–42.

Zelazny, B., Autar, M.L., Singh, R. and Malone, L.A. (1988) *Papuana uninodis*, a new host for the baculovirus of *Oryctes*. *Journal of Invertebrate Patholology* **51**, 157–160.

6 Virus production

INTRODUCTION

The production of BVs in the insect host can be easily conducted as a small-scale laboratory process or industrially to provide inoculum to treat large areas of crops. Standardization of the process is greatly assisted by culture of host insects on artificial diets of precisely controlled composition, and under appropriate micro-climatic and axenic conditions in a properly designed facility (Chapter 24).

The alternative method for propagating BVs, namely in cell culture, continues to receive increasing attention (Shuler *et al.*, 1995). The development of cell lines with improved susceptibility to virus, and increased productivity, has raised hopes that virus can be produced at an economic cost and in sufficient quantities to compete successfully with that produced in insects.

This chapter includes a discussion of the relationship between insect host growth and virus productivity (kinetics), the various factors affecting the processes involved (diet composition, pH, temperature etc.), and the ideal facility design for both small- and large-scale productions of BVs. The microbiological examination of virus produced *in vivo* is described followed by a brief introduction to insect tissue culture procedures.

PRODUCTION IN INSECTS

Notable reviews on production of virus in insects include those of Shapiro (1986) and Shieh (1989). An especially useful account of the development of the production process is given by Shapiro, Bell and Owens (1981) for the NPV of *Lymantria dispar* while Shieh (1989) covers the commercial production of *Helicoverpa zea* NPV.

The design of virus production facilities is less well covered than is insect production, but Shieh (1989) gives consideration to commercial facilities, including the housing of the large-capacity equipment required and the automation of processes.

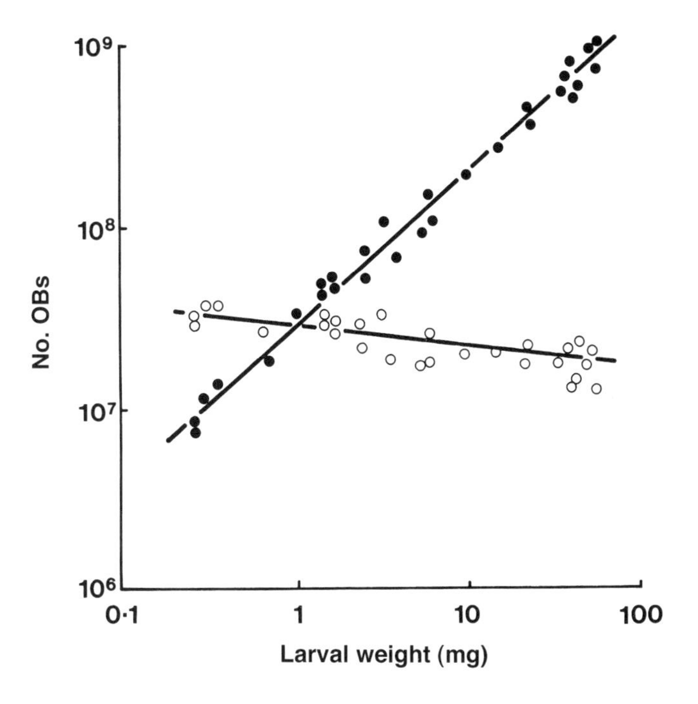

Figure 6.1 The numbers of NPV OBs at death in larvae of different sizes of winter moth, *Operophtera brumata*: data as OBs per larva (●) and OBs per mg larval body weight (○). (After Wigley, 1976, with permission.)

QUANTITATIVE RELATIONSHIPS AND KINETICS OF VIRUS INCREASE

Virus productivity in relation to host weight

There is a close and constant relationship between host weight at death and virus OB productivity; this is applicable throughout the larval period (Figure 6.1). OB production per mg of final body weight varies between 9.2×10^6 and 4.3×10^7 (NPVs in Lepidoptera), 2×10^5 and 5×10^6 (NPVs in diprionid sawflies) and seems to be about 2×10^7 for GVs in Lepidoptera (Entwistle and Evans, 1985). For this reason, the growth size of the host at the end of the logarithmic phase of OB multiplication is one of several considerations involved in the development of efficient virus production processes.

Recognizing this relationship, attempts have been made to maximize the size of the host larvae so that yield per individual insect can be increased. There have been two approaches. In lymantriid moths, fully grown female larvae tend to be heavier than males: they occupy six rather than the more usual five instars. While it may prove impracticable to separate male from female larvae, and this anyway might be wasteful of the insect production process, selection of moth strains with a sex ratio skewed in favour of females might be considered. Experimentally, the value of such an approach has been demonstrated for *L. dispar* (Shapiro *et al.*, 1981). The other approach involves feeding juvenile hormone. For instance, a juvenile hormone mimic, Altosoid, was tested on NPV production in *H. zea* by diet incorporation at 0.4%: at 7 days post-infection there was an increase in larval weight by 28%, accompanied by an increase in OB production of 15.8 % (Shieh, 1989).

Kinetics of occlusion body increase

It is only possible to decide on the optimal time to harvest infected larvae by obtaining some understanding of the kinetics of BV OB multiplication for the particular virus and host species concerned.

OB production involves logarithmic growth over a short period of time. For instance, in *H. zea* this begins 4 days after NPV inoculation and occupies a period of only 24–36 h (Figure 6.2) (Shieh, 1989). Over the following 3 to 4 days (and until mortality was complete) the curve of multiplication rapidly flattens. A closely similar pattern has been found for NPV in *Mamestra brassicae* (Evans, Lomer and Kelly, 1981), *L. dispar* (Shapiro *et al.*, 1981), *Spodoptera exigua* (Smits and Vlak, 1988) and *Helicoverpa armigera* (Teakle and Byrne, 1989). The period occupied by the logarithmic phase varies in the different species but because temperatures at which observations were made are not always stated it is difficult to make close interspecies comparisons. The early phase of OB multiplication has proved difficult

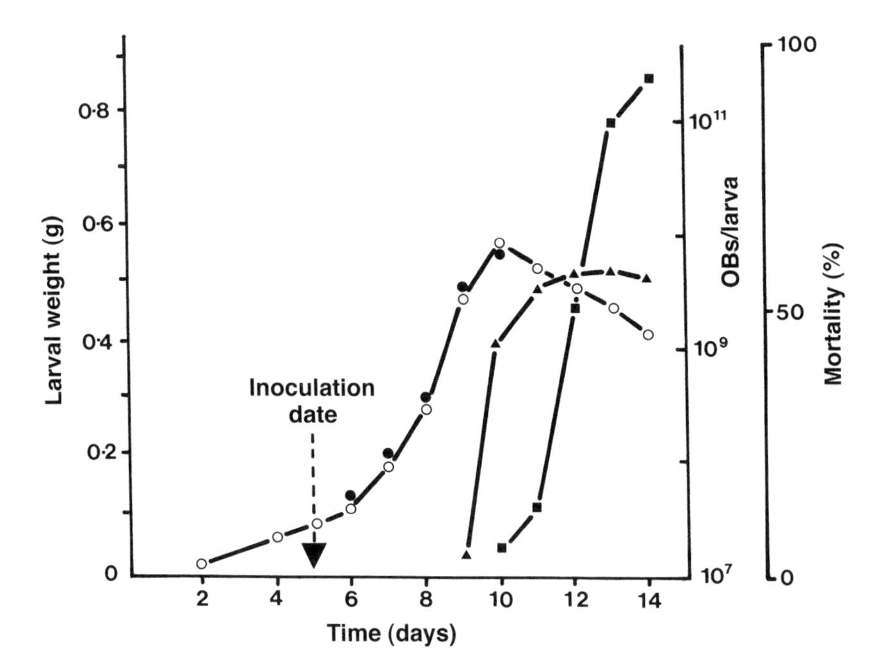

Figure 6.2 Kinetics of NPV production in larvae of *H. zea* on an artificial diet: ●, weight of uninfected larvae; ○, weight of infected larvae; ▲, growth in number of OBs per individual larva; ■, mortality of infected larvae. (From Shieh (1989) *Advances in Virus Research*, reproduced by permission of Academic Press, Inc.)

to quantify using the optical microscope since values below about 1×10^7 OB/ml cannot be assessed accurately. (If necessary, the early phase can be 'illuminated' by employing ELISA with antibodies to OB protein (Evans *et al.*, 1981).

For the limited amount of information available it is interesting to compare curves of larval mortality and of OB multiplication. For *H. zea* (Shieh, 1989) and for *L. dispar* (Shapiro *et al.*, 1981), 50% mortality coincides approximately with the virtual cessation of OB multiplication (Figure 6.2).

The kinetics of OB multiplication clearly vary with the species. Though the literature has not been analysed in detail it seems likely that for GVs the period between infection and logarithmic OB increase may prove to fall into two classes: those where this phase is relatively short and similar to that of NPVs (e.g. *Pieris brassicae*, *Cydia pomonella*) and those where it is abnormally long (e.g. *Agrotis segetum*, *Adoxophyes orana, Spodoptera littoralis*). The actual time span of logarithmic multiplication, however, seems likely to be similar in all species, though this has not yet been clearly established

The relationship between post-infection larval growth and the logarithmic phase of OB replication varies. For those species where infection involves considerable midgut dysfunction, the rate of host growth tends rapidly to decline. This is especially true of the sawflies (Hymenoptera: Symphyta) in which NPVs cause diseases solely of the midgut, and for the GV of *Harrisina metallica* (*brillians*) (Lepidoptera: Zygaenidae). However, in the majority of Lepidoptera, whilst midgut infections occur, they seem usually slight and hence larval growth may continue unaffected. This situation is illustrated for *H. zea* in Figure 6.2 where it appears host growth continued until the close of the phase of logarithmic multiplication of NPV OB.

PRODUCTION IN ALTERNATIVE HOST SPECIES

Production of BV in alternative host species is often desirable where the homologous host species is difficult to culture, to mass produce or is physically small so that there are problems associated with low yields per larva and the extra time needed for handling larger numbers of insects. Certain problems may arise in the use of an alternative host some of which are outlined below.

Excitation of latent infections

Feeding the heterologous virus may stimulate latent viral infections so that these are produced instead (e.g. *Spodoptera* spp.).

Reduced productivity ratio (PR)

Yield in the more convenient host may be less than that of the homologous virus for that host, e.g. yields of *Panolis flammea* NPV were less than those of *M. brassicae* NPV in *M. brassicae* larvae (Table 6.1).

However, some BVs can be successfully produced in alternative host species. For example, AcMNPV is routinely replicated in larvae of *Trichoplusia ni* as is the NPV of *Anagrapha falcifera*; and *P. flammea* NPV is routinely propagated in *M. brassicae*.

Table 6.1 Productivity ratios (PRs) for some BVs

BV[a]/host species	PR[b]	Reference
Lymantriidae		
Euproctis chrysorrhoea	1×10^3	Kelly *et al.*, (1989)
Lymantria dispar	4×10^3	Shapiro *et al.* (1981)
Orgyia pseudotsugata	2.15×10^3	Martignoni (1978)
Noctuidae		
Helicoverpa zea	1×10^5	Shieh (1989)
Helicoverpa armigera NPV in *Heliothis virescens*	2.43×10^5	Bell (1991)
Mamestra brassicae (mean of fourth and fifth instars)	1.28×10^3	Evans *et al.* (1981)
M. brassicae	2.2×10^3	Kelly and Entwistle (1988)
Panolis flammea NPV in *M. brassicae*	8.4×10^2	Kelly and Entwistle (1988)
Spodoptera exigua	1.2×10^6	Smits and Vlak (1988)[c]

[a] The homologous NPV unless otherwise stated.
[b] Where diet surface contamination is used to inoculate, the value for virus (OB/mm^2) is employed in calculating the PR.
[c] These authors showed a quadratic relationship between inoculum rate and yield.

Microclimate

Temperature

Some general comments on the response of the infection cycle to temperature are to be found in Chapter 4. Clearly, inhibition of replication may set in before temperature begins to curtail larval growth: the situation is very variable for different virus–host systems.

The response of *M. brassicae*, in terms of the weight of dead, NPV-infected larvae harvested, was investigated by Kelly and Entwistle (1988). An approximately linear relationship was detected with the highest yields (18.0 g) at 20°C and the lowest at 30°C (3.9 g). The percentage of harvestable (dead) larvae closely followed this trend. At 30°C, many larvae died before infection developed, possibly through desiccation, which also increased with temperature. However, the moisture content of harvested larvae (*c.* 79.4 ± 1.6%) stayed constant.

For *L. dispar* NPV, over the range 23–29°C, productivity of OBs per larva was constant at *c.* 1.1×10^9 but at 32°C it fell to 3.5×10^8 (a threefold reduction factor). The LT_{50} decreased with temperature to 29°C (Shapiro *et al.*, 1981). These results are in general agreement with those of the work on *Lymantria monacha* NPV (Yavada, 1970).

Synchronization of larval growth during incubation of viral diseases can be very sensitive to temperature fluctuations. For *H. zea*, Shieh (1989) found that a deviation of just over 1.0°C could result in 'as much as 50% (larval size) deviation in a 7-day incubation period'. Clearly, regulation of virus production facility temperatures is very important and especially so if a condition of an adopted protocol concerns harvesting before larval mortality is complete, as is done for *H. zea*, *L. dispar* and *S. littoralis* (Jones, 1994).

Humidity

Here it is important to draw a clear distinction between the epizootiological role of humidity in the field and humidity levels to which inoculated larvae are exposed in the virus production facility. In the latter situation, Yavada (1970) and Shapiro *et al.* (1981) found no notable effect of humidity, an observation supported by Kelly and Entwistle (1988). However, a combination of high temperature and low humidity can lead to larval mortality, possibly because of diet desiccation. Under such conditions (*c.* 30–35% relative humidity (RH)) diet replacement may be necessary whilst at 45–60% RH this will usually be unnecessary.

Photoperiod

Very little information is available on the effect of light availability, but for *L. dispar* larvae infected as neonates and held at light–dark regimes of 16:8, 12:12 and 8:16 h, no differences were observed in virus-induced mortality or yield (Shapiro *et al.*, 1981). A light period of 16 h seems common in insectaries.

Plant food, artificial diets and virus production

That the nature of natural insect food influences the infectivity of BVs is well known. Often the process appears to be adversely affected by tannins and polyphenols (Chapter 4). Where natural foods are used (e.g. foliage) in rearing insects for virus production, it is possible these and other adverse nutritional influences can be involved. Another aspect of natural foods is that the content of adverse substances (usually secondary plant metabolites) may vary seasonally, with the result that maintenance of consistency in inoculum effectiveness is difficult.

The nature of artificial diet constitution can also alter the effectiveness of virus production. For instance adjusting the pH range of *L. dispar* diet with 4 M KOH and 1 M HCl showed NPV production to be optimal at pH 5.7 but little less at 5.0 and 4.0. However, production declined towards pH 8.0 (by *c.* 30%), a process accompanied, though not necessarily connected with, an increase in microbial contamination (Shapiro *et al.*, 1981).

A 10-fold increase in a vitamin mix concentration increased *L. dispar* NPV production by *c.* 30% (Shapiro *et al.*, 1981).

The nature and source of protein in the diet should also be considered. For instance, it was found for *L. dispar* NPV that many alternative sources could act as a substitute for the expensive casein. Torula yeast, an inexpensive by-product of the paper pulp industry, is a likely candidate. By increasing the proportion of wheat germ, casein can be eliminated as a diet ingredient (Shapiro *et al.*, 1981). A high wheat germ diet made production of *Euproctis chrysorrhoea* possible in the laboratory: the low productivity ratio (PR) obtained (Table 6.1) might or might not reflect a need to reconsider dietary requirements for this virus–host system (Kelly, Speight and Entwistle, 1989).

Presentation of infecting inoculum

The method of presentation of inoculum may affect the inoculum dose required to cause a high level of infection (usually > 95% is required) and so can have considerable effects on the productivity ratio. Where rearing on artificial diet is employed, there are three main methods of inoculum presentation.

Controlled doses to individual larve

Feeding highly controlled doses on small 'plugs' of diet to individually constrained larvae is often used in bioassay (Chapter 26) but is not practicable for large-scale production. It is useful in precise investigation of production kinetics and productivity ratios (e.g. Evans *et al.*, 1981).

Incorporation into artificial diet

Incorporation of virus into an artificial diet is seldom now considered. The OB bodies must be incorporated before the diet gels (i.e. 50°C for agar) and the possibility of some degree of heat inactivation exists. Such a method is impossible where carageenans, with their higher gelling temperatures, are a diet component. Incorporation of virus into the diet of some pest and insects of stored products, where diet may be mixed at ambient temperatures, seems feasible. However, incorporation in the diet may be desirable for insects that burrow into their food.

Application to the diet surface

At a large-scale or industrial level, application of virus to the diet surface is the usual method employed and is effected by spraying an OB suspension at a pre-calculated concentration and volume per container. However, where larvae are reared in individual cells (e.g. cannibalistic species such as *H. zea*), inoculum is usually injected into each cell by syringe. This process has been automated (Shieh, 1989). An alternative method employed with *S. littoralis* is to spray diet sheets and then impress on these a honeycomb material (Aeroweb) and to insert one larva into each cell before covering the whole (Jones, 1994).

In the now comparatively rare instances where virus production relies on rearing insects on natural food plants, the inoculum may be presented by spray or by dipping foliage (or other), employing a concentration of inoculum pre-determined in tests. The application of this method to production of *Neodiprion sertifer* NPV in the field is particularly interesting because of the interrelationship of host larval population density, amount of infecting inoculum and larval death. A sprayed dose level can be selected appropriate for maximum larval death and this will decline with increasing host density. The dose selected will be appropriate for the instars to be infected. In this way the productivity ratio can be maximized.

The productivity ratio

The usual objective in virus production is to achieve maximum larval size by the time the brief logarithmic multiplication phase of OBs is complete. This can be attained, on the one hand, by feeding a low dose some considerable period before this or, on the other hand, by feeding a much larger dose of inoculum with a consequently briefer interval before maximum OB production occurs.

The difference between the size of the infecting inoculum and the per-larval OB yield is expressed as the **productivity ratio** (PR):

$$PR = \text{No. OBs yielded/larva} \div \text{No. infecting OBs}$$

Clearly, in the last instance described above, the productivity ratio will be low. It will be optimized by presentation of that minimal dose which will provide maximum yield. It was shown by Evans *et al.* (1981) that between the first and fifth instars of *M. brassicae*, the productivity ratio declined from 83 500 to 1280 (a mean value for fourth and fifth instars). However, because of general inconvenience, few workers would choose to utilize early instars for virus production.

There seems to be no precise guide to predicting the optimal productivity ratio and it seems necessary to conduct an investigation for each virus–host species system of interest. A particularly useful example of such an investigation was described by Bell (1991) for *H. armigera* NPV production in *Heliothis virescens* larvae. It was shown that the exact time for optimal harvesting can be critical. Under Bell's conditions, harvesting at day 7 could increase yield by *c.* 1.35-fold. Lower doses of inoculum actually provided higher yields. Bell tested inoculum doses (per square millimetre artificial diet surface) between 54 and 2708 OBs. Plotting log yield against log inoculum for day 7 (Figure 6.3) indicated a linear

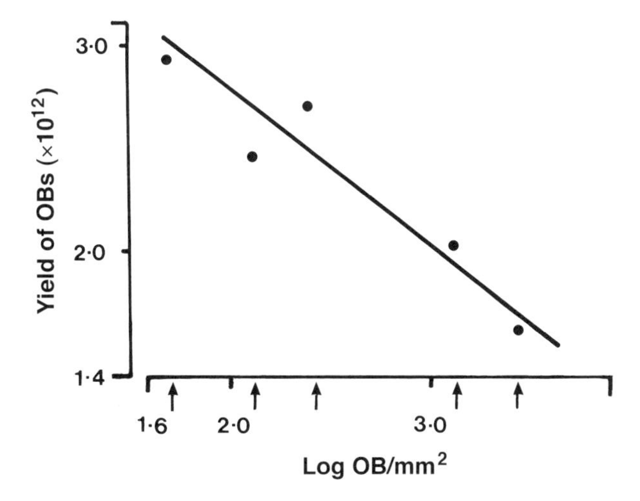

Figure 6.3 Yield of OBs of *H. armigera* NPV in *H. virescens* larvae. The inoculum was presented on the surface of an artificial diet at five concentrations (arrows on abscissa) and the yield assessments were made 7 days post-inoculation (Bell (1991) *Journal of Entomological Science*; reproduced by permission of the Georgia Entomological Society Inc., M. R. Bell.)

decline with increasing inoculum size. Bell concluded that for this (laboratory) system a dose of 54 OB/mm^2 would be optimal.

Values for productivity ratio published in the literature vary greatly; some are shown in Table 6.1.

Time of virus harvest and biological activity

Reference has been made elsewhere in this book (p. 33) to observed differences in the biological activity of virus in OBs harvested from living or harvested from dead larvae. This was true for NPV of *H. zea* (Ignoffo and Shapiro, 1978) and for *H. armigera* NPV produced in *H. virescens* (Bell, 1991). For *H. armigera* NPV it was shown that the infectivity was significantly less in larvae dosed at 7 days of age than for those dosed when younger, notably 6 and 5 days (Teakle and Byrne, 1989).

This situation also applied to *S. exigua* NPV, where virus from dead larvae was about 50% more active (Smits and Vlak, 1988).

A possible further dimension was added by Teakle and Byrne (1989), who noted that the infectivity per NPV polyhedron from *H. armigera* larvae infected at daily intervals from 0 to 7 days peaked at day 6 and then declined sharply at day 7.

Genetic drift in host and virus during production

Both the host stock and the virus stock will have been selected for optimal productivity whilst in other respects remaining representative of relevant wild genotypes. In the insectary, deviation from these ideals can occur by inbreeding of stock, especially if this is already of a restricted genetic base. Breeding from survivors of virus-exposed populations is very undesirable as it could possibly lead to selection for a degree of resistance, with a consequent adverse impact on productivity.

Changes in virulence during passage

Loss of infectivity for the homologous host has been reported following repeated passage of an NPV of *Bombyx mori* through *Galleria mellonella* and *Chilo suppressalis*. Its virulence for *C. suppressalis*, however, increased markedly (Aizawa, 1975). Similarly, changes in virulence have been noted for various NPVs passaged in homologous and non-homologus hosts (Tompkins *et al.*, 1981, 1988).

The interaction of factors relevant to virus production

The more important factors, correct manipulation of which should lead to maximization of OB production of the most biologically active virus and to maintenance of virus genetic character, are summarized in Figure 6.4. Interactions between the eight parameters included in this figure have been inadequately investigated and so it is possible useful information remains to be uncovered.

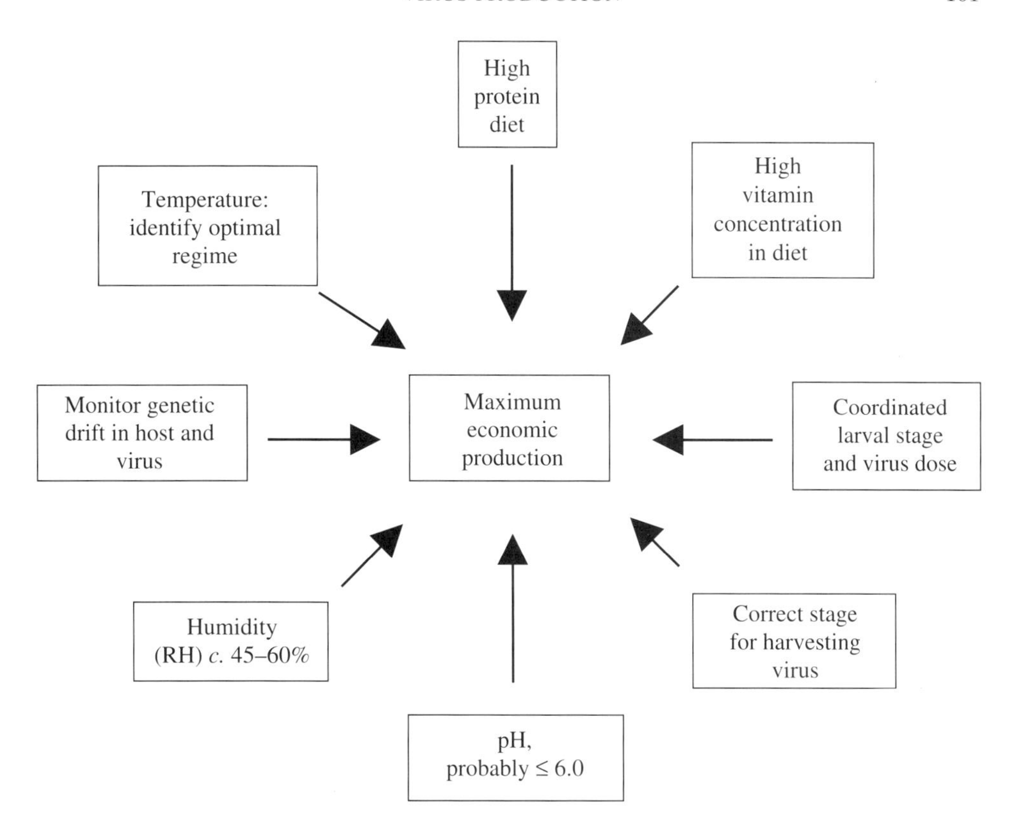

Figure 6.4 Salient factors relevant to efficient, economic production of BVs in insect hosts.

VIRUS PRODUCTION PROCESSES

Small-scale virus production

Small-scale production, as described here, will be adequate to provide supplies of virus for laboratory studies (e.g. genomic characterization, serology, bioassay) and for small-scale field trials. At the laboratory level we have, for instance, produced enough NPV to spray aerially over 500 ha at around 2×10^{11} OB/ha.

The actual practical methods described here are flexible, adaptation being possible to accommodate to immediate circumstances, for instance choice of containers, some aspects of artificial diet composition, methods of harvesting infected larvae, etc. However, basic stipulations such as those summarized in Figure 6.4, above, are mandatory for efficiency, and attention to them will ensure a high level of success.

A suggested protocol is illustrated in Figure 6.5. **The insect production facility must be kept strictly microbiologically isolated from all aspects of the virus production process**. Separate buildings are an ideal but if housed in the one building a positive air pressure system will minimize airborne contamination of the insectary. Basic plans for the design of insect rearing and virus production units are given in

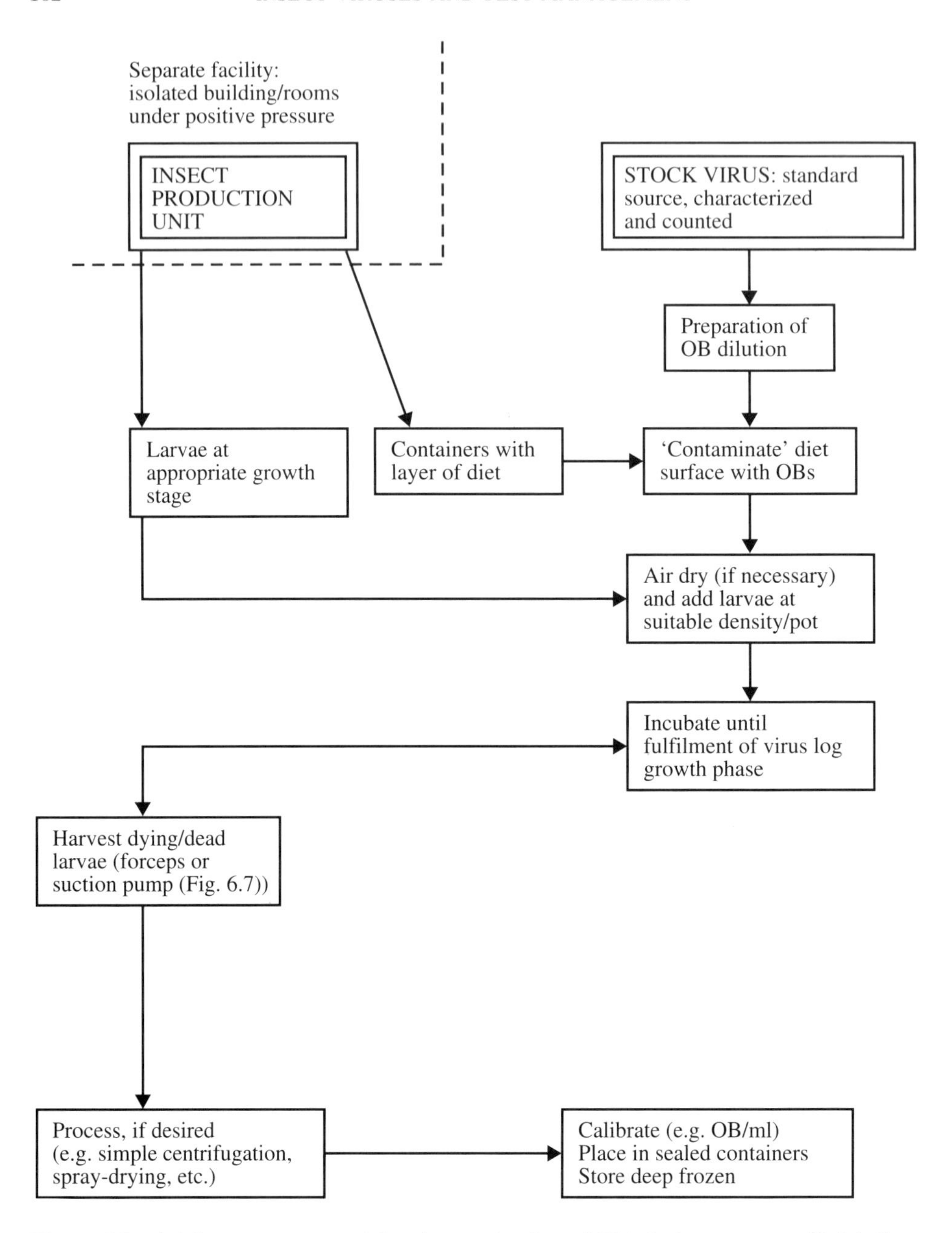

Figure 6.5 A laboratory protocol for the production of BVs in insects on artificial diet.

Figures 24.1 (p. 376) and 25.3 (p. 405), respectively. To ensure maintenance of the viral genotype, dilutions to provide inoculum for infection of successive batches of larvae should be drawn from a single frozen, pre-counted, bulk source. Figure 6.5 is self-explanatory and the process can be conducted from a very small to a modestly large level. Not being automated, it is labour intensive.

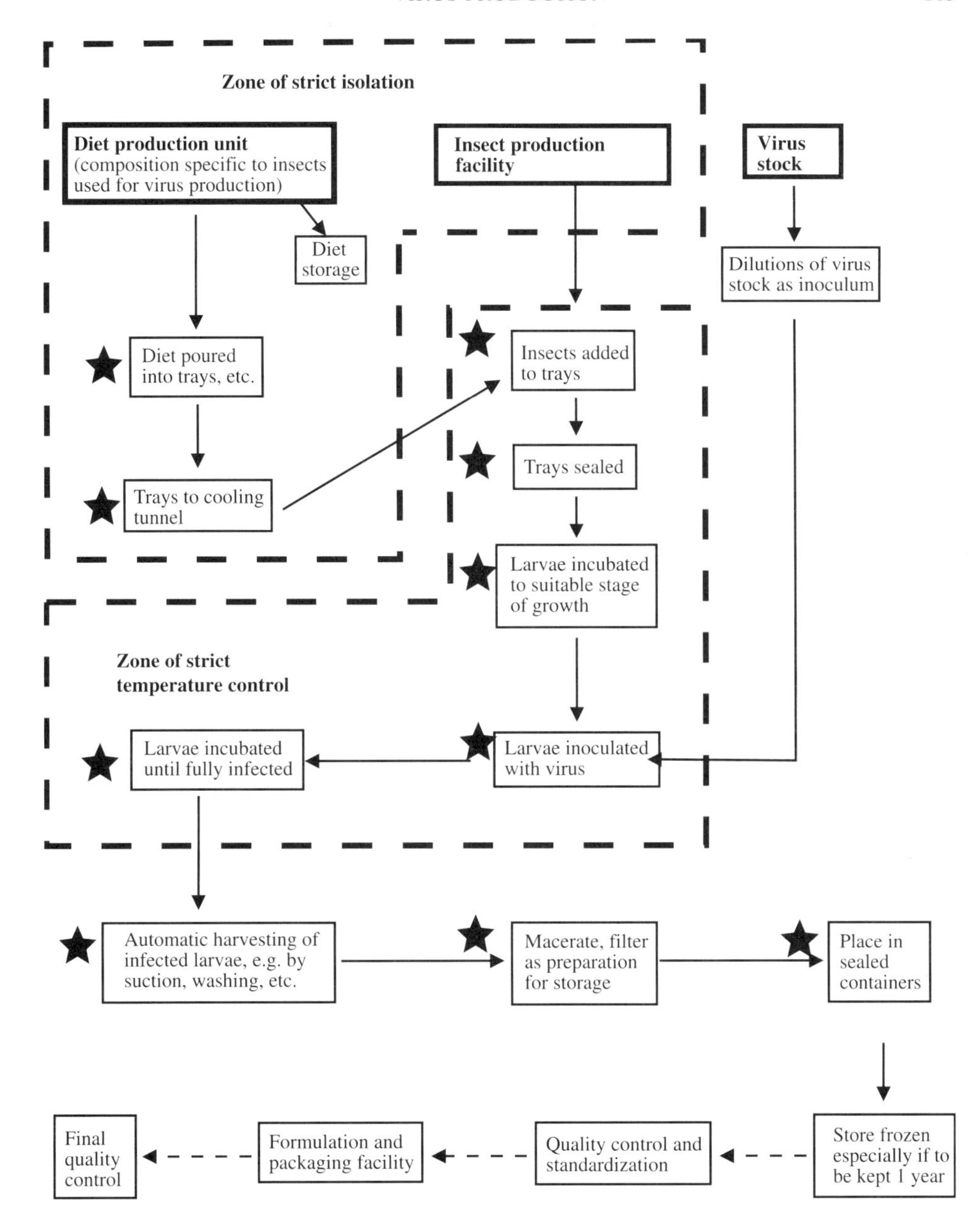

Figure 6.6 A commercial protocol for the production of BVs in insects on artificial diet (★: automatable process).

Large-scale production

Some insect viruses are, or have been, produced on a large scale commercially. Information concerning their current status is given in the chapters in Part Two and in the tables in Chapter 11. The protocol shown in Figure 6.6 contains elements drawn from several sources.

Pilot plant mass-production facilities can be inexpensively developed, in the first instance, by use of a series of interconnected temporary buildings or even mobile homes (caravans). Because of its flexibility, such an approach can aid the further development of organizational concepts. As a general rule, the costs of a production facility should be recouped during the first 3 years from sales of products. This is clearly more likely to be achieved with a simple facility but will be less attainable with custom-made facilities. A guiding principle here is that, to be economic, the latter must aim at a much larger market.

Another major cost component is employee remuneration. The extent to which this is problematic will depend on local wage conditions. In developing nations, labour-intensive operations are often commercially viable. Indeed it could be considered that the mass production of viral insecticides, with an added attraction of export potential, could be ideally conducted where such labour conditions pertain. In technically developed nations, automation is obligatory if the higher wage costs are to be accommodated. Aspects of this are described by Shieh (1989) in an account that provides a good basis for elaboration of technological and automational concepts and advances.

COLLECTING, PROCESSING AND STORING INFECTED HOST MATERIAL

Collection of infected larvae

In small-scale production, infected larvae, provided they have not first become too fragile, can be easily collected using forceps, but this would be tedious and labour intensive at any larger scale. With human operators a suction system can be employed, as shown in Figure 6.7. Here a small vacuum pump provides suction so that larvae can be collected rapidly into a pre-sterilized flask. It is important that the suction pump exhaust air, which will be contaminated with virus, be directed through filters or otherwise exhausted safely. In industrial-scale production, which may well be robotized, it is possible that infected larvae could be washed from their diet substrate, though this may result in some contamination with residual diet and faeces. The amount of diet remaining, however, can be minimized by providing just sufficient to support larval growth until collection.

Processing collected material before storage

There is no reason why some processing of infected larvae should not be done before storage. For instance, material may be macerated, filtered to remove large contaminant particulates (but being careful to avoid loss of appreciable amounts of virus in the process) and, in some species such as *E. chrysorrhoea* and *L. dispar*, the urticaceous larval hairs may be removed (p. 123).

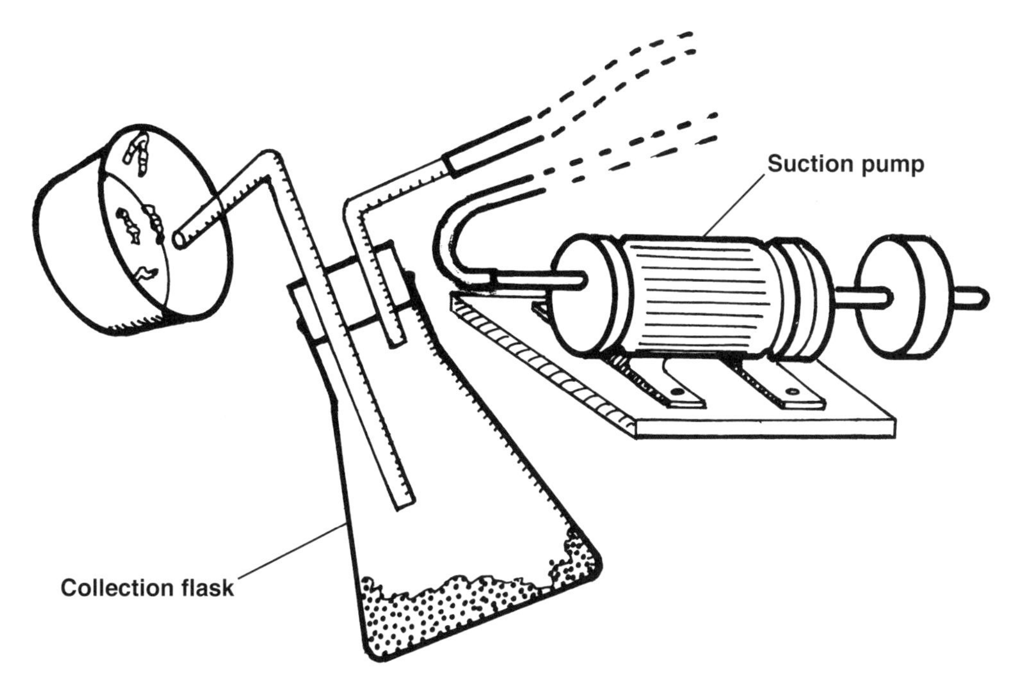

Figure 6.7 Simple small-scale suction apparatus to collect infected larvae.

Storage of collected material

Some virus production systems, e.g. with NPV-infected *S. littoralis*, involve removing infected larvae from the incubator to a lower temperature (<14°C) for 7 days before final storage. This permits the full development of infection while minimizing multiplication of microbial contaminants (McKinley *et al.*, 1989; Anon., 1991). Whether such a system is practised or not, when virus replication has attained the desired level, infected larvae can be stored at a low temperature. At *c.* 4°C the infectivity of BVs will be retained almost indefinitely. Some micro-organisms will, however, continue to multiply so that the material may be so severely contaminated that when formulated it may fail to reach registration standards. Hence, if storage is to be for more than a short time, it is best achieved at –20 or –70°C. It is important that subsequent thawing be achieved rapidly, otherwise there may be some loss of infectivity. For the same reason, repeated cycles of freezing and thawing should be avoided. McKinley *et al.* (1989) freeze-dried and formulated NPV-infected *S. littoralis* larvae and achieved a product that did not need refrigeration during storage or the addition of anti-microbial agents. The question of microbial suppression in the final formulated virus product, which will almost always be required to withstand storage at reasonable temperatures, is addressed in Chapter 7.

PRODUCTION IN INSECTS NOT ADAPTED TO ARTIFICIAL DIETS

Diprionid sawflies and NPVs

Most diprionid sawflies (Hymenoptera, Symphyta Dipronidae), all of which are conifer feeders, cannot easily be cultured in the laboratory or insectary. *Gilpinia hercyniae*, European spruce sawfly, is an exception, but its culture is onerous and the productivity of larvae is low. No sawfly species has been cultured using an artificial diet.

The NPV of *N. sertifer*, the European pine sawfly, may be obtained by hand collection of infected larvae in the field, either in natural disease outbreaks, which are frequent, or from areas of young forest sprayed with the virus. Alternatively Bird (1950) sprayed virus onto field-collected larvae placed on foliage in trays. The larvae of *N. sertifer* feed gregariously and, given adequate food, tend to stay in the trays. If the tray base is wire netting many dead larvae will fall through and can easily be collected. Others must be tediously picked from the foliage and twigs. Rollinson, Hubbard and Lewis (1970) collected larvae that were half to two-thirds developed and held them on pine foliage in stacked cardboard shoe boxes at about 100 larval colonies per box. A suspension of 1×10^7 OB/ml was sprayed on the foliage to 'a point of near dripping'. Larvae began to die at 7 days, peaking at 8.

These methods will also be applicable to the biologically closely similar *Neodiprion lecontei*.

Smirnoff (1964) constructed 'carousels' for *Neodiprion swainei*, the Jack pine sawfly. Indoors these consisted of wooden racks with wire mesh bases, holding in all 1200 pine sprigs with 75 000 larvae. Each rack was sprayed daily with 50 ml at 1×10^6 OB/ml. At 25–30°C, death occurred at 10–12 days. An outdoor 'carousel' had eight arms pinned with sprigs carrying up to 5×10^5 larvae. It was rotated frequently to prevent larvae aggregating on the sunny side. Screen (mesh) trays below caught fallen larvae but permitted frass to pass through. Because of temperature control, indoors production was easier but the outdoor method was less expensive.

Oryctes rhinoceros virus

Rhinoceros beetle is an insect that has not so far been reared on an artificial diet in the conventional sense. However, it can be mass cultured and this is done especially to rear adults that will be infected orally with the virus and later released into palm plantations to initiate economically favourable epizootics. Infection of adult beetles and the method of harvesting midguts and processing them to yield stable virus preparations is described in Chapter 23. As the yield of inoculum from a single adult is sufficient to infect a further 1000 adults, and as rhinoceros beetle is a classic low-population-density pest, it is not at present regarded as necessary to elaborate methods for larger scale production.

MICROBIOLOGIAL EXAMINATION OF VIRUS PRODUCED *IN VIVO*[†]

Preparations of virus propagated *in vivo* are invariably contaminated with a variety of bacteria and fungi (Podgewaite, Bruen and Shapiro, 1983). In order to ensure the safety of the final formulation it is important to exclude the possibility of contamination with human or veterinary pathogens. Safety regulations for BV products may include a limit on the total number of contaminating microorganisms and stipulate the absence of particular groups of bacteria such as coliforms, regardless of species or proven pathogenicity to humans.

The scope of the microbiological screening programme adopted will depend on the objectives to be achieved. For obtaining safety clearance, the minimum requirements will include screening the final product for pathogens and estimating the total bacterial load. It is worthwhile, however, making a detailed microbiological examination of all stages involved, particularly when establishing a new production system. By doing so it should be possible to detect and eliminate the presence of any insect pathogens that are likely to affect adversely the vigour of the rearing colony and hence the yield and quality of product. Such data may be required, also, for registration purposes.

Published data concerning the bacterial flora of microbial pesticides produced from insects are limited (Podgewaite *et al.* 1983), but a number of main subgroups may be distinguished as outlined below.

Gut microflora

The gut microflora of insectary-reared individuals comprises a more limited range of species than that of wild insects, although the actual numbers of organisms present in the former are greater. Enterococci (e.g. *Enterococcus faecalis* and *E. faecium*) dominate. Microaerophilic or anaerobic species may also be present, but as techniques for their isolation are more difficult and rarely used little is known of them at present. Yeasts are found regularly but in fewer numbers than bacteria. The larger number of species identified in the gut microflora of larvae collected from the wild may reflect the greater diversity of microorganisms on natural foods.

Oportunistic saprophytes

Opportunistic saprophytes are found on diet, frass and colonizing larval corpses. They include various members of the genera *Bacillus*, *Alcaligenes*, *Flavobacterium*, *Micrococcus* and *Pseudomonas*. Amongst these, *Bacillus cereus* and *B. sphericus* are the most commonly reported and numerically abundant. All are common in the environment and may gain access via airborne dust, contaminated diet and equipment. *Bacillus* spp. are difficult to eliminate as they form spores that will withstand temperatures commonly used for preparing diet and washing equipment. They cannot, however, withstand autoclaving at 15 psi (121°C) for 30 min.

[†] (contributed by David Grzywacz)

Bacteria infecting humans

Species of bacteria normally restricted to human skin may be recovered from microbial pesticides and probably originate from the personnel handling and processing the product. While many of these pose no hazard, their presence indicates an unsatisfactory level of hygiene.

It is stressed in the published literature concerning microbiological standards for virus pesticides that there should be a complete absence of species of *Shigella, Salmonella* and strains of *Staphylococcus aureus*. There may also be a requirement to report numbers of coliforms (lactose-fermenting bacteria, e.g. *Escherichia, Klebsiella, Citrobacter* and *Enterobacter* spp.). Therefore, screening is directed primarily against these target species.

The product must also be tested for the presence of *B. cereus*. This species is reported to be a common contaminant of NPV produced *in vivo*. Even strains that are not pathogenic to humans are toxic when injected intraperitoneally into mice in large quantities. This poses a particular problem in the interpretation of the mouse toxicity test, which is designed to establish the absence of mammalian pathogens or bacterial toxins from the product.

The procedure for screening a product for bacterial contamination is summarized in Figure 6.8. Details of the tests to be performed are given in Chapter 28.

INSECT TISSUE CULTURE TECHNIQUES[†]

Since Grace (1962) developed the first continuous insect cell lines, many others have been reported (Hink, 1972, 1976, 1980; Hink and Hall, 1989). They have been widely used particularly in the study of insect pathology, insect physiology and in biotechnology. The number of cell lines available, however, is still very limited, particularly considering the number of insect species. In spite of improvements in technique and culture media, it is still difficult to obtain proliferating cell populations from primary cultures of insect tissues. Cells from some species are particularly intractable in spite of repeated attempts to culture them. This lack of success may result from a failure to understand their nutritional requirements and, therefore, to provide them with a suitable medium. The recent development of cell lines with improved susceptibility to virus, and greater productivity, has raised hopes of the possibility of propagating sufficient quantities of BVs for field application at a competitive cost (Granados *et al.*, 1994; Davis and Granados, 1995; Granados and McKenna, 1995; Shuler *et al.*, 1995).

Equipment

Since all cultures are maintained under strictly sterile conditions, a clean room or a laminar airflow cabinet is necessary for handling all sterile materials. Glass and metal equipment is sterilized by dry heat, rubber and heat-stable plastic materials by wet heat (Chapter 23), and media and heat-labile salt solutions by membrane

[†] (contributed by Jun Mitsuhashi)

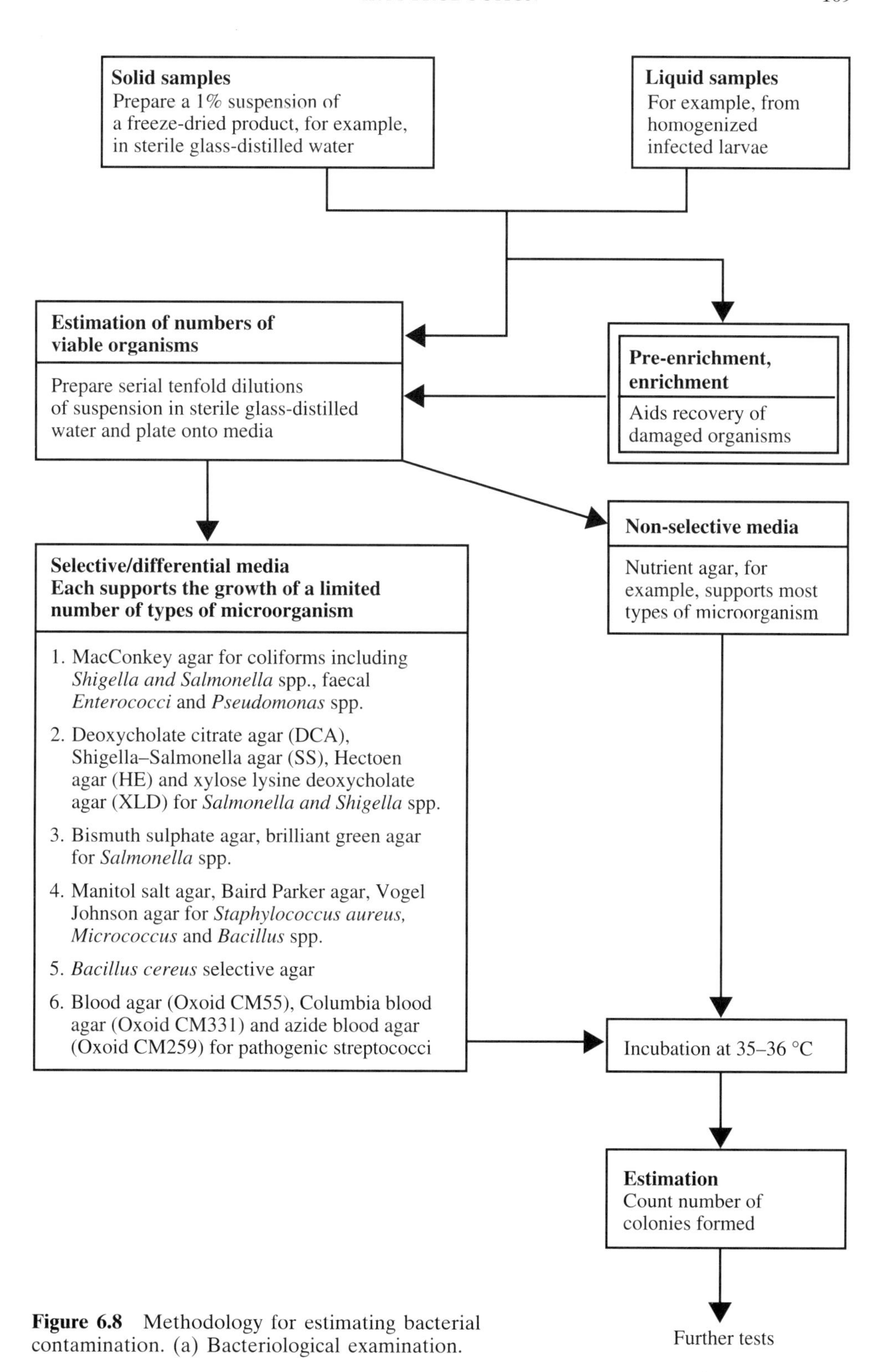

Figure 6.8 Methodology for estimating bacterial contamination. (a) Bacteriological examination.

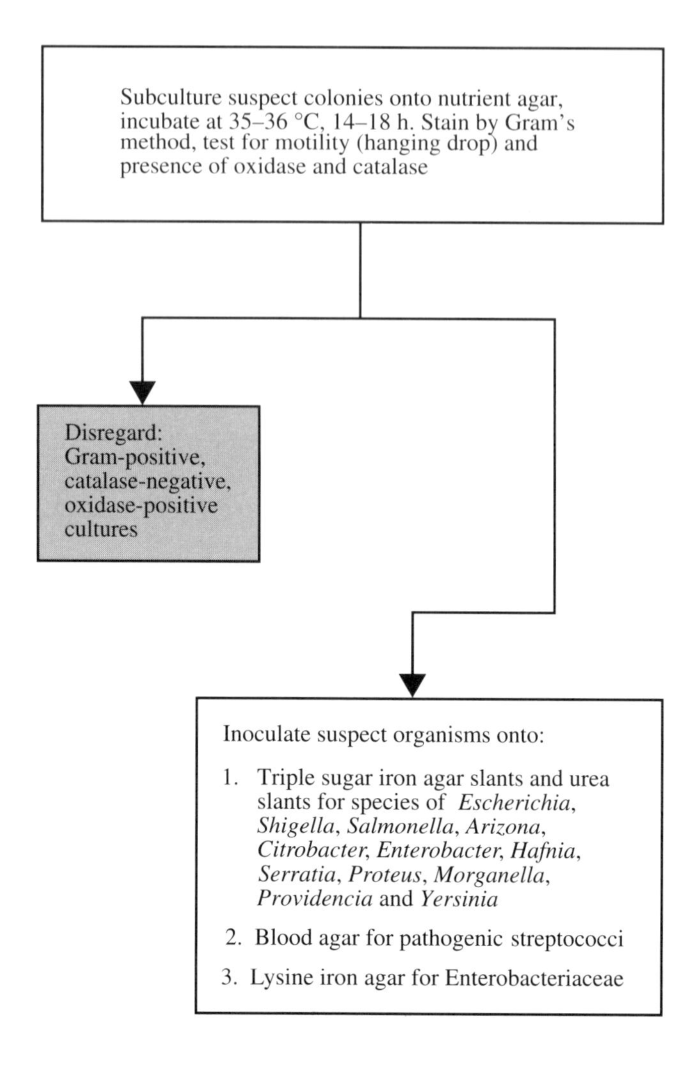

Figure 6.8 (*cont.*) (b) Further tests.

filtration (pore size 0.2 μm), (Chapter 23). Setting up of primary cultures requires the use of a dissecting microscope, opthalmologist's scissors, watchmaker's forceps, fine mounted needles and a dissecting tray for dissecting small insects. Such cultures are best made in small glass or plastic tissue culture vessels (bottles, tubes, multi-well plates or Petri dishes). Instruments are stored and sterilized by immersion in 70% ethanol followed by flaming or rinsing in sterile distilled water immediately before use. An inverted microscope is essential for viewing growing cultures.

Media and salt solutions

Various media have been formulated for insect cell cultures (Mitsuhashi, 1982). Many are chemically defined but most require the addition of vertebrate serum, more particularly fetal bovine serum (5–10% for established cell lines or 15–20% for primary cultures) or insect haemolymph. The high cost of serum has resulted in the recent development of serum-free media (mostly chemically undefined) that are effective for culturing cells and propagating virus (McIntosh, Grasela and Goodman, 1995). No one medium can support the growth of all types of insect cell culture, but some can be used for a range of species. Many, including some that are serum-free, are commercially available. They include Grace's (1962) and TC100 (general purpose) (Gardiner and Stockdale, 1975), Mitsuhashi and Maramorosch's (1964) (general, especially good for mosquito cells), Schneider's (*Drosophila*) (Mitsuhashi, 1982), Mark's M-14 and M-20 (cockroaches) (Mitsuhashi, 1982), Chiu and Black's (1967) (leafhoppers) and Goodwin's IPL-52, IPL-52B and IPL-76 media (lepidopterans) (Mitsuhashi, 1982). IPL-4 is recommended for growing cells in spinner cultures (King and Possee, 1992) and IPL-41 for large-scale production of virus in *S. frugiperda* cells (Weiss *et al.*, 1981). Examples of serum-free media that are available commercially include CPSR-3 (Sigma), EX-CELL 400 (JRH Biosciences, Lenexa, KS, USA) and SF900 and IPL-41 (Gibco BRL Life Technologies).

Some physiological salt solutions are used frequently during the establishment of primary cell cultures. These include Ringer solution and its modifications. Among them, Carlson's solution (Carlson, 1946) is excellent for maintaining tissues of various insect species for short periods of time although it was designed for culturing grasshopper neuroblasts (see Chapter 27 for composition). Cell culture medium can be substituted for such salt solutions.

Primary cultures

Organ cultures

Organs removed from the insect are maintained in such a way that they continue to function normally for as long as possible. Cell migration and multiplication is not desirable. For short-term culture, a simple glucose-containing salt solution can be used. For long-term cultures, however, more nutrient-rich media, usually consisting of inorganic salts, sugars, amino acids, vitamins and other ingredients, such as those used for propagating established cell cultures, are necessary.

Cell cultures

In general, cultures are initiated from chopped tissues excised from living insects, or from suspensions of individual cells obtained by treating chopped tissues with enzymes such as trypsin. Digestion by enzyme may be halted by addition of serum or dilution with medium or salt solution.

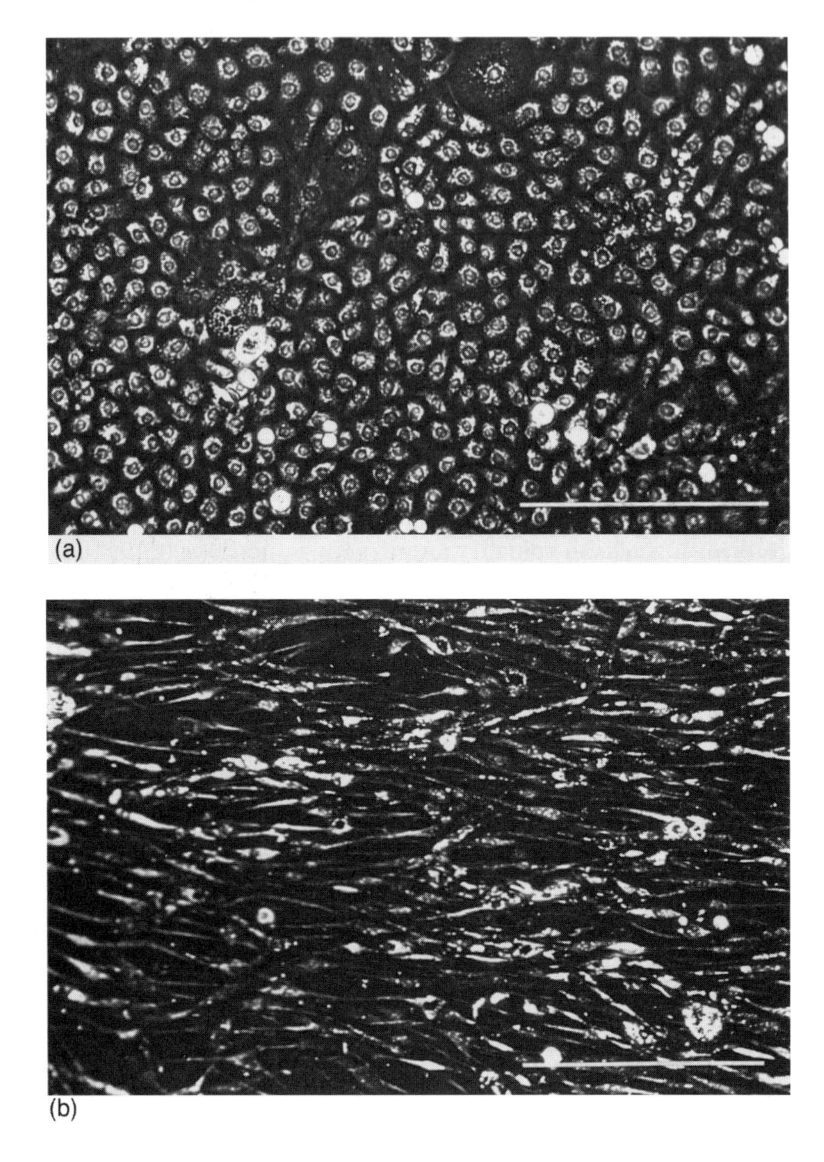

Figure 6.9 Primary explant cultures. (a) Cells sheet from *M. brassicae* pupal ovary (bar = 200 μm). (b) Network developed from *S. litura* larval fat bodies (bar = 200 μm).

In explant cultures (pieces of chopped tissue), cells will first migrate from the tissue and will then increase in number by mitoses. Three types of cell are usually recognizable: epithelioid cells which form compact sheets (Figure 6.9a), fibroblast-like cells that form networks (Figure 6.9b) and individually dispersed haemocyte-like cells (Figure 6.9c,d). Primary cultures of haemocytes consist of cells scattered over the bottom of the culture vessel. Some may multiply by mitosis. Growth of cells is very slow in primary culture. It may take many months for migrated cells to multiply sufficiently to be subcultured.

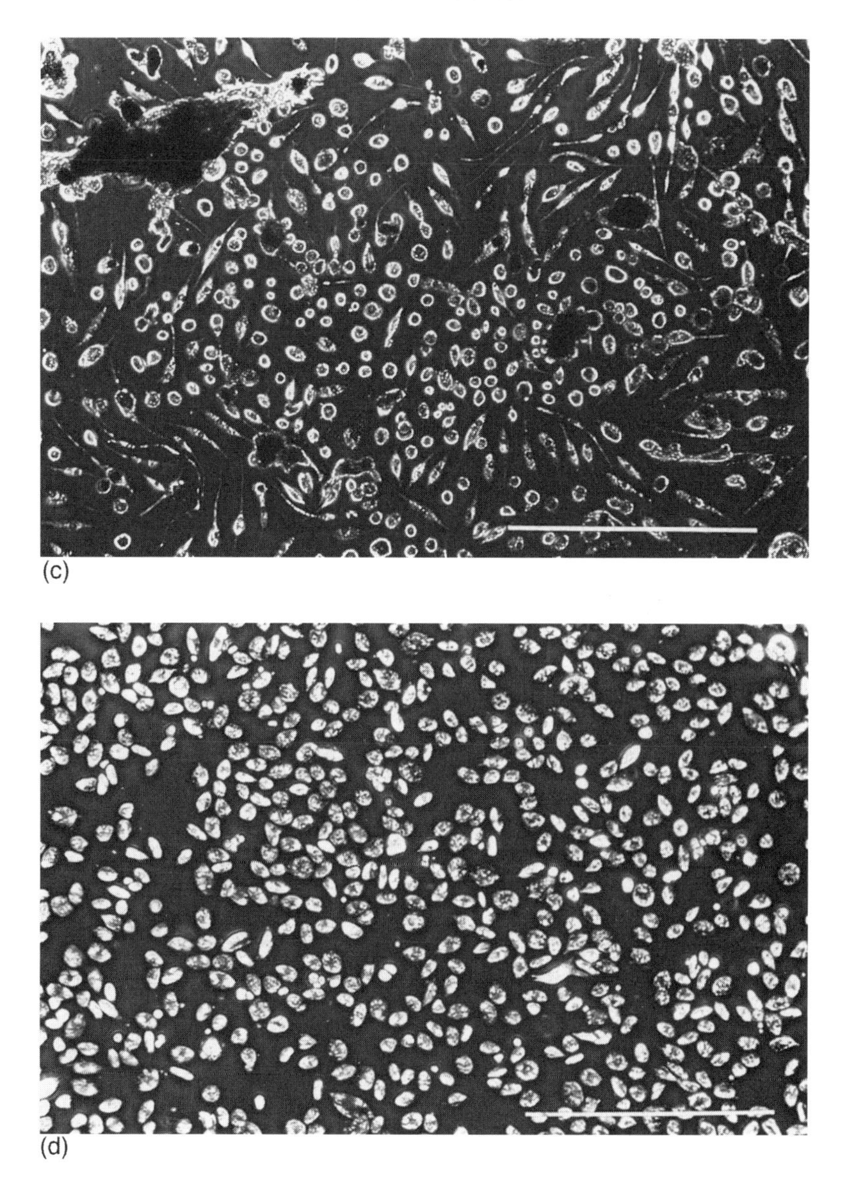

Figure 6.9 (*cont.*) (c) Free cells migrated from larval fat bodies of *D. spectablis* (bar = 200 μm). (d) Haemocytes of *Aromala rufocuprea* (bar = 100 μm).

Subculturing cells and continuous cell lines

In primary culture, cells increase in numbers by mitosis and may eventually cover the whole area of the bottom of the culture vessel. Unlike mammalian cells, contact inhibition is not usual so that even when cells touch each other they continue to grow and multiply, piling up on each other. Some cells attach firmly to the substrate while others attach loosely.

When the cell density is considered adequate, the culture is passaged by dislodging the cells, diluting and resuspending them in fresh medium and dispensing the diluted cell suspension in culture vessels. Loosely attached cells can be resuspended by pipetting up and down gently to prevent damaging them. Firmly attached cells will require the use of a rubber policeman (Figure 27.4, p. 503) to detach them or possibly treatment, for as short a period of time as practicable, with an enzyme such as trypsin (0.25% trypsin and 0.1% EDTA in Rinaldini's salt solution). Cells, such as lepidopteran cells, that are sensitive to trypsin should be treated with pancreatine (0.003%).

Cells should be passaged routinely. The growth rate of early passages is usually slow so that freshly subcultured cells may take a month to recover cell density. With successive passages, the growth rate will increase until it reaches a constant value. When this stage is reached the cells are considered to constitute a continuous cell line and to be immortal.

NPVs are relatively easy to propagate in cell culture while GVs have proved more recalcitrant. To date two GVs, those of *C. pomonella* and *T. ni*, can be propagated satisfactorily *in vitro* (McIntosh, 1994). Consequently, knowledge of the molecular biology of the NPVs, particularly that of the cabbage looper *Autographa californica*, is far in advance of that of the GVs.

REFERENCES

Aizawa, K. (1975) Selection and strain improvement of insect pathogenic microorganisms for microbial control. In *Approaches to Biological Control*. K. Yasumatsu and H. Mori (eds). Tokyo: University of Tokyo Press, pp. 99–105.

Anon. (1991) *NRI Report on Operational Programmes: 1989–91*. Chatham Maritime, UK: Natural Resources Institute, pp. 162–166.

Bell, M.R. (1991) *In vivo* production of a nuclear polyhedrosis virus utilizing tobacco budworm and a multicellular larval rearing container. *Journal of Entomological Science* **26**, 69–75.

Bird, F.T. (1950). *Bi-monthly Progress Report of the Canadian Department of Agriculture*, Vol. 6, pp. 2–3.

Carlson, J.G. (1946) Protoplasmic viscosity changes in different regions of the grasshopper neuroblast during mitosis. *Biological Bulletin*, **90**, 109–121.

Chiu, R.J. and Black, L.M. (1967) Monolayer cultures of insect cell lines and their inoculation with a plant virus. *Nature* **215**, 1076–1078.

Davis, T.R. and Granados, R.R. (1995) Development and evaluation of host insect cells. In *Baculovirus Expression Systems and Biopesticides*. M.L. Shuler, R.R. Granados, D.A. Hammer and H.A. Hood (eds). New York: Wiley, pp. 121–130.

Entwistle, P.F. and Evans, H.F. (1985) Viral control. In *Comprehensive Insect Physiology, Biochemistry and Pharmacology*, Vol. 12. G.A. Kerkut and L.I. Gilbert (eds). Oxford: Pergamon Press, pp. 347–412.

Evans, H.F., Lomer, C.J. and Kelly, D.C. (1981) Growth of nuclear polyhedrosis virus in larvae of the cabbage moth, *Mamestra brassicae* L. *Archives of Virology* **70**, 207–214.

Gardiner, G.R. and Stockdale, H. (1975) Two tissue culture media for production of lepidopteran cells and polyhedrosis virus. *Journal of Invertebrate Pathology* **25**, 363–370.

Grace, T.D.C. (1962) Establishment of four strains of cells from insect tissue culture grown *in vitro*. *Nature* **195**, 788–789.

Granados, R.R. and McKenna, K.A. (1995) Insect cell culture methods and their use in virus research. In *Baculovirus Expression Systems and Biopesticides*. M.L. Shuler, R.R. Granados, D.A. Hammer and H.A. Hood (eds). New York: Wiley, pp. 13–39.

Granados, R.R., Guoxun, L., Derksen, A.C.G., McKenna, K.A. (1994) A new insect cell line from *Trichoplusia ni* (BTI Tn-5B1–4) susceptible to *Trichoplusia ni* single enveloped nuclear polyhedrosis virus. *Journal of Invertebrate Pathology* **64**, 260–266.

Hink, W.F. (1972) A catalogue of invertebrate cell lines. In *Invertebrate Tissue Culture*, Vol. II. C. Vago (ed.). New York: Academic Press, pp. 363–387.

Hink, W.F. (1976) A compilation of invertebrate cell lines and culture media. In *Invertebrate Tissue Culture, Research Applications*. K. Maramorosch (ed). New York: Academic Press, pp. 319–369.

Hink, W.F. (1980) The 1979 compilation of invertebrate cell lines and culture media. In *Invertebrate Systems* In Vitro. E. Kurstak, K. Maramorosch and A. Dübendorfer (eds). Amsterdam: Elsevier/North-Holland Biomedical Press, pp. 553–585.

Hink, W.F. and Hall, R.L. (1989) Recently established invertebrate cell lines In *Invertebrate Cell System Applications*, Vol. II. J. Mitsuhashi (ed.). Boca Raton, FL: CRC Press, pp. 269–293.

Ignoffo, C.M. (1966) Insect viruses. In *Insect Colonization and Mass Production*. C.N. Smith (ed.). London: Academic Press, pp. 501–530.

Ignaffo, C.M. and Shapiro, M. (1978) Characteristics of baculovirus preparations processed from living and dead larvae. *Journal of Economic Entomology* **71**, 186–188.

Jones, K.A. (1994) Use of baculoviruses for cotton pest control. In *Insect Pests of Cotton*. G.A. Matthews and J. Tunstall (eds). Wallingford, UK: CAB International, pp. 445–467.

Kelly, P.M. and Entwistle, P.F. (1988) *In vivo* mass production in the cabbage moth (*Mamestra brassicae*) of a heterologous (*Panolis*) and a homologous (*Mamestra*) nuclear polyhedrosis virus. *Journal of Virological Methods* **19**, 249–256.

Kelly, P.M., Speight, M.R. and Entwistle, P.F. (1989) Mass production and purification of *Euproctis chrysorroea* (L.) nuclear polyhedrosis virus. *Journal of Virological Methods* **25**, 93–100.

King, L.A. and Possee, R.D. (1992) *The Baculovirus Expression System. A Laboratory Guide*. London: Chapman & Hall.

Martignoni, M.E. (1978) Production, activity and safety. In *Technical Bulletin, 1585. The Douglas-Fir Tussock Moth: A Synthesis*. M.H. Brooks and R.W. Campbell (eds). Washington, DC: USDA Forest Service, pp. 140–147.

McKinley, D.J., Moawad, G.M., Jones, K.A., Grzywacz, D. and Turner, C. (1989) The development of nuclear polyhedrosis virus for control of *Spodoptera littoralis* (Boisd.) in cotton. In *Pest Management in Cotton*. M.B. Green and D.J. de B Lyon (eds). Chichester, UK: Ellis Harwood, pp. 93–100.

McIntosh, A.H. (1994) Specificity of baculoviruses. In *Insect Cell Biotechnology*. K. Maramorosch and A.H. McIntosh (eds). Boca Ratan, FL: CRC Press, pp. 57–69.

McIntosh, A.H., Grasela, J.J. and Goodman, C.L. (1995) Replication of *Helicoverpa zea* nuclear polyhedrosis virus in homologous cell lines grown in serum-free media. *Journal of Invertebrate Pathology* **66**, 121–124.

Mitsuhashi, J. (1982) Media for insect cell cultures. In *Advances in Cell Culture*. K. Maramorsch (ed.). New York: Academic Press, pp. 133–196.

Mitsuhashi, J. and Maramorosch, K. (1964) Leafhopper tissue culture: embryonic, nymphal and imaginal tissues from aseptic insects. *Contributions to the Boyce Thompson Institute* **22**, 435–460.

Podgwaite, J.D., Bruen, R.B. and Shapiro, M. (1983) Microorganisms associated with production lots of the nuclear polyhedrosis virus of the gypsy moth *Lymantria dispar*. *Entomophaga* **28**, 9–16.

Rollinson, W.D., Hubbard, H.B. and Lewis, F.B. (1970) Mass rearing of the European pine sawfly for production of the nuclear polyhedrosis virus. *Journal of Economic Entomology* **63**: 343–344.

Shapiro, M. (1986) *In vivo* production of baculoviruses. In *The Biology of Baculoviruses*. Vol. II. R.R. Granados and B.F. Federici (eds). Boca Raton, FL: CRC Press, pp. 31–61.

Shapiro, M., Bell, R.A. and Owens, C.D. (1981) *In vivo* mass production of gypsy moth nucleopolyhedrosis virus. In *The Gypsy Moth: Research Toward Integrated Pest*

Management. C.C. Doane and M.L. McManus (eds). Washington, DC: USDA, Forest Service, pp. 633–655.

Shieh, T.R. (1989) Industrial production of viral pesticides. *Advances in Virus Research* **36**, 315–343.

Shuler, M.L., Granados, R.R., Hammer, D.A. and Wood, H.A. (1995) Overview of the baculovirus–insect cell system. In *Baculovirus Expression Systems and Biopesticides*. M.L. Shuler, R.R. Granados, D.A. Hammer and H.A. Hood (eds). New York: Wiley, pp. 1–11.

Smirnoff, W.A. (1964) *Forest Chronicle* **40**, 187–194.

Smits, P.H. and Vlak, J.M. (1988) Quantitive and qualitative aspects in the production of a nuclear polyhedrosis virus in *Spodoptera exigua* larvae. *Annals of Applied Biology* **112**, 249–257.

Teakle, R.E. and Byrne, V.S. (1989) Nuclear polyhedrosis virus production in *Helicoverpa armigera* infected at different larval ages. *Journal of Invertebrate Pathology* **53**, 21–24.

Tompkins, G.J., Vaughn, J.L., Adams, J.R. and Reichelderfer, C.F. (1981) Effects of propagating *Autographa californica* nuclear polyhedrosis virus and its *Trichoplusia ni* variant in different hosts. *Environmental Entomology* **10**, 801–806.

Tompkins, G.J., Dougherty, E.M., Adams, J.R. and Diggs, D. (1988) Changes in the virulence of nuclear polyhedrosis viruses when propagated in alternate noctuid (Lepidoptera: Noctuidae) cell lines and hosts. *Journal of Economic Entomology*, **81**, 1028–1032.

Weiss, S.A., Smith, G.C., Kalter, S.S. and Vaughn, J.L. (1981) Improved method for the production of insect cell cultures in large volumes. *In Vitro* **17**, 495–502.

Wigley, P.J. (1976) The epizootiology of a nuclear polyhedrosis virus disease of the Winter moth, *Operophtera brumata* L., at Wistman's Wood, Dartmoor. D Phil thesis, University of Oxford, UK.

Yavada, R.L. (1970) Studien uber den Einfluss von Temperatur und relativerluftfeuchtikeit auf die Entwicklung der Kern polyedrose der None (*Lymantria monacha* L.) und des Schwamm-spinners (*L. dispar* L.). *Zeitschrift für Angewandte Entomologie* **65**, 167–174.

RECOMMENDED READING

Chapman, R.F. (1982) *The Insects, Structure and Function*, 3rd edn. London: Hodder & Stoughton.

Maramorosch, K. and McIntosh, A.H. (eds) (1994) *Insect Cell Biotechnology*. Boca Raton, FL: CRC Press.

Vlak, J.M., de Gooijer, C.D., Tramper, J. and Miltenburger, H.G. (eds) (1996) *Insect Cell Cultures: Fundamental and Applied Aspects*. Reprinted from *Cytotechnology*, Dordrecht: Kluwer.

Wigglesworth, V.B. (1965) *The Principles of Insect Physiology*, 6th edn. London: Methuen.

7 Formulation

INTRODUCTION

Like other biological insecticides (notably *Bacillus thuringiensis* (*B.t.*)) and, especially, chemical insecticides, the need to protect commercial interests has greatly restricted publication of information on formulation in the scientific and technical presses (Yearian and Young, 1982). This is understandable but is not necessarily too serious as enough is known to permit development of a range of satisfactory BV formulations. Burges and Jones (1997) have recently reviewed the available information on formulation of BVs and *B.t.*

Formulation requirements for microbial pesticides have often been more perceived than actually necessary and this has probably arisen by extrapolation from practices with chemical pesticides. It should be born in mind that formulation of chemical insecticides is essentially the art of handling soluble toxicants whereas formulation of microbial pesticides involves the suitable preparation of particulates.

The need for formulation encompasses two main areas: shelf and tank (finished spray) formulations.

Shelf formulations

Commercial practice for pesticides requires a shelf stability of at least 18 months, a period which allows adequate time for product distribution, marketing and reasonable retention by the end-user prior to actual application. With BVs this can be achieved either by retaining the active ingredient frozen until the time of use or by developing a formulation that confers stability under a reasonable range of shelf storage conditions. The first of these two options is of much less commercial convenience, though it should not be dismissed entirely. Formulations will be designed to retain biological activity of the active ingredient, to prevent the in-storage replication of any contaminant microorganisms (largely bacteria and fungi) and to ease handling of the product. The species composition and count (numbers per unit volume) of such contaminants is usually subject to registration regulation.

Tank (finished spray) formulations

Tank formulations are made more or less immediately before product application and are determined by the physical method of application and the rate of decay of the BV once it is deposited at its target site. Some tank additives, such as UV protectants and stickers, may be added either at this stage or previously to the shelf formulation. Others, such as anti-evaporant carriers, including spray oils, will often be added to the tank at field mixing.

Few workers have referred to the fairly obvious need to avoid the use of chlorinated (tap) water. Exceptions are Jaques (1973) and Podgwaite *et al.* (1986) and the promoters of BVs in Vector NPO, Novosibirsk, FSU (Former Soviet Union). Conversely, a warning against the use of 'carbonated' water is given only by Ignoffo and Montoya (1966). The extent to which these two aspects of mixing water are of practical importance requires further examination. Some water authorities, for example in the UK, will provide information on the degree of chlorination. The risk can be minimized by leaving the water to stand for some time before addition of the BV formulation and by making up the spray suspension, as needed, just prior to application.

SHELF FORMULATION TYPES

With insecticides in general, formulation fashions change with time, a process that is essentially driven by solution of field persistence problems, developments in application equipment and methodologies, and costs. The increasing need of the farmer to apply two or more pesticides simultaneously can also have an effect on the nature of BV formulations. For instance, joint application with any alkaline chemical (e.g. some fungicides) probably dictates a need to protect the BV by some form of micro-encapsulation.

With the current level of usage of BV pesticides, it is on the whole preferable that their application be compatible with the most commonly available types of spray equipment. Any major departures from this attitude are probably either for the future or for situations where it has already been decided BVs are the preferred option and their most effective use indicates adoption of special machinery design. Environmentally sensitive areas like some types of forest would fall into this category.

On the whole, however, formulations are designed to 'suit' existing equipment. At present formulation tends towards development of suspension concentrates and water-dispersible granules and emulsions in water, whilst emulsifiable concentrates and wettable powders are becoming less popular. Such current trends have been reviewed by Seaman (1990).

DRY PREPARATIONS

Given storage under acceptable temperature conditions and in containers sealed to avoid ambient fluctuations in humidity and, ideally, to maintain moisture content below 5%, some dry preparations may retain viability.

Air-dried infected host insects

The internal environment of the host (dried at temperatures probably below 30°C) is known to favour retention of activity. Material of this type requires pulverization and filtration before use in a spray.

Freeze drying, or lyophilization

As with many other microorganisms, stability is retained after freeze drying. It involves freezing aqueous suspensions of BV OBs (the degree of prior purification is optional) under vacuum for complete removal of moisture by volatilization. The resultant powder requires grinding and the addition of formulants before storage in tightly sealed containers. The process is often regarded as too expensive for commercial use though, nevertheless, it has been used as a step in the formulation of *Lymantria dispar* NPV as Gypchek (Shapiro, 1982), *Orgyia pseudotsugata* NPV as TM Biocontrol-1 (Martignoni, 1978) and *Spodoptera littoralis* NPV (McKinley *et al.*, 1989). Difficulties in resuspension may to some extent be overcome by lyophilization in the presence of lactose as a paste with the virus (Ignoffo, 1964) (see below).

Spray drying

The spraying of droplets of BV suspensions into a column of dry, warm, air results in dried particles of product. The 'crumb' size is related to that of the initial spray droplet and the concentration of solids within it and is thus subject to process control. This method is very widely used commercially for the production of 'instant' coffee and tea, etc. Whilst NPVs of *Autographa californica* (Sandoz 404), *Helicoverpa zea* (Elcar) and *Trichoplusia ni* (Sandoz 405) have been successfully prepared in this way, *Choristoneura fumiferana* NPV (Cunningham *et al.*, 1978) and *Cydia pomonella* GV (Sandoz 406) (Young and Yearian, 1986) were shown to be less stable.

Co-precipitation with lactose

The addition of acetone to an aqueous suspension of BV OBs in the presence of dissolved lactose (4–6%) results in precipitation of both BV and lactose. The precipitate, recovered by filtration, may be air-dried, ground and stored. This process was originally developed by Dulmage, Correa and Martinez (1970a) for recovery of spores and crystals of *B.t.* and was later adapted for use with NPVs (Dulmage, Martinez and Correa, 1970b) and GVs (Hunter, Collier and Hoffman, 1973). Whilst this formulation resuspends rather easily in water, the storage stability seems to be inadequate especially at higher temperatures (e.g. 35°C) (McGaughey, 1975; Hunter, Collier and Hoffman, 1977; Ignoffo and Shapiro, 1978).

Micro-encapsulation

The finished micro-capsule product may usually be stored dry (see Chapter 28 for details of micro-encapsulation processes).

Dusts and granules

Dust and granules are terms used to imply preparations to be 'field-applied' in their dry forms, mostly without further dilution by other (dry) materials. As has been remarked by Lewis (1982), the main difference between dusts and granules is that of particle or crumb size. However, it must be added that while dusts will seldom incorporate a binder (such as glycerol), in granules some additive is usually necessary to ensure cohesion of the ingredients. Very often such an adhesive can also help to make granules water repellent and so increase their field life.

More interest in the use of dusts and granules has been shown in relation to *B.t.* and fungi than to BVs. Dusts are not now favoured for reasons of their storage bulk, problems of handling and airborne particles and because of a general shortage of application equipment. Granules tend to have specialist applications. They are valuable where, for instance, the leaf axil morphology, as in maize, collects granules by funnelling and makes them especially accessible to such major pests as *Spodoptera frugiperda*, fall armyworm, and *Ostrinia nubilalis*, European corn borer. Baits are rather easily incorporated into dusts and granules and, in the case of brans of various types, have the advantage that they can sometimes act simultaneously as the filling or bulking agent. The opacity of granules in particular assists greatly in protection of BVs from solar UV. Granules also tend to be favoured for soil application against cutworms (*Agrotis* and *Euxoa* spp.). Granules also avoid the handling problems encountered with dusts.

The development of granules based on bran and hominy (hulled maize, which may also, by definition, be crushed) for *H. zea* NPV, as Elcar, was abandoned because it was less effective than a sprayable formulation (Hostetter and Pinnell, 1983).

WET PREPARATIONS

Wet preparations are usually water-based concentrates that, on dilution in the tank, should be capable of efficient dispersal or emulsion in the presence of spray oils. Very often their preparation involves one of the processes listed above, e.g. lyophilization or spray-drying. They are essentially of two types: flowable and settling concentrates.

Flowable concentrates

In flowable concentrates, the BV is maintained in suspension. Ensuring a 'permanent' suspension of BV OBs during shelf storage means that the concentrate does not require agitation before dilution into the spray tank: it also assists considerably in preventing OB aggregation, which can result in very uneven distribution among spray droplets and hence in erratic coverage of the target crop. Addition of a dispersant (see Table 7.4, below) may assist further. Suspension is achieved by inclusion either of thickeners or thixotropic agents as discussed below and has been very successfully used, for instance, for *Neodiprion sertifer* NPV in the commercially registered product Virox.

Concentrates which settle

In almost all aqueous suspensions, BV OBs will settle to the bottom of the container during storage. Provided settling is not associated with aggregation, shaking rapidly produces a suspension suitable again for tank mixing. The use of a dispersing agent may be very important, e.g. Darvan No 1 or No 2 (Table 7.4, below), which appears to act by putting like charges on particles and so causing natural repulsion. Settling, especially of particles less than 10 μm, is delayed. *Pieris rapae* GV as formulated in Wuhan, the People's Republic of China, is an example of an easily resuspendable and very effective BV insecticide. It appears that aggregation in aqueous concentrates of NPVs can sometimes be rectified by agitation with spray oils prior to application, probably because of the presence of surfactants. This seemed to take place with *Panolis flammea* and *Mamestra brassicae* NPVs in the presence of Actipron (Ulvapron) but it should always be checked in advance of actual spraying. The greater difficulties in producing flowable concentrates of NPVs than of GVs is discussed by Young and Yearian (1986).

BAIT PREPARATIONS

BV preparations may be baited to promote insect feeding and also to mask innate phagodeterrent qualities in some formulations (Table 7.8, below). The literature on baiting and insect attractant materials is very large so that only an outline is provided in this book.

ESTABLISHMENT AND MAINTENANCE OF LOW MICROBIAL IN-STORE CONTAMINATION

The presence of a complex of saprophytic microorganisms developing in BV-diseased and BV-killed insects appears to have no adverse effect on occluded virus (Jaques and Huston, 1969). In resisting the action of proteases, the polyhedral envelope may be involved in this type of stability (Gipson and Scott, 1975).

Registration authorities, however, have requirements on the types and levels of microbial contaminants of shelf-storage preparations of BVs (Chapter 6 and 28). At formulation there should ideally be lower contaminant levels than actually demanded and it is also essential that there be no significant changes during a reasonable period of storage (≥18 months). For BVs produced in insects grown under controlled conditions, usually on sterilizable semi-synthetic diets, general hygiene greatly assists conformation with an acceptable microbial species complex and low cell density. However, testing for microbial contaminants is still an essential procedure.

If virus is produced in insects cultured on plant tissue or collected as cadavers from the wild, the microbial load can be minimized through the method of selective virus recovery (e.g. centrifugation) and the extent of dilution in the shelf-storage preparation. Such an end has been achieved, for instance, for *N. sertifer* NPV, with commercial products being registered in Finland, UK, Canada and the USA. High levels of purity of BV preparations can, of course, be achieved through centrifugation, a process which is not necessarily commercially too expensive. *N. sertifer* NPV has been prepared by continuous zonal rotor centrifugation (Mazzone, Breillatt and Anderson, 1970; Breillatt *et al.*, 1972). The employment of certain chemical disinfectants is also possible (e.g. sodium omidine); these must, however, be washed out before product storage or use, which involves a further process step (Dubois, 1976).

Maintenance of unaltered contamination levels is automatic in dried preparations in sealed containers, e.g. Elcar, TM Biocontrol-1 etc., and foil-packed *P. rapae* GV in The People's Republic of China.

In liquid formulations, maintenance of an acceptable concentration of cells of contaminant microorganisms is more difficult. Available methods of stabilization are of two types. A low pH (≤4.0) suppresses replication of many microorganisms without loss of BV activity. The pH may be adjusted with 1–5 N H_2SO_4 or can be achieved with buffers. In addition, certain substances often widely employed in the food industry, e.g. sorbic acid and potassium sorbate, may be included in the shelf formulation (Table 7.1). Room exists for much further investigation along these lines.

Table 7.1 Suppression of contaminant microorganisms during shelf storage

Substance	Concentration	Function	References
EDTA (ethylene diaminetetraacetic acid) disodium salt	5 mM	Enzyme inactivation	
Potassium sorbate	0.1%	Mould and yeast inhibitor	
Sorbic acid	0.1%	As K sorbate	
Tween 20	0.02%	Inhibit germination of some fungal spores	Boyette *et al.* (1991)
Tween 80	0.02%	As Tween 20	
Adjustment of pH		Low pH inhibits bacterial growth	
	0.04%		
Multifilm Buffer-X			Ignoffo and Montoya 1966)
Phosphate buffers	–		P.F. Entwistle, unpublished data
Saline buffer	–		Johnson and Lewis (1982)
Shade[a]	2.0–2.5%		Ignoffo *et al.* (1976a,b), Smith *et al.* (1978), Yendol *et al.* (1977), Vail *et al.* (1977)
Sorba Spray Zip	0.053%		Vail *et al.* (1977)
Sulphuric acid	1–5 N		Smirnoff *et al.* (1962) etc.
Tris buffers	–		Huber and Dickler (1977)

[a] UV protectant additive that also reduces pH.

In designing shelf-storage formulae, it should be born in mind that while there may be adequate information on the BV itself, all registration authorities will require testing on the full storage formulation.

It will always be an advantage if formulations, no matter of what type, can be stored at a low temperature, e.g. at +4°C, or less.

Urticaceous hairs

Some larval Lepidoptera bear setae that are urticaceous. The consequences of undue contact with these vary in intensity with the individual person but include dermal urticaria, conjunctivitis and some respiratory conditions. The problem is mainly associated with nettle caterpillars within the Limacodidae, some species of Lymantriidae, some processionary caterpillars in the Thaumetopoeidae and some *Dendrolimus* species in the Lasiocampidae.

The majority of such hairs can be removed by initial filtration of larval macerates and, if necessary, the level can be further greatly reduced by a simple two-step low-speed centrifugation procedure (Kelly, Speight and Entwistle, 1989). Full removal will seldom be essential but can be achieved by density gradient centrifugation as for *L. dispar* (Podgwaite and Mazzone, 1981).

Dermal, conjunctival and respiratory observations will indicate if a product is fit to go forward to a full range of safety tests.

FORMULATION ADDITIVES

Terminology

Some of the terms employed in formulation literature can be misleading in themselves and may overlap in usage with others. Furthermore, some of the terminology relating to pesticides in general is confusing and can have misleading implications for microorganisms such as insect viruses. Therefore, the following notes may be helpful.

Anti-evaporant/humectants

Anti-evaporant is often used in the separate senses of a substance that (a) actually retards evaporation of water (usually) from the spray droplet or (b) by being of very low volatility itself, preserves a droplet mass adequate to ensure carriage of the active ingredient to the spray target and for penetration of the droplet through any boundary air layer surrounding the target (p. 584). As shown by Payne (1983), incorporation of emulsifiable oils as anti-evaporants does not prevent loss of water and, indeed, can actually lead to an increase in the rate of loss of water from a spray droplet surface, leaving only the oil component of the mixture.

Humectant

Humectants are formulants employed to provide an aqueous environment for the pathogen; this term is also sometimes used to imply anti-evaporant qualities in relation to the preservation of droplet mass.

Anti-drift agent

The term anti-drift agent is misleading since there are no substances, *per se*, that prevent or reduce spray drift. Essentially anti-drift agents are anti-evaporants. Behind the use of this term is the belief that as droplet size decreases (e.g. by evaporation of water) distance of drift increases. This is an oversimplification because the extent of drift is controlled by such factors as wind speed, crop canopy roughness (both of which are involved in turbulence creation), droplet size and crop boundary–air-layer stability. Because of this complexity, the term anti-drift agent is best avoided. It probably had its origin in commercial sales literature.

Thickeners

Though thickeners are employed to delay OB settling in shelf-storage formulations, the term thickener has also been used in an anti-drift agent sense.

Stickers/adhesives

Probably because of the particulate nature of BVs, it has often been perceived that stickers are essential in formulations. Though they may be desirable, the surface of some plants retain BV (and CPV) OBs very strongly. For instance, despite rain falling immediately after spraying, the biological impact of BV sprays tends often not to be diminished. Formal laboratory demonstrations of very pronounced rain-fastness have been made, e.g. of PbGV on cabbage leaves (David and Gardiner, 1966), *C. pomonella* GV on apple leaves (Burgerjon and Sureau, 1985) and *Thaumetopoea pityocampa* CPV on pine needles (Burgerjon and Grison, 1965). However, some leaves, possibly those with markedly hydrophilic surfaces, may not naturally retain OBs. From a formulation point of view, the matter is further complicated by possible innate differences in the adhesion of OBs and particulate formulants such as some UV protectants. Clearly, in the absence of a sticker the value of a rain-labile UV protectant may be compromised.

Wetting agents

The use of wetting agents, or surfactants, with BV OB-containing sprays is a complex area in which misconceptions are possible. BV OBs seem to be negatively charged at pH values higher than 3.0–4.0. The cationic detergent CTAB can permanently reduce electrophoretic mobility as can the chelating agent EDTA. Such sensitivity suggests an impact from quite small changes in formulation, for instance on attachment to plant surfaces (Small and Moore, 1987). Usage of a wetter has generally been based on intuition rather than experience. However, Smith, Hostetter and Ignoffo (1978) found *H. zea* NPV in water to be more active in the absence of the wetter Triton C5–7, which tends to agree with experience with several other surfactants (Hostetter *et al.*, 1982). The following are some situations in which wetters may or may not be useful.

With reflectant surfaces Surfactants are useful where reflectant surfaces are involved because small spray droplets may be deflected from naturally reflective surfaces unless the correct proportion of surfactant has been added. This quantity is one that often greatly exceeds the critical micelle concentration (that concentration which provides a monomolecular surface coverage to the droplet) (Wirth, Storp and Jacobsen, 1991).

Penetration of stomatal cavities and hydathodes Addition of surfactants may aid penetration of stomatal cavities and hydathodes (Zidack, Backman and Shaw, 1992) especially by GVs and possibly also by NPVs. This may be especially important for protection of the virus from UV and also for the control of leaf-mining insects such as potato tuber moth, *Phthorimaea opercullela*, which would otherwise ingest very little virus (Reed and Springett, 1971). To deliver a liquid into leaf stomata it has been shown the liquid must have a critical oil–water surface tension of ≤30 dyne/cm (Schonherr and Bukovac, 1972).

With apolar surfaces Surfactants are not useful with apolar surfaces of leaves because these are rather easily wetted; incorporation of a surfactant may lead to more rapid spray droplet spread, confluence, and hence spray fluid run-off from the leaf. Whether this necessarily involves coincident loss of BV OBs should be questioned in view of the acquisition by some leaves of NPVs from spray droplets rolling rapidly across their surface (P. F. Entwistle, unpublished data). In any event, the proper function of any particulate UV protectant could be compromised. Spray droplets may bounce more easily from a moisture film than from a dry leaf surface (Wirth *et al.*, 1991) so that the promotion of droplet confluence induced by added wetters may accelerate a process of spray droplet 'rejection'. These are probably understudied areas the clarification of which may lead to a more constructive approach to formulation.

The concentration of baculovirus in spray droplets

Evidence exists to indicate various degrees of pathogen loss from sprays (Chapter 8) and some formulation work has been directed towards this problem (Smith, Hostetter and Pinnell, 1980). The influence on BV concentration in the spray droplet of any alteration or addition to a formulation should, therefore, be checked.

FORMULATING MATERIALS

We consider here the principles and practical aspects of formulation. Some of the substances mentioned appear not yet to have been employed with BVs but, from their known capacities and performance, have potentially useful characteristics. Micro-encapsulating agents and UV protectants are discussed below.

Extensive lists of formulating materials are given by Ignoffo and Montoya (1966), Angus and Luthy (1971), Couch and Ignoffo (1981), Entwistle and Evans (1985) and Hostetter *et al.* (1982); these will be found useful for careful evaluation of a variety of formulating materials. For reviews on dusts and granules, see Lewis (1982) and for baits, Henry (1982).

Spreaders, wetters and emulsifiers

Spreaders, wetters and emulsifiers are all terms embraced within the general term surfactant. In addition, adhesiveness is often involved. Chapter 32 discusses the nature of plant surfaces in relation to wettability. It is by no means certain that it is always necessary for a spray fluid to have the capacity to spread (see above). For instance, in droplets that remain discrete on the target surface, any UV-protectant materials will be held as closely as possible to the virus, but as a droplet spreads the value of UV protectants may decrease. Furthermore, the tendency for the leaf to become thoroughly wetted, through droplet confluence, and then to reject further spray droplets (Wirth *et al.*, 1991) will be diminished. However, spray fluid may assist entry of OBs to stomatal cavities and a certain level of surfactant presence is necessary for droplet collection by hydrophobic surfaces (see above). Whether the proportion of surfactant present in emulsions of water with proprietary spray oils is generally adequate for droplet retention by water-reflective surfaces seems not yet to have been studied. However, there seemed to be no hindrance to collection of a very high proportion of small (40–120 μm diameter) droplets of water with 20% Actipron by the hydrophobic needles of *Pinus contorta*, Lodgepole pine.

Where an emulsion of water with a spray oil is required, a surfactant is essential. Most workers have relied upon commercial spray oils in which surfactants are present but Smirnoff, Fettes and Haliburton (1962) emulsified 80% No. 2 Fuel oil with Span 80 whilst Prior, Jollands and Le Patourel (1988) emulsified equal proportions of water and coconut oil with 0.01% Tween-80 (as a carrier for desiccation-susceptible fungal spores).

Table 7.2 lists surfactants employed in 'biological' spray fluids and the concentrations at which they have been employed. In general, the source literature does not provide information on the reasons for selection of these concentrations. Where a surfactant has been employed specifically as a wetter to ensure spray droplet capture by a hydrophobic plant surface, it seems important to be sure that the critical micelle concentration is exceeded (Wirth *et al.*, 1991).

Stickers

Some stickers also appear to have spreading, and presumably wetting, capacity. Table 7.3 lists those most commonly used with BVs and in other microbial pesticide formulations. As discussed above, the need for stickers should be questioned. It almost certainly is dictated by the nature of the target surface. In tests on the rain-fastness of some chemical insecticides on pine foliage, Bio-Film was found to give best results (Nord and Pepper, 1991) but, as far as is known, this has yet to be tested with BVs.

Table 7.2 Surfactants (spreaders, wetters and emulsifiers)

Name	Concentration (%)	Source	References
Agral NN	0.05–0.10	ICI	Smits *et al.* (1988), Bourner *et al.* (1992)
Arlacel 'C'	1.00	Atlas Powder Co., USA	Clark and Reiner (1956)
Chevron	0.05–2.34	Ring-Around Products Inc., USA	Yendol *et al.* (1977), Wollam *et al.* (1978)
Colloidal X-77	0.025	Colloidal Products Inc., USA	Elmore (1961), Tanada and Reiner (1962), Tanada (1964)
FloMo			Smith *et al.* (1978)
Lissapol NS (lauryl alcohol type)	20.00		Ahmed *et al.* (1973)
Rhoplex B60A	2.00	Rohm and Haas Co., USA	McEwan and Hervey (1959), Podgwaite *et al.* (1986, 1991)
Shell 'Wetting Agent'	0.009		Teakle *et al.* (1985)
Silwet L-77[a]	0.20	Union Carbide, USA	Zidack *et al.* (1992)
Span 80		Atlas Powder Co., USA	Clark and Reiner (1956), Smirnoff *et al.* (1962)
Teepol[b]	0.05–0.50		Topper *et al.* (1984), Easwaramoorthy and Jayaraj (1991)
Tergitol TMN6[c]		Union Carbide, USA	Wirth *et al.* (1991)
Triton B-1956	0.03–1.00	Union Carbide, USA	Tanada (1956), Glass (1958), Genung (1960), Wolfenbarger (1965), Ignoffo and Montoya (1966), Schuster and Clark (1977), Ives and Muldrew (1978)
Triton CS-7	0.01	Union Carbide, USA	Smith *et al.* (1978, 1982), Hostetter *et al.* (1982)
Triton X-100	0.03–0.53	Union Carbide, USA	Ignoffo *et al.* (1965), Allen *et al.* (1966), Falcon *et al.* (1968), Vail *et al.* (1971), Rabindra *et al.* (1988), Sopp *et al.* (1989)
Triton X-114	0.20	Union Carbide, USA	
Triton X-152		Union Carbide, USA	Ignoffo and Montoya (1966)
Triton X-172		Union Carbide, USA	Ignoffo and Montoya (1966)
Tween 20[d]	0.005–0.02	Atlas Chemical Industries Inc, USA	Boyette *et al.* (1991)
Tween 80	0.004–0.33	Atlas Chemical Industries Inc, USA	Clark and Reiner (1956), Huber and Dickler (1977), Vail *et al.* (1977, 1991), Bell and Kanavel (1978), Bell and Romine (1980), Hostetter *et al.* (1982), Prior *et al.* (1988), Bell (1991)
Plyac	0.125–0.150		Jaques *et al.* (1981)

[a] Polyalkyleneoxide-modified polydimethyl siloxene, used for bacterial penetration of leaf stomata etc.
[b] Sodium secondary alkyl sulphonate.
[c] 2,6,8-Trimethylnonaol hexaglycolether, concentration dependent on droplet size.
[d] Polyoxyethylene sorbitan monolaiate.

Table 7.3 Stickers[a]

Name	Concentration (%)	Source	References
Agral NN	0.05–0.10	ICI	Smits *et al.* (1988)
Chevron[b]	0.05–2.34	Chevron Chemical Co., USA	Smirnoff (1977), Wollam *et al.* (1978), Hostetter *et al.* (1982), Shepherd *et al.* (1984), Nord and Pepper (1991)
Flo-Mo			Smith *et al.* (1978)
Geon 652B[c]		Goodrich Chemical Co., USA	Smirnoff *et al.* (1962)
Hyvis 150[d]	0.50	British Petroleum Co.	Prior and Ryder (1987)
Lovo[e]	1.00–2.50		Morris (1963), Rollinson *et al.* (1965)
Methyl and hydroxy-methyl cellulose	0.20–5.00	Dow Chemicals of Canada Ltd	Angus (1954), Clark and Reiner (1956)
Milk (skim)	0.60–3.80		Huber and Dickler (1977), Glen and Payne (1984)
Nu-film[c]	0.10–1.25	Miller Chemical and Fertilizer Corp., USA	Andrews *et al.* (1975), Nord and Pepper (1991)
Pinolene	0.40		Zethner (1976)
Plyac	0.015–1.240	Allied Chemical Co., USA	Fernandez *et al.* (1969), Jaques (1971, 1973), Jaques *et al.* (1977, 1981), Stacey *et al.* (1977b), Livingston *et al.* (1980), Pritchett *et al.* (1980), Nord and Pepper (1991)
Polyvinyl alcohol (PVA)	0.50		Smith *et al.* (1978, 1980), Hostetter *et al.* (1982)
Polyvinyl chloride			Angus and Luthy (1971)
Polyvinyl-pyrrolodone (PVP)	1.00	Chemical Develop-ments of Canada	
Rhoplex B6OA	0.19–2.00	Rohm and Haas Co., USA	McEwan and Hervey (1959), Podgwaite *et al.* (1986, 1991), Webb *et al.* (1990)
Triton B-1956	0.03–1.00	Rohm and Haas Co., USA	Clark and Reiner (1956), Tanada (1956), Ignoffo and Montoya (1966)
Triton CS-7		Rohm and Haas Co., USA	Smith *et al.* (1978)
Triton X-45		Rohm and Haas Co., USA	McEwan and Hervey (1958)
Triton X-100	0.01–0.53	Rohm and Haas Co., USA	Clark and Reiner (1956), Ignoffo *et al.* (1965), Ignoffo and Garcia (1966)

[a] Where a substance has a dual function of wetter and sticker it is sometimes not clear from the literature if the given concentration relates specifically to one or the other function.
[b] Alkyl aromatic polymers 45% + 55% inert ingredients.
[c] A latex.
[d] Iso-polybutanes.
[e] Amine stearates.

Thickeners and thixotropic agents

Thickeners are normally water-soluble substances employed to give a good homogenous quality to flowable shelf formulations and to maintain the active ingredient and any other particulate formulants in uniform suspension during storage. Among them, an algin (under the trade names Keltose and Kelzan) has thixotropic properties, i.e. whilst a gel when stationary, it rapidly converts to a flowable condition on agitation (Table 7.4). A combination of VanGel B, an industrial thickening suspending agent, with xanthan gum provides particle stability and dispersion without the formulation being thixotropic (product literature).

Binders

Binders are essential in granular formulations to attach the active ingredient to the carrier, along, of course, with any other formulants (Table 7.4).

Anti-desiccants, humectants and anti-evaporants

As an aqueous environment is generally unnecessary during storage or in tank mixes, a substance with water-retaining properties is probably useful only to preserve spray droplet mass during flight from the atomizer to the target. (With some bacteria, especially where a spore or crystalline endotoxin is not the important biocidal entity, preservation of moisture may be essential, e.g. *Pseudomonas aeruginosa* cells (Stephens, 1959)).

Adjustment of pH

In general, the pH tolerance of occluded BVs lies between 4.0 and 9.0. Below 4.0 activity rapidly deteriorates, e.g. *Heliothis* NPV was essentially inactivated at pH 1.2 (Ignoffo and Garcia, 1966) and greatly reduced when buffered at pH 2.0: the concentration of phosphate buffer had little influence (Gudauskas and Canerday, 1968). Above pH 9.0, the matrix protein of the OB breaks down and virions are liberated. Hence storage above pH 9.0 results in rapid deterioration (for information on the dissolution characteristics of a range of BV OBs see Griffith (1982)). Storage at low pH has the benefit of inhibiting the replication of many microorganisms, notably bacteria, and hence is of value in complying with registration regulations on contamination.

A non-occluded rod-shaped virus of *Panonychus citri*, citrus red mite, was most infective at pH 5.0–7.0 with an optimum of 5.6–6.6: buffer salt concentration appeared to have little influence (Chambers, 1968). Some workers have formulated BVs in buffers, e.g. *C. pomonella* GV in tris-buffer at pH 8.0 (Huber and Dickler, 1977) and, of course, this approach has also been employed in attempts to counteract the alkaline condition of cotton leaf surfaces (Falcon, 1971; Young and Yearian, 1986). pH may be easily and safely adjusted with 1–5 N H_2SO_4 (Table 7.1) while mixing or agitating.

The pH of rainwater is often acid both because of dissolved CO_2 and, as a pollutant, H_2SO_4. In industrial areas, values of 4.3 are common (Charlson and Rodhe, 1982).

Table 7.4 Miscellaneous formulation components

Name	Concentration (%)	Source	References
Thickeners and thixotropic agents (viscosity-modifying agents)			
Keltose (a thixotropic, algin)	0.50	Kelco Co., USA	Smith *et al.* (1978, 1980)
Carboxymethylcellulose	1.0–10.0	Union Carbide, USA	Couch and Ignoffo (1981)
Hydroxyethylcellulose (Cellulosize QP4400)	2.00 (pH 5.4–5.8)	Union Carbide, USA	McLaughlin (1967), Andrews *et al.* (1975)
Kelzan gum[a]	0.10–1.1	Kelco Inc., USA	Couch and Ignoffo (1981), Smith *et al.* (1982)
VanGel B	0.40	R.T. Vandebilt, Co., USA	
Veegum[b]	0.40–0.50	K and K Grief Ltd, USA	
Xanthan gum[a]	0.05–0.10	Kelco Inc., USA	
Anti-desiccants, anti-evaporants and humectants			
Biofilm CMC			Nord and Pepper (1991)
Casein–mucin–sucrose mix			Stephens (1959)
Glycerol	1.0–4.0		Dowden and Girth (1953)
Molasses	12.50–25.00		Yendol *et al.* (1977), Wollam *et al.* (1978), Pritchett *et al.* (1980), Shepherd *et al.* (1984)
Nalco-trol[c]			Pfrimmer (1979)
Sorbitol			
Binders			
Gelatine	2.00		Hostetter and Pinnell (1983)
Glycerol	1.00–20.00		McLaughlin *et al.* (1971), Ahmed *et al.* (1973)
Decagin	0.50–1.00	Diamond Shamrock Co., USA	McLaughlin *et al.* (1971), Andrews *et al.* (1975), Bell and Kanavel (1975), Hostetter and Pinnell (1983)
Hydroxyethylcellulose	0.20–0.50		McLaughlin *et al.* (1971), Andrews *et al.* (1975), Bell and Kanavel (1975)
Hydroxymethylcellulose	0.20		Henry (1971), Henry *et al.* (1978)
Paraffin, liquid	20.00		Ahmed *et al.* (1973)
Dispersing agent			
Darvan[d]	2.00	Air Products and Chemicals Inc., USA	
Spray marker dyes			
Rhodamine B	0.04		Shepherd *et al.* (1984), Otvos *et al.* (1987)
Nigrosine (water soluble)	0.50		
Brilliant Sulphaflavine	0.14	Organic Dyestuffs, USA	Yendol *et al.* (1990)

[a] Polyheterosaccharide.
[b] Colloidal magnesium aluminum silicate derived from natural clays.
[c] Polyvinyl substance.

UV PROTECTION

In common with other microorganisms, e.g. protozoa (Teetor and Kramer, 1977), bacteria (Cohen *et al.*, 1991), nematodes (Gaugler and Boush, 1979) and fungi employed as insect pest control agents, BVs can be damaged by solar radiation. It is mostly assumed that the damage is caused to the virus DNA (p. 572) but there is also the possibility of alteration in virus proteins, which has not yet been investigated.

Damage has been thought to be largely a result of UV-B radiation but work by Shapiro and Robertson (1992) suggests that there is also a strong effect from UV-A. These authors examined the protectant capacities of 79 dyes and found that where UV-B absorption was equal, those which also absorbed successfully in UV-A provided the best protection. Previous studies employing monochromators to permit exposure of virus samples to narrow increments of the UV waveband have suggested that UV-A is of little consequence in BV inactivation (Chapter 25). It, thus, seems probable that, in addition to the known damaging impact of UV-B itself, there may be an adverse interaction of UV-B with UV-A. Jones *et al.* (1993) observed that there was some deleterious effect of wavelengths between 320 and 400 nm and above 665 nm. However, with exposure to wavelengths between 400 and 665 nm, in addition to wavelengths above 665 nm, no deleterious effects were noted. The resolution of these issues will have an important bearing on future studies on the protection of insect viruses from solar radiation.

In this section we consider the use of UV-protected formulation for liquid sprays of BV and also the role of protectants in the usually less liquid bait formulations. The former are by far the most important at present.

Damage by solar UV probably commences from the moment of spray droplet formation at the spray nozzle or rotary atomizing disc and continues while the droplet is in flight as well as after it has impacted on the plant target. The significance of events occurring during droplet flight have not been specifically studied for BVs or other insect pathogens but two aspects seem of importance. Firstly, the surface area–volume ratio for very small droplets (≤100 μm diameter), as employed in the increasingly adopted ultra-low-volume (ULV) spraying techniques, is relatively greater than for the more conventional larger droplets generated in spray applications at higher volume; therefore, the exposure of contained virus will be greater. Secondly, being relatively little under the influence of gravity, the flight time of smaller droplets is much greater than that of larger ones. It is determined by factors such as wind speed and air turbulence over the target crop. Under the influence of light winds, small droplets may carry for over 200 m before coming to rest; counter-intuitively, greater wind speeds result in capture by crops in shorter distances than this. In so far as their formulation involves a volatile component (e.g. water), droplet size will decrease during flight, and the rate of decrease can be calculated taking into account atmospheric RH and temperature (Matthews, 1992). The impact of UV will, therefore, potentially increase with this process. Other aspects of droplet size and protectants are commented on by Killick (1987, 1990). It would not be difficult to assess any decrease in BV infectivity occurring during droplet flight. Spray droplets, especially those of small size, could be collected in an aerosol-sampling device and numbers and infectivity (by bioassay) of OBs could be assessed in comparison with a standard preparation of virus. In selecting treatments for inclusion in such bioassays, the possible impact of formulating materials on infectivity should not be overlooked.

Droplets impacting on target surfaces usually spread, and the extent of spread will have a bearing on the subsequent biological impact of solar UV. For instance, a droplet that immediately before impact has a diameter of 50 μm will at the same time have a surface area of $7.854 \times 10^3 \mu m^2$. If on impact the droplet spreads by a factor of three times its original diameter, the surface area then exposed to UV will approximate to 1.767×10^4 μm^2, i.e. *c.* 2.25 times greater than the same droplet in terminal flight. Up to about a spread factor of ×5, the relationship between free-flight surface area and deposited surface area approximates to linearity: droplets are unlikely to spread more than this. The greater the spread factor, the thinner will be the final deposit and the less will be the thickness of UV protectant between virus and incident solar radiation. The physical position of OBs in droplets at rest on plant surfaces is probably also of importance. There is some evidence that in oily carriers they may be very close to or on the droplet surface and often they may have a distribution within the 'resting' droplet that is different from that for particulate protectants. For this and the foregoing reasons, it seems clear that spread of spray droplets on impact should be minimized as much as possible. Pathogen acquisition by pest insects during feeding is likely to be little influenced by the spread factor of very small droplets because the micro-scale of events is unlikely to influence the rate of droplet encounter.

It cannot, incidentally, be assumed that the exposure of the in-flight droplet to UV radiation is equal to only half of its total surface area. The extent of UV irradiation will, in fact, be determined by the proportion of direct and diffuse UV at the time. For further discussion of this topic the reader is referred to the work of Cox (1987) on the aerial fate of microorganisms.

Comments above on the significance of droplet spread on the spray target almost certainly become less precise where larger droplets are concerned. According to the angle of presentation of the target surface (a leaf, for instance) and the hydrophobicity of that surface there will be run-off of spray fluid. It has often been said that this will involve loss to a crop of pathogens included in the spray; this is not necessarily true. It has been demonstrated, for instance, that a droplet of an aqueous suspension of NPV OBs hitting an inclined cabbage leaf and running off very rapidly leaves a deposit of OBs in its track (P. F. Entwistle and W. R. Carruthers, unpublished data). These have been very efficiently acquired from the spray fluid probably by the action of a complex of micro-forces (see Small, 1985; Small, Moore and Entwistle, 1986; Small and Moore, 1987). Under such circumstances, the fate of UV protectant materials in the spray formulation will be problematic; almost certainly they will provide less protection than might, in theory, be anticipated. This whole topic requires further investigation but it does point strongly to disadvantages in the employment of large spray droplets in the context of UV protection.

The technology of UV protectants and their inclusion into sprays of insect pathogens is very much a developing subject area. For instance, the value of chromophores in protection has been only recently appreciated. The joint capacities of strong UV protection and infectivity enhancement of BVs by some optical brighteners is also a recently developing field study. In terms of their mode of action, UV protectant materials appear to fall into four, or possibly five, main classes, with the action of some materials still unclear. Known protectants are listed in Table 7.5 and can be grouped by a functionally based classification.

Reflectants

Some materials that are both opaque and reflectant have been used as UV protectants. They are usually bright substances such as metallic aluminium (powdered), aluminium oxide (AlO_2) and titanium dioxide (TiO_2) though use of the last may be impermissible as titanium is now considered an environmental pollutant. Zinc oxide, often incorporated into anti-sunburn creams as a sun 'blocker', seems not to have been tested unless it is an undeclared component of proprietary protectants. In addition, 'day-glow' pigments or optical brighteners such as the Tinopals (stilbene derivatives) are valuable reflectants and have the advantage of being supplied in a very finely divided form.

General absorbents

Some black materials absorb from the infrared through the visible to the UV wavelengths, carbon being the prime example. Carbon is usually employed in a very finely divided form such as activated charcoal/carbon, which is used industrially for clarifying, deodorizing and filtering and is widely marketed as Norit, Carboraffin, Opocarbyl and Ultracarbon (in addition to the names quoted in Table 7.5). India ink has been successfully employed but the extent to which ingredients other than carbon (e.g. tannins, which are a traditional ink component) contributed to spray effectiveness has not been determined. Activated charcoal/carbon is prepared from wood and other vegetable matter (e.g. coconut shell) and can be easily and locally produced by most developing nations. Other black substances such as Naphthalene Black (Buffalo Black) appear promising (Bull *et al.*, 1976). Molasses has also been shown to be highly effective (Jones, 1988). It is worth noting that the insect debris, protein, etc. in unpurified BV preparations provide considerable protection from UV (Jones, 1988).

Selective absorbents

Some substance with strong absorbance in UV-B have been selected and tested as UV protectants. Such materials are often wholly or largely transparent to the visible wavelengths. For example, various natural flavinoids protect plants from solar UV while permitting photosynthesis because they are transparent to visible light. Thus Quericetrin, a plant-derived protectant, absorbs strongly at 291–389 nm. Other examples are Congo Red, folic acid and to a lesser extent *p*-aminobenzoic acid. A study of the protectant capacity of 79 dyes showed 18 to be effective of which six were especially promising. Congo Red emerged as the most successful agent (Shapiro, 1989; Shapiro and Robertson, 1992).

Table 7.5 Substances with some capacity for protection of BVs from the inactivating effects of solar UV-B radiation

Protectants	Composition	Quantity/ concentration	Source of supply	References
Reflectants[a]				
Aluminium	Powdered Al	NPV : Al = 1 : 49 in micro-capsules		Ignoffo and Batzer (1971)
Aluminium oxide	AlO_2			Ignoffo and Batzer (1971)
Blancophor BBH	Stilbene sulphonic acid	0.02–1.0%	Burlington Chemicals	Shapiro (1992), Shapiro and Robertson (1992), Dougherty *et al.* (1996), Webb *et al.* (1996)
Leucophor BS and BSB, Phorwite AR, Tinopal LPW	Stilbene sulphonic acid	0.1–1.0%	Chemical Development of Canada	Shapiro (1992), Shapiro and Robertson (1992), Dougherty *et al.* (1996), Webb *et al.* (1996)
Tinopal CBS	Naphthotriozole stilbene derivative			Ignoffo *et al.* (1991)
Tinopal CBS-X	Distryl biphenyl	1.0%	Ceiba-Geigy Corp.	Martignoni and Iwai (1985)
Tinopal DCS	Distryl biphenyl	5.0%		Martignoni and Iwai (1985)
Tinopal RBS 200	Distryl biphenyl	0.001%		Martignoni and Iwai (1985)
Titanium dioxide	TiO_2			Topper *et al.* (1984), Bull *et al.* (1976)
General absorbants				
Carbon	Activated high-porosity carbon type RB	1 kg/ha	Pittsburgh Activated Carbon Co.	Ignoffo *et al.* (1991)
	Carbon black, Carbo-jet black	1–5%	Supa Specta, Cities Services Co.	Bull *et al.* (1976)
	India ink	1–5%		Krieg *et al.* (1980)
Naphthalene Black (Buffalo Black (BB))		NPV : BB : oil = 1 : 1 : 48		Ignoffo and Batzer (1971)
Selective absorbants				
Amelozan		0.05%		Injac (1977)
p-Aminobenzoic acid		≥5.0%		Dunkle and Shasha (1989)
Benzilidine sulphonic acid	A sun screen-agent			Shapiro (1985)
2-Hydroxy-4-methoxyl benzophenone	A benzophenone derivative			Shapiro *et al.* (1983)
Benzyl cinnamate		3%		

Brilliant Yellow (3-(4-methylbenzyliden)-camphor	A camphor derivative	0.1%		Krieg *et al.* (1980)
Chevron		0.5–2.4%	Chevron Chemical Co., USA	Vasiljevic and Injac (1975), Injac (1977)
Congo Red (Direct Red 28)		0.5–1.0%	Aldrich Biochemicals, USA	Dunkle and Shasha (1989), Shapiro (1989), Ignoffo *et al.* (1991)
Folic acid	A reduced pteridine nucleus with a *p*-amino-benzoic glutamic acid side chain	1.0%	Aldrich Chemical Co.	Shapiro (1985), Dunkle and Shasha (1988)
Lignin sulphate		8.5%	ITT Rayonier Inc. USA	Martignoni and Iwai (1985)
Orzan LS-50	Probably lignin sulphate	12.0%	ITT Rayonier Inc. USA	Podgwaite *et al.* (1991)
Pantothenic acid		1.0%		Shapiro (1985)
Pyridoxine		1.0%		Shapiro (1985)
Quercetrin (flavin yellow shade)	A bark extract	6.3%	ITT Rayonier Inc. USA	Martignoni and Iwai (1985)
Raymix powder	A lignosulphonate	0.5–1.0%	e.g. BDH Ltd, UK	Shapiro (1985)
Riboflavin				
SAN-240 wp			Sandoz Inc. USA	Jaques (1977)
SNA-285			Sandoz Inc. USA	
SHADE (IMC90001)	A polyflavinoid, catechin leucocyaniden co-polymer	0.25–6.0% (1–12 kg/ha)	Sandoz Inc. USA	Vasiljevic and Injac (1975), Injac (1977), Shapiro *et al.* (1983), Martignoni and Iwai (1985)
Uric acid		1.0%		Shapiro (1984)
Xanthopterin	A pterin	1.0%		Shapiro (1985)
Chromophores				
Acriflavin	3,6-Diamino-10-methyl acridinium		Fluka, Buchs, Switzerland	Cohen *et al.* (1991)
Ethyl Green (probably a chromophore)				
Methyl Green			Fluka, Buchs, Switzerland	Cohen *et al.* (1991)
Rhodamine B			Merck, Darmstadt, Germany	Cohen *et al.* (1991)
Free-radical scavengers				
Carbon/charcoal	Probably best activated			
n-Propyl gallate		>0.01 mg/ml	Sigma St Louis, USA	Ignoffo and Garcia (1994)
Catalase	Oxidaline enzyme	>0.4 mg/ml	Sigma St Louis, USA	Ignoffo and Garcia (1994)

Table 7.5 (*Continued*)

Protectants	Composition	Quantity/ concentration	Source of supply	References
Miscellaneous substances				
Egg albumin		3.0%	e.g. Nutritional Biochemicals Co., USA	Jaques (1971, 1972)
Milk	Powdered/skimmed peptonized	1.0–5.0%	e.g. Difco Laboratories	Jaques (1971)
Soy hydrolysate		5.0%		Jaques (1971)
Yeast	Brewer's/extract	3.0–5.0%		Jaques (1971)
Lovo 192		0.05%	Fison Chemical Co., UK	Vasiljevic and Injac (1975)
Nufilm P, Nufilm 17		0.1%	Miller Chemical and Fertilizer Corp., USA	Injac (1977)
Molasses	Animal feed/Das molasses, etc.	10.0–25.0%		Shapiro *et al.* (1983), Topper *et al.* (1984), Martignoni and Iwai (1985)
Spray oil		0.1%		Injac (1977)
Spray oil universal DS49	1.0%			
Host-derived materials				
Extract of various life stages				Shapiro (1984), and see Teetor and Kramer (1977), Gaugler and Boush (1979)
Extract of faeces				
Impure BV preparations	Body material from diseased host present e.g. PbGV, NPV of *Mythimna separata*, *Spodoptera littoralis*, *Trichoplusia ni*, etc.			David and Gardiner (1966), Jacques (1971, 1972), Manjunath and Mathad (1978), Elnager *et al.* (1980)
Combinations				
Brewer's yeast + charcoal				Jacques (1971, 1972)
Egg albumin + charcoal				Jacques (1971, 1972)
Skim milk + charcoal				Jacques (1971, 1972)
Raymix + molasses				Martignoni and Iwai (1985)

[a] Optical brighteners (stilbene derivatives) absorb UV and emit visible light.

Optical brighteners in virus protection and enhancement of activity

It has been known for some time that optical brighteners (fluorescent brighteners) can provide protection against UV light (Topper *et al.*, 1984). These absorb UV radiation and emit light in the blue region of the spectrum. More recently, it has been discovered that some optical brighteners also enhance or potentiate the infectivity of insect viruses.

In screening for their UV-protectant capacity, Shapiro (1992) tested 23 brighteners belonging to various chemical groups (stilbene, oxazole, pyrazole, naphthalic acid, lactone and coumarin) and found that a number of these, all stilbene disulphonic acids, provided the greatest protection. These were Leucophor BS and BSB, Phorwite AR and Tinopal LPW. Their effects were dose dependent and when Phorwite AR and Tinopal LPW, for instance, were presented with *L. dispar* NPV, there was 15% OAR at 0.001%; 72–84% OAR at 0.01%; 97–100% OAR at 0.1% and 100% OAR at 1.0% concentration of brightener.

Impressive potentiation has been found in several host–virus systems. For *L. dispar* NPV, the addition of stilbene brighteners reduced the LC_{50} in laboratory assays from 18 000 OB/ml to 10–44 OB/ml and the LT_{50} also fell (Shapiro and Robertson, 1992); further work by Dougherty *et al.* (1996) demonstrated an enhancement of 214-fold. Successful results were also obtained in field trials, allowing a considerable reduction in virus dosage (Webb *et al.*, 1994b; Cunningham *et al.*, 1997), while brightener–NPV combinations could be used to broaden the window of application to include third and fourth instar *L. dispar* larvae (Webb *et al.*, 1994a). Enhancement also occurred in other systems; for instance the activity of AcMNPV in *T. ni* was increased 41-fold (Dougherty *et al.*, 1996) and *Anagrapha falcifera* MNPV in *H. zea*, *Spodoptera exigua* and *T. ni* was enhanced 2.6–13.6-fold in LC_{50} tests and 3.7–16-fold in LC_{90} tests, accompanied by a 2.1-fold decrease in LT_{50} (Vail *et al.*, 1996). Enhancement of *S. frugiperda* NPV applied to maize in the field has also been demonstrated along with a study of interactions between a braconid parasitoid and a naturally occurring ascovirus (Hamm, Chandler and Sumner, 1994). The activities of GV in *S. frugiperda* and of *Helicoverpa* iridovirus in *H. zea* were also increased (Shapiro, Hamm and Dougherty, 1992). The zone for active potentiation was found to lie between 0.25 and 1.0% (w/v) (e.g. Vail *et al.*, 1996; Webb *et al.*, 1996); Topper *et al.* (1984) used a much lower concentration, which probably explains why they did not detect an effect using Tinopal.

The discoveries of the use of optical brighteners resulted in the award of a US Patent with licensing of the technology to both American Cyanamid and Sandoz-biosys (Shapiro *et al.*, 1992).

Enhancement has also been demonstrated for other groups of insect pathogenic viruses. For instance, the LC_{50} for *L. dispar* CPV fell 864-fold and its LT_{50} declined from 13.2 to 8.4 days in the presence of Phorwite AR (Shapiro and Dougherty, 1994). These authors demonstrated that addition of Phorwite AR changed *L. dispar* from non-susceptible to susceptible when challenged by AcMNPV or *Amsacta* EPV.

The precise means by which stilbene disulphonic acids mediate the altered susceptibilities to homologous and heterologous viruses in Lepidoptera is not yet fully understood. However, stilbene brighteners may compromise the physical structure of the peritrophic membrane, which is considered to be one of the main mechanical barriers protecting midgut cells from microorganisms. This effect may result from the known interference with chitin fibrilogenesis. The possibility that stilbene brighteners can act as anion transport inhibitors may also be a factor (Shapiro, 1995).

Chromophores

Chromophores function through fast energy transfer from highly absorbent and, therefore, UV susceptible molecules (e.g. DNA) to the energy-accepting molecules or chromophores before damage is significant. The potential of chromophores in plant protection was first demonstrated by Margulies *et al.* (Margulies, Rozen and Cohen, 1985; Margulies, Cohen and Rozen, 1987), who showed that the UV-labile, but otherwise very effective, contact pyrethroid insecticide bioresmethrin could be protected by joint adsorption with a chromopore (e.g. methyl green) to an appropriate surface. The surface used was the finely divided clay montmorillonite, which served to bring the donor and receptor molecules close enough together to permit adequate energy transfer. This method of protection was also shown to stabilize a UV-labile insecticide, a nitromethylene hetercycle (Margulies, Rozen and Cohen, 1988). The value of this approach to microbiological control was next demonstrated with the insect-toxic δ-endotoxin protein of *B.t.* (Cohen *et al.*, 1991). There is no direct evidence that DNA would be similarly protected but it seems likely. In the People's Republic of China, ethyl green has been found to protect PbGV efficiently from inactivation. Ethyl green has a chemical structure similar to that of methyl green and may also turn out to be a chromophore. The molecular structures of methyl green, acroflavin and rhodamine, all effective chromophores, are given by Cohen *et al.* (1991).

Free radical scavengers

Free radicals are defined as 'any (molecular) species capable of independent existence that contains one or more unpaired electrons' (i.e. electrons in unshared rings) (Halliwell and Gutteridge, 1991). This condition sometimes makes the species highly reactive. For instance, UV radiation can lead to the generation of hydroxyl radicals ($OH^{\bullet}$), which are responsible for a large part of the damage done to cellular DNA. The attack of hydroxyl radicals on DNA generates a series of modified purine and pyrimidine bases. Preferential acquisition of such a free-radical (scavenging is the term often used) by another substance would be of benefit to the maintenance of DNA integrity. Since free-radical formation consequent on UV irradiation is likely to be rapid, the presence of scavengers that will be adequately active from the onset of spray droplet formation might be critical. Carbon is said to act as a free-radical scavenger and this capacity will be enhanced when the carbon is very finely divided so facilitating reactions by reducing intermolecular spacing. The role of free radicals in BV inactivation and, *ipso facto* protection by scavenging substances, needs further investigation. The role of free radicals in biology and medicine is the subject of a book by Halliwell and Gutteridge (1991). Experimental evidence supporting the free-radical theory was provided by Ignoffo and Garcia (1994) when they demonstrated the protective action of anti-oxidants and oxidative enzymes, the former to scavenge and the latter to degrade reactive radicals.

Miscellaneous substances

Various substances have been found to offer some protection from UV-B but their mode of action is not necessarily clear. In addition (as reviewed in Jaques (1985)), there is evidence that some materials work well in concert, for example some proteinaceous complexes coupled with absorbants such as carbon.

The importance of the physical proximity of BVs with photoprotectants has been referred to above. It seems likely to be generally true that this will favour the enhancement of some beneficial reactions, e.g. with chromophores (Margulies *et al.*, 1985, 1987, 1988; Cohen *et al.*, 1991). The function of micro-encapsulation, in which the BV is held in close proximity with protectant materials, would in theory favour such reactions as well as effecting an economy in use of protectants. However, this is not necessarily so and Bull *et al.* (1976) found no evidence that, for equal quantities of carbon, a micro-encapsulated formulation was better than a simple joint suspension of carbon with OBs of an NPV.

Earlier processes of micro-capsule formation involved the use of various organic solvents some of which caused a degree of virus inactivation. The use of such solvents may not now be necessary. For instance, sunlight protectants and *Heliothis* (*sens. lat.*) NPV have been starch-encapsulated (Ignoffo, Shasha and Shapiro, 1991) employing a technique first applied to *B.t.* (Dunkle and Shasha 1988, 1989). The micro-capsule size achieved, about 100 μm, is probably compatible with applications involving conventional larger droplet sprays but for ULV, controlled droplet applications (CDA) systems it may be necessary to develop production of a much smaller size capsule.

While attempts have been made to achieve very close proximity of BV and photoprotectant, actual staining of the OB with a dark dye or with a selective absorber of UV has not been reported. This seems worthwhile investigating, as Ignoffo and Garcia (1992) found greater resistance to simulated sunlight in hyphomycetous fungal spores with naturally high levels of dark pigmentation.

It is a notable aspect of many published accounts of UV-B protection of BVs that substances for which a protective capacity has been claimed have been inadequately titrated, a process essential to determine the optimal concentration and the characteristics of the relevant curve relationship. One exception to this is the work of Bull *et al.* (1976), who titrated carbon black (80% of particles in the 10–30 μm size range) concentration against proportion of residual infectivity of *H. zea* NPV following exposure to UV (using a germicidal lamp with 87% emission at 240–260 nm) for 1, 8, 24 and 48 h. For all periods of exposure, protection tended to plateau at 50–60% carbon. A concentration dependence was found for extracts of gypsy moth faeces and for uric acid as NPV protectants (Shapiro, 1984).

Solid and semi-solid formulations

We should not assume that UV cannot penetrate solid or semi-solid formulations or otherwise have an effect, though such possibilities are likely to be less than for liquid formulations. The impact of UV will vary with the composition of the formulation and with the size of 'droplets' or 'particles'. Owing to the nature of the relationship between surface area and droplet volume, such effects will be inversely related to droplet size. A penetration of 5 μm deep into a deposited 'droplet' (assumed to be hemispherical) of 200 μm resting diameter will reach >7.0% of volume, assuming, for the sake of argument, no decrease in UV intensity with depth of penetration.

UV formulation and safety

All materials included in BV formulations, whether for protection from solar UV or for other purposes, should be considered for human, veterinary and environmental safety. While information on the safety of individual substances may already be available, the possibility of adverse interactions involving otherwise benign materials should be considered.

There may necessarily be a trade-off between formulation costs, including UV protectants, and the level of pest control achieved. Again, as mentioned in Chapter 4, this points to the basic necessity to determine the population density to which a particular pest insect must be reduced to achieve an acceptable economic return.

MICRO-ENCAPSULATION

Various techniques of micro-encapsulation are available and have been studied for a wide range of purposes (Table 7.6). Not all have been applied to insect viruses and there is consequently much scope for developmental work.

The main attraction of micro-encapsulation of BVs is that the virus and selected formulating materials can be held in very close physical proximity. This is of especial value where close molecular spacing is an advantage: for instance, between BVs and free radical scavengers and chromophores employed as UV protectants. It is also seen as a possible means of economical use of spray additives.

To be successful four main qualities are required of micro-encapsulation.

1. The process must not adversely affect the effectiveness of the virus.
2. The capsular material should be readily degradable in insect gut.
3. Technical control of micro-capsule size should be possible. This is because capsules of different sizes may be required for different droplet application procedures. Ultra-low-volume, controlled droplet spraying, for example, may well require micro-capsules of ≤50 μm diameter.
4. The cost–benefit ratio must be advantageous. It is desirable that the technology involved should not be complex and does not require heavy capital investment.

Table 7.6 summarizes the major encapsulation processes; each process consists essentially of three main steps.

1. The encapsulating material is dissolved in the appropriate solvent.
2. The active ingredient (BV for the present purposes) is added together with any other formulants, such as UV protectants, and is mixed thoroughly.
3. The material is solidified and the micro-capsules are produced.

Table 7.6 Micro-encapsulation: materials, processes and applications

Capsular materials	Solvents	Capsule formation process	Capsule sizes so far produced (μm)	Recorded applications	References
Not disclosed	Not disclosed	Not disclosed	201	Chemical insecticides *Bacillus thuringiensis*	Raun and Jackson (1966) Raun (1968)
Ethylcellulose gelatin 'polymer'	Toluene, methylethyl ketone, petroleum distillate	Not disclosed, but possible as below	Range 10–100	*Heliothis* (*sens. lat.*) NPV	Ignoffo and Batzer (1971)
Styrene maleic anhydride half ester (SMA-2625A)	Ethyl acetate	High-velocity rotating disc	10–30 (>80%) and none >50	*Heliothis* (*sens. lat.*) NPV	Bull *et al.* (1976)
Calcium alginate	Water	Inject into 100 mM $CaCl_2.2H_2O$ or 50 mM $CaCl_2$	Not disclosed	Vertebrate cells, plant embryos, nematodes (of insects), bacteria, fungi etc. (for soil releases)	Lim and Moss (1981), Redenbaugh *et al.* (1986), Kaya and Nelson (1985), Trevors (1991), Lackey *et al.* (1993)
Carob gum, xanthan gum and mixtures	Water (≅ 1 g/l)			Bacteria	Jung *et al.* (1982), Mugnier and Jung (1985)
κ-Carrageenan	Water (held at 42°C during solubilization)	Inject into 0.3 M KCl		Bacteria, fungi, yeasts	Wada *et al.* (1980), Nasri *et al.* (1987); see Trevors (1991)
Starch–borate, starch–xanthate	Water at alkaline pH	Involves H_2O_2 and H_2SO_4		Chemical insecticides	Shasha (1980), Trimnell *et al.* (1982), Wing and Otey 1983), Shasha *et al.* (1984)
Starch	Water	Dry and then blend, grind or otherwise pulverize	750–1550 utilized >106 μm	*Bacillus thuringiensis* *Heliothis* (*sens. lat.*) NPV	Dunkle and Shasha (1988, 1989), Ignoffo *et al.* (1991), Moguire (1992)

The method for producing the micro-capsules varies.

1. Where organic solvents are employed, solidification is achieved by evaporation from small droplets, for example, those formed when a liquid mix is delivered to a rapidly rotating disc. Small droplets are produced from which the solvent evaporates during their flight. Droplet and, therefore, micro-capsule size will probably be to some extent controllable through adjustments to flow rate to the disc and to its rate of revolution.
2. Encapsulating materials such as calcium alginate and κ-carrageenan can be injected into solutions of $CaCl_2$ or KCl, respectively, to achieve solidification (it will be necessary to check on the effect of such solutions on BVs).
3. If encapsulation uses starch dissolved in water, the mixture is dried and then milled or otherwise pulverized down to an appropriate particle size.

Employment of certain organic solvents may to some extent adversely affect BV infectivity, but ethyl acetate does not seem to fall into this category (Ignoffo *et al.*, 1991).

Alginate capsules may be coated with a biopolymer, providing a semi-permeable membrane (Posillico, 1986) or a lipid membrane (Kaya and Nelsen, 1985).

The use of micro-capsules for spray application to cotton may provide BVs with some protection from the high pH values often encountered on that crop, as was indicated by Bull *et al.* (1976) who found NPV incorporated into SMA micro-capsules only to be adversely affected above pH 9.0.

On the surface of many plants, OBs have strong adherance. The investigator should always check whether this is also true of micro-capsules.

CARRIERS FOR SPRAYS, DUSTS AND GRANULES

Sprays

The nature of the spray fluid chosen will be related to the linked requirements of the spray application methodology employed and the physicochemical nature of the target surface (Chapter 32). Most shelf formulations are water based and water is also often the main carrier fluid at the spray tank mix level. Where a spray of large droplet size is selected, which almost always implies medium- to high-volume spraying, it will usually be unnecessary to add emulsifiable spray oils, which, anyway, would probably be too expensive at this high level of usage. Where small droplets are selected, as in low-volume or ultra-low-volume work, it is almost certain that a spray oil or other 'anti-evaporant' will be required. This is because the rate of evaporation from small water droplets is so rapid (high surface : volume ratio) that the droplet may otherwise be reduced to an inefficient size before closing on the target. The volatility of spray oils themselves is too low to be a consideration during spraying. The proportion of oil to water required will vary with the mean size of spray droplet selected and the atmospheric conditions (temperature and humidity, pp. 123 and 161). Spray oil carriers are listed in Table 7.7.

As far as is known most, if not all, spray oils are not injurious to occluded BVs and, *ipso facto*, nor are the associated surfactants. Spray oils are often also used

Table 7.7 Spray and dust carrier materials

Name	Concentration (%)	Source	References
Spray carriers			
Actipron (Ulvapron)	≤50.00	British Petroleum, UK	Topper *et al.* (1984), Entwistle and Evans (1985), Smits *et al.* (1988)
Coconut oil	50.00		Prior *et al.* (1988)
Codacide oil	90.00	Microcide Ltd, UK	Chadd (1986), Helyer (1993)
Cotton oil	45.00		Bell and Kanavel (1978), Topper *et al.* (1984), Bell (1991)
Diesel oil			Clark and Reiner (1956)
Emulsifiable oil	25.00		Otvos *et al.* (1987)
Flo-Mo			Smith *et al.* (1978)
Fuel Oil, No. 2	80.00		Smirnoff *et al.* (1962)
Naphthalene oils			Wolfenbarger (1964)
Top oil	45.00		Smith *et al.* (1978)
ULV oil		Hoechst Co., Holland	Smith *et al.* (1978)
Dust Carriers[a]			
Attapulgite (attaclay)			Montoya *et al.* (1966), Ahmed *et al.* (1973), Bell (1991)
Brans (e.g. wheat, rice hulls)			Henry (1971), Hostetter and Pinnell (1983)
Flours (e.g. wheat, rice)			Bell and Romine (1980), Pritchett *et al.* (1980)
Starches (e.g. corn, potato)			Rorer (1910), Hostetter *et al.* (1982)
Talc			Rabindra *et al.* (1988)
Lactose			Rabindra *et al.* (1988)

[a] Others, largely used with *B.t.*, are listed by Angus and Luthy (1971) and Couch and Ignoffo (1981).

for *B.t.*, while recent studies on control of locusts involve applications of fungi in coconut oil : water systems with Tween-80 to aid emulsification (Prior *et al.*, 1988). Close study of the physical behaviour of BV OBs and particulate UV-protectant materials in oily droplets may prove to be worthwhile. Actipron–water droplets that were collected on very clean glass surfaces (Chapter 29) showed quite different distributions for an NPV and carbon particles: OBs tended to raft centrally and carbon tended to be peripheral. The relevance of this type of effect to the quality of UV protection needs to be studied.

Oils (e.g. paraffinic oils) have also been used as binders and, presumably, as waterproofing agents, in 'dry' granular formulations of insect pathogens.

Spray oils in general, commercial or otherwise, seem not to be phytotoxic. In addition they do not necessarily directly affect target insects, e.g. *Heliothis* larvae were not affected by spray oils on corn silks (Wolfenbarger, 1964) though Codacide oil killed spider mites (Helyer, 1993). However, their possible adverse impact on beneficial insects should be borne in mind. In some countries, diesel oil is banned as a spray oil because of potential human health hazards.

Most spray oils have a very high flash point and so appear not to present a fire or explosion risk, which could be a problem with organic substances of lower volatility, especially in aerosol form. However, the possibility of such a risk existing should always be considered. Wolfenbarger (1964) discussed various oils in relation to NPV control of *Heliothis* while Wrigley (1973) considered mineral oils in general for spraying.

Care must be given to the use of BVs in electrostatic sprayers. The Electrodyn machine employs its own proprietary oil carrier. By comparison the Rothamsted electrostatic sprayer will impart a charge to droplets of a wide variety of spray fluids including water–commercial spray oil emulsions (Arnold and Pye, 1981). *S. littoralis* NPV has been successfully applied with both sprayers (Jones, 1994).

Dusts

Dusts may pose storage problems because of their bulk and need to be kept dry. An appropriate particle size will often have to be achieved by milling and sieving. Some dust formulations involve a 'wet' phase during their production so that drying then becomes an additional industrial process.

Dust carriers (alias extenders or fillers) are listed in Table 7.7. As far as is known there is no one generally preferable dust, some being more appropriate for a particular pest or set of economic circumstances than another. Abrasive dusts may scarify insect cuticle resulting in severe desiccation and can be said to have innate insecticidal value (David and Gardiner, 1950). The capacity of such abrasive dusts (e.g. carborundum) to injure peritrophic membrane and so to possibly influence the BV infection process has not been studied.

Granules

Formulations of BVs as granules are unusual. With other insect pathogens, e.g. *B.t.*, the fillers often are those also employed in dusts, e.g. attapulgite. It may be noted that fillers such as talc have also sometimes been used to 'extend' sprayable NPV formulations (Rabindra, Muthaiah and Jayaraj, 1988).

Adjuvants, including the concepts of baiting and gustatory stimulants

The effectiveness of BVs may be very considerably improved by application with adjuvants. The term adjuvant is generally employed to exclude solar UV protectants (though a sun-screen may be incorporated in some proprietary adjuvants) but apparently to cover such concepts as larval feeding attractants (baits), phagostimulants in the sense of increasing innate feeding rates, and, possibly, enhancement of BV effectiveness (Table 7.8). The adjuvant capacity of some optical brighteners, mentioned above, should be noted also. As most of the work on adjuvants has been empirical, it is seldom possible to determine precise modes of action, but the possibility of two or more principles operating together must be considered.

Substances identified on the basis of laboratory assays as having potential adjuvant value will not necessarily exhibit such activity when applied to the insects' host plant (see Bartelt, McGuire and Black, 1990). Therefore, the logical sequence of adjuvant assessment would be laboratory assays followed by individual plant tests (e.g. in greenhouses) followed by small-scale field trials.

Baiting can have a number of distinct purposes, not necessarily mutually exclusive:

- presenting highly concentrated inoculum, thus ensuring a shorter disease incubation period (e.g. a reduction from 9.1 to 3.3 days with a cotton-derived bait against *H. virescens* (Bell and Kanavel, 1978)
- probably economizing on inoculum quantity
- favouring inoculum accumulation sites, e.g. granules will funnel into corn leaf whorls for *S. frugiperda* control
- providing ancillary UV protection by nature of the density of some baits
- altering pest larval behaviour to favour control; even in the absence of *B.t.* or BV, baits may provide some control, e.g. of *Pectinophora gossypiella* and of *Heliothis* spp., the latter by delaying boll and bud penetration and so perhaps favouring natural enemy activity (Henry, 1982).

Some successful baits have been developed using extracts of the crop to which they are then applied, e.g. cotton and soybean oil and flour-based baits, examples of proprietary forms of which are Coax and Gustol. Such baits may be valuable for the same and sometimes different insects on the same or different crop plants; for example, cotton boll weevil, *Anthonomus grandis*, bait is effective with *Heliothis* species on cotton (Andrews *et al.*, 1975) whilst Coax, developed for *Heliothis* on cotton, is successful for *Ostrinia nubilalis*, European corn borer, on maize (Bartelt *et al.*, 1990).

In general, the efficiency of unbaited sprays of pathogens is increased by increasing droplet density on the crop, probably simply because droplet encounter by a pest is a chance event. However, the reverse seems possible with baited preparations because larvae are attracted to droplet landing sites. Work with baited sprays of *H. zea* NPV seems to support this concept: sprays of larger droplets containing more OBs were most successful (Stacey, Young and Yearian, 1977b) while with granular baits against *O. nubilalis* on corn, *B.t.* concentration in baits was of greater importance than bait volume, which did not itself significantly affect control (Lynch *et al.*, 1977).

It has also been noted that the attractiveness of a bait may be less to larvae that have first fed on the host plant. For example, a bait of 'diet ingredients' was less effective than Coax when *O. nubilalis* larvae had first fed on maize for 24 h (Bartelt *et al.*, 1990).

The ideal, or essential, nature of baits was defined by Bartelt *et al.* (1990) on the basis of work with *O. nubilalis*. They considered there was important ingredient interaction and that the presence of sugar, lipid and protein was required for good stimulant activity. This concept was compatible with the demonstrated success of Coax or of a mixture of *O. nubilalis* semi-synthetic diet ingredients.

Most BV-related studies on baits have been with cotton insect pests, especially *Helicoverpa* and *Heliothis* spp. and *A. grandis*, as is reflected in Table 7.8, which lists the most important adjuvant ingredients tested. The most successful baiting

Table 7.8 Adjuvant agents[a] for BV preparations, including those registered

Name	Pest	Type of test[b]	References
Water extracts			
Green beans	*Helicoverpa zea*	L	Montoya *et al.* (1966)
Green tomatoes	*H. zea*	L	Montoya *et al.* (1966)
Soft dough sorghum	*H. zea*	L	Montoya *et al.* (1966)
Green okra	*H. zea*	L	Montoya *et al.* (1966)
Fresh corn	*Heliothis virescens*	L, F	Montoya *et al.* (1966)
Corn silks	*H. zea*	L, F	Starks *et al.* (1965), Guerra and Shaver (1969)
Corn kernels	*H. zea* and *H. virescens*	L, G, F	Allen and Pate (1966), Guerra and Shaver (1969), Bell and Kanavel (1978)
Corn meal	*H. zea*	L, F	Stacey *et al.* (1977a)
Corn seed	*H. zea*	L, F	Stacey *et al.* (1977a)
Clover seed	*H. zea*	L, F	Stacey *et al.* (1977a)
Wheat	*H. zea*	L, F	Stacey *et al.* (1977a)
Organic solvent extract			
Cottonseed (50% ethanol, chloroform, methanol)	*Heliothis* spp., *Pectinophora gossypiella*	L, F	Bell and Kanavel (1975, 1977)
Wheat germ (hexane)	*Ostrinia nubilalis*	L	Bartelt *et al.* (1990)
Flours and pulps			
Soybean flour	*H. virescens*	L, G	Bell and Kanavel (1978), Hostetter *et al.* (1982)
Cottonseed flour	*H. zea*	L	Hostetter *et al.* (1982)
Citrus pulp	*H. zea*	L	Smith *et al.* (1982)
Proprietary preparations[c]			
Boll Weevil Bait	*Heliothis* spp.	L, F	McLaughlin (1967), McLaughlin *et al.* (1971), Andrews *et al.* (1975)
Coax (Trader Oil Mill Co., USA)	*Heliothis* spp.	L, F	Bell and Kanavel (1978), Bell and Romine (1980), Johnson (1982), Lutterell *et al.* (1982a, 1983)
	O. nubilalis	L, G	Bartelt *et al.* (1990)
Gustol (Sandoz Inc., USA)	*Heliothis* spp.	L, F	Johnson (1982), Lutterell *et al.* (1982a, 1983)
IMC 607	*Heliothis* spp.	F	Andrews *et al.* (1975)
SAN-285 (sticker, UV screen and feeding adjuvant)	*Heliothis* spp.	F	Pfrimmer (1979)
SAN-285-WP66	*Cydia pomonella*	F	Jaques *et al.* (1981)
Sandoz Liquid Adjuvant-1975	*H. zea*	L, F	Ignoffo *et al.* (1976b), Stacey *et al.* (1977a), Yearian *et al.* (1980)

Table 7.8 (*Continued*)

Name	Pest	Type of test[b]	References
Sandoz Viral Adjuvant (may be the same as the above)	*Lymantria dispar*	F	Wollam *et al.* (1978)
Brans and grits			
Corn bran	*H. zea*	L	Hostetter and Pinnell (1983)
Wheat bran	*H. zea*	L	Hostetter and Pinnell (1983), Bourner *et al.* (1992)
	Agrotis ipsilon	F	Salama *et al.* (1990)
Corn hominy grits	*H. zea*	L	Hostetter and Pinnell (1983)
Oils			
Corn	*H. zea*	L	Hostetter *et al.* (1982)
Cotton seed	*H. zea*	L	Bell and Kanavel (1978), Hostetter *et al.* (1982)
Soybean	*H. zea*	L	Hostetter *et al.* (1982)
Optical brighteners			
Leucophar BS and BSB, Phorwite AR and RKH, Tinopal LPW	Various Lepidoptera	L, F	Shapiro (1995)
Sugars			
Invert (dextrose and levulose 1 : 1)	*H. zea*	L, F	Stacey *et al.* (1977a)
Molasses and citrus molasses	*L. dispar*	F	Wollam *et al.* (1978)
	A. ipsilon	F	Salama *et al.* (1990)
	Heliothis spp.	L, F	Teakle *et al.* (1985), Bell and Kanavel (1978)
Sucrose	*H. virescens*	L, G	Bell and Kanavel (1978)

[a] Very many substances have been tested. Only those with a good response level are indicated here. For the effect of substances in combination consult text, original papers and Table 7.6. No attempt is made to classify into baits, phagostimulants and synergists, as the literature is generally unclear.
[b] L, laboratory; G, greenhouse; F, field.
[c] Response to two of Staley's baits appears to have been nil (Montoya *et al.*, 1966).

materials are plant derivatives, usually concentrated extracts of the host crop itself. It has generally been found that extracts of plant seeds are more attractive than foliage (McMillan and Starks, 1966). Water extracts are adequate, but organic solvent extraction may be less successful, as in the case of maize (Starks *et al.*, 1965).

The concept, or technology, of baiting has been under-exploited, though already shown to be of great value. *Heliothis* and *Pectinophora* control on cotton has been considerably improved. Starch-encapsulated *B.t.* was virtually non-attractive to *O. nubilalis* in the absence of baiting materials (Bartelt *et al.*, 1990).

Table 7.9 Some successful baits

Ingredients	Bait types (% composition)		
	Boll weevil (%)[a]	Coax (%)[b]	Bait described in Smith *et al.* (1982)
Sucrose	27.3	25.0	10.0
Cotton seed	55.9	12.3	–
Soybean oil	–	–	5.0
Cotton seed flour[c]	–	63.3	–
Soy flour[c]	–	–	80.0
Tween 80	–	0.33	1.0
Triton CS-7	–	–	–
Glycerol	10.9	–	–
Thixin	3.6	–	–
Dacagin	3.6	–	–
Hydroxyethylcellulose	1.36	–	–
Original target species	*A. grandis*	*Heliothis* spp.	*H. zea*

[a] Formulated in water, 60%.
[b] As defined and developed by Bell and Kanavel (1978).
[c] In preparing baits with cotton seed or soybean flour, 7.5% represented the upper limit of mixability in water (Hostetter *et al.*, 1982).

Some baits, and the insects against which they have been developed, are compared in Table 7.9. They have great similarities. Unless incorporated in granules or in micro-capsules, baits, especially proprietary baits, or adjuvants (e.g. Coax, Gustol) would normally be added to the BV at the tank or finished spray stage of pre-spray preparation. The rate of application of a bait, for instance in a spray formulation, cannot be given here precisely. As a guide Coax, to which the response is dose related, has been used at 3.36 kg/ha cotton with AcMNPV against a *Heliothis* population, predominantly *H. virescens* (Bell and Romine, 1980). Coax and Gustol elevated *Heliothis* control on cotton at 1.1 kg in 46.8 l spray fluid/ha, but the adjuvant dose needed further study (Lutterell *et al.*, 1982b). In an investigative study, Lutterell, Yearian and Young (1983) examined the mode of action of Coax and Gustol. They found neither to have a direct effect on *Heliothis* larval mortality or, at 2.5%, to screen against UV. However, in laboratory tests, third instar larvae ate more (leaf discs) and the concentration was more important than the rate per hectare. Therefore, the amount of 'adjuvant' ideally required will be related to the selected spray droplet size and density (number of droplets/unit area) considered in relation to the response of the pest species concerned (Chapter 9).

It appears that adjuvants may enhance, or even synergize, BV action, as seemed to occur with cotton seed flour extract, which is also a gustatory stimulant (Hostetter *et al.*, 1982). The active principle has not been determined. However, following work by Tanada (1985) and others, lecithin appears to synergize *H. armigera* NPV (Tuan and Hou, 1988). Lecithin is commercially extracted from soybean but no connection has yet been made between this observation and the action of soybean-related adjuvants or adjuvants of any other lecithin-containing plant source.

REGISTRATION

The requirements of registration authorities are an essential adjunct to formulation development. Requirements for registration are dealt with in Chapter 30.

REFERENCES

Ahmed, S.M., Nagamma, M.V. and Majumdan, S.I. (1973). Studies in granular formulations of *Bacillus thuringiensis* Berliner. *Pesticide Sci.* **4**: 19–23.

Allen, G.E. and Pate, T.L. (1966). A potential role of a feeding stimulant used in combination with a nuclear polyhedrosis virus of *Heliothis*. *J. Invert. Pathol.* **8**, 129–131.

Allen, G.E., Gregory, B.G. and Brazzel, J.R. (1966). Integration of the *Heliothis* nuclear polyhedrosis virus into a biological control programme for cotton. *J. Econ. Entomol.* **59**: 1333–1336.

Andrews, G.L., Harris, F.A., Sikorowski, P.P. and McLaughlin, R.E. (1975). Evaluation of *Heliothis* nuclear polyhedrosis virus in a cotton seed oil bait for control of *Heliothis virescens* and *H. zea* on cotton. *J. Econ. Entomol.* **68**: 87–90.

Angus, T.A. (1954). Use of methyl cellulose in laboratory tests of bacterial pathogens of insects. *Can. Entomol.* **86**: 203.

Angus, T.A. and Luthy, P. (1971). Formulation of microbial insecticides. In *Microbial Control of Insects and Mites*. H.D. Burges and N.W. Hussey (eds). London: Academic Press, pp. 623–638.

Arnold, A.J. and Pye, B.J. (1981). In *Proc. 1981 Br. Crop Protection Conf. Pests and Diseases*. Croydon: BCPC, pp. 661–666.

Bartelt, R.J., McGuire, M.R. and Black, D.A. (1990). Feeding stimulants for the European corn borer (Lepidoptera: Pyralidae): additives to a starch-based formulation for *Bacillus thuringiensis*. *Environ. Entomol.* **19**: 182–189.

Bell, M.R. (1991). Effectiveness of microbial control of *Heliothis* spp. developing on early season wild geraniums: field and cage tests. *J. Econ. Entomol.* **84**: 851–854.

Bell, M.R. and Kanavel, R.F. (1975). Potential of bait formulations to increase effectiveness of nuclear polyhedrosis virus against the pink bollworm. *J. Econ. Entomol.* **70**: 389–391.

Bell, M.R. and Kanavel, R.F. (1977). Field tests of a nuclear polyhedrosis virus in a bait formulation for control of pink bollworm and *Heliothis* spp. in cotton in Arizona. *J. Econ. Entomol.* **70**: 625–629.

Bell, M.R. and Kanavel, R.F. (1978). Tobacco budworm: development of a spray adjuvant to increase effectiveness of a nuclear polyhedrosis virus. *J. Econ. Entomol.* **71**: 350–352.

Bell, M.R. and Romine, C.L. (1980). Tobacco budworm field evaluation of microbial control in cotton using *Bacillus thuringiensis* and a nuclear polyhedrosis virus with a feeding adjuvant. *J. Econ. Entomol.* **73**: 427–430.

Bourner, T.C., Vargas-Osuna, E., Williams, T., Santiago-Alvarez, C. and Cory, J.S. (1992). A comparison of the efficiency of nuclear polyhedrosis and granulosis viruses in spray and bait formulations for the control of *Agrotis segetum* (Lepidoptera: Noctuidae) in maize. *Biocontrol Sci. Technol.* **2**: 315–326.

Boyette, C.D., Quimby, P.C., Connick, W.J., Daigle, D.J. and Fulgham, F.E. (1991). Progress in the production, formulation and application of mycoherbicides. In *Microbial Control of Weeds*. D.O. TeBeest (ed.). New York: Chapman & Hall, pp. 209–222.

Breillatt, J.P., Brantley, J.N., Mazzone, H.M., Martignoni, M.E., Franklin, J.E. and Anderson, N.G. (1972). Mass purification of nucleopolyhedrosis virus inclusion bodies in the K-series centrifuge. *Appl. Microbiol.* **23**: 923–930.

Bull, D.L., Ridgway, R.L., House, V.S. and Pryor, N.W. (1976). Improved formulations of the *Heliothis* nuclear polyhedrosis virus. *J. Econ. Entomol.* **69**, 731–736.

Burgerjon, A. and Grison, P. (1965). Adhesiveness of preparations of *Smithiavirus pityocampa vago* on pine foliage. *J. Invert. Pathol.* **7**: 281–284.

Burgerjon, A. and Sureau, F. (1985). La 'Carpovirusine', bio-insecticide experimental a base du baculovirus de la granulose pour lutter contre le carpocapse (*Cydia pomonella* L.). *C.R. 5 5eme Coll. Recherches Fruitieres*, Bordeaux, INRA/CTIFL, pp. 53–56.

Burges, H.D. and Jones, K.A. (1997). Formulation of bacteria, protozoa and viruses. In *Formulation of Beneficial Microorganisms*. H.D. Burges (ed.). London: Chapman & Hall, in press.

Chadd, E. (1986). *Evaluation of vegetable oils*. Report prepared for Microcide Ltd, Stanton Bury St Edmunds, UK.

Chambers, D.L. (1968). Effect of ionic concentration on the infectivity of a virus of the citrus red mite, *Panonychus citri*. *J. Invert. Pathol.* **10**: 245–251.

Charlson, R.J. and Rodhe, H. (1982). Factors controlling the acidity of natural rainwater. *Nature* **295**: 683–685.

Clark, E.C. and Reiner, O.E. (1956). The availability of certain proprietary adjuvants for use with the polyhedrosis viruses of insects. *J. Econ. Entomol.* **49**: 703–704.

Cohen, E., Rozen, H., Joseph, T., Braun, S. and Margulies, L. (1991). Photoprotection of *Bacillus thuringiensis kurstaki* from ultraviolet irradiation. *J. Invert. Pathol.* **57**: 343–351

Couch, T.L. and Ignoffo, C.M. (1981). Formulation of insect pathogens. In *Microbial Control of Pests and Plant Diseases 1970–1980*. H.D. Burges (ed.). London: Academic Press, pp. 621–634.

Cox, C.S. (1987). *The Aerobiological Pathway of Microorganisms*. Chichester: John Wiley.

Cunningham, J.C., Kaupp, W.J., House, G.M., McPhee, J.R. and de Groot, P. (1978). Aerial application of spruce budworm baculovirus: tests of virus strains, dosages and formulations in 1977. *Can. For. Serv. Inf. Rep. FPM-X-3.*

Cunningham, J.C., Brown, K.W., Payne, N.J., Mickle, R.E., Grant, G.G., Fleming, R.A., Robinson, A., Curry, R.D., Langevin, D. and Burns, T. (1997). Aerial spray trials in 1992 and 1993 against gypsy moth, *Lymantria dispar* (Lepidoptera: Lymantriidae), using nuclear polyhedrosis virus with and without an optical brightener compared to *Bacillus thuringiensis*. *Crop Protect.* **16**: 15–23.

David, W.A.L. and Gardiner, B.O.C. (1950). Particle size and adherence of dusts. *Bull. Entomol. Res.* **41**: 1–61.

David, W.A.L. and Gardiner, B.O.C. (1966). Persistence of a granulosis virus of *Pieris brassicae* applied to cabbage leaves. *J. Invert. Pathol.* **8**: 496–501.

Dougherty, E.M., Guthrie, K.P. and Shapiro, M. (1996). Optical brighteners provide baculovirus activity enhancement and uv radiation protection. *Biol. Control* **7**: 71–74.

Dowden, P.B. and Girth, H.B. (1953). Use of a virus to control European pine sawfly. *J. Econ. Entomol.* **46**: 525–526.

Dubois, N. (1976). Effectiveness of chemically decontaminated *Neodiprion sertifer* polyhedral inclusion bodies. *J. Econ. Entomol.* **69**: 93–95.

Dulmage, H.T., Correa, J.A. and Martinez, A.J. (1970a). Coprecipitation with lactose as a means of recovering the spore-crystal complex of *Bacillus thuringiensis*. *J. Invert. Pathol.* **15**: 15–20.

Dulmage, H.T., Martinez, A.J. and Correa, J.A. (1970b). Recovery of the nuclear polyhedrosis virus and the cabbage looper, *Trichoplusia ni*, by coprecipitation with lactose. *J. Invert. Pathol.* **16**: 80–83.

Dunkle, R.L. and Shasha, B.S. (1988). Starch-encapsulated *Bacillus thuringiensis*: a potential new method for increasing environmental stability of entomopathogens. *Environ. Entomol.* **17**: 120–126.

Dunkle, R.L. and Shasha, B.S. (1989). Response of starch-encapsulated *Bacillus thuringiensis* containing ultraviolet screens to sunlight. *Environ. Entomol.* **18**: 1035–1041.

Easwaramoorthy, S. and Jayaraj, S. (1991). Influence of spray fluid volume and purity on the effectiveness of a granulosis virus infecting sugar-cane shoot borer, *Chilo infuscatellus* Sneller. *Trop. Pest Manag.* **37**: 134–137.

Elmore, J.C. (1961). Control of the cabbage looper with a nuclear polyhedrosis virus disease. *J. Econ. Entomol.* **54**: 47–50.

El-Nagar, S., Abul-Nasr, S. and Nasr, S.A. (1980). Effect of direct sunlight on the virulence of NPV (nuclear polyhedrosis virus) of the cotton leafworm, *Spodoptera littoralis* (Boisd.) *Z. Angew. Entomol.* **90**: 75–80.

Entwistle, P.F. and Evans, H.F. (1985). Viral control. In *Comprehensive Insect Physiology, Biochemistry and Pharmacology*. G.A. Kerkut and L.I. Gilbert (eds). Oxford: Pergamon Press, pp. 347–412.

Falcon, L.A. (1971). Microbial control as a tool in integrated control. In *Biological Control*. C.B. Huffaker (ed.). New York: Plenum Press, pp. 346–364.

Falcon, L.A., Kane, W.R. and Bethell, R.S. (1968). Preliminary evaluation of a granulosis virus for control of the codling moth. *J. Econ. Entomol.* **61**: 1208–1213.

Fernandez, A.T., Graham, H.M., Lukefahr, M.J., Bullock, H.R. and Hernandez, N.S. (1969). A field test comparing resistant varieties plus applications of polyhedral virus with insecticides for control of *Heliothis* spp. and other pests of cotton. *J. Econ. Entomol.* **62**: 173–177.

Gaugler, R. and Boush, G.M. (1979). Laboratory tests on ultraviolet protectants of an entomogenous nematode. *Environ. Entomol.* **8**: 810–813.

Genung, W.G. (1960). Comparison of insecticides, insect pathogens and insecticide–pathogen combinations for control of cabbage looper, *Trichoplusia ni* (Hübner). *Florida Entomol.* **43**: 65–68.

Gipson, I. and Scott, N.A. (1975). An electron microscope study of effects of various fixatives and thin-section enzyme treatments of a nuclear polyhedrosis virus. *J. Invert. Pathol.* **26**: 171–179.

Glass, E.H. (1958). Laboratory and field tests with the granulosis of the red-banded leafroller. *J. Econ. Entomol.* **51**: 454–457.

Glen, D.M. and Payne, C.C. (1984). Production and field evaluation of codling moth granulosis virus for control of *Cydia pomonella* in the United Kingdom. *Ann. Appl. Biol.* **104**: 87–98.

Griffith, I.P. (1982). A new approach to the problem of identifying baculoviruses. In *Microbial and Viral Pesticides*. E. Kurstak (ed.). New York: Marcel Dekker, pp. 507–527.

Gudauskas, R.T. and Canerday, D. (1968). The effect of heat, buffer salt and H-ion concentrations, and ultraviolet light on the infectivity of *Heliothis* and *Trichoplusia* nuclear-polyhedrosis viruses. *J. Invert. Pathol.* **12**: 405–411.

Guerra, A.A. and Shaver, T.N. (1969). Feeding stimulants from plants for larvae of the tobacco budworm and bollworm. *J. Econ. Entomol.* **62**: 98–100.

Halliwell, B. and Gutteridge, J.M.C. (1991). *Free Radicals in Biology and Medicine*, 2nd edn. Oxford: Clarendon Press.

Hamm, J.J., Chandler, L.D. and Sumner, H.R. (1994). Optical brighteners provide baculovirus activity enhancement and UV radiation protection. *Florida Entomol.* **77**, 425–437.

Helyer, N. (1993). *Verticillium lecanii* for control of aphids and thrips on cucumber. *IOBC/WPRS Working Group: Integrated Control in Glasshouses*. Pacific Grove, USA, pp. 63–66.

Henry, J.E. (1971). Experimental application of *Nosema locustae* for control of grasshoppers. *J. Invert. Pathol.* **18**: 389–394.

Henry, J.E. (1982). Use of baits in microbial control of insects. *Proc. 3rd Int. Coll. Invert. Pathol.* and *XVth Annu. Meet. Soc. Invert. Pathol.*, Brighton, UK, pp. 45–48.

Henry, J.E., Oma, E.A. and Onsager, J.A. (1978). Relative effectiveness of ULV spray applications of spores of *Nosema locustae* against grasshoppers. *J. Econ. Entomol.* **71**: 629–632.

Hostetter, D.L. and Pinnell, R.E. (1983). Laboratory evaluation of plant-derived granules for bollworm control with virus. *J. Georgia Ent. Soc.* **18**: 155–159.

Hostetter, D.L., Smith, D.B., Pinnell, R.E., Ignoffo, C.M. and McKibben, G.H. (1982). Laboratory evaluation of adjuvants for use with *Baculovirus heliothis* virus. *J. Econ. Entomol.* **75**: 1114–1119.

Huber, J. and Dickler, E. (1977). Codling moth granulosis virus: its efficiency in the field in comparison with organophosphorus insecticides. *J. Econ. Entomol.* **70**: 557–561.

Hunter, D.K., Collier, S.J. and Hoffman, D.F. (1973). Effectiveness of a granulosis virus of the Indian meal moth as a protectant for stored inshell nuts: preliminary observations. *J. Invert. Pathol.* **22**: 481.

Hunter, D.K., Collier, S.J. and Hoffman, D.F. (1977). Granulosis virus of the Indian meal moth as a protectant for stored in-shell almonds. *J. Econ. Entomol.* **70**: 493–494.

Ignoffo, C.M. (1964). Production and virulence of a nuclear polyhedrosis virus from larvae of *Trichoplusia ni* (Hübner) reared on a semi-synthetic diet. *J. Insect. Pathol.* **6**: 318–326.

Ignoffo, C.M. and Batzer, O.F. (1971). Microencapsulation and ultraviolet protectants to increase sunlight stability of an insect virus. *J. Econ. Entomol.* **64**: 850–853.

Ignoffo, C.M. and Garcia, C. (1966). The relationship of pH to the activity of inclusion bodies of a *Heliothis* nuclear polyhedrosis. *J. Invert. Pathol.* **8**: 426–427.

Ignoffo, C.M. and Garcia, C. (1992). Influence of conidial color on inactivation of several entomogenous fungi (Hyphomycetes) by simulated sunlight. *Environ. Entomol.* **21**: 913–917.

Ignoffio, C.M. and Garcia, C. (1994). Antioxidant and oxidative enzyme effects on the inactivation of inclusion bodies of the *Heliothis* baculovirus by simulated sunlight-UV. *Environ. Entomol.* **23**: 1025–1029.

Ignoffo, C.M. and Montoya, E.L. (1966). The effects of chemical insecticides and insecticidal adjuvants of *Heliothis* nuclear polyhedrosis virus. *J. Invert. Pathol.* **8**: 409–412.

Ignoffo, C.M. and Shapiro, M. (1978). Characteristics of baculovirus preparations processed from living and dead larvae. *J. Econ. Entomol.* **71**: 186–188.

Ignoffo, C.M., Chapman, A. and Martin, A. (1965). The nuclear polyhedrosis virus of *Heliothis zea* (Boddie) and *Heliothis virescens* (Fabricius). III. Effectiveness of the virus against field populations of *Heliothis* on cotton, corn and grain sorghum. *J. Invert. Pathol.* **7**: 227–235.

Ignoffo, C.M., Hostetter, D.L. and Smith, D.B. (1976a). Gustatory stimulant, sunlight protectant, evaporation retardant: three characteristics of a microbial adjuvant. *J. Econ. Entomol.* **69**: 207–210.

Ignoffo, C.M., Yearian, W.C., Young, S.Y., Hostetter, D.L. and Bull, D.L. (1976b). Laboratory and field persistence of new commercial formulations of the *Heliothis* nucleopolyhedrosis virus, *Baculovirus heliothis*. *J. Econ. Entomol.* **69**: 233–236.

Ignoffo, C.M., Shasha, B.S. and Shapiro, M. (1991). Sunlight ultraviolet protection of the *Heliothis* nuclear polyhedrosis virus through starch-encapsulation technology. *J. Invert. Pathol.* **57**: 134–136.

Injac, M. (1977). Protection of the granulosis virus (Baculovirus) of the fall webworm (*Hyphantria cunea* Drury) from ultraviolet light. *Zastita Bilja* **28**: 311–317.

Ives, W.G.H. and Muldrew, J.A. (1978). Preliminary evaluations of the effectiveness of nucleopolyhedrosis virus sprays to control the forest tent caterpillar in Alberta. *Information Report NOR-X-204*, Northern Forest Research Centre, Canadian Forestry Service.

Jaques, R.P. (1971). Tests on protectants for foliar deposits of the polyhedrosis virus. *J. Invert. Pathol.* **17**: 9–16

Jaques, R.P. (1972). The inactivation of foliar deposits of viruses of *Trichoplusia ni* (Lepidoptera: Pieridae) and tests on protectant additives. *Can. Entomol.* **104**: 1985–1994.

Jaques, R.P. (1973). Tests on microbial and chemical insecticides for control of *Trichoplusia ni* and *Pieris rapae* on cabbage. *Can. Entomol.* **105**: 21–27.

Jaques, R.P. (1977). Stability of entomopathic viruses. In *Environmental Stability of Microbial Insecticides. Misc. Publ. Entomol. Soc. Am.* **10**: 99–116.

Jaques, R.P. (1985). Stability of insect viruses in the environment. In *Viral Insecticides for Biological Control.* K. Maramorosch and K.E. Sherman (eds). Orlando, FL: Academic Press, pp. 285–360.

Jaques, R.P. and Huston, F. (1969). Tests on microbial decomposition of polyhedra of the nuclear-polyhedrosis virus of the cabbage looper, *Trichoplusia ni*. *J. Invert. Pathol.* **14**: 289–290

Jaques, R.P., MacLellan, C.R., Sandford, K.H., Proverbs, M.D. and Hagley, E.A.C. (1977). Preliminary orchard tests on control of codling moth larvae by granulosis virus. *Can. Entomol.* **109**: 1079–1081.

Jaques, R.P., Laing, J.E., MacLellan, C.R., Proverbs, M.D., Sanford, K.H. and Trottier, R. (1981). Apple orchard tests on the efficacy of the granulosis virus of the codling moth, *Laspeyresia pomonella* (Lep.: Olethreutidae). *Entomophaga* **26**: 111–118.

Johnson, D.R. (1982). Suppression of *Heliothis* spp. on cotton by using *Bacillus thuringiensis*, *Baculovirus heliothis* and two feeding adjuvants. *J. Econ. Entomol.* **75**: 207–210.

Johnson, T.B. and Lewis, L.C. (1982). Evaluation of *Rachiplusia ou* and *Autographa californica* nuclear polyhedrosis viruses in suppressing black cutworm damage to seedling corn in the greenhouse and field. *J. Econ. Entomol.* **75**: 401–404.

Jones, K.A. (1988). Studies on the persistence of *Spodoptera littoralis* nuclear polyhedrosis virus on cotton in Egypt. PhD Thesis, University of Reading, UK.

Jones, K.A. (1994). Use of baculoviruses for cotton pest control. In *Insect Pests of Cotton*. G.A. Matthews and J. Tunstall (eds). Wallingford, UK: CAB Int., pp. 445–467.

Jones, K.A., Moawad, G. McKinley, D.J. and Grzywacz, D. (1993). The effect of natural sunlight on *Spodoptera littoralis* nuclear polyhedrosis virus. *Biocontrol Sci. Technol.* **3**, 189–197.

Jung, G., Mugnier, J., Diem, H.G. and Dommergues, Y.R. (1982). Polymer-entrapped *Rhizobium* as an inoculant for legumes. *Plant and Soil* **65**: 219–231.

Kaya. H.K. and Nelsen, C.E. (1985). Encapsulation of steinernematid and heterorhabditid nematodes with calcium alginate: a new approach for insect control and other applications. *Environ. Entomol.* **14**: 572–574.

Kelly, P.M., Speight, M.R. and Entwistle, P.F. (1989). Mass production of *Euproctis chrysorrhoea* (L.) nuclear polyhedrosis virus. *J. Virol. Meth.* **25**: 93–100.

Killick, H.J. (1987). Ultraviolet light and *Panolis* nuclear polyhedrosis virus: a non-problem? In *Bulletin 67: Population Biology and Control of the Pine Beauty Moth*. S.R. Leather, J.T. Stoakley and H.F. Evans (eds). Edinburgh: Forestry Commission, pp. 69–75.

Killick, H.J. (1990). Influence of droplet size, solar ultraviolet light and protectants, and other factors on the efficacy of baculovirus spray against *Panolis flammea* (Schiff.) (Lepidoptera; Noctuidae). *Crop Protection* **9**: 21–28.

Krieg, A., Groner, A., Huber, J. and Matter, M. (1980). The effect of medium- and long-wave ultraviolet rays (UV-B and UV-A) on insect-pathogenic bacteria and viruses and their influence by UV-protectants. *Nachr. Deutsch. Pflanzen.* **32**: 100–105.

Lackey, B.A., Muldoon, A.E. and Jaffee, B.A. (1993). Alginate pellet formulation of *Hirsutella rhossiliensis* for biological control of plant-parasitic nematodes. *Biol. Control* **3**: 155–160.

Lewis, L.C. (1982). Use of granules and dusts to disseminate insect pathogens. In *Proc. 3rd Int. Coll. Invert. Pathol.* and *XVth Annu. Meet. Soc. Invert. Pathol.*, Brighton, UK, pp. 66–70.

Lim, F. and Moss, R.D. (1981). Microencapsulation of living cells and tissues. *J. Pharmacol. Sci.* **70**: 351–354.

Livingstone, J.M., McLeod, P.J., Yearian, W.C. and Young, S.Y. (1980). Laboratory and field evaluation of a nuclear polyhedrosis virus of the soybean looper, *Pseudoplusia includens*. *J. Georgian Entomol. Soc.* **15**: 194–199.

Lutterell, R.G., Yearian, W.C. and Young, S.Y. (1982a). Mortality of *Heliothis* spp. larvae treated with *Heliothis zea* nuclear polyhedrosis virus spray adjuvant combinations on cotton and soybean. *J. Georgian Entomol. Soc.* **17**: 447–453.

Lutterell, R.G., Young, S.Y., Yearian, W.C. and Horton, D.L. (1982b). Evaluation of *Bacillus thuringiensis*–spray adjuvant–viral insecticide combinations against *Heliothis* spp. (Lepidoptera: Noctuidae). *Environ. Entomol.* **11**: 783–787.

Lutterell, R.G., Yearian, W.C. and Young, S.Y. (1983). Effect of spraying adjuvants on *Heliothis zea* (Lepidoptera: Noctuidae) nuclear polyhedrosis virus efficacy. *J. Econ. Entomol.* **76**: 162–167.

Lynch, R.E., Lewis, L.C., Berry, E.C. and Robinson, J.F. (1977). European corn borer: granular formulations of *Bacillus thuringiensis* for control. *J. Econ. Entomol.* **70**: 389–391.

Manjunath, D. and Mathad, S.B. (1978). Temperature tolerance, thermal inactivation and ultraviolet light resistance of nuclear polyhedrosis virus of the armyworm, *Mythimna separata* (Wlk.) (Lepid., Noctuidae). *Z. Angew. Entomol.* **87**; 82–90.

Margulies, L., Rozen, H. and Cohen, E. (1985). Energy transfer at the surface of clays and protection of pesticides from photoinactivation. *Nature* (London) **315**: 658–659.

Margulies, L., Cohen, E. and Rozen, H. (1987). Photostabilization of bioresmethrin by organic cations on a clay surface. *Pest. Sci.* **18**: 79–87.

Margulies, L., Rozen, H. and Cohen, E. (1988). Photostabilization of a nitromethylene heterocycle insecticide on the surface of montmorillonite. *Clays and Clay Minerals* **36**: 159–164.

Martignoni, M.E. (1978). Virus in biological control: production, activity and safety. In *The Douglas-fir Tussock Moth: a Synthesis*. M.J.H. Brooks, R.W. Stark and R.W. Campbell (eds). Washington, DC: USDA, pp. 140–147.

Martignoni, M.E. and Iwai, P.J. (1985). Laboratory evaluation of new ultraviolet absorbers for protection of Douglas-fir tussock moth (Lepidoptera: Lymantriidae) Baculovirus. *J. Econ. Entomol.* **78**: 982–987.

Matthews, G.A. (1992). *Pesticide Application Methods*. London: Longman.

Mazzone, H.M., Breillatt, J.P. and Anderson, N.G. (1970). Zonal rotor purification and properties of a nuclear polyhedrosis virus of the European pine sawfly, (*Neodiprion sertifer*, Geoffroy). In *Proc. IVth Int. Coll. Invert. Pathol.*, College Park, MD, pp. 371–379.

McEwan, F.L. and Hervey, G.E.R. (1958). Control of the cabbage looper with a virus disease. *J. Econ. Entomol.* **51**: 626–631.

McEwan, F.L. and Hervey, G.E.R. (1959). Microbial control of two cabbage insects. *J. Insect Pathol.* **1**: 86–94.

McGaughey, W.H. (1975). A granulosis virus for Indian meal moth control in stored wheat and corn. *J. Econ. Entomol.* **68**: 346–348.

McKinley, D.J., Moawad, G., Jones, K., Grzywacz, D. and Turner, C. (1989). The development of nuclear polyhedrosis virus for control of *Spodoptera littoralis* (Boisd.) in cotton. In *Pest Management in Cotton*. M.B. Green and D.J. de B. Lyon (eds). Chichester: Ellis Horwood, pp. 93–100.

McLaughlin, R.E. (1967). Development of the basic principle for boll-weevil control II. Field-cage tests with a feeding stimulant and the protozoan *Mattesia grandis*. *J. Invert. Pathol.* **9**, 70–77.

McLaughlin, R.E., Andrews, G. and Bell, M.R. (1971). Field tests for control of *Heliothis* spp. with a nuclear polyhedrosis virus included in a Boll weevil bait. *J. Invert. Pathol.* **18**: 304–305.

McMillan, W.W. and Starks, K.J. (1966). Feeding responses of some Noctuidae larvae to plant extracts. *Ann. Entomol. Soc. Am.* **59**: 516–519.

Moguire, M.R. (1992). Starch encapsulation of microbial pesticides. In *Abst. XXV Annu. Meet. Soc. Invert. Pathol.*, Heidelberg, Germany, p. 263.

Montoya, E.L., Ignoffo, C.M. and McGarr, R.L. (1966). A feeding stimulant to increase effectiveness of, and a field test with, a nuclear-polyhedrosis virus of *Heliothis*. *J. Invert. Pathol.* **8**: 320–324.

Morris, O.N. (1963). The natural and artificial control of the Douglas-fir tussock moth *Orgyia pseudotsugata* McDunnough, by a nuclear polyhedrosis virus. *J. Insect Pathol.* **5**: 401–414.

Mugnier, J. and Jung, G. (1985). Survival of bacteria and fungi in relation to water activity and the solvent properties of water in biopolymer gels. *Appl. Environ. Microbiol.* **50**: 108–114.

Nasri, M., Syadi, S., Barbotin, J.N. and Thomas, D. (1987). The use of immobilization of whole living cells to increase stability of recombinant plasmids in *Escherichia coli*. *J. Biotechnol.* **6**: 147–157.

Nord, J.C. and Pepper, W.D. (1991). Rain fastness of insecticide deposits on loblolly pine foliage and the efficacy of adjuvants in preventing wash off. *J. Entomol. Sci.* **26**: 287–298.

Otvos, I.S., Cunningham, J.C. and Friskie, L.M. (1987). Aerial application of nuclear polyhedrosis virus against Douglas-fir tussock moth, *Orgyia pseudotsugata* (McDunnough)

(Lepidoptera: Lymantriidae): I. Impact in the year of application. *Can. Entomol.* **119**: 697–706.
Payne, N.J. (1983). A quantification of turbulent dispersal and deposition of coarse aerosol droplets over a wheat field. PhD thesis, Cranfield Institute of Technology, UK.
Pfrimmer, T.R. (1979). *Heliothis* spp.: control on cotton with pyrethroids, carbamates, organophosphates, and biological insecticides. *J. Econ. Entomol.* **72**: 593–598.
Podgwaite, J.D. and Mazzone, H.M. (1981). Development of insect viruses as pesticides: the case of the gypsy moth (*Lymantria dispar*, L.) in North America. *Protect. Ecol.* **3**: 219–227.
Podgwaite, J.D., Rush, P., Hall, D. and Walton, G.S. (1986). Field evaluation of a nuclepolyhedrosis virus for control of redheaded pine sawfly (Hymenoptera: Diprionidae). *J. Econ. Entomol.* **79**: 1648–1652.
Podgwaite, J.D., Reardon, R.C., Kolodny-Hirsch, D.M. and Walton, G.S. (1991). Efficacy of ground application of the gypsy moth (Lepidoptera: Lymantriidae) nucleopolyhedrosis virus product, Gypchek. *J. Econ. Entomol.* **84**: 440–444.
Posillico, E.G. (1986). Microencapsulation technology for large-scale antibody production. *BioTechnology* **4**: 114–117.
Prior, C. and Ryder, K. (1987). Effect of low volume copper sprays with polisobutene sticker on mango blossom blight (*Glomerella cingulata*) in Dominica. *Trop. Pest Manag.* **33**: 350–352.
Prior, C., Jollands, P. and Le Patourel, G. (1988). Infectivity of oil and water formulations of *Beauveria bassiana* (Deuteromycotina: Hyphomycetes) to the cocoa weevil pest *Pantorhytes plutus* (Coleoptera: Curculionidae). *J. Invert. Pathol.* **52**: 66–72.
Pritchett, D.W., Young, S.Y. and Yearian, W.C. (1980). Efficacy of baculoviruses against field populations of fall webworm, *Hyphantria cunea* (Drury). *J. Georgian Entomol. Soc.* **15**: 332–336.
Rabindra, R.J., Muthaiah, C. and Jayaraj, S. (1988). Laboratory evaluation of the CDA formulation of nuclear polyhedrosis virus against *Heliothis armigera* (Hbn.). *J. Entomol. Res.* **12**: 166–168.
Raun, E.S. (1968). Microencapsulation. *Crop Soils Magazine* **20**: 16–18.
Raun, E.S. and Jackson, R.D. (1966). Encapsulation as a technique for formulating microbial and chemical insecticides. *J. Econ. Entomol.* **59**: 620–622.
Redenbaugh, K., Paasch, B.D., Nichol, J.W., Kossler, M.E., Viss, P.R. and Walker, K.A. (1986). Somatic seeds: encapsulation of asexual plant embryos. *BioTechnology* **4**: 797–801.
Reed, E.M. and Springett, B.P. (1971). Large-scale field testing of a granulosis virus for control of the potato moth (*Phthorimaea opercullela* (Zell.) (Lep.: Gelachiidae)). *Bull. Entomol. Res.* **61**: 223–233.
Rollinson, W.D., Lewis, F.B. and Waters, W.E. (1965). The successful use of a nuclear-polyhedrosis virus against the gypsy moth. *J. Invert. Pathol.* **7**: 515–517.
Rorer, J.B. (1910). The green muscardine of froghoppers. *Proc. Agric. Soc. Trin.* **10**: 467–482.
Salama, H.S., Moawad, S., Salah, R. and Ragaei, M. (1990). Field tests on the efficacy of baits based on *Bacillus thuringiensis* and chemical insecticides against the greasy cutworm *Agrotis ypsilon* Redf. in Egypt. *Anz. Schad. Pflanzen. Umwelt.* **63**: 33–36.
Schonherr, J. and Bukovac, M.J. (1972). Penetration of stomata by liquids: dependence on surface tension, wettability, and stomatal morphology. *Plant Physiol.* **49**: 813–189.
Schuster, D.J. and Clark, R.K. (1977). Cabbage looper: control on cabbage with formulations of *Bacillus thuringiensis* and synthetic pyrethroids. *J. Econ. Entomol.* **70**: 366–368.
Seaman, D. (1990). Trends in the formulation of pesticides – an overview. *Pesticide Sci.* **29**: 437–449.
Shapiro, M. (1982). *In vivo* mass production of insect viruses. In *Microbial and Viral Pesticides*. E. Kurstak (ed.). New York: Marcel Dekker, pp. 463–492.
Shapiro, M. (1984). Host tissue and metabolic products as ultraviolet screens for the gypsy moth (Lepidoptera: Lymantriidae) nuclear polyhedrosis virus. *Environ. Entomol.* **13**: 1131–1134.

Shapiro, M. (1985). Effectiveness of B vitamins as UV screens for the gypsy moth (Lepidoptera: Lymantriidae) nuclear polyhedrosis virus. *Environ. Entomol.* **14**: 705–708.
Shapiro, M. (1989). Congo red as an ultraviolet protectant for the gypsy moth (Lepidoptera: Lymantriidae) nuclear polyhedrosis virus. *J. Econ. Entomol.* **82**, 548–550.
Shapiro, M. (1992). Use of optical brighteners as radiation protectants for gypsy moth (Lepidoptera: Lymantriidae) nuclear polyhedrosis virus. *J. Econ. Entomol.* **85**, 1682–1686.
Shapiro, M. (1995). Radiation protection and activity enhancement of viruses. In *Biorational Pest Control Agents: Formulation and Delivery, Am. Chem. Soc. Symp. 595*. F.R. Hall and J.W. Barry (eds). Washington, DC: American Chemical Society, pp. 153–164.
Shapiro, M. and Dougherty, E.M. (1994). Enhancement in activity of homologous and heterologous viruses against the gypsy moth (Lepidoptera, Lymantriidae) by an optical brightener. *J. Econ. Entomol.* **87**, 361–365.
Shapiro, M. and Robertson, J.L. (1992). Enhancement of gypsy moth (Lepidoptera: Lymantriidae) baculovirus activity by optical brighteners. *J. Econ. Entomol.* **84**, 1120–1124.
Shapiro, M., Agin, P.P. and Bell, R.A. (1983). Ultraviolet protectants of the gypsy moth (Lepidoptera: Lymantriidae) nuclear polyhedrosis virus. *J. Econ. Entomol.* **12**, 982–985.
Shapiro, M., Hamm, J.J. and Dougherty, E.M. (1992). Compositions and methods for biocontrol using fluorescent brighteners. US Patent No. 5,124,149.
Shasha, B.S. (1980). Starch and other polyols as encapsulation matrices for pesticides. In *Controlled Release Technologies: Methods, Theory and Applications*, Vol. 2. A.F. Kydoniens (ed.). Boca Raton, FL: CRC Press, pp. 207–224.
Shasha, B.S., Trimnell, D. and Otey, F.H. (1984). Starch–borate complexes for EPTC encapsulation. *J. Appl. Polym. Sci.* **29**: 67–73.
Shepherd, R.F., Otvos, I.S., Chorney, R.J. and Cunningham, J.C. (1984). Pest management of Douglas-fir tussock moth (Lepidoptera: Lymantriidae): prevention of a Douglas-fir tussock moth outbreak through early treatment with a nuclear polyhedrosis virus by ground and aerial applications. *Can. Entomol.* **116**: 1533–1542.
Small, D.A. (1985). Aspects of the attachment of a nuclear polyhedrosis virus from the cabbage looper *Trichoplusia ni* to the leaf surface of cabbage (*Brassica oleracea*). DPhil thesis, University of Oxford, UK.
Small, D.A. and Moore, N.F. (1987). Measurement of surface charge of baculovirus polyhedra. *Appl. Environ. Microbiol.* **53**: 598–602.
Small, D.A. and Moore, N.F. and Entwistle, P.F. (1986). Hydrophobic interactions involved in attachment of a baculovirus to hydrophobic surfaces. *Appl. Environ. Microbiol.* **52**: 220–223.
Smirnoff, W.A. (1977). Confirmations experimentales du potential du complex *Bacillus thuringiensis* et chitinase pour la repression de la Tordeuse des bourgeons de l'epinette *Choristoneura fumiferana* (Lepidoptera: Tortricidae). *Can. Entomol.* **109**: 351–358.
Smirnoff, W.A., Fettes, J.J. and Haliburton, W. (1962). A virus disease of Swaine's jack pine sawfly, *Neodiprion swainei* Midd. sprayed from an aircraft. *Can. Entomol.* **94**: 477–486.
Smith, D.B., Hostetter, D.L. and Ignoffo, C.M. (1978). Formulation and equipment effects on application of a viral (*Baculovirus Heliothis*) insecticide. *J. Econ. Entomol.* **71**: 814–817.
Smith, D.B., Hostetter, D.L. and Pinnell, R.E. (1980). Laboratory formulation comparisons for a bacterial (*Bacillus thuringiensis*) and a viral (*Baculovirus heliothis*) insecticide. *J. Econ. Entomol.* **73**: 18–21.
Smith, D.B., Hostetter, D.L., Pinnell, R.E. and Ignoffo, C.M. (1982). Laboratory studies of aerial adjuvants: formulation development. *J. Econ. Entomol.* **75**: 16–20.
Smits, P.H., Rietstrra, I.P. and Vlak, J.M. (1988). Influence of application techniques on the control of beet armyworm larvae (Lepidoptera: Noctuidae) with nuclear polyhedrosis virus. *J. Econ. Entomol.* **81**: 470–475.
Sopp, P.I., Gillespie, A.T. and Palmer, A. (1989). Application of *Verticillium lecanii* for the control of *Aphis gossypii* by a low volume electrostatic rotary atomiser and a high volume hydraulic sprayer. *Entomophaga* **34**: 417–428.
Stacey, A.L., Yearian, W.C. and Young, S.Y. (1977a). Evaluation of *Baculovirus heliothis* with feeding stimulants for control of *Heliothis* larvae on cotton. *J. Econ. Entomol.* **70**: 779–784.

Stacey, A.L., Young, S.Y. and Yearian W.C. (1977b). *Baculovirus heliothis*: effect of selective placement of *Heliothis* on mortality and efficacy in directed sprays on cotton. *J. Georgian Entomol. Soc.* **12**: 167–173.
Starks, K.J., McMIllian, W.W., Sekul, A.A. and Cox, H.C. (1965). Corn earworm larvae feeding response to corn silk and kernel extracts. *Ann. Entomol. Soc. Am.* **58**: 74–76.
Stephens, J.M. (1959). Mucin as an agent promoting infection by *Pseudomonas aeruginosa* (Schroeter) Migula in grasshoppers. *Can. J. Microbiol.* **5**: 73–77.
Tanada, Y. (1956). Microbial control of some lepidopterous pests of crucifers. *J. Econ. Entomol.* **49**: 320–329.
Tanada, Y. (1964). A granulosis virus of the codling moth, *Carpocapsa pomonella* (Linnaeus) (Olethreutidae, Lepidoptera). *J. Insect Pathol.* **6**: 378–380.
Tanada, Y. (1985). A synopsis of studies on the synergistic property of an insect baculovirus: a tribute to Edward A. Steinhaus. *J. Invert. Pathol.* **45**: 125–138.
Tanada, Y. and Reiner, C. (1962). The use of pathogens in the control of the corn earworm, *Heliothis zea* (Boddie). *J. Insect Pathol.* **4**: 139–154.
Teakle, R.E., Jensen, J.M. and Mulder, J.C. (1985). Susceptibility of *Heliothis armigera* (Lepidoptera: Noctuidae) on sorghum to nuclear polyhedrosis virus. *J. Econ. Entomol.* **78**: 1373–1378.
Teetor, G.E. and Kramer, J.P. (1977) Effect of ultraviolet radiation on the microsporidian *Octosporea muscaedomesticae* with reference to protectants provided by the host *Phormia regina*. *J. Invert. Pathol.* **30**: 348–353.
Topper, C., Moawad, G., McKinley, D., Hosny, M., Jones, K., Cooper, J., El-Nagar, S., El-Sheik, M., Nagar, S.E. and Sheik, M.E. (1984). Field trials with a nuclear polyhedrosis virus against *Spodoptera littoralis* on cotton in Egypt. *Trop. Pest Manag.* **30**, 372–378.
Trevors, J.T. (1991). Respiratory activity of alginate-encapsulated *Pseudomonas fluorescens* cells introduced into soil. *Appl. Microbiol. Biotechnol.* **35**: 416–419.
Trimnell, D., Shasha, B.S., Wing, R.E. and Otey, F.H. (1982). Pesticide encapsulation using a starch–borate complex as wall material. *J. Appl. Polym. Sci.* **27**: 3919–3928.
Tuan, S.J. and Hou, R.F. (1988). Enhancement of nuclear polyhedrosis virus infection by lecithin in the corn earworm *Heliothis armigera*. *J. Invert. Pathol.* **52**, 180–182.
Vail, P.V., Whitaker, T., Toba, H. and Kishaba, A.N. (1971). Field and cage tests with polyhedrosis virus for control of the cabbage looper. *J. Econ. Entomol.* **64**: 1132–1136.
Vail, P.V., Henneberry, T.J. and Bell, M.R. (1977). Cotton leaf perforator: effect of a nuclear polyhedrosis virus on field populations. *J. Econ. Entomol.* **70**: 727–728.
Vail, P.V., Barnett, W., Cowan, D.C., Sibbett, S., Beede, R. and Tebbets, J.S. (1991). Codling moth (Lepidoptera: Tortricidae) control on commercial walnuts with a granulosis virus. *J. Econ. Entomol.* **84**: 1448–1453.
Vail, P.V., Hoffmann, D.F. and Tebbets, J.S. (1996). Effects of a fluorescent brightener on the activity of *Anagrapha falcifera* (lepidoptera: Noctuidae) nuclear polyhedrosis virus to four noctuid pests. *Biol. Control* **7**: 121–125.
Vasiljevic, L. and Injac, M. (1975). Investigation of protection of the nuclear polyhedrosis viruses of the gypsy moth (*Lymantria dispar* L.) against ultraviolet radiation. *Zastita Bilja* **26**: 353–363.
Wada, M., Kato, M.J. and Chibata, I. (1980). Continuous culture of ethanol using immobilized growing yeast cells. *Eur. J. Appl. Microbiol.* **10**: 275–287.
Webb, R.E., Podgwaite, J.D., Shapiro, M., Taman, K.M. and Douglas, L.W. (1990). Hydraulic spray application of Gypchek as a homeowner control tactic against gypsy moth (Lepidoptera: Lymantriidae). *J. Entomol. Sci.* **25**: 383–393.
Webb, R.E., Dill, N.H., Podgwaite, J.D., Shapiro, M., Ridgway, R.L., Vaughn, J.L., Venables, L. and Argauer, R.J. (1994a). Control of third and fourth instar gypsy moth (Lepidoptera: Lymantriidae) with Gypchek combined with stilbene disulfonic acid additive on individual shade trees. *J. Econ. Entomol.* **29**: 82–91.
Webb, R.E., Shapiro, M., Podgwaite, J.D., Ridgway, R.L., Venables, L., White, G.B., Argauer, R.J., Cohen, D.L., Witcosky, J., Kester, K.M. and Thorpe, K.W. (1994b). Effect

of optical brighteners on the efficacy of gypsy moth (Lepidoptera, Lymantriidae) nuclear polyhedrosis virus in forest plots with high or low levels of natural virus. *J. Econ. Entomol.* **87**: 134–143.

Webb, R.E., Dill, N.H., McLaughlin, J.M., Kershaw, L.S., Podgwaite, J.D., Cook, S.P., Thorpe, K.W., Farrar, R.R., Ridgway, R.L., Fuester, R.W., Shapiro, M., Argauer, R.J., Venables, L. and White, G.B. (1996) Blankophor BBH as an enhancer of nuclear polyhedrosis virus in arborist treatment against the gypsy moth (Lepidoptera: Lymantriidae). *J. Econ. Entomol.* **89**: 957–962.

Wing, R.E. and Otey, F.H. (1983). Determination of reaction variables for starch xanthide encapsulation of pesticides. *J. Polym. Sci., Polym. Chem. Ed.* **21**: 121–140.

Wirth, W., Storp, S. and Jacobsen, W. (1991). Mechanisms controlling leaf retention of agricultural spray solutions. *Pesticide Sci.* **33**: 411–420.

Wolfenbarger, D.A. (1964). Parafinic and naphthenic oil fractions in combinations with DDT and a *Heliothis* virus for corn earworm control. *J. Econ. Entomol.* **57**: 732–735.

Wolfenbarger, D.A. (1965). Polyhedrosis-virus–surfactant and insecticide combinations, and *Bacillus thuringiensis*–surfactant combinations for cabbage looper control. *J. Invert. Pathol.* **7**: 33–38.

Wollam, J.D., Yendol. W.G. and Lewis, F.B. (1978). Evaluation of aerially-applied nuclear polyhedrosis virus for suppression of the gypsy moth, *Lymantria dispar* L. *For. Ser. Res. Paper NE-396*. Broomall, PA: USDA, Northeastern Forest Experiment Station.

Wrigley, G. (1973). Mineral oils as carriers for ultra-low-volume (ULV) spraying. *PANS* **19**: 54–61.

Yearian, W.C. and Young, S.Y. (1982). Control of insect pests of agricultural importance by viral insecticides. In *Microbial and Viral Pesticides*. E. Kurstak (ed.). New York: Marcel Dekker, pp. 387–423.

Yearian, W.C., Luttrell, R.G., Stacy, A.L. and Young, S.Y. (1980). Efficacy of *Bacillus thuringiensis* and *Baculovirus heliothis*–chlordimeform spray mixtures against *Heliothis* spp. on cotton. *J. Georgian Entomol. Sci.* **15**: 260–271.

Yendol, W.G., Hedlund, R.C. and Lewis, F.B. (1977). Field investigations of a baculovirus of the gypsy moth. *J. Econ. Entomol.* **70**: 598–602.

Yendol, W.G., Bryant, J.E. and McManus, M.L. (1990). Penetration of oak canopies by a commercial preparation of *Bacillus thuringiensis* applied by air. *J. Econ. Entomol.* **83**: 173–179.

Young, S.Y. and Yearian, W.C. (1986). Formulation and application of baculovirus. In *The Biology of Baculoviruses*, Vol. II. R.R. Granados and B.A. Federici (eds). Boca Raton, FL: CRC Press, pp. 157–179.

Zethner, O. (1976). Control experiments on the nun moth (*Lymantria monacha* L.) by nuclear polyhedrosis virus in Danish coniferous forests. *Z. Ang. Entomol.* **81**: 192–207.

Zidack,N.K., Backman, P.A. and Shaw, J.J. (1992) Promotion of bacterial infection of leaves by an organosilicone surfactant: implications for biological weed control. *Biol. Control.* **2**: 111–117.

8 Spray application of baculoviruses

INTRODUCTION

Spraying is by far the most frequent method of presentation of BVs to pest insects and is necessary for most crop-protection uses because of the need to ensure rapid control through good coverage of the crop at an optimal stage in the life cycle of the pest. Although there is wide experience and suitable equipment available for chemical pesticides, this technology may not be directly suitable for application of viral insecticides. It is, therefore, important to consider the specific attributes of the latter agents.

1. Viral insecticides consist of solid particles (the OBs) in liquid suspension (less commonly as dusts or granules); by comparison, most chemical insecticides are soluble substances. This affects the dilution characteristics of the two. For instance, where a chemical is concerned even the smallest of spray droplets, in any spectrum of spray droplet size, will contain some active ingredient but, for any given concentration of BV, droplets below a certain size will be unlikely to contain any active ingredient. For a given dose (in terms of OBs/unit area of crop) this relationship places a constraint on sensible droplet number (see Figure 8.1, Table 8.1): it will be futile to generate more droplets than there are OBs, except in the instance of simultaneous application of two (or more) viruses where the required dosage levels are very different (Doyle and Entwistle, 1988).
2. Sprayed viruses act following ingestion and replication in the host, but the majority of chemical insecticides combine both contact and ingestion routes. Coupled with the incubation period, viruses inevitably act more slowly than chemical insecticides.
3. Viruses are very specific pesticides in that they have narrow host ranges and have no adverse effects on beneficial animals (parasitoids and predators, whether vertebrate or invertebrate). This is in total contrast to the majority of chemical insecticides where use may also kill beneficial arthropods.
4. Because viruses replicate in the pest insects that they ultimately kill, there is the possibility of a follow-on effect in which secondary inoculum will contribute to prolonged protection of the crop. This is often a very notable effect (Chapter 9).

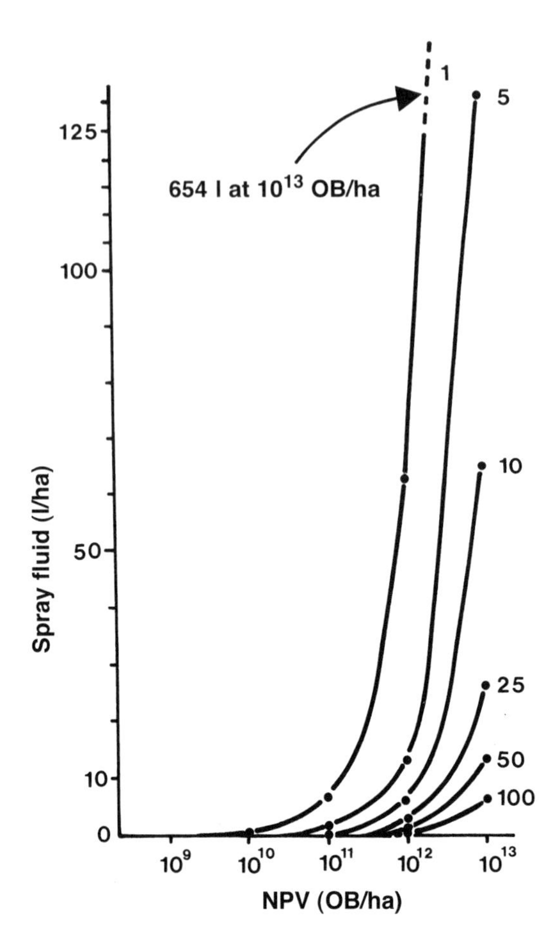

Figure 8.1 The relationship between the number of NPV OBs per hectare and the volume of spray fluid at varying OB densites per 50 μm diameter droplet (1, 5, 25, 50 and 100 OB/droplet). For any given droplet size at desired rates of OBs per hectare, and OBs per droplet, the volume of spray fluid can be calculated from the following formula where 1×10^{15} μm^3 = 1 litre:

$$\text{Vol./ha} = \frac{\text{Droplet vol.}\left(\dfrac{\text{OB/ha}}{\text{OB/droplet}}\right)}{1 \times 10^{15}}$$

Table 8.1 gives values for the number of OBs per droplet volume.

5. Current environmental concerns have put pressure on the manufacturers of chemical insecticides for products that, though effective in the short term, rapidly decay so as not to enter food chains significantly or to have other undesirable environmental consequences. Increasing the persistence or durability of viral deposits on crops is, however, a desirable trend, one which will increase the effectiveness of this type of microbiological control without prejudice to the environment. There are, however, caveats on the persistence of genetically modified viruses.

Table 8.1 The volume of NPV OB material as a percentage of the volume of droplet fluid for a 50 μm diameter droplet

OB diameter (μm)	Percentage of droplet volume occupied by OBs for various numbers of OBs per droplet				
	1	5	10	100	1000[a]
1.0	0.0008	0.004	0.008	0.08	0.8 (0.1)
2.5	0.0125	0.062	0.125	1.25	12.5 (1.56)
5.0	0.1	0.5	1.0	10.0	100.0 (12.5)

For example, in row 1 where OB diameter is 1.0 μm, the OB comprises 0.0008% of a droplet containing 1 OB and 0.004% of the droplet volume for 5 OB/droplet.
[a] Figures in parentheses in the final column are for a 100 μm droplet.

6. Change of response with pest age is much greater with viruses than with chemical insecticides. Because of this, timing of applications is more critical for viral than for chemical treatments. However, sometimes too much has been made of this point: it is important to appreciate that irrespective of the nature of a pesticide it is necessary to take account of application timing if damage to the crop is to be avoided.

The principle of spray application to field crops and forests, and the nature of the associated equipment, is a very large subject that is, in general, beyond the scope of the present volume. The reader is referred to some key works on this subject: Matthews (1992) on pesticide application methods and equipment, Quantick (1985) for a very comprehensive account of aerial applications and Bache and Johnstone (1992) for a detailed and analytical account of spray application in relation to micro-climate.

The only available review specifically devoted to machinery and application factors associated with the use of entomopathogens is that of Smith and Bouse (1981). These authors make a serious and very important point that is still essentially valid, that the majority of spray-control studies employing insect pathogens either report only biological aspects or report very inadequately the characteristics of spray application employed and of spray deposition achieved. Again, as Smith and Bouse state, the multiplicity of factors associated with spray formulation (e.g. viscosity, surface tension, pathogen-loading ratios), spray machinery (e.g. exact type, pressure of spray fluid, speed of rotation where spinning disc generators are used, flow rates of spray fluids) and environmental factors (e.g. adiabatic conditions, wind speeds, temperature, humidity) make it generally difficult to assess the conditions under which a given set of control results were achieved or to reconcile the accounts of different experiments by different, or even the same, authors; hence it is often impossible to derive general principles for pathogen application from the literature.

Disciplined, analytical studies are still greatly needed in order for the optimal parameters for the spray application of pathogens to be defined, and, indeed to allow consideration of the possible need to design specific application equipment.

Table 8.2 Studies on particle loss and particle distribution in spray droplets

Particle type	Estimated loss (%)	Interdroplet distribution measured?	Method of droplet generation	Reference
Fluorescent particles	40	?	?	Himel *et al.* (1965)
B.t.[a]	97	No	Micronair spinning cage	Fast (1976), Morris (1977)
B.t.	11[b]	Yes	Hydraulic, TX-1 nozzle	Smith *et al.* (1978)
Fluorescent particles and NPV	90	No	Spinning disc	Payne, (1983)
GV	91	Yes	Spinning disc	Richards (1984)
Fluorescent particles and *B.t.*	70-80	Yes	Spinning disc	Aston (1989)
Fluorescent particles	?	Yes	?	Fraser (in Aston, 1989)
NPV	Not calculated	Yes	Spinning disc	P. F. Entwistle and H. F. Evans, unpublished data

[a] *Bacillus thuringiensis* spores and especially crystals.
[b] A relative difference between two levels of hydraulic pressure: 552 and 138 kPa.

These parameters will, of course, differ according to the situation and according to insect habit and sensitivity to a given pathogen.

PARTICLE LOSS FROM SPRAY DROPLETS AND THE RELATIONSHIP BETWEEN DROPLET SIZE AND PARTICLE CONTENT

There is a growing body of evidence that indicates that there may be a severe loss of solid particles (e.g. *B.t.* spores and crystals, BV OBs and fluorescent particles) from spray droplets in the period between the finished formulation in the spray tank and the arrival of the spray droplet at its target (Table 8.2). Clearly, the validity of such a conclusion depends on the reliability of the methods employed to determine droplet size and particle content. However, the finding is particularly disturbing because the same general conclusion has been reached using quite a wide range of methods and by the activities of several workers over a period of 15 years.

Another general conclusion seems to be that smaller droplets tend to contain a disproportionately high number of particles. This may not always be the case. Richards (1984), for instance, states that the difference between spray deposition was 'greater when measured by Uvitex determination than by bioassay' and that 'this difference increased with decreasing droplet size, with an apparent "loss" of 91% of infective virus between the spray tank and the leaf surface when virus was applied at 50 l/ha by a spinning disc sprayer (119 μm vmd)'. Richards considered this result, obtained with CpGV, to have been largely uninfluenced by other factors, e.g. ULV formulants, UV decay, Uvitex or OAF (open air factor).

The calculations of Fast (1976) were based on information on counts of *B.t.* crystals in droplets caught on targets on the ground during forest spraying (Morris, 1977). It was observed that 100 μm diameter droplets contained only 144 crystals instead of the 4580 anticipated on the basis of tank concentration. While it is not disputed here that this implies a severe loss, the picture is nevertheless incomplete. Smaller droplets would be unlikely to have reached the targets on the ground and may have contained disproportionately more *B.t.* It is, therefore, very important to measure the content of a representative range of spray droplet sizes in order to assess fully the biological implications of these phenomena.

An association of loss with spray fluid pressure (using a TX-1 nozzle) was noted by Smith, Hostetter and Ignoffo (1978). The *B.t.* content of droplets was about 11% less at 552 kPa than at 138 kPa. This value is for relative loss, and there may have been an unmeasured background loss from droplets at either or both pressures.

Fluorescent particles are particularly useful in studies of this kind. Not only can they be obtained in sizes approximating closely to microbiological agents (e.g. 1–5 μm) but droplets containing fluorescent particles can be visualized under UV light and the particle content of droplets can be estimated using image analysis equipment (Aston, 1989). However, while fluorescent particles may be useful in investigating the distribution of solid particles in droplets, they may accumulate within some sprayers (e.g. in the disc of spinning disc droplet generators) and this may preclude their use in studies on absolute particle content. Other problems with the use of fluorescent particles are discussed in Chapter 29.

Probably the most comprehensive comparison of methodologies for investigating particle numbers in spray droplets was that conducted by Aston (1989) using scanning electron microscopy, optical microscopy and computerized image analysis.

The processes involved in particle loss and in the counter-intuitive uneven distribution of particles between droplets of different sizes are not yet understood. On the whole, it is considered that loss is unlikely to occur during droplet flight as surface tension would prevent particles 'falling out' of droplets. However, Payne (1983) postulated that, as water evaporates from a spray oil–water formulation, the particles (in this case the polyhedral OBs of an NPV) will remain on the surface of oil droplets and will then be removed as the droplet is accelerated and decelerated in atmospheric turbulence before reaching its target. Also, as the conclusion that loss seems independent of method of droplet capture (e.g. on solid surfaces (Morris, 1973, 1977; Fast, 1976) or as aqueous droplets into open dishes of oil (Aston, 1989)), it seems unlikely that particles are thrown out at impact.

A clear distinction should be made between the questions of unexpected particle distributions and actual particle absence as it is entirely possible that they have different causes. For instance, Aston (1989) discussed events occurring between delivery of spray fluid to a spinning disc and the formation of droplets by 'fracture' of fluid ligules developed at the teeth on the disc margin. Through the action of friction, the fluid moves more slowly close to the disc surface and much more rapidly towards the fluid sheet free surface. The fluid is thus subject to a strong shear force and it is suggested that particles may be redistributed in accordance with the velocity profile and will tend to move into the 'fast lane'. If this happens, it may help to explain the deviation between the anticipated Poisson distribution and measured values for particle content in relation to droplet size observed by Aston. This worker commented that 'the number of fp's/drop was [found to be] a function of the droplet diameter^2 and not the diameter^3'.

At present, it is not clear how loss from spray droplets can be counteracted. It has been suggested (Smith and Bouse, 1981) that it may be advantageous to use thickeners rather than surfactants in microbial spray formulations. However, it must be considered that this suggestion may conflict with indications for the need for surfactants to prevent droplet bounce from some types of plant surface.

It is clear that particle loss must be considered in the context of droplet formation methods and formulation functions.

REFERENCES

Aston, R.P. (1989) The use of *Bacillus thuringiensis* (Berliner) for the control of *Heliothis armigera* (Hubner) (Lepidoptera: Noctuidae) on cotton. PhD thesis, Cranfield Institute of Technology, UK.

Bache, D.H. and Johnstone, D.R. (1992) *Environmental Management Science and Technology Series: Microclimate and Spray Dispersion*. Chichester: Ellis Horwood. p. 239.

Doyle, C.J. and Entwistle, P.F. (1988) Aerial application of mixed virus formulations to control joint infestations of *Panolis flammea* and *Neodiprion sertifer* on lodgepole pine. *Annals of Applied Biology* **113**, 119–127.

Fast, P.G. (1976) Further calculations relevant to field application of *Bacillus thuringinensis*. *Bi-Monthly Research Notes* **32**, 27.

Himel, C.M., Vaughn, L., Miscus, R.P. and Moore, R.P. (1965) A new method for spray deposit assessment. *US Forest Research Note PSW 87*.

Matthews, G.A. (1992) *Pesticide Application Methods*. London: Longman, pp.1–405.

Morris, O.N. (1973) A method of visualizing and assessing deposits of aerially sprayed insect microbes. *Journal of Invertebrate Pathology* **22**, 115–121.

Morris, O.N. (1977) Relationship between microbial numbers and droplet size in aerial spray applications. *Canadian Entomologist* **109**, 1319–1323.

Payne, N.M. (1983) A quantification of turbulent dispersal and deposition of coarse aerosol droplets over a wheat field. PhD thesis, Cranfield Institute of Technnology, UK.

Quantick, H.R. (1985) *Aviation in Crop Protection, Pollution and Insect Control*. London: Collins.

Richards, M.G. (1984) The use of a granulosis virus for control of codling moth, *Cydia pomonella*: application methods and field persistence. PhD thesis, University of London.

Smith, D.B. and Bouse, L.F. (1981) Machinery and factors that affect the application of pathogens. In *Microbial Control of Pests and Plant Diseases 1970–1980*. H.D. Burges (ed.). London: Academic Press, pp. 635–653.

Smith, B.D., Hostetter, D.L. and Ignoffo, C.M. (1978) Performance specifications for two microbial insecticides. In *Miscellaneous Publications of the Entomological Society of America. Vol. 10: Formulation and Application of Microbial Insecticides*. C.M. Ignoffo and L.A. Falcon (eds). Washington, DC: Entomological Society of America, pp. 44–66.

9 Conduct and recording of field control trials

The general history of field control trials of insects and mites with BVs suggests that many have had a strong empirical element in their design. At best, trials have been designed only on the basis of previous experience, and only in terms of a selection of dosage of pathogen, its formulation and application system. This has often resulted in tedious and protracted developmental programmes.

Field trials will probably always be a necessary preliminary to deciding the optimal parameters of virus use for refined economic application. However, a detailed knowledge of target pest biology and investigations in both the laboratory and field should make it possible to design field trials to test hypotheses, thus substantially reducing the *ad hoc* nature of many previous trials. Such an approach, apart from being more scientifically enjoyable, will undoubtedly be more economical in effort and materials and should lead more rapidly to firm practical recommendations. Such a framework can also lead to the development of predictive models that may be of practical value to other workers who wish to use BVs in the field.

DEVELOPING THE INITIAL FIELD CONTROL TRIAL PLAN

To develop a preliminary BV control trial plan it is necessary to obtain at least approximate values for a small number of key parameters.

Identification and distribution of virus acquisition sites

Information on virus acquisition sites is especially relevant to first, second and, perhaps, third instar larvae, but occasionally, where early instars have concealed feeding habits, the behaviour of a later-stage larvae may be important, e.g. as in spruce budworm, *Choristoneura fumiferana*, which does not feed on foliage treatable with spray droplets until the fourth instar, having previously fed under cover in a silk tent. The feeding sites of older larvae, the control of which may now be made possible by the use of substances effective both as UV protectants and enhancers of virus activity (Chapter 7), should also be considered. Several

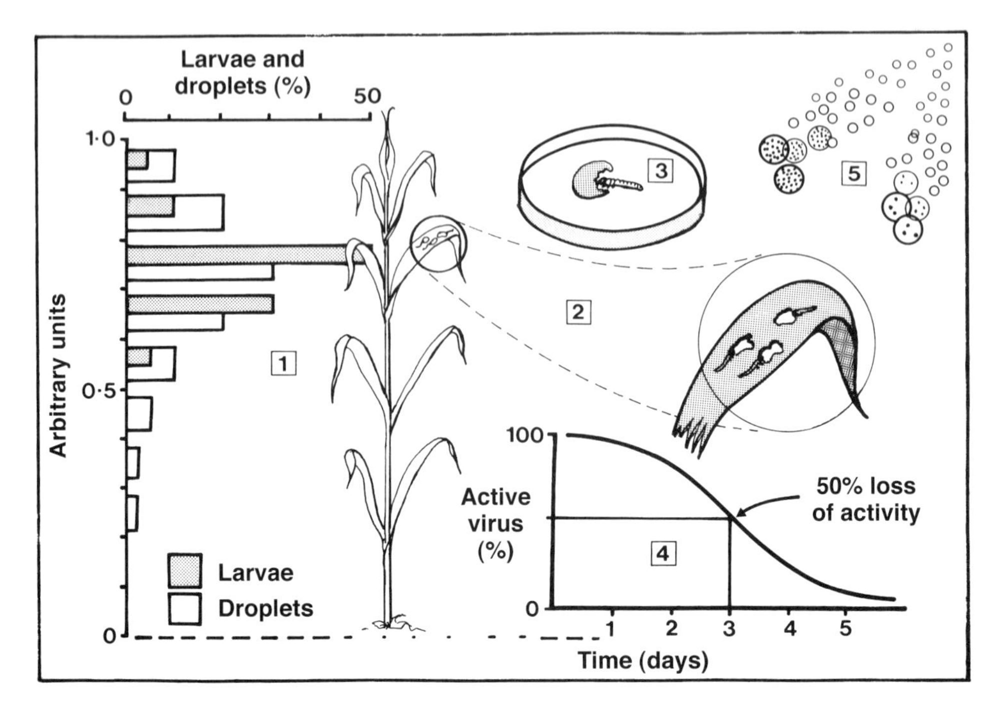

Figure 9.1 A guide to development of a first field trial (see text for details) (After Entwistle *et al.*, 1990).

instances are already known of virus acquisition from sites where there is no obvious feeding by larvae, so this aspect of larval behaviour should always be investigated. This type of information should be quantitatively comparable to deposition patterns of spray droplets (Figure 9.1, step 1). The development process can be divided into five steps, illustrated in Figure 9.1.

1. Determine the statistical distribution of virus acquisition sites (these are not necessarily entirely coincident with overt feeding sites on the crop) and then select droplet size and application methodology appropriate to best match of spray deposition (Figure 9.1, step 1).
2. Measure feeding rate (leaf surface area consumed in unit time say, 24–48 h) of the most susceptible life stages to which spray droplets are accessible: usually, but not always, the first two instars.
3. Conduct laboratory bioassays, preferable with inoculum in a realistic number of appropriately sized spray droplets, on leaf discs of an area directed by results from step 1 and 3, to determine LD_{90-95} values over a selected time period (e.g. 24–48 h).
4. By field-exposure studies determine the total inactivation picture in the time interval employed in step 3 so that dose/droplet density can be corrected for decay or formulation can be designed to counteract decay.
5. Make test batch of 'final' spray formulation, apply under strictly operational conditions and inspect statistically reliable sample of trapped droplets to see if they contain the expected number of virus OBs.

Susceptibility of the main target instars

It is essential that virus dosage–mortality relationships should be investigated in the laboratory. From this a dose that will provide, in theory, a high level of field mortality can be selected, e.g. an LD_{90} or LD_{95}, or higher. The methodology employed should relate as closely as possible to field conditions. It is suggested a leaf-disc technique be used (Chapter 26). This is more realistic than conducting the assay on BV-contaminated semi-synthetic diet because the larval feeding rate and the interaction of the plant tissue with the BV may differ from that on diet. The area of the leaf disc on which doses are presented should be selected to mimic the area consumed in a 'convenient' period of time, e.g. 12–24 h, in relation to both virus degradation and decreasing target susceptibility with increasing larval age.

Determination of droplet size

An essential requirement is to determine the droplet spectrum (in terms of the volume median diameter (VMD) and the VMD to NMD (number median diameter) ratio and span) that gives that crop coverage which best matches the distribution of virus acquisition sites by larvae, as discussed above. This can be assessed by laboratory and field observations of the relationships between droplet size and crop (vertical) penetration using an identical formulation and delivery rate to that likely to be used in routine field programmes selected.

The nature of the crop surface

The likely response of the crop surface to spray droplets of the size and composition selected must also be determined. The surfaces of some crops are very reflective to spray droplets not incorporating a suitable proportion of surfactant (Chapter 7)

Virus degradation

Solar UV irradiation and, less commonly, the chemical nature of leaf surfaces constitute the major causes of biological degradation of BV spray deposits. In addition, deposits may be physically eroded by windborne dust and by rain. These losses can be reduced by formulation. It is necessary to determine the nature of the virus degradation curve in the field and to relate this to the selected virus acquisition period. If a long acquisition period is necessary, then attrition may be significant. The selection of appropriate protectants can be made on this sort of basis.

Spray droplet number

The number of spray droplets that must be generated to achieve the selected droplet density at the virus acquisition sites depends on two factors:

- the proportion of all droplets lost to the crop by drift outside the target area
- the proportion of the droplets captured by areas of the crop of little or no interest in terms of virus acquisition by the pest.

These data will enable both the spray volume required per hectare and the number of BV OBs in that volume to be calculated. A worked example of these two simple processes is provided below.

The virus dose per hectare can be calculated from the following data; the values given are arbitrary but in the realms of possibility.

Larval unit feeding area in, say 48 h	10 mm^2
LD_{95} (leaf disc method)	50 OBs
Including a factor to cover decay, (e.g. ×2 over 48 h)	100 OBs
Correct for only 75% of drops falling on targeted zone of crop	133 OBs
OB/mm^2 leaf tissue	13.3
mm^2/ha	1×10^{10}
Foliar index	$\times 3$
Dose/ha	$13.3(3 \times 10^{10}) = 4 \times 10^{11}$ OBs

Similarly spray fluid volume/hectare can be calculated.

Deliver, say, 4 droplets/10 mm^2	1.2×10^{10} droplets/ha
Only 75% droplets, say, in target area	1.6×10^{10} droplets/ha
Ideal, selected, droplet is 80 μm diameter	2.68×10^5 μm^3
Spray volume/ha	$\dfrac{(1.6 \times 10)(2.68 \times 10^5)}{1 \times 10^5} = 4.29$ l

From these calculations it can be seen that employing a droplet of 80 μm VMD would require 4×10^{11} OBs applied in 4.29 l/ha. The above calculations assume an even vertical distribution of spray droplets on the crop. However, droplet size will have been selected to give a skewed droplet deposition pattern favouring larval feeding zones. Taking this into account could lead to a reduction in dose and volume on the above calculations.

The complete formulation

It is now possible to design a tentative tank formulation that will minimally include some, or all, of the following components.

1. The active ingredient (virus)
2. Water
3. Anti-evaporant, especially where small droplets have been selected and/or atmospheric conditions will lead to a high rate of evaporation
4. Surfactant, to counteract any plant surface droplet reflectivity
5. UV protectant and/or any other substance necessary to prevent deposit decay.

The occlusion body content of droplets at the target

At present, the conditions under which there is loss of solid particles (such as OBs and, possibly, particulate UV protectants such as carbon) between the spray tank and the spray droplet target are not understood (Chapter 28) and hence it is not possible to predict loss or to design formulations to prevent its occurrence. It is, therefore, desirable to make up a trial batch of the selected formulation and to test this for loss of solid particles using exactly the spray machinery and settings selected for the forthcoming field control experiments. Techniques for inspecting the OB content of captured droplets are discussed in Chapter 29. Such an investigation may indicate a need to modify the spray formulation or the spray application equipment and its settings (or both) but it is much better to know about this at an early stage in programme development than to discover the problem later on. The general steps discussed above are summarized in Figure 9.1.

FIELD DESIGN AND STATISTICAL TREATMENT

The statistical basis of field design for a control experiment employing BV treatments will be very similar to that for chemical insecticides. In any event, field experiments will often incorporate comparisons of biological (viral, bacterial, fungal) and chemical insecticides so that a design will be required to satisfy the nature of all treatments. Standard texts on field trial design and analysis should be followed.

In any field control trial in which it is desired to establish a virus dose–mortality relationship, the selection of dose levels in treatments is very important. Field doses that result in 0 or 100% mortalities are of very limited value in establishing such a relationship. A minimum of four dosage levels providing mortality responses within the range 5–95% are necessary for satisfactory analysis (e.g. log dose–probit mortality), a position which, of course, is equally applicable to laboratory LD_{50} studies (Chapter 26).

In any spray experiments, the buffer zones between plots should be wide enough to allow for spray drift and so to prevent treatment interactions. With BVs, an even wider buffer zone may be selected because of the possibility of BV transport by beneficial animals (parasitoids and predators) from one plot to another. Figure 9.2 illustrates actual extra-plot NPV disease that developed following two methods of aerial spray application; spread was considerable, especially downwind. If it is required to record an experiment for more than one host generation, then even greater consideration should be given to buffer zone dimensions.

Where trials are to be conducted by spray drifting, as in ULV applications, it is necessary to ensure a commercially realistic pattern of droplet deposition. This will usually involve spraying along several upwind spray lanes in addition to spraying directly over the plot itself (Chapter 8). As plots will almost certainly be demarcated before the day of spraying, this will necessitate a wide boundary zone on all sides of the plot because the direction of the wind on the spray day can seldom be reliably predicted. Such boundary zones must probably be wider for aerial than for ground applications.

In some insect species, natural BV infections will appear in the field. Usually, but not always, these will develop more slowly than experimentally induced infections and so will be particularly evident in control plots; however they will tend to be pre-empted and masked in the BV treatments. Where control plot infection reaches a high level the actual degree of crop protection (e.g. yield, growth and

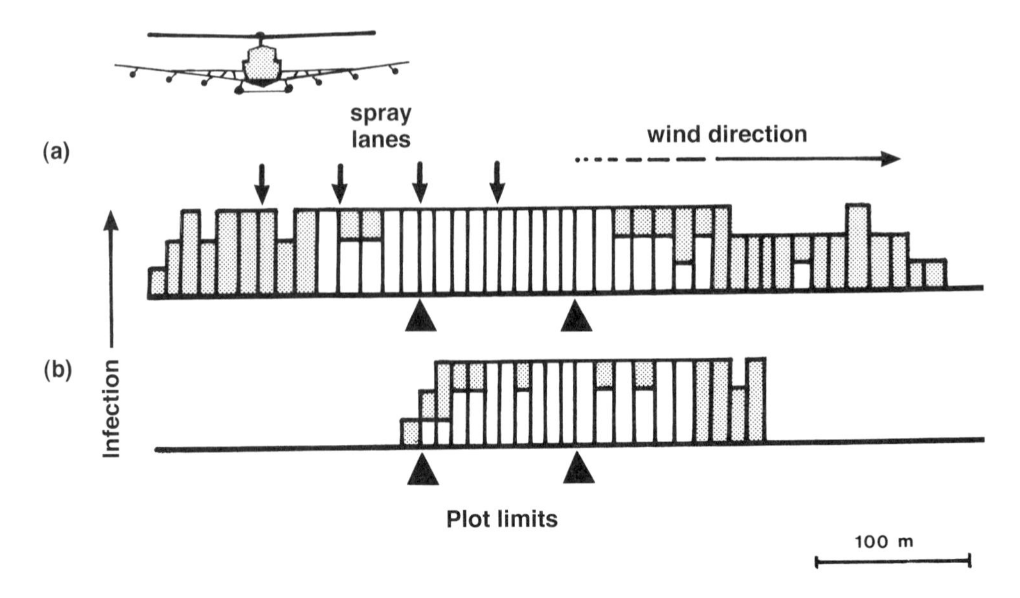

Figure 9.2 Appearance of disease in NPV-sprayed European pine sawfly, *Neodiprion sertifer* (Hymenoptera: Diprionidae), infestations on young pine trees. The figure contrasts the degree of development of disease inside and outside two plots aerially (helicopter) sprayed by two different methods but with the same dose of NPV (5×10^9 OB/ha). The height of the bars indicates the degree of infection. (a) Down-wind drifting of spray from flying along four lanes with 50 m separation (droplet size 110 μm diameter). (b) 'Direct' spraying with hydraulic boom and nozzle equipment with spray lane width equal to boom span (droplet size 220 μm diameter). The plots were both sprayed on 14 June, 1983 and the infection patterns depicted are for 24 days (open bars) and 38 days (shaded bars) later.

cosmetic protection) may prove to be a better parameter for assessing treatment impact than relative pest mortality levels.

BVs may accumulate and persist in soil (Chapters 4 and 5), resulting in background infection, so that where a succession of trials is planned it may be best to use a series of fresh field sites.

It is often the fate of entomologists planning field experiments to find that natural pest populations are unusually low, but the vagaries of natural population densities of insects need not necessarily prevent the conduct of field control trials. It is generally possible to maintain laboratory cultures of insects and to 'seed' test crops. While the population density at which such artificial infestations are feasible may not be high enough to induce expected levels of economic damage, information on control *per se* can still be obtained.

One of the benefits of using selective pesticides, such as BVs, with very restricted host ranges is the lack of direct effects on populations of beneficial insects. It is easier to demonstrate this component of overall pest control with larger than with smaller plots, especially in situations where previous use of chemical insecticides has severely depressed the general levels of beneficial insects. Ultimately, the true value of selective biopesticides of this type can only be demonstrated in very large plots where the chances of beneficial insect suppression resulting from the effects of neighbouring chemical applications are strictly minimized. Therefore, large-scale, or extension, trials should always be a part of the BV testing and developmental process in areas where chemical insecticides have been the norm.

RECORDING BACULOVIRUS CONTROL TRIALS

In insect pest control trials the usual economic measure of response to insecticide applications is yield and growth of the crop itself. However, in experiments where it is often desirable to obtain a more rapid assessment of impact, response may be measured in terms of the proportion of insects killed, as calculated from pre- and post-spray population counts and comparisons with control plot populations. BVs have an incubation period during which a variable amount of larval feeding may take place. While the incubation period may be relatively long, the amount of feeding within it may be very little. For instance in the spruce sawfly, *Gilpinia hercyniae*, at 5 days after NPV infection of the third instar, mortality levels were similar to the controls but feeding was 97% less (Figure 9.3). Here an assessment of the comparative impact of a chemical insecticide and a NPV spray during this period would have misrepresented the impact of the NPV in real terms of crop protection.

In most BV control schemes, spray application is timed deliberately to coincide with the very early larval period. There are two reasons for this. Firstly, sensitivity to BVs declines as larvae increase in size so that early sprays are likely to be the most successful. Secondly, the amount of feeding during the first and second

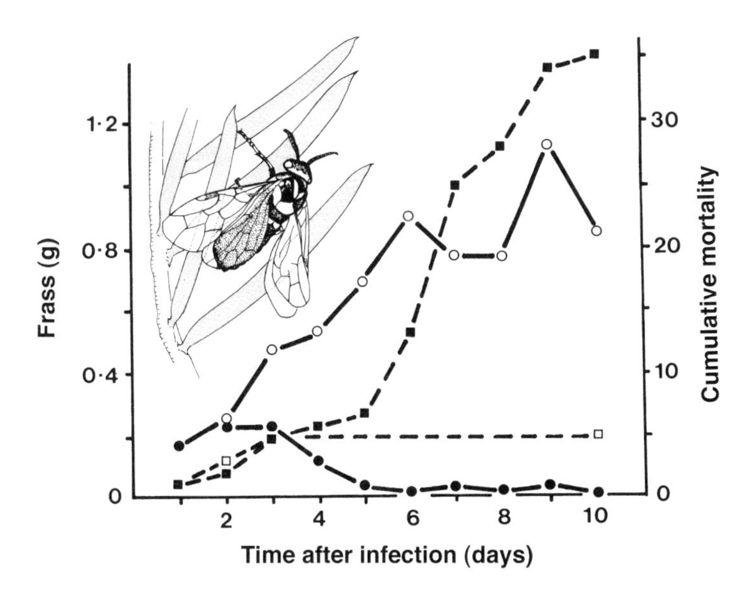

Figure 9.3 Effect of NPV infection on feeding and mortality of third instar larvae of European spruce sawfly, *Gilpinia hercyniae* (Hymenoptera: Diprionide). Frass production: ●, diseased larvae; ○, healthy larvae. Mortality: ■, diseased larvae; □, healthy larvae.

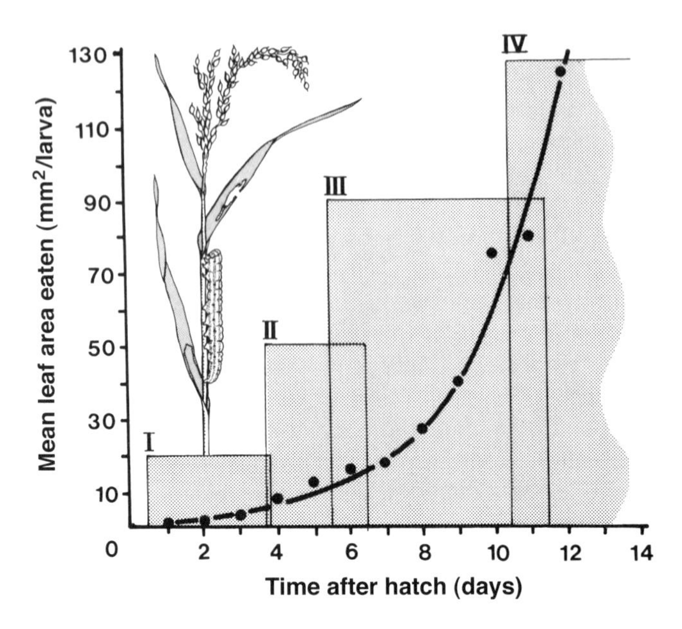

Figure 9.4 The increasing feeding rate with age (instar I–IV) of larvae of the rice armyworm, *Mythimna separata* (Lepodoptera: Noctuidae), on rice leaf. (Bar width shows instar durations but bar height is arbitrary.)

instars is dramatically less than in older larvae, so that control at this stage minimizes crop damage. A typical example is rice leaf area consumption by first instar larvae of the rice armyworm, *Mythimna separata*, which is only about 1% of that of fourth instars (Figure 9.4). As most crop plants will tolerate some damage before yield is affected, for properly timed spray applications the crop yields obtained by BVs are often no different from those resulting from applications of chemical insecticides. For instance while the rate of establishment of control by chemical insecticides and *B.t.* is faster than for NPV when these are sprayed against the European skipper butterfly (the 'Essex' skipper in the UK), *Thymelicus lineola*, yields of timothy grass are similar (Figure 9.5).

As a corollary to this, it should be pointed out that where a BV is employed for control, the economic threshold for treatment may be set at a lower pest density than for a more rapidly acting chemical insecticide. The difference in threshold levels for implementation of spraying will depend on the type of insect feeding damage. For leaf-eating insects, the difference may be slight because of the relative degree of independence that may exist between leaf area and crop yield. Where young larvae penetrate buds or fruiting structures (e.g. bollworms of cotton, *Heliothis* spp., *Diparopsis* spp., *Pectinophora gossypiella*; codling moth, *Cydia pomonella*, in apples) or where trivial damage is deemed economically important because crop appearance affects marketability (cosmetic damage), threshold values may be wider. For instance, the question of action thresholds for *Pieris rapae* on cole crops, for which cosmetic damage to wrapper leaves is important, was investigated by Webb and Shelton (1991), who showed that GV control equivalent to permethrin is possible.

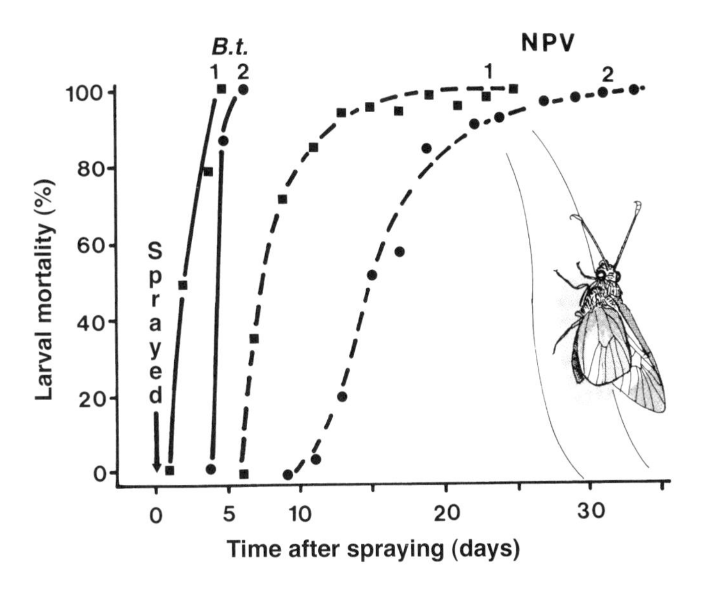

Figure 9.5 Two field trials (1 and 2) comparing the use of a variety of chemical insecticides (not show here – most caused 100% mortality within 3 days) with *B.t.* and NPV for control of the European skipper butterfly, *Thymelicus lineola* (Lepidoptera: Hesperidae) on timothy grass. Yield increase from NPV was 425% and from the range of chemicals was 409–541%. (From Thompson (1977) *Journal of Economic Entomology*; reproduced by permission of L. S. Thompson, Entomological Society of America.)

Estimations of infection

With some insect species, infection estimates can be made with comparative reliability directly in the field. For instance, when infected, sawflies of gregarious larval habit, such as *Neodiprion lecontei* and *N. sertifer*, show a degree of colony disruption, cessation of feeding and, close to and following death, most larvae adhere to the foliage and stems. Here the incidence of colonies infected or of colonies dead can be a reliable and rapidly measured spray impact assessment parameter. However, for most Lepidoptera, accurate assessment of infection depends upon collection of samples of larvae: each larva must then be separately diagnosed (Chapter 25). Methods of diagnosis suitable to the processing of large samples usually only detect infection once a certain amount of virus has been generated during the infection process. Therefore, the value obtained for the proportion of the sample infected will generally understate the real situation. When, for instance, a comparison was made between an immediate diagnosis (larvae were frozen within a few hours of collection and at a later date tissue smears were Giemsa's stained and microscopically inspected for OBs of NPV) and larvae collected directly into separate sterile containers and reared until full development or death on surface-sterilized foliage in individual isolation, a relationship between 'instant' and 'real' infection was derived (Figure 9.6a). This showed that low levels of 'instant' infection characteristic of the period shortly after spraying were especially prone to distort the 'real' level; for instance, 20% 'instant' infection appeared to indicate about 50% 'real' infection. The implications of this for an actual field trial are shown in Figure 9.6b. This, or a

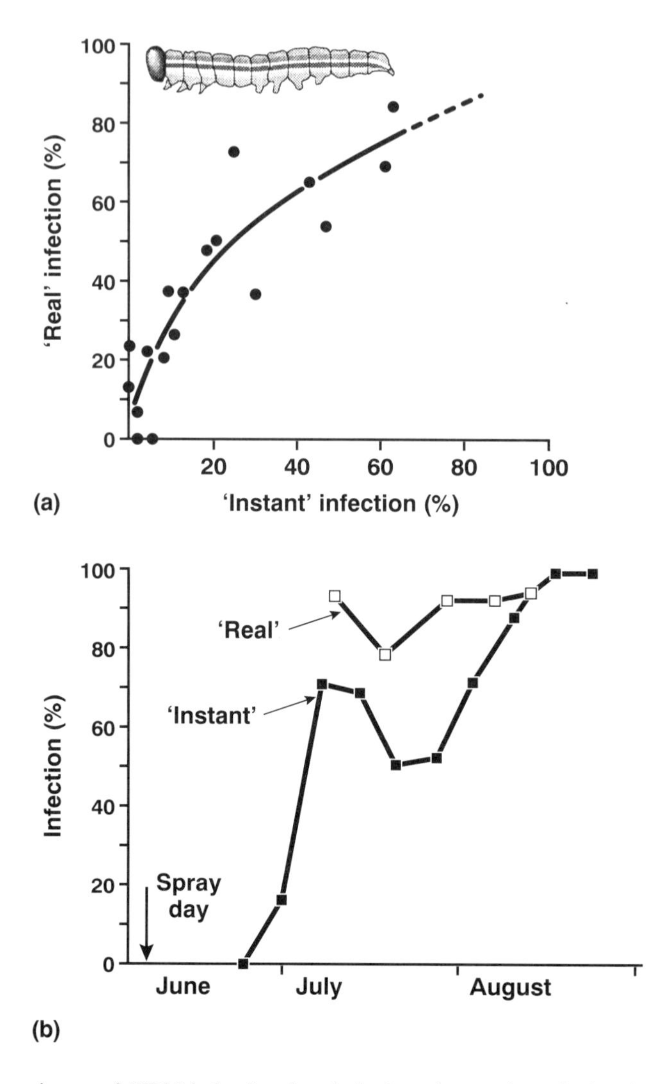

Figure 9.6 Comparisons of NPV infection levels in larval samples of pine beauty moth, *Panolis flammea* (Lepidoptera: Noctuidae), estimated from specimens frozen very soon after collection ('instant' infection) or held in individual isolation on sterile foliage until maturation or death from disease ('real' infection). (a) Relationship of 'instant' (x) with 'real' (y) infection from data collected in 1982: $y = 1.7 + 9.1 \sqrt{x}$. (b) Development of infection in a 1 ha trial plot sprayed with NPV in 1985, measured in terms of 'instant' and 'real' infection; the latter was observed only for the five intermediate sample periods.

similar type of transformation, should always be employed to achieve a realistic interpretation of virus impact over time.

Recording over time

While the impact of modern chemical insecticides (which tend not to be notably persistent) and also to a large extent of *B.t.* is seen only in the population actually

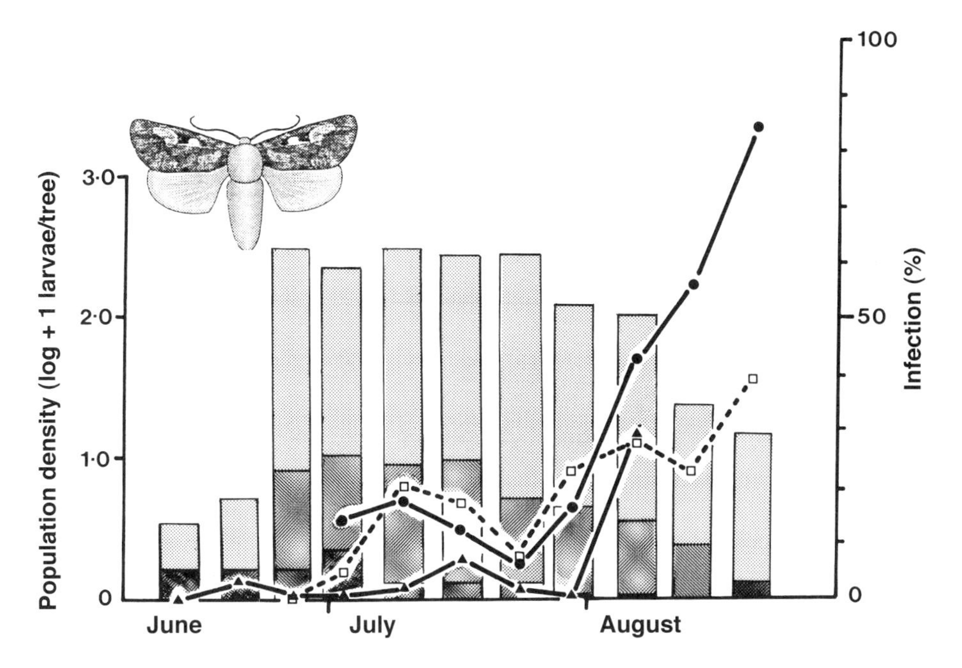

Figure 9.7 Continuation of control in 2 years, 1987 and 1988, following spraying of pine beauty moth, *Panolis flammea* with NPV on 18 June, 1986. Disease incidence was measured as 'instant' infection in all years: 1986 (■); 1987 (□) and 1988 (▲). Pupae with NPV disease are not shown. Larval population density is expressed in logarithmic terms and shading of bars in the histogram follow the declining sequence 1986, 1987, 1988 for lightest to darkest shading. To obtain large enough numbers of larvae in 1987 and 1988 to permit a realistic estimation of infection levels, tree samples sizes had greatly increased. (P.F. Entwistle and J.S. Cory, unpublished data.)

treated, the impact of BVs may be also seen in subsequent pest generations. This is usually the result of generation of secondary inoculum rather than of persistence of the original sprayed virus. For the collection of comprehensive control data it is, therefore, desirable to record experimental plots for two or more pest generations after spraying. The pest regulatory impact of secondary inoculum may be especially apparent in forests (Figure 9.7) but the impact on field crops may also be appreciable. This can occur through accumulation of virus in soil or on the plant (Chapter 4). For instance, NPV epizootics caused by the mobilization by rain of inoculum persisting on leaves are common in *Trichoplusia ni* in cabbages in North America (Hofmaster and Ditman, 1961; and discussed in Entwistle, 1986).

Recording over space

A contribution from spray drift and the propensity of BV diseases in some host species to spread makes it worthwhile to record disease incidence outside the boundaries of treated plots. In addition to demonstrating the benefits of this type of pest control, information is obtained relevant to selection of the plot spacing necessary to prevent experimental treatment interactions. Figure 9.2 compares the distribution of NPV disease (measured by visual assessment as described above) in *N. sertifer* larval populations at two dates after spraying by two different methods. The results clearly indicate the value of extra-plot records. A more full study would also have involved recording over more than 1 year, especially as *N. sertifer* is an univoltine species.

THE CONDITION OF THE VIRUS INOCULUM

It is critically important to the conduct of field trials and to an adequate interpretation of their results that the condition of the virus material applied (by spray or by any other method) has not been compromised up to the moment of application. Four main concerns should be considered.

1. The inoculum should have been held under optimal storage conditions.
2. It should be demonstrated during previous developmental studies that virus activity is not adversely affected by any of the formulating substances (surfactants, anti-microbials, thixotropic agents, UV protectants, etc., Chapter 7) or by the bulk materials employed in actual field delivery and often as mixed 'on site' (carrier oils: not forgetting the material used by the manufacturers in their formulations, dusts or bait ingredients). The quality of water used, for instance its pH value, presence of contaminating chemicals, etc., should also be considered.
3. There should be adequate mixing in the spray tank to ensure proper dispersion of virus in the spray fluid. This is important to ensure that all spray droplets contain the expected, or calculated, number of OBs.
4. Activity assessments should be made. On the day of the field trial applications, two virus samples should be retained for subsequent laboratory assessments of activity: a subsample of the virus concentrate itself before dilution and a subsample from the field tank mix after mixing and immediately before application. (Ideally, spray tank residues will also be sampled immediately after applications; if their virus concentration exceeds the expected value, then clearly the dose applied in the field will have been less than intended and may indicate settling in the tank. The settling rate is likely to be aggravated by any clumping of OBs.) These subsamples should be bioassayed in comparison with the laboratory standard isolate of virus and may require collection and transportation on ice if ambient temperatures are high. Such comparative bioassays should be conducted with minimum delay.

The origin of any appreciable reduction in the potency of the virus preparation applied should then be traced back along the line of developmental studies, and rectified.

REFERENCES

Entwistle, P.F. (1986) Epizootiology and strategies of microbial control. In *Biological Plant and Health Protection.* Franz JM (ed.). Stuttgart: Fischer Verlag, pp. 257–278.

Entwistle, P.F., Evans, H.F., Cory, J.S. and Doyle, C. (1990) Questions on the aerial application of microbial pesticides to forests. In *Proceedings of the Vth International Colloquium on Invertebrate Pathology and Microbial Control*, Adelaide, Australia, pp. 159–163.

Hofmaster, R.N. and Ditman, L.P. (1961) Utilisation of a nuclear polyhedrosis virus to control the cabbage looper on cole crops in Virginia. *Journal of Economic Entomology* **54**, 921–923.

Thompson, L.S. (1977) Field tests with chemical and biological insecticides for control of *Thymelicus lineola* on timothy. *Journal of Economic Entomology* **70**, 324–326.

Webb, S.E. and Shelton, A.M. (1991) A simple action threshold for timing applications of a granulosis virus to control *Pieris rapae* (Lep.: Pieridae). *Entomophaga* **36**, 379–389.

10 Future developments

VIRUS IMPROVEMENT

Strain selection

It is likely that most insect species are susceptible to a range of virus diseases and we have isolated only a tiny proportion of the viruses that exist in insect populations. The situation with insect virus isolations today is probably similar to the situation with *B.t.* in the early 1980s when about 1000 isolates were known and new isolates were found from time to time in a rather sporadic and haphazard way. When a new method of isolating *B.t.* from the environment was discovered in the mid 1980s, the number of new strains isolated rose dramatically to the many tens of thousands of strains known today. This has provided a previously undreamt of diversity of strains that has fuelled the huge commercial development of *B.t.* insecticides. Unfortunately, it is unlikely that any method will be developed for screening large numbers of samples for the presence of insect viruses. Nevertheless, new isolates of viruses will be made and will continue to increase the range of insect species that can be controlled. An interesting example of a recently discovered virus strain that appears to have benefits for commercial exploitation was the NPV from the celery looper *Anagrapha falcifera*. Although this virus seems to be quite closely related to AcNPV, it is more infectious for some other species and may, therefore, be suitable for use on a wider range of major pests. There is, therefore, still plenty of scope for finding new viruses or new strains of viruses with improved characteristics as insecticides.

Genetic engineering

One of the disadvantages of BVs is that they are slower to kill insects than chemical control agents and crop damage may continue for several days following virus application. Although some BVs take much longer to kill their host than others, it is unlikely that natural strains will be discovered that kill significantly faster than many current strains. Several laboratories are, therefore, now looking to genetic engineering to produce more pathogenic virus strains. A variety of different genes has been inserted into BVs in attempts to enhance pathogenicity; those showing the most promise are genes for insect toxins obtained from a mite,

Pyemotes tritici, and a scorpion, *Androctonus australis*. Recombinant strains of AcMNPV expressing these toxins have been shown to kill significantly faster than the wild-type strain. Moreover, paralysis occurs prior to death so that feeding ceases even earlier, thus further reducing crop damage. In fact, it may be preferable to produce viruses that rapidly paralyse their host rather than kill prematurely since a virus which kills very quickly is unlikely to replicate to the same level as a normal virus and it may, therefore, be difficult to mass produce such a virus in larvae economically.

Another recombinant AcMNPV that has been shown to kill insects more rapidly than the wild-type virus is an *egt* deletion virus, and there is already commercial interest in using this virus. The *egt* gene has been identified in several other BVs, including GVs as well as NPVs. The effect of deleting these genes is being investigated and may lead to recombinant strains of other viruses with enhanced speed of kill. The advantage of this strategy is that release of deletion mutants may be considered to pose a lower risk than other genetically engineered viruses, since no new genetic material is being introduced into the environment.

Factors other than speed of kill may be modified by genetic engineering. In a few cases, the host range or infectivity of viruses has already been modified either by repeated passaging in a host that is initially of very low susceptibility or by co-transfection using DNAs from two viruses. In some of these cases, the genetic changes that take place have not been characterized, but in one case the genetic changes have been characterized in great detail. The host range of AcMNPV can be modified to included *Bombyx mori* by changing just a small region of the AcMNPV helicase gene to be like the region found in the helicase of BmNPV. The extent to which the helicase gene is responsible for determining the host range in other viruses and with other hosts is unknown but as other genes that affect host range are characterized, there will be increasing possibilities for deliberately modifying host range by genetic engineering. Eventually, it may be possible to produce viruses with host ranges that are more suited to particular pest complexes.

VIRUS PRODUCTION

In vivo production

In vivo production is likely to remain the only feasible method for large-scale production of most BVs for the foreseeable future. Although it is perceived as a relatively 'low-tech' procedure, there is plenty of scope for improvement, leading to more efficient systems, and some new methods for high-density insect rearing have recently been patented. The major objectives are to rear, infect and harvest larvae at very high densities with minimal inputs of materials and labour. It is also essential to maintain high standards of hygiene both to prevent contamination of stock insects and to minimize microbial contamination of the product. Automation of the insect rearing process is achievable, as has been shown with silkworm production. Unfortunately, many insect species are less amenable to intensive production, particularly where larvae are antagonistic to one another.

Some improvement can be obtained by careful selection of stock insects. For instance, we found considerable differences between stocks of *Cydia pomonella* obtained from different sources. With one stock, larvae were highly antagonistic and could be reared only in isolation. With another, it was possible to rear larvae together and final larval size was also greater. Selective breeding has allowed great variation and many desirable features can be obtained in a wide variety of animals. Owing to the very short life cycle of insects, significant improvements may be achievable within short periods of time.

An alternative approach, where the normal host insect is difficult to rear or has other disadvantages such as a hairy body, is to find an alternative host in which to produce the virus. For example, a process has been patented in which CpGV is produced in *Cryptophlebia leucotreta.* Although this species is smaller than the natural host, it can be reared to very high densities, resulting in a much greater biomass than can be achieved with *C. pomonella.* CpGV is about 1000 times less infectious for this alternative host, but the amount of inoculum virus required is still an insignificant cost. In cases where a suitable host is not known, it may be possible to adapt the virus to multiply in a host that is normally non-permissive. For example, although *Orgyia pseudotsugata* MNPV (OpMNPV) was produced on a large scale in *O. pseudotsugata* for many years, the larvae are slow to develop and are very hairy. In contrast, *Trichoplusia ni* is an insect that is easy to mass produce but is not susceptible to OpMNPV. However, by blind passaging material from *T. ni* inoculated with OpMNPV for several passages, a strain of OpMNPV was isolated that gradually became increasingly virulent to *T. ni.* Eventually a virus was obtained that not only multiplied well in *T.ni* but also proved to have enhanced infectivity for *O. pseudotsugata.* The extent to which other BVs in general could be adapted to new host species is unknown, but the success of the example given above suggests that this approach may be worth pursuing when the normal host is not readily amenable to mass rearing.

In vitro production

Production of viruses in large-scale cell cultures has many potential advantages over production in larvae. In particular, the process can be fully automated, which is a great attraction to industry. It would result in a good quality product with virtually no microbial contamination and, therefore, improved storage properties. The problems with *in vitro* production are that there are relatively few BVs that can be propagated in cell culture and of those that can, only AcMNPV and its variants can be produced at a level that is anywhere near economic. Much effort is, however, being put into reducing the cost of the medium required. Unfortunately, AcMNPV, though by far the most widely studied BV, has so far had very little success as a pest control agent (Part two). One factor that may change this situation is the advent of genetically engineered strains of AcMNPV with improved insecticidal properties, such as those described above. These improvements may confer a sufficient advantage on the virus to make its use in the field more attractive. Patented, genetically improved strains of a virus are likely to attract a price premium that may enable the virus to be profitably produced *in vitro.*

FORMULATION AND APPLICATION

Frequent sprays at low doses

Control application methodology has been dominated by the idea of single, or widely spaced, sprays each designed to give a high level of infection/death. There are some suggestions in the literature (FSU, Guatemala, Netherlands) that more frequent, lower dose, applications may be more effective. In terms of total consumption of OBs per crop season, this approach may even be more economical than more infrequent larger dosing, so that the cost of extra spraying rounds could be offset. Of course, it is possible that if the grower is a small-scale farmer working for himself, additional applications may have no real monetary costs.

An analysis of the voluminous USA literature on *Heliothis* control on cotton (Entwistle and Evans, 1985) suggested that, above a certain total OB threshold, frequent applications were more important than the actual dosage of OBs. It is surprising that no attempts seem to have been made to test the findings of this analysis experimentally. Cotton is an especially good crop for frequent sprayings since growers everywhere are accustomed to making 10–20 (or more!) applications per crop season.

The concept of 'less virus more frequently' may not be tenable where the physical application costs are high, e.g. the use of expensive helicopters. However, costs may be ameliorated if applications can be made with other substances which anyway have to be applied for other purposes, e.g. fungicides (beware possible adverse effects on viruses), aphicides, etc.

The behaviour of particulates in sprays

The behaviour of particulates in sprays is an area requiring considerable research. Indeed, some recent papers suggest that much remains to be understood about the behaviour of spray droplets themselves! An example of a fairly recent and fundamental finding has been that unless an appropriate amount of surfactant is included in water-based sprays droplets will be reflected from those plant surfaces that by their chemical nature (e.g. waxiness) or physical micro-structure are unwettable. However, we know that as droplets run off such hydrophobic surfaces they can leave behind OBs! Questions that need to be considered include what happens to contained OBs when droplets hit and are deflected by hydrophobic surfaces. Are some OBs left on the plant? And what happens to particulate, insoluble, UV protectants (e.g. carbon, tinopal, etc.)? The problem of the loss of solid particles (e.g. entomopathogens and solid UV protectants) from spray droplets before they reach their target remains to be adequately investigated. There seems to be very little doubt loss can occur but the causal factors are far from being adequately explained and all we have at present is a body of empirical knowledge. Examination of this area would ideally require the co-operative work of a chemist–physicist–biologist grouping.

The approach to UV protection of entomopathogens has had a tendency to be *ad hoc*, with progress being made on the basis of empirical experience. Consideration should be given to factors that either could adversely affect or could protect virus infectivity, both while airborne and after hitting the plant surface. The formation of free radicals, which go on to damage DNA, may begin in the airborne droplet, while the most significant period of exposure to UV probably takes place at the plant surface; as a result, the resting juxtaposition of OBs and particulate UV protectants in the 'flat' dried spray droplet could be very important. A protectant could be innately valuable but useless as employed.

ECOLOGICAL ASPECTS

Virus reservoirs

The value of managing, and even creating, virus reservoirs in the soil has probably been under-exploited as a form of prophylactic control. In this book, it has been referred to in relation to grass pest control in New Zealand and to at least three species of *Colias* butterflies attacking clover/alfalfa. Its most obvious extension would be to the various moth species that may collectively be called cutworms because the larvae live in soil attacking roots or, often nocturnally, emerging to attack aerial parts of crops. We know *Agrotis* larvae can acquire NPV purely from the soil, and no doubt this is true for other such species. The actual incorporation of BVs into soil, with appropriate modification of cultural practices, including ones designed to disperse soil-located virus onto plants (e.g. by overhead irrigation splash or by deliberate soil-surface disturbance such as shallow harrowing in dry weather), seems worthwhile investigating.

Out-of-season pest reservoirs

Outside the main crop season, some insect pests, especially in warmer climates, retreat to other plant species or to more restricted areas of the main crop species. It is possible that they could be more efficiently treated here than on the main crop itself. Recently, attention is being paid in the USA to the virus treatment of *Heliothis* spp. on field geranium weeds outside the cotton season and, because of the mobility of adult moths, such experimentation has been carried out on a very wide scale. In Egypt, there seems to have been reluctance to take such an approach in attacking *Spodoptera littoralis* on berseem in the winter. Berseem is the agricultural 'fuel' in Egypt, being grown as fodder for draft animals. The importance of *S. littoralis* reservoirs on berseem is recognised to the extent that, in theory, there is a closed period when it is illegal to grow either crop. In this way, it is calculated the pest will be disadvantaged. Of course, this ban is often flouted. NPV treatment of berseem towards the end of the winter might be very valuable.

This type of approach can also be employed with the deliberate planting of trap crops. In Central America it was found that small plots of cotton grown early attracted hibernating boll weevils, which could then be destroyed before they infested the main crop. Trap crops for Lepidoptera could equally be treated with viruses. Volunteer crop plants also often represent very localized centres of persistent pest populations, e.g. volunteer potatoes support potato tuber moth (PTM) and could easily be treated.

NEW STRATEGIES

Synergy

A lot has been written on the question of the interaction of viruses with chemical insecticides, *B.t.* and other pathogens. Is there synergy, additivity or antagonism? Most of the studies concern the simultaneous presentation of virus with entity 'X' to host insects; only in a minority of studies have test entities been presented in lag phase. A notable example of such an intelligent enquiry is that of Savanurmath and Mathad (1981), investigating virus–chemical insecticide control of *Mythimna separata* in India. How an entity (virus or other) may predispose an insect to react to a subsequently presented different entity is an essential part of any study seeking to identify beneficial interactions (or indeed adverse interactions, so that they can be avoided in the field).

Clearly, the whole topic of virus interactions with other control methods, especially where these involve sprayable chemicals and microbiologicals, should be opened up and some science needs to be applied.

OPTIMIZING EFFICIENCY

Modelling control

It is worthwhile drawing attention to comments in the text of this book and elsewhere in the literature on the feasibility of designing field experiments that are not of a time-wasting, *ad hoc*, nature. This is achievable for almost any insect pest species for which a little is known of the larval biology and virus dose–mortality relationships. Given these, and given that spraying is not the imprecise uncontrollable process that so many people assume it to be, field experiments can become more a process of testing theoretical conclusions rather than a chance experience. Not only is this vastly more satisfying but it can also save a lot of time and money. We should not really be very far from the stage at which the control 'experiment' can be computer modelled in advance. The model could be fine tuned to yield maximum pest control by some optimal really practical strategy. Development of such a modelling approach is definitely to be encouraged.

SAFETY AND REGISTRATION REQUIREMENTS

Are present safety test schedules necessary? As viruses tend to have smaller markets than do chemical pesticides (they are, for instance, more host specific) developmental costs are relatively more disadvantageous. This applies particularly to the present registration situation. Substantial saving could be made if it was not necessary for every single BV to be put through the same, expensive, safety-testing protocol. Because of the host taxa with which they have evolved, we do not expect BVs to be intrinsically hazardous. In addition, so many have now been formally safety tested that it might be possible to make a sound case for curtailed ('fast-track') testing in the future. Possibly an approach route could be via a very detailed review and analysis of what has already been done in terms of testing and what were the results.

THE USE OF VIRUSES OTHER THAN BACULOVIRUSES

The emphasis of this book has been heavily on the BVs for which the majority of fundamental and applied investigations have occurred. There have been rather few examples of successful control work with viruses in other groups. As seen at present, some virus groups (e.g. ascoviruses, entomopoxviruses, iridoviruses) appear to offer little promise in comparison with BVs. CPVs have shown some success but their rather slow rate of action is generally disadvantageous. However, there have been some notably successful demonstrations of the employment of *Nudaurelia* β viruses, parvoviruses and picornaviruses, especially for the control of Lepidoptera in the tropics (*Gonometa* and *Nudaurelia* spp. and Limacodidae (Chapter 18); Limacodidae (Chapter 15); *Acharia* and *Eacles* spp. (Chapter 22) and *Dacus* spp. (Chapter 12)). The potential for stability in the field and in refrigeration has been demonstrated. It seems very likely that viruses belonging to these families are especially prevalent in the tropics; most have been isolated there despite the fact that the majority of insect pathologists are located in the temperate zones of the world. Viruses are undoubtedly important in the regulation of natural populations of some tropical Lepidoptera, e.g. in Papua New Guinea, *Lymantria ninayi* is affected by a picornavirus, a *Nudaurelia* β virus, a nodavirus and an unidentified, pleiomorphic virus in addition to an NPV; in decimating epizootics in first and second instar larvae, a picornavirus has been identified as the key virus. Many epizootics in Limacodidae have been shown to be driven by non-occluded viruses. The barrier to the practical use of such viruses is the perception that some may be infective in humans and other vertebrates. Balanced against this concern is the fact that there are no known instances of human infection even where people have been heavily exposed. Nevertheless, an essential prerequisite to serious investigation of their potential as control agents is a detailed assessment of their medical, veterinary and environmental hazard potential. Perhaps because such viruses represent a potentially valuable pest control resource for developing nations, it is especially important that such studies be undertaken. A measure of the interest in such viruses is that, without hazard studies having been made, many have already been employed widely in the tropics.

THE CONTROL BY VIRUSES OF ARTHROPODS OF MEDICAL AND VETERINARY IMPORTANCE

While viruses belonging to several of the major families associated with phytophagous insects are also represented in haematophagous arthropods associated with vertebrates, there have been few instances of these viruses receiving fundamental scrutiny or being considered as candidates for bio-pest control. (This is in distinction to other pathogen taxa; for instance, there have been well-developed studies on the bacterial control of mosquitoes and blackfly, especially with *Bacillus thuringiensis* var *israelensis* (*B.t.i.*), and of mosquitoes with fungi, *Coelomomyces* spp., and a nematode, *Romanomermis culicivorax*. From these studies *B.t.i.* has emerged as a highly commercializable product.) The most notable example of a serious candidate virus is that of tsetse fly, *Glossina* spp. (Chapter 18). Provided that this virus proves to be benign to vertebrates, it seems almost certainly assured of an important role in the manipulated control of tsetse. This is extremely encouraging as the biology of *Glossina* suggests it to be fundamentally inimical to colonization by a pathogenic microorganism, especially where this has a narrow host range: it has no free-living larval stage (adult females are viviparous) and it is a naturally low-density taxon. Indeed it has been described as a K-strategist to which pathogens need to adapt very closely for success (p. 85). An essential prerequisite to any beneficial outcome is a careful study of the biology and ecology of viral pathogens. An example of such work is that of Linley and Nielsen (1968a,b) on an iridovirus of the mosquito *Aedes taeniorrhynchus*; adults carry a latent infection that kills larvae; other larvae acquire virus by cannibalism, their progeny dying in the next generation. It takes little imagination to see that a stable iridovirus-contaminated bait might be valuable in elevating infection levels and assisting in population suppression: it is encouraging that iridoviruses may be easy to mass produce in other hosts (e.g. *Tipula* iridovirus by injection in *Pieris brassicae* larvae) and *in vitro*.

Another approach could be through genetic engineering. For instance, the viral gene(s) responsible for reduced reproductive efficiency in *Glossina* spp. (p. 285) or in *Merodon equestris* (p. 18) might be of use for insertion into viruses that are otherwise unsatisfactory in biocontrol of other vector arthropod species. As has already been shown in studies on the genetic 'improvement' of BVs as pesticides, there must be many such possibilities; these should be promoted to extend the present scope of viral control of pest insects and also, specifically, so that the biocontrol of mosquitoes and blackflies does not continue to rely largely on bacteria, notably *B.t.i.*

REFERENCES

Entwistle, P.F. and Evans, H.F. (1985) Viral control. In *Comprehensive Insect Physiology, Biochemistry, and Pharmacology*. I. Kerkut and L.I. Gilbert (eds). Oxford: Pergamon Press, pp. 347–412.

Linley, J.R. and Nielsen, H.T. (1968a) Transmission of a mosquito iridescent virus in *Aedes taeniorrhynchus*. I Laboratory experiments. *Journal of Invertebrate Pathology* **12**, 7–16.

Linley, J.R. and Nielsen, H.T. (1968b) Transmission of a mosquito iridescent virus in *Aedes taeniorrhynchus*. II Experiments related to transmission in nature. *Journal of Invertebrate Pathology* **12**, 17–24.

Savanurmath, C.J. and Mathad, S.B. (1981) Efficacy of fenitrothion and nuclear polyhedrosis virus combinations against the armyworm *Mythimna* (*Pseudaletia*) *separata* (Wlk.) (Lepidoptera: Noctuidae). *Zeitschrift für Angewandte Entomologie* **91**, 464–474.

PART TWO

WORLD SURVEY

Figure 11.1 The geographical regions selected for separate treatment in Part two; they are close to conventional biogeographical zonation. 1. Western Europe; 2. Eastern Europe and the former Soviet Union; 3. Indian subcontinent; 4. South-east Asia and the Western Pacific; 5. The People's Republic of China; 6. Japan; 7. Africa, the Near and Middle Easts; 8. Australasia; 9. North America; 10. Central America and the Caribbean; and 11. South America.

11 A world survey of virus control of insect pests

PHILIP F. ENTWISTLE

INTRODUCTION

The survey which follows is in general organized according to widely accepted zoogeographical regions (Figure 11.1). It presents information on both current control practices and on research and technical studies aimed at the development of control methodologies (but not necessarily yet brought to completion). In any one zoogeographical region, the report on an insect pest refers, in general, to work done or in progress in that region only. For instance, for a particular species and its virus, safety tests may have been mentioned in one region but not in another where less progress has so far been made on that topic or, indeed, where it may not be a necessary developmental or legal requirement. For each region a selected bibliography is provided.

A separate progress check table is provided for each region and for ease of comparison these are placed together at the beginning of this section (Tables 11.1–11.11). The objective of these tables is to enable the reader to obtain rapidly a picture of the pest/virus situations of interest in each zoogeographical area. For this purpose, a series of thirteen aspects of the control development process have been adopted. Progress in each of these has been assessed for each pest species on a rating of +, ++ and +++ (mostly completed), which can, obviously, be only a broad guide to development. In these tables the terms 'commercialization' and 'sales' should not be too strictly interpreted: together they could indicate a level of practical usage as, for instance, in the People's Republic of China, where 'sales' perhaps cannot be interpreted in the capitalist sense. These terms are also used to indicate self-help initiatives, for example, where farmers produce and apply viruses without involvement in the commercial world.

The reader could easily amend the Tables 11.1–11.11 to keep a running record of advances.

Table 11.1 Progress check table for Western Europe

	Strain search	Character-ization	Infectivity testing	Ecology	Safety tests	Environmental impact	Formu-lation	Field trials	Extension trials	Production development	Registration	Commercial-ization	Sales
Adoxophyes orana GV	++	++	++	+	+++	+	++	+++	++	++	+++	++	++
Agrotis segetum GV	++	++	+++	++	+++	+	++	++	+	++	+++	+	+
Cnephasia spp. GV								++					
Cydia pomonella GV	+	+++	+++	++	+++	++	++	+++	+++	+++	+++	+++	+++
Euproctis chrysorrhoea NPV	+	++	+++	++	+	++	++	+++	+	++	–	–	–
Hyphantria cunea GV and NPV	+	+	++	++	–	+	+	++	+	+	–	–	–
Lacanobia oleracea NPV		+	+					+					
Leucoma salicis NPV								+					
Lymantria dispar NPV	+++	+++	+++	++	++	+	++	+++	++	+	–	–	–
Lymantria monacha NPV	+	+	++	+	–	+	+	++	+	+	–	–	–
Neodiprion sertifer NPV	+	++	++	+++	+++	++	++	+++	+++	++	++	+++	++
Mamestra brassicae NPV	++	+++	+++	++	+++	+	++	+++	++	++	+++	+	+
Panolis flammea NPV	++	+++	++	++	+++	++	++	+++	+++	++	+	–	–
Spodoptera exigua NPV	++	+++	+++	++	+++	+	++	+++	+	++	+++	+	+
Thaumetopoea pityocampa CPV	+	++	++	++	–	+	++	+++	++	++	–	–	–
Yponomeuta spp. NPV								+					
Zeiraphera diniana GV	+	+	++	++	–	+	+	++	–	+	–	–	–

For commercial development of *Spodoptera littoralis* NPV see Africa.
Progress is assessed on an increasing scale from + (some) to +++ (mostly complete).

Table 11.2 Progress check table for Eastern Europe and the Former Soviet Union

	Strain search	Character-ization	Infectivity testing	Ecology	Safety tests	Environmental impact	Formu-lation	Field trials	Extension trials	Production development	Registration	Commercial-ization	Sales
Agrotis segetum GV and NPV	+	+	++	–	+++	–	++	++	++	++	+++	++	++
Cydia pomonella GV	+	–	++	–	–	–	+	++	–	++	+++	+	+
Dendrolimus sibiricus GV	+	+	+	–	–	–	+	++	+	++	+++	+	+
Helicoverpa armigera NPV	++	+	+	–	+++	–	++	+++	++	++	+++	++	++
Hyphantria cunea NPV and GV	+	–	++	–	–	–	+	++	+	++	+++	+	+
Hyponomeuta malinellus NPV	+	–	+	–	–	–	+	+	–	+	–	–	–
Leucoma salicis NPV	++	+	++	+	+++	+	++	++	+	+++	+++	+	+
Lymantria dispar NPV	+++	++	+++	+++	+++	+	++	+++	+++	++	+++	++	++
Lymantria monacha NPV	++	++	+	–	–	–	+	+	–	+	–	–	–
Mamestra brassicae NPV	++	+	+++	–	+++	+	++	+++	+++	++	+++	++	+
Malacosoma neustria NPV	+	+	+	–	+++	–	+	++	–	+	+++	+	+
Neodiprion sertifer NPV	+	–	+	+	+	–	+	++	+	++	+++	+	+
Pieris brassicae GV	+	–	+	–	–	–	+	++	–	–	–	–	–
Pieris rapae GV	+	–	+	–	–	–	+	++	–	–	–	–	–
Xestia c-nigrum GV and CPV	+	+	++	–	–	–	+	++	–	+	–	–	–

Progress is assessed on an increasing scale from + (some) to +++ (mostly complete).

Table 11.3 Progress check table for the Indian subcontinent

	Strain search	Character-ization	Infectivity testing	Ecology	Safety tests	Environmental impact	Formu-lation	Field trials	Extension trials	Production development	Registration	Commercial-ization	Sales
Agrotis ipsilon GV	–	–	+	–	–	–	+	–	–	–	–	–	–
Amsacta spp. NPV	–	–	–	+	+	+	–	++	–	+	–	–	–
Chilo infuscatella GV	–	++	++	+	+	+	–	+	–	+	–	–	–
Helicoverpa armigera NPV	+	–	++	++	++	+	+++	+++	++	–	–	–	–
Mythimna separata NPV	–	–	++	++	++	+	+	++	+	–	–	–	–
Oryctes rhinoceros OrV	–	–	+	–	–	–	+	+++	+++	–	–	+++	–
Spodoptera litura NPV	–	–	++	++	–	+	++	++	–	–	–	–	–

Progress is assessed on an increasing scale from + (some) to +++ (mostly complete).

Table 11.4 Progress check table for south-east Asia and the western Pacific

	Strain search	Character-ization	Infectivity testing	Ecology	Safety tests	Environmental impact	Formu-lation	Field trials	Extension trials	Production development	Registration	Commercial-ization	Sales
Crocidolomia binotalis NPV	[a]	[a]	+	–	[a]	–	–	?	–	–	–	–	–
Helicoverpa armigera NPV	++	+++	++	–	++	–	++	++	+++	++	++	++[b]	–
Limacodidae (various viruses)	+	+	+	+	–	–	+	++	+++	+	[b]	N.A.	N.A.
Mythimna separata NPV	++	++	+	–	–	–	+	+	–	+	–	–	–
Oryctes rhinoceros OrV	+++	+++	+++	+++	++	++	+++	+++	+++	+++	[b]	[c]	N.A.
Penaeus monodon BV	+	–	–	–	–	–	N.A.	N.A.	N.A.	N.A.	N.A.	N.A.	N.A.
Pieris rapae NPV	+	+++	+++	+	–	+	++	++	–	++	–	–	–
Plutella xylostella GV and NPV	++	++[a]	+++	+	–[a]	–	++	++	–	+	–	–	–
Spodoptera exigua NPV	++	+++	++	+	–	–	+	++	+	++	++[b]	++	++
Spodoptera litura NPV	++	++	++	+	–	–	+	+	–	+	–	–	–

Progress is assessed on an increasing scale from + (some) to +++ (mostly complete).
N.A. Information not available
[a] As part of the AcMNPV 'complex' this virus has received very considerable attention.
[b] Starter samples of stock virus distributed free.
[c] Because stock virus is distributed free, sales seem unlikely.

Table 11.5 Progress check table for the People's Republic of China

	Strain search	Character-ization	Infectivity testing	Ecology	Safety tests	Environmental impact	Formu-lation	Field trials	Extension trials	Production development	Registration	Commercial-ization	Sales
Adoxophyes orana GV	–	–	+	–	–	–	–	?	–	–	–	–	–
Anomis flava NPV	–	–	+	–	–	–	–	++	–	–	–	–	–
Buzura spp. NPVs	–	–	++	–	++	++	++	++	–	++	–	–	–
Cnaphalocrocis medinalis GV	–	–	+	–	–	–	–	–	–	–	–	–	–
Dendrolimus punctatus NPV and CPV	–	–	+	–	–	–	–	++	++?	+?	–	–	–
Ectropis obliqua NPV	–	–	+	–	++	++	++	+++	++	++	–	+?	–
Euproctis pseudo-conspersa NPV	–	–	+	–	–	–	–	++	+	–	–	–	–
Euproctis similis NPV	–	–	++	–	–	++	–	++	+	–	–	–	–
Gynaephora ruorgensis NPV	–	–	++	–	–	–	++	++	+++	+++	–	++	–
Helicoverpa armigera NPV	–	–	+++	–	–	–	++	+++	++	++	++	++	–
Mythimna separata NPV	++	++	++	–	++	++	?	+	–	–	–	–	–
Pieris rapae GV	++	++	++	+	+++	++	++	+++	++	++	++?	++	++
Plutella xylostella GV	–	–	++	–	–	++	–	++	–	–	–	–	–
Spodoptera litura NPV	++	–	++	–	–	++	++	–	–	–	+	++	+

Progress is assessed on an increasing scale from + (some) to +++ (mostly complete).

Table 11.6 Progress check table for Japan

	Strain search	Character-ization	Infectivity testing	Ecology	Safety tests	Environmental impact	Formu-lation	Field trials	Extension trials	Production development	Registration	Commercial-ization	Sales
Adoxophyes orana fasciata GV	++	+	+	+	+	+	+	++	++	++	–	–	–
Dendrolimus spectabilis CPV	+	+	++	++	+++	+	++	++	++	++	+++	++	++
Hyphantria cunea NPV	+	+	++	++	++	+	++	++	+++	++	–	–	–
Lymantria fumida NPV and CPV	+	+	+	++	–	++	+	++	++	–	–	–	–
Spodoptera litura NPV	+	+	++	–	+	–	+	++	++	++	–	–	–
Thanatarctia imparilis NPV	+	+	++	+++	++	++	++	++	++	++	–	–	–

Progress is assessed on an increasing scale from + (some) to +++ (mostly complete).

Table 11.7 Progress check table for Africa

	Strain search	Character-ization	Infectivity testing	Ecology	Safety tests	Environmental impact	Formu-lation	Field trials	Extension trials	Production development	Registration	Commercial-ization	Sales
Casphalia extranea Parvov.	+	+	+	–	–	–	–	+	–	–	–	–	–
Chrysodeixes spp. NPV	+	–	++	–	–	–		++	++	++	–	+	–
Cryptophlebia leucotreta GV	++	+	++	+	–	–	–	++	–	+	–	–	–
Glossina spp. DNA indet.	+	+	++	++	–	–	–	–	–	–	–	–	–
Gonometa podocarpi picornavirus	+	++	+	+	–	–	–	+	+	–	–	–	–
Helicoverpa armigera NPV	+	+	++	–	+++	–	+	++	–	–	–	–	–
Kotochalia junodi NPV	++	+	++	+	–	–	+	++	++	++	–	+	–
Oryctes spp. OrV	++	++	++	+	–	–	+	++	+	–	–	++	–
Parasa viridissima picornavirus	+	+	+	–	–	–	–	+	–	–	–	–	–
Phthorimaea opercullela GV	+	+	–	+	–	–	+	++	–	+	–	–	–
Spodoptera exempta NPV	+	+	++	++	+++	–	–	–	–	–	–	–	–
Spodoptera littoralis NPV	++	++	+++	+	+++	+	++	+++	+++	+++	++	?	?
Thysanoplusia orichalcea NPV	+	++	++	–	–	–	–	++	++	++	–	+	–

Progress is assessed on an increasing scale from + (some) to +++ (mostly complete).

Table 11.8 Progress check table for Australasia

	Strain search	Characterization	Infectivity testing	Ecology	Safety tests	Environmental impact	Formulation	Field trials	Extension trials	Production development	Registration	Commercialization	Sales
Cydia pomonella GV	–	–	–	–	–	–	–	++	–	+	–	–	–
Epiphyas postvittana NPV	–	–	+	–	–	–	+	++	–	+	–	–	–
Helicoverpa armigera and *punctigera* NPV	–	–	++	–	–	–	++	+++	++	++	++	+	+
Lymantria ninayi NPV	–	–	+	+	–	–	+	+	–	–	–	–	–
Oryctes rhinoceros OrV	++	++	–	–	–	–	–	++	+++	++	–	–	–
Phthorimaea opercullela GV	++	–	++	++	–	–	+	+++	++	++	–	–	–
Pieris rapae GV	–	–	–	–	–	–	–	++	–	+	–	–	–
Scapanes australis OrV	–	–	++	–	–	–	+	++	+	++	–	–	–
Tiracola plagiata NPV	–	–	+	–	–	–	–	+	–	–	–	–	–
Wiseana spp. NPV	++	++	++	+++	–	++	–	+++	++	–	–	–	–

Progress is assessed on an increasing scale from + (some) to +++ (mostly complete).

Table 11.9 Progress check table for North America

	Strain search	Character-ization	Infectivity testing	Ecology	Safety tests	Environmental impact	Formu-lation	Field trials	Extension trials	Production development	Registration	Commercial-ization	Sales
Anagrapha falcifera NPV	++	++	+++	+	++	–	++	++	–	++	+++	+++	–
Anticarsia gemmatalis NPV	+	++	+++	++	–	–	+	+++	–	–	–	–	–
Autographa californica NPV	++	+++	+++	++	+++	++	++	+++	–	+++	+++	+++	–
Choristoneura fumiferana NPV	++	+++	++	++	++	+	++	+++	–	+	++	–	–
Cydia pomonella GV	+	++	+++	++	+++	+	+++	+++	+++	++	+++	++	–
Harrisina metallica GV	++	++	+++	+++	–	–	–	++	–	–	–	–	–
Heliothis spp. NPV	+++	+++	+++	+++	+++	++	+++	+++	+++	+++	+++	+++	+
Lymantria dispar NPV	+++	+++	+++	+++	+++	+++	+++	+++	+++	+++	+++	+	–
Melanopus sanguinipes EPV	+++	+++	++	+++	+++	+++	+++	+++	–	–	–	–	–
Neodiprion lecontei NPV	–	++	++	+++	+++	++	++	+++	+++	++	+++	–	–
Neodiprion sertifer NPV	–	++	++	+++	+++	++	++	+++	+++	++	+++	–	–
Orgyia pseudo-tsugata NPV	++	+++	+++	++	+++	++	++	+++	+++	+++	+++	+	–
Pieris rapae GV	–	+	++	+	–	–	+	++	–	–	–	–	–
Plodia interpunctella	–	++	+++	–	–	+++	+++	+++	–	+	+++	+	–
Spodoptera exigua NPV	+	++	+++	++	+++	++	+++	+++	+++	+++	+++	+++	–
Trichoplusia ni NPV	+	++	++	++	–	–	–	+++	–	–	+++	–	–

Progress is assessed on an increasing scale from + (some) to +++ (mostly complete).

Table 11.10 Progress check table for Central America and the Caribbean

	Strain search	Character-ization	Infectivity testing	Ecology	Safety tests	Environmental impact	Formu-lation	Field trials	Extension trials	Production development	Registration	Commercial-ization	Sales
Diaphania hyalinata SaNPV[a]	–	–	++	–	–	–	++	+	–	–	–	++	++
Estigmene acraea NPV	++	+++	+++	–	–	–	+	++	–	+	–	–	–
Heliothis zea NPV	–	–	+	–	–	–	–	–	–	++	–	–	–
Hellula phidilealis AcNPV[b]	–	–	++	–	–	–	++	+	–	–	–	–	–
Plutella xylostella AcNPV	–	–	++	–	–	–	++	++	++	+	+	+	+
Pseudoplusia includens NPV	++	+++	+++	–	–	–	+	+++	++	++	+	+	+
Spodoptera albula NPV	++	+++	+++	+	–	–	+	+++	++	++	+++	++	++
Spodoptera exigua NPV	++	+++	+++	–	–	–	+	+++	++	++	+	++	++
Spodoptera frugiperda NPV	++	+++	+++	+	–	–	+	+++	–	++	–	–	–
Trichoplusia ni NPV	++	+++	+++	–	–	–	+	++	++	+	+	++	++

Progress is assessed on an increasing scale from + (some) to +++ (mostly complete).

[a] *Spodoptera albula* NPV (SaNPV) is registered in Guatemala as VPN–82. It is employed against *Diaphania hyalinata, Spodoptera albula* and *Spodoptera exigua.*

[b] AcNPV is registered as VPN–80 in Guatemala, where it is employed against *Spodoptera exigua* and *Trichoplusia ni.*

Table 11.11 Progress check table for South America

	Strain search	Character-ization	Infectivity testing	Ecology	Safety tests	Environmental impact	Formu-lation	Field trials	Extension trials	Production development	Registration	Commercial-ization	Sales
Acharia spp. PV	–	–	+	+	–	+	+	++	–	–	–	–	+
Anticarsia gemmatalis NPV	++	++	+++	++	++	++	++	+++	+++	+++	++	++	+++
Chrysodeixis includens AcNPV	–	–	+	–	–	–	+	+	–	–	–	–	–
Cydia pomonella GV	–	–	+	–	–	–	–	++	–	–	–	–	–
Diatraea saccharalis GV	+	+	++	+	–	+	–	++	–	–	–	–	–
Dione spp. NPV	–	+	++	+	–	–	+	++	–	–	–	–	–
Eacles imperialis picornavirus and CPV	–	–	+	+	–	–	+	+	–	–	–	–	–
Erinnyis ello GV	+	+	++	+	–	+	+	++	++	+	–	+	–
Heliothis spp.[a] NPV	–	–	+	–	–	–	–	+	–	–	–	–	–
Phthorimaea opercullela GV	++	++	++	+	–	+	++	+++	++	++	?	+	++
Rachiplusia nu NPV	+	–	++	–	–	–	+	++	+	+	–	–	–
Scrobipalpula absoluta GV	–	–	+	–	–	–	–	+	–	–	–	–	–
Spodoptera frugiperda NPV	++	+	+++	++	–	++	++	+++	+++	++	+	++	++

Progress is assessed on an increasing scale from + (some) to +++ (mostly complete).
[a] *sensu lato*, i.e. includes *Helicoverpa*.

12 Western Europe

JURG HUBER

PERSPECTIVE

Franz and Huber in 1979 presented a comprehensive, tabular review of field tests using insect pathogenic viruses in Europe. They listed 18 insects of importance in agriculture and forestry against which biological control methods using insect viruses had been tested in the field. To that list, only five new species can be added since that time. The review presented here essentially overlaps with and builds on Franz and Huber's (1979) paper, in which much of the earlier literature is cited.

Insect control with viruses in Western Europe has been dominated by forest pests, notably the lepidopterans *Lymantria dispar, Lymantria monacha, Panolis flammea* and *Thaumetopoea pityocampa* and the diprionid sawfly *Neodiprion sertifer*. The forest-pest–virus associations tend to have been rather well explored and virus control is an available technique dependent only upon acquiring or producing the necessary supply of virus. However, there has also been very considerable development of the control of lepidopterous pests of orchards, notably of codling moth, *Cydia pomonella*, for which viruses are currently available as genuinely practical alternatives to chemical insecticides. Lepidopterous pests of greenhouse (especially *Spodoptera* spp.) and field crops, notably *Agrotis segetum*, have also been very well explored and, again, some viruses are available.

The unification of safety test standards and registration of biological pesticides through the European Union (Chapter 30) should considerably ease the transfer of virus control technologies from one member state to another. The more significant investigations in Western Europe are listed in Table 12.1.

VIRUS PRODUCTION AND FORMULATION

Much of the more recent and significant work in Europe has related to viruses destined for registration and commercialization. Therefore, those production and formulation processes that have been well developed tend to represent publically unavailable information. An exception to this is production work at the well-known

Table 12.1 Virus control of insect pests in Western Europe: significant investigations

	Belgium	Denmark	Finland	France	Germany	Italy	The Netherlands	Norway	Spain	Sweden	Switzerland	United Kingdom	Former Yugoslavia
Adoxophyes orana											+		
Agrotis segetum		+							+	+		+	
Cnephesia spp.					+								
Cydia nigricana					+								
C. pomonella				+	+	+	+?		+		+	+	
Euproctis chrysorrhoea												+	+?
Hyphantria cunea													+
Lacanobia oleracea												+	
Leucoma salicis	+						+						
Lymantria dispar				+									+
L. monacha		+					+			+			
Neodiprion sertifer			+		+	+		+		+		+	+
Mamestra brassicae				+								+	
Panolis flammea												+	
Spodoptera exigua							+		+?				
S. littoralis				+					+				
Thaumetopoea pityocampa				+		+							
Yponomeuta spp.					+								
Zeiraphera diniana											+?		

I.N.R.A. Laboratoire de Lutte Biologique at Surgeres. Here, the NPVs of *Mamestra brassicae* and *Spodoptera littoralis* have been mass produced (de Coninck and Biache, 1988). The technology employed, however, dates back to the 1970s.

Safety tests and registration

The insect viruses listed in Table 12.2 have mostly been safety tested within member states of the European Union. However, *N. sertifer* NPV (NsNPV) safety test data, courteously made available by the US Department of Agriculture Forest Services, was accepted as a valid part of a registration package for Virox submitted to the UK Ministry of Agriculture Fisheries and Food. Safety tests on another NsNPV preparation, Kemira Sertifvirus, were conducted in Finland, which has recently joined the EU. At least one other virus not yet commercialized, the NPV of *P. flammea*, has been subjected to formal laboratory safety tests and has also been the subject of some environmental impact observations (Institute of Virology and Environmental Microbiology, Oxford, UK: unpublished). Attention is also drawn to environmental impact studies made in the UK on CpGV, especially as these relate to small mammals (Bailey and Hunter-Fujita, 1987).

Table 12.2 summarizes viruses that have been commercialized in Western Europe.

FIELD EXPERIMENTATION

Adoxophyes orana (GV and NPV)

An interesting separation of two strains of the *Adoxophyes orana* GV replicating in either the nucleus or the cytoplasm has been made, but it is not known which of these has been employed in field tests. Though slow in action, a GV was developed to a commercial product, Capex, and was registered in Switzerland in 1989 (Andermatt, 1991). Between 1979 and 1983 the NPV was field tested in several European countries in the framework of a 'CEC-Programme on Integrated and Biological Control'. Good efficacy was demonstrated (Dickler and Huber, 1983; Peters *et al.*, 1984). The reciprocal activation of infections in *A. orana* and *M. (Barathra) brassicae* by feeding heterologous NPVs was convincingly demonstrated by Jurkovicova (1979): the possible pest control implications appear not to have been pursued.

Agrotis segetum (GV and NPV)

The turnip moth (*A. segetum*) is also known as the common cutworm. The larvae spend much of their life in the soil where they attack the roots of a wide range of vegetable crops in addition to wheat and maize and some cultivated flowers. The successful use of a GV (AsGV) has been reported on rootcrops in Denmark (Zethner, 1980), on maize in Spain (Caballero *et al.*, 1990, 1991) and lettuces and dwarf asters in Germany (Fritzsche, Geissler and Schliephake, 1991). Work has also been conducted in Pakistan and the former Soviet Union (pp. 221, 235). The superiority of

Table 12.2 Virus preparations commercialized in Europe

Pets insect	Virus type	Product trade name	Company[a]	Registered country	Registered date
Adoxophyes orana	GV	Capex	Andermatt-Biocontrol AG, Grossdietwil (R+P)	Switzerland	12/89
Agrotis segetum	GV	Agrovir	Saturnia, Copenhagen (R+P)	Denmark	[b]
Cydia pomonella	GV	Madex	Andermatt-Biocontrol AG, Grossdietwil (R+P)	Switzerland	12/87
		Granupom	Hoechst A.G., Frankfurt (R) PROBIS, Pforzheim (P)	Germany	03/89
		Carpovirusine	Calliope SA, Beziers (R+P)	France	10/92
Mamestra brassicae	NPV	Mamestrin	Calliope SA, Beziers (R+P)	France	07/93
Neodiprion sertifer	NPV	Monisärmiövirus (Kemira Sertifvirus)	Kemira Oy, Espoo (R+P)	Finland	05/83
		Virox	Microbial Resources Ltd (R+P)[c], Oxford Virology Ltd (P)	UK	1984
Spodoptera exigua	NPV	Spod-X	Brinkmann BV, s'Gravenzande (R), Crop Genetics Int., Columbia USA (P)	The Netherlands	12/93
Spodoptera littoralis	NPV	Spodopterin	Calliope SA, Beziers (R+P)	France	[d]

[a] R, registration; P, production.
[b] Notification only, up to July 1993 no registration necessary.
[c] No longer in business.
[d] Applied for registration.

control over parathion in carrots has been noted in Denmark (Zethner, 1980). Until July 1993, no registration was required for microbial preparations in Denmark and the AsGV product Agrovir was sold on a limited scale. However, as such a small market cannot pay for impending registration costs, the product is likely to be withdrawn (L. Ogaard, personal communication). In any case, more recent studies have concentrated on a very promising NPV (de Oliveira, 1988; de Oliveira and Entwistle, 1990) and on a comparison of this with the GV (Bourner *et al.*, 1992). AsGV isolates from Denmark and Spain were indistinguishable. An NPV has been isolated in Poland (Lipa, Ziemnicka and Gudz-Gorban, 1971) and in the UK (Sherlock, 1983). In a maize trial in Spain, AsGV (4×10^{13} OB/ha) and AsNPV (4×10^{12} OB/ha) were compared by delivery in a spray or a bran bait. One day after application, infections for NPV sprays and bait, respectively, were 87.5% and 91.0% and for GV 12.5% and 55.0%; at 4 days the values were 78.0% and 100% for NPV and 13.0% and 6.0% for GV. Evidently, NPV in a bait formulation was the most effective treatment. Further trials on maize in Spain and on beetroot in the UK again showed NPV to act more rapidly than GV: mixed inoculum gave an intermediate mortality response (Bourner, Cory and Popay, 1994). However, cost-effectiveness studies involving spray and bait comparisons have yet to be conducted (Bourner *et al.*, 1992).

Cnephasia spp. (GV)

In field-collected *Cnephasia* spp., Glas (1991) found a GV that infects both *C. longana* and *C. pumicana*. In the wild, this can lead to collapse of larval populations of these tortricid moths. These two species of *Cnephasia* are pests of cereals. Eggs are not laid directly on the crop but on the trunk of neighbouring trees. After hatching, the first instar larvae hibernate there and only in spring do they migrate or are blown into the adjacent fields. Glas (1991) showed that it is possible to spread the virus disease by spraying GV suspensions on the bark of trees. A dose of 2×10^{10} granules per trunk was sufficient to induce more than 90% larval mortality and to lead to complete breakdown of the pest population. In a field population where the virus was already present, spraying of the eggs on the trees was still useful since the population then collapsed 3 weeks earlier than in the untreated plots.

Cydia nigricana (CpGV)

Field trials with CpGV (see below), as the commercial formulation Granupom, against pea moth gave nearly 72% control while control with pyrethroid and organophosphate insecticides gave 87.5 and 24.6% control, respectively. The result with the virus, and the fact that laboratory storage for 3 years showed no significant loss of activity, suggests Granupom to be an effective pea moth control agent (Geissler, 1994).

Cydia pomonella (GV)

The virus of codling moth, which is a worldwide pest of pome crops and of walnuts, was identified by Tanada (1964) from larvae collected by L. E. Caltagirone in northern Mexico. There have been very few subsequent 'original' isolations and epizootics have never been detected. Nevertheless, CpGV has been shown to be a potent agent for control. The use of CpGV is especially straightforward in regions where codling moth is univoltine, as in much of Europe. Where there is more than one annual generation, e.g. California, and especially where such generations have not an exactly synchronized phenology, the situation is more difficult. The use of pheromone trapping to reveal adult emergence patterns permits improved accuracy of spray timing. Developmental programmes have been conducted in many countries: outside Europe in Australia, Canada, Chile, the former Soviet Union, New Zealand and the USA and inside Europe in Austria, Germany, the Netherlands, United Kingdom and Switzerland. Under an IOBC/WPRS working group, the latter group of nations, together with Hungary, in 1976–7 tested the virus in more than a dozen orchards using material produced in Darmstadt, Germany (Huber, 1981).

A new phase of research began in 1979 under an EC programme on 'Biological Control in Apple Orchards', a programme that included studies on the NPV control of *A. orana*. This project involved France, Germany, Greece, the Netherlands and the UK and resulted in the accumulation of a vast amount of information on production, standardization, formulation and application (Cavalloro and Piavaux, 1983, 1984). In France, INRA began a multiinstitute apple orchard IPM programme based on the use of CpGV. A standardized product, Carpovirusine, was developed jointly between INRA and Solvay SA and this was tested in a dozen orchards in France, Israel and Italy (Burgerjon and Sureau, 1985). Later on, under a new contract, the development of Carpovirusine was carried forward jointly by INRA and Calliope SA. Meanwhile in Switzerland a small enterprise, Andermatt-Biocontrol AG, has developed CpGV under the trade name of Madex and this was registered in 1987. Commencing in 1984 in Germany, Hoechst AG produced CpGV for field trials and a product, Granupom, was registered in 1989. Virus for this Hoechst product is actually produced by the Institut für Umweltanalytik (IFU), Pforzheim. Granupom appeared on the market in 1991. In the Netherlands, a Hoechst CpGV product under the name Granusal was tested and found to be as effective as a chemical regime. It also proved advantageous to split the first recommended dose between two applications (i.e. one half dose CpGV each), this being especially economical in the event of an early fungicidal spray in which one of the half doses could be included (given compatibility of GV and fungicide) (Helsen, Blommers and Vaal, 1992). Field comparisons of Carpovirusine and Granupom in Italy suggested both were as effective as standard insecticides, though Granupom required more treatments than Carpovirusine (Pasqualini, Antropoli and Faccioli, 1994). Despite the present high price of CpGV products, their use is attractive within IPM programmes and also to the organic grower who can charge a premium on his produce. Studies on the non-chemical control of *C. pomonella* in France over a 10 year period were reviewed by Andemard (1986), who concluded mating disruption and CpGV were the most promising directions. The story of the worldwide commercial development of CpGV is given by Huber (1990) who also (1986) discussed production questions. Further work in Germany (Helsen *et al.*, 1992) confirmed the efficacy of another commercial formulation, Granusal, which prevent deep entry of larvae by about 90%.

The host range of CpGV includes the pea moth, *Cydia nigricana* (Payne, 1981), preliminary field trials on which showed promise (Huber, Bathon and Gillich, 1985).

Dacus oleae (picornavirus)

Cricket paralysis virus (CrPV) was found to replicate well in adult olive fly and infection brought about a pronounced shortening of longevity. In addition, the virus is transmissible by faecal contamination (Manousis and Moore, 1987). In view of the difficult nature of olive fly control, further investigation of a virus approach seems desirable.

Euproctis chrysorrhoea (NPV)

Brown-tail moth is a common European defoliator principally of woody rosaceous plants, in addition to which the severely urticaceous hairs of the larvae pose a public health problem. NPV control was developed in England, where *E. chrysorrhoea* is at the edge of its geographical range, using a locally isolated virus (Sterling, 1989). No other host has been found for this virus but the host is also susceptible to an NPV from *Euproctis similis*. There are two windows for successful NPV spray application: during the first two instars in the late summer–early autumn and post-hibernation of early instars in the following spring. Successful applications were made using a fan-assisted ULV/CDA machine (Turbair-Fox) at 1×10^9 OB/m (Kelly *et al.*, 1990; Speight *et al.*, 1992). (This dosage form is employed because the experiments involved linear hedgerow 'plots' rather than ones of a conventional, isodiametric shape.) The virus preparation involved dehairing by two cycles of low-speed centrifugation and the OBs were then suspended in the bacteriostatic fluid employed for PfNPV.

Hyphantria cunea (GV and NPV)

Fall webworm, a native of North America, has spread to various parts of the world, e.g. Japan and the eastern Adriatic region. It is a pest of many amenity trees and also walnuts (*Juglans regia*); as the larval hairs are irritant it is also a minor public health problem.

Lacanobia oleracea (GV)

Limited greenhouse trials indicated that *L. oleracea* GV has a strong potential as a pest control agent on tomato (Crook and Brown, 1982).

Leucoma salicis (NPV)

Virus control of the satin moth, to which much attention has been paid in former Eastern Block nations, has also been studied in Belgium (Nef, 1971) and in the Netherlands (Lameris *et al.*, 1985). The NPV gave very promising results.

Lymantria dispar (NPV)

NPV control of the gypsy moth has been widely studied in North America and the former Eastern Block and various commercial products have appeared. At present, an attempt is also being made at commercialization in Germany, where there have been several mass outbreaks in the 1990s. Different virus preparations from Canada, USA, Russia and the Czech Republic have been compared in field trials. In 1994, the virus was applied to about 100 ha (Huber and Langenbruch, 1994). In the Mediterranean, where gypsy moth is jeopardizing stands of valuable cork oak (*Quercus suber*), field trials with the virus have been continued (Magnoler, 1985).

Lymantria monacha (LmNPV)

Most work on the NPV control of nun moth has been in the former Eastern Block, though Zethner (1976) demonstrated conclusively the value of applications of this virus in Denmark. The efficacy of LmNPV was compared with Dimilin in aerial applications in the Netherlands and the impact of treatments on *L. monacha* parasitoids was recorded (Lameris *et al.*, 1986). In Germany, Schonherr and Ketterer (1979) reported the success of a mixed application of NPV at a normal rate with a sublethal level of *B.t.* Also in Germany, this was taken further when, for the first time, a mixture of a very low dosage of NPV and *B.t.* (1/10 of the normal concentration) was tested in the field with astonishingly good results (Altenkirch, Huber and Krieg, 1986).

Malacosoma neustria (NPV)

Larvae of the lackey moth are polyphagous on broadleaved woody plants and form tents much as do the larvae of *E. chrysorrhoea.* Treating cork oak trees in nurseries in Sardinia with a motor blower employing 8.8×10^{10} to 3.2×10^{11} OB/ha gave mortality from 93% to 100% when applied to instars two and three (Magnoler, cited by Franz and Huber, 1979; Magnoler, 1985).

Mamestra brassicae (MbNPV)

In a series of field trials conducted between 1977 and 1982 in Germany, the integration of MbNPV with *B.t.* and pirimicarb for the respective control of *M. brassicae, Plutella xylostella* and aphids was shown to be a realistic approach to pest control on cabbages (Langenbruch, Hommes and Groner, 1986). While the LD_{50} of MbNPV on brassica leaves for *P. xylostella* is very high, considerable reductions (10 times) can be achieved by use with 25% of the normal dose of cypermethrin (Biache, Severini and Injac, 1989). Based on field studies performed mostly in France and Yugoslavia in the 1970s (Bues *et al.*, 1981), INRA, France developed the NPV of the cabbage moth to the commercial product Mamestrin, which was registered in France in July, 1993 (de Coninck and Biache, 1988). This virus was not, in the first place, intended for control of cabbage moth in Europe, but for control of several other noctuids, such as *Heliothis* spp. and *Spodoptera* spp., in the tropics. It has, however, the potential to control several different insect species in Europe. For instance, trials in the former German Democratic Republic demonstrated successful control of *M. brassicae* on greenhouse roses (Fritzsche *et al.*, 1991). An early formulation 'Mamestrin' from Calliope SA, France and another from the FSU (Virin-KSH) both showed considerable promise in the control of *P. flammea* on pines in the UK.

Neodiprion sertifer (NsNPV)

European pine sawfly is principally a pest of pre-canopy-closure pine plantations in the British Isles; elsewhere in Europe it appears to be a serious pest of all pine growth stages. It occurs widely in the Palaearctic and has been accidentally introduced into eastern North America. Bird (1950) using an NPV obtained from Sweden demonstrated good control. This virus was imported from Canada into Scotland where it again proved itself in field trials. It has now been widely and successfully tested throughout Europe and the former Soviet Union and has been used in operational control in the UK, Scandinavia, FSU, etc. NsNPV work was reviewed in detail by Cunningham and Entwistle (1981). In Europe, two commercial preparations have been available: a semi-purified preparation from the Kemira Oy Chemical Co., Finland and a stabilized unpurified emulsifiable preparation, Virox, in the UK. The last preparation is not available at present. Application of from 2×10^9 to 5×10^9 OB/ha either with hydraulic equipment or by ULV/CDA has given very good control. Using the latter approach with droplets of around 50 μm VMD, virus can be delivered in a little over 1 l/ha using a spray fluid of water with 20% anti-evaporant oil (Actipron, BP). Application should ideally be made to populations in the first to third instars (Entwistle *et al.*, 1985). In a field comparison in Italy, Virox acted more rapidly than Neochek-S (a North American commercial preparation of NsNPV) but both ultimately gave good control (Baronio, Faccioli and Antopoli, 1987). Joint applications with PfNPV are possible to control mixed infestations with *P. flammea* (Doyle and Entwistle, 1988).

As work is no longer in an experimental stage, research has now concentrated on epizootiology and field persistence studies (Olofsson, 1988).

Panolis flammea (PfNPV)

Pine beauty moth is a severe defoliator of pines (*Pinus* spp.) in Central and Northern Europe. Early instar larvae feed internally in very young needles and then browse needles of any age. Heavy attacks on *Pinus contorta*, lodgepole pine, in the British Isles have resulted in tree death over very large areas. Control employing an NPV isolated in Sutherland, Scotland and closely related to MbNPV was developed during the 1980s. PfNPV OBs were semi-purified and a working concentrate prepared in a simple surfactant medium of low pH designed to inhibit microbial replication. Owing to the limited surface feeding of the susceptible, early instars, a high spray droplet density was required. This was calculated at 1×10^{10} to 2×10^{10} droplets/ha and was obtained by development of spinning-disc equipment ('X-15', Micron Sprayers Ltd).

Small droplets containing 20% anti-evaporant oil (Actipron, BP) were drifted onto the crop from helicopters flying cross-wind. A single dose of 2×10^{11} OB/ha delivered in about 9 l applied at about 90% egg hatch (Cory and Entwistle, 1990) gave over 95% control. Persistent virus residues on the crop gave further control with population suppression extending at least into the third year (Entwistle, 1986; Entwistle and Evans, 1987). Control is also possible with MbNPV, a Russian formulation of which (Virin-KSH) was found to be acceptable. Joint infestations with *N. sertifer* were controllable by a single application of a mixture of NsNPV and PfNPV (Doyle and Entwistle, 1988). Birds have been shown to be involved in dispersal of PfNPV (Entwistle *et al.*, 1993).

Spodoptera exigua (NPV)

Beet armyworm is polyphagous and especially notable in North and Central America, Africa, southern Asia, Australia and southern Europe. In northern Europe, it is an accidental introduction with sustained populations in glasshouses, notably in the Netherlands. This population is probably of North American origin. Smits (Smits, 1987; Smits, van der Vrie and Vlak, 1987) obtained three MNPV isolates from a single larva from a Netherlands glasshouse, which, because of their close similarity to MNPVs from *M. brassicae*, he named MbMNPV-NL8O, MbMNPV-NL82 and MbMNPV-NL83. The virus has also been isolated in North America, while an NPV from *Spodoptera albula* (registered as VPN-82 in Guatemala) in Central America is cross-infective, as is AcMNPV. Differences in infectivity of these viruses are small. Values for LD_{50} for the first four larval instars are low (4, 3, 39 and 102 OB) and these differences are to a very large degree compensated for by the increasing, age-related, larval feeding rate. Another comparison of isolates, from the USA, Thailand and Spain, revealed distinct but closely related genotypes. The Thai isolate was the most infective in the second instar *S. exigua* (Caballero *et al.*, 1992b). It was found (Smits, 1987; Smits *et al.*, 1987) that on chrysanthemums control by NPV (two applications at 1×10^{12} OB/ha) was cheaper than diflubenzuron (four applications necessary) or methomyl (eight applications necessary), these being, owing to the development of resistance, the only chemicals available at the time. Assiduous promotion led, by the end of 1993, to a commercial product, SPOD-X, and its subsequent registration in the Netherlands.

NPV epizootics have been observed in sunflower fields and in vegetables in glasshouses in Spain (Caballero *et al.*, 1992a). A modelling approach was taken by de Moed, van der Werf and Smits (1990) to the question of inoculative introductions for creation of economically viable epizootics in glasshouse chrysanthemums. They concluded that a stable polyhedral population is a factor crucial to success.

Spodoptera littoralis (NPV)

While much developmental and commercialization activity for *S. littoralis* NPV has been centred on France (Calliope SA) it has largely been directed towards use in other geographical regions as *S. littoralis* is not a notable pest in Europe (see Chapter 18).

Thaumetopoea pityocampae (CPV)

The pine processionary moth, made famous by the observations of Jean Henri Fabre, is a severely urticaceous pest. Control of this pest was one of the earliest of European ventures and was the first example of application of a CPV for biological control. A mixture of CPV with bentonite dust was applied on about 320 ha by helicopter in France (Grison, Maury and Vago, 1959; Grison, 1960). The virus was mass produced in larvae fed on natural foliage. Handling these urticaceous larvae undoubtedly inhibited production of this virus.

Yponomeuta spp. (NPV)

Feemers (1986a) isolated a NPV from *Y. evonymellus*, which was also infective for *Y. padellus*. The LD_{50} for *Y. padellus* third instar larvae was only 22 OBs. In a small field trial, a virus concentration of 1×10^7 OB/ml, applied to run-off on *Prunus spinosa* bushes, was able to prevent defoliation by both species. Untreated bushes with the same larval density (four nests per bush) were completely defoliated (Feemers, 1986b).

Zeiraphera diniana (GV)

The subalpine larch (*Larix* spp.) forests of Austria, France, Italy and Switzerland suffer cyclical outbreaks of larch budmoth (Baltensweiler *et al.*, 1977) often with associated GV epizootics. Attempts to use the GV for control have not been very successful. For instance when 12 ha was helicopter-sprayed at 5×10^{11} to 8×10^{11} OB/ha only 30–50% control was achieved (Benz, 1976). Similar treatment of 17 ha at 1.4×10^{13} OBs in 40 l/ha gave 0–30% control (Schmid, 1973; Benz, 1976).

REFERENCES

Altenkirch,W., Huber, J. and Krieg, A. (1986) Versuche zur biologischen Bekamfung der Nonne (*Lymantria monacha* L.). *Zeitschrift für Pflanzenkrankheiten und Pflanzenschutz* **93**, 479–493.

Andemard, M. (1986) Lutte biologique contre le carpocapse (*Cydia pomonella* L.). *Colloques de l'INRA* **34**, 15–28.

Andermatt, M. (1991) Field efficacy and environmental persistence of the granulosis viruses of the summer fruit tortrix, *Adoxophyes orana*. *IOBC/WPRS Bulletin* **14**, 73–74.

Bailey, M.J. and Hunter-Fujita, F.R. (1987) Specific immunological response against the granulosis virus of the codling moth (*Cydia pomonella*) in woodmice (*Apodemus sylvaticus*): field observations. *Annals of Applied Biology* **111**, 649–660.

Baltensweiler, W., Benz, G., Bovey, P., and Delucchi, Y. (1977) Dynamics of larch bud moth populations. *Annual Review of Entomology* **22**, 79–100.

Baronio, P., Faccioli, G. and Antopoli, A. (1987) Utilizzazione di una nucleopoliedrosi virale specifico nella lotta contro *Neodiprion sertifer* (Geoffr.) (Hym., Diprionidae) confronto tra due preparati. *Bollettino dell Instituto di Entomologia 'Guido Grande' della Universita degli Studi di Bologna* **41**, 233–240.

Benz, G. (1976) Current microbial control programs in Western Europe exclusive of the United Kingdom and Ireland. *Proceedings of the 1st International Colloquium of Invertebrate Patholology*, Kingston, Canada, pp. 52–58.

Biache, G., Severini, M. and Injac, M. (1989) Sensibilite de *Plutella xylostella* a une associacion de baculovirus de la polyedrose de *Mamestra brassicae* et d'un pyrethroide de synthese. *Mededelingen van de Faculteit Landbouwwetenschappen, Gent* **54,** 917–921.

Bird, F.T. (1950). The dissemination and propagation of a virus disease affecting the European pine sawfly, *Neodiprion sertifer* (Geoff.). *Bi-monthly Progress Report of the Canadian Department of Agriculture*, **6**, 2–3.

Bourner, T.C., Vargas-Osuna, E., Williams, T., Santiago-Alvarez, C. and Cory, J.S. (1992) A comparison of the efficacy of nuclear polyhedrosis and granulosis viruses in spray and bait formulations for the control of *Agrotis segetum* (Lepidoptera: Noctuidae) in maize. *Biocontrol Science and Technology* **2**, 315–326.

Bourner, T.C., Cory, J.S. and Popay, A.J. (1994) Nuclear polyhedrosis and granulosis viruses for the control of the common cutworm, *Agrotis segetum* (Lepidoptera: Noctuidae). *Proceedings of the 47th New Zealand Plant Protection Conference*, Waitangi Hotel, pp. 159–162.

Bues, R., Poitout, H.S., Injac, M. and Burgerjon, A. (1981) Trois annees d'experimentation du virus de la polyedrose nucleaire pour la lutte contre *Mamestra brassicae* (Lep., Noctuidae) en culture de choux-fleurs. *Defense Vegetale* **210**, 283–293.

Burgerjon, A. and Sureau, F. (1985) La 'carpovirusine', bio-insecticide experimental a base du baculovirus de la granulose pour lutter contre le carpocapse (*Cydia pomonella* L.). *Dèbats de la 5ième Colloque sur les Recherches Fruitieres*, Bordeaux, 1985, pp. 53–66.

Caballero, P., Vargas-Osuna, E. and Santiago-Alvarez, C. (1990) Aplicacion en campo del virus de la granulosis de *Agrotis segetum* Schiff. (Lepidoptera: Noctuidae). *Boletin de Sanidad Vegetal Plagas* **16**, 333–337.

Caballero, P., Vargas-Osuna, E. and Santiago-Alvarez, C. (1991) Efficacy of a Spanish strain of *Agrotis segetum* granulosis virus (Baculoviridae) against *Agrotis segetum* Schiff. (Lep.: Noctuidae) on corn. *Journal of Applied Entomology* **112**, 59–64.

Caballero, P., Aldebis, H.K., Vargas-Osuna, E. and Santiago-Alvarez, C. (1992a) Epizootics caused by a nuclear polyhedrosis virus in populations of *Spodoptera exigua* in southern Spain. *Biocontrol Science and Technology* **2**, 35–38.

Caballero, P., Zuidema, D., Santiago-Alvarez, C. and Vlak, J.M. (1992b) Biochemical and biological characterization of four isolates of *Spodoptera exigua* nuclear polyhedrosis virus. *Biocontrol Science and Technology* **2**, 145–157.

Cavalloro, R. and Piavaux, A. (eds) (1983) *C.E.C. Programme on Integrated and Biological Control – Progress Report 1979–1981. (EUR 8273 EN)*. Luxembourg: Office for Official Publications of the European Communities.

Cavalloro, R. and Piavaux, A. (eds.) (1984) *C.E.C. Programme on Integrated and Biological Control – Final Report 1979–1983. (EUR 8689)*. Luxembourg: Office for Official Publications of the European Communities.

Cory, J.S. and Entwistle, P.F. (1990) The effect of time of spray application on infection of the pine beauty moth, *Panolis flammea* (Den. & Schiff.) (Lep., Noctuidae), with nuclear polyhedrosis virus. *Journal of Applied Entomology* **110**, 235–241.

Crook, N.E. and Brown, J.D. (1982) Isolation and characterization of a granulosis virus from the tomato moth, *Lacanobia oleracea*, and its potential as a control agent. *Journal of Invertebrate Pathology* **40**, 221–227.

Cunningham, J.C. and Entwistle, P.F. (1981) Control of sawflies by baculovirus. In *Microbial Control of Pests and Plant Diseases 1970–1980*. H.D. Burges (ed.). London: Academic Press, pp. 379–407.

de Coninck, P. and Biache, G. (1988) The production of baculovirus preparations at the Biological Control Laboratory of Le Magneraud. In *Production and Application of Viral Bio-pesticides in Orchards and Vegetables (EUR 10424-EN-FR)*. H. Audemard and R. Cavalloro (eds). Luxembourg: Directorate-General for Agriculture, Commission of the European Communities, pp. 127–132.

de Moed, G.H., van der Werf, W. and Smits, P.H. (1990) Modelling the epizootiology of *Spodoptera exigua* nuclear polyhedrosis virus in a spatially distributed population of *Spodoptera exigua* in greenhouse chrysanthemums. *SROP/WPRS Bulletin* **XIII**, 135–141.

de Oliveira, M.R.V. (1988) The ecology of *Agrotis segetum*, turnip moth (Lepidoptera:

Noctuidae) nuclear polyhedrosis virus and its use as a control agent. MSc Thesis, University of Oxford, UK.
de Oliveira, M.R.V. and Entwistle, P.F. (1990) *Agrotis segetum* nuclear polyhedrosis virus as control agent for the cutworm, *Agrotis segetum* (Schiff.) (Lepidoptera: Noctuidae), and assessment of plant damage. In *Proceedings and Abstracts Vth International Colloquium on Invertebrate Pathology and Microbial Control*, Adelaide, Australia. p. 493.
Dickler, E. and Huber, J. (1983) Microbial control of *Adoxophyes orana* in combination with granulosis virus control of codling moth. In *C.E.C. Programme on Integrated and Biological Control. Progress Report 1979–1981* R. Cavalloro and A. Piavaux (eds). Luxembourg: Office for Official Publications of the European Community, pp. 16–21.
Doyle, C.J. and Entwistle, P.F. (1988) Aerial application of mixed virus formulations to control joint infestations of *Panolis flammea* and *Neodiprion sertifer* on lodgepole pine. *Annals of Applied Biology* **113**, 119–127.
Entwistle, P.F. (1986) Spray droplet deposition patterns and loading of spray droplets with NPV inclusion bodies in the control of *Panolis flammea* in pine forests. In *The 4th International Colloquium of Invertebrate Pathology: Fundamental and Applied Aspects of Invertebrate Pathology*, Wageningen, the Netherlands. R.A. Samson, J.M. Vlak and D. Peters (eds). pp. 613–615.
Entwistle, P.F. and Evans, H.F. (1987) Trials on the control of *Panolis flammea* with a nuclear polyhedrosis virus. In *Forestry Commission Bulletin 67: Population Biology and Control of the Pine Beauty Moth in Scotland.* S.R. Leather, J.T. Stoakley and H.F. Evans (eds). London: HMSO, pp. 61–68
Entwistle, P.F., Evans, H.F., Harrap, K.A. and Robertson, J.S. (1985) Control of European pine sawfly *Neodiprion sertifer* (Geoffr.) with its nuclear polyhedrosis virus in Scotland. In *Site Characteristics and Population Dynamics of Lepidopteran and Hymenopteran Forest Pests. Forestry Commission Research and Development Paper 135*, D. Bevan and J.T. Stoakley (eds). London: HMSO, pp. 36–46.
Entwistle, P.F., Forkner, A.C., Green, B.M. and Cory, J.S. (1993) Avian dispersal of nuclear polyhedrosis viruses after induced epizootics in the pine beauty moth, *Panolis flammea* (Lepidoptera: Noctuidae). *Biological Control* **3,** 61–69.
Feemers, M. (1986a) Untersuchungen uber ein Kernpolyeder-Virus aus *Yponomeuta evonymellus* L. (Lep.,Yponomeutidae) und seine Wirkung auf verschiedene *Yponomeuta*-Arten. 1. Morphologie und Pathogenese. *Journal of Applied Entomology* **101**, 89–100.
Feemers, M. (1986b) Untersuchungen uber ein Kernpolyeder-Virus aus *Yponomeuta evonymellus* L. (Lep., Yponomeutidae) und seine Wirkung auf verschiedene *Yponomeuta*-Arten. 2. Wirkung und Umweltabhangigkeit des Virus, *Journal of Applied Entomology* **101**, 425–444.
Franz, J.M. and Huber, J. (1979) Feldversuche mit Insektenpathogenen Viren in Europa. *Entomophaga* **24**, 333–343.
Fritzsche, R., Geissler, K. and Schliephake, E. (1991) Results of experiments with granulosis and nuclear polyhedrosis virus preparations in fruits, vegetables and ornamental plants. *Nichrichtenblatt des Deutschen Pflanzenschutzdienstes* **43**, 92–95.
Geissler, K. (1994) Eingung des Granulose-Virus des Apfelwickers (*Cydia pomonella* L.) zur Bekamfung des Erbsenwicklers (*Cydia nigricana* Steph.). *Archives of Phytopathology and Plant Protection* **29**, 191–194.
Glas, M. (1991) Tortricids in cereals. In *World Crop Pests, 5. Tortricid Pests. Their Biology, Natural Enemies and Control* L.P.S. van der Geest and H.H. Evenhuis (eds). Amsterdam: Elsevier, pp 553–562.
Grison, P. (1960) Utilisation en foret d'une preparation a base de virus specifique contre *Thaumatopoea pityocampa* Schiff. *Zeitschrift für Angewandte Entomologie* **47**, 24–31.
Grison, P., Maury, R. and Vago, C. (1959) *Revue de Forestiere Française* **5**, 353–370.
Helsen, H., Blommers, L. and Vaal, F. (1992) Efficacy and implementation of granulosis virus against codling moth in orchard IPM. *Mededelingen van de Faculteit Landbouwwetenschappen Gent.* **57**, 569–573.
Huber, J. (1981) Apfelwickler-Granulosevirus: Produktion und Biotests. *Mitteilungen der Deutschen Gesellschaft für Allgemeine und Angewandte Entomologie* **2**, 141–145.

Huber, J. (1986) In vivo production and standardization. In *The 4th International Colloquium of Invertebrate Pathology: Fundamental and Applied Aspects of Invertebrate Pathology*, Wageningen, the Netherlands. R.A. Samson, J.M. Vlak and D. Peters (eds). Wageningen, The Netherlands, pp. 87–90.

Huber, J. (1990) History of the CpGV as a biological control agent – its long way to a commercial viral pesticide. In *Proceedings and Abstracts Vth International Colloquium on Invertebrate Pathology and Microbial Control*, Adelaide, Australia, pp. 424–427.

Huber, J. and Langenbruch, G.A. (1994) Use of microbial pathogens for control of forest insect pests in Germany. In *Proceedings of the VIth International Colloquium on Invertebrate Pathology and Microbial Control*, Montpellier, pp. 388–392.

Huber, J., Bathon, H. and Gillich, H. (1985) Erste Versuche zur Bekamfung des Erbenswicklers mit dem Apfelwickler-Granulosevirus. *Jahresbericht 1985*, Biologische Bundesanstalt für Landwirtschaft und Forstwirtschaft, Brunschweig, p. 75.

Jurkovicova, M. (1979) Activation of latent virus infections in larvae of *Adoxophyes orana* (Lepidoptera: Tortricidae) and *Barathra brassicae* (Lepidoptera: Noctuidae) by foreign polyhedra. *Journal of Invertebrate Pathology* **34**, 213–223.

Kelly, P.M., Entwistle, P.F., Sterling, P.H., Speight, M.M. and Laport, R.F. (1990) Virus control of the brown-tail moth, *Euproctis chrysorrhoea*. In *USDA Technical Report NE-123: The Lymantriidae: a Comparison of New and Old World Tussock Moths*. W.E. Walner and K.A. McManus (eds). Washington, DC: USDA, pp. 427–438.

Lameris, A.M.C., Ziemnicka, J., Peters, D., Grijpma, P. and Vlak, J.M. (1985) Potential of baculoviruses for control of the satin moth *Leucoma salicis* L. (Lepidoptera: Lymantriidae). *Mededelingen van de Faculteit Landbouwwetenschappen Gent*. **50**, 431–439.

Lameris, A.M.C., Laport, R.F., Grijpma, P. and Vlak, J.M. (1986) Aerial application of nuclear polyhedrosis virus with Dimilin for control of the nun moth, *Lymantria monacha* L. (Lep., Lymantriidae). In *The VIth International Colloquium on Invertebrate Pathology: Fundamental and Applied Aspects of Invertebrate Pathology*, Wageningen, the Netherlands. R.A. Samson, J.M. Vlak and D. Peters (eds). p. 134.

Langenbruch, G.A., Hommes, M. and Groner, A. (1986) Feldversuche mit dem Kernpolyedervirus der Kohleule (*Mamestra brassicae*). *Zeitschrift für Pflanzenkrankheiten und Pflanzenschutz* **93**, 72–86.

Lipa, J.J., Ziemnicka, J. and Gudz-Gorban, A.P. (1971) Electron microscopy of nuclear polyhedrosis virus from *Agrotis segetum* Schiff. and *A. excamationis* L. (Lepidoptera, Noctuidae). *Acta Microbiologica Polonica Series B*. **3**, 55–61.

Magnoler, A. (1985) L'impiego di due baculovirus nella lotto contro *Malacosoma neustria* L. e *Lymantria dispar* L. nei querceti della Sardegna. *Difesa delle Piante* **8**, 451–462.

Manousis, T. and Moore, N.F. (1987) Cricket paralysis virus, a potential control agent for the olive fruit fly, *Dacus oleae* Gmel. *Applied and Experimental Microbiology* **53**, 142–148.

Nef, L. (1971) Influence des traitements insecticides chimiques et microbiens sure une population de *Stilpnotia* (=*Leucoma*) *salicis* L. et sur ses parasites. *Journal of Applied Entomology* **69**, 357–367.

Olofsson, E. (1988). Persistence and dispersal of the nuclear polyhedrosis virus of *Neodiprion sertifer* (Geoffroy) (Hymenoptera: Diprionidae) in a virus-free lodgepole pine-plantation in Sweden. *Canadian Entomologist* **120**, 887–892.

Pasqualini, E., Antropoli, A. and Faccioli, G. (1994) Performance tests of granulosis virus against *Cydia pomonella* L. (Lepidoptera, Olethreutidae). *Bulletin OILB-SROP* **17**, 113–119.

Payne, C.C. (1981) The susceptibility of the pea moth, *Cydia nigricana*, to infection by the granulosis virus of the codling moth, *Cydia pomonella*. *Journal of Invertebrate Pathology* **38**, 71–77.

Peters, D., Wiebenga, J, van Maanen, H.J., Vanwetswinkel, G. and Blommers, L.H.M. (1984) The control of summer fruit tortrix moth (*Adoxophyes orana*) with a nuclear polyhedrosis virus in orchards. In *C.E.C. Programme on Integrated and Biological Control. Final Report 1979–1983*. R. Cavarollo and A. Piavaux (eds). Luxembourg: Office for Official Publications of the European Community, pp. 69–78.

Schmid, A. (1973) PhD Thesis No. 5045, ETH Zurich, Switzerland.
Schonherr, J. and Ketterer, R. (1979) Zur Frage der kombinierten Anwendung von Polyedervirus und *Bacillus thuringiensis* bei der Nonne, *Lymantria monacha* L. (Lepidoptera). *Zeitschrift für Pflanzenkrankheiten und Pflanzenschutz* **86**, 483–488.
Sherlock, P.L. (1983) The natural incidence of disease in the cutworm *Agrotis segetum* in England and Wales. *Annals of Applied Biology* **102**, 49–56.
Smits, P.H. (1987) Nuclear polyhedrosis virus as biological control agent of *Spodoptera exigua*. PhD thesis, Wageningen, the Netherlands.
Smits, P.H., van der Vrie, M. and Vlak, J.M. (1987) Nuclear polyhedrosis virus for control of *Spodoptera exigua* larvae on glasshouse crops. *Entomologia Experimentalis et Applicata* **43**, 73–80.
Speight, M.R., Kelly, P.M., Sterling, P.H. and Entwistle P.F. (1992) Field application of a nuclear polyhedrosis virus against the brown-tail moth, *Euproctis chrysorrhoea* (L.) (Lep., Lymantriidae). *Journal of Applied Entomology* **113**, 295–306.
Sterling, P. (1989). Natural mortalities of *Euproctis chrysorrhoea* (L.) and the use of its baculovirus in biocontrol. DPhil Thesis, University of Oxford, UK.
Tanada, Y. (1964) A granulosis virus of the codling moth, *Carpocapsa pomonella* (Linnaeus) (Olethreutidae). *Journal of Insect Pathology* **6,** 378–380.
Zethner, O. (1976) Control experiments on the nun moth (*Lymantria monacha* L.) by nuclear-polyhedrosis virus in Danish coniferous forests. *Zeitschrift für Angewandte Entomologie* **81**, 192–207.
Zethner, O. (1980) Control of *Agrotis segetum* (Lep.: Noctuidae) [on] root crops by granulosis virus. *Entomophaga* **25**, 27–35.

13 Eastern Europe and the former Soviet Union

JERZY J. LIPA

PERSPECTIVE

Zoogeographically, the argument for treating Eastern Europe and the former Soviet Union (FSU) with Europe as the Western Palaearctic is compelling. However, the direction, progress and attainments of applied insect virology in the old Eastern Block are all notably different from those of Western Europe and so merit separate treatment.

There has been a large-scale effort in the development, production and utilization of viruses for insect control in this area, most of which has been in the FSU. Unfortunately, the results of both basic and applied studies are mostly published in slavic languages and in conference proceedings with very limited circulation so that they are largely inaccessible outside Central and Eastern Europe (CEE) .

Recent reviews (Lipa, 1990, 1991) have presented general information on microbial control of insects in the CEE region. Practical use has been preceded by intensive fundamental research, on which there is a vast literature. Readers interested in this literature are advised to consult specialized books (Weiser *et al.*, 1966; Kok, 1973; Vorobeva, 1974, 1976; Lipa, 1975; Tarasevich, 1975; Kok, Skuratovskaya and Strokovskaya, 1980; Chukhrii, 1988; Gulii and Rybina, 1988). Those who are interested in applied research should consult the books, conference proceedings and reviews for each country: Bulgaria (Videnova, Velichkova and Encheva, 1981), Hungary (Kozar *et al.*, 1989), Romania (Petre, 1981), Poland (Lipa, 1975, 1990, 1991), Czechoslovakia (Weiser *et al.*, 1966), FSU ((Vorobeva, 1974, 1976; Tarasevich, 1975, 1985; Orlovskaya, 1976, 1984, 1989; Sikura, 1976; Gulii and Golosova, 1977; Aleshina, 1980; Beglyarov, Sikura and Orlovskaya, 1981; Gulii, Ivanov and Shternshis, 1982a,b; Tarasevich and Gulii, 1985; Gulii and Rybina, 1988; Samersov, 1990; Urakov, 1990).

Scientific, technological, organizational, economic and safety problems associated with virus insecticides are discussed in various conference proceedings, which also provide information on new formulations (translated into English, these are: *Viruses of Insects and Perspectives of their Use in Plant Protection against Pests*

Table 13.1 Virus control of insect pests in Eastern Europe and the FSU: significant investigations

Pest species	Bulgaria	FSU	Hungary	Latvia	Poland	Romania
Agrotis segetum		+			+	
Cydia pomonella		+	+	+	+	
Dendrolimus sibiricus		+				
Helicoverpa armigera		+				
Hyphantria cunea	+	+		+		+
Hyponomeuta malinellus		+				
Leucoma salicis		+			+	
Lymantria dispar		+			+	+
Lymantria monacha		+			+	
Malacosoma neustria		+		+	+	
Mamestra brassicae	+	+				
Neodiprion sertifer		+		+	+	
Pieris spp.		+		+		
Xestia c-nigrum		+			+	

in Member Countries EPS/IOBC (Beglyarov *et al.*, 1981); *Results and Perspectives of Production and Use of Virus Preparations in Agriculture and Forestry* (Orlovskaya, 1984); *Problems of Development and Use of Microbiological Means of Plant Protection* – Part II (Borovik, 1989); *2nd Symposium of COMECON Countries on Microbial Pesticides* (Urakov, 1990)).

Practical attempts to use viruses in insect control have been made in all CEE countries (see Table 13.1 for a list of all significant investigations) but industrial production of virus insecticides has been organized only in the FSU. It is only in that area that there has been sufficient demand and an appropriate specialized scientific and industrial infrastructure. At first, efforts were centred on the All-Union Scientific Research Institute of Microbial Means for Plant Protection and Bacterial Preparations in Moscow (Aleshina, 1980). Later the development of virus insecticide production was the responsibility of the All-Union Scientific Research Institute of Molecular Biology, Novosibirsk and of the Scientific and Production Association 'Vektor', both connected with the Ministry of Medical and Microbiological Industry, which, having over 130 factories with almost 200 000 employees, was in a position to produce all kinds of microbial pesticides on a large scale (Bozhko, 1988). At present, after the dissolution of the FSU, there is no leading scientific centre working for the biopesticide needs of the present FSU. For Russia, such a leading role is performed by the All-Russian Institute of Biological Plant Protection (VNIIBZR) and the All-Russian Institute of Applied Microbiology (VNIIPM), and in Moldavia by the Institute of Biological Plant Protection (VNIIBZ). However previously, and at present, because of the large scale of such industry there is no interest in small-scale operations and, therefore, the production of preparations such as Virin-KSh, Virin-Diprion and Virin-EKS (see Table 13.2), for which there is a limited demand, is by regional biofactories or the research institutes that developed such products. In countries other than the FSU, notably Bulgaria, Poland and Roumania, viruses, even if given a brand name as occurs in Bulgaria, are produced by research institutes.

Table 13.2 BV insecticides developed in Central and Eastern Europe and the FSU

Trade names (target insects)	Virus type	Formulations (titre)	Crops protected	Dose and number of treatments/season or generation[a]	Country	References
Hifantrin (*Hyphantria cunea*)	GV	50% glycerine/water (1×10^9 OB/ml)	Orchards and parks	Dose not given: 1 treatment	Bulgaria	Weiser *et al.* (1986)
Mamestrin (*Mamestra brassicae*, *Lacanobia oleracea*, *Plusia gamma*)	NPV	WP (9×10^9 OB/g)	Vegetables	1×10^6 to 5×10^6 OB/ml at 600–700 l/ha: 1–2 treatments/generation	Bulgaria	Weiser *et al.* (1986); Velichkhova-Kozhukharova (1988)
Virin-ABB, Virin-ABB-3 (*Hyphantria cunea*)	NPV + GV	50% glycerine/water (1×10^8–2×10^9 OB/ml)	Orchards, parks, forests	0.1–0.2 kg/ha: 1 treatment	USSR	Sikura and Smetnik (1980)
Virin-Diprion (*Neodiprion sertifer*)	NPV	50% glycerine/water (1×10^9 OB/ml)	Pine forests	Dose not given: 1 treatment	Poland, USSR	Glowacka-Pilot *et al.* (1987); Zarinsh (1980)
Viris EKS (*M. brassicae*)	NPV	WP or 50% glycerine/water	Cabbage, pea, sugar beet	0.1–0.15 kg/ha: 2 treatments at 8–10 day intervals on L1–L2 of each generation	USSR	Kravtsov and Golyshin (1989); Rutskaya (1990)
Virin-ENSh (*Lymantria dispar*)	NPV	50% glycerine/water (1×10^9 OB/ml)	Orchards, parks, forests	100 ml/ha against L2–L3; for egg treatment on 10–15% trees, 0.2 ml/ha when <2 egg masses/tree; 2 ml/ha when 0.5–2.0 egg masses/tree	USSR	Kravtsov and Golyshin (1989)
Virin-GSSh (*Dendrolimus sibiricus*)	GV	?	Pine forests	Dose not given: 1 spray in autumn against L1–L2, L4	USSR	Orlovskaya (1989)
Virin-GYaP (*Cydia pomonella*)	GV	50% glycerine/water (3×10^9 OB/ml)	Apple orchards: in regions with one generation and when <13% fruits damaged	0.3 kg/ha; 1st application at beginning of egg hatch, 2nd at maximum hatch	USSR	Kravtsov and Golyshin (1989); Kanapatskaya *et al.* (1990)
Virin-KhS (*Helicoverpa armigera*)	NPV	WP (7×10^9 OB/g)	Cotton	0.3 kg/ha 1–2 applications/generation	USSR	Kravtsov and Golyshin (1989); Simonova *et al.* (1984)

Virin-KSh (*Malacosoma neustria*)	NPV	50% glycerine/water (1×10^9 OB/ml)	Orchards, parks	0.2 kg/ha; 1 treatment against L1–L3	USSR	Zarinsh (1984)
Virin-LS (*Leucoma salicis*)	NPV	WP (?)	Parks and urban trees	0.1 kg/ha; 2 treatments in spring against L2–L3, in summer against L1–L2	Poland, USSR	Ziemnicka *et al.* (1991a,b)
Virin-PShM (*Lymantria monacha*)	NPV	50% glycerine/water (1×10^9 OB/ml)	Forest	Dose not given: 1 treatment against L1–L2	USSR	Orlovskaya (1989)
Virin-OS (*Agrotis segetum*)	GV	WP (3×10^9 OB/g)	Cotton	0.3 kg/ha; 1–2 treatments against L1–L2 in mid-Asia	USSR	Kravtsov and Golyshin (1989)
Virin-YaM (*Yponomeuta malinellus*)	NPV	?	Orchards	?	USSR	Simonova *et al.* (1980)

WP, wettable powder.
[a] L1–L4 is larval stages first to fourth instar.

A general view of the situation in the microbiological industry and of microbial insecticide production in the USSR until 1988 was given by Rimmington (1989). In Table 13.2 a list of 14 virus insecticides is given together with their 'trade names' and major features. Further information on these is available in Anon. (1994) and Kravtsov and Golyshin (1989).

In the CEE area, the philosophy of virus utilization for control has followed two main routes. Firstly, viruses are mainly used as sprayable biological insecticides. Secondly, however, they are employed to generate epizootics with the objective of long-term and widespread suppression of pest populations. This 'focal (epizootic) technology' was pioneered by Orlovskaya (1961) employing *Lymantria dispar* and its NPV as a research system (see below). Of course, this approach is most successful where there is a natural capacity for epizoosis that can be triggered by artificial increases in inoculum mass. This is done by limited and spatially dispersed point introductions of virus. The approach works particularly well for insects with a gregarious larval habit, e.g. *L. dispar*, *Malacosoma neustria* and *Neodiprion sertifer*, but it is inappropriate for some pest behaviour, for example with *Mamestra brassicae*, the larvae of which are often hidden within cabbages where they cannot be reached by viruses. Based on such an approach, Shekhurina (1985) developed a method of predicting epizootics in populations of the grey grain cutworm, *Apamea anceps*, with the purpose of obviating the unnecessary use of chemical insecticides.

Few data are available on the economic aspects of virus insecticides since in the centrally planned economy, prices of products (chemical and biological) were heavily subsidized at different rates so that price comparisons may be quite artificial.

The autocratic policy typical of a socialistic economy, and the limited use of modern technologies, negatively affected the volume and quality of microbial pesticides, their distribution and scale of use (Rimmington, 1989). Despite these hindrances, the programmes of biological control in the CCE region are very impressive. The scientific, technological and political integration of Eastern and Western countries, which is occurring in the 1990s, will undoubtedly have a great positive influence on ideas and future development of biological control in the whole of Europe.

VIRUS PRODUCTION

All virus production is *in vivo*, utilizing where possible inexpensive semi-synthetic diet, but such a simple technology could clearly be extended to some insect species at present reared on leaves. Literature providing more detailed process descriptions is cited above. There is little information on cost and the volume of production. While discussing the cost of virus preparations, Chukrii *et al.* (1989b) indicated that the cost of rearing of one third instar larva of *Helicoverpa (Heliothis) armigera* was 0.20–0.25 roubles, of *M. brassicae* 0.15–0.2 roubles and *Agrotis segetum* 0.10–0.15 roubles (at the time of that evaluation US $1 = 0.62 roubles, while the exchange rate in the mid-1990s was much higher, US $1 ~ 3000 roubles). This determines the high cost of products for which it is necessary to use 1500–2000 larvae to produce 1 kg. In the case of *Hyphantria cunea* (Virin-ABB) and *L. dispar*

(Virin-ENSh), the larvae of which are reared on leaves, the cost of 1 kg of virus insecticide can be 45–50 roubles.

At the Atakskaya biofactory, Moldavia, 4 kg of Virin-KS (MbMNPV) was produced in 1988 and 40 kg in 1989 and this was used on 56 and 500 ha, respectively, against cabbage moth *M. brassicae* (Rutskaya, 1990). However, this virus has been exported to the UK for field trials and limited commercial use against pine beauty moth, *Panolis flammea*. About 20 kg of Virin-OS (*A. segetum* NPV) was produced during 1987–8 and used in Uzbekistan and Moldavia (Kitik, 1989). While discussing the industrial production of viral insecticides, Kolyganov, Samojlenko and Dumanova (1984) provided the data that at the Ungenska Factory, Ungenska, Moldavia, during 1971–82 the volume of production of Virin ENSh increased about 55 times, of Virin-EKS 80 times and of Virin-ABB 120 times.

Formulation

Most products are formulated at two levels of concentration and frequently as wettable powders and in liquid form. The least concentrated are intended for higher volume applications while those with a higher concentration, usually suffixed 'C' (e.g. Virin-KSC) are for low and ULV spraying: Virin-KS contains 5×10^9 OB/g while Virin-KSC contains 5×10^{10} OB/g. All preparations are stated to contain 'a filler with stabilizing and anti-oxidizing components on the basis of aerosil or silicic acid'. The bacterial titre is given as 1×10^7 colonies/g and the pH of a water-glycerine suspension is 7.0–7.2. The material is supplied in hermetically sealed plastic bags and should be stored at 0–20°C in a dark, dry place, when it will have a shelf life of 18 months (information from NPV Vektor, Novosibirsk). The filler was originally kaolin, but as this frequently blocked sprayer nozzles poor plant coverage was a common feature. Work is, therefore, in progress to replace the old product Virin-ENSh (1×10^9 OB/ml in liquid form with kaolin) with, for instance, new wettable powders elaborated by the 'Vektor' Association. These are Virin-ENSh-1 (5×10^9 OB/g with zeolite) and Virin-ENSh-2 (5×10^9 OB/g with silicic acid). Both have performed well in field trials against *L. dispar* (respectively, 85.1–93.9 and 77.2–96.9% control) (Yatsenko, 1984; Yatsenko *et al.*, 1990). The old liquid formulation, Virin-Gyap (3×10^9 OB/ml in glycerine) is now being replaced by the wettable powder formulation (Kanapatskaya, Korol and Zarinsh, 1990). New formulations of Virin-ABB-2 (Chukhrii, Voloshchuk and Neshcharet, 1989a) and Virin-KhS-2 (Chukhrii *et al.*, 1990) have also been developed.

FIELD EXPERIMENTATION

Agrotis segetum (GV and NPV)

Working on improvement of Virin-OS (a GV preparation), Kitik (1989) during 1987–9 produced about 20 kg, which was used in Uzbekistan and Moldavia to control winter cutworm on cotton, sugar beet, vegetables and winter cereals. A mixture of GV and NPV has also been tested with reasonable success in the FSU (Table 13.3).

Table 13.3 Virus effectiveness against salient pest insects in the CEE region

Pest	Virus type	Virus dose or concentration	Volume (l/ha)	Efficacy (% population reduction)	Country
Agrotis segetum	GV	1×10^7 OB/ml	500	65–85	USSR
	GV + NPV	5×10^6 OB/ml	100–500	77.5	USSR
Cydia pomonella	GV	9×10^{10}–2.9×10^{11} OB/tree	4–8 l/tree	74–88	Poland, USSR
		7.7×10^7 OB/ml	1.5 l/tree	77–95	Hungary, USSR
Dendrolimus sibiricus	GV	1×10^9 OB/ml	40	74–90	USSR
Helicoverpa armigera	NPV	1×10^{10}–1×10^{11} OB/ha	20–200	90–100	USSR
Lymantria dispar	NPV	2.7×10^8–5×10^{13} OB/ha	10–1000	60–100	Romania, USSR
Lymantria monacha	NPV	6.8×10^{11} OB/ha	80	90	Poland, USSR
Malacosoma neustria	NPV	8.8×10^{10}–3.2×10^{11} OB/ha	200	88–100	Latvia USSR
Mamestra brassicae	NPV	1×10^{11}–1×10^{13} OB/ha	400–1000	51–100	USSR
	NPV + GV	1×10^6 OB/ml	600	84–90	USSR
Neodiprion sertifer	NPV	1×10^5–1.5×10^{12} OB/ha	10–80	88–100	Poland, USSR
Pieris brassicae	GV	1×10^6 OB/ml	600	72–95	USSR
Pieris rapae	GV	1×10^6 OB/ml	600	89–94	USSR
Xestia c-nigrum	GV + NPV	5×10^6 OB/ml	600	85	USSR

From Tarasevich (1985): modified and enlarged.

Cydia (Carpocapsa) pomonella (GV)

Virus control of codling moth has been globally very widely tested. In CEE it was tested in Hungary, for instance, during 1976–7 (together with tests in Austria, Germany, the Netherlands, United Kingdom and Switzerland) within the framework of an IOBC/WPRS working group on 'Genetic Control of Codling Moth and Adoxophyes', which was extended to include CpGV (Dickler, 1978; Cranham *et al.*, 1980). The virus has also been tested in Poland (Ziemnicka *et al.*, 1989), where it is soon to be registered as Virin-Gyap, and the FSU. Recommended application rates in the latter area are quite modest: 9×10^{11} OB/ha (0.3 kg/ha) in two–three treatments, one at the commencement of hatching and another at full hatch (but only in regions with a single generation a year) (Kravtsov and Golyshin, 1989; Kanapatskaya *et al.*, 1990; Anon., 1994). Other findings, however (Table 13.3), indicate much higher dosage levels may be required.

Dendrolimus sibiricus (GV and CPV)

Mass outbreaks of the lasiocampid moth, *Dendrolimus sibiricus*, and of *Dendrolimus pini* lead to defoliation of pines (*Pinus* spp.) in the FSU. The epizootiology of a naturally occurring GV of *D. sibiricus* was studied by Baranovsky and Litvina (1978) and subsequently this virus was commercialized. Applications are best made in the autumn against early instars (Orlovskaya, 1989) (Tables 13.1 and 13.2). Both *D. pini* and *D. sibiricus* are susceptible to a CPV of *Dendrolimus spectabilis* imported from Japan for test purposes (Golosova and Chkhubianishvili, 1978) but this virus has not been used commercially in the FSU. Slizynski and Lipa (1975), in a thorough study of *D. pini* CPV in Poland, evaluated its infectivity to five lymantriid species including *L. dispar* and *Euproctis chrysorrhoea*.

Helicoverpa (Heliothis) armigera (NPV)

Despite its importance in warmer regions of the FSU, there appears to be very little published information on the *H. armigera* NPV, though it is a registered product (Virin-KhS). On cotton, one to two applications per generation at about 2×10^{12} OB/ha are recommended (Anon., 1993, 1994; Kravtsov and Golyshin, 1989; Simonova, Mitrofanov and Mamedov, 1984). Trials conducted in 1988 on tomato in Tajikistan gave 72% control (3.0×10^{12} to 4.8×10^{12} OB/ha, hand sprayer) and 89% control (6.0×10^{12} to 8.0×10^{12} OB/ha, Mini-Ulva (a ULV/CDA sprayer)). The high infectivity of MbMNPV, as noted elsewhere in Europe (e.g. Doyle *et al.*, 1990), may question the need to produce Virin-KhS.

Hyphantria cunea (NPV and GV)

H. cunea is a serious defoliator of poplar and other deciduous trees in Ukraina, Bulgaria, Moldavia, Slovakia and Romania. Experimental virus insecticides were developed and tested in Bulgaria (Hifantrin), Russia and Ukraina (Virin-ABB) (Table 13.2). Satisfactory results were reported from Ukraina (Sikura and Smetnik, 1980; Krasnitskaya, 1984). At present the only production (as Virin-ABB-3) is in Moldavia (Chukrii, 1993).

Leucoma (Stilpnotia) salicis (NPV)

The satin moth, widespread in the Palaearctic, is a defoliator of broadleaved trees, especially poplars and willows (*Populus* and *Salix* spp.). A CPV is known while NPVs seem to be widely present; a Polish NPV isolate was about seven times more infectious than one from Yugoslavia (Lameris *et al.*, 1985; Ziemnicka, 1981). A new insecticide, Virin-LS, has been developed and patented (Zeimnicka *et al.*, 1991a,b) and is produced by Moldavia (Chukrii, 1993).

Lymantria dispar (NPV)

The gypsy moth is an extremely important Palaearctic defoliator of broadleaved trees that has also spread extensively in the Nearctic. Protection of forests, shelter belts and orchards is required. Amongst the insect viruses produced in the FSU *L. dispar* NPV (LdMNPV) is the most important. This is registered as Virin-ENSh (and, for a higher concentration product, Virin-ENShC) and its development has in part been the fruit of a USA/USSR working team (Ignoffo, Martignoni and

Table 13.4 The areas sprayed and the effectiveness of Virin-ENSh in the control of gypsy moth, *L. dispar*, during 1972–8 in the USSR (after Orlovskaya, 1980)

Year	Area treated (ha)	Population reduction (%)
1972	750	66–96
1973	2 000	90
1974	8 800	62–88
1975	21 500	66–100
1976	25 000	99–100
1977	39 000	74–94
1978	53 000	70–100

Vaughn, 1980, 1983). In a broad comparative study it was found that Gypchek (the USA LdMNPV product) contained approximately twice as many virions per polyhedron as did Virin-ENSh. However, Virin-ENSh contained approximately twice as many nucleocapsids per virion. Restriction endonuclease digest patterns revealed significant biochemical differences between the two NPVs. In addition, Virin-ENSh is infectious for the Nearctic lymantriid forest pest *Orgyia pseudotsugata* while Gypchek is not. Surprisingly, Virin-ENSh was not infectious in *Leucoma (Stilpnotia) salicis*, although, according to Orlovskaya (1989), the virus strain used for production of Virin-ENSh was originally isolated from *L. salicis* and adapted to *L. dispar*. It may, therefore, be concluded that these two commercial products contain different BVs.

Virin-ENSh represents the largest volume of production in the FSU and also the product used to treat the largest area. Beginning in 1959, Orlovskaya (1961, 1980) pioneered the use of LdNPV and already by 1960 it had been used on 1200 ha. Usage data between 1972 and 1978 are given in Table 13.4. Simonova *et al.* (1989) reported on a new formulation which they used on 1745 ha of taiga forest in the Tatarian Autonomic Republic, Siberia; Gninenko (1984) reported that Virin-ENSh was used on 800 ha of forest in the Kazachstan Republic.

As mentioned, above, the focal (epizootic) approach to the use of BVs in pest control in the FSU was first developed by Orlovskaya (1961) with *L. dispar*. Such an approach allows dosages of Virin-ENSh as low as 0.2–2.0 ml/ha, corresponding to 0.4–4.0 LE/ha (Orlovskaya, 1976). The contamination of *L. dispar* egg batches in the spring of 1978 with 1.0% Virin-ENSh (1×10^6 OB/ml) on 50% of poplar trees reduced pest density in the same year by 65.2%, in 1979 by 96.1% and in 1980 by 99.5%; further treatment was unnecessary until the spring of 1982 (Yatsenko, 1984) (Figure 13.1). Current recommendations for implementation of this strategy of control are treatment of egg masses on 10–15% of trees. An egg mass occurrence of 2 egg masses/tree is treated with 0.2 ml (2×10^8 OBs) per ha; where the density is higher 2 ml (2×10^9 OBs) per ha is required. For overall spraying of larval infestations 100 ml (1×10^{11} OBs) per ha should be applied on instars two and three.

Yatsenko (1984) compared the cost of using Virin-ENSh with kerosene brushing of the egg masses. Using the virus to treat some egg masses (0.02 kg/ha) to create epizootic foci cost 20.26 roubles/ha while overall spraying (0.275 kg/ha) cost 50.08 roubles. Use of kerosene to brush all the egg masses (100 l/ha) in a pome orchard

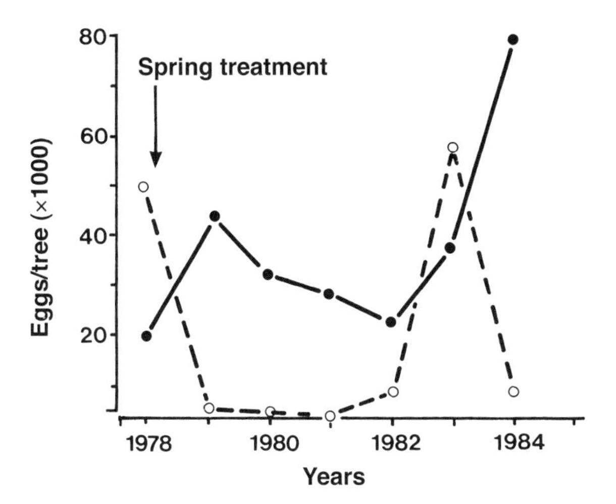

Figure 13.1 The long-term impact of local (focal) application of NPV (as Virin-ENSh) to contaminate egg masses of gypsy moth, *L. dispar*, in orchards: ●, untreated trees; ○, NPV-treated trees. (After Yatsenko, 1984.)

cost 78.33 roubles. Thus treatment of a portion of the egg masses resulted in a saving of 58.07 roubles/ha. Gninenko (1984) also reported considerable savings compared with the use of *B.t.* in the treatment of 800 ha of birch forest. Virin-ENSh cost 0.56 roubles/ha compared with 11.51 roubles for Gomelin, the *B.t.* preparation. Benefits of this order will no doubt remain as long as labour is inexpensive so that the economic balance lies in favour of treatment from the ground with brush applicators, hand-operated injectors and ground sprayers and making use of low rates of LdNPV; by contrast, application from the air is, *per se*, expensive and also requires a minimum of 50 times more virus.

During 1979–81 Likhovidov (1984) studied the impact of Virin-ENSh applications in relation to *L. dispar* outbreak phases. He employed dosages and timings in accordance with general recommendations. On an area of 1400 ha forest, he treated egg masses in areas with low, intermediate and very high populations and he also made an aerial application on 24 ha. Based on the results, he drew the following conclusions: (a) treatments were not effective against depressed populations; (b) at intermediate, increasing densities, virus application resulted in a short-term population stabilization; and (c) at peak densities virus treatments were very effective in achieving population reduction but did not prevent defoliation. From these conclusions, Likhovidov suggested several improvements in use technology, dosages, etc. and he also emphasized that lack of consideration of such factors as the population development phase may explain why some workers (e.g. Orlovskaya, 1980) reported a lack of effectiveness in virus control of *L. dispar*.

Lymantria monacha (NPV)

Known as the nun moth in Western Europe, *L. monacha* is a widespread Palaearctic species attacking trees in many coniferous genera. Outbreaks tend to be at long intervals but are commonly catastrophic. An NPV has always been a key mortality factor during both past and recent moth outbreaks (Glowacka-Pilot, 1988, 1989) throughout Europe. In the FSU, one application (dose not stated) of the NPV as Virin-PShM was recommended against instars one to two (Orlovskaya, 1989).

Malacosoma neustria (NPV)

M. neustria is a tent caterpillar polyphagous on broadleaved trees and shrubs in the Palaearctic. The first control studies were in Latvia. Applications of 1×10^7 OB/ml gave good control. Results were improved further by applying NPV at 1×10^4 OB/ml with 0.02% Dipterex or 0.05% Entobacterin (a *B.t.* preparation). Currently in the FSU, where the NPV is registered as Virin-KSh, one application of 2×10^{11} OB/ha on instars one to three is recommended (Zarinsh, 1984; Anon., 1993, 1994). About 20 kg of the commercial preparation was produced per year (Zarinsh, 1980, 1984). Virin-KSh is also used in Latvia where a formulation with lignosulphonate has been found to be especially effective (Eglite and Zarinsh, 1987; Zarinsh and Eglite, 1993).

Mamestra brassicae (NPV)

Because of the economic importance of *M. brassicae*, the NPV has been extensively studied as a microbial control agent. Wettable powder and liquid formulations of Virin-EKS (at 1×10^9 OB/ml (or per g)) have been registered and commercialized in the FSU and, presently, in Russia. On cabbage, two treatments are recommended at rates of 0.1–0.15 l/ha or kg/ha (1.8×10^{10}–3.6×10^{10} OB/ha) (Weiser *et al.*, 1986; Anon., 1993, 1994) (Tables 13.1 and 13.2). Field trial data for 1989–90 from the Voronezh and Novosibirsk regions of Russia indicate about 86% mortality when applied at 1.5×10^{11} OB/ha by tractor sprayer and about 97% mortality when applied in an aerosol at 1×10^{11} OB/ha. Two treatments are generally required. In Bulgaria, an experimental product, Mamestrin, was studied but not commercialized (Weiser *et al.*, 1986; Velichkhova-Kozhukharova, 1988).

Neodiprion sertifer (NPV)

N. sertifer is one of the simplest insects to control with virus sprays. Use of the virus has been widely explored throughout the Holarctic. In Poland Glowacka-Pilot *et al.* (1987) conducted satisfactory ground application trials but the virus is not in operational use in that country. In the FSU, studies have been in progress since at least 1970 when successful aerial applications were made (Gulii and Zimerikin, 1971). Zarinsh (1980) reported on an FSU commercial preparation (Tables 13.2 and 13.3).

Pieris brassicae and *P. rapae* (GV)

Cabbage and other cruciferous vegetables are quite important crops in the CEE region and they are effectively protected against *Pieris* spp. with a number of *B.t.*

products (Lipa, 1991). This partly explains why in this region there is relatively little research on virus use against these pests. However, Rituma (1990) reported on evaluation of an experimental product, Virin-KB, in Latvia, which gave 70–96% control and increased the yield value by 150 roubles/ha. No data on titre, formulation types and dosages were given. Vorobeva (1976) gave results of 1970–1 field experiments with an unnamed experimental GV product based on virus strains isolated from *P. brassicae* and *P. rapae*. The pest density on cabbages was 3.6–6.2 larvae/plant and the plants were sprayed using 600–1200 l/ha at a virus concentration of 1×10^6 OB/ml. Efficacy in various experiments was high and the pest population was reduced by 72.1–94.6%. Vorobeva also used a mixture of MbMNPV and *Pieris* spp. GVs at a total concentration of 1×10^6, 1×10^7 and 1×10^8 OB/ml. Such a combination provided protection of cabbage against a complex of pierid and noctuid pests.

Xestia (Amathes) c-nigrum (GV, NPV and CPV)

From field-collected *X. c-nigrum* larvae, Lipa and Ziemnicka (1971) isolated a GV which was also infectious to many other noctuids. *X. c-nigrum* larvae were not susceptible to NPVs isolated from *A. segetum* and *A. exclamationis* (Lipa, Ziemnicka and Gudz-Gorban, 1971) but they were susceptible to CPV isolated from *Noctua (Triphaena) pronuba* (Lipa, 1970). None of these viruses were field evaluated or commercialized in the CEE or FSU.

Yponomeuta (Hyponomeuta) malinellus (NPV)

Laboratory tests and 1978 summer field trials showed moderate effectiveness of Virin-YaM, an NPV preparation against third instar larvae of *Y. malinellus* (Simonova, Kazanskaya and Nikitina, 1980). After 2–3 years of storage, a wettable powder or fluid formulation showed great variations (0.0001–0.03) in LC_{80} values.

REFERENCES

Aleshina, O.A. (1980) Study of entomopathogenic viruses in the USSR. In *Characterization, Production and Utilization of Entompathogenic Viruses. Proceedings of the Second Conference of Project V, Microbiological Control of Insect Pests of the US/USSR*. C.M. Ignoffo, M.E. Martignoni and J.L. Vaughn (eds). Washington, DC: Joint Working Group on the Production of Substances by Microbiological Means. pp. 1–16.

Anon. (1993) The list of chemical and biological means for pests, diseases and weed control, growth regulators and pheromones permitted for usage in agriculture, including farms, forest and the municipal economy during 1992–1996 years. *Bio-Preparations. Zashchita Rastenii* **3**, 68–71 (in Russian).

Anon. (1994) *Preparaty Dlya Zashchity Rastenii Spisok khimicheskikh i biologicheskikh sredstv borby s vreditelyami, bolexnyami rastenii i sornyakami, regulyatorov rosta rastenii i feromonov, razreshennykh dlya primeneniya v selskom, v tom chisle fermerskom, lesnom i kommunalnom khozyaistvakh na 1992–1996*. Moskva: Kolos.

Baranovsky, V.I. and Litvina, L.A. (1978) Some epizootiological observations and gel electrophoresis of larvae of *Dendrolimus sibiricus* infected with granulosis virus. *International Colloquium on Invertebrate Pathology, Progress in Invertebrate Pathology, 1958–1978*. September, 1978, Prague, pp. 13–14.

Beglyarov, G.A., Sikura, A.I. and Orlovskaya, E.V. (eds) (1981) *Virusy Nasekomykh i Perspektiva ikh Prakticheskogo Ispolzovaniya v Zashchite Rastenii ot Vreditelei v Stranakh-Chienakh VPS/MOBB* (Moskva, 19–21 November, 1980). Moskva.

Borovik, R.V. (ed.) (1989) *Problemy Sozdaniya i Primeneniya Mikrobiologicheskikh Sredstv Zashchity Rastenii* (Velegozh, 16–18 May, 1989), Moskva.

Bozhko, N.A. (1988) Razrabotka i primenenie v narodnom khozaystve entomopatogennykh virusnykh preparatov: problemy i perspektivy. *Biotekhnoliya* **4**, 267–272.

Chukhrii, M.G. (1988) *Biologiya Bakulovirusov i Virusov Isitoplazmaticheskogo Poliedroza.* Shtinitsa: Kishinev.

Chukhrii, M.G. (1993) Priglashaem k sotrudnichestvu. *Zashchita Rastenii* **5**, 3–4.

Chukhrii, M.G., Voloshchuk, L.F. and Neshcharet, A.S. (1989a) Sokhranenie aktivnosti Virin-ABB-3 na rasteniyakh. In *Problemy Sozdaniya i Premeneniya Mikrobiologicheskikh Sredstv Zashchity Rastenii.* R.V. Borovik (ed.). Moskva, p. 193.

Chukhrii, M.G., Voloshchuk, L.F., Gyrlya, V.I. and Kozhokary, A.D. (1989b) Sostoyanie i perspektiva proizvodstva virusnykh insektitsidov v usloviakh proizvodstvennykh biolaboratorii. In *Problemy Sozdaniya i Primeneniya Mikrobiologicheskikh Sredstv Zashchity Rastenii.* R.V. Borovik (ed.). Moskva, p. 172.

Chukhrii, M.G., Voloshchuk, L.F., Gyrlya, V.I., Rusnak, A.F., Foka, M.I. and Khegaj, E.I. (1990) Teckhnologicheskie aspekty proizvodstva virusnogo inektitsida Virin-KhS-2 na osnove massovogo razvedeniya khlopkovoj sovki. In *II Simpozium Stran-Chlenov SEV po Mikrobnym Pestitsidam, Protvino, SSSR.* N.N. Urakov (ed.). p. 21.

Cranham, J.E., Gruys, P., Steiner, H. and Wildbolz, T. (eds) (1980) Biological control in orchards/biology and control of codling moth. *IOBS/WPRS Bulletin* **3**, 88.

Dickler, E. (ed.) (1978) The use of integrated control and the sterile insect technique for control of the codling moth. *Mitteilungen Biologie Bundesanstande Landbouwwetenschappen Forstwirtschaft* (Berlin-Dahlem) **180**, 120.

Doyle, C.J., Hirst, M.L., Cory, J.S. and Entwistle, P.F. (1990) Risk assessment studies: detailed host range testing of wild-type cabbage moth, *Mamestra brassicae* (Lepidoptera: Noctuidae), nuclear polyhedrosis virus. *Applied and Environmental Microbiology* **56**, 2704–2710.

Eglite, G. and Zarinsh, I. (1987) Effectiveness and properties of the virus preparation Virin-KSh with different carrier materials. *Trudy Latviiskoi Selskokhozyaistnennei Akademii* **236**, 48–53.

Glowacka-Pilot, B. (1988) Pathogenic viruses and bacteria of the nun moth (*Lymantria monacha* L.) during the outbreak 1978–84 in Poland. In *USDA Technical Report NE-123, Lymantriidae: A Comparison of Features of New and Old World Tussock Moths.* Washington, DC: USDA, pp. 401–415.

Glowacka-Pilot, B. (1989) Epizootic diseases of the nun moth (*Lymantria monacha* L.) and possibilities of its microbial control. *Prace Instytutu Badawczego Lestnictwa* **691**, 1–71.

Glowacka-Pilot, B., Ziemnicka, J., Lipa, J.J. and Chlodny, J. (1987) Laboratory and field efficacy of Virox in control of the European pine sawfly (*Neodiprion sertifer* Geoffr.) (Hymenoptera: Diprionidae). *Prace Naukowe IOR Poznan* **28**, 399–407.

Gninenko, Y.I. (1984) Effektivnost Virin-ENSh dlya zashchity ot gusenits neparnogo shelkopryda. In *Perspektivy Proizvodstva i Primeneniya Virusnykh v Selskom i Lesnom Khozyastve.* E.V. Orlovskaya (ed.). Moskva, pp. 86–91.

Golosova, M.A. and Chkhubianshvili, T.A. (1978) A cytoplasmic polyhedrosis virus of *Dendrolimus pini* and *Dendrolimus sibiricus*. In *International Colloquium on Invertebrate Pathology: Progress in Invertebrate Pathology, 1958–1978*, September, 1978, Prague, pp. 73–74.

Gulii, V.V. and Golosova, M.A. (1975) *Virusy v Zashchite Lesa ot Vrednukh Nasekomykh.* Moskva: Lesnaya Promyshlennost.

Gulii, V.V. and Rybina, S.J. (1988) *Virusnye Bolezni Nasekomykh i ikh Diagnostika.* Kishinev: Shtinitsa.

Gulii,V.V. and Zimerikin, U.N. (1971) Opyt aviatsionnoi virusologicheskoi borby s rizhim sosnovym pililshchikom. *Lesovedenie* **3**, 87–89.

Gulii, V.V., Ivanov, G.M. and Shternshis, M.V. (1982a) *Mikrobiologicheskaya Borba s Vrednymi Organizmami*. Moskva: Kolos.
Gulii, V.V., Ivanov, G.M. and Shternshis, M.V. (1982b) *Mikroorganizmy Poleznye Biometoda*. Novosibirsk: Nauka.
Ignoffo, C.M., Martignoni, M.E. and Vaughn, J.L. (eds) (1980) *Proceedings of the Second Conference of Project V, Microbiological Control of Insect Pests, of the US/USSR: Characterization, Production and Utilization of Entomopathogenic Viruses*. Clearwater Beach, FL. Washington, DC: Joint Working Group on the Production of Substances by Microbiological Means.
Ignoffo, C.M., Martignoni, M.E. and Vaughn, J.L. (eds) (1983) *A Comparison of the US (Gypchek) and USSR (Virin-ENSH) Preparations of the Nuclear Polyhedrosis Virus of the Gypsy Moth,* Lymantria dispar. Results of Research Conducted Under Project V-01.0705, Microbiological Control of Insect Pests, of the US/USSR. Washington, DC: Joint Working Group on the Production of Substances by Microbiological Means.
Kanapatskaya, V.A., Korol, I.T. and Zarinsh, I.A. (1990) Novaya forma virusnogo preparata Virin-GYaP. In *Biologicheskii Metod Zashoty Rastenij Tezisy Dokladov Nauchno-Proizvodstvennoj Konferencii*, Minsk 18–19 April, 1990. V.F. Samersov (ed.).
Kitik, U.S. (1989) Usovershenstvovanie biotekhnologii polucheniya virusnogo insektitsida Virin-OS. In *Problemy Sozdaniya i Primeneniya Mikrobiologheskikh Sredstv Zashchitu Rastenij*, 16–18 May 1989 Velegozh. R.V. Borovik (ed.). Moskva: 182 pp.
Kok, I.P. (1973) *Nukeleinovye Kisloty Nasekomykh I Virusov*. Kiev: Naukova Dumka.
Kok, I.P., Skuratovskaya, I.I. and Strokovskaya, L.I. (1980) *Molekularnye Osnovy Reproduktsii Bakulovruso*. Kiev: Naukova Dumka.
Kolyganov, Y.D., Samojlenko, L.D. and Dumanova, O.D. (1984) Osvoenie tekhnologii proizvodstva preparatov Virin-ENSh, Virin-EKS i Virin-ABB. In *Itoqi i Perspektiva Proizvodstva i Primeneniya Virusnykh Preparatov v Selskom i Lesnom Khozyastve*. E.V. Orlovskaya (ed.). pp. 15–18.
Kozar, F., Buday, E., Shimon, E. and Ilovay, Z. (1989) Istoricheskij obzor razvitiya biologicheskoj zashchity rastenii v Vengrii. *EPS/IOBC Bulletin* **25**, 7–17.
Krasnitskaya, R.S. (1984) Ispolzovaniye Virin-ABB protiv amerykanskoj beloi babochki na Ukraine. In *Itogi I Perspektivy Proizvodstva i Primeneniya Virusnykh Preparatov v Selskom i Lesnom Khozyaistve*. E.V. Orlovskaya (ed.). Moskva, pp. 106–110.
Kravtsov, A.A. and Golyshin, N.M. (1989) *Khimicheskie i Biologisheskie Sredstva Zashchitu Rasenii*. Moskva: Agropromizdat.
Lameris, A.M.C., Ziemnicka, J., Peters, D., Grijpma, P. and Vlak, J.M. (1985) Potential of baculoviruses for control of the satin moth, *Leucoma salicis* L. (Lepidoptera: Lymantriidae). *Mededelingen van de Faculteit Landbouwwetenschappen Rijksuniversiteit Gent* **50**, 431–439.
Likhovidov, V.E. (1984) Opyt primeneniya Virin-ENSh v borbe s neparnym shelkopryadom v lesakh yugozapadnoj chasti SSSR. In *Mikroorganizmy v Zashchite Rastenij*. V.V. Gulii (ed.). pp. 15–20.
Lipa, J.J. (1970) A cytoplasmic polyhedrosis virus of *Triphaena pronuba* L. (Lepidoptera, Noctuidae). *Acta Microbiologica Polonica, Series B* **2**, 237–242.
Lipa, J.J. (1975) *An Outline of Insect Pathology*. Warsaw: F.S.C. P.D.
Lipa, J.J. (1989) Lotta biologica nell'Europa orientale. In *Applicazioni Alternativa Nella Difesa delle Pianti*. Forli: Agro-Bio-Frut, pp. 139–148.
Lipa, J.J. (1990) Update: microbial pest control in Eastern Europe. *IPM Practitioner* **12**, 1–5.
Lipa, J.J. (1991) Microbial pesticides and their use in the EPRS/IOBC region (Eastern Europe). *IOBC/WPRS Bulletin*, **14**, 23–32.
Lipa, J.J. and Ziemnicka, J. (1971) Studies on the granulosis virus of cutworms *Agrotis* spp. (Lepidoptera, Noctuidae). *Acta Microbiologica Polonica, Series B* **3**, 155–162.
Lipa, J.J., Ziemnicka, J. and Gudz-Gorban, A.P. (1971) Electron microscopy of nuclear polyhedrosis virus from *Agrotis segetum* Schiff. and *A. exclamationis* L. (Lepidoptera, Noctuidae). *Acta Microbiologica Polonica, Series B* **3**, 55–61.

Orlovskaya, E.V. (1961) Rezultaty polevykh ispytanii polyedrennykh virusov protiv neparnogo shelkopryada. *Byul. Vses. Inst. Zash. Rastenij* **3–4**, 54–57.

Orlovskaya, E.V. (1976) Aktualnye voprosy proizvodstva i primeneniya entomopatogennukh virusnykh preparatov. In *Mikrobiologicheskie Metody Zashchity Rastenii; Tezisy Dokladov 1-j Vsesoyuznoj Nauchnoj Konferencii, Kishinev*, 4–7 October 1976. A.I. Sikura (ed.). pp. 187–178.

Orlovskaya, E.V. (1980) Production of preparations for controlling the cabbage moth (*Mamestra brassicae*) and gypsy moth (*Lymantria dispar* L.) based on nuclear polyhedrosis viruses. In *Characterization, Production and Utilization of Entomopathic Viruses. Proceedings of the Second Conference of Project V, Microbiological Control of Insect Pests of the US/USSR*. C.M. Ignoffo, M.E. Martignoni and J.L. Vaughn (eds). Washington, DC: Joint Working Group on the Production of Substances by Microbiological Means. pp. 54–76.

Orlovskaya, E.V. (ed.) (1984) *Itogi i Perspektivy Proizvodstva i Primeneniya Virusnykh Preparatov v Selskom i Lesnom Khozyajstve*. Moskva.

Orlovskaya, E.V. (1989) Ispolzovanie virusov v borbe s khvoelistogrizuschimi nasekomymi v USSR. *IOBC/EPRS Bulletin* **27**, 69–72.

Petre, Z. (1981) Issledovaniya entomopatogennykh virusov, isolirovannykh v Rumunii. In *Virusy Nasekomykh i Perspektivy ikh Prakticheskogo Ispolzovaniya v Zashchite Rastenii ot Vreditelei y Stranakh-Chienakh VPS/MOBB*, Moskva, 19–21 November 1980. G.A. Beglyarov, A.I. Sikura and E.V. Orlovskaya (eds). pp. 21–31.

Rimmington, A. (1989) The production and use of microbial pesticides in the USSR. *International Industrial Biotechnology* **9**, 10–14.

Rituma, I.A. (1990) Ispolzovanie bakulovirusov v borbe s vreditelyami selskokhozyastvennykh kultur. In *II Simpozium Stran-Chlenov SEV Po Mikrobnym Pestitsidam, 15–19 October 1990, Protivno SSRR*, Moskva: Tezisy Dokladov.

Rutskaya, V.I. (1990) Tekhnologicheskij protsess proizvodstva virusnogo preparata Virin-KhS. In *II Simpozium Stran-Chlenov SEV po Mikrobnym Pestitsidam*, 15–19 October 1990. N.N. Urakov (ed.). Moskva: Protivno SSRR, Tezisy Dokladov.

Samersov, V.F. (ed.) (1990) *Biologicheskij Metod Zashchity Rastenij. Tezisy Dokladov Nauchno-Proizvodstvennoj Konferencii*. Minsk 18–19 April 1990, Minsk.

Shekhurina, T.A. (1985) Prognozirovanie virusnykh epizootii u vrednykh nasekomykh s tselju sokrashcheniya khimicheskikh obrabotok. *IOBC/EPRS Bulletin* **12**, 69–74.

Sikura, A.I. (ed.) (1976) *Mikrobiologicheskie Metody Zashchity Rastenii; Tezisy Dokaladov 1-j Vsesoyuznoj Nauchnoj Konferencii*. (Kishinev 4–7 October 1976).

Sikura, A.I. and Smetnik, A.I. (1980) Use of viruses against the fall webworm (*Hyphantria cunea* Drury). In *Characterization, Production, and Utilization of Entomopathogenic Viruses. Proceedings of the Second Conference of Project V, Microbiological Control of Insect Pests of the US/USSR*. C.M. Ignoffo, M.E. Martignoni and J.L. Vaughn (eds). Washington, DC: Joint Working Group on the Production of Substances by Microbiological Means. pp. 203–215.

Simonova, E.Z., Kazanskaya, V.A. and Nikitina, N.I. (1980) Otsenka virusnogo insektitsida Virin-YaM. In *Biologicheskij Metod Borbu s Vrednymi Nasekhomymi i Kleshchami*. Riga: Zinatis, pp. 62–64.

Simonova, E.Z., Mitrofanov, V.B. and Mamedov, Z.M. (1984) Rezultaty issledovanya virusnogo preparata Virin-KhS v borbe s khlopkovoi sovkoi. In *Itogi i Perspektivy Proizvodstvai i Primeneniya Virusnykh Preparatov v Selskom i Lesnom Khozaystve*. E.V. Orlovskaya (ed.). pp. 97–100.

Simonova, E.Z., Novikova, L.K., Nikitina, N.I. and Konstantinov, Y.N. (1989) Virusnyi insektitsid dlya zaschchity lesa. In *Problemy Sozdaniya i Primeneniya Mikrobiologischeskikh Sredstv Zashchity Rastenii*. R.V. Borovik (ed.). Moskva, pp. 196.

Slizynski, K. and Lipa, J.J. (1975) A cytoplasmic polyhedrosis virus of pine moth *Dendrolimus pini* L. (Lepidoptera: Lasiocampidae). *Prace Naukowe I.O.R. Poznan* **17**, 29–60.

Taresevich, L.M. (1975) *Virusy Nasekomykh*. Moskva: Nauka.

Taresevich, L.M. (1985) *Virusy Nasekomykh Sluzhit Cheloveku*. Moskva: Nauka.

Taresevich, L.M. and Guliij, V.V. (1985) Sostoyanie i perspektivy primeneniya entomopatogennykh virusov v zashchite rastenii. *IOBC/EPRS Bulletin* **12**, 64–69.

Urakov, N.N. (ed.) (1990) *II Simpozium Stran-Chlenov SEV Po Mikrobnym Pestitsidam*, 15–19 October 1990. Moskva: Protivno SSRR, Tezisy Dokladov.

Velichkova-Kozhkharova, M. (1988) Mamestrin-virusnyi preparat dlya borby s sovkami *Mamestra brassicae* L., *Mamestra oleracea* L., *Phytometra gamma* L. In *Pervyi Bolgaro-Sovetskii Simpozium s Mezhdunarodnum Uchastiem Po Mikrobialnum Pestitsidam – Programma Rezyume* 24–26 October 1988. Plovdiv.

Videnova, E., Velichkova, M. and Encheva, L. (1981) Izuchenie raspostraneniya virusnykh boleznei nasekomykh v Bulgarii i ikh rol v reguliovanii chislennosti vrediteley rastenii. In *Virusy Nasekomykh i Perspektivy ikh Prakticheskogo Ispolzovaniya v Zashchite Rastenii ot Vrediteley v Stranakh-Chlenakh VPS-MOBB*, Moscova 19–21 November 1980. G.V. Beglyarov, A.I. Sikura and E.V. Orlovskaya (eds). pp. 15–20.

Vorobeva, N.N. (ed.) (1974) *Virusy Nassekomykh*. Novosibirsk: Nauka.

Vorobeva, N.N. (1976) *Entomopatogennye Virusy*. Novosibirsk: Nauka.

Weiser, J., Videnova, E., Kandybin, N.V. and Smirnov, O.V. (1986) Tekhnicheskaya kharakteristika i standartizatsiya mikrobnykh entmotsidnykh preparatov. *IOBC/EPRS Bulletin* **16**, 44–52.

Yatsenko, V.G. (1984) Primenenie virusnogo preparat Virin-ENSh v borbe s neparnym shelkopryadom v sadakh lesostepnoi zony Ukrainy. In *Itogi i Perspektivy Proizvodstva i Primeneniya Virusnykh Preparatov v Selskom i Lesnom Khozyastve*. E.V. Orlovskaya (ed.). pp. 91–96.

Yatsenko, V.G., Rudnev, A.G., Vasileva, V.L., Trusov, V.I., Bozhko, N.A. and Karavaev, V.S. (1990) Effektivnost i bezopastnost novykh preparativnykh form Virin-NSh protiv neparnogo shelkopryada v sadakh. In *II Simpozium Stran-Chlenov SEV po Mikrobnym Pestitsidam* , 15–19 October 1990. N.N. Urakov (ed.). Moskva: Protivno SSRR, Tezisy Dokladov, pp. 126–127.

Zarinsh, I.A. (1980) Use of entomopathogenic viruses. In *Characterization, Production and Utilization of Entomopathic Viruses. Proceedings of the Second Conference of Project V, Microbiological Control of Insect Pests of the US/USSR*. C.M. Ignoffo, M.E. Martignoni and J.L. Vaughn (eds). Washington, DC: Joint Working Group on the Production of Substances by Microbiological Means. pp. 77–90.

Zarinsh, I.A. (1984) Rezultaty ispytaniya virusnogo preparata Virin-KSh raznykh form v 1981–1982 v usloviyakh Pribaltiki. In *Itoqi i Perspektivy Proizvodstva i Primeneniya Virusnykh Preparatov v Selskom i Lesnom Khozyaystve*. E.V. Orolovskaya (ed.). Moskva, pp. 78–81.

Zarinsh, I. and Eglite, G. (1993) Investigations of entomopathogenous viruses in Latvia and their potential as pest control agents. *Latvijas Zinatnu Akademijas Vestas B. ala, Dabaszinatnis* **12**, 49–53.

Ziemnicka, J. (1981) Studies on nuclear and cytoplasmic polyhedrosis viruses of the satin moth (*Stilpnotia salicis* L.) (Lepidoptera: Lymantriidae). *Prace Naukowe I.O.R. Poznan* **13**, 75–142.

Ziemnicka, J., Konopacka, W., Lipa, J.J., Korol, I.T. and Kozlowski, J. (1989) Infectivity of Virin-Gjap for Polish population of codling moth. *Materialy XXIX Sesji Naukowej I.O.R. Poznan* Cz. 2 – Postery, pp. 39–43.

Ziemnicka, J., Lipa, J.J., Chukhrii, M.G., Voloshchuk, L.F. and Gyrlya, V.I (1991a) Method of production of insecticidal preparation: Virin-LS. Patent No. 289570. Poland.

Ziemnicka, J., Lipa, J.J., Chukhrii, M.G., Voloshchuk, L.F. and Gyrla, V.I. (1991b) Insecticidal preparation: Virin-LS. Patent No. 289571. Poland.

14 Indian subcontinent

S. EASWARAMOORTHY

PERSPECTIVE

In the Indian subcontinent (India, Pakistan, Bangladesh, Sri Lanka, Andaman and Maldive Islands) virus control studies have been concentrated on:

Amsacta albistriga	peanut red hairy caterpillar
Chilo infuscatella	sugarcane shoot borer
Helicoverpa armigera	gram pod borer
Mythimna separata	oriental armyworm
Oryctes rhinoceros	coconut rhinoceros beetle
Spodoptera litura	tobacco cutworm.

Some attention has been paid to *Agrotis ipsilon* in Pakistan and to *Plocaederus ferrugineus* (Coleoptera; Lamiidae) in Sri Lanka. As far as is known there has been no virus pest control work in Bangladesh, though undoubtedly work elsewhere in the region will be relevant to the agricultural problems of this country.

BVs have been detected in 33 pest Lepidoptera and CPVs in three.

The most extensive studies have been concentrated on *H. armigera* (on chickpea, pigeonpea, field bean, sunflower, cotton and tomato), *S. litura* (on cotton, tobacco, banana, castor, blackgram and cauliflower), *A. albistriga* (groundnut) and *M. separata* (sorghum). Pigeonpea pests and their viral and chemical control have been reviewed (Anon., 1986).

A particular feature of work in this region has been the rather large number of laboratory and field studies on the joint use of BVs with chemical insecticides, either in simultaneous applications or in lag-phase applications, notably in *H. armigera* and *S. litura* control. Such an approach has been stimulated by the need to preserve beneficial insects (e.g. by seeking to identify synergistic associations of BVs with sublethal concentrations of chemical insecticides) or by the need to treat multiple pest infestations that include some insect species not responsive to BVs.

As an aid to development of control methods, there have been some studies on the environmental relationships of BVs, for example, in terms of temperature and solar radiation tolerances, persistence in soils, and interactions with parasitoidal insects and avian insectivores. Other studies are more overtly related to the

Table 14.1 Yield of BV OBs from final instar larvae of selected species

Species	OB dose/ larva	OB yield/ larva	Productivity ratio	Reference
NPV				
A. albistriga	8.5×10^6	7.1×10^9	8.4×10^2	Narayanan *et al.* (1978a)
H. armigera	1.1×10^4	2.6×10^9	2.4×10^5	P.J. Rabindra (unpublished data)
S. litura	1.0×10^7	5.3×10^9	5.3×10^2	Santharam (1986)
GV				
Chilo infuscatellus	1.0×10^5	2.1×10^9	2.1×10^4	S. Easwaramoorthy and G. Santhalaksh (unpublished data)

development of control registration packages, for example non-target host susceptibility tests, formulation, including the use of a quite wide range of potential adjuvants, and compatibility with chemical insecticides and fungicides.

However, in this geographical region no virus has yet been developed to the point of registration and commercial or other form of widespread use, though deliberate releases of OrV for rhinoceros beetle control on coconut palms in Kerala, the Maldive Islands and the Andaman Islands do represent an example of the geographically widespread use of an insect virus.

VIRUS PRODUCTION

Small-scale production of virus was achieved by field collection of *A. albistriga* and *C. infuscatellus* larvae or by rearing on a semi-synthetic diet for *H. armigera, M. separata* and *S. litura* employing oral, foliage or diet contamination with BV inoculum. Table 14.1 states the initial doses, yields and productivity ratios for final instar larvae of some pest species.

Formulation

Experimental dust formulations of *H. armigera* and *S. litura* NPVs have incorporated talc or kaolin (china clay) as fillers together with a dispersing agent. For *H. armigera* it was found that there was no advantage in a wettable powder formulation compared with unformulated virus. A dust formulation appeared to inhibit larval feeding and so had reduced efficiency (Ethiraju, Rabindra and Jayaraj, 1988). A ULV formulation also had no advantage over unformulated virus (Muthiah, 1988). All formulations studied rapidly reduced stability compared with unformulated virus.

The viruses were invariably used with such surfactants as Tween 20, Triton X-100 or Teepol. Tests with other adjuvants have provided conflicting results and probably require further evaluation. For instance, Rao, Rao and Chandra (1987) found increased efficiency of *S. litura* NPV with tannic acid (0.25%) or boric acid (0.35%) as did Chaudhari (1992). Addition of boric acid to *H. armigera* NPV in one study did not improve performance (Muthiah, 1988) but in another appeared to be advantageous (Chundarwar, Pawar and More, 1990). Addition of other substances gave advantage either in the laboratory or in field tests: crude sugar (0.5%), cotton seed kernel powder (1.0%), peanut oil cake (3.0%), chickpea flour (1.0%) (Rabindra and Jayaraj, 1988a), larval extracts of *H. armigera*, *S. litura* and *Corcyra cephalonica* (4.0%), whole milk (20.0%), whole egg homogenate (10.0%), egg yolk (10.0%), egg white (10.0%) and coconut milk (20.0%) (Rabindra and Jayaraj, 1988b; Rabindra *et al.*, 1989). Baits containing extracts of *H. armigera* larvae with either chickpea or maize flour were more effective than several other substances (soybean flour, sugars, cotton seed kernel flour) (Dhandapani, Jayaraj and Rabindra, 1994).

In contrast, no significant improvement of *H. armigera* NPV performance was observed when high volume mixes included crude sugar, cotton seed kernel extract, whole egg homogenate, whole milk and, as whitening agents, Ranipal and Robin blue or when ULV mixes included milk powder, sucrose, lactose or boric acid (Muthiah, 1988).

THE USE OF VIRUSES WITH PESTICIDES

No adverse effects were observed when *S. litura* NPV was mixed with the fungicides mancozed (0.15%) or copper oxychloride (0.125%) immediately before spraying, but 0.025% carbendizan reduced infectivity. However, for all these fungicides, suspension for 24 h greatly reduced infectivity (Sachithanandam, 1988).

In laboratory bioassays, *H. armigera* NPV combined with 5 ppm DDT or 10, 20 or 50 ppm pyrethrin produced supplemental effects. However, potentiation occurred with pyrethrin at 5 ppm. Gamma-HCH (BHC) produced an additive effect only at 50 ppm while malathion was antagonistic to the virus (Komolpith and Ramakrishnan, 1978). Among 12 insecticides tested, chlorinated hydrocarbons and carbamates were synergistic to the virus while some organophosphates (but not diazinon) were antagonistic (Chaudhari and Ramakrishnan, 1983). With *S. mauritia* NPV, sublethal doses of fenthion and endosulfan produced potentiation, permethrin produced a subadditive action while phosphamidon and fenitrothion showed antagonism (Mathai, 1982). The action of the insecticide–virus mixture varied with both the concentration of the insecticide and the virus; it also varied with larval age (Table 14.2) (Santharam, 1986).

For comments on the synergistic benefits of applying viruses and some chemical insecticides in temporal lag phase, see Savanurmath and Mathad (1981).

SAFETY TESTS

Extensive safety tests of *Spilosoma obliqua* NPV on rats included acute oral and dermal toxicity, eye and primary skin irritation, acute inhalation, allergenicity and subacute dietary administration (Battu and Ramakrishnan, 1983). Acute toxicity tests with *H. armigera* NPV unformulated (Narayanan, 1979) and formulated (Muthiah, 1988), *M. separata* NPV (Vijayakumar and Mathad, 1984a,b) and GVs of *C. infuscatellus* and *C. saccariphagus indicus* (Easwaramoorthy, 1984) on rats

Table 14.2 Interaction between chemical insecticides and *S. litura* NPV (Santharam 1986)

NPV dose	Insecticide and dose	Interaction type[a]	
		Fourth instar	Fifth instar
LC_{25}	Chlorpyridos LC_{25}	S	S
	Monocrotophos	A	S
	Fenverlate	S(M)	S
	Endosulfan	S(M)	S
LC_{25}	Chlorpyridos LC_{50}	AN	S(M)
	Monocrotophos	AN	AN
	Fenverlate	AN	AN
	Endosulfan	S(M)	AN
LC_{50}	Chlorpyridos LC_{25}	S(M)	S
	Monocrotophos	S	S
	Fenverlate	AN	S
	Endosulfan	S	AN

[a] A, additive; AN, antagonistic; S, synergistic; M, moderate.

and of *A. albistriga* NPV on mice showed no adverse effects. Nor were rats allergic to dermal exposure to GVs or NPV (Vijayakumar and Mathad, 1978). NPVs of *H. armigera* (Narayanan, 1979), *S. litura* (Regupathy *et al.*, 1978) and *A. albistriga* (Narayanan *et al.*, 1978b) were safe to chickens in acute feeding tests. The NPVs of *A. albistriga* and *H. armigera* were also tested on fish and found to be harmless (Narayanan *et al.*, 1977; Narayanan, 1979).

Fairly extensive tests have shown a lack of response to the NPVs and GVs mentioned here by a range of beneficial insects and mites. In addition, the NPVs of *A. albistriga*, *H. armigera*, *M. separata*, *S. litura* and *S. obliqua* and GVs of *Chilo* spp. have been variously tested on a range of economic species of silkmoth (*Antheraea cerana indica, Antheraea mylitta, Bombyx mori* and *Samia (Philosamia) ricini*), all with negative results.

FIELD EXPERIMENTATION

Agrotis ipsilon (GV)

In a joint project between Denmark and Pakistan, *A. ipsilon* (greasy cutworm) control on tobacco was studied using a Danish *A. segetum* GV strain. Spraying tobacco seedlings with concentrations from 5×10^7 to 1×10^9 OB/ml at 60 to 250 ml/m^{-2} reduced second instar larval damage by 72–100% (Chaudhry, Shah and Gul, 1976; Shah *et al.*, 1979; Zethner *et al.*, 1983). Addition of activated charcoal was not an improvement (Shah *et al.*, 1979).

Amsacta albistriga (NPV)

A. albistriga is a considerable pest of millets and field legumes. In India, in areas not receiving chemical insecticidal treatment, *A. albistriga* has a high rate of parasitism, especially by Braconidae (Hymenoptera), Tachinidae and a member of the Phoridae (Diptera). An homologous NPV gives very effective control (Jayaraj *et al.*, 1977; Chandramohan and Kumaraswami, 1979; Gunathilagaraj and Babu, 1987).

Amsacta atkinsoni (NPV)

Application of virus at 250 LE/ha (LE, larval equivalent) significantly reduced the larval population of *A. atkinsoni* inside field bean pods but not on the foliage. A better result was obtained when 125 LE/ha was sprayed with 0.035% endosulfan twice at a 15 day interval. There was a significant decrease in population and pod damage together with a yield increase (Narayanan, 1987).

Chilo infuscatellus (GV)

In sugarcane, application of GV at 1×10^7 to 1×10^9 OB/ml on days 35, 50, 65 and 80 after planting reduced *C. infuscatellus* larval incidence below the economic injury level of 20.0% (Easwaramoorthy, 1984; Easwaramoorthy and Santhalakshmi, 1988). Application at high volume was more effective than at low volume. Addition of charcoal did not improve the efficacy of the virus.

Helicoverpa armigera (NPV)

Field experiments have been conducted on chickpea (*Cier arietinum*), pigeonpea (*Cajanus cajan*), field bean (*Vicia faba*), sunflower (*Helianthus annuus*), cotton (*Gossypium* spp.) and tomato (*Lycopersicon esculentum*) (Jayaraj, Rabindra and Narayanan, 1989).

Chickpea

Reductions in pod damage by *H. armigera* following virus application have been reported (Makode, 1978; Bakwad, 1979; Mistry *et al.*, 1984; Jayaraj, Rabindra and Santharam, 1987). Application thrice weekly in the evening against early instars at 125 or 250 LE/ha (1 LE = 6×10^9 OBs) significantly reduced larval populations, giving a yield increase equal to that obtained from 0.07% endosulfan (Narayanan, 1979). ULV/CDA application in 20.0% crude sugar was as effective as when used at high volume or as 0.07% endosulfan (Rabindra and Jayaraj, 1987). NPV at 250 LE/ha with 262 g a.i./ha (a.i., active ingredient) endosulfan given as a combined application or given as a dose of virus alone followed by the dose of endosulfan 5 days later were as effective as 524 g a.i. endosulfan/ha (Rabindra and Jarayaj, 1987). In another study, NPV at 375 LE/ha was comparable with 0.05% carbaryl (Santharam and Balasubramanian, 1982). These promising results are probably due to the exposed nature of early larval feeding on chickpea being conducive to acquisition of a high dose and, to a lesser extent, of the older larvae feeding also on leaves, in addition to flower buds and pods where they acquire less virus (Rabindra and Jayaraj, 1987).

Table 14.3 Comparative efficacy of NPV and endosulfan, separately and in combination, against *H. armigera* on pigeonpea at five Indian locations (Jayaraj *et al.*, 1989)

Treatments	Mean pod damage (%) at				
	Madurai	Coimbatore	Bangalore	Ludhiana	Hyderabad
NPV 250 LE/ha	9.8(b)	33.3(b)	24.1(b)	30.4(b)	15.4(b)
NPV 250 LE/ha + endosulfan 0.035%	–	24.5(a)	15.4(a)	20.5(a)	7.4(a)
Endosulfan 0.07%	6.0(a)	24.4(a)	15.4(a)	19.6(a)	7.7(a)
Control	19.1(c)	51.6(c)	28.5(c)	25.3(b)	38.8(c)

In columns, means bearing same letter are not significantly different.

Pigeonpea

H. armigera has been reported to be less effective on pigeonpea infestations than on chickpea (Chelliah *et al.*, 1978) even at 375 LE/ha (Santharam, Balasubramanian and Chelliah, 1981). However, very extensive trials at five locations in India demonstrated the value of application of NPV at 250 LE/ha in combination with a reduced dose (0.035%) of endosulfan (Jayaraj *et al.*, 1989) (Table 14.3). In a comparison of NPV in a wettable powder, NPV in a ULV application and endosulfan there were no treatment differences (Muthiah and Rabindra, 1991). However, the fluid economy of ULV applications will often be found valuable.

Joint application of NPV with the insecticides chlorpyriphos and cypermethrin were also effective (Srinivas, 1987).

Field bean

In *H. armigera* on field beans, a combination of 125 LE NPV/ha with 0.035% endosulfan twice at a weekly interval (Jayaraj, 1981; Jayaraj *et al.*, 1987) or NPV with 0.006% cypermethrin (Srinivas, 1987) gave appreciable protection of pods.

Cotton

The best protection of squares and bolls of cotton was achieved with two high-volume applications of *H. armigera* NPV at 125 LE NPV/ha with 0.035% endosulfan at 1 week's interval. Four rounds of ULV applications of 450 LE NPV/ha with 350 g a.i. endosulfan/ha with a variety of adjuvants (4.0% larval extract or 10.0% whole milk plus 15.0% crude sugar or 2.5% cotton seed kernel powder plus 17.5% crude sugar) were as effective as 700 g a.i. endosulfan/ha in reducing damage and increasing yield (Dhandapani, Jayaraj and Rabindra, 1987).

Sunflower

A single application of *H. armigera* NPV to the flower heads of sunflowers at 250 LE NPV/ha with 0.035% endosulfan 3 days later was comparable with two applications of the virus at the same rate (Rabindra, Jayaraj and Balasubramanian, 1985).

Tomato

NPV at 100 LE/ha caused heavy *H. armigera* larval mortality on tomatoes (Mistry *et al.*, 1984).

Mythimna separata (NPV)

NPVs of *M. separata* from China, India and Japan were characterized biochemically and by infectivity (Hatfield and Entwistle, 1988), the Indian strain probably being that employed in laboratory and field work by Neelgund (1977). His trials in Karnataka on sorghum (*Sorghum bicolor*) achieved a field mortality of 43.9% at 2.5×10^6 OB/ml and 71.4% at 1×10^7 OB/ml.

Oryctes rhinoceros (OrV)

In Minicoy Island (northern Maldives) OrV was employed by releasing infected adults into an apparently disease-free population. Releases were made in April, 1983 and observations 2 years later showed a larval population reduction of >90% and this was subsequently maintained at around 86%. There was an associated diminution in damage (Mohan, Jayapal and Pillai, 1986). In 1987, the virus was introduced into Androth (Lakshadweep), the Andaman and Nicobar Islands and, a year later, into parts of Kerala State (G.B. Pillai, personal communication). There was also an extensive introduction campaign in the Maldive Islands during 1984 (FAO, 1986). The Kerala isolate of OrV was initially released at four locations along the chain of the Andaman Islands; at all places it reduced coconut palm damage by around 90% in 43 months. The rate of spread of the virus was about 1 km/month (Chapter 15) and beetle populations have remained at a low level (Jacob, 1996).

Plocaederus ferrugineous

P. ferrugineous is a root and stem borer of cashew (*Anacardium occidentale*) against which OrV was tested in Sri Lanka by releasing infected *P. ferrugineus* adults or by placing a mixture of virus with sawdust at the base of infested stems. The latter gave the best results (Rajapakse and Jeevaratnam, 1982).

Spodoptera litura (NPV)

Control of *S. litura* has been studied on banana (*Musa* spp.), blackgram (*Vigna mungo*), castor, cauliflower (*Brassica oleracea* v. *botrytis*), cotton and tobacco using the NPV alone or with a low concentration of insecticide (Table 14.4). Larval virus mortality varied between 50 and 90% depending on dose and age. In some experiments, reduction in plant damage was assessed. The cotton leaf damage index (0–5 scale) was reduced from 3.84 in the control to 2.03 following NPV spraying at 250 LE/ha and the number of larvae per five plants fell from 75.7 to 21.6 (Jayaraj *et al.*, 1981). On tobacco, application of 9.3×10^7 OB/ml, or half this dose with 0.02% carbaryl, significantly reduced damage to seedlings

Table 14.4 Efficiency of NPV against *S. litura* on various crops in India

Crop	Dosage (NPV/ha)	Mortality of larvae (%)	Reference
Cotton			
Instar II and III	250–500 LE	80.6–89.7	Jayaraj *et al.* (1981)
Instar IV and V	250 LE	58.5	Jayaraj *et al.* (1981)
Combined dose with metho-midophos (1.25 l) or with molasses (5%)	125 LE	64.7–72.0	Jayaraj *et al.* (1981)
Tobacco	125 LE	62.8	Santharam and Balasubramanian (1980)
	250 LE	86.4	Santharam and Balasubramanian (1980)
Banana	1.2×10^{12}, 2.4×10^{12} or 3.6×10^{12} OB	80.1–90.4	Santharam *et al.* (1978)
Castor	1.3×10^{12} OB	83.2	Mahadevan (1978)
Blackgram	1.55×10^{12} OB	87.1	Mahadevan and Kumaraswami (1980)

(Ramakrishnan, 1976; Ramakrishnan *et al.*, 1981). In later studies (Chari, Bharpoda and Patel, 1985) the use of virus at 100 LE/ha, along with castor as a trap crop on the plot borders, was suggested as a tobacco nursery management practice. Virus application on cauliflower reduced the leaf area damage from 64.0 cm^2 (control) to 37.9 cm^2: addition of activated charcoal resulted in a further decrease in leaf damage to 22.8 cm^2 (Dhandapani and Jayaraj, 1989). Improved control was achieved when the virus was combined with low doses of insecticides (Mahadevan, 1978; Santharam and Balasubramanian, 1980). However, on cauliflower a combination of NPV and 50 ppm endosulfan did not increase foliar protection over NPV alone while protection was only slightly improved when 10 ppm DDT was added to NPV (Chaudhari and Ramakrishnan, 1980). The LC_{50} of *S. litura* NPV was reduced about 10 times by incorporation of 0.5% boric acid and about 20 times with 1.0% boric acid, which also reduced the LT_{50} by about 40% (Chaudhari, 1992).

FUTURE DEVELOPMENT

Decisions need to be made on whether viruses are to be produced on a smaller scale at local levels (e.g. farmer cooperatives) or on a more central, larger scale, or both. Development of appropriate production and formulation protocols can then be addressed. For some pest species, time and effort may be saved by adopting, and possibly adapting, technological findings from other geographical regions of the world; for example, studies of *H. armigera* can make use of North American advances on *H. virescens/zea* NPV, work in The People's Republic of China on *M. separata* and *S. litura* and, by analogy with the very closely related *S. littoralis*, the very complete and appropriate NPV production methods being practised in Egypt and the *Agrotis* spp. development in Europe and Brazil. Little, if any, further work should be necessary on OrV control of coconut rhinoceros beetle, as this has been investigated and successfully practised in a very wide geographic context.

REFERENCES

Anon. (1986) *ICRISAT Annual Report, 1985, Pigeonpea: Insect Pests*. pp. 190–197.

Bakwad, D.G. (1979) Studies on nuclear polyhedrosis virus infections in *Heliothis armigera* Hübner and *Anomis sabulifera* (Guenee). MSc thesis, Marathwada Agricultural University, Parbhani, India.

Battu, G.S. and Ramakrishnan, N. (1983) *Safety Testing of the Nuclear Polyhedrosis Virus of* Diacrisia obliqua *(Walker)* (Technical Bulletin). New Delhi: Division of Entomology, Indian Agricultural Research Institute.

Chandramohan, N. and Kumaraswami, T. (1979) Comparative efficiency of chemical and nuclear polyhedrosis virus in the control of groundnut red hairy caterpillar, *Amsacta albistriga* (Walker). *Science and Culture* **45**, 202–204.

Chari, M.S., Bharpoda, T.M. and Patel, S.N. (1985) Studies on integrated management of *Spodoptera litura* Fb., in tobacco nursery. *Tobacco Research* **11**, 93–98.

Chaudhari, S. (1992) Formulation of nuclear polyhedrosis virus of *Spodoptera littoralis* with boric acid. *Indian Journal of Entomology* **54**, 202–206.

Chaudhari, S and Ramakrishnan, N. (1980) Field efficacy of baculovirus and its combination with sub-lethal dose of DDT and endosulfan on cauliflower against tobacco caterpillar (*Spodoptera litura* (Fabricius)). *Indian Journal of Entomology* **42**, 592–596.

Chaudhari, S. and Ramakrishnan, N. (1983) Effect of insecticides on the activity of nuclear polyhedrosis virus of *Spodoptera litura* (Fabricius) in laboratory bioassay tests. *Journal of Entomological Research* **7**, 173–179.

Chaudhey, M.I., Shah, B.H. and Gul, H. (1976) Preliminary studies on the biology of greasy cutworm *Agrotis ipsilon* Roth, and its susceptibility to granulosis virus of *A. segetum* D&S from Denmark. *Pakistan Journal of Forestry* **26**, 140–144.

Chelliah, S., Surulivelu, T., Balasubramanian, G. and Vasudeva Menon, P.P. (1978) Studies on the control of redgram (*Cajanus cajan* L.) pod borers. *Madras Agricultural Journal* **65**, 183–187.

Chundarwar, P.M., Pawar, V.M. and More, M.R. (1990) Efficacy of nuclear polyhedrosis virus in combination with boric acid and tannic acid against *Helicoverpa armigera* on chickpea. *International Chickpea Newsletter* **27**, 17–18.

Dhandapani, N. and Jayaraj, S. (1989) Efficacy of nuclear polyhedrosis virus formulation for the control of *Spodoptera litura* (Fab.) on chillies. *Journal of Biological Control* **3**, 47–49.

Dhandapani, N., Jayaraj, S. and Rabindra, R.J. (1987) Efficacy of ULV application of nuclear polyhedrosis virus with certain adjuvants for the control of *Heliothis armigera* (Hbn.) on cotton. *Journal of Biological Control* **1**, 111–117.

Dhandapani, N., Jayaraj, S. and Rabindra, R.J. (1994) Nuclear polyhedrosis virus bait formulations against larvae and adults of *Heliothis armigera* (Hbn.) *Indian Journal of Experimental Biology* **32**, 294–295.

Easwaramoorthy, S. (1984) Studies on the granulosis viruses of sugarcane shoot borer, *Chilo infuscatellus* Snellen and internode borer, *C. sacchariphagus indicus* (Kapur). PhD thesis, Tamil Nadu Agricultural University, Coimbatore, India.

Easwaramoorthy, S. and Santhalakshmi, G. (1988) Efficacy of granulosis virus in the control of shoot borer, *Chilo infuscatellus* Snellen. *Journal of Biological Control* **2,** 26–28.

Ethiraju, S., Rabindra, R.J. and Jayaraj, S. (1988) Laboratory evaluation of certain formulations of nuclear polyhedrosis virus against the larvae of *Heliothis armigera*. *Journal of Biological Control* **2**, 21–25.

FAO, Maldives (1986) FAO project statement on biological control of rhinoceros beetle. *Quarterly Newsletter, Asian and Pacific Plant Protection Commission* **29**, 46–49.

Gunathilagaraj, K. and Babu, P.C.S. (1987) *Amsacta albistrigata* Wlk. on groundnuts and other crops. *FAO Plant Protection Bulletin* **35**, 63–64.
Hatfield, P.F. and Entwistle, P.F. (1988) Biological and biochemical comparison of nuclear polyhedrosis virus isolates pathogenic for the oriental armyworm, *Mythimna separata* (Lepidoptera: Noctuidae). *Journal of Invertebrate Pathology* **52**, 168–176.
Jacob, T.K. (1996) Introduction and establishment of baculovirus for the control of rhinoceros beetle, *Oryctes rhinoceros* (Coleoptera: Scarabaeidae) in the Andaman Islands (India). *Bulletin of Entomological Research* **86**, 257–262.
Jayaraj, S. (1981) Control of gram caterpillar on lab-lab and bengalgram with nuclear polyhedrosis virus. *Tamil Nadu Agricultural University Newsletter* **11**, 1.
Jayaraj, S., Sundaramurthy, V.T., Mahadevan, N.R. and Swamiappan, M. (1977) Relative efficacy of nuclear polyhedrosis virus and *Bacillus thuringiensis* Berliner in the control of groundnut red hairy caterpillar *Amsacta albistriga* (Walker). *Madras Agricultural Journal* **64**, 130–131.
Jayaraj, S., Santharam, G., Narayanan, K., Sundararajan, K. and Balagurunathan, R. (1981) Effectiveness of the nuclear polyhedrosis virus against field populations of the tobacco caterpillar, *Spodoptera litura*, on cotton. *Andhra Agricultural Journal* **27**, 26–29.
Jayaraj, S., Rabindra, R.J. and Santharam, G. (1987) Control of *Heliothis armigera* (Hübner) on chickpea and lablab bean by nuclear polyhedrosis virus. *Indian Journal of Agricultural Science* **57**, 738–741.
Jayaraj, S., Rabindra, R.J. and Narayanan, K. (1989) Development and use of microbial agents for control of *Heliothis* spp. (Lep.: Noctuidae) in India. *Proceedings of the Workshop on Biological Control of* Heliothis: *Increasing the Effectiveness of Natural Enemies*. New Delhi: IOBC Heliothis work group, pp. 483–503.
Komolpith, U. and Ramakrishnan, N. (1978) Joint action of a baculovirus of *Spodoptera litura* (Fabricius) and insecticides. *Journal of Entomological Research* **2**, 15–19.
Mahadevan, N.R. (1978) Studies on the biology and control of the tobacco caterpillar, *Spodoptera litura* (F.) (Noctuidae: Lepidoptera) with nuclear polyhedrosis virus. MSc (Ag.) thesis, Tamil Nadu Agricultural University, Madurai, India.
Mahadevan, N.R. and Kumaraswami, T. (1980) A note on the effectiveness of nuclear polyhedrosis virus against *Spodoptera litura* (F.) on blackgram, *Phaseolus mungo* L. *Madras Agricultural Journal* **67**, 138–140.
Makode, D.L. (1978) Relative efficacy of synthetic insecticides and the NPV against gram pod borer *Heliothis armigera* Hübner. MSc thesis, Marathwada Agricultural University, Parabhani, India.
Mathai, S. (1982) Joint action of insect pathogen–insecticide mixtures in the control of crop pests. PhD thesis, Kerala University, Trivandrum, India.
Mistry, A.S., Yadav, D.A., Patel, R.C. and Pawar, B.S. (1984) Field evaluation of nuclear polyhedrosis virus against *Heliothis armigera* Hübner (Lepidoptera: Noctuidae) in Gujarat. *Indian Journal of Plant Protection* **12**, 31–33.
Mohan, K.S., Jayapal, S.P. and Pillai, G.B. (1986) Biological suppression of coconut Rhinoceros beetle *Oryctes rhinoceros* (L.) in Minicoy, Lakshadweep by *Oryctes* baculovirus – impact on pest population and damage. *Journal of Plantation Crops* **16** (Suppl.), 163–170.
Muthiah, C. (1988) Studies on the nuclear polyhedrosis virus of *Heliothis armigera* (Hbn.) and its formulations. MSc (Ag) thesis, Tamil Nadu Agricultural University, Coimbatore, India.
Muthiah, C. and Rabindra, R.J. (1991) Control of gram-pod borer (*Heliothis armigera*) on pigeonpea (*Cajanus cajan*) with controlled droplet application of nuclear polyhedrosis virus and effect of oral feeding on mulberry silkworm (*Bombyx mori*). *Indian Journal of Agricultural Sciences* **61**, 449–452.
Naryanan, K. (1979) Studies on the nuclear polyhedrosis virus of gram pod borer, *Heliothis armigera* Hübner (Noctuidae: Lepidoptera). PhD thesis, Tamil Nadu Agricultural University, Coimbatore, India.
Narayanan, K. (1987) Field efficacy of nuclear polyhedrosis virus of *Adisura atkinsoni* Moore on field beans. *Journal of Biological Control* **1**, 73–74.

Narayanan, K., Govindarajan, R., Jayaraj, S., Paulraj, S. and Kutty, K.N. (1977) Non-susceptibility of the common carp, *Caprinus carpio* L. to nuclear polyhedrosis virus *Baculovirus amsacta* of ground-nut red hairy caterpillar. *Madras Agricultural Journal* **64**, 411–412.

Narayanan, K. Santharam, G., Easwaramoorthy, S. and Jayaraj, S. (1978a) Influence of larval age and dosage of virus on the recovery of polyhedral inclusion bodies of the nuclear polyhedrosis virus of groundnut red hairy caterpillar *Amsacta albistriga* (W.). *Indian Journal of Experimental Biology* **16**, 1324–1325.

Narayanan, K., Santharam, G., Easwaramoorthy, S. and Jayaraj, S. (1978b) Lack of susceptibility of poultry birds to nuclear ployhedrosis virus of groundnut red hairy caterpillar, *Amsacta albistriga*. *Indian Journal of Experimental Biology* **16**, 1322–1324.

Neelgund, Y.F. (1977) Studies on nuclear polyhedrosis of the armyworm, *Mythimna (Pseudaletia) separata* Walker. *Research Publication Series*, No. 31. Karnataka University, Dharwad, India.

Rabindra, R.J. and Jayaraj, S. (1987) Control of *Heliothis armigera* (Hbn.) on chickpea with controlled droplet application of nuclear polyhedrosis virus in combination with endosulfan and boric acid. *Journal of Biological Control* **1**, 122–125.

Rabindra, R.J. and Jayaraj, S. (1988a) Evaluation of certain adjuvants for nuclear polyhedrosis virus (NPV) of *Heliothis armigera* (Hbn.) on chickpea. *Indian Journal of Experimental Biology* **26**, 60–62.

Rabindra, R.J. and Jayaraj, S. (1988b) Larval extracts and other adjuvants for increased efficiency of nuclear polyhedrosis virus against *Heliothis armigera* larvae. *Journal of Biological Control* **2**, 102–105.

Rabindra, R.J., Jayaraj, S. and Balasubramanian, M. (1985) Efficacy of nuclear polyhedrosis virus for the control of *Heliothis armigera* (Hübner) infesting sunflower. *Journal of Entomological Research* **9**, 246–248.

Rabindra, R.J., Sathiah, N., Muthiah, C. and Jayaraj, S. (1989) Controlled droplet application of nuclear polyhedrosis virus with adjuvants and UV protectants for the control of *Heliothis armigera* Hbn. on chickpea. *Journal of Biological Control* **3**, 37–39.

Rajapakse, R.H.S. and Jeevaratnam, K. (1982) Use of a virus against the root and stem borer *Plocaederus ferrugineus* L. (Coleoptera: Cerambycidae) of the cashew. *Insect Science and its Application* **3,** 49–51.

Ramakrishnan, N. (1976) Development of microbial control agents: nuclear polyhedrosis virus of *Spodoptera litura* (Fabricius). *Proceedings of the National Academy of Sciences, India* **B46**, 110–116.

Ramakrishnan, N., Chaudhari, S., Kumar, S. and Rao, R.S.N. (1981) Field efficacy of nuclear polyhedrosis virus against tobacco caterpillar, *Spodoptera litura* (F.), on tobacco. *Tobacco Research* **7**, 129–134.

Rao, R.S.N., Rao, S.G. and Chandra, I.J. (1987) Biochemical potentiation of nuclear polyhedrosis virus of *Spodoptera litura* (F.). *Journal of Biological Control* **1**, 36–39.

Regupathy, A., Santharam, G., Easwaramoorthy, S., Jayaraj, S. and Kannan, P. (1978) A note on the effect of oral feeding of nuclear polyhedrosis virus of *Spodoptera litura* (Fabricius) on poultry. *Indian Journal of Animal Science* **48**, 242–244.

Sachithanandam, S. (1988) Studies on the nutritional management of pests and diseases and NPV control of *Spodoptera litura* (Fabricius) on groundnut (*Arachis hypogea* L.). MSc thesis, Tamil Nadu Agricultural University, Coimbatore, India.

Santharam, G. (1986) Studies on the nuclear polyhedrosis virus of tobacco cutworm, *Spodoptera litura* (Fabricius) (Noctuidae: Lepidoptera). PhD thesis, Tamil Nadu Agricultural University, Coimbatore, India.

Santharam, G. and Balasubramanian, M. (1980) Note on the control of *Spodoptera litura* (F.) (Lepidoptera: Noctuidae) on tobacco with NPV and diflubenzuron. *Indian Journal of Agricultural Science* **50**, 726–727.

Santharam, G. and Balasubramanian, M. (1982) Effect of nuclear polyhedrosis virus (NPV) used alone and in combination with insecticides in controlling *Heliothis armigera* (Hübner) on bengalgram. *Journal of Entomological Research* **6**, 417–420.

Santharam, G., Regupathy, A., Easwaramoorthy, S. and Jaraj, S. (1978) Effectiveness of nuclear polyhedrosis virus against field populations of *Spodoptera litura* (F.) in banana. *Indian Journal of Agricultural Science* **48**, 676–678.

Santharam, G., Balasubramanian, M. and Chelliah, S. (1981) Control of *Heliothis armigera* (Hübner) on redgram (*Cajanus cajan* L.) with a nuclear polyhedrosis virus and insecticides. *Madras Agricultural Journal* **68**, 417–420.

Savanurmath, C.J. and Mathad, S.B. (1981) Efficacy of fenitrothion and nuclear polyhedrosis virus combinations against the armyworm *Mythimna separata* (Lepidoptera: Noctuidae). *Zeitschrift für Angewandte Entomologie* **91**, 464–474.

Shah, B.H., Zethner, O., Gul, H. and Chaudhry, M.I. (1979) Control experiments using *Agrotis segetum* granulosis virus against *Agrotis ipsilon* (Lep.: Noctuidae) on tobacco seedlings in northern Pakistan. *Entomophaga* **24**, 393–401.

Srinivas, P.R. (1987) Studies on the ecology and integrated management of *Heliothis armigera* (Hübner). PhD thesis, Tamil Nadu Agricultural University, Coimbatore, India.

Vijayakumar, M. and Mathad, S.B. (1978) Dermal exposure of albino rats to the nuclear polyhedrosis virus of the armyworm, *Mythimna (Pseudaletia) separata*. *Current Science* **47**, 124–125.

Vijayakumar, M. and Mathad, S.B. (1984a) Insusceptibility of albino rats to per os administration of *Mythimna (Pseudaletia) separata* nuclear polyhedrosis virus. *Journal of Karnataka University, Science* **29**, 59–64

Vijakumar, G. and Mathad, S.B. (1984b) Lack of toxicity/pathogenicity of orally administered *Baculovirus mythimna* of *Mythimna (Pseudaletia) separata* in albino rats. *Journal of Karnataka University, Science* **29**, 65–67.

Zethner, O., Chaudhry, M.I., Gul, H. and Khan, B.H. (1983) Prospects of the use of *Agrotis* granulosis virus for the control of cutworms in the N.W.F.P. Pakistan. *Bulletin of Zoology* **1**, 143–147.

15 South-east Asia and the western Pacific

KEITH. A. JONES, BERNHARD ZELAZNY,
UTHAI KETUNUTI, ANDREW CHERRY
and DAVID GRZYWACZ

PERSPECTIVE

The south-east Asian region (Thailand, Malaysia, Burma (Myanmar), Indonesia, Laos, Vietnam, Taiwan and Korea) comprises an area of high population and rapid economic growth. This has resulted in an increasingly intensive agricultural system for both home consumption and export. Throughout the region, a wide variety of food and fibre crops are grown and losses owing to insect attack are a major concern. Since the 1980s, very high levels of insecticide use have been recorded – both in amount applied and frequency sprayed. This has resulted in considerable problems of pesticide poisonings, environmental pollution and insecticide resistance. For example, in Thailand Chinese kale is sprayed up to ten times per crop, with up to six crops per season (Rumakom and Prachuabmoh, 1992), most sprays being directed against insect pests. Similarly, cotton is treated with 18 to 25 sprays per crop. Examples of pesticide poisonings and insecticide resistance have been reported from many countries (Rengam, 1992; Telekar, 1992; Jones *et al.*, 1993).

These problems have led to increasing pressure for alternative control measures and south-east Asia is at the forefront of implementation of Integrated Pest Management (IPM) programmes (Ooi *et al.*, 1992). The use of BVs as one part of an IPM programme has been considered for a number of countries in the south-east Asian region (Table 15.1). Stages of development range from laboratory research to commercialization of a BV insecticide (Table 11.4, p. 193). Activities in the smaller and geographically dispersed nations of the Western Pacific have essentially concentrated on a common problem, that of attack by rhinoceros beetle, *Oryctes rhinoceros*, to coconut palms. This problem is, of course, shared by most coconut-producing areas outside the New World and in this book is also referred to in the context of Australasia (especially Papua New Guinea), India and Africa.

Table 15.1 Virus control of insect pests in south-east Asia and the western Pacific

Pest Species	Indonesia	Korea	Malaysia	Philippines	Taiwan	Thailand	Western Pacific
Crocidolomia binotalis			+				
Helicoverpa armigera	+				+	+	
Limacodidae[a]	+		+			+	
Mythimna separata		+					
Oryctes rhinoceros	+		+	+			+
Penaeus spp.	+			+	+	+	
Pieris rapae					+		
Plutella xylostella			+		+	+	
Spodoptera exigua					+	+	
Spodoptera litura		+	+				

[a] *Darna catenatus*, *Darna furva*, *Darna trima*, *Parasa lepida*, *Quasithosea* sp., *Setora nitens* and *Setothosea asigna*.

This chapter provides a brief summary of the current status of applied insect virus – mainly BV – research and development in south-east Asia and the western Pacific. It does not intend to summarize all activities; because of difficulties in obtaining information from some countries, this is impossible, so the chapter provides an indication of the advances that have taken place and how the products are utilized. Within the area under discussion, two examples of very considerable progress stand out. They are the advanced work in Thailand on NPV control of *Helicoverpa armigera* and *Spodoptera exigua* and the geographically very widespread and successful virus control of *O. rhinoceros*, which has evolved from studies that commenced in Western Samoa in 1967 (Marschall, 1970).

VIRUS PRODUCTION

Apart from work with *O. rhinoceros*, most virus production development studies have been with the NPVs of *H. armigera* and *S. exigua* in Thailand. These viruses are produced *in vivo* in their homologous host insect. Larvae are reared on artificial diet under laboratory conditions until 7 days old. At this point they are infected by surface contamination of the artificial diet, which has been poured as a thin layer in the bottom of 300 g plastic tubs. The tubs are then stored at room temperature (28°C). From the sixth day after dosing, the tubs are inspected daily and virus-killed insects collected with the aid of a suction pump. The resultant crude virus is stored frozen as a suspension in distilled water adjusted to 1×10^9 OB/ml. This basic production method has been improved by introduction of the techniques used for *Spodoptera littoralis* NPV in Egypt. In brief, the tubs

Table 15.2 Production criteria for *H. armigera* (HaNPV) and *S. exigua* (SeNPV) NPVs in Thailand in a mass rearing system

	Dose[a] (OB/larva)	Larval weight at dosing (mg)	Incubation period (days)	Larval weight at harvesting (mg)	Mean yield (OB/larva)
HaNPV	1.25×10^5	40–50	6	257	5.9×10^9
SeNPV	8×10^4	30–40	5	86	1×10^9

[a] The dose is much higher than is needed to cause infection *per se*.

are replaced by large trays divided into cells for individual larvae, this being achieved by pressing Aeroweb honeycomb into diet poured into locally obtained trays. A single larva is placed in each cell and a ventilated perspex lid placed over the tray. The larvae are incubated at 28°C until a point just prior to death, when they are harvested. Harvesting alive results in less bacterial contamination (a mean of 1.1×10^4 to 4.6×10^4 colony-forming units (cfu)/larva compared with 0.9×10^6 to 1.4×10^6 cfu/larva for larvae harvested after death). The age of the larva at dosing, virus dose applied and length of the incubation period used for each species, as determined by experimentation in Thailand, are given in Table 15.2.

The viral suspension is not purified before use as this would be prohibitively expensive. The types of contaminant bacteria present in unpurified production batches have been identified (Table 15.3) and this confirmed that no primary human pathogens are present. Unpurified suspensions are kept frozen until required and the virus is often formulated in the field just prior to use.

The Departments of Agriculture and Agricultural Extension produced small bottles of seed inoculum of *S. exigua* NPV, which was distributed to farmers along with instructions on use and virus multiplication (Huber, 1986; Jones *et al.*, 1993). Farmers are advised to collect virus-killed insects after spraying, to store the larvae in a cool place, filter through cloth and apply again much as in some other parts of the world, for instance with *Anticarsia gemmatalis* NPV in Brazil. This has proved to be very successful, but there is a need for a mechanism of maintaining some sort of quality control. The most practical solution is to ensure an adequate supply of seed inoculum and to recommend only limited recycling. Isolates of

Table 15.3 Contaminant bacteria isolated from production batches of *H. armigera* and *S. exigua* NPVs in Thailand

Spore-forming bacteria	Vegetative bacteria
Bacillus cereus	*Acinetobacter lwoffi*
Bacillus coagulans	*Enterobacter georgoviae*
Bacillus mycoides	*Micrococcus* spp.
Bacillus pantotheticus	*Proteus mirabilis*
Bacillus sphericus	*Proteus vulgaris*
	Pseudomonas cepecia
	Pseudomonas fluorescens
	Streptococcus spp.

S. exigua NPVs from Spain, the USA and from Thailand were shown by restriction endonuclease (REN) analysis to contain distinct, but closely related, genotypes (variants) (Caballero *et al.*, 1992). Bioassays of the isolates in second instar *S. exigua* larvae showed differences in biological activity to be small but significant, with the Thai strain the most potent. Values for LD_{50} ranged from 1.5 OBs for the Thai isolate to 5.8 for the Spanish isolate.

The demonstrable success of BVs in Thailand has led to interest in commercial production of both the above viruses. The UK Department of Overseas Aid (DOA) has constructed a pilot plant for virus production in Bangkok (and Thailand has been designated by the UNDP-sponsored Regional Network for Pesticide Producers as the regional centre for training in production and use of insect viruses). A local cotton seed company has also constructed a small factory for virus production (Jones *et al.*, 1993). However, production seems now to have been discontinued. International agrochemical companies have also shown an interest in marketing viruses in Thailand and this has led to such organizations pursuing commercial registration of BVs in the region.

The dynamics of *S. exigua* NPV production have also been studied in Taiwan (Huang and Kao, 1994). Optimal yields were obtained by presenting fourth instar larvae with artificial diet sprayed with 2×10^5 OB/cm^2 and then rearing at 30°C for 5–6 days, when yields peak at 1.3×10^9 OB/larva. Under these conditions, microbial contamination was 1×10^8 cfu/larva. Mass culture of Limacodidae for virus production has not been attempted. It has so far proved sufficient to collect dead, naturally infected larvae opportunistically from palms and store them frozen until required.

An effective method for the production of OrV has been described in detail by Bedford (1981). Field-collected or reared larvae (100–150) are placed in plastic boxes containing sterile sawdust mixed with eight virus-killed larvae. Live larvae are allowed to feed for 5 to 7 days before being transferred to boxes containing sawdust alone. Larvae dying from virus move to the sawdust surface and are collected daily. The dead larvae are ground and mixed with water for use in the field. However, this method has to some extent been replaced by a method which uses virus that is provided as room temperature-stable preparations by a low-technology processing technique using virus produced in the adult gut. This permits successful transport from one region to another. This method is described in detail in Chapter 25.

Crocidolomia binotalis (NPV)

In some parts of the region the pyralid moth *C. binotalis*, sometimes known as the cabbage cluster caterpillar, is a very serious pest of cruciferous crops. Early laboratory studies (P.F. Entwistle, unpublished) employing a *C. binotalis* stock from Papua New Guinea showed it to be amenable to culture on a semi-synthetic diet and to be highly susceptible to AcMNPV. More recently, a closely related virus, *Galleria melonella* MNPV (GmMNPV), has also been found to be infective (Kadir, 1992). There is, therefore, the possibility of dual control with *Plutella xylostella* (see below) using GmMNPV.

H. armigera and *S. exigua* (NPVs)

Because work on *H. armigera* and *S. exigua* has been very much concentrated in Thailand, in the context of the same project and because of crop host plant overlap, they are treated here together.

Field testing of both viruses has been undertaken over a number of years. The trials have been briefly reported (Ketunuti and Prathomrut, 1989; Ketunuti and Tantichodok, 1990; Jones, Ketunuti and Grzywacz, 1994) but no details have been published. *S. exigua* NPV has been successfully used for control on cabbage, grape vine, okra, asparagus, cotton and shallot while *H. armigera* has been successfully controlled by NPV on tomato, okra, asparagus, sorghum, yardlong bean, cotton and tangerine. The application rate for both viruses is normally 9.375×10^{11} OB/ha on crops other than cotton, though 2×10^{12} OB/ha has been used on tomato. Most trials have used unpurified suspensions that have been diluted in the field. A wettable powder formulation (freeze-dried, virus-infected larvae, 100 g; bentonite/china clay, 60 g; neosyl containing 50% Etocas 30 DG, 40 g) has been tested on tomato, cotton and grape vine: initial results have indicated no difference in activity from unpurified suspensions. Although there is some indication that the wettable powder persists longer than the unpurified suspension, both persist longer than purified suspensions (K. A. Jones and U. Ketunuti, unpublished results). Commercial wetting agents (Pitsulin or Triton) are added to all formulations prior to application. A standard volume application rate using a motorized or hand-lever knapsack is 90 l/ha. Recent trials have also shown that oil-based, ULV applications can give as good control (K. A. Jones, A. J. Cherry and U. Ketunuti, unpublished results).

The effectiveness of a number of chemicals as UV protectants for *S. exigua* NPV showed that stability was improved by 1% xanthine, uric acid-activated carbon or folic acid.

Since 1979, farmers have adopted the use of NPV in some areas, most particularly in onion and shallot, where chemical control of *S. exigua* has become extremely difficult. Good control is achieved by applying the virus every 5 days, a frequency that is necessary because larvae feed within the onion leaf. Eggs, however, are laid on the leaf surface and, therefore, the young larva ingests a lethal dose during hatching and penetrating the leaf.

H. armigera NPV is used as an important part of the cotton IPM programme in Thailand. IPM demonstration plots have shown that virus can replace four to six of the chemical sprays to the crop (Ketunuti and Prathomrut, 1989). Application doses in this case ranged from 1.3×10^{12} to 1.95×10^{12} OB/ha. Recently, very good control has been demonstrated on tangerine, some trials showing twice the yield in virus-treated plots (1.87×10^{12} to 2.5×10^{12} OB/ha), compared with the standard insecticide treatments (U. Ketunuti, unpublished results). This is partly because it is necessary to spray the tangerine blossom and chemical pesticides have a detrimental effect on pollinating insects.

H. armigera NPV is also recommended as part of the IPM programme for cotton in Indonesia, where demonstration plots have been established in each cotton-growing area (Ruchijat and Sukmaraganda, 1992). In Taiwan, the commercial product Elcar, now out of production, was reported to give good control of *H. armigera* in field tests on maize (Yen, 1988).

Limacodidae

Slug and nettle caterpillars are recorded widely as pests in the tropics, especially of palm crops. In south-east Asia they are known as frequent defoliators of oil palm and, rather less often, of coconuts. While attacks are seldom lethal, the slow rate of frond replacement in palms usually means that a single defoliation results in a long period of yield suppression. Outbreaks occur naturally but can also be induced by indiscriminate use of non-selective chemicals. About 40 species of limacodids collectively host a remarkable assemblage of viruses (Entwistle, 1987). An account of the viral diseases of the five dominant limacodids in Indonesia is given by Desmier de Chenon *et al.* (1988). Over half the pathogens found are small, non-occluded viruses (densoviruses, picornaviruses and *Nudaurelia* β) and the rest are divided between BVs and reoviruses. As is generally common, these viruses are often associated in the same host species, making identification of the disease-causing agent difficult.

The practical use of viruses for control of Limacodidae in south-east Asia is summarized in Table 15.4. Application methodology is diverse, with the use of motor-powered knapsack sprayers the most frequent. In all instances, inoculum is prepared from field-collected diseased larvae, which are macerated, filtered and diluted in water. Control is usually very effective and inoculum persisting on the palms following a control exercise (or a natural epizootic) can provide quite long-term protection. Tiong and Munroe (1976) showed that resurgence to pest levels of *Darna trima* followed more slowly when virus, rather than a chemical insecticide, was the control agent applied. In Indonesia, a BV has been isolated from *Parasa lepida*, a major pest of coconut palms in southern Sumatra (Ginting and Chenon, 1987).

It is difficult to state precisely the level of use of this valuable and simple technology. In a survey of the status of nettle caterpillars by the Palm Oil Research

Table 15.4 Cases of successful use of virus to control Limacodidae in south-east Asia

Pest	Area	Virus	Dosage (larva/ha)	Reference
Darna catenatus	Central Sulawesi	?	*c.* 20 larvae/l[a]	
Darna furva	Thailand	?	?	U. Ketunuti, personal communication (1966)
Darna trima	Sabah	GV + *Nudaurelia* β	300–400	Tiong and Munroe (1976); Tiong (1982)
Quasithosea spp.	Thailand	Non-occluded virus (undetermined spp.)	50	U. Ketunuti, personal communication (1995)
Setora nitens	North Sumatra	*Nudaurelia* β	300 g[b]	Desmier de Chenon *et al.* (1988); Sipayung *et al.* (1989)
Setothosea asigna	North Sumatra and Sarawak	*Nudaurelia* β + CPV	300 g[b]	Tiong (1982); Desmier de Cheonon *et al.* (1988); Sipayung *et al.* (1989)

[a] Volume spray fluid/ha not known.
[b] Dosage as grams larvae per hectare.

Institute of Malaysia (PORIM) covering the period 1981–90, the spraying of virus suspensions was found to be practiced in only four oil palm estates in Sabah and Sarawak, despite the majority of estates regarding biological control as a suitable method for long-term control of these pests (Kamurudin and Wahid, 1992). It has also been stated (Salman Shah, personal communication) that the frequency of *D. trima* outbreaks in Sabah has declined and viruses are now rarely used. PORIM is assessing the potential of BVs for control of Limacodidae on oil palm, initially through a search for local BVs isolates.

Despite the great promise of viruses for limacodid control, there may be difficulties (from the regulatory point of view) of registering a product containing several different viruses that have not all been safety tested. Under a subcontract from the Food and Agriculture Organization of the United Nations, two limacodid viruses are being safety tested in the University of Otago, New Zealand.

Mythimna (Pseudaletia) separata (NPVs)

In Korea 21 NPV isolates from the armyworm *Mythimna separata* and other Lepidoptera have been assayed in the laboratory. When dosed on leaf surfaces, three local isolates from the homologous host were most effective, giving LC_{50} values of of 4.8×10^2 to 1.8×10^3 OB/ml in first instar larvae (Park and Muneo, 1989).

O. rhinoceros (OrV)

Studies on the virus control of *O. rhinoceros* commenced with the discovery of the virus in Malaysia (Huger, 1966), rapidly progressing to its release into many islands (Table 15.5) to which the pest had spread during the period 1909–63 (Lever, 1969). Invariably the disease became established easily after contaminating breeding sites or releasing virus-infected adult beetles.

Releasing adult beetles was eventually found to be more convenient than contaminating breeding sites with virus-killed larvae. Agricultural workers infect

Table 15.5 Records of introductions of the OrV into *O. rhinoceros* populations (including locations outside south-east Asia and the western Pacific)

Country	Year	Release method	Reference
W. Samoa	1967	Contaminating breeding sites	Marschall (1970)
Tonga	1970	Contaminating breeding sites	Young (1974)
Fiji	1970	As above and release of infected adult beetles	Bedford (1976)
Wallis Island	1970	Contaminating breeding sites	Hammes and Monsarrat (1974)
Tokelau Islands	1970	Contaminating breeding sites	Swan (1974)
Mauritius	1970	Contaminating breeding sites	Monty (1974)
Palau	1970	Contaminating breeding sites	Swan (1974)
American Samoa	1972	Release of infected adult beetles	Swan (1974)
Papua New Guinea	1978	Release of infected adult beetles	Gorick (1980)
Maldives	1984	Release of infected adult beetles	Zelazny *et al.* (1989b)

beetles collected by farmers using the virus preparation produced by the method described on p. 429. A small drop of the inoculum, mixed with sucrose to promote ingestion, is placed on the mouth of the beetles, which are then allowed to fly away at dusk. Adults can also be infected by immersion in a virus suspension (10 min in a 10% suspension of ground infected larvae). As few as 10 inoculated beetles are sufficient to establish the disease on one island.

OrV appears to be endemic on the mainland of India, Sri Lanka, Malaysia, Indonesia and the Philippines. However, *O. rhinoceros* can still be a pest in such areas; considerable efforts have been made to understand how the disease spreads, how it reduces the pest population and whether it could be used to prevent or control outbreaks in areas where it is already established.

Transmission and spread

For both adults and larvae, probably the only natural route of infection is by mouth; transovarial passage and spread through pupae appear not to occur (Zelazny, 1976). The main field routes of infection are between adult and larva in breeding sites (decaying wood, trash and compost heaps) and during mating. The frequency of transmission in breeding sites is very dependent on such conditions as the density and type of decaying coconut trunk available for breeding. If many trunks are available (e.g. during replanting) contact between adults and larvae is rare and disease prevalence low. Dead, standing palms are the preferred breeding site and at a density of five per hectare these can promote disease spread effectively; many will contain both virus-infected larvae and adults (Zelazny and Alfilier, 1986; Zelazny, Alfiler and Lolong, 1989a). The first mating seems to take place in coconut palms, the later ones in breeding sites. The fact that the virus is often transmitted during mating explains its capacity to survive at low pest densities and contributes to its success as a biological control agent.

Further information on disease biology, including effects on the host insect and dispersal characteristics within the coconut environment can be found in Chapter 5 and in Huger (1973), Young (1974), Bedford (1976); Monsarrat and Veyrunes (1976), Zelazny (1977a,b) and Young and Longworth (1981).

Viral control in areas of viral endemicity

Comments here refer also to areas where the virus has been deliberately established. Disease may decline to very low levels and this acts as a trigger to pest outbreaks. The most common accompanying reason for an outbreak with low disease is the presence of many lying tree trunks during replanting. This provides a sudden abundance of potential breeding sites and, because in such a situation adults and larvae seldom meet, virtually arrests transmission. To promote increased disease spread, the following practice is, therefore, suggested:

1. Fell palms as close to the ground as possible
2. Pile the trunks and plant a cover crop round the piles (overgrown logs are unattractive as breeding sites (Wood, 1968)
3. Leave five dead palms standing per hectare.

This results in more contact between beetles and in accelerated disease transmission. The system is currently under test in Indonesia. It seems probable that release of infected adult beetles would hasten control establishment under these conditions.

Detection of infection in the field

Infection cannot be detected in adults except by dissection and examination of the midgut: this is thin and semi-translucent brown when healthy and opaque white and enlarged when infected. At an advanced phase of infection, larvae become rather translucent and flaccid. However, in neither life stage can such visual methods be described as sensitive. A more reliable dot-blot assay, using a non-radioactive detection system based on DNA hybridization (distributed by Boehringer, Mannheim), has been found suitable for field use in Indonesia (Crawford, 1988). Small spots are prepared with a piece of infected tissue (e.g. adult beetle midgut) on nitrocellulose membrane in the field. The membranes, which can be stored in glass bottles over silica gel for up to 3 months without loss of sensitivity, are sent to a suitable laboratory for processing.

In laboratory tests, some evidence was found for development of resistance to the virus in adults, but not in larvae, in the Philippines. The consequences of this, if any, for control have yet to be established.

Cross-infectivity

The susceptibility of *Papuana uninodis*, a dynastid pest of root crops of Papua New Guinea origin but now causing serious damage to root crops in Fiji, has been demonstrated (Zelazny *et al.*, 1988). We are unaware of any field trials. For information on the question of *Scapanes* susceptibility, see Chapter 19.

Safety tests

Extensive tests conducted in France have failed to show that OrV can develop in vertebrates or in vertebrate cell cultures (Gourreau *et al.*, 1979; Gourreau, Kaiser and Monsarrat, 1981, 1982).

Papuana uninodis

see *O. rhinoceros* above.

Penaeus monodon (BV)

A BV affecting the cultured giant tiger prawn was reported as early as 1981 in prawns originating from Taiwan. It was subsequently found to be widespread in shrimps from Taiwan, the Philippines and Sumatra. Its exact role in the mass deaths of cultured shrimp is not clear as it is apparently widespread and, by itself, of low pathogenicity. However, when prawns are overcrowded both its prevalence and lethality increase to a serious level, as does its tendency to facilitate secondary infections by other pathogens such as bacteria, which may be the direct causes of mass mortality.

Pieris (Artogeia) rapae (GV)

The effectiveness of spraying the homologous GV of *P. rapae* alone or in combination with *B.t. kurstaki*, has been reported in Taiwan (Su, 1986, 1991). Control was effective for 14 days in the field with the persistence of GV exceeding that of *B.t.*

Plutella xylostella (GV and NPV)

Much work has been conducted on the virus control of the diamondback moth, *Plutella xylostella*, in Malaysia, especially in Malaya and Taiwan. Kadir (1992) has assessed the potential of three viruses. Laboratory studies undertaken in the UK showed a specific GV (PxGV) to be highly potent in neonate larvae (1.9–5 OB/larva) but *P. xylostella* was also found to be susceptible to *G. melonella* NPV (GmMNPV) and AcMNPV. Values for LD_{50} were similar for both NPVs (21.1–24 and 26 OB/larva, respectively). The NPVs kill more rapidly than the GV (LT_{50} values: AcMNPV 3.8–6.3, GmMNPV 2.5–4.3, PxGV 4.9–5.4 days). As GmMNPV is also infective in *C. binotalis*, this is the virus of choice. However, in Malaysia, field tests comparing PxGV and GmMNPV have not yet been reported. In greenhouse tests, a combination of 9×10^{12} OB/ha with molasses in an aqueous formulation gave superior control to a formulation of 9×10^{13} OB/ha without molasses. It is considered that in these greenhouse trials, where glass screened out solar UV, the benefit conferred by molasses probably resulted from its capacities as a sticker, anti-evaporant and, possibly, as a feeding stimulant (Jones and McKinley, 1986; Jones, 1990). Wang and Rose (1978) and Su (1988, 1990, 1991) have also tested PxGV in Taiwan, concluding that the virus had good potential as a control agent; these studies included evaluation of PxGV/*B.t.* mixtures, which were also effective.

PxGV was rapidly inactivated on choy sum (*Brassica rapa*) plants sprayed at 10 LE/l. Larval mortality reduced from 100% to about 50% after 6 h exposure to sunlight. Telekar (1992) also reported that PxGV was rapidly inactivated in the field; although it had previously been demonstrated (Kao and Rose, 1976) that activated carbon or India ink provided good UV protection, this was not reflected in the field. Currently, IPM programmes for diamondback moth are concentrating more on the use of *B.t.* and parasitoids, as well as the careful use of chemical insecticides (e.g. Anon., 1991; Chin *et al.*, 1992). However, the development of resistance in this insect to *B.t.* (Zoebelein, 1990; Jones *et al.* 1993; Shelton *et al.*, 1993) highlights the need to develop further alternative control agents such as BVs.

Spodoptera litura (NPV)

The use of NPV for control of *S. litura*, tobacco cutworm, has been studied in Korea (Im *et al.*, 1990a,b). The values for LC_{50} against third and fifth instar larvae were 1.32×10^3 and 1.09×10^5 OB/ml, respectively. Virus activity persisted for more than 10 days on the underside of soybean leaf but decreased after 3 days on leaf surfaces exposed to sunlight. Formulations with increased persistence of activity were prepared using sucrose as a feeding attractant, polyvinyl alcohol as a suspending (anti-precipitant) agent, Triton X-100 as a wetting agent and carbon white as a UV protectant. In field evaluations in soybean (Im *et al.*, 1990c) pest mortality reached 93% in 14 days in NPV treatments, but mortality commenced 7 days later than in chemical treatments. In Malaysia, an NPV has also been isolated from S. *litura* attacking young acacia plantations (Sajap and Wahab, 1994) and this virus is currently under development for controlling this pest on acacia and tobacco (Norowi, personal communication, 1995).

REFERENCES

Anon. (1991) Demonstration of IPM of diamondback moth on farmers', fields in the lowlands. *Progress Report: Asian Vegetable Research and Development Center, 21–25.* AVRDC, Taipai, Taiwan.

Bedford, G.O. (1976) Use of a virus against the coconut palm rhinoceros beetle in Fiji. *Pest Articles and News Summaries* A, **22**, 11–25.

Bedford, G.O. (1981) Control of rhinoceros beetle by baculovirus. In *Microbial Control of Pests and Plant Diseases 1970–1980.* H.D. Burges (ed.). London: Academic Press, pp. 409–426.

Caballero, P., Zuidema, D., Santiago-Alvarez, C. and Vlak, J.M. (1992) Biochemical and biological characterization of four isolates of *Spodoptera exigua* nuclear polyhedrosis virus. *Biocontrol Science and Technology* **2**, 145–157.

Chin, H., Othman, Y., Loke, W.H. and Rahman, S.A. (1992) National integrated pest management in Malaysia. In *International Pest Management in the Asia–Pacific Region.* P.A.C. Ooi, G.S. Lim, T.H. Ho, P.L. Manalo and J. Waage (eds). Wallingford, UK: CAB International, pp. 191–209.

Crawford, A.M. (1988) Detection of baculovirus infection in rhinoceros beetle (*Oryctes rhinoceros*) and the purification and identification of virus strains. *Integrated Coconut Pest Control Project:, Annual Report 1988.* Manado, North Sulawesi, Indonesia: Coconut Research Institute, pp. 120–141.

Desmier de Chenon, R., Mariau, D., Monsarrat, P., Fediere, G. and Sipayung, A. (1988) Recherches sur les agents entomopathogenes d'origine virale chez les lepidopteres defoliateurs du palmier a huile et du cocotier. *Olagineaux* **43**, 107–117.

Entwistle, P.F. (1987) Virus diseases of Limacodidae. In *Slug and Nettle Caterpillars in South East Asia.* M.J.W. Cock, C.H. Godfray and J.D. Holloway (eds). Wallingford, UK: CAB International, pp. 213–221.

Ginting, C. and Chenon, R. (1987) New biological prospects for controlling a major coconut pest in Indonesia: *Parasa lepida* Cramer, Limacodidae, by the use of viruses. *Olagineaux* **42**, 107–118.

Gorick, B.D. (1980) Release and establishment of the baculovirus disease of *Oryctes rhinoceros* (L.) (Coleoptera: Scarabaeidae) in Papua New Guinea. *Bulletin of Entomological Research* **70**, 445–453.

Gourreau, J.M., Kaiser, C., Lahellec, M., Chevrier, L. and Monsarrat, P. (1979) Etude de l'action pathogene eventuelle du *Baculovirus* d'*Oryctes* pour le porc. *Entomophaga* **24**, 213–219.

Gourreau, J.M., Kaiser, C. and Monsarrat, P. (1981) Etude de faction pathogene eventuelle de baculovirus de oryctes sur cultures cellulaires de vertebres en ligne continue. *Annales de Virologie (Institute Pasteur)* **132E**, 347–355.

Gourreau, J.M., Kaiser, C. and Monsarrat, P. (1982) Study of the possible pathogenic action of the *Oryctes baculovirus* in the white mouse. *Annales de Virologie (Institute Pasteur)* **133E**, 423–428.

Hammes, C. and Monsarrat, P. (1974) Recherches sur *Oryctes rhinoceros* L. Cah. *ORSTOM Series in Biology* **22**, 43–111.

Huang, L.H. and Kao, S.S. (1994) Production of *Spodoptera exigua* nuclear polyhedrosis virus in larvae. *Chinese Journal of Entomology* **14**, 343–352.

Huber, J. (1986) Use of baculoviruses in pest management programs. In *The Biology of Baculoviruses*, Vol. II, *Practical Application for Insect Control*. R.R. Granados and B.A. Federici (eds). Boca Raton, FL: CRC Press, pp. 181–202.

Huger, A.M. (1966) A virus disease of the Indian rhinoceros beetle *Oryctes rhinoceros* (L.) caused by a new type of insect virus, *Rhabdionvirus oryctes* gen. n., sp. n.. *Journal of Invertebrate Pathology* **8**, 38–51.

Huger, A.M. (1973) Grundlgen zur biologischen Bekamfung des Indischen Nashornkafers, *Oryctes rhinoceros* (L.), mit *Rhabdionvirus oryctes*: Histopathologie der Virose bei Kafern. *Zeitschrift für Angewandte Entomologie* **72**, 309–319.

Im, D.J., Jin, B.R., Choi, K.M. and Kang, S.K. (1990a) Microbial control of the tobacco cutworm, *Spodoptera litura* (Fab.) using *S. litura* nuclear polyhedrosis virus. I. The effect of spray on soybean leaves, temperature, storage and sunlight on the pathogenicity of the virus. *Korean Journal of Applied Entomology* **29**, 184–189.

Im, D.J., Jin, B.R., Choi, K.M. and Kang, S.K. (1990b) Microbial control of the tobacco cutworm, *Spodoptera litura* (Fab.), using *S. litura* nuclear polyhedrosis virus as viral insecticides. *Korean Journal of Applied Entomology* **29**, 244–251.

Im, D.J., Jin, B.R., Choi, K.M. and Kang, S.K. (1990c) Microbial control of the tobacco cutworm, *Spodoptera litura* (Fab.), using *S. litura* nuclear polyhedrosis virus. III. Field evaluation of the viral insecticides. *Korean Journal of Applied Entomology* **29**, 252–256.

Jones, K.A. (1990) Control of *Spodoptera littoralis* in Crete and Egypt with NPV. In *Pesticides and Alternatives: Innovative Approaches to Pest Control*. J.E. Casida (ed.). Amsterdam: Elsevier, pp. 131–142.

Jones, K.A. and McKinley, D.J. (1986) UV inactivation of *Spodoptera littoralis* nuclear polyhedrosis virus in Egypt: assessment and protection. In *Fourth Colloquium of Invertebrate Pathology: Fundamental and Applied Aspects of Invertebrate Pathology*, Wageningen, the Netherlands. R.A. Samson, J.M. Vlak and D. Peters (eds), p. 5.

Jones, K.A., Westby, A., Reilly, P.J.A. and Jeger, M.J. (1993) The exploitation of micro-organisms in the developing countries of the tropics. In *Exploitation of Micro-organisms*. D.G. Jones (ed.). London: Chapman & Hall, pp. 343–370.

Jones, K.A., Ketunuti, U. and Grzywacz, D. (1994) Production and use of NPV to control *Helicoverpa armigera* and *Spodoptera litura* in Thailand. *Abstracts, VIth International Colloquium on Invertebrate Pathology and Microbial Control*, Montpellier, France, 28 August to 2 September 1994. Society for Invertebrate Pathology, p. 177.

Kadir, H.A. (1992) Potential of several baculoviruses for the control of diamondback moth and *Crocidolomia binotalis* on cabbages. In *Diamondback Moth and other Crucifer Pests*. N.S. Talekar (ed.). Tainan, Taiwan: Asian Vegetable Research and Development Center, pp. 185–192.

Kamarudin, N.H.J. and Wahid, M.B. (1992) A survey of current status and control of nettle caterpillars (Lepidoptera: Limacodidae) in Malaysia (1981–1990). *Palm Oil Research Institute of Malaysia Occasional Paper* No. 27.

Kao, H.W. and Rose, R.I. (1976) Effect of sunlight on the virulence of granulosis virus of the diamondback moth and the evaluation of some protective adjuvants. *Plant Protection Bulletin Taiwan* **18**, 391–395.

Ketunuti, U. and Prathomrut, S. (1989) Cotton bollworm larvae control by *Heliothis armigera* nuclear polyhedrosis virus. *Abstracts of the First Asia-Pacific Conference of Entomology*, 8–13 November 1989, Chang Mai, Thailand. Bangkok: The Secretariat, APCE, p. 16.

Ketunuti, U. and Tantichodok, A. (1990) The use of *Heliothis armigera* nuclear polyhedrosis virus to control *Heliothis armigera* (Hübner) on okra. In *Proceedings of the Vth International Colloquium on Invertebrate Pathology and Microbial Control*, Adelaide, Australia, 20–24 August 1990. Society for Invertebrate Pathology, p. 257.

Lever, R.J.A.W. (1969) Pests of the coconut palm. *FAO Agricultural Studies*, No. 77. Rome: FAO.

Marschall, K. (1970) Introduction of a new virus disease of the coconut rhinoceros beetle in Western Samoa. *Nature (London)* **225**, 288–289.

Monsarrat, P. and Veyrunes, J.-C. (1976) Evidence of *Oryctes* virus in adult feces and new data for virus characterization. *Journal of Invertebrate Pathology* **27**, 387–389.

Monty, J. (1974) Teratological effects of the virus *Rhabdionvirus oryctes* on *Oryctes rhinoceros* (L.) (Coleoptera: Scarabaidae). *Bulletin of Entomological Research* **64**, 633–636.

Ooi, P.A.C., Lim, G.S., Ho, T.H., Manalo, P.L. and Waage, J. (eds) (1992) *International Pest Management in the Asia-Pacific Region.* Wallingford, UK: CAB International.

Park, S.D. and Muneo, O. (1989) Cross infectivity of nuclear polyhedrosis virus to the common armyworm, *Pseudaletia separata. Korean Journal of Applied Entomology* **28**, 10–15.

Rengam, S.V. (1992) IPM: the role of governments and citizens action groups. In *International Pest Management in the Asia-Pacific Region.* P.A.C. Ooi, G.S. Lim, T.H. Ho, P.L. Manalo and J. Waage (eds). Wallingford, UK: CAB International, pp. 13–19.

Ruchijat, E. and Sukmaraganda, T. (1992) National integrated pest management in Indonesia: its successes and challenges. In *International Pest Management in the Asia-Pacific Region.* P.A.C. Ooi, G.S. Lim, T.H. Ho, P.L. Manalo and J. Waage (eds). Wallingford, UK: CAB International, pp. 329–347.

Rumakom, M. and Prachuabmoh, O. (1992) National IPM in Thailand. In *International Pest Management in the Asia-Pacific Region.* P.A.C. Ooi, G.S. Lim, T.H. Ho, P.L. Manalo and J. Waage (eds). Wallingford, UK: CAB International, pp. 211–236.

Sajap, A.S. and Wahab, Y.A. (1994) Natural enemies associated with some insect pests of forest plantations in peninsular Malaysia. In *BIOTROP Special Publication* No. 53, *Forest Pest and Disease Management.* pp. 167–169.

Sipayung, A., Desmier de Chenon, R. and Sudharto, P.S. (1989) Recent work with viruses in the biological control of leaf-eating caterpillars in North Sumatra, Indonesia. *International Oil Palm Conference*, PORIM, Kuala Lumpur, Malaysia, 5–9 September, 1989.

Shelton, A.M., Robertson, J.L., Tang, J.D., Perez, C., Eigenbrode, H.K., Wilsey, W.T. and Cooley, R.J. (1993) Resistance development of diamondback moth (Lepidoptera, Plutellidae) to *Bacillus thuringiensis* subspecies in the field. *Journal of Economic Entomology* **86**, 697–705.

Su, C.Y. (1986) Field efficiency of granulosis virus (GV) for control of the small cabbage white butterfly, *Artogeia rapae. Chinese Journal of Entomology* **6**, 79–82.

Su, C.Y. (1988) Utilization of *Plutella xylostella* granulosis virus for control of *Plutella xylostella. Chinese Journal of Entomology* **8**, 161–163.

Su, C.Y. (1990) Microbial control of the diamondback moth, *Plutella xylostella*, using *Bacillus thuringiensis* and granulosis virus. *Plant Protection Bulletin* **32**, 10–32.

Su, C.Y. (1991) Field trials of a granulosis virus and *Bacillus thuringiensis* for control of *Plutella xylostella* and *Artogeia rapae. Chinese Journal of Entomology* **11**, 174–178.

Swan, D.I. (1974) *A Review of the Work on Predators, Parasites and Pathogens for the Control of* Oryctes rhinoceros *(L.) (Coleoptera: Scarabaeidae) in the Pacific Area* (*Miscellaneous Publications* No. 27). London, UK: Commonwealth Institute of Biological Control, Commonwealth Agricultural Bureaux.

Telekar, N.S. (1992) Integrated management of diamondback moth: a collaborative approach in southeast Asia. In *International Pest Management in the Asia-Pacific Region.* P.A.C. Ooi, G.S. Lim, T.H. Ho, P.L. Manalo and J. Waage (eds). Wallingford, UK: CAB International, pp. 37–49.

Tiong, R.H.C. (1982) Oil palm pests in Sarawak and the use of natural enemies to control them. In *Proceedings of an International Conference on Plant Protection in the Tropics*, pp. 362–372.

Tiong, R.H.C. and Munroe, D.D. (1976) Microbial control of an outbreak of *Darna trima* (Moore) on oil palm (*Elaeis guineensis* Jacq.) in Sarawak (Malaysian Borneo). In *Proceedings of the International Agriculture Oil Palm Conference*, 1976, Paper 41.

Wang, C.L. and Rose, R.I. (1978) Control of the imported cabbageworm, *Pieris rapae crucivora* Boisduval, with granulosis virus in the field. *Plant Protection Bulletin, Taiwan* **20**, 16–20.

Wood, B.J. (1968) Studies on the effect of ground vegetation on infestations of *Oryctes rhinoceros* (L.) (Col. Dynastidae) in young oil palm replanting in Malaysia. *Bulletin of Entomological Research* **59**, 85–96.

Yen, F. (1988) Studies on the integrated control of key pests of supersweet corn. *Research Bulletin, Tainan District Agricultural Improvement Station* **22**, 25–37.

Young, E.C. (1974) The epizootiology of two pathogens of the coconut palm rhinoceros beetle. *Journal of Invertebrate Pathology* **24**, 82–92.

Young, E.C. and Longworth, J.F. (1981) The epizootiology of the baculovirus of the coconut palm rhinoceros beetle (*Oryctes rhinoceros*) in Tonga. *Journal of Invertebrate Pathology* **38**, 362–369.

Zelazny, B. (1976) Transmission of a baculovirus in populations of *Oryctes rhinoceros*. *Journal of Invertebrate Pathology* **27**, 221–227.

Zelazny, B. (1977a) *Oryctes rhinoceros* populations and behavior influenced by a baculovirus. *Journal of Invertebrate Pathology* **29**, 210–215.

Zelazny, B. (1977b) Occurrence of the baculovirus disease of the coconut palm rhinoceros beetle in the Philippines and in Indonesia. *FAO Plant Protection Bulletin* **25**, 73–77.

Zelazny, B. and Alfiler, A. (1986) *Oryctes rhinoceros* (Coleoptera: Scarabaeidae) larva abundance and mortality factors in the Philippines. *Environmental Entomology* **15**, 84–87.

Zelazny, B., Autar, M.L., Singh, R. and Malone, L.A. (1988) *Papuana uninodis*, a new host for the baculovirus of oryctes. *Journal of Invertebrate Pathology* **51**, 157–160.

Zelazny, B., Alfiler, A.R. and Lolong, A. (1989a) Possibility of resistance to a baculovirus in populations of the coconut rhinoceros beetle (*Oryctes rhinoceros*). *FAO Plant Protection Bulletin* **37**, 77–82.

Zelazny, B., Lolong, A. and Crawford, A.M. (1989b) Introduction and field comparison of baculovirus strains against *Oryctes rhinoceros* (Coleoptera: Scarabaeidae) in the Maldives. *Environmental Entomology* **19**, 1115–1121.

Zoebelein, G. (1990) Twenty-three-year surveillance of development of insecticide resistance in diamondback moth from Thailand (*Plutella xylostella* L., Lepidoptera, Plutellidae). *Mededelingen van de Faculteit Landbouwwetenschappen Rijksuniversiteit, Ghent* **55**, 313–322.

16 People's Republic of China

PHILIP F. ENTWISTLE

PERSPECTIVE

In the People's Republic of China the attitude to insect viruses is unique in that they are consciously regarded as a national resource to be isolated, catalogued and, when promising, developed and exploited as pest control agents (Anon., 1984a; Liang *et al*, 1986).

Insect virology in China goes back at least to the 1950s but there has been an especially marked surge of activity since the early 1970s. Language difficulties present the West with particular problems of access to information so that some comments on Chinese insect virus literature may be helpful. Some key sources are *Studies on the Polyhedrosis of Insect Pests* (Lee, 1980), *Collected Papers on Insect Viruses, Research of the Department of Virology* (Wuhan University) (Anon., 1984a) and *Microbial Control of Insects* (Yu, 1989). In addition, *Acta Virologica Sinica* and *Acta Entomologica Sinica* are rich sources of information. Individual universities and specific commodity research institutes (e.g. Institute of Tea Research (Hangchow), Forest Research Institutes (Kirin Province and Beijing)) also often publish relevant material. Reviews in English of insect pathology in the People's Republic of China have appeared, notably Su (1982) and Hussey and Tinsley (1981), the latter being based on a UK Royal Society Delegation visit of 1977 to study Chinese biological control (Risbeth, 1978).

In China, as with most other nations, applications of insect virology have found their place mostly in agriculture and forestry with only limited attention elsewhere, for example mosquitoes. Agricultural problem solving in China has largely been directed against pest species that are well known elsewhere in the world. Notable examples are *Helicoverpa armigera*, *Pieris rapae*, *Plutella xylostella* and *Spodoptera litura*. However, as usual, pests of forests and tree crops have a tendency to be more geographically localized, e.g. *Buzura thibetaria*, *Dendrolimus punctatus*, *Ectropis obliqua* and *Euproctis pseudoconspersa*, and the same is true of the pasture pest *Gynaephora ruoergensis*, though there are such notably more widespread exceptions as *Adoxophyes orana* and *Buzura suppressaria*.

VIRUS PRODUCTION

Not much information is available on the details of those viruses that are intended for large-scale use, though a precise production system has been described for *E. obliqua* NPV (Yin, 1988) and *P. rapae* GV (Liang, Chang and Tsai, 1979). In insect semi-synthetic diets, the emphasis is on controlling ingredient costs. Therefore, for NPV production *Mythimna separata* diet employs chopped leaf to replace the use of more expensive constituents (F. Zeng, personal communication). The GV of *P. rapae* has been in production for several years, both as dried and suspension concentrates, and is regarded very much as a success story. More recently, *H. armigera* NPV production has begun at the Jianghu Virus Insecticide Experiment Factory, Hubei. *S. litura* NPV production commenced in China in the early 1990s on a modest scale at Zhongshan University.

FORMULATION

Details of formulation are often not available; for instance, the constitution of the *E. obliqua* NPV aqueous flowable concentrate appears not to have been published.

For practical reasons and for effectiveness, unpurified preparations are used, often with the addition of 0.01–0.1% washing powder, presumably to improve physical performance. Protection from solar radiation is mainly by addition of activated charcoal, but a *P. rapae* GV liquid formulation contains 2.0% Fast Green (Malachite Green). However, for PbGV, 0.001% nigrosin or 0.012 'black dye' were suggested as UV protectants (Hu and Wan, 1986).

Safety tests

Very careful consideration is given to the safety of insect virus preparations. At least three of the economically more important viruses have been subjected to high levels of conventional safety tests on standard laboratory and domestic animals. In addition, infectivity tests have been conducted on the silkmoths *Bombyx mori* and *Samia* (*Philosamia*) *cynthia* x *ricini*, while other cross-infectivity tests on non-target economic Lepidoptera have given some indications of the likely breadth of host ranges.

FIELD EXPERIMENTATION AND IMPLEMENTATION

Adoxophyes orana (GV)

A. orana is a widespread, Eurasian pest of fruit trees with, in Japan, a race attacking tea, from which a GV was first reported (Aizawa and Nakazato, 1963), which was successfully transmitted to the apple race (Yamada and Oho, 1973). Later a GV was isolated in China (Liang *et al.*, 1986), while an NPV and a CPV are known in Europe. Jurkovikova (1979) provided evidence that the NPV could exist as a latent infection.

It is not known if field control trials have been conducted in China. There have been quite extensive field studies in Japan, where a mass production method employing semi-synthetic diet has been described (Yamada and Oho, 1973).

Anomis flava (NPV)

The cotton looper, *A. flava* (actually a semi-looper), is known especially as a very widespread Old World cotton defoliator that also attacks kenaf or okra (*Hibiscus esculentus*), e.g. in Taiwan. Eighty per cent control was obtained by spraying its NPV at 3×10^8 OB/ml and 60.0% control at 2×10^6 OB/ml (Liu *et al.*, 1984a,b).

Buzura spp. (NPV)

Buzura suppressaria and *B. thibetaria* are geometrid defoliators of tea for which NPV preparations have been produced, providing effective control (Peng *et al.*, 1992; Anon., 1993).

Cnaphalocrocis medinalis (GV)

Cnaphalocrocis medinalis is a rice leafroller with a wide distribution from Pakistan through south-east Asia to the Eastern Archipelago, and it is a common and major pest in China. At times and in places (e.g. the Solomon Islands and in Fiji) it has been confused with *Marasmia patnalis*.

A GV was isolated in China but its infectivity was very low until fed simultaneously with *S. litura* NPV. In such mixed infections, the viruses usually, but not always, replicate in separate cells (Hu and Liu, 1984). There do not appear to have been field trials.

Dendrolimus punctatus (NPV and CPV)

Members of the genus *Dendrolimus* are often pests of coniferous trees in Siberia, China and Japan. In southern China, *D. punctatus* is one of the most destructive defoliators of pines, especially *Pinus massoniana*: 1.6–2.0 million hectares are damaged annually, with timber loss approaching five million cubic metres. The larvae have irritant hairs and constitute a public health problem.

BV infections are known in five species. Virus control has been studied in China (and see *D. spectabilis*, p. 272). The NPV LC_{50} was given as 8.2×10^6 OB/ml in second instar larvae (Pu, 1978; Hsao, 1981). It is not clear what type of virus has been employed in *D. punctatus* control, as the polyhedrosis first isolated in 1973 in Kwangchow and subsequently field tested could have been a CPV, an NPV or a mixture. Whatever the case, field mortality was 89–99% compared with 22% in control areas. However, in a field trial, *D. spectabilis* CPV was sprayed against *D. punctatus* at 1×10^7 OB/ml giving 75% mortality. An aerial spray on 133 h achieved 51–71% mortality (Xiao, 1985).

***Ectropis (Boarmia) obliqua* (NPV)**

Several species of the Old World genus *Ectropis* are defoliators of trees and shrubs such as cinchona, cocoa, gambir and tea. Interest in China centres on *E. obliqua* as a tea pest.

The NPV of *E. obliqua* has been the subject of applied research at the Tea Research Institute, Chinese Academy of Agricultural Science, Hongzhou. An experimental product produced at Shi-zu-pu Tea Plantation, Anhui was tested in Zhejiang and Anhui. As an aqueous flowable it was applied at 7.5×10^{10} and 3.0×10^{11} OB/ha in 75–300 l/ha. Control was good and one application a year was adequate. The virus both persists and spreads (Yin, 1982). Commencing with only 1 ha in 1980–2, treatment increased to 160 ha during 1983–6. There is no information for more recent years.

Basic toxicity tests have been conducted and there were no infections in non-target vertebrates and invertebrates (K.-S. Yin, personal communication, 1991). A production system has been described in detail (Yin, 1988).

***Euproctis pseudoconspersa* (NPV)**

Euproctis is a very large Old World genus containing many pest species. In China, 78 of the 270 lymantriids belong to *Euproctis* (Xilin, 1989) and several are pests of woody crops. NPVs have been recorded in nine *Euproctis* species in China. In addition to attacking tea, *E. pseudoconspersa*, tea tussock moth, causes serious urticaria in tea-pickers.

NPV control on tea is practised by staff of the Tea Research Institute, Hangchow in Kweichow Province (Hussey and Tinsley, 1981). The virus seems first to have been recorded in 1976 (Su, 1982) and a CPV has also been noted (Lee, 1980).

***Euproctis similis* (NPV)**

E. similis has a wide host plant range and a broad Old World distribution. In China and Japan, it is of especial concern as a pest of mulberry (mulberry tussock moth). It causes urticaria in workers gathering mulberry leaves to feed silkworms, *B. mori*, and also constitutes a minor threat to the foliage itself. *E. similis*, therefore, requires to be controlled on mulberry but, because of the risk to silkworms, chemical insecticides cannot be used.

A multiply enveloped NPV was isolated in China in 1974 (Chu *et al.*, 1975). (Work in Japan (Watanabe and Aratake, 1974) indicated the possibility of two types of EsMNPV distinguishable by polyhedron shape.)

Much of the work in China has been by the Control Department of the Institute of Entomology, Shanghai. EsMNPV has been applied at concentrations of 1.5×10^{3} OB/ml to third instar larvae in quantities dependent on larval population density. At 7.7×10^{10} OB/ha, larvae died 5–14 days after spraying and there was evidence of a bimodal infection curve, similar to that noted for NPV disease in *Euproctis chrysorrhoea* in the UK (Sterling *et al.*, 1988), indicating an impact of progeny virus on the residual host population.

All cross-infectivity tests (e.g. the silkmoths *B. mori* and *Philosamia cynthia ricini* and also *Diacrisia subcarnea*, *Galleria mellonella* and *M. separata*) have been negative.

Gynaephora ruoergensis (NPV)

G. ruoergensis is a lymantriid pest of grasslands on the Chuan Qing Zang plateau. A pesticide consisting of the homologous NPV, *B.t.*, Ca^{2+}, a small quantity of chemical pesticide and a UV-protectant agent was applied successfully over a very large area (Liu *et al.*, 1993).

Helicoverpa armigera (NPV)

Commonly known as cotton bollworm, *H. armigera* is possibly the most important pest of agriculture in the Old World, where it is present everywhere apart from the cooler regions. It attacks many crops: in the People's Republic of China it is a major pest of cotton on which it may account for 40% of all insect-attributed yield loss. Maize, tomatoes, tobacco and several types of bean are also attacked.

In 1974, two NPV isolates were independently discovered in Hupeh, Wuhan (Anon., 1975) and in Kiangsu (Fudan University, 1975), both being highly virulent. The Hepeh virus came from the Chinchow district and was designated VHA-273 (Lee, 1980). This is a multiply enveloped strain: later a singly enveloped NPV was isolated by Fudan University (Yu and Chen, 1979) and a second SNPV designated P-NPV was isolated in 1982 (Cai, Ding and Wang, 1986). A GV and a CPV have also been found in China. An NPV has also been found in *Helicoverpa assulta*, tobacco budworm (Tsai and Ding, 1982); this insect has a wide Old World distribution but is seldom an important pest. VHA-273 is cross-infectious to *H. assulta*.

In the laboratory, an LT_{50} of 2.4–3.4 days followed exposure to 1×10^6 OB/ml and there was a 98% final mortality (Hussey and Tinsley, 1981). In another assay the same concentrations yielded 85 and 80% mortalities in first and second instar larvae at 6 and 8 days, respectively. Since 1975, extensive field trials have been conducted in Kiangsu and Hupeh Provinces where 2.27×10^{11} to 4.55×10^{11} OB/ha (approximately 105–150 LE) resulted in 60–70% control on cotton. This was as effective as chemical insecticides (DDT + compound 1605). Combinations of NPV with 'dilute sevin' (carbaryl, a carbamate insecticide) or *B.t.* gave results better, or as effective, as the NPV or sevin alone (Su, 1982).

'Several thousand mu [1 mu = 0.09 ha] cotton have been treated' with 71.6–91.6% control (Wu, 1986). A wettable powder formulation of *H. armigera* NPV has been produced by the Jianghu Virus Insecticide Experiment Factory, Hubei. It is quoted as useful for *H. armigera* control on cotton, tobacco, pepper, tomato, corn and sorghum. Recommendations are that 303–606 g of the powder be applied in 606–909 l water/ha. Such a spray should be applied two to three times during one pest generation at intervals of 3 days beginning with peak egg hatch. Effectiveness is said to be improved by aerial application, but the type of spray equipment and the droplet sizes generated are not indicated.

With the addition of small amounts of copper sulphate and ferrosulphate, synergism is said to occur. Washing powder (0.1%) and approximately 380 g activated carbon/ha are added, the latter as a UV protectant. The use of other particulate additives is considered by Xia and Xu (1983).

Mythimna (Leucania) separata (NPV)

M. separata is an economically prominent member of a genus essentially associated with Graminae and occurs from Pakistan to New Zealand. While it is fairly important on rice, corn, sorghum and barley in China, it is particularly known as a pest of wheat. It is said to be responsible for a loss of 100 million kg crops annually in China. Breeding populations survive the winter in south China, and with the advent of warmer weather commence to migrate northwards and northwestwards as far as Inner Mongolia. Individual moths fly up to 1000 km (Li, Wong and Woo, 1964) and, as with other migratory pests (e.g. *Spodoptera exempta* in East Africa), the first warning to farmers is often an apparently sudden outbreak of larvae (Hirai, 1984). In the People's Republic of China, the movement of *M. separata* in relation to weather patterns has been studied since the early 1970s. To provide guidance in crop pest control, a predictive system is in place (reviewed by Sharma and Davies, 1983).

NPVs of *M. separata* (MsNPVs) have been collected in India and Japan while in China there have been over 50 isolations. Some isolates from these three main areas, together with the NPV of *Mythimna (Pseudaletia) unipuncta*, have been compared in terms of infectivity and REN (Hatfield and Entwistle, 1988). Genomically an Indian and a Chinese isolate were closely similar but an isolate from Japan appeared more closely related to MbNPV. However, a German isolate of MbNPV was of low infectivity to a Japanese strain of *M. separata* (Doyle *et al.*, 1990). MsNPV is said to be cross-transmissible to *Helicoverpa assulta* in China (Tsai, 1965). A genomic comparison of two MsNPV isolates ('794' and 'Fuyang') from China showed differences in *Eco*R1 digests but revealed widely different molecular weights (75.2×10^6 and 93.3×10^6) (Li, 1984); polypeptides of these two isolates were compared by Mao (1984).

The infectivity of MsNPV isolates has been measured in terms of LD_{50} and LC_{50}. Values for LD_{50} were obtained using a laboratory host culture obtained from Japan:

Chinese MsNPV	209, 371 and 565 OB
Japanese MsNPV	413 and 934 OB
Indian MsNPV	380 OB
M. unipuncta NPV	68 and 118 OB

The figures are for replicate tests with second instar larvae (Hatfield and Entwistle, 1988). For a Chinese isolate, the LC_{50} at 20°C was 3.69×10^6 to 5.35×10^6 OB/ml but the test instar is not known (F. Zeng, personal communication). (Neelgund (1977) in India gave tables of dose-related mortality but did not calculate LD_{50} values.)

Figures provided by Liu *et al.* (1984a) indicate a 30% activity loss following exposure to about 8 h natural sunlight, and almost total inactivation after 3 days.

No field trials data are available for China, but there has been extensive work on the Indian subcontinent, where conventional safety tests have also been conducted.

Pieris (Artogeia) rapae (GV)

P. rapae is a western Old World species which now occurs in China, New Zealand and North America. It is often known as the cabbage butterfly but is a general pest of cruciferous crops and in China is especially important on Chinese cabbage (*Brassica chinensis*).

GVs have been isolated from *P. rapae*, and other pest *Pieris* species, in various parts of the world. In China, PrGV was first found in the 1960s, more than a decade before it was formally described (Su, 1982) and it was again isolated in Canton Province in 1977. Four Chinese isolates compared by *Eco*R1 digests showed slight differences (Yuan and Yin, 1984).

At doses between 9.8×10^2 and 9.8×10^4 OB/ml, the LT_{50} was 5.5–6.7 days. The LC_{50}, following 4 days of exposure of single larvae, was 1 OB/mm^2 but this increased to 89.4 OBs when larvae were exposed in groups.

Much of the developmental work has centred on the Institute of Virology, Wuhan University, Hubei Province. Two sprayable commercial-type preparations are available: a dried foil-packed powder and a liquid OB suspension, both semi-purified. The latter incorprates 2% Fast Green (Malachite Green) as a UV protectant. Fast Green is inexpensive in China.

Using 1 g of the powdered GV product in 30 l water, presumably sprayed to reasonable droplet coverage, gave 90% control on cabbage. The results of further spray trials are referred to under *P. xylostella*, below.

The addition of activated charcoal to a PrGV bait formulation enhanced effectiveness (Su, 1982). Synergism with various chemicals has been studied and was especially marked with acephate (Huang and Dai, 1991).

Formal safety testing of PrGV in China employed pigs, cows, sheep, mice, rabbits, chickens (and other birds), frogs, fish and shrimps. Tests were also conducted with human embryonic lung cells and rabbit and chicken embryonic cells. All these tests gave negative results (Anon., 1984b). Using ELISA on human sera, 2.1% of 241 samples taken in Wuhan reacted, but there was no associated illness (Jiang and Qi, 1984). PrGV was also tested (26×10^{-3} mg/ml) on 22 lepidopterous species, including the silkmoths *B. mori* and *P. cynthia* x *ricini*. Most species were not susceptible but 2% infections were found in *P. xylostella* and *A. flava* (Anon., 1984b).

A PrGV production flow chart is available (Liang *et al.*, 1979). When stored at –7°C for 15 months, PrGV lost about 4% of its activity, at 4°C it lost 12.5% and at room temperature it lost 18.3%.

Plutella xylostella (GV)

Though probably of European origin, *P. xylostella*, diamondback moth, now occurs globally as a pest of cruciferous crops. A GV has been isolated several times in south-east Asia and a naturally occurring NPV is known. (In addition AcNPV and *Anagrapha falcifera* NPV (AfNPV), wide host-spectrum viruses, are cross-infective.) In China, GV was first isolated in Canton Province and later in Hubei Province.

In China, as in many other parts of the world, *P. xylostella* and *P. rapae* may simultaneously infest brassica crops, so attention has been paid to the joint application of their GVs. In one trial, after 11 days there was 100% *P. rapae* and 66.5–89.5% *P. xylostella* control, but in another trial, in which a joint application was compared with the use of the two GVs separately, results were poorer (Liu *et al.*, 1979; Anon., 1984a).

Spodoptera (Prodenia) litura (NPV)

S. litura is an important pest species occurring in most of the warm regions of the Old World with the exception of Africa and western parts of the Middle East, where it is replaced by the closely allied *S. littoralis*. In China, *S. litura* is a notable defoliator of cotton and also attacks rice, tomato, tobacco and brassica (cole) crops.

In China, an NPV of *S. litura* was first recorded near Kwangchow (Canton), Kwangtung Province in 1960 (Tai, 1973). (Further isolations have been made from India to Japan and in the Eastern Archipelago.) A GV virus isolated in Quangzhou Region (Dai, Shi and Yian, 1982) appears not to have been tested as a control agent.

The third instar LD_{50} is reported as 105.3 OB/ml, but this should probably be referred to as an LC_{50}. SlNPV has been field tested on water spinach, taro (*Colocasia* sp.) and cotton. Washing powder (0.01%), presumably as a wetter/spreader, was added to a PIB suspension. Three to four days after spraying, the mean mortality was 80%. In another field test on cotton, a spray of 1×10^7 OB/ml (volume not quoted) gave 75% mortality after 6 days. Tai (1980) obtained best results by spraying before the third instar.

A wettable powder formulation is produced at the Institute of Entomology, Zhongshan University, Guangzhou. Though this is not yet widely licensed it has been used in Central and South China, including Guangdong, Guangxi, Hunan, Hupei and Zhangi Provinces. Applications have been at 4.5×10^{12} to 9.0×10^{12} OB/ha, yielding population reductions between 83 and 90%. The virus has been applied at about 170 l/ha from the air as well as from the ground. During 1987–9 180 ha were treated each year and in 1990 this rose to 360 ha.

SlNPV does not infect the silkmoths *B. mori* and *P. cynthia ricini* or a parasitoid (not named) of *S. litura* (Q.-X. Long, personal communication, 1991). In addition it was not cross-transmissible to *Agrotis ipsilon*, *Agrotis segetum* or *M. separata* (Tai, 1973).

REFERENCES

Aizawa, K. and Nakazato, Y. (1963) Diagnosis of diseases in insects 1959–62. *Mushi* **37**, 155–158.

Anon. (1975) In *Plant Protection Soil and Fertilizer in Hupei*. Chinchow Microbiology Experimental Station and Huachung Teachers College, pp. 29–33.

Anon. (1984a) *Collected papers on Insect Virus Research of the Department of Virology*. T. Lee (ed.). Hubei: Wuhan University.

Anon. (1984b) Safety tests of a granulosis virus insecticide against cabbage butterfly, *Pieris rapae*. In *Collected Papers on Insect Virus Research of the Department of Virology*. T. Lee (ed.). Hubei: Wuhan University. pp. 82–87.

Anon. (1993) Investigation on the nuclear polyhedrosis virus of the cloud geometrid (*Buzura thibetaria*). *Tea in Guizhou* **1**, 19–22.

Cai, X.-Y., Ding, C. and Wang, M. (1986) Isolation and infectivity of a new isolate of nuclear-polyhedrosis virus from the cotton bollworm, *Heliothis armigera*. *Sinozoologia* **4**, 31–36.

Chu, K.-K., Hseih, Y.-T., Chang, H.-E. and Fang, C.-C. (1975) *Acta Microbiologia Sinica* **15**, 93–100.

Dai, G.-Q, Shi, M.-B. and Xian, B.-C. (1982) Preliminary observations on the granulosis virus of the noctuid moth *Prodenia litura* F. in Guangzhou Region. *Acta Virologica Sinica* **2**, 119–123.

Doyle, C.J., Hirst, M.L., Cory, J.S. and Entwistle, P.F (1990) Risk assessment studies. Detailed host range testing of a wild-type cabbage moth, *Mamestra brassicae* (Lepidoptera: Noctuidae), nuclear polyhedrosis virus. *Applied Environmental Microbiology* **56**, 2704–2710.

Fudan University (1975) In *Biological Control*. Shanghai: The New Agricultural Techniques Office, pp. 100–103.

Hatfield, P.R. and Entwistle, P.F. (1988) Biological and biochemical comparison of nuclear polyhedrosis virus isolates pathogenic for the Oriental armyworm, *Mythimna separata* (Lepidoptera: Noctuidae). *Journal of Invertebrate Pathology* **52**, 168–176.

Hirai, K. (1984) Migration of *Pseudaletia separata* Walker (Lepidoptera: Noctuidae): considerations of factors affecting time of taking-off and flight period. *Applied Entomology and Zoology* **19**, 422–429.

Hsao, K.-J. (1981) The use of biological agents for the control of the pine defoliator, *Dendrolimus punctatus* (Lepidoptera, Lasiocampidae), in China. *Protection Ecology* **2**, 297–303.

Hu, C. and Wan, X.S. (1986) Selection of protectants for granulosis virus of *Pieris brassicae*. *Natural Enemies of Insects*, *Kunchang Tiandi*, **8**, 109–112.

Hu, Y. and Liu, N. (1984) Mixed infections of *Cnaphalocrocis medinalis* (Guenee) granulosis virus *Prodenia litura* nuclear polyhedrosis virus against riceleaf rollers and the morphogenesis of these two viruses. In *Collected Papers on Insect Virus Research of the Department of Virology*. T. Lee (ed.). Hubei: Wuhan University, pp. 181–189.

Huang, H. and Dai, G. (1991) Studies on synergism of *Pieris rapae* granulosis virus with insecticides. *Journal of the South China Agricultural University* **12**, 96–103.

Hussey, N.W. and Tinsley, T.W. (1981) Impressions of insect pathology in the People's Republic of China. In *Microbial Control of Pests and Plant Diseases*, H.D. Burges (ed.). London: Academic Press, pp. 785–795.

Hsao, K.-J. (1981) The use of biological agents for the control of the pine defoliator, *Dendrolimus punctatus* (Lepidoptera, Lasiocampidae), in China. *Protection Ecology* **2**, 297–303.

Jiang, C. and Qi, Y. (1984) The detection of *Pieris rapae* granulosis virus natural antibodies in human sera. In *Collected Papers on Insect Virus Research of the Department of Virology*. T. Lee (ed.). Hubei: Wuhan University, pp. 123–125.

Jurkovikova, M. (1979) Activation of latent virus infections in larvae of *Adoxophyes orana* (Lepidoptera: Tortricidae) and *Barathra brassicae* (Lepidoptera: Noctuidae) by foreign polyhedra. *Journal of Invertebrate Pathology* **34**, 213–233.

Lee, T.-C. (1980) *Studies on the Polyhedrosis of Insect Pests*. Wu-Han, Hupei, China: Huazhong Teachers College.

Li, K.-P., Wong, H.-H. and Woo, W.-S. (1964) Route of the seasonal migration of the Oriental armyworm in eastern part of China as indicated by a three year result of releasing and recapturing marked moths *Acta Phytologica Sinica* **3**, 101–110.

Li, S. (1984) Characterization of *Leucania separata* NPV-DNA. In *Collected Papers on Insect Virus Research of the Department of Virology*. T. Lee (ed.). Hubei: Wuhan University, pp. 196–203.
Liang, D, Cai, Y., Liu, D., Zang, Q., Hu, Y.-Y., He, H. and Zhao, K. (1986) *The Atlas of Insect Viruses in China*. Hubei: Department of Virology, Wuhan University.
Liang, T.-Z., Chang, C.-L. and Tsai, Y.-N. (1979) Production of *Pieris rapae* granulosis virus. *Journal of Wuhan University* **1979**, 97–104.
Liu, N.Z. *et al.* (1979) In *Chinese Collected Papers of Virology*, pp. 72–74.
Liu, N., Liang, D., Chang, Q., He D.-Q., Chao, C. and Yang, F. (1984a) Isolation and virulence of polyhedrosis virus against cotton looper. In *Collected Papers on Insect Virus Research of the Department of Virology*. T. Lee (ed.). Hubei: Wuhan University, pp. 41–42.
Liu, N., Liang, D., Chang, Q., Chao, C. and Yang, F. (1984b) Effect of a nuclear polyhedrosis virus against *Anomis flava* (Fabricius). In *Collected Papers on Insect Virus Research of the Department of Virology*. T. Lee (ed.). Hubei: Wuhan University, pp. 43–46.
Liu, S.G., Yang, Z.R., Wu, T.Q., Wang, Z.G., Zhou, S., Liu, S.G., Yang, Z.R., Wu, T.Q., Wang, Z.G., and Zhou, S. (1993) The production and application in large areas of GrNPV insecticide. *Acta Pratacultura Sinica* **2**, 47–50.
Mao, P. (1984) Analysis of polypeptides of two *Leucania separata* nuclear polyhedrosis virus isolates. In *Collected Papers on Insect Virus Research of the Department of Virology*. T. Lee (ed.). Hubei: Wuhan University, pp. 204–211.
Neelgund, Y.F. (1977) Studies on nuclear polyhedrosis of the armyworm, *Mythimna (Pseudaletia) separata* Walker. Dharwad, India: Extension Service and Publications, Karnatak University.
Peng, H.Y., Xie, T.N., Jing, F., Zhang, Y.L. and Liu, Y (1992) Study on new viral pesticide with high effect and without environmental pollution in China. *Biochemical and Biophysical Research Communications* **189**, 680–683.
Pu, G.-L. (ed.) (1978) *The Principles and Methods of the Biological Control of Insect Pests*. Peking: Academic Press.
Risbeth, J. (1978) *The Royal Society Delegation on Biological Control to China*, Ch. 3. London: The Royal Society, p. 77.
Sharma, H.C. and Davies, J.C. (1983) *Miscellaneous Report No. 59: The Oriental Armyworm*, Mythimna separata *(Wlk.): Distribution, Biology and Control: a Literature Review*. London: Centre for Overseas Pest Research.
Sterling, P.H., Kelly, P.M., Speight, M.R. and Entwistle, P.F. (1988) The generation of secondary infection cycles following the introduction of nuclear polyhedrosis virus to a population of the brown-tail moth (*Euproctis chrysorrhoea* (L.)) (Lepidoptera: Lymantriidae). *Journal of Applied Entomology* **106**, 302–311.
Su, T.-M. (1982) Use of bacteria and other pathogens to control insect pests in China. In *Microbial and Viral Pesticides*, E. Kurstak (ed.). New York: Marcel Dekker, pp. 317–332.
Tai, G.-C. (1973) A preliminary study of the polyhedrosis on the cotton leafworm, *Prodenia litura* Fab., in Canton area. *Acta Entomologica Sinica* **16**, 89–90.
Tai, G.-C. (1980) *Journal of Huanan Agricultural Institute* **1**, 128–135.
Tsai, S.-Y. (1965) Studies on a nuclear polyhedrosis of *Pseudaletia separata* (Walk.). *Acta Entomologica Sinica* **14**, 534–550.
Tsai, S.-Y. and Ding, T. (1982) Some insect viruses discovered in China. *Acta Entomologica Sinica* **25**, 413–415.
Watanabe, T. and Aratake, Y. (1974) Two isolates of a nuclear polyhedrosis virus of the brown tail moth, *Euproctis similis*, exhibiting different occlusion body shapes. *Journal of Invertebrate Pathology* **24**, 383–386.
Wu, Z.-Q. (1986) Virus study of *Heliothis armigera* in China. In *Proceedings of IVth International Colloquium of Invertebrate Pathology: Fundamental and Applied Aspects of Invertebrate Pathology*, the Netherland. R.A. Samson, J.M. Vlak and D. Peters (eds). p. 138.
Xia, B.Y. and Xu, B.C. (1983) Influence of the effect of NPV (nuclear polyhedrosis virus) of *Heliothis* with different solid diluents against corn earworm, *Heliothis armigera* (Hübner). *Natural Enemies of Insects, Kunchong Tiandi*, **5**, 146–149.

Xiao, G. (1985) Status of biological control of forest insect pests in China. *Chinese Journal of Biological Control* **1**, 25–35.

Xilin, S. (1989) Lymantriid forest pests in China. In *General Technical Report NE-123 Lymantriidae: A Comparison of Features of New and Old World Tussock Moths*. W.E. Wallner and K.A. McManus (eds). Washington, DC: Forestry Service, pp. 51–64.

Yamada, H. and Oho, N. (1973) A granulosis virus, possible biological agent for control of *Adoxophyes orana* (Lepidoptera: Tortricidae) in apple orchards. I: Mass Production. *Journal of Invertebrate Pathology* **21**, 144–148.

Yin, K.-S. (1982) Study and application of *Boarmia obliqua* NPV. *Report of the Tea Research Institute*, pp. 39–54.

Yin, K.-S. (1988) Production of *Ectropis obliqua* nuclear polyhedrosis virus. *China Tea* **10**, 4–6.

Yu, Y.-X. and Chen, M.-C. (1979) In *The Meeting of the Horticultural Society*, Shanghai, China.

Yu, Z. (ed.) (1989) *Microbial Control of Insects*. Chinese Agricultural Sciences Press.

Yuan, L. and Yin, C. (1984) Comparison of structural polypeptides and genomes of four isolates of PrGV. In *Collected Papers on Insect Virus Research of the Department of Virology*. T. Lee (ed.). Hubei: Wuhan University, pp. 113–115.

17 Japan

YASUHISA KUNIMI

PERSPECTIVE

Historically in Japan, studies on the diseases of the silkworm, *Bombyx mori*, as a factor affecting the well-being of the silk industry, have dominated insect pathology. However, work by Hidaka (1933) demonstrating the possibility of controlling the pine caterpillar, *Dendrolimus spectabilis*, with the fungus *Beauveria bassiana*. Until the early 1960s, fungi were the entomopathogens of choice as biopesticides. Although eclipsed for a while by synthetic organic insecticides interest has recently resurged (e.g. Taketsune, 1983; Kawakami, 1986; Kawakami and Shimane, 1986; Nose, 1990; Shibata *et al.*, 1991; Shimazu *et al.*, 1992; Tsutsumi and Yamanaka, 1996, 1997). A formulation of *Beauvaria brongniartii* for the control of *Psacolhea hilaris* on mulberry and of *Anoplophora malasiaca* on citrus trees was registered in 1995 (Baiorisa-Kamikiri, Nitto Denko Co.).

While *Bacillus moritai*, active against the house fly, *Musca domestica*, and a CPV of *D. spectabilis* were registered in 1962 and 1974, respectively, the only widely used microbial pesticides in Japan are *B.t.* preparations.

A total of 237 insect–virus associations have been described from Japan with 44% being identified as NPVs, 20% as CPVs, 15% as GVs and 8% as iridoviruses (IVs) (Kunimi, 1990, 1993).

As with the People's Republic of China, India and other sericultural countries, a key aspect of the development of microbial pesticides is that they should be demonstrably safe for *B. mori*. Awareness of this potential problem area has for a long time influenced the course of applied investigations in Japan (e.g. Aizawa and Fujiyoshi, 1968) with a sense of caution.

VIRUS PRODUCTION

The CPV/NPV mixture employed experimentally to spray *Lymantria fumida* outbreaks was obtained by collection of larvae in the field and their infection in the fifth and sixth instars in the laboratory. *D. spectabilis* CPV was deliberately multiplied in the field by spraying pine branches *in situ* with a semi-purified suspension of 1×10^6 OB/ml and containing seventh instar larvae on the branches in gauze sleeves. After 2 to 3 weeks, dead and infected larvae were collected, homogenized and filtered through cheese cloth. Each infected larva yielded approximately 5×10^8 OBs.

Virus production in the other species referred to in Table 11.6 (p. 195) was in larvae cultured on semi-synthetic diets. A mass production method for *Adoxophyes orana* GV was developed by Yamada and Oho (1973) and modified by Sato (1984). Virus propagation is in an alternative *Adoxophyes* species. Egg masses are surface sterilized by soaking in 1% $HgCl_2$ for 3 min followed by aqueous rinsing and dipping in a suspension of 1 full-grown virus-infected larva/10 ml distilled water. These egg masses are placed in groups of seven or eight in polypropylene containers ($28 \times 20 \times 6$ cm) with 140 g artificial diet flakes (Tamaki, 1966). They are incubated for 3 weeks at 25°C with a photoperiod of 16 h. An average of 250 diseased larvae can be reared per container. The production cost is about 4.4 yen per infected larva or the equivalent of 21 000 yen/ha. Sixth instar larvae of *Hyphantria cunea* are fed diet containing 2×10^6 NPV OB/g diet for 3 days before transferrence in groups of 200 to wooden boxes containing virus-free diet, where they are incubated for a further 10 days. They are then hand-collected, homogenized in water and filtered. *Lemyra* (*Spilosoma*, *Thanatarctia*) *imparilis* seventh instar larvae are fed on *B. mori* diet containing 1×10^8 NPV OB/g for 72 h at 25°C followed by a further 10–14 days on virus-free diet. For the latter stage, larvae are incubated in groups of 250 in disposable paper boxes from which they are hand-harvested, homogenized, filtered and semi-purified by three cycles of centrifugation at $1000 \times g$. Production of 1×10^9 OBs is estimated at 3.9 yen, 73% of which is attributable to labour costs. Larvae of *Spodoptera litura* are reared on an artificial diet containing pinto beans and bran powders as the principle ingredients (Okada, 1987). For maximum yield of NPV OBs, middle fifth instar larvae are fed diet containing 5×10^7 OB/g for 2 days: they are then reared individually for a further 10 days on virus-free diet. Infected larvae are harvested by aspiration through a glass tip connected to a vacuum line, homogenized in distilled water, centrifuged and resuspended before lyophilization for storage. On average a sixth instar larva yields 8×10^9 OBs at a cost of 10 yen (more than 50 000 larvae were produced by two workers in 5 weeks).

Formulation

Comparatively little attention has been paid to formulation. Several preparations have been purified to varying extents by centrifugation. Lyophilization is a common pre-storage treatment. The final commercial *D. spectabilis* CPV powder (Matsukemin) contained 1×10^8 OB/g; white carbon was added as a UV protectant. White carbon was also used, at 1% final concentration, with *H. cunea* NPV. Following the final round of centrifugation, the OB pellet of *L. imparilis* NPV was resuspended in 0.2% benzalkonium chloride for 60 min to kill contaminant bacteria, a treatment which did not adversely affect viral infectivity.

SAFETY TESTS AND REGISTRATION

There has been a quite intensive level of safety testing of entomoviruses of commercial potential. Tests on *D. spectabilis* CPV are an especially valuable experience as the safety of very few viruses in this group has been studied. The tests included oral administration and intracerebral, intramuscular and intraperitoneal injection of OBs to rats, mice, rabbits, hamsters, carp and killifish. No adverse effects were observed. Honey bees (*Apis mellifera*) were unaffected by the virus but first instar *B. mori* larvae proved susceptible.

Three NPVs, those of *H. cunea*, *L. imparilis* and *S. litura*, have been subjected to varying levels of safety testing. These include intraperitoneal and subcutaneous injections in mice and rats and exposure of carp and chickens. For *Lemyra* NPV, test inocula included a preparation of virions liberated from OBs; exposure of *B. mori* and spiders was examined in addition to the usual laboratory test vertebrates. Apart from a reddening of the eye of mice following *H. cunea* NPV drops, there were no ill effects.

D. spectabilis CPV was produced commercially as Matsukemin by Chugai Pharmaceutical Co. Ltd following registration in 1974. Since 1973, the Tokyo Metropolitan Sericultural Centre has produced a total of 8×10^{13} OB/year of *H. cunea* NPV for distribution to the City Office for use in control of *H. cunea* on roadside trees. Since 1984, the Department of Plant Protection, the Ministry of Agriculture, Forestry and Fisheries has coordinated a programme for NPV control of *S. litura*, tobacco cutworm, and for GV control of *Adoxophyes* sp. and *Homona magnanima*. Several prefectural governments, national and local government offices subsidize the use of *S. litura* NPV. For example, *S. litura* NPV and GVs of *Adoxophyes* sp. and *H. magnanima* formulations prepared by the Kagoshima Agricultural Experiment Station are being distributed to farmers by the Extension Service Branch (Makino *et al.*, 1987; Nishi and Nonaka, 1996).

FIELD EXPERIMENTATION AND USAGE

Adoxophyes orana fasciata (GV)

The nominative species *Adoxophyes orana* is a pest of tea but the subspecies *fasciata* is the summer fruit tortrix. It is an important pest of apple in Japan and has four generations a year. Dependence on chemical insecticides, such as methomyl and salithion, with a broad spectrum of activity has caused residue problems, destruction of summer fruit tortrix natural enemies and insecticide resistance. There is, therefore, great demand for a selective method of control.

A GV of *A. orana* was first isolated in Japan (Aizawa and Nakazato, 1963). (For the history of further virus isolations from *A. orana*, see Chapter 16). Field trials were conducted in Japan during 1970–2 with progressively more apple trees being treated each year (Shiga *et al.*, 1973; Oho, 1975; Shiga and Nakazato, 1980). In the final year, the area treated had increased to four orchards (900–2500 m^2), each tree being sprayed with a macerate of 200 full grown diseased larvae in 30 l of water. Results were promising: in 1971, for instance, there was 74% mortality in treated plots compared with 6% in control plots. Virus infections recurred in the tortrix population for at least three consecutive generations. However, the percentage parasitism by *Apanteles adoxophyesi*, *A. orana* and *Bracon adoxophyesi* (Braconidae) and *Pristomerus* sp. (Ichneumonidae) during the first and second generations after spraying was significantly lower than in the control plots. This difference vanished in the third generation post-treatment (Shiga *et al.*, 1973).

Adoxophyes sp. and *Homona magnanima* (GVs)

Known respectively as the smaller tea tortrix and the oriental tea tortrix in Japan, *Adoxophyes* sp. and *H. magnanima* are serious pests of tea production, the control of which has depended on the use of chemical insecticides.

A GV of *H. magnanima* was isolated by Sato, Oho and Kodomari (1980) and with *A. orana* GV was tested for the joint control of these two pests. Tea was sprayed with a GV mixture (*Adoxophyes* : *Homona* as 2 : 1 LEs) at the rate of 1500 LE/ha in 2000 l of water In the treated generation, *Adoxophyes* was reduced 50–88% and *Homona* 50–79%, with considerable resultant mortality in the second and even the third generations. In six of ten plots, *Adoxophyes* control was better than with chemicals and for *Homona* it was better in all plots (Kodomari, 1987). A CPV and an EPV are also known from *Adoxophyes* sp. (Watanabe, 1973; Ishikawa, Shimamura and Watanabe, 1983).

Bombyx mori (various viruses)

B. mori is the classic silkworm species, which is of very considerable economic importance in Japan. Much attention has been paid to its diseases, including those of viral aetiology. On an applied level, studies have mainly been directed towards control of silkworm virus diseases. However, it has also been perceived that viruses (and other taxa of pathogens) that have potential for control of pest insects should ideally not be infective in *B. mori*. In addition, it has also been considered that through cross-infectivity *B. mori* diseases could have populations in other, wild insects that could constitute a source from which silkworms could become reinfected. This can be a real and immediate problem: for instance, *Diaphania* (*Glyphodes*) *pyloalis*, the mulberry pyralid, has been described as an habitual host of non-occluded viruses pathogenic to *B. mori* (Watanabe, Kurihara and Wang, 1988). Conversely, *H. cunea* was found to be resistant to *B. mori* CPV (Yamaguchi, 1976); more extensive studies by Aratake and Kayamura (1973a) showed appreciable, and differing, host ranges for two types of *B. mori* CPV. Aratake and Kayamura (1973b) found *B. mori* NPV to be infective *per os* in 8 out of 19 species of Lepidoptera tested. There have been a number of other such studies that collectively expanded the understanding of preservation of health in silkworm rearing.

Dendrolimus spectabilis (CPV)

The pine caterpillar, *D. spectabilis*, is one of the major defoliators of *Pinus* spp. in Japan. Adults appear in July and August and the larvae moult three to four times before winter hibernation.

D. spectabilis CPV was first isolated from overwintering larvae in the Kanto district (Koyama, 1958) and has since been tested in state forests under the control of the Government Forest Experimental Station. Helicopter spray trials were conducted in 1974 and 1975 in Ishikawa Prefecture, when 450 ha were treated at 1×10^{11} OBs in 60 l water. In the year of application, there was an average infection rate of 34% but in the following year no larvae could be found (Itaya, 1975; Itaya and Sato, 1975). The total cost, including virus production, labour and use of the helicopter was estimated at 21 597 yen/ha compared with 9212 yen for a chemical insecticide. For satisfactory control it was found necessary to treat *D. spectabilis* larval populations at intermediate densities, i.e. before infestations became so dense that serious economic damage had been inflicted on the trees.

There are practical benefits to spraying a mixture of *B.t.* and CPV (a rapid mortality response to *B.t.* while CPV kills many of the survivors overwinter) (Katagiri *et al.*, 1977, 1978).

Hyphantria cunea (NPV)

The fall webworm was accidentally introduced into Japan from the USA just after World War II. *H. cunea* is bivoltine and attacks many kinds of deciduous tree including cherry (*Prunus* spp.), planes, mulberry and persimmon (*Diospyros kaki*). It is especially well known as a street tree pest in urban areas and, in connection with sericulture, on mulberry trees. Attacks of this kind in the Kanto district and the Tokyo area in 1963 led to a decision by the Kanagawa and Tokyo Sericultural Experimental Station to investigate the utilization of *H. cunea* NPV (HcNPV) as an alternative to chemical insecticides, which, of course, are difficult to apply to mulberry plantations during the silkworm rearing season.

The virus was originally isolated from larvae collected in Tokyo (Aruga *et al.*, 1960) and was first applied to 25 ha mulberry in 1967 by helicopter at 3.6×10^{12} OB/ha in 60 l water. More than 90% larval mortality was observed after 15 days: there was no pre-spray infection in this population. In the following year, 20 ha of mulberry was treated at the lower rate of 1×10^{11} OB/ha, again in 60 l water. Average mortalitics in first, second and third instar larvae 15 days after spraying were 90.0, 74.5 and 67.8%, respectively (Dohi *et al.*, 1967).

Lymantria fumida (CPV and NPV)

The red belly tussock moth, *L. fumida*, is one of the main defoliators of the Japanese fir, *Abies firma*, and the Japanese larch, *Larix leptolepis*, throughout Japan, except Hokkaido where it becomes univoltine. In Tokyo, *L. fumida* outbreaks have occurred at regular intervals of 6 to 7 years, with NPV epizootics being observed 2 to 3 years after the peak of each outbreak (Koyama and Katagiri, 1959; Katagiri, 1977). Although important in reducing *L. fumida* populations, epizootics occur too late to prevent either infestation or damage.

Infestation of over 60 ha fir forest in 1965 resulted in the Forest Agency (Ministry of Agriculture and Forestry) applying a mixture of NPV and CPV by helicopter on 64 ha at a rate of 3×10^{10} OB/ha in 60 l water. Spraying was timed to coincide with the peak presence of second instar larvae. Maximum infection was noted 2 weeks later and was followed by a secondary infection peak after a further 2 weeks. Total larval mortality was estimated to be 84%, of which the greater part was caused by NPV disease. In the following generation, the density of egg masses was reduced by about 90% and the number of eggs per mass by one half.

Mythimna (Pseudaletia) separata (NPV)

M. separata is essentially a pest of graminaceous crops and in Japan assumes particular significance on rice. Much investigation of virus control has been conducted in the People's Republic of China and in India. A proteinaceous factor associated with the spheroid (OB) of *M. separata* entomopoxvirus synergizes the action of the MsNPV (Xu and Hukuhara, 1992) but no attempt has so far been made to exploit this relationship for pest control.

Lemyra (Spilosoma, Thanatarctia) imparilis (NPV)

The mulberry tiger moth, *L. imparilis*, occurs throughout Japan on a limited number of trees and shrubs, e.g. *Fagar ailoanthoides*, *Mallotus japonicus* and *Morus kagayamae*. It is univoltine and, following overwinter hibernation, the larvae become highly polyphagous, attacking low-growing plants and periodically causing damage to mulberry, citrus and to foliage plants such as *Phoenix roebelens* and *Monstera beliciosa*.

An NPV was first isolated in 1981 from overwintering larvae on Hachijo Island, about 300 km from Tokyo (Kunimi, 1986). In field populations, the incidence of NPV disease in pre-overwintering larvae is low (*c.* 1%) but becomes high when they have overwintered. This phenomenon seems attributable to the changing larval habits with age: younger, pre-overwintering larvae are unlikely to encounter virus as there is little possibility of it being splashed from the soil to leaves several metres above the ground; later on the larvae hibernate in the contaminated soil and also may feed on low-growing plants. The resultant epizootics occur too late to prevent damage to foliage plants and so it was decided to test the feasibility of initiating an artificial epizootic in pre-overwintering larvae. Trials commenced in 1982 and, as a result of observations made during the first 2 years, a protocol was developed that has been used on 120 ha on Hachijo Island since 1984. Virus (2×10^9 OB/ha per 420 l) is sprayed on first and second instar larvae and has resulted in an 85% population reduction and a complete absence of damage. The data indicate that this type of treatment results in virus persistence in the population for 3 years (Kunimi, 1987). Application of virus has not been observed to affect rates of parasitism either by *Compsilura concinnata* and *Exorista japonica* (Diptera: Tachinidae) or by *Meteorus pulchricornis* (Hymenoptera: Braconidae).

Spodoptera litura (NPV)

The so-called 'tobacco' cutworm, *S. litura*, is one of the most serious agricultural pests in south-western Japan. The larvae attack over 80 species of plants including sweet potato, soybean, yam, strawberry and cabbage.

An NPV was first isolated in Japan in Kagoshima prefecture (Okada, 1968). It was quickly shown (Okada, 1977) that *S. litura* could be successfully controlled by an aqueous spray of 1×10^{12} OB/ha directed against early-stage larvae using an ULV sprayer applying 5 l/ha. The residue on soybean was said to be still infective to the next generation of larvae 30 days after spraying, probably at least in part because of the presence of secondary inoculum. Further small-scale field trials have been undertaken, notably in Ehime (Yamasaki and Yashioka, 1987), Aichi (Asayama and Takimoto, 1987) and Saitama Prefectures (Nemoto and Okada, 1987), but the largest and most detailed study was in Kagoshima (Makino *et al.*, 1987; S. Makino, personal communication). In a series of experiments conducted on soybean during 1984–7, it was found that application of NPV by mist spray at 2×10^{12} OB/ha in 1.5 l/ha caused greater mortality (87–93%) than application of the same amount of virus sprayed from a helicopter in 100 l/ha (54–83% mortality). In Kagoshima Province, application of NPV on soybean was most effective between early and mid September. Field trials have also shown excellent control on taro, yam and red field beans (Okada, 1987) and on strawberries in the greenhouses (Nemoto and Okada, 1987, 1990).

Other insect pests

These are detailed in Table 17.1.

FUTURE DEVELOPMENTS

So far, only one viral pesticide, *D. spectabilis* CPV, has been registered as a biopesticide in Japan. This is probably in part because of the lack of guidelines for the registration of microbial pesticides but also because of the reluctance of farmers to abandon wide-spectrum pesticides for the less familiar but more target specific biologicals. There are signs that in the 1990s this attitude has been changing, however, as a result of the detection of chemical insecticidal residues not only in crops but also in the environment. Another appreciable factor in altering attitudes is the serious problem of increasing pest resistance to chemical insecticides. Therefore, utilization of virus and other selective microbial pesticides is likely to progress rapidly in Japan in the near future.

Table 17.1 Ground spray trials of entomopathogenic viruses to control some insect pests in Japan

Pest species	Larval instar	Plant or crop	Virus	Year	Total area (a)	Type of spray equipment	Dosage (PIB/a)	Volume (l/a)	Population reduction (%)	Reference
Euproctis subflava	13	*Quercus* spp.	NPV	1969	0	Hand sprayer	2×10^8	3	80	Tsurumi *et al.* (1969)
	15	*Quercus* spp.	NPV	1969	25	Power sprayer	9×10^7	0.9	95	Tsurumi *et al.* (1969)
	Old	–	NPV	1972	–	Power sprayer	10^6/ml	–	74–93	Sudo *et al.* (1972)
	Young	–	NPV	1972	–	Power sprayer	10^6/ml	–	10	Sudo *et al.* (1972)
Euproctis similis	5–6	Mulberry	NPV	1971	4	Power sprayer	2×10^{11}	7	95	Tomita *et al.* (1972)
Dendrolimus superans	4–5	Hemlock	CPV + *B.t.*	1976	40	Power sprayer	3×10^9	3	82	Iwata (1977)
Pieris rapae	1–3	Cabbage	GV	1968	5	Power sprayer	8 infected larvae	10	80	Akutsu (1971)
Mamestra brassicae	–	Cabbage	NPV	1962	–	Hand sprayer	7×10^7	9.9	83	Akutsu (1967)
	–	Cabbage	NPV	1962	–	Hand sprayer	2×10^8	9.9	100	Akutsu (1967)
Plutella xylostella	–	Cabbage	GV	1978	0.21	Hand sprayer	1 infected larvae/100 ml		50	Akutsu (1979)
Homona magnanima	Young	Tea plant	GV	1977	1	Power sprayer	5 infected larvae/1 l	20	74	Sato *et al.* (1986)
Adoxophyes sp.	Young	Tea plant	GV	1977	1	Power sprayer	10 infected larvae/1 l		61	Sato *et al.* (1986)

Area: 1 a = 0.01 ha.

REFERENCES

Aizawa, K. and Fujiyoshi, N. (1968) Selection and breeding of bacteria for control of insect pests in the sericultural countries. In *Proceedings of the US–Japan Seminar on Microbial Control*, Fukuoka, pp. 79–83.
Aizawa, K. and Nakazato, Y. (1963) Diagnosis of diseases in insects, 1959–62. *Mushi* **37**, 155–158.
Akutsu, K. (1967) The use of viruses for control of cabbage armyworm, *Mamestra brassicae* (Linnaeus), and common cabbage worm, *Pieris rapae crucivora* (Boisduval). In *Proceedings of the US–Japan Seminar on Microbial Control of Insect Pests*, pp. 43–49.
Akutsu, K. (1971) Control of the common cabbage worm *Pieris rapae crucivora* Boisduval by a granulosis virus. *Japan Applied Entomology and Zoology* **15**, 56–62 (in Japanese with English summary).
Akutsu, K. (1979) A granulosis virus of the diamondback moth, *Plutella xylostella* (Lepidoptera: Plutellidae). *Bulletin of the Tokyo Agricultural Experimental Station* **12**, 19–24 (in Japanese with English summary).
Aratake, Y. and Kayamura, T. (1973a) Pathogenicity of a cytoplasmic-polyhedrosis virus of the silkworm, *Bombyx mori*, for a number of lepidopterous insects. *Japanese Journal of Applied Entomology and Zoology* **17**, 101–106 (in Japanese with English summary).
Aratake, Y. and Kayamura, T. (1973b) Pathogenicity of a nuclear-polyhedrosis virus of the silkworm, *Bombyx mori*, for a number of lepidopterous insects. *Japanese Journal of Applied Entomology and Zoology* **17**, 121–126 (in Japanese with English summary).
Aruga, H., Yoshitake, N., Watanabe, H. and Hukuhara, T. (1960) Studies on nuclear polyhedroses and their inductions in some Lepidoptera. *Japanese Journal of Applied Entomology and Zoology* **4**, 51–56 (Japanese with English summary).
Asayama, T. and Takimoto, M. (1987) Use of nuclear polyhedrosis virus for the control of the tobacco cutworm, *Spodoptera litura*, in Aichi Prefecture, Japan. *Food and Fertilizer Technology Center Extension Bulletin* **257**, 18.
Dohi, K., Ohta, Y., Morii, K. and Imai, T. (1967) Studies on the control of the fall webworm, *Hyphantria cunea*. *Technical Bulletin, Kanagawa Prefecture* (in Japanese).
Hidaka, Y. (1933) Utilization of natural enemies for control of the pine caterpillar. *Journal of the Japanese Forestry Society* **15**, 1221–1231 (in Japanese).
Ishikawa, I., Shimamura, A. and Watanabe, H. (1983) A new entomopoxvirus disease of the smaller tea tortrix, *Adoxophyes* sp. (Lepidoptera: Tortricidae). *Japanese Journal of Applied Entomology and Zoology* **27**, 300–303 (in Japanese with English summary).
Itaya, Y. (1975) Utilization of microbial insecticides for control of the pine caterpillar in the Noto area. *Forest Pests* **24**, 54–58 (in Japanese).
Itaya, Y. and Sato, K. (1975) Control of the pine caterpillar using microbial insectidides. *Forest Pests* **24**, 84–87 (in Japanese).
Iwata, Z. (1977) Notes on outbreak of the hemlock caterpillar, *Dendrolimus superans* (Butler), in Hokkaido, Japan, *Forest Pests* **26**, 199–202.
Katagiri, K. (1977) Epizootiological studies on the nuclear and cytoplasmic polyhedrosis of the red belly tussock moth, *Lymantria fumida* Butler (Lepidoptera: Lymantriidae). *Bulletin of the Government Forestry Experiment Station* **294**, 85–135 (in Japanese with English summary).
Katagiri, K., Itawa, Z., Kushida, T., Fukuizumi, Y. and Ishizuka, H. (1977) Effects of application of *Bt*, CPV and a mixture of *Bt* and CPV on the survival rates in populations of the Pine caterpillar, *Dendrolimus spectabilis*. *Journal of the Japanese Forestry Society* **59**, 442–448.
Katagiri, K., Iwata, Z., Ochi, K. and Kobayashi, F. (1978) Aerial application of a mixture of CPV and *Bacillus thuringiensis* for the control of the pine caterpillar, *Dendrolimus spectabilis*. *Journal of the Japanese Forestry Society* **60**, 94–99 (in Japanese with English summary).
Kawakami, K. (1986) Notes on the new method for control of the yellow-spotted longicorn beetle, *Psacothea hilaris*. *Konetu no Noyaku* **30**, 59–64 (in Japanese).
Kawakami, K. and Shimane, T. (1986) Microbial control of the yellow-spotted longicorn beetle, *Psacothea hilaris* Pascoe (Coleoptera: Cerambycidae), by an entomogenous

fungus, *Beauveria tenella*. *Journal of Sericultural Science, Japan* **55**, 227–234 (in Japanese with English summary).
Kodomari, S. (1987) Use of granulosis viruses for the control of two tea tortricid moths. *Food and Fertilizer Technology Center Extension Bulletin* **257**, 22–23.
Koyama, R. (1958) Yellow muscardine on *Dendrolimus spectabilis*. *Ringyo-shinchishiki* **53**, 8–9 (in Japanese).
Koyama, R. and Katagiri, K. (1959) On the virus disease of *Lymantria fumida* Butler. *Journal of the Japanese Forestry Society* **41**, 4–18 (in Japanese with English summary).
Kunimi, Y. (1986) Epizootiological studies on a nuclear polyhedrosis of the mulberry tiger moth, *Spilosoma imparilis* Butler (Lepidoptera: Arctiidae). *Bulletin of the Tokyo Metropolitan Sericultural Center* **2**, 1–93 (in Japanese with English summary).
Kunimi, Y. (1987) The use of nuclear polyhedrosis virus for the control of the mulberry tiger moth, *Spilosoma imparilis*. *Food and Fertilizer Technology Center Extension Bulletin* **257**, 35–36.
Kunimi, Y. (1990) A list of viral diseases of Japanese insects. *Plant Protection* **44**, 1–8 (in Japanese).
Kunimi, Y. (1993) A list of insect diseases of Japanese insects. In *Manual for the Researches on Insect Pathogens*. H. Iwahana, M. Okada, Y. Kunimi and M. Shimazu (eds). Tokyo: Nihon Shokubutuboeki-kyoukai, (in Japanese).
Makino, S., Fukamachi, S., Yamashita, S., Horikiri, M., Tanaka, A. and Okada, M. (1987) Ground and aerial applications of nuclear polyhedrosis viruses for the control of the tobacco cutworm, *Spodoptera litura*, in Kagoshima Prefecture, Japan. *Food and Fertilizer Technology Center Extension Bulletin* **257**, 14–15.
Nemoto, H. and Okada, M. (1987) Microbial control of the tobacco cutworm, *Spodoptera litura*, on strawberry grown in the greenhouse. *Food and Fertilizer Technology Centre Extension Bulletin* **257**, 19–20.
Nemoto, H. and Okada, M. (1990) Pest management for strawberries grown in the greenhouse: microbial control of the tobacco cutworm, *Spodoptera litura*, and agricultural, chemical and physical control of aphids. *SROP/WPRS Bulletin* **XIII**/5, 149–152.
Nishi, Y. and Nonaka, T. (1996) Biological control of the tea tortrix: Using granulosis virus in the tea field. *Agrochemicals Japan* **69**, 7–10.
Nose, T. (1990) Utilization of *Beauveria bassiana* for control of planthoppers and leafhoppers in paddy fields. Ms. Thesis, Tokyo University of Agriculture and Technology. (in Japanese).
Oho, N. (1975) Possible utilization of a granulosis virus for control of *Adoxophyes orana* Fischer von Roslerstamm (Lepidoptera: Tortricidae) in apple orchards. In *Approaches to Biological Control*. K. Yasumatsu and H. Mori (eds). Tokyo: University of Tokyo Press, pp. 61–68.
Okada, M. (1968) Isolation of entomopathogens from the tobacco cutworm. *Bulletin of Chugoku-branch of Japanese Society of Applied Entomology and Zoology* **10**, 36–39 (in Japanese).
Okada, M. (1977) Studies on the utilization and mass production of *Spodoptera litura* nuclear polyhedrosis virus for control of the tobacco cutworm, *Spodoptera litura* Fabricius. *Bulletin of the Chugoku National Agricultural Experiment Station* **E**(12), 1–66 (in Japanese with English summary).
Okada, M. (1987) Utilization and mass production of nuclear polyhedrosis viruses for the control of some noctuid larvae. *Food and Fertilizer Technology Center Extension Bulletin* **257**, 11–14.
Sato, T. (1984) Mass production methods of baculoviruses of *Adoxophyes* spp. *Plant Protection* **38**, 366–399 (in Japanese).
Sato, T., Oho, N. and Kodomari, S. (1980) A granulosis virus of the tea tortrix, *Homona magnanima* Diakonoff: its pathogenicity and mass-production method. *Applied Entomology and Zoology* **15**, 409–415.
Sato, T., Oho, N. and Kodomori, S. (1986) Utilization of granulosis viruses for controlling leafrollers in tea fields. *Japanese Agricultural Research Quarterly* **19**, 171–175.
Shibata, E., Yoneda, Y., Higuchi, T. and Ichinose, H. (1991) Control method of the adult sugi borer, *Semanotus japonicus* Lacordaire (Coleoptera: Cerambycidae) using non-

woven fabric sheet with an entomogenous fungus, *Beauveria brongniartii* (Sacc.) Petch, in Japanese cedar, *Cryptomeria japonica* D. Don, Stand. *Applied Entomology and Zoology* **26**, 587–590.

Shiga, M. and Nakazato, H. (1980) Use of granulosis virus for the control of *Adoxophyes orana*. *Plant Protection* **34**, 128–132 (in Japanese).

Shiga, M., Yamada, H., Oho, N, Nakazawa, H. and Ito, Y. (1973) A granulosis virus, possible biological agent for control of *Adoxophyes orana* (Lepidoptera: Tortricidae) in apple orchard. II Semipersistent effect of artificial dissemination into an apple orchard. *Journal of Invertebrate Pathology* **21**, 149–157.

Shimazu, M., Kushida, T., Tsuchiya, D. and Mitsuhashi, W. (1992) Microbial control of *Monochamus alternatus* Hope (Coleoptera: Cerambycidae) by implanting wheat-bran pellets with *Beauveria bassiana* in infected tree trunks. *Journal of the Japanese Forestry Society* **74**, 325–330.

Sudo, C., Funabashi, T., Takahoko, K., Ito, A. and Asayama, T. (1972) Studies on biological control of the tussock moth, *Euproctis subflava* (Bremer). *Bulletin of Nagoya Institute of Health* **19**, 81–90 (in Japanese).

Taketsune, A. (1983) Studies on the population dynamics of the pine sawyer, *Monochamus alternatus* Hope (Coleoptera: Cerambycidae), and its control by some pathogens. *Bulletin of the Hiroshima Forest Experiment Station* **18**, 39–62 (in Japanese with English summary).

Tamaki, Y. (1966) Mass rearing of the smaller tea tortrix, *Adoxophyes orana* Fischer von Roslerstamm, on a simplified artificial diet for successive generations (Lepidoptera: Tortricidae). *Applied Entomology and Zoology* **1**, 120–124.

Tomita, I., Kojima, Y. and Tanabe, H. (1972) Studies on microbial control of the injurious insect of mulberry field. I Infectivity of polyhedrosis virus to larvae of *Leuproctis similis* and its control. *Bulletin of Aichi Agricultural Experimental Station* **D(3)**, 69–76 (in Japanese with English summary).

Tsurumi, M., Saida, I., Fujii, R., Kusaya, S., Dosi, M., Asanuma, K., Inoue, M. and Hanabusa, T. (1969) Studies on biological control of the oriental tussock moth, *Euproctis subflava*. *Technical Bulletin*, Okayama Prefecture (in Japanese).

Tsutsumi, T. and Yamanaka, M. (1996) Effects of non-woven fabric sheet containing entomogenous fungus, *Beauveria brongniartii* (Sacc.) Petch GSES, on adult yellow spotted longicorn beetle, *Psacothea hilaris* (Pascoe) (Coleoptera: Cerambycidae) on fig tree. *Japanese Journal of Applied Entomology and Zoology* **40**, 145–151 (in Japanese with English summary).

Tsutsumi, T. and Yamanaka, M. (1997) Infection by entomogenous fungus, *Beauveria brongniartii* (Sacc.) Petch GSES, of adult yellow spotted longicorn beetle, *Psacothea hilaris* (Pascoe) (Coleoptera: Cerambycidae) by dispersing conidia from non-woven fabric sheet containing fungus. *Japanese Journal of Applied Entomology and Zoology* **41**, 45–49 (in Japanese with English summary).

Watanabe, H. (1973) On a cytoplasmic polyhedrosis of the smaller tea tortrix, *Adoxophyes fasciata* Walsingham. *Japanese Journal of Applied Entomology and Zoology* **17**, 1–4 (in Japanese with English summary).

Watanabe, H., Kurihara, Y. and Wang, Y.-X. (1988) Mulberry pyralid, *Glyphodes pyloalis*: habitual host of nonoccluded viruses pathogenic to the silkworm, *Bombyx mori*. *Journal of Invertebrate Pathology* **52**, 401–408.

Yamada, H. and Ono, N. (1973) A granulosis virus, possible biological agent for control of *Adoxophyes orana* (Lepidoptera: Tortricidae) in apple orchard. I Mass production. *Journal of Invertebrate Pathology* **21**, 144–148.

Yamaguchi, K. (1976) Resistance of the fall webworm, *Hyphantria cunea*, to the cytoplasmic polyhedrosis virus of the silkworm, *Bombyx mori*. *Journal of Sericultural Science, Japan* **45**, 377–378 (in Japanese).

Yamasaki, Y. and Yoshioka, K. (1987) The use of nuclear polyhedrosis virus for control of the tobacco cutworm, *Spodoptera litura*, in soybean fields in Ehime Prefecture, Japan. *Food and Fertilizer Technology Center Extension Bulletin* **257**, 16–17.

Xu, J. and Hukuhara, T. (1992) Enhanced infection of a nuclear polyhedrosis virus in larvae of the armyworm, *Pseudaletia separata*, by a factor in the spheroids of an entomopoxvirus. *Journal of Invertebrate Pathology* **60**, 259–264.

18 Africa, the Near and Middle East

EDNA KUNJEKU, KEITH A. JONES and GALAL M. MOAWAD

PERSPECTIVE

Work in Africa is notable for some relatively early ventures into virus control of pests. Especially worthwhile singling out are the pioneering studies of Ossowski (1957) on *Kotochalia junodi*, wattle bagworm, in South Africa, Coaker (1958) on *Helicoverpa armigera*, Old World cotton bollworm, in Uganda and Abul-Nasr in Egypt on *Spodoptera littoralis*, Egyptian cotton leafworm, on cotton and some other crops (Abul-Nasr, 1959a,b). At present, control of the cotton leafworm is achieved mainly by hand collection of egg masses, a method that has been in use since at least 1905 (Willcocks, 1905), having been first recommended by a government committee in 1885 (Hafez, 1985). Also the early season infestation of cotton is reduced by restricting the irrigation of berseem (Egyptian clover, *Trifolium alexandrinum*), which is a source of the first annual generation of cotton leafworm. This tradition of using alternative methods of control is indicative of the wish in Egypt to minimize the use of chemical insecticides.

By far the most research on BVs in Egypt, and in the northern part of the region under discussion, has been devoted towards cotton leafworm. Here development of BV has been pursued in a thorough, coordinated and comprehensive manner. In the 1990s, it has reached the position of a fully available technology in which those aspects so often neglected in other programmes, notably safety tests, environmental impact assessments and production methods leading to a microbiologically non-hazardous final sprayable inoculum by methods appropriate to farmers' capabilities, have been adequately addressed.

H. armigera is a notable pest, which has been studied over a long period of time and in at least five African nations. However, work has been fragmentary and uncoordinated and there is now a great need to make further progress. The problem of *H. armigera* breaks down into two main areas. Firstly, control on sorghum, which has been shown to be exceptionally successful in Africa, a situation that is well supported by results obtained on *Helicoverpa* and *Heliothis* spp. on sorghum in North America and Australia. There is almost no technical reason why control on this crop should not now be pressed forward: success also seems likely on various vegetable crops and on soybean. Secondly, control on cotton, where the problem with NPV control of *H. armigera* is twofold. Control experiments have not been notably successful but there has been a little work that takes adequate cognisance of the intensive developments in *Heliothis* and *Helicoverpa* spp. control on cotton in North America and, to a lesser but useful extent, in Australia. An analysis of field control trials in the USA (Entwistle and Evans, 1985) suggests that one of the main keys to successful *Helicoverpa zea* control lies in frequent application of fairly low doses, e.g. 2×10^{11} OB/ha in perhaps as many as 18 sprays, with a seasonal virus total usage of about 3.6×10^{12} OBs, the latter figure being similar to some single application doses! Formulation also requires more attention in Africa. The other problem with cotton in Africa, compared with other major growing areas such as the USA and Australia, is the taxonomic diversity of bollworm species, which includes *Pectinophora gossypiella*, *Diparopsis* and *Earias* spp. and, occasionally, *Cryptophlebia leucotreta*. There are however, real possibilities of both *H. armigera* and *Diparopsis watersi* control with MbNPV; and the susceptibility of other bollworms to MbNPV apparently remains to be reported. Should MbNPV become widely available, its use could be considered for *H. armigera* control on other crops. It seems surprising that BVs have not yet been found in some other notorious pests of cotton, namely *Earias biplaga*, *E. insulana* and *P. gossypiella*, though non-occluded viruses are known in the last species (Monsarrat, Abot-Ela and Zeddam, 1994).

In general, virus control in Africa has been neglected; for instance, at a Southern African Development Cooordination Conference (SADCC) workshop (Namponya, 1989) no mention is made of entomopathic viruses in Angola, Botswana, Lesotho, Malawi or Swaziland. In several instances, such as Botswana, studies on entomopathic viruses have been carried out by expatriate workers and on completion of their contracts, work stopped. But real possibilities lie everywhere, for instance in forestry, subsistence grain crops such as maize and sorghum, vegetables and such important cash crops as cotton and oil palm. The fairly recent studies on a virus inducing sterility in tsetse fly, *Glossina* spp., raise the real possibility of control by sterile male release (p. 285).

Significant virus control investigations in Africa, the Near and Middle Easts are shown in Table 18.1.

VIRUS PRODUCTION

By far the most intensive development has been concentrated on *S. littoralis* NPV in Egypt and the UK. This research has been entirely directed toward *in vivo* NPV production. (Some research on *in vitro* production in Israel has been directed towards the use of BVs as expression vectors (Neutra, Levi and Shoham, 1992)). Early results indicated clearly that production of purified NPV suspensions is uneconomic (Topper *et al.*, 1984; Jones, 1988a). Techniques were, therefore, developed for the production of an acceptable preparation of unpurified NPV. The method evolved requires that infected insects be harvested while still alive in order to minimize bacterial contamination (McKinley *et al.*, 1989). Briefly, the host insect is infected and reared on an artificial diet in individual cells (this obviates losses through cannibalism) made by pressing an aluminium honeycomb (Aeroweb) into

Table 18.1 Virus control of insect pests in Africa, the Near and Middle Easts: significant investigations

Pest species	Botswana	Cape Verde Islands	Cameroon	Egypt	Ivory Coast	Kenya	Malawi	Nigeria	Oman	South Africa	Tanzania	Tchad	Tunisia	Uganda	Zimbabwe
Casphalia extranea	–	–	–	–	+	–	–	–	–	–	–	–	–	–	–
Chrysodeixes spp.	–	–	–	–	–	–	–	–	–	–	–	–	–	–	+
Colias electo	–	–	–	–	–	–	–	–	–	–	–	–	–	–	–
Cryptophlebia leucotreta	–	+	–	–	–	–	–	–	–	–	–	–	–	–	–
Glossina spp.	–	–	–	–	+	+	–	–	–	+	–	–	–	–	–
Gonometa podocarpi	–	–	–	–	–	–	–	–	–	–	–	–	–	+	–
Helicoverpa armigera	+	–	+	–	+	–	+	+	–	–	–	+	–	+	–
Kotochalia junodi	–	–	–	–	–	–	–	–	–	+	–	–	–	–	–
Oryctes spp.	–	–	–	–	+	–	–	–	+	–	+	–	–	–	–
Parasa viridissima	–	–	–	–	+	–	–	–	–	–	–	–	–	–	–
Phthorimaea opercullela	–	–	–	+	–	–	–	–	–	–	–	–	+	–	–
Spodoptera exempta	–	–	–	–	–	+	–	–	–	+	–	–	–	–	–
Spodoptera littoralis	–	–	–	+	–	–	–	–	–	–	–	–	–	–	–
Thysanoplusia orichalcea	–	–	–	–	–	–	–	–	–	–	–	–	–	–	+

large trays containg a layer of the diet 1 cm in depth. The diet surface is sprayed with virus inoculum (equivalent to approximately 1×10^6 OB/larva) prior to insertion of the honeycomb. A single 7 day-old larva is manually inserted into each cell and after placement of a lid on the tray the larvae are incubated at 25°C for 7 days. The live larvae are then collected and subjected to low temperature incubation at <14°C for 7 days to increase virus production to an optimum while limiting the growth of bacteria (McKinley *et al.*, 1989; Anon., 1991). Although a maximum yield of 5.4×10^9 OB/larva can be obtained, under operational conditions in Egypt the average yield was 7×10^8 OB/larva (Jones, 1994). The harvested larvae are freeze-dried, a process which appears to have no adverse effect on *S. littoralis* NPV.

In Zimbabwe, the Plant Protection Research Institute, Ministry of Agriculture uses the semi-loopers *Thysanoplusia orichalcea* and *Chrysodeixes chalcites* to produce rather modest quantities of NPV for distribution as 1 g samples to farmers who then infect larvae that are to be placed live in soybean fields to initiate epizootics. Stock virus partial purification and recovery is by the lactose co-precipitation method of Dulmage, Martinez and Correa (1970). The resultant powder is frozen or refrigerated pending distribution. Farmers are instructed to resuspend virus in water with a drop or two of household detergent and to dip and dry soybean leaves on which larvae of about 1 cm in length are then fed. Eventually the larvae turn yellowish and sluggish, at which stage they are placed in heavy infestation foci in the field. Farmers are encouraged to keep diseased larvae from the fields at the end of the growing season and to store them frozen for use next season. For small farmers without facilities, a fresh supply of virus is issued each season. Recommendations have been made for a central facility to produce virus under controlled conditions to ensure adequate monitoring and restriction of undesirable bacterial types and to maintain a continuous check on the identity of the virus genome.

Formulation

In Egypt, freeze-dried NPV-infected *S. littoralis* larvae are formulated to give a wettable powder as this results in a product that does not need to be refrigerated during storage or require the addition of anti-microbial agents to limit the multiplication of unwanted bacteria or other microorganisms. The formulation has been described by McKinley *et al.* (1989) and consists of 100 g freeze-dried NPV-infected larvae, 60 g Speswhite china clay and 40 g Neosyl synthetic silica onto which a wetting agent, Etocas 30, has been adsorbed at a ratio of 50:50. The final product is milled and sieved. The china clay improves flowability and Etocas promotes wettability. When field tested, this formulation was shown to be effective in controlling *S. littoralis* (Jones *et al.*, 1994). The excellence of its environmental persistence, as estimated from field application rates, is indicated by a drop in mortality of neonate larvae from 92% to 80% between 8 and 82 days (Jones, 1988b). Field tank additives are mentioned below. The suitability of *S. littoralis* NPV for formulation in oil, with a view to ULV application, was tested by exposure to six vegetable and 12 mineral oils during 18 months' storage. Arachis oil seemed most suitable as it was associated with least potency degradation (although rapid degradation occurred in this oil after this period) and did not have anti-feedant properties. Further work is in progress (Cherry *et al.*, 1994).

The use of viruses with pesticides

The integration of use of MbNPV (as Mamestrin, Calliope SA, France) with several chemical insecticides has been inspected in trials on *H. armigera* control on cotton in Cameroun and the Ivory Coast (Calliope SA, 1985).

Safety tests and registration

Comprehensive conventional laboratory safety test of the NPVs of *H. armigera*, *S. exempta* and *S. littoralis* have been conducted by the Institute of Virology and Environmenal Microbiology, Oxford, UK in a UK Overseas Development Agency-funded project. There was no evidence of the virus being harmful to mammals (Carey and Harrap, 1980; McKinley, 1980a).

FIELD EXPERIMENTATION AND USAGE

Agrotis ipsilon and *A. segetum* (NPVs)

In Egypt, an NPV has been described in the black cutworm, *A. ipsilon*, which, in combination with chemical insecticides, gave an improved level of mortality (Salama and Moawad, 1988). Also in Egypt, laboratory tests showed *A. segetum* to be susceptible to an homologous NPV (which has not been isolated in Egypt) and to *H. armigera* NPV. *A. ipsilon* was susceptible to AcMNPV (S. Khattab, personal communication).

Casphalia extranea

C. extranea is discussed under Limacodidae, below.

Chrysodeixes acuta and *C. chalcites*

C. acuta and *C. chalcites* are discussed under semi-loopers, below.

Colias electo (NPV)

Epizootics in the South African lucerne butterfly, *C. electo*, can be initiated by use of overhead irrigation to splash inoculum from soil to foliage or even by dragging a branch through the fields (Polson and Tripconey, 1970).

Cryptophlebia leucotreta (GV)

The false codling moth is a pest of several fruit crops, maize and cotton in tropical and southern Africa, the larvae mainly feeding internally. GV isolates have been obtained from the Cape Verde Islands, Ivory Coast and South Africa. The GV has been shown to be very infective, at a level similar to CpGV in *Cydia pomonella*; these GVs are closely similar (Fritsch and Huber, 1986). Persistence on citrus leaves was investigated in the Cape Verde Islands and the field half life was found to be about 2 days (Fritsch and Huber, 1989). Field control trials in the Cape Verde Islands showed applications at concentrations of 1×10^8 and 1×10^9 OB/ml with skimmed milk and a wetting agent reduced damage by 77% on citrus and 65% on peppers (*Capsicum*) (Fritsch, 1988).

In the Ivory Coast, Angelini and Vandamme (1972) isolated an NPV from *Argyroploce* sp. on cotton (possibly a *Cryptophlebia* species). An attempt to transmit this NPV to the cocoa flush leaf defoliator, *Anomis leona*, resulted in high mortality in the first three instars but with symptoms atypical for NPV disease (Lavabre, Ban and Vandamme, 1966).

Ctenoplusia limberina

C. limberina is discussed under semi-loopers, below.

Diparopsis watersi (NPV)

An NPV that was shown by REN analysis to be closely related to AcMNPV was isolated from the African cotton bollworm *D. watersi* in Cameroun (Croizier *et al.*, 1980). Despite its high infectivity, even in late instars, there appear to have been no field trials. MbNPV, as Mamestrin (Calliope SA) is said to be more infective to *D. watersi* than to the homologous host.

A gut-infecting virus, possibly a CPV, was recorded in the Ivory Coast (Angelini and Vandamme, 1964).

Forest insects

Gonometa podocarpi *(picornavirus)*

Gonometa podocarpi is a lasiocampid moth that attacks many tree species in East Africa, notably *Pinus patula*. The virus caused decimating epizootics in Uganda (Austara, 1971) and was sprayed as a very effective control agent (Brown, 1965). It has been classified as a picornavirus (Moore, Reavy and King, 1985).

Nudaurelia cytherea capensis *(Nudaurelia β virus)*

The pine tree emperor moth causes severe defoliation of pines, particularly *Pinus radiata*, in the Cape Province of South Africa. Nudaurelia β virus is the type name for this virus group (Tripconey, 1970; Moore *et al.*, 1985) and may be involved, with at least four other viruses (Juckes, Longworth and Reinganum, 1973), in decimating epizootics. Epizootics of 'wilt' disease also occur in *N. cytherea cytherea* in Natal, Transvaal and Swaziland (Hepburn, 1964) but no attempts appear yet to have been made to develop virus control for either pest subspecies.

Glossina *spp. (unclassified virus)*

Hyperplasia of the salivary glands (h.s.g) of both male and female adults of the tsetse fly *G. pallidipes* was first observed in South Africa (Whitnall, 1934). The association of this condition with a virus was noted in 1978 (Jaenson, 1978) and further explored in 1980 (Otieno *et al.*, 1980). The virus was described as a rod approximately 50 nm in diameter and often exceeding 1000 nm in length and containing dsDNA (Odindo, Payne and Crook, 1984). Infection is associated with gonadal lesions in both sexes (Jura *et al.*, 1988), the males being sterile (Odindo, 1988). The complete biology of the virus is probably not yet known but it has been shown that adults can be infected *per os* and by micro-injection. No similar pathology has been found in other *Glossina* spp. in East and Southern Africa, but in the Ivory Coast, West Africa, hyperplasia of the salivery glands was observed in *G. palpalis*, *G. pallicera* and *G. nigrofusca* (Gouteux, 1987). When third instar larvae of *G. morsitans morsitans* were inoculated with the virus, resultant adult males were sterile but reproductively active and the post-coital receptivity of their mates for healthy males was greatly reduced, thus decreasing the possibility of subsequent fertilization (Jura and Davies-Cole, 1992). The collective implication of the studies is that the *G. pallidipes* virus may, through sterile male release, offer the possibility of biological control of tsetse and that such control may be effective for other species, e.g. *G. m. morsitans* and some West African species, at the very least. Until this virus has been more precisely characterized and until its biological properties are better understood, it is recommended that it should be treated with caution.

Helicoverpa armigera **(NPVs and GV)**

The African 'cotton bollworm' is possibly the most important Old World insect pest. In Africa, *H. armigera* is an especially notable pest of sorghum, maize, cotton, chickpeas and tomatoes. Virus control studies have been limited to exploration of the use of NPVs on cotton and sorghum. The infectivity of a commercial *B.t.* preparation and an NPV isolate from Botswana were compared by Daoust and Roome (1974). Calculations based on LC_{90} and LD_{90} suggested that concentrations for control of *H.armigera* on sorghum would be rather low and more economic than for the chemicals then in use. The NPV isolate appeared to be slightly more potent than that tested on *H. zea* by Ignoffo (1965). Field trials on sorghum in Botswana demonstrated that a single application, at 100 LE/ha with 0.6% molasses (200 LE/ha if molasses is omitted) gave season-long control (Roome, 1975). Neither the commercial *H.zea* NPV nor *B.t.* was as effective as the local virus. A high NPV activity was detected up to 30 days after application, with considerable survival reported well beyond this (Roome and Daoust, 1976). NPV has also been isolated in Egypt, e.g. from a laboratory culture (Moawad *et al.*, 1986; Salama, Moawad and Magahed, 1986) with LD_{50} values of 7.82×10^2 OBs for second instar larvae and 1.47×10^5 OBs for third instar larvae (Moawad *et al.*, 1986). These isolates have not been characterized or field tested.

However, the efficacy of NPV control on cotton has been poorer. In Uganda, Coaker's early study (1958) showed that application of 5×10^8 OBs in 20 ml or 1.5×10^9 in 60 ml per plant (approximately equal to 1.25×10^{12} to 3.35×10^{12} OB/ha) resulted in 56.6 and 70.0% control, respectively. McKinley (1971) compared a local NPV isolate with Biotrol VHF (Nutrilite Products Inc.), a commercial

H. zea NPV preparation, and, despite good laboratory performance, obtained poor field responses. Similarly Roome (1975) in Botswana obtained much poorer results on cotton than on sorghum. Work in the Ivory Coast using an isolate from Bouake was successful but at the expense of a dose of 5000 LE/ha (probably $>5 \times 10^{12}$ OBs) (Angellini and Vandamme, 1972). The first field trials in Tchad employed a local isolate and were not successful (Atger, 1969). Later (Cadou and Soubrier, 1974) a local *H. armigera* NPV isolate was compared with Viron H (International Minerals Corp., USA), another *H. zea* NPV commercial preparation, and also gave inadequate responses. Trials using Mamestrin (MbNPV) were reported by Calliope SA (1985). It is difficult to analyse the data presented: in an Ivory Coast trial there appeared to be no control treatment and NPV (at 1×10^{13} OB/ha per treatment) was not separately tested from combinations with chemical insecticides. In Cameroun, application of the same Mamestrin dose every 3 or 7 days with triazophos at 250 g/ha a.i. every 14 days considerably elevated yields. Eventually, it was shown in Cameroun that a combination of MbNPV (1×10^{13} OB/ha) as Mamestrin and cypermethrin at a very low dose (5 g/ha) gave results as good as those obtained with pyrethroids alone at normal dose (Renou, 1987). The addition to this combination of the phagostimulant GustolR (1kg/ha) was recommended by Montaldo (1991). Virus control with *H. zea* NPV (as Elcar) of *H. armigera* on tomato was studied in comparison with *B.t.* (as Dipel and Thuricide) in Nigeria. No interactive effects were noted in mixed applications, each substance separately providing promising control (Lutwama and Matanmi, 1988). In any event, in the light of successes in virus control of *Heliothis* on cotton in North America (the reward of much effort), and the successful control of *H. armigera* on several crops in Thailand (Chapter 15), the subject of *H. armigera* NPV control *per se* should be reopened in Africa.

Heliothis NPV (as two commercial preparations, Elcar and Viron-H) at 6×10^{11} OB/ha was compared with monocrotophos (as Nuvacron) at 380 g/ha a.i. for *H. armigera* control of chickpea (*Cicer arietinum*) in the field in Syria. The chemical insecticide was more effective than either virus preparation, but none of the treatments significantly affected pod damage or seed yield, possibly because the *Heliothis* population before treatment was below the economic threshold level and possibly also because of the presence of other pests not susceptible to the virus (Abdally, Makkouk and Cardona, 1987). A GV seems first to have been isolated in South Africa, where its persistence in bird faeces was shown (Gitay and Polson, 1971). Joint infections of an NPV and a GV, isolated in South Africa (Whitlock, 1974), resulted in interference, indicating the importance of prevention, through good-quality control, of GV contamination during *H. armigera* NPV production (Whitlock, 1977a). Mortality patterns induced by the GV (Whitlock, 1977b) suggest it may not be a suitable virus for field control.

Kotochalia junodi (NPV)

Very few psychid moths are pests, but *K. junodi*, wattle bagworm, a polyphagous widespread African species, is a regular defoliator in South Africa of black wattle, *Acacia mearnsii*, the bark of which is a major source of commercial tannins.

Virus control of *K. junodi* was studied by L. L. J. Ossowski at the Wattle Research Institute in South Africa. His work is historically notable both because it was one of the very earliest essays of its kind and because it had a successful outcome. Ossowski was probably also the first in applying an insect virus from the air. The results of pilot experiments (Ossowski, 1957) showed that application of 100 ml per tree of an OB suspension of 1×10^7 OB/ml with a hand atomizer caused newly hatched larvae to become moribund in 3 to 4 days and to die 2 days later. Applications at a higher rate conferred no extra benefit. In the following season, larger-scale experiments employed lower concentrations (1×10^6 to 1×10^7 OB/ml) at about 68 l/ha (6.8×10^{11} to 6.8×10^{12} OB/ha) with a motorized sprayer. Mortality was very high (Ossowski, 1958). Large-scale trials were conducted in 1957, when over 500 acres (200 ha) were treated from the air in various parts of the wattle area. In treating first instar populations, concentrations of 2.5×10^5 to 1×10^6 OBs in 57 l/ha (1.43×10^{10} to 5.7×10^{10} OB/ha) were used. A rate of 2.85×10^{10} OB/ha gave a high mortality while at 1.43×10^{10} OB/ha the trees remained green. However, at the higher doses the bagworm population returned to an endemic density in the succeeding year while at the lowest dose this process was more gradual (Ossowski, 1962). Ossowski's investigations were wide reaching and included observations and experiments showing that virus introduced from another area had a higher virulence than endemic virus (Ossowski, 1960).

Twenty-five full-grown virus-killed larvae yielded sufficient inoculum to treat a hectare of 8 m tall trees.

Limacodidae

Casphalia extranea *(parvovirus)*

Experimental spraying in the Ivory Coast of a parvovirus was conducted on two 10 ha oil palm plots at 50 and 100 LE/ha. One additional 10 ha plot served as a control and 50 ha were treated with deltamethrin by helicopter. Both virus treatments gave 92% control of *C. extranea* after 2 weeks. By week four, disease spread had brought about a gradual population decline in the adjacent control plot. Kill by deltamethrin was rapid but the chemical was expensive and is said to have been 'inconvenient' (Fediere *et al.*, 1986a).

Parasa viridissima *(picornavirus)*

The nettle caterpillar *P. viridissima* is widely known in West Africa. A life table study in Nigeria showed early larval mortality, which included that from diseases and parasites, to be the key factor in natural regulation of populations on oil and coconut palms. In the Ivory Coast, the virus was applied aerially at 425, 1902 and 3704 g dead larvae/ha on 4–5 m tall oil palms. Two weeks later, mortalities were 82.5, 96.6 and 97.2%, respectively. Larval populations on adjacent control trees also had a high mortality, possibly owing to disease spread. During subsequent generations *P. viridissima* was absent or at very low densities in treated areas (Fediere *et al.*, 1986b).

Oryctes monoceros, O. boas and *O. rhinoceros* (OrV)

Oryctes monoceros and *O. boas* are endemic African rhinoceros beetle species whilst *O. rhinoceros* is the well-known rhinoceros beetle of the Pacific and south-east Asian spheres, finding its westermost hold in Oman. All attack coconut and oil palms. *O. boas* breeds only in manure and sludge and so is largely restricted to inhabited areas, but *O. monoceros* and *O. rhinoceros* breed in many forms of decaying plant material. In Africa, *O. monoceros* is the most important species. No virus disease has been recorded from either African species. However, *O. boas* seems to be very susceptible to OrV (Julia and Mariau, 1976). The susceptibility of *O. monoceros* to OrV varies throughout its range. The Seychelle Islands population is sufficiently susceptible for control by OrV to be a realistic proposition (Lomer, 1986), but populations in the Ivory Coast (Julia and Mariau, 1976) are of lesser susceptibility. The Tanzanian population was considered sufficiently susceptible to proceed as far as pilot field control trials (Paul, 1985). However, Purini (1989) conducted laboratory tests on *O. monoceros* susceptibilty to three OrV isolates in Tanzania. Susceptibility to isolates from the Phillipines (PV), West Samoa (WS) or the Seychelles (SEY) was low but co-infection with PV + WS resulted in very much higher mortality, which did not occur for other combinations of these isolates. In control trials, Purini released adults infected with single OrV isolates and the PV + WS combination, but the comparatively high infection levels subsequently recorded in the field suggested prior presence of the virus, perhaps as a result of earlier releases by Paul. Surprisingly, in post-release monitoring Purini found only adults, and not larvae, to be infected.

An outbreak of *O. rhinoceros* in the Salalah region of Oman was the object of control work with OrV. In 1989, 667 adult beetles were collected, infected and released, the population having been first screened and found to be free of the virus. A survey of nine sites the following year showed 50% of beetles to be infected, while damage to coconut fronds had decreased from about 75% to less than 10% (Anon., 1989, 1990).

Parasa viridissima

Parasa viridissima is discussed under Limacodidae.

Phthorimaea opercullela (GV)

Of South American origin, *Phthorimaea opercullela,* the potato tuber moth, is now almost uniformly distributed through the warmer climatic areas. It is a pest of potatoes (and some other solonaceous crops), both the green parts of the plant and the tubers in field and, especially, in store, in North Africa, East Africa and South Africa. Although originally isolated in Sri Lanka, a GV has been found in Tunisia, Yemen, Egypt and elsewhere in Africa and several isolates have been characterized (Vickers, Cory and Entwistle, 1991). GV production methodologies developed in South America have been found highly suitable and so have been adopted in Tunisia (Alcazar, 1996) and Egypt, where facilities capable of producing enough virus to treat 10 000 tonnes of tubers are planned (Abol-Ela *et al.*, 1996).

In Tunisia, trials were conducted to attempt to reduce infestation in tubers at harvest, and in store thereafter, by field applications to the soil surface before harvest. Two treatments were compared, a spray application and a powder employing magnesium silicate as a carrier for the GV. The spray reduced field infestation of tubers by 73% and the dust by 35%. In-store infestations failed to develop from tubers subjected to either field treatment. On grounds of better efficiency, the spray was recommended for areas with a Mediterranean climate (Salah and Albu, 1992). Earlier work in Tunisia had concentrated on in-store protection. At the time, there was a move away from chemical pesticides in general and, particularly, from the most hazardous. Therefore, parathion had been replaced by the less-hazardous pyrethroid Deltamethrin (von Arx *et al.*, 1988) and experiments were conducted evaluating single applications of GV (at 0.002, 0.2 and 2.0 LE/kg potatoes) and a standard application of *B.t.* With *B.t.* only 0.4% of moths survived from egg to adult emergence and no reproduction occurred, while with GV survival was 34.7, 11.1 and 0.8%, respectively (control 32.5%), with fecundity reduced only at the highest concentration. The possible longer term benefits of GV, with its greater capacity for replication, were not discussed (von Arx and Gebhardt, 1990).

Trials in Egypt with a GV identical to that used in Tunisia demonstrated a much better rate of post-harvest protection (95%) using 2 LE in 10 ml water with 0.001% Tween 20 per kg potatoes. Contrary to the Tunisian experience, a dust formulation was found preferable for pre-harvest use (Abol-Ela *et al.*, 1996).

A nucleic acid non-radioactive probe has been developed in Egypt to screen for the GV (Zeddam *et al.*, 1994).

For further information, see Chapters 19 and 22.

Semi-loopers (NPV)

In Zimbabwe, considerable attention has been paid to NPV control of a complex of semi-looper caterpillars attacking soybean. In approximate order of importance these are *Thysanoplusia orichalcea*, *Chyrsodeixes chalcites*, *Chrysodeixes acuta* and *Ctenoplusia limberina*. The virus, isolated from natural field infections, is infective to all these species. It seems possible the inoculum in use is a mixture of three or four NPVs and studies to resolve this question are in progress.

Following a study of the economic threshold of semi-loopers on soybean (Taylor and Kunjeku, 1983), farmers have been advised not to use chemical insecticides on soybeans. This is in order to have larval populations that would support the virus. Most farmers have complied and are willing to sustain the non-economic defoliation of soybean during its vegetative stage. Control is initiated by farmers placing infected larvae, at the sluggish stage, in the field in patches where semi-looper infestation is heavy (see above for the method employed in producing larvae in this state). Epizootics result and most farmers have reported good control and have no need of chemical insecticides. In high rainfall areas (over 1000 mm per year) most farmers report natural virus epizootics every year and have no need for artificial infection. This occurs irrespective of the cropping pattern. A study of how the virus persists and survives the dry season has been started in these areas. However, in some seasons, some farmers have had to use insecticides for control of *H. armigera*, which seems to be gaining status as a soybean pod feeder.

Following the initial issue of virus starter stock, those farmers who will continue to produce virus for further seasons, rather than relying on annual issue from a central facility, are advised to collect infected larvae in the field at the point of death rather than decaying insects. This is to minimize any possible microbial hazard to farmers. For the same reason, farmers are also advised to spread infected insects in the field rather than spray. If farmers do spray the crude virus preparation, they are advised to use high volume sprays to minimize inhalation of the small droplets produced in low volume spraying. When spraying, the use of protective clothing and the safe practices employed with chemical insecticides have to be followed.

Spodoptera exempta (NPV)

The African armyworm is a pest of grasses including maize, rice and sorghum. The strongly migratory behaviour of adults results in the 'sudden' development of larval outbreaks, often not noticed until much damage has been inflicted. A long-term study, based on an East African network of light traps, has led to a well publicized warning system. There has, however, been no effective marriage between this and the potentially rewarding prospect of NPV control.

An NPV discovered in Kenya (Brown and Swaine, 1965) is highly infective (Odindo, 1981) and has considerable epizootic potential (Odindo, 1983). Strong evidence for NPV latency and its activation by feeding heterologous *Spodoptera* NPVs to larvae has been produced (McKinley *et al.*, 1981). Production of *S. exempta* NPV has been optimized at NRI (Natural Resources Institute), UK, using a Kenyan isolate of the virus, and initial studies on infectivity, production and environmental persistence have been undertaken in Kenya (A.J. Cherry and K.A. Jones, personal communication).

Spodoptera littoralis (NPV)

The Egyptian cotton leaf worm is in fact a polyphagous noctuid species that is widespread in Africa and the Mediterrancan. A presumed polyhedrosis was first positively reported in 1937 from Egypt (Willcocks and Bahgat, 1937) and there have been several subsequent NPV isolations. The polyhedral proteins and polypeptides of three isolates have been compared (Merdan *et al.*, 1977). Virus from Egypt was also purified and characterized by Harrap, Payne and Robertson (1977) and found to be of the multiply enveloped type. This is now designated as strain type B, which is apparently present also in Morocco and Madagascar (Croizier, Boukhoudmi-Amari and Crozier, 1986). Strain type A, which is singly enveloped, is known from Israel. Several genotypic variants of *S. littoralis* NPV have been isolated from field-collected larvae, for instance by Abol-Ela *et al.* (1988), who found variation between isolates not to be large, though they did not compare biological activity. Hunter-Fujita *et al.* (1990) isolated six genetic variants and here there were significant differences in activity and one isolate appeared to be cross-infective to *S. exempta*. The histology of infection has been described (Abul-Nasr, 1956). It is apparently not known if types A and B have differing histopathologies. The NPV used in the control studies described below is of a wild type, which is probably a mixture of genotypes of strain B.

Table 18.2 Infectivity of *S. littoralis* NPV to *S. littoralis* larvae

Instar	Infectivity	
	LD_{50}[a]	OBs/larva[b]
I	5.1 ± 3.89	–
II	17.7 ± 5.97	175
III	48.7 ± 24.32	3 700
IV	1017.3 ± 667.58	18 000
V	616.6 ± 390.35	140 000
VI	2957.3	–

[a] From McKinley (1985)
[b] From El-Nagar *et al.* (1983).

Laboratory studies of the infectivity (LD_{50}) of purified NPV in Egypt (El-Nagar, Tawfik and Abdelrahman, 1983) and in the UK (McKinley, 1985) to different instars yielded widely different results (Table 18.2). As the virus used in each study had a common source, these results probably reflect variable susceptibility in *S. littoralis* itself or in bioassay conditions, a situation further demonstrated in Egypt (A. Farghaly, personal communication, 1995) and in Israel (Klein and Podoler, 1978). Abul-Nasr (1956) assayed NPVs from two different sources, Egypt and Morocco, against *S. littoralis* and found no significant difference in potency. No profiles were obtained in this study to indicate whether there was any genotypic variation between the two virus samples.

As increased temperature reduces the LT_{50} without affecting overall mortality (Hassan, Tawfik and Marei, 1979; Moawad, 1986) it has been concluded that field temperatures will not adversely affect disease development. McKinley (1985) and Jones (1988b) also found that field temperatures would not inactivate the virus. There is a dose-related decrease in the amount of food consumed by NPV-infected larvae (Eid *et al.*, 1985), while larvae that survive NPV exposure can die of infection as pupae (McKinley, 1985) or may give rise to atypically light-weight pupae yielding adults with 22–48% reduction in fecundity (Abul-Nasr, Ammar and Abol-Ela, 1979). Larvae can be infected by surface contamination of egg masses or of adult moths (especially females), while low levels of infection occurred in the generation following the treatment of pupae with NPV. There is no evidence for transovarial transmission (El-Nagar, Tawfik and Abdelrahman, 1982, 1985).

The first small-scale field trials were conducted in Egypt, where doses, as far as can be inferred, were between 1×10^{11} and 1×10^{12} OB/ha equivalent of unpurified material. When very young larvae were treated, high levels of mortality were later observed, especially in the third instar (Abul-Nasr, 1959a). Further small-scale field tests (Abul-Nasr, 1959b) used the same virus suspension, diluted to an estimated 2×10^{5} OB/ml (dose per ha not stated), on cotton and sweet potato infested with second and third instar larvae and on maize that had been invaded by fifth and sixth instar larvae from alfalfa fields. Control on cotton was in the region of 88% and on maize about 84%. Spraying late instars on maize was ineffectual.

Initial studies on the use of NPV to control *S. littoralis* on lucerne in Crete (McKinley, 1980b) were followed by a more comprehensive cooperative programme between Egyptian universities, the Egyptian Plant Protection Research Institute and the UK Tropical Development and Research Institute (now NRI). Highly purified virus was used and the trials were undertaken to determine the best formulation and application technique (Topper *et al.*, 1984). The most effective formulation proved to be an aqueous NPV suspension containing 10% molasses as a sticker/UV protectant, 0.001% Tinopal RBS200 (Ciba-Geigy Inc.) as a UV protectant and 0.05% Teepol as a wetting agent. Effective application was identified as essential to good control, as the virus must be ingested by the first to third instar larvae, which are normally on the underside of leaves. Several sprayers were tested both for physical criteria of coverage and biological efficacy. Volume application rates ranged from 10 to 140 l/ha. A knapsack sprayer fitted with a cotton tailboom, which directs the spray upwards from a low level so that the underside of leaves are targeted, at 140 l/ha was identified as the best combination. The NPV was shown to be as effective as either the hand collection of egg masses or the use of chemical insecticides. More recently, trials with the wettable powder (p. 283) have demonstrated effective control with 1×10^{12} OB/ha and 90 l/ha (Jones *et al.*, 1994) with increased cotton yields (A. Farghaly, personal communication, 1995). An economic assessment showed the cost of production and application of NPV was less than the current alternative: chemical pesticides or the hand collection of egg masses (Jones, 1994). The commercial product, Spodopterin produced by Calliope SA in France (Guillon, 1987), has been field tested in Egypt but the results have yet to be published. Farghaly also personally reports the successful control of *S. littoralis* on berseem and tomato.

Directly related to field use has been extensive research into persistence of *S. littoralis* NPV, especially on cotton plants. It was demonstrated at Fayoum that two factors determine the persistence of virus on cotton in Egypt: inactivation by sunlight and physical loss of polyhedra (Jones and McKinley, 1987; Jones, 1988a; El-Sheik, 1984). Growth of the plant also acts to dilute the amount of virus per unit area (Jones and McKinley, 1987). In the field, the majority of sunlight inactivation is by wavelengths between 300 and 320 nm (Jones *et al.*, 1993), the importance of these wavelengths also being confirmed in the laboratory by El-Nagar (1985). The perception of early studies that persistence on the whole leaf was a basic criterion for control success tended to magnify the problem. For example, El-Sheik (1984) determined the field persistence on cotton of several formulations (Table 18.3). All formulations had an LT_{50} of about 2 to 3 days (representing a drop in mortality amongst neonate larvae from $60.2 \pm 5.4\%$ to $8.5 \pm 3.0\%$ in 7 days) and none was significantly better than unformulated virus.

However, it has been shown (Jones, 1988b; McKinley *et al.*, 1989) that the persistence of purified, unformulated NPV depended on where the virus was placed on the plant. Significant amounts of virus were calculated to remain after several weeks of exposure on the underside of leaves and on shaded parts of the plant – these being exactly the areas where target insects for control are located. Therefore, most of the virus inactivation observed in the study by El-Sheik was probably on the upper surface of the leaf, which does not play an important role in field control.

Several additives tested in the field, (e.g. 1–10% molasses, 0.1–1.0% Indigo Carmine, 0.1–1.0% riboflavin and 0.5–5.0% clay) were found to give significant UV protection. However, none of the formulations, with the exception of that

Table 18.3 LT_{50} values of *S. littoralis* on cotton in Egypt

NPV formulation[a]	LT_{50} (days)	
	High volume	Low volume
1. Actipron	2.68	2.42
Molasses	1.97	2.62
Coax	2.06	1.47
2. Virus alone	2.83	2.42
Molasses	2.29	2.01
Coax	2.54	3.48

From El-Sheik (1984).
Actipron, an emulsifiable spray oil; molasses, considered to function as a sticker, UV protectant and feeding stimulant (phagostimulant); Coax, a phagostimulant derived from cotton seed.

containing 5 or 10% molasses over long exposure periods, were significantly more effective than unpurified virus. Unpurified virus was up to 100 times more resistant to inactivation by sunlight than purified virus. None of a range of gums and stickers tested over a 14 day period was completely effective and unpurified virus was shown to persist physically as long as any of the formulations tested. It was concluded that, in terms of ease of formulation and cost, unpurified preparations of NPV were the most effective (Jones, 1988a). This was also demonstrated in the field by El-Nagar and Abul-Nasr (1980).

A number of studies have been carried out on the compatibility of *S. littoralis* NPV with other control agents. A mixture of 38.4% chlorpyrifos and 2.4% diflubenzuron with NPV was more effective than the sum of the individual treatments, but a decreased level of mortality was observed with the combined treatment when the amount of diflubenzuron was increased to 4.8% (Moawad and Elnabrawy, 1987). The importance of dose on the effect of virus–insecticide combinations was also found in studies by K. A. Jones (unpublished, see Table 18.4). Benz (1971) has pointed out that the type of interaction that occurs between chemical insecticides and microorganisms is dependent on the insecticide dose. In contrast to the results shown in Table 18.4, Ferron, Biache and Aspirot (1983), working in France, showed that synergism occurred when low NPV doses and deltamethrin were fed to first instar larval *S. littoralis*. It seems likely, therefore, that the effect of a microbial–insecticide combination may also differ when applied to a range of instars. An effect such as this was observed by Matter and Zohdy (1981) in Egypt, who found that combinations of *Heliothis* NPV and *B.t.* were additive against higher instars but antagonistic against lower instars.

Some inter-microorganism studies have been made. For instance, it was found that a combination of *S. littoralis* NPV with *B.t. entomocidus* was additive but that with *B.t. galleriae* it was antagonistic (Salama, Moawad and Zaki, 1987). A combination with the microsporidium *Variomorpha ephestiae* was near to the predicted mortality when the microorganisms were dosed simultaneously or when the microsporidium was dosed after the virus, but antagonism was displayed when the virus was offered after the microsporidium (G. Moawad, personal communication). It can be concluded that the combination of NPV with other control agents gives

Table 18.4 The effects on fourth instar larvae of *S. littoralis* of combinations of its NPV with deltamethrin

Virus concentration (OB/larva)	Mortality[a] at varying delthamethrin concentrations (ng/larva)		
	0.05	0.1	0.5
1×10^4	–8.01 ± 0.09	–	–
2×10^3	–18.01 ± 7.92	+7.51 ± 6.01	–1.81 ± 5.84

[a] Mortality as the difference between the measured percentage mortality of the combination treatment and the estimated value from treatment with the individual components. Estimated mortality of combined treatment determined by: [(% NPV mortality + deltamethrin mortality)/100] × (100 – %NPV mortality)]. The insecticide was topically applied 24 h after ingestion of NPV dose.

very variable results and field control effects cannot be easily predicted from laboratory studies. Combinations would be better tested in field trials. Field tests in which a mixture of dichrotophos and endrin with *S. littoralis* NPV (at 6×10^{11} OB/ha) was applied showed control of *S. littoralis* with this combination to be superior to the insecticide alone (Hafez *et al.*, 1970). In West Africa, field tests have shown that a one tenth dosage of both NPV and a pyrethroid (Deltamethrin) gave an equivalent effect to that of the normal full dose of the latter (Guillon, 1987).

The effect of *S. littoralis* NPV on beneficial insects has been investigated. In laboratory tests, the virus showed no adverse effects on the predators *Orius albidipennis*, *Chrysopa carnea* and *Labidura* spp. (Abbas, 1988). In field trials the numbers of the predatory *Coccinella undecimpunctata*, *Scymnus* spp., *Paederus afieri*, *C. carnea* and *Orius* spp. were not greatly reduced after application of *S. littoralis* NPV whereas a very large drop followed the application of chemical insecticides (Topper, 1984). When *S. littoralis* eggs were treated with the *S. littoralis* NPV prior to exposure to the egg parasitoid *Trichogramma evanescens*, only 14% gave rise to *Trichogramma* adults and the remainder contained dead larvae and pupae. A similar result was obtained when *H. armigera* eggs were treated with *Heliothis* NPV (as Elcar) (Abbas, 1986). In these studies, it was not determined whether the virus infected the parasitoids and, therefore, it is not clear whether death was caused by the virus or by some other agent present in the virus dose. In the same study, it was found that development of the ectoparasitic *Bracon brevicornis* was unaffected when *H. armigera* was infected with Elcar 24 h before parasitization, whereas all parasitoids died when the host larva was parasitized 3 to 6 days after infection. Although it is not clear from the report, it is likely that in the former case the parasitoids had time to develop before the host could die of the virus infection, but in the latter the host succumbed to infection before the parasitoides were fully developed. Hence it can be concluded that in this instance the virus had no direct effect on the parasitoid. Results from field trails in Crete also indicate that the overall level of parasitism of *S. littoralis* larvae is not adversely affected by application of NPV (Jones, 1990).

Some studies have taken place to see if *S. littoralis* NPV could control insect pest species outside the Lepidoptera. There was a small but significant reduction in hatch following treatment of *Ceratitis capitata* (Diptera: Tephritidae) eggs with the virus. There was also a reduction in subsequent pupation, both effects apparently being dose related. The cause is unclear as virus was not reisolated (Moawad and Shehata, 1987). Testing the termite *Kalotermes flavicollis* (Isoptera) with unpurified virus resulted in variable mortality (between 8 and 81%), which was not dose related. However, NPV-like polyhedra were found in termite cadavers and in fluid defecated prior to death (Al-Fazairy and Hassan, 1988). In Israel, Bensimon *et al.* (1987) reported that *S. littoralis* NPV could be cross-transmitted to two species of locust (Orthoptera: Acrididae): *Locusta migratoria migratoriodes* and *Schistocerca gregaria*. A lethal condition, 'dark cheeks disease', resulted and virus recovered from the locusts was confirmed by REN analysis to be *S. littoralis* NPV type B. However, preliminary attempts to cross-infect *S. gregaria* at NRI, University of Greenwich, UK resulted in neither disease nor premature death (M. Brown personal communication, 1995).

A GV was isolated from laboratory and field populations of *S. littoralis* in Egypt (Hunter-Fujita *et al.*, 1990). Laboratory studies showed the GV to be highly potent but slow acting, extending the larval period to 8 weeks before any mortality. It was also found to interact antagonistically with *S. littoralis* NPV (Hunter-Fujita *et al.*, 1992).

Thysanoplusia orichalcea

Thysanoplusia orichalcea is discussed under semi-loopers.

REFERENCES

Anon. (1989) *Commonwealth Institute for Biological Control Annual Report 1989*. Ascot, UK: International Institute of Biological Control.

Anon. (1990) *IOBC Annual Report 1990*. Ascot, UK: International Institute of Biological Control.

Anon. (1991) *NRI Report on Operational Programmes 1989–91*. Chatam, UK: Natural Resources Institute, pp. 162–166.

Abbas, M.S.T. (1986) Interaction between host, egg and larval parasites, and nuclear polyhedrosis virus. In *Proceedings of the 4th International Colloquium of Invertebrate Pathology: Fundamental and Applied Aspects of Invertebrate Pathology*, Wageningen, the Netherlands. R.A. Sampson, J.M. Vlak and D. Peters (eds). p. 132.

Abbas, M.S.T. (1988) Interaction between nuclear polyhedrosis virus, host and predators. *Zeitschrift für Pflanzenkrankheit und Pflanzenschutz* **95**, 606–610.

Abdally, A., Makkouk, K.M. and Cardona, C. (1987) Control of *Heliothis* spp. on chickpea by insect pathogenic nuclear polyhedrosis virus. *Arab Journal of Plant Protection* **5**, 78–80.

Abol-Ela, S.M., Remillet, M., Khamis, O.A. and Croizier, G. (1988) Genomic variation at the molecular level among viral strains isolated from larve of *Spodoptera littoralis* (Boisd.) from different regions in Egypt. *Bulletin of the Faculty of Agriculture, University of Cairo* **39**, 655–668.

Abol-Ela, S., El-Bolboi, H., Monsarrat, A. and Giannotti, J. (1996) Improving the production and application of the potato tuber moth granulosis virus in Egypt. In *IOBC/WRPS Bulletin* **19**, 106.

Abul-Nasr, S. (1956) Field tests on the use of a polyhedrosis virus disease for control of cotton leaf-worm, *Prodenia litura* F. *Bulletin Société Entomologique d'Egypte* **40**, 321–332.

Abul-Nasr, S. (1959a) Field tests on the use of a polyhedrosis virus disease for the control of the cotton leaf-worm, *Prodenia litura* F. *Bulletin Société Entomologique d'Egypte* **43**, 231–243.

Abul-Nasr, S. (1959b) Further tests on the use of a polyhedrosis virus in the control of the cotton leafworm *Prodenia litura* Fabricius. *Journal of Insect Pathology* **1**, 112–120.

Abul-Nasr, S.E., Ammar, E.D. and Abol-Ela, S.M. (1979) Effects of nuclear polyhedrosis virus on various developmental stages of cotton leafworm *Spodoptera littoralis*. *Zeitschrift für Angewandte Entomologie* **88**, 181–187.

Alcazar, J. (1996) Use of granulosis virus to control potato tuber moth, *Phthorimaea operculella* (Zeller). *IOBC/WRPS Bulletin* **19**, 40.

Al-Fazairy, A.A. and Hassan, F.A. (1988) Infection of termites by *Spodoptera littoralis* NPV. *Insect Science and its Application* **9**, 37–39.

Angelini, A. and Vandamme, P. (1964) Une virose intestinale chez *Diparopsis watersi* (Lepidoptera-Noctuidae). *Coton et Fibres Tropicales* **19**, 265–270.

Angelini, A. and Vandamme, P. (1972) Les moyens de lutte biologique contre certains ravageurs du contonnier et une perspective sur la lutte integree en Cote d'Ivoire. *Coton et Fibres Tropicales* **27**, 283–289.

Atger, P. (1969) Observations sur la polyhedrose nucleaire d'*Heliothis armigera* (Hbn.) au Tchad. *Coton et Fibres Tropicales* **24**, 243–244.

Austara, O. (1971) *Gonometa podocarpi* Aur. (Lepidoptera: Lasiocampidae): a defoliator of exotic softwoods in East Africa. The biology and life cycle at Muko, Kigezi District in Uganda. *East African Agricultural and Forestry Journal* **36**, 275–289.

Bensimon, A., Zinger, S., Gerassi, E., Hauschner, A., Harpaz, I. and Sela, I. (1987) 'Dark cheeks', a lethal disease of locusts provoked by a lepidopterous baculovirus. *Journal of Invertebrate Pathology* **50**, 254–260.

Benz, G. (1971) Synergism of micro-organisms and chemical insecticides. In *Microbial Control of Insect Pests and Mites*. H.D. Burges and N.W. Hussey (eds). London: Academic Press, pp. 327–355.

Brown, E.S. and Swaine, G. (1965) Virus disease of the African armyworm, *Spodoptera exempta* (Wlk.). *Bulletin of Entomological Research* **56**, 95–116.

Brown, K.W. (1965) *Quarterly Report of the Forest Entomologist*, Uganda, July–September. (Cyclostyled, 2 pp.).

Cadou, J. and Soubrier, G. (1974) Utilisation d'une polyedrose nucleaire dans la lutte contre *Heliothis armigera* (Hb.) (Lep. Noct.) en culture cotonniere au Tchad. *Coton et Fibres Tropicales* **29**, 357–365.

Calliope SA (1985) Mamestrin®, a new biological insecticide. Trade documentation. Beziers, France.

Carey, D. and Harrap, K.A. (1980) Safety tests on the NPVs of *Spodoptera littoralis* and *Spodoptera exempta*. In *Invertebrate Systems* in vitro. E. Kurstak, K. Maramorosch and A. Dubendorfer (eds). Amsterdam: Elsevier, pp. 441–450.

Cherry, A.J., Parnell, M.A., Smith, D. and Jones, K.A. (1994) Oil formulations of insect viruses. *IOBC/WPRS Bulletin* **17**, 254–257.

Coaker, T.H. (1958) Experiments with a virus disease of the cotton bollworm *Heliothis armigera*. *Annals of Applied Biology* **46**, 536–541.

Croizier, G., Amargier, A., Godse, D.-B., Jaquemard, P. and Duthoit, J.-L. (1980) Un virus de polyedrose nucleaire decouvert chez le lepidoptere Noctuidae *Diparopsis watersi* (Roth.) nouveau variant du baculovirus d'*Autographa californica* (Speyer). *Coton et Fibres Tropicales* **35**, 415–423.

Croizier, G., Boukhoudmi-Amari, K. and Croizier, L. (1986) Physical map of the DNA of the *Spodoptera littoralis* nuclear polyhedrosis virus of type B. Genetic comparison of the *S. littoralis* and *Spodoptera litura* Fabr. baculoviruses. In *Proceedings of the IVth International Colloquium of Invertebrate Pathology: Fundamental and Applied Aspects of Invertebrate Pathology*, Wageningen, the Netherlands. R.A. Sampson, J.M. Vlak and D. Peters (eds). p. 104.

Daoust, R.A. and Roome, R.E. (1974) Bioassay of a nuclear-polyhedrosis virus and *Bacillus thuringiensis* against the American (sic) bollworm, *Heliothis armigera*, in Botswana. *Journal of Invertebrate Pathology* **23**, 318–324.

Dulmage, H.T., Martinez, A.J. and Correa, J.A. (1970) Recovery of the nuclear polyhedrosis virus of the cabbage looper, *Trichoplusia ni*, by coprecipitation with lactose. *Journal of Invertebrate Pathology* **16**, 80–83.

Eid, M.A.A., El-Nagar, S., Salem, M.S. and Badawy, E. (1985) Effect of nuclear polyhedrosis virus ingestion on consumption and utilization of food by *Spodoptera littoralis* larvae. *Bulletin of the Entomological Society of Egypt, Economic Series* **13**, 67–74.

El-Nagar, S. (1985) The inactivation of a nuclear polyhedrosis virus by radiation. *Bulletin of the Entomological Society of Egypt, Economic Series* **13**, 171–174.

El-Nagar, S. and Abul-Nasr, S. (1980) Effect of direct sunlight on the virulence of nuclear polyhedrosis virus of the cotton leafworm *Spodoptera littoralis*. *Zeitschrift für Angewandte Entomologie* **90**, 75–80.

El-Nagar, S., Tawfik, M.W.S. and Abdelrahman, T.A. (1982) Transmission of the nuclear polyhedrosis virus disease of *Spodoptera littoralis* via exposure to the virus. *Zeitschrift für Angewandte Entomologie* **94**, 152–156.

El-Nagar, S., Tawfik, M.W.S. and Abdelrahman, T.A. (1983) The susceptibility to nuclear polyhedrosis virus amongst populations of *Spodoptera littoralis*. *Zeitschrift für Angewandte Entomologie* **96**, 459–463.

El-Nagar, S., Tawfik, M.W.S. and Abdelrahman, T.A. (1985) Transmission of the nuclear polyhedrosis virus disease of *Spodoptera littoralis* via egg and pupal exposure to the virus. *Bulletin of the Entomological Society of Egypt, Economic Series* **13**, 31–38.

El-Sheikh, M.O.A. (1984) The nuclear polyhedrosis virus of *Spodoptera littoralis* (Boisd.): an evaluation of its role for pest management in Egypt. PhD Thesis, Cairo University.

Entwistle, P.F. and Evans, H.F. (1985) Viral control. In *Comprehensive Insect Physiology, Biochemistry and Pharmacology*, Vol. 12. Oxford: Pergamon Press, pp. 347–412.

Fediere, G., Monsarrat, P., Mariau, D. and Bergoin, M. (1986a) Lutte microbiologique par virus entomopathogenes contre deux lepidopteres Limacodidae ravageures du palmier a huile et du cocotier. *Congres sur la Protection de la Sante Humain et des Cultures en Lilieu Tropical*, Marseilles, France, July 2–4 1986, pp. 363–368.

Fediere, G., Monsarrat, P. Mariau, D. and Bergoin, M. (1986b) A densovirus of *Casphalia extranea* (Lepidoptera: Limacodidae): characterization and use for biological control. *Proceedings 4th International Congress Invertebrate Pathology*, Eindhoven, the Netherlands, p. 705.

Ferron, P., Biache, G. and Aspirot, J. (1983) Pathologie animale – synergisme entre baculovirus a polyedres nucleares de Lepidopteres Noctuidae en doses reduites de pyrethrinoides photostables. *Compte Rendus de l'Academiè des Science, Paris III*, **296**, 511–514.

Fritsche, E. (1988) Biologische Bekamfung des Falschen Apfelwickers, *Cryptophlebia leucotreta* (Meyrick) (Lep., Tortricidae), mit Granuloseviren. *Mitteilungen der Deutschen Gesellschaft für Allgemeine und Angewandte Entomologie* **6**, 280–283.

Fritsch, E. and Huber, J. (1986) A granulosis virus of the false codling moth, *Cryptophlebia leucotreta* (Meyr.). In *Proceedings of the 4th International Colloquium of Invertebrate Pathology: Fundamental and Applied Aspects of Invertebrate Pathology*, Wageningen, the Netherlands. R.A. Sampson, J.M. Vlak and D. Peters (eds). p. 12.

Fritsch, E. and Huber, J. (1989) Comparative field persistence of granulosis viruses under tropical and European conditions. In *Study Group: Insect Pathogens and Insect–Parasitic Nematodes*. C.C. Payne (ed.). Versailles, France: IOBC/WPRS, pp. 84–87.

Gitay, H. and Polson, A. (1971) Isolation of granulosis virus from *Heliothis armigera* and its persistence in avian feces. *Journal of Invertebrate Pathology* **17**, 288–290.

Gouteux, J.-P. (1987) Prevalence of enlarged salivary glands in *Glossina pallipes*, *G. pallicera* and *G. nigrofusca* (Diptera: Glossinidae) from the Vavoua area, Ivory Coast. *Journal of Medical Entomology* **24**, 268.

Guillon, M. (1987) Marketing of baculovirus-based biological insecticides, application to *Mamestra brassicae* and *Spodoptera littoralis* NPV. *Mededelingen van de Faculteit Landbouwwetenschapen Rijksuniversiteit Gent* **52**, 147–153.

Hafez, M. (1985) Synopsis of entomological research in Egypt. *AAIS Bulletin of African Insect Science* **9**, 10–15.

Hafez, M., Kamel, A.A.M., Mostafa, T.H. and Omar, E.E. (1970) Field test of combinations of polyhedrosis virus suspensions and certain chemical insecticides for control of

the cotton leaf worm, *Spodoptera littoralis* (Boisd.) (Lepidoptera: Noctuidae). *Bulletin of the Entomological Society of Egypt, Economic Series* **4**, 65–69.
Harrap, K.A., Payne, C.C. and Robertson, J.S. (1977) The properties of three baculoviruses from closely related hosts. *Virology* **79**, 14–31.
Hassan, S.M., Tawfik, M.F.S. and Marei, S. (1979) Effect of temperature on pathogenicity of polyhedrosis virus disease in the cotton leafworm, *Spodoptera littoralis* (Lepidoptera Noctuidae). *Bulletin of the Entomological Society of Egypt, Economic Series* **9**, 1–6.
Hepburn, G.A. (1964) The status of forest insects in the Republic of South Africa. *FAO/IUFRO Symposium on Internationally Dangerous Forest Diseases and Insects*, Oxford, Meetings II/III, pp. ii, 1–4. FAO of UN.
Hunter-Fujita, F.R., Radley, E., Smith, I.R.L. and Adiku, T.K. (1990) Characterization of baculoviruses isolated from cotton fields in Egypt. *Proceedings of the 5th International Colloquium on Invertebrate Pathology and Microbial Control*, Adelaide, Australia. p. 255.
Hunter-Fujita, F.R., Dodsworth, J., Jones, K.A., Cherry, A., Smith, I.R.L., Boner, W. and Moawad, G.M. (1992) Observations on the relationship between a granulosis virus and a nuclear polyhedrosis virus infecting Egyptian cotton leafworm. In *Abstracts of the 25th Annual Meeting of the Society for Invertebrate Pathology*, Heidelberg, Germany.
Ignoffo, C.M. (1965) The nuclear polyhedrosis virus of *Heliothis zea* (Boddie) and *Heliothis virescens* (Fabricius). IV. Bioassay of virus activity. *Journal of Invertebrate Pathology* **7**, 315–319.
Jaenson, T.G.T. (1978) Virus-like rods associated with salivary gland hyperplasia in tsetse *Glossina pallidipes*. *Transactions of the Royal Society of Tropical Medicine and Hygiene* **72**, 234–238.
Jones, K.A. (1988a) The use of insect viruses for pest control in developing countries. *Aspects of Applied Biology* **17**, 425–433.
Jones, K.A. (1988b) Studies on the persistence of *Spodoptera littoralis* nuclear polyhedrosis virus on cotton in Egypt. PhD thesis, University of Reading, UK.
Jones, K.A. (1990) Use of a nuclear polyhedrosis virus to control *Spodoptera littoralis* in Crete and Egypt. In *Pesticides and Alternatives*. J.E. Casida (ed.). Amsterdam: Elsevier, pp. 131–142.
Jones, KA (1994) Use of baculoviruses for cotton pest control. In *Insect Pests of Cotton*. G.A. Matthews (ed.). Wallingford, UK: CAB International, pp. 445–467.
Jones, K.A. and McKinley, D.J. (1987) Persistence of *Spodoptera littoralis* nuclear polyhedrosis virus on cotton in Egypt. *Aspects of Applied Biology* **14**, 323–334.
Jones, K.A., Moawad, G., McKinley, D.J. and Grzywacz, D. (1993) The effect of natural sunlight on *Spodoptera littoralis* nuclear polyhedrosis virus. *Biocontrol Science and Technology* **3**, 189–197.
Jones, K.A., Irving, N.S., Grzywacz, D., Moawad, G.M., Hussein, A.H. and Fargahly, A. (1994) Application rate trials with a nuclear polyhedrosis virus to control *Spodoptera littoralis* (Boisd.) on cotton in Egypt. *Crop Protection* **13**, 337–340.
Juckes, I.R.M., Longworth, J.F. and Reinganum, C. (1973) A serological comparison of some non-occluded insect viruses. *Journal of Invertebrate Pathology* **21**, 119–120.
Julia, J.F. and Mariau, D. (1976) Recherches sur l'*Oryctes monoceros* Ol. en Cote-d'Ivoire II. – Essai de lutte biologique avec le virus *Rhabdionvirus oryctes*. *Oleagineux* **31**, 113–117.
Jura, W.G.Z.O. and Davies-Cole, J.O.A. (1992) Some aspects of mating behaviour of *Glossina morsitans morsitans* males infected with a DNA virus. *Biological Control* **2**, 188–192.
Jura, W.G.Z.O., Odhiambo, T.R., Otieno, L.H. and Tabu, N.O. (1988) Gonadal lesions in virus-infected male and female tsetse, *Glossina pallidipes* (Diptera: Glossinidae). *Journal of Invertebrate Pathology* **52**, 1–8.
Klein, M. and Podoler H. (1978) Studies on the application of a nuclear polyhedrosis virus to control populations of the Egyptian cottonworm, *Spodoptera littoralis*. *Journal of Invertebrate Pathology* **32**, 244–248.
Lavabre, E.M., Ban, J. and Vandamme, P. (1966) Etude preliminaire de la transmission de viroses nucleaires et de bacterioses aux chenilles des cacaoyers. *Café, Cacao, Thé* **10**, 336–341.

Lomer, C.J. (1986) Release of *Baculovirus oryctes* into *Oryctes monoceros* populations in the Seychelles. *Journal of Invertebrate Pathology* **47**, 237–246.

Lutwama, J.J. and Matanmi, B.A. (1988) Efficacy of *Bacillus thuringiensis* subsp. *kurstaki* and *Baculovirus heliothis* foliar applications for suppression of *Heliothis armigera* (Hübner). *Bulletin of Entomological Research* **78**, 173–179.

Matter, M.M. and Zohdy, N.Z.M. (1981) Biotic efficiency of *Bacillus thuringiensis* Berl. and a nuclear polyhedrosis virus on larvae of the American bollworm, *Heliothis armigera* Hbn. (Lepid., Noctuidae). *Zeitschrift für Angewandte Entomologie* **2**, 336–343.

McKinley, D.J. (1971) Nuclear polyhedrosis virus of the cotton bollworm in Central Africa. *Cotton Growing Review* **48**, 297–303.

McKinley, D.J. (1980a) The use of viruses in the control of *Spodoptera* species and prior safety testing. In *Environmental Protection and Biological Forms of Control of Pest Organisms*. B. Lundholm and M. Stackerud (eds). *Ecological Bulletin* (Stockholm) **31**, 75–80.

McKinley, D.J. (1980b) Evaluation of nuclear polyhedrosis viruses by the Centre for Overseas Pest Research. *Project Report 1974–79*. Overseas Development Administration Research Scheme R2870.

McKinley, D.J. (1985) Nuclear polyhedrosis virus of *Spodoptera littoralis* Boisd. (Lepidoptera: Noctuidae) as an effective agent in its host and related insects. PhD Thesis, University of London.

McKinley, D.J., Brown, D.A., Payne, C.C. and Harrap, K.A. (1981) Cross-infectivity and activation studies with four baculoviruses. *Entomophaga* **26**, 79–90.

McKinley, D.J., Moawad, G., Jones, K., Grzywacz, D. and Turner, C. (1989) The development of nuclear polyhedrosis virus for control of *Spodoptera littoralis* (Boisd.) in cotton. In *Pest Management in Cotton*. M.B. Green and D.J. de B. Lyon (eds). Chichester: Ellis Horwood, pp. 93–100.

Merdan, A., Croizier, L, Veyrunes, J.-C. and Croizier, G. (1977) Etude comparee des proteines des polyedres et des virions de trois isolats de baculovirus de *Spodoptera littoralis*. *Entomophaga* **22**, 413–420.

Moawad, G.M. (1986) The influence of temperature and relative humidity on the interaction of nuclear polyhedrosis virus and the cotton leafworm, *Spodoptera littoralis* (Boisd.). *Agricultural Research Review* **61**, 203–209.

Moawad, G.M. and Elnabrawy, I.M. (1987) Effectiveness of nuclear polyhedrosis virus and insecticides against the cotton leafworm *Spodoptera littoralis* Boisd. *Insect Science and its Applications* **8**, 89–94.

Moawad, G.M. and Shehata, N.F. (1987) Effects of the nuclear polyhedrosis virus of the Egyptian cotton leafworm *Spodoptera littoralis* Boisd. on the Mediterranean fruit fly *Ceratitis capitata* Wied. *Insect Science and its Applications* **8**, 365–368.

Moawad, G.M., Abdeen, S.A.O., Saleh, W.S. and Gadallah, A.I. (1986) Relative susceptibility of the bollworm *Heliothis armigera* Hbn. to a nuclear polyhedrosis virus and its biochemical effects. *Bulletin of the Entomological Society of Egypt, Economic Series* **14**, 385–398.

Monsarrat, A., Abol-Ela, S. and Zeddam, J.-L. (1994) Prevalence of a new picorna-like virus among *Pectinophora gossypiella* natural populations in Egyptian cotton fields. In *The 6th International Colloquium on Invertebrate Pathology and Microbial Control*, Montpellier, France. Society for Invertebrate Pathology, p. 241.

Montaldo, T. (1991) La lutte microbiologique en culture cotonnière au Nord Cameroun: synthèse de l'expérimentation menée de 1979 à 1988. *Coton et Fibres Tropicales* **46**, 217–229.

Moore, N.F., Reavy, B. and King, L.A. (1985) General characteristics, gene organization and expression of small RNA viruses of insects. *Journal of General Virology* **66**, 647–659.

Namponya, C.R. (eds) (1989) *Strategy for Integrated Pest Management and Weed Control in SADCC Countries*. Mbabane, Swaziland.

Neutra, R., Levi, B.Z. and Shoham, Y. (1992) Optimization of protein production by the baculovirus expression vector system in shake flasks. *Applied Microbiology and Biotechnology* **37**, 74–78.

Odindo, M.O. (1981) Dosage–mortality and time–mortality responses of the armyworm, *Spodoptera exempta* to a nuclear polyhedrosis virus. *Journal of Invertebrate Pathology* **38**, 251–255.

Odindo, M.O. (1983) Epizootiological observations on a nuclear polyhedrosis of the African armyworm *Spodoptera exempta* (Walk.). *Insect Science and its Application* **4**, 291–298.

Odindo, M.O. (1988) *Glossina pallidipes* virus: its potential for use in biological control of tsetse. *Insect Science and its Application* **9**, 399–403.

Odindo, M., Payne, C.C. and Crook, N.E. (1984) A novel virus isolated from the tsetse fly *Glossina pallidipes. Abstracts of the Sixth International Congress of Virology* p. 336.

Ossowski. L.L.J. (1957) The biological control of the wattle bagworm, *Kotochalia junodi* (Heyl.), by a virus disease. I. Small-scale experiments. *Annals of Applied Biology* **45**, 81–89.

Ossowski, L.L.J. (1958) Erfahrungen mit PolyederViren gegen den Akaziensackwurm, *Kotochalia junodi* (Heyl.) – Psychidae. In *Transactions of the 1st International Conference on Insect Pathology and Biological Control*, Prague, pp. 247–253.

Ossowski, L.L.J. (1960) Variation in virulence of a wattle bagworm nuclear polyhedrosis virus. *Journal of Insect Pathology* **2**, 35–43.

Ossowski, L.L.J. (1962) A polyhedral virus in wattle bagworm control. In *Der 11te Internationaler Kongress für Entomologie, Wien* 1960, Vol. II, pp. 810–814.

Otieno, L.H., Kokwaro, E.D., Chimtawi, M. and Onyango, P. (1980) Prevalence of enlarged salivary glands in wild populations of *Glossina pallidipes* in Kenya, with a note on the ultrastructure of the affected organs. *Journal of Invertebrate Pathology* **36**, 113–118.

Paul, W.D. (1985) *Integrierte Bekamfung von Palmenschadlingen in Tanzania.* Aichtal, German Federal Republic: Verlag Josef Margraf (Monographs on agriculture and ecology of warmer climates, Vol. I).

Polson, A. and Tripconey, D. (1970) A virus disease of the lucerne caterpillar, *Colias electo* Linn. *Phytophylactica* **2**, 17–20.

Purini, K. (1989) *Baculovirus oryctes* release into *Oryctes monoceros* population in Tanzania, with special reference to the interaction of virus isolates used in our laboratory infection experiments. *Journal of Invertebrate Pathology* **53**, 285–300.

Renou, A. (1987) Les acquis en lutte biologique contre *Heliothis armigera* (Hbn.), ravageur de la culture cotonniere au Nord Cameroun (36th International Symposium in Crop Protection). *Mededelingen van de Faculteit Landbouwwetenschappen, Rijksuniversiteit, Gent* **52**, 311–318.

Roome, R.E. (1975) Field trials with a nuclear polyhedrosis virus and *Bacillus thuringiensis* against larvae of *Heliothis armigera* (Hb.) (Lepidoptera, Noctuidae) on sorghum and cotton in Botswana. *Bulletin of Entomological Research* **65**, 507–514.

Roome, R.E. and Daoust, R.A. (1976) Survival of the nuclear polyhedrosis of *Heliothis armigera* on crops and in soil in Botswana. *Journal of Invertebrate Pathology* **27**, 7–12.

Salah, H.B. and Albu, R. (1992) Field use of granulosis virus to reduce initial storage infestation of the potato tuber moth, *Phthorimaea opercullela* (Zeller), in North Africa. *Agriculture Ecosystems and Environment* **38**, 119–126.

Salama, H.S. and Moawad, S.M. (1988) Joint action of nuclear polyhedrosis virus and chemical insecticides against the black cutworm *Agrotis ipsilon* Hufn. *Acta Biologica Hungaricae* **39**, 99–108.

Salama, H.S., Moawad, S.M. and Magahed, M.I. (1986) Effect of nuclear polyhedrosis virus on the cotton bollworm *Heliothis armigera. Journal of Applied Entomology* **102**, 123–130.

Salama, H.S., Moawad, S.M. and Zaki, F.N. (1987) Effects of nuclear polyhedrosis virus *Bacillus thuringiensis* combinations on *Spodoptera littoralis* Boisd. *Journal of Applied Entomology* **194**, 23–27.

Taylor, D.E. and Kunjeku, E. (1983) Development of an economic threshold for semiloopers (Lepidoptera: Noctuidae) on soyabeans in Zimbabwe. *Journal of Agricultural Research* **21**, 89–100.

Topper, C. (1984) Report on research and development of nuclear polyhedrosis virus of *Spodoptera littoralis* 1979–1983. *Unpublished Report ODA and Egyptian Academy of Sciences* (3 vols). London: Overseas Development Administration.

Topper, C., Moawad, G., McKinley, D.J., Hosny, M., Jones, K., Cooper, J., El-Nagar, S. and El-Sheik, M. (1984) Field trials with a nuclear polyhedrosis virus against *Spodoptera littoralis* on cotton in Egypt. *Tropical Pest Management* **30**, 372–378.

Tripconey, D. (1970) Studies on a nonoccluded virus of the pine tree emperor moth. *Journal of InvertebratePathology* **15**, 268–275.

Vickers, J.M., Cory, J.S. and Entwistle, P.F. (1991) DNA characterisation of eight geographic isolates of granulosis virus of the potato tuber moth, *Phthorimaea opercullela*. *Journal of Invertebrate Pathology* **57**, 334–342.

von Arx, R. and Gebhardt, F. (1990) Effects of a granulosis virus and *Bacillus thuringiensis* on life-table parameters of the potato tubermoth, *Phthorimaea opercullela*. *Entomophaga* **35**, 151–159.

von Arx, R. Ewell, P.T., Goueder, J., Essamet, M., Cheikh, M. and Ben Temine, A. (1988) *Management of the Potato Tuber Moth by Tunisian Farmers*. International Potato Center (CIP) and Institut National de la Recherche Agronomique de Tunisie (INRAT).

Whitlock, V.H. (1974) Symptomatology of two viruses infecting *Heliothis armigera*. *Journal of Invertebrate Pathology* **23**, 70–75.

Whitlock, V.H. (1977a) Simultaneous treatments of *Heliothis armigera* with a nuclear polyhedrosis and a granulosis virus. *Journal of Invertebrate Pathology* **29**, 297–303.

Whitlock, V.H. (1977b) Effect of larval maturation on mortality induced by nuclear polyhedrosis and granulosis virus infections of *Heliothis armigera*. *Journal of Invertebrate Pathology* **30**, 80–86.

Whitnall, A.B.M. (1934) The trypanosome infections of *Glossina pallidipes* in the Umfolosi Game Reserve, Zululand. *Onderstepoort Journal of Veterinary and Animal Industry* **11**, 7–21.

Willcocks, F.C. (1905) The Egyptian cotton leafworm. In *Yearbook of the Khedivial Agricultural Society*, Egypt: Khedivial Agricultural Society.

Willcocks, F.C. and Bahgat, S. (1937) The insects and related pests of Egypt. *Royal Agricultural Society of Egypt*, **I**, 591.

Zeddam, J.L., El Bolbol, H., El Guindy, N., Lagnaoui, A., Al-Abssi, G., Fediere, G., Lery, X., Monsarrat, A., Abol-Ela, S. and Gianotti, J. (1994). Use of non-radioactive nucleic acid probes for epidemiological survey of potato tuber moth granulosis virus. *IOBC/WRPS Bulletin* **17**, 285.

19 Australasia

ROBERT. E. TEAKLE

PERSPECTIVE

This chapter covers New Zealand, Australia and the island of New Guinea, which is divided into Irianjaya, Indonesia to the west and Papua New Guinea to the east. The latter nation includes the offshore islands of New Britain and New Ireland. Because they share with the New Guinea area the problem of *Scapanes* attack on coconuts, the Solomon Islands are included here rather than in Oceanea.

Insect viruses have been considered for the control of agricultural, forestry and pasture pests in Australia and New Zealand and an inventory of 74 insect-derived viruses containing 50 BVs, 6 entomopoxviruses and 10 small isometric viruses has been published (Anon., 1981). Eight lepidopteran hosts of BVs are of economic importance in New Zealand: *Cydia pomonella*, *Chrysodeixis eriosoma*, *Epiphyas postvittana*, *Helicoverpa armigera*, *Mythimna separata*, *Phthorimaea opercullela*, *Spodoptera litura* and *Wiseana cervinata* but none was considered of sufficient importance to warrant a major development programme (Longworth, 1980). Additional important BV hosts in Australia are *Chrysodeixis argentifera, Helicoverpa punctigera, Herpetogramma licarsisalis, Mythimna convecta, Merophyas divulsana* and *Spodoptera mauritia.*

Outbreaks of forest defoliators occur sporadically, but production of viruses for their control would rarely be feasible, despite the high economic thresholds. However, while most native insect species are subject to heavy natural control, they could be very destructive in overseas plantings of, for example, Australian *Eucalyptus* species. In this event, individual BV isolates could play an important role in the control of their expatriate hosts (Teakle, 1969). The geometrid moth *Pseudocoremia (Selidosema) suavis* is the only real forest insect pest in New Zealand (Longworth, 1980) but control employing a recorded small RNA virus and a CPV (Dugdale, 1964) has not been attempted.

Pasture is an extremely important component of the agricultural economy in New Zealand, where small RNA viruses of the grass grub *Costelytra zealandica*, the black beetle *Heteronychus arator* and the field cricket *Teleogryllus commodus* may have potential for control. Ecological studies have established the value of accumulations of NPV in topsoil in the natural suppression of *Wiseana* spp. in

New Zealand (Kalmakoff and Crawford, 1982) and *H. punctigera* in lucerne in Australia (Cooper, 1979). The use of selective BVs for the control of *H. punctigera* and *M. divulsana* on lucerne (alfalfa) could conserve natural predators and imported aphid parasites (Teakle, 1978).

In Australia and New Zealand, laboratory and field assessments have been made of four locally produced BV insecticides. Although promising results were obtained, these assessments have not progressed beyond the experimental phase to commercial development and registration. One imported virus formulation, Elcar, based on a NPV from *Helicoverpa zea*, has been registered for use on cotton and sorghum in Australia, but it has not been used commercially to any appreciable extent. Workers in New Zealand have made very substantial contributions to an understanding of the non-occluded baculovirus-like pathogen (OrV) of *Oryctes rhinoceros*. These include development of a permissive insect cell line (Crawford, 1982) (using cell culture technology largely pioneered by Grace (1967) in Australia), exploration of host range (Crawford *et al.*, 1985), elucidation of genotypic variation in geographical isolates, involvement in studies leading to production of highly stabilized virus preparations (Zelazny, Alfiler and Crawford, 1897) and, through a key study in the Tonga Islands (Young, 1974), an understanding of the dynamics of spread of disease in previously healthy rhinoceros beetle populations. Virus produced *in vitro* and exported from New Zealand has aided the rhinoceros beetle control programme in the Maldive Islands.

The most important work in Papua New Guinea and the Solomon Islands has been in the virus control of *O. rhinoceros* and, using the same virus, in exploration of the possibility of control of *Scapanes australis*, a dynastid pest of coconut palms restricted to this immediate region. In addition, the possibility of control of species of the locally important genus *Papuana*, also a dynastid, with OrV virus has been considered (Zelazny *et al.*, 1988). Studies on the complex of viruses infecting *Lymantria ninayi*, a pest of introduced pines in the highlands of Papua New Guinea, have established a basis for control strategy. Work by van Velsen (1967) on control of *Tiracola plagiata* by sprays of an NPV represents one of the very few applications of this technology in cocoa, an extremely valuable tropical cash crop.

Significant regional control investigations are listed country by country in Table 19.1.

In Australia and New Zealand, a number of factors contributed to the virtual cessation of the assessment of locally produced viruses in the 1970s and their failure to achieve commercial usage. Perhaps the choice of viruses was unfortunate in that, with the exception of *Pieris rapae* GV, the host insects tend to feed in concealed locations. This makes effective placement of the virus and mortality assessment difficult. Furthermore, the *P. rapae* GV is unduly specific in that it does not control other lepidopteran pests of crucifers. In Australia, its replacement by the less specific AcNPV would be hampered by the infectivity of the latter for lepidopteran *Cactoblastis cactorum*, an important biocontrol agent of the weed *Opuntia* (Vail, Vail and Summers, 1984).

All of the initial assessments were carried out by Government research organizations, but none included safety testing. This lack would have constituted a considerable financial disincentive to further development, except in the case of

Table 19.1 Virus control of insect pests in Australasia: significant investigations

Pest species	Australia	New Zealand	Papua New Guinea	Solomon Islands
Cydia pomonella	+	+		
Epiphyas postvittana	+	+		
Helicoverpa armigera and *punctigera*	+			
Lymantria ninayi			+	
Oryctes rhinoceros			+	+
Phthorimaea opercullela	+			
Pieris rapae	+	+		
Scapanes australis			+	+
Tiracola plagiata			+	
Wiseana cervinata and other spp.		+		

H. zea NPV. A major dampener to interest in virus-based insecticides was undoubtedly the advent of the cheap and effective synthetic pyrethroids. The timing of the initial importation of Elcar into Australia in 1977 was unfortunate in that it coincided with the introduction of the pyrethroids. Consequently, there was little incentive to examine its potential fully. A renewal of interest in the viruses should emerge as the synthic pyrethroids lose their appeal, largely through the development of resistance. However, despite the discovery of resistance to these insecticides in *H. armigera* on cotton in 1983 (Gunning *et al.*, 1984), Elcar was withdrawn from the market in 1984.

Since then, a reappraisal of the potential of the virus has been made. This resulted in a decision to modify genetically a local *Helicoverpa* NPV isolate by the insertion of a gene coding for a toxin, thereby causing rapid disruption to feeding (Christian and Oakeshott, 1989). Unfortunately, this virus is not expected to be available commercially as a microbial insecticide before the year 2000 (P.D. Christian, personal communication). Such a combination of new technology and long-term commitment could prove the key to the successful development of this and further virus-based insecticides in Australia.

VIRUS PRODUCTION

The technology for production of *Heliothis* NPV was imported from Sandoz, USA, the owners of Elcar. Production of *P. opercullela* GV was achieved in super-infesting plots of field potatoes by release of laboratory-reared adults. The potatoes were grown under stress (low nitrogen and water) to minimize the amount of foliage to be searched for larvae. Virus was applied at 6250 LE/ha in 500 l water with 0.2% Triton X114. When the proportion of larvae visibly infected reached 50%, foliage with larvae was put into large polythene bags, which were sealed and placed in the sun for 2 h. Larvae emerged and were separated by washing (Mattheissen *et al.*, 1978).

E. postvittana NPV was produced initially in larvae reared on broad bean, *Vicia faba*: late second to early third instars were sprayed with 1×10^5 OB/ml. Larvae were harvested after 10–12 days. This method was more convenient than two semi-synthetic diets tested (MacCollom and Reed, 1971).

OrV was produced in continuous cultures of a black beetle, *Heteronychus arator*, cell line (Crawford and Sheehan, 1984).

Formulation

A Queensland-based firm, Biocontrol Ltd, started to develop a local sunlight-stable formulation of *Heliothis* NPV in the early 1980s but discontinued owing to a change in priorities. Preliminary studies on the solar inactivation of *E. postvittana* NPV indicated no benefit from inclusion of 0.15% charcoal or Fire Orange fluorescent pigment, or a combination of these (MacCollom and Reed, 1971).

The addition of a wetter seems very desirable to aid penetrance of *P. operculella* GV into leaf stomatal cavities, which in potato are on both leaf surfaces, from where it can best be acquired by the leaf-mining larvae (Reed, 1971).

FIELD EXPERIMENTATION

Cydia pomonella (GV)

Studies on *C. pomonella* GV were made as part of a Cooperative Research Programme on Pest Management in Pome Fruit Orchards of South-eastern Australia (Geier *et al.*, 1971). This involved four state Departments of Agriculture and the CSIRO Division of Entomology. The aim was to develop a programme that allowed the beneficial insects to continue to operate while selectively controlling codling moth.

Weekly schedules of three concentrations of virus (0.3, 0.1 and 0.025 μg/ml) applied to neglected apple trees resulted in proportions of codling moth 'stings' developing into deep entries ranging from 1.4 to 5.7%. It was concluded that the acceptable level of codling moth damage in terms of deep entries lay within the concentration range. A problem associated with the use of the virus was that virus-infected larvae damaged the surface of fruit, even though they finally succumbed to the infection.

Epiphyas postvittana (NPV)

The study of *E. postvittana* was done as part of the above Cooperative Research Programme and followed laboratory studies by MacCollom and Reed (1971). Assessments in Tasmania indicated no effect on fruit damage and yield, although a proportion of the larval population became infected and died (A. Terauds, personal communication, 1989). By contrast, Longworth (1980) reported that both this virus and *C. pomonella* GV gave adequate control with a reduction in fruit damage in New Zealand. However, control did not meet export market standards.

Helicoverpa spp. (NPV)

The commercially produced virus from *H. zea* is highly infectious for the major pest species *H. armigera* and *H. punctigera* (Teakle, 1979) even though *H. zea* does not occur naturally in Australasia. Two imported commercial formulations have been used.

Viron H

Viron H is a commercial formulation that was used experimentally against *H. armigera* on sorghum in Western Australia, where it was effective provided timing of application coincided with an appropriate stage of panicle development (Michael, 1973).

Elcar

Elcar was an improved formulation (Shieh, 1978); it was assessed in eastern Australia on cotton, sorghum, navy bean and sweet corn.

Cotton Elcar was used as the first option for *Helicoverpa* spp. control in an experimental, computer-based pest management programme for cotton called 'Fly'. In the first two seasons, 1976–77 and 1977–78, yields were maintained with fewer spray applications than with conventional insecticides (Room, 1979). In central Queensland, Waite (1980) concluded that Elcar did not perform well under heavy population pressure but that it might be useful for mid-season control when pest infestations were diminished and beneficial insects were most abundant. The virus formulation was unable to compete with the newly introduced synthetic pyrethroids and consequently there was little commercial usage. In 1981, this resulted in Elcar being omitted from the commercial cotton pest management programme 'SIRATAC' (Hearn *et al.*, 1981), the successor to 'Fly'.

Sorghum This was a highly favourable crop on which to control *Helicoverpa* with the virus. Preliminary assessments in southern and central Queensland indicated that high levels of infection could be achieved without disrupting the often considerable natural control on this crop (Teakle *et al.*, 1983). Taking advantage of the high level of synchrony between sorghum head and *Helicoverpa* larval development and the predelection of young larvae for the pollen sacs, it was shown that applications of virus to sorghum at 100% anthesis could result in high levels of secondary infection (Teakle, Jensen and Mulder, 1985).

Interest in the use of the virus on sorghum has suffered because of its failure to control the economically important pest sorghum midge, *Contarinia sorghicola*. However, the increasing availability of midge-resistant sorghum varieties may overcome this problem. A further difficulty is that the optimal timing of virus application (at approximately 100% anthesis) is too early for the effectiveness of natural control to be able to be reliably established. Consequently, farmers may have to use the virus as an insurance treatment before it can be established that it is really required.

Navy bean Elcar and fenvalerate applied as single treatments to navy bean, *Phaseolus vulgaris*, were compared for effectiveness (Rogers, Teakle and Brier, 1983). No significant differences in seed yield or stained beans were found in the respective treatments. Elcar was most effective against first to early-third instar larvae, in which >90% mortality was recorded.

Sweet corn Elcar was applied at tasselling with 60% anthesis. Sampled larvae and larvae hatching from sampled eggs both showed about 50% virus infection after transfer to artificial diet. It was concluded that treatment early enough for virus to be released from larvae at early silking should achieve a significant level of control (Hamilton, 1980).

Lymantria ninayi (NPV, picornavirus, Nudaurelia β virus, nodavirus, etc.)

Lymantri ninayi is an indigenous New Guinea species feeding inoffensively on native deciduous trees but with the capacity to become epidemic on introduced pines, especially *Pinus patula* (Roberts, 1987). Its biology is very similar to that of gypsy moth, *Lymantria dispar*, and like the western forms of this species the adult female is almost flightless (Roberts, 1979). Periodic decimating epizootics of virus diseases, usually apparently with a strong involvement of NPV disease, are especially found after rain in areas where disease has been noted in at least the previous 6 years. A simple epizootic prediction system based on three factors – time since last epizootic, host larval population density at that time and whether rain has fallen on the present infestation during its egg/early larval period – will indicate whether or not control measures should be instituted (Entwistle, 1986). A control trial near Goroka, Papua New Guinea, in which helicopter applications of NPV were made was unassessable. The population was sprayed during the first two instars, but after a day or two, heavy virus mortality occurred in all plots, including the control. It was concluded a general epizootic, involving NPV and picornavirus infections, had already commenced (Entwistle, 1983).

Oryctes rhinoceros (OrV)

In the area concerned, coconut rhinoceros beetle is at present absent from the mainland of New Guinea but is present in Manus Island, the Gazelle Peninsula of New Britain and New Ireland, all offshore regions of Papua New Guinea. Much basic work on the dynastid pests of coconuts, *Oryctes* and *Scapanes*, and their virus relationships was conducted in New Britain by G.O. Bedford in the 1970s (e.g. Bedford, 1976). Field releases of marked adults of *O. rhinoceros* pre-infected with a Western Samoa virus isolate were made during 1978 and 1979 on Manus Island, at four sites on New Ireland and at 12 sites on the Gazelle Peninsula (Gorick, 1980). The virus became established at most of the release sites and a spread rate of 1 km/month was indicated. Post-release monitoring was not reported on long enough to define adequately the long-term impact of the virus.

Phthorimaea opercullela (GV)

Following preliminary studies on GV control of *P. opercullela*, potato tuber moth (PTM) (Reed, 1969, 1971), large-scale field trials were conducted on potato in Western Australia (Reed and Springett, 1971). These were the classic field trials, rather than in-store control on tubers, which opened the way to international investigation of virus control of PTM. These studies indicated that single, early applications of virus could achieve control and that extensive spread of the virus to untreated areas occurred. Subsquent studies showed that field populations differed in their susceptibility to the virus and indicated that a single dominant autosomal gene, which segregated according to simple Mendelian ratios, controlled resistance in a laboratory strain (Briese, 1982).

Pieris rapae (GV)

P. rapae GV was the first virus to be assessed for microbial control; preliminary trials were conducted in both New Zealand and Australia. Kelsey (1957) applied the virus (750 infected final instar larvae were macerated in 150 ml water and a 0.11% dilution of this was sprayed at about 190 ml per plant) to cabbages in the field and concluded that aqueous suspensions could be of value for the control of *P. rapae* in New Zealand. However, Kelsey (1960) also stated that under farm-crop conditions, virus application would seldom be necessary as *P. rapae* was usually in a state of equilibrium with its parasites, *Apanteles glomeratus* and *Pteromalus puparum*, and the GV. In home gardens, however, some form of control is usually necessary. In Australia, Wilson (1960) noted that epizootics resulting from aqueous sprays of the virus to cabbage persisted for at least 3 weeks. The resultant populations in the sprayed plots were lower than those in the control plots and the numbers of larvae reaching the later instars were much lower.

Scapanes australis (OrV)

Several subspecies of *S. australis*, a very large dynastid beetle, occur in the New Guinea–Solomon Islands area not, apparently, suffering from any endemic virus condition. Bedford (1973) showed that larvae from the first to early-third instar (there are only three larval instars in dynastids) of subspecies *grossepunctatus* in New Britain were susceptible. Further studies by B.D. Gorick (unpublished data) failed to establish adult susceptibility to the virus. The fate of field releases of the virus on Guadalcanal, Solomon Islands remain unknown.

Tiracola plagiata (NPV)

T. plagiata is an opportunist in tropical forest clearings, often increasing very rapidly on weeds and shade plants and then infesting other crops such as cocoa, *Theobroma cacao* (Entwistle, 1972). In Sabah, larval population collapse was thought to be caused by an NPV (Conway, 1971). In Papua New Guinea, van Velsen (1967) conducted a preliminary control trial using a homologous, local, NPV. Larvae were most susceptible during the first 18 days of life.

Wiseana spp. (NPV)

Porina, which appears to consist of several species dominated in the field by *Wiseana cervinata*, is a major pasture pest in New Zealand. In undisturbed fields, it seems normally to be in equilibrium with its virus disease but in short-term leys serious outbreaks may occur, apparently because conventional, deep cultivation dilutes virus that had accumulated towards the top of the soil where most porina larval feeding takes place. Management of pastures by shallow cultivation during reseeding appears to allow maintenance of a favourable host–virus balance and avoidance of infestation development (Kalmakoff and Crawford, 1982).

REFERENCES

Anon. (1981) Invertebrate Pathology Resources Inventory. *Australasian Invertebrate Pathology Working Group Newsletter* **2**, 11–32.

Bedford, G.O. (1973) Experiments with the virus *Rhabdionvirus oryctes* against the coconut palm rhinoceros beetles *Oryctes rhinoceros* and *Scapanes australis grossepunctatus* in New Guinea. *Journal of Invertebrate Pathology* **22**, 70–74.

Bedford, G.O. (1976) Rhinoceros beetles in Papua New Guinea. *South Pacific Bulletin* **26**, 38–41.

Briese, D.T. (1982) Genetic basis for resistance to a granulosis virus in the potato moth, *Phthorimaea opercullela*. *Journal of Invertebrate Pathology* **39**, 215–218.

Christian, P.D. and Oakeshott, J.E. (1989) The potential for genetically engineered NPV in *Heliothis* control. *Australian Cotttongrower* **10**, 77–78.

Crawford, A.M. (1982) A coleopteran cell line derived from *Heteronychus arator* (Coleoptera: Scarabaeidae). *In Vitro* **18**, 813–816.

Crawford, A.M. and Sheehan, C. (1984) An *Oryctes rhinoceros* (L.) (Coleoptera: Scarabaeidae) baculovirus inoculum derived from tissue culture. *Journal of Economic Entomology* **77**, 1610–1611.

Crawford, A.M., Sheehan, C.M., King, P.D. and Meekings, J. (1985) *Oryctes* baculovirus infectivity for New Zealand scarabs. In *Proceedings 4th Australasian Conference on Grassland Invertebrate Ecology*, Lincoln College, Canterbury, 13–17 May. R.B. Chapman (ed.). Canterbury, New Zealand: Caxton Press.

Conway, R.G. (1971) *Pests of Cocoa in Sabah and their Control, with a List of the Cocoa Fauna*. Sabah: Department of Agriculture.

Cooper, D.J. (1979) The pathogens of *Heliothis punctigera* Wallengren. PhD thesis, University of Adelaide, South Australia.

Dugdale, J.S. (1964) Polyhedral virus and the 1960–62 *Selidosema suavis* epidemic at Eyrewell S.F. *Symposium on Insect Pathology and Annual Conference*, May, Rotorua, NZ. Entomological Society.

Entwistle, P.F. (1972) *Pests of Cocoa*. London: Longman.

Entwistle, P.F. (1983) Viral spray trials for control of insect defoliation of pine trees in the highlands of Papua New Guinea. *EEC Consultancy Project 5604.31.61.019*. Luxembourg: Office for Official Publications of the European Communities.

Entwistle, P.F. (1986) Epizootiology and strategies of microbial control. In *Biological Plant and Health Protection*. J.M. Franz (ed.). Stuttgart: Gustav Fischer Verlag, pp. 257–278.

Geier, P.W. *et al.* (1971) A co-operative program of research into the management of pest insects in pome-fruit orchards of south-eastern Australia. In *CSIRO, Division of Entomology, Progress Reports 1969–1971*. Canberra: CSIRO.

Gorick, B.D. (1980) Release and establishment of the baculovirus disease of *Oryctes rhinoceros* (L.) (Coleoptera: Scarabaeidae) in Papua New Guinea. *Bulletin of Entomological Research* **70**, 445–453.

Grace, T.D.C. (1967) Insect cell culture and virus research. *In Vitro* **III**, 104–117.

Gunning, R.V., Easton, C.S., Greenup, L.R. and Edge, V.E. (1984) Pyrethroid resistance in *Heliothis armigera* (Hübner) (Lepidoptera: Noctuidae) in Australia. *Journal of Economic Entomology* **77**, 1283–1287.

Hamilton, J.T. (1980) Control of *Heliothis armigera* in sweet corn and maize with NPV. *Proceedings of a Workshop on Biological Control of* Heliothis *spp.*, 23–25 Sept. 1980, Toowoomba, Queensland, Australia, pp. 93–94.

Hearn, A.M., Ives, P.M., Room, P.M., Thomson, N.J. and Wilson, L.T. (1981) Computer-based cotton pest management in Australia. *Field Crops Research* **4**, 321–332.

Kalmakoff, J. and Crawford, A.M. (1982) Enzootic virus control of *Wiseana* spp. in the pasture environment. In *Microbial and Viral Pesticides*. E. Kurstak (ed.). New York: Marcel Dekker, pp. 435–448.

Kelsey, J.M. (1957) Virus sprays for control of *Pieris rapae* L. *New Zealand Journal of Science and Technology* **38**, 644–646.

Kelsey, J.M. (1960) Interaction of virus and insect parasites of *Pieris rapae* L. *Proceedings of the 11th International Congress of Entomology*, Vol. 2, pp. 790–796.

Longworth, J.F. (1980) Insect pathology in New Zealand. *Australasian Invertebrate Pathology Working Group Newsletter* **1**, 19–23.

MacCollom, G.B. and Reed, E.M. (1971) A nuclear polyhedrosis virus of the light brown apple moth, *Epiphyas postvittana*. *Journal of Invertebrate Pathology* **18**, 337–343.

Matthiessen, J.N., Christian, R.L., Grace, T.D.C. and Filshie, B.K. (1978) Large-scale field propagation and the purification of the granulosis virus of the potato moth, *Phthorimaea opercullela* (Zeller) (Lepidoptera: Gelechiidae). *Bulletin of Entomological Research* **68**, 385–391.

Michael, P.J. (1973) Biological control of *Heliothis* in sorghum. 2. Trials with virus sprays. *Journal of Agriculture of Western Australia, (Series 4)* **14**, 223–224.

Reed, E.M. (1969) A granulosis virus of potato moth. *Australian Journal of Science* **31**, 300.

Reed, E.M. (1971) Factors affecting the status of a virus as a control agent for the potato moth (*Phthorimaea opercullela* (Zell.) (Lep., Gelechiidae)). *Bulletin of Entomological Research* **61**, 207–222.

Reed, E.N. and Springett, B.P. (1971) Large-scale field-testing of a granulosis virus for the control of the potato moth (*Phthorimaea opercullela* (Zell.) (Lep., Gelechiidae)). *Bulletin of Entomological Research* **61**, 223–233.

Roberts, H. (1979) *Lymantria ninayi* B. Br. (Lep., Fam. Lymantriidae) a potential danger to *Pinus* defoliation in the highlands of Papua New Guinea. *Tropical Forestry Research Note*, *SR*, Vol. **37**.

Roberts, H. (1987) Forest insect pests of Papua New Guinea. 4. Defoliators of *Pinus* (Pines) in the highlands. *Entomology Bulletin* **48**, 103–108.

Rogers, D.J., Teakle, R.E. and Brier, H.B. (1983) Evaluation of *Heliothis* nuclear polyhedrosis virus for control of *Heliothis armigera* on navy beans in Queensland, Australia. *General and Applied Entomology* **15**, 31–34.

Room, P.M. (1979) A prototype 'on-line' system for management of cotton pests in the Namoi Valley, New South Wales. *Protection Ecology* **1**, 245–264.

Shieh, T.R. (1978) Characteristics of a viral pesticide Elcar®. In *Proceedings of the 2nd International Colloquium on Invertebrate Pathology and 11th Annual Meeting of the Society for Invertebrate Pathology*, Prague, September 11–17, pp. 191–193.

Teakle, R.E. (1969) A nuclear-polyhedrosis virus from *Anthela varia* (Lepidoptera: Anthelidae). *Journal of Invertebrate Pathology* **14**, 18–27.

Teakle, R.E. (1978) Possibilities for the control of non-aphid insects on lucerne. Paper delivered to a *Workshop on Lucerne Aphids*, Tamworth, N.S.W., Australia, November 1978, pp. 122–126.

Teakle, R.E. (1979) Relative pathogenicity of nuclear polyhedrosis viruses from *Heliothis punctiger* and *Heliothis zea* for larvae of *Heliothis armiger* and *Heliothis punctiger*. *Journal of Invertebrate Pathology* **34**, 231–237.

Teakle, R.E., Page, F.D., Sabine, B.N.E. and Giles, J.E. (1983) Evaluation of *Heliothis* nuclear polyhedrosis virus for control of *Heliothis armiger* on sorghum in Queensland. *General and Applied Entomology* **15**, 11–18.

Teakle, R.E., Jensen, J.M. and Mulder, J.C. (1985) Susceptibility of *Heliothis armigera* (Lepidoptera: Noctuidae) on sorghum to a nuclear polyhedrosis virus. *Journal of Economic Entomology* **78**, 1373–1378.

Vail, P.V., Vail, S.S. and Summers, M.D. (1984) Response of *Cactoblastis cactorum* (Lepidoptera: Phycitidae) to the nuclear polyhedrosis virus isolated from *Autographa californica* (Lepidoptera: Noctuidae). *Environmental Entomology* **13**, 1241–1244.

van Velsen, R.J. (1967) A nuclear polyhedral virus disease affecting the larvae of *Tiracola plagiata* Walk. *Papua New Guinea Agricultural Journal* **18**, 134–136.

Waite, G.K. (1980) Elcar in cotton pest management. *Proceedings of a Workshop on Biological Control of* Heliothis *spp.*, 23–25 September 1980, Toowoomba, Queensland, Australia. pp. 96–98.

Wilson, F. (1960) The effectiveness of a granulosis virus applied to field populations of *Pieris rapae* (Lepidoptera). *Australian Journal of Agricultural Research* **11**, 485–497.

Young, E.C. (1974) The epizootiology of two pathogens of the coconut palm rhinoceros beetle. *Journal of Invertebrate Pathology* **24**, 82–92.

Zelazny, B., Alfiler, A.R. and Crawford, A.M. (1987) Preparation of a baculovirus inoculum for use by coconut farmers to control rhinoceros beetle (*Oryctes rhinoceros*). *FAO Plant Protection Bulletin*, **35**, 36–42.

Zelazny, B., Autar, M.L., Singh, R. and Malone, L.A. (1988) *Papuana uninodis*, a new host for the baculovirus of *Oryctes*. *Journal of Invertebrate Pathology* **51**, 157–160.

20 North America

JOHN T. CUNNINGHAM

PERSPECTIVE

Despite the long history of research on insect viruses in both Canada and the USA and prolific publications on laboratory and field trials, the current practical use of viral insecticides is still very limited. Viruses for insect pest control may be technical successes in several instances, but a technical success does not necessarily ensure a commercial success (Bohmfalk, 1982). There are two viral insecticides currently registered for forestry use in the USA and three for agricultural use. A further two have reached the point of being granted Experimental Use Permits. In Canada, three virus insecticides are registered for two forest pest species. There are no commercial products currently available in the USA or Canada although several companies have expressed interest in marketing viral insecticides in the near future. Viruses in which a commercial interest has been shown, together with registered trade names, are listed in Table 20.1.

The high degree of host specificity of viruses has, to some degree, discouraged commercialization and not many companies wish to face the problems involved with *in vivo* production in insect larvae. Several problems have to be addressed before *in vitro* production of viral insecticides can become a commercial reality. Formerly, the costs and efficacy of insect viruses were measured against chemical insecticides. In the 1980s, *B.t.* has become a major competitor to insect viruses, particularly in the North American forestry market, and it is now the yardstick for comparative studies. In fact, use of viral insecticides is insignificant compared with the use of *B.t.* In some instances, viruses do have advantages over *B.t.* particularly when they initiate and sustain epizootics in insect populations. *B.t.*, which affects a wide range of species of Lepidoptera, has been considered too broad spectrum in some ecologically sensitive areas where Lepidoptera are important in food chains, include endangered species or are considered aesthetically attractive. Presently, *B.t.* is the only biological option available for control of many species. Reliance on a single biological alternative to chemical insecticides is not a sound management strategy and development of resistance to *B.t.* has already been observed in some pest species (Marrone and MacIntosh, 1993).

Table 20.1 Virus products registered in North America and their current status

Host (virus)	Trade name	Licencee	Country	Current status
Anagrapha falcifera (NPV)	–	7	USA	In development
Autographa californica (NPV)	SAN-404	5	USA	Discontinued
Cydia pomonella (GV)	Decyde TM	4	USA	Discontinued
	SAN-406	5	USA	Discontinued
	Specific-T-1	6	USA	On going
	UCB 87	1	USA	On going
Heliothis spp. (NPV)	Biotrol-VHZ	3	USA	Discontinued
	Elcar	5	USA	Revived interest
	Viron H	2	USA	Sold to Sandoz
Lymantria dispar (NPV)	Disparvirus	1, 9	Canada	In development
	Gypchek	8	USA	Routine use
Neodiprion lecontei (NPV)	Lecontvirus	9	Canada	Routine use
Neodiprion sertifer (NPV)	Neochek-S	8	USA	Reregistration pending
Orgyia pseudotsugata (NPV)	TM BioControl-1	8, 9	USA and Canada	In periodic use
	Virtuss	9	Canada	In periodic use
Plodia interpunctella (GV)	–	8	USA	In development
Trichoplusia ni (NPV)	Biotrol VTN	4	USA	Discontinued
	SAN-405	6	USA	Discontinued
	Viron T	3	USA	Discontinued

1, Association for Sensible Pest Control Inc. (ASPCI); 2, International Minerals and Chemical Corporation Inc. (IMC); 3. Nutrilite Products Inc.; 4, Microgenesis Inc.; 5, Sandoz Inc.; 6, University of California; 7, United States Department of Agriculture (USDA); 8, United States Department of Agriculture, Forest Services (USDA, For. Ser.); 9, Canadian Forest Service.

Forestry accounts for about 5% of all insecticides used in Canada and this figure is probably even less in the USA. However, pest management in forests is well controlled and strongly politically motivated. Many forests are public lands and environmentalists have a voice in determining which pest control agent is applied. The public is bitterly opposed to aerial application of chemical pesticides and supportive of biological control options. The situation in agriculture is much more complex, with most farmers opting for short-term gain by using fast-acting chemical pesticides to obtain maximum yield in that year and ignoring the longer-term ecological and environmental effects. There is a small group of organic farmers that obtains higher prices for its crops by using no chemical pesticides and selling its products to a discriminating, health-conscious sector of the community. Since the middle of the 19th century, apple orchards have been particularly well studied by economic entomologists and excellent pest management strategies have been developed to conserve natural enemies and reduce dependence on chemical insecticides in this ecosystem.

VIRUS PRODUCTION

All virus production, except on a very small scale mainly for laboratory purposes, is *in vivo*. The majority of the host species discussed here can be grown on artificial diets. However, the moth *Harrisina metallica* (*brillians*) and the two diprionid sawfly species, *Neodiprion lecontei* and *Neodiprion sertifer*, are exceptions. A method of rearing and infecting *N. sertifer* larvae on cut foliage in boxes was described by Rollinson, Hubbard and Lewis (1970). Over a 2 year period, 1.35 million virus-infected larvae were harvested at a cost of 1 cent per larva and a yield of 1×10^8 OB/larva. *Neodiprion* viruses are also produced by spraying heavily infested plantations when larvae reach the fourth instar and harvesting diseased and dead colonies (Cunningham and McPhee, 1986).

Production of EPV in the grasshopper *Melanopus sanguinipes* is unusual in that the optimal procedure is to inject virions into nymphs, which are frozen after 21 days (Oma and Streett, 1993).

FORMULATION

There has been a great deal of work in North America on the formulation of insect viruses, very large sections of which relate to the employment of UV protectants and gustatory stimulants and baits. Much of this work has been centred on the NPV of *Helicoverpa/Heliothis* spp. and the lepidopterous pests of cole crops and a comprehensive discussion is provided in Chapter 7. Impressive and fairly recent advances resulting from the recognition that at least some optical brighteners have not only a UV-protectant capacity but also enhance the activity of entomopathic viruses are also discussed in Chapter 7.

THE USE OF VIRUSES WITH PESTICIDES

Much work has been devoted to study of the interactions of chemical insecticides with insect viruses, especially seeking to employ reduced dosages of both. Studies by Jaques (1988), Jaques and Laing (1978) and Jaques, Laing and Maw (1989) on cole crop Lepidoptera are notable. The principles involved in analysing such interactions are discussed in Chapter 5.

REGISTRATION

In the USA, the Environmental Protection Agency (EPA) regulates the use of pesticides under the Federal Insecticide, Fungicide and Rodenticide Act (FIFRA). Guidelines for testing safety of microbial pesticides, including viruses, are contained in *Pesticide Assessment Guidelines – Subdivision M – Biorational Pesticides* (Anon., 1983). Registration requirements, including product analysis, toxicology and environmental effects, are described and discussed by Betz (1986). Both toxicological and environmental tests are arranged in a three tier system. If the material shows no adverse effects in tier I, testing in tier II and tier III is not required.

Tier I toxicology tests consist of short-term studies in several mammalian species, such as rats, rabbits, guinea pigs and hamsters, and include oral, dermal, inhalation, ocular and injection routes of exposure. Also irritancy, hypersensitivity and tests of effects on the immune system are required as well as cell culture studies in a variety of mammalian cell lines. Tier I environmental tests include avian oral, avian injection, wild mammal, freshwater fish, aquatic invertebrate, esturine and marine animal, plant and non-target insect. BVs with their limited host range have not required testing past tier I. In fact, it may be possible to obtain waivers for some tests because of the well-established safety record of BVs. The potential hazard in viral insecticides is not from the BV, but from insect material and extraneous microorganisms in the product.

In Canada, pesticides are regulated under the Pest Control Products Act. This act has been administered by Agriculture and Agri-Food Canada with Health, Environment and the Canadian Forest Service (for forestry products) as advisors. In a recent restructuring, the lead role has gone to Health Canada. *Registration Guidelines for Microbial Pest Control Agents* (Anon., 1993) is still in the form of a regulatory proposal. These guidelines cover all microbial control products and it is probable than many waivers and exceptions can be obtained with appropriate documentation for BVs. The Canadian guidelines are similar to the American, but involve five ecozones with more stringent environmental requirements for non-indigenous microorganisms than for those which are indigenous.

FIELD EXPERIMENTATION

Anagrapha falcifera (NPV)

Fairly recently discovered, the celery looper *Anagrapha falcifera* NPV (AfNPV) has a wide host range (Hostetter and Putler, 1991) of more than 31 species of Lepidoptera in ten families. This host range is similar to alfalfa looper NPV, but it is much more active against *Helicoverpa zea*. The host range includes many other major lepidopterous pests such as the *Spodoptera* complex, European cornborer (*Ostrinia nubilalis*), black army cutworm (*Ochropleura fennica*) and various stored products and vegetable pests. Field trials on cotton and vegetables were conducted in 1991 (Vail *et al.*, 1993a,b). AfNPV is closely related to AcMNPV and could be considered a variant of the same virus (P. V. Vail, personal communication). AfNPV is being developed for commercial use by Sandoz Agro, Inc. as biosis and a registration petition has been submitted to the EPA.

Anticarsia gemmatalis (NPV)

A velvetbean caterpillar NPV was isolated from larvae in Brazil (Allen and Knell, 1977; Carner and Turnipseed, 1977). Field trials conducted in Florida indicated that a dosage of about 1×10^{11} OB/ha suppresses larval populations to below damaging levels (Moscardi, Allen and Greene, 1981; Moscardi and Ferreira, 1985). Use in Florida and South Carolina is still experimental and the recommended dosage is 1.25×10^{10} OB/ha in 235 l. The areas treated have been very small. Results varied depending on the size of larvae and population density at the time of application. When there is significant suppression, yields of soybean are similar to those obtained with the best chemical insecticide treatments. The virus causes epizootics and persists in the upper layers of the soil.

Readers should consult Chapter 22, which discusses the NPV control of *A. gemmatalis*, which has been developed on a huge scale in South America.

Autographa californica (NPV)

Alfalfa looper NPV (AcNPV) was discovered in a single larva collected on alfalfa in the southern desert of California. It was found to have a broad host range (Vail *et al.*, 1971; Vail and Jay, 1973), which has been listed (Payne, 1986) as 43 species of Lepidoptera in 11 families and keeps expanding. Many of the hosts are of economic importance and there has been considerable interest in developing the virus. Sandoz Inc. had an experimental product, SAN 404, which was field tested in California between 1979 and 1981 under an Experimental Use Permit against several pests but it was discontinued by the company (Falcon, 1982). In 1978, a registration petition was sent to the EPA by the USDA Agricultural Research Service but no action was taken and since then the regulations have changed considerably.

AcMNPV has been tested on a wide range of pests on several different crops. Tobacco budworm, *Heliothis virescens*, is more susceptible to AcMNPV than to *Heliothis* NPV, but the cotton bollworm, *H. zea*, is less susceptible (Vail *et al.*, 1978; Vail and Collier, 1982). Dosages have ranged from 2.5×10^{11} to 2.5×10^{12} OB/ha in volumes from 47 to 564 l, with two to eight applications seasonally. In several field trials on cotton, AcMNPV gave no better control of cabbage looper, *Trichoplusia ni*, than did the NPV of *T. ni* itself. AcMNPV, though infective to cotton leaf perforator, *Bucculatrix thurberiella*, did not reduce populations in the field. Against pink bollworm of cotton, *Pectinophora gossypiella*, its effectiveness was markedly improved when combined with *B.t.* and crude cotton seed oil, but it was no better than the *B.t.* alone. Nor were populations of beet armyworm, *Spodoptera exigua*, western yellow-striped armyworm, *Spodoptera praefica*, or saltmarsh caterpillar, *Estigmene acrea*, affected when these were present (Falcon, 1982).

In Canada, AcNPV was used at 1.5×10^{12} OB/ha with two to six applications per year on small plots infested with *T. ni.* Virus persisted in the soil but this did not result in adequate crop protection in succeeding years (Jaques, 1975). Mixtures of half the dosage of AcMNPV and half the dosage of certain recommended chemical insecticides provided crop protection exceeding that of either component used alone. AcMNPV has usually been applied to cruciferous crops in combination with imported cabbage worm, *Pieris rapae*, GV or permethrin (Jaques, 1977, 1988; Jaques and Laing, 1978). Some degree of synergy appears to occur between AcMNPV and permethrin (Jaques *et al.*, 1989).

Experimentally AcMNPV could be successfully disseminated by transovum transmission into tobacco, but this did not result in crop protection (Nordin *et al.*, 1990).

AcMNPV is the BV most intensively studied by molecular biologists and it was the first BV to be used for the expression of foreign genes. It grows well in both *T. ni* and *Spodoptera frugiperda* cell lines. Over 140 exogenous genes have now been expressed in this system but little of this research has been conducted from the standpoint of enhancing the virus as an insecticide. In the USA, a test was conducted on the persistence of AcMNPV from which the polyhedrin gene had been deleted and which had been co-occluded with wild-type virus (Hamblin *et al.*, 1990; H.A. Wood, personal communication). It was tested on *T. ni*, with 0.0625 ha treated, using three applications of 7.5×10^{12} OB/ha in 284 l water. When cells are co-infected, the wild-type virus provides inclusion body protein and so the persistence of this engineered virus is determined by the probability of such cellular co-infection; laboratory data indicate that the strain lacking polyhedrin cannot persist under natural conditions. This is supported by similar studies in the UK (Bishop, 1986). A next step is to field test AcMNPV transformed with an insecticidal gene inserted in the strain lacking polyhedrin and to provide protection by co-infection with the wild type. If such an approach results in an acceptable level of control, the fundamental instability of the genetically modified component should contribute to its environmental acceptability.

AcMNPV was registered by the US EPA in 1994 under the trade name Gusano TM, based on data submitted by Crop Genetics International. There is currently no product for sale in the USA.

Choristoneura fumiferana (NPV, GV and entomopoxvirus)

Spruce budworm, *Choristoneura fumiferana*, is Canada's most important forest insect pest and virtually every biological control agent has been tested at some time (Cunningham, 1985a). Extensive field trials were conducted with viruses during 1971–83; a total of 65 plots with a combined area of 2656 ha was treated (Cunningham, 1985b). Mainly NPV was applied but an entomopoxvirus and a GV were also tested. Viruses that infect *C. fumiferana* also infect western spruce budworm, *Choristoneura occidentalis*, and jack pine budworm, *Choristoneura pinus*. Between 1976 and 1982, six plots with a combined area of 424 ha were treated with NPV or GV in British Columbia to control *C. occidentalis* and one 50 ha plot was treated in Ontario to control *C. pinus*.

Aerial sprays, whether biological or chemical, are usually applied on spruce budworm at bud flush when larvae are exposed to the deposit. By then, they have reached their fourth instar. Infected larvae die at the onset of pupation, no foliage is saved and there is no opportunity for horizontal virus transmission. Naturally occurring epizootics have never been observed to terminate a budworm outbreak. Following a virus spray application, vertical transmission does occur but the incidence of infection is insufficient to regulate the pest population. Since the mid 1980s use of *B.t.* for operational budworm control has steadily increased in eastern Canada (van Frankenhuyzen, 1990) and field testing of viruses has been discontinued until some major advancements are made in terms of reduced production costs and effectiveness. Because of the economic importance of spruce budworm, genetic manipulation and enhancement of viruses for its control are research priorities for the Canadian government and two networks are actively involved in this endeavour.

Cydia pomonella (GV)

A GV found in codling moth *Carpocapsa pomonella* (CpGV) in Mexico (Tanada, 1964) showed considerable promise in field tests on apple and pear trees and there has been interest in the development and registration of this virus in both North America and Europe. Sandoz Inc. developed an experimental product, SAN 406, which was granted an Experimental Use Permit by the EPA in 1981, but the project was terminated in 1982. Microgenesis, in Connecticut, made a brief incursion into the viral insecticide market in the mid-1980s and obtained an Experimental Use Permit for a product named Decyde TM. Behringwerke AG, Germany, registered CpGV as Granusul TM in 1990.

In collaboration with a growers' cooperative, the Association for Sensible Pest Control Inc. (ASPCI), L.A. Falcon at the University of California has been working on CpGV, designated UCB 87, with additional funding from the State of California and the USDA. About 20 ha were treated in 1987, 50 ha in 1988 and 200 ha in 1989 and 1990. The dosage was 2.5×10^{13} OB/ha applied in 950 to 3800 l/ha using conventional orchard spray equipment. The cost was US $62–75/ha treatment and the virus was applied every 4 to 7 days during the period from 2 to 98% egg hatch. Control is purely insecticidal as the virus does not initiate epizootics. In 1990, about 2400 ha of treatment coverage was applied to almost 200 ha of apple, pear and walnut trees (i.e. each area had approximately 10 treatments) by over 40 growers in California, Washington, Colorado and Oregon under an Experimental Use Permit that exempted the crops from residue tolerance requirements. The results, ranging from very good to poor, have been greatly influenced by spray timing and coverage quality. As potential registrants, the Regents, University of California made submission to the EPA in 1992 for a product, Specific-T-1, to be available to the private sector under licence by the University (L. A. Falcon, personal communication).

In Canada, CpGV has been tested in Ontario, Nova Scotia and British Columbia. Prior to 1986 material was produced by R. P. Jaques, during 1986–9 an unregistered French product, Carpovirusine (Solvay), was tested and in 1990–1 material was obtained from L. A. Falcon. Dosages from 1×10^{13} to 1×10^{14} OB/ha in 100 l/ha were applied to small apple trees. In Ontario and British Columbia where there are two generations per year, four to eight applications were made, but in Nova Scotia where codling moth is univoltine, only two to four applications were made though, in some cases, one sufficed. Plot sizes were about 0.5 ha and studies were carried out under an Agriculture Canada Research Permit. Fruit protection equalled that by organophosphate insecticides, azinophos methyl or, in some cases, phosmet (Jaques *et al.*, 1977, 1981, 1987, 1994; R.P. Jaques, personal communication).

Integrated pest management practices have reached a high degree of refinement in orchards. Use of chemical insecticides destroys predators and parasites of codling moth and this results in outbreaks of other arthropod pests such as the red spider mite. In many areas, codling moth is the only major pest that requires chemical insecticides for its control, and organic growers in North America are enthusiastic about the commercialization of its GV as a viable pest management option.

Crop Genetics International is developing a codling moth GV product called Cyd-X; a registration petition has been submitted to EPA and it should be available commercially soon.

Gilpinia hercyniae (NPV)

Control of *Gilpinia hercyniae* the European spruce sawfly, is the largest and most dramatic biological control programme in Canada. This sawfly was a serious defoliator of spruce trees in eastern Canada and the adjacent USA. Between 1933 and 1951, 27 species of parasites from Europe and Japan were released for its control (McGugan and Coppel, 1962). An NPV appeared in the sawfly population in the late 1930s and acted as a key factor in regulating this pest (Balch and Bird, 1944). It is believed that the virus was introduced with parasites and that its rapid spread could have resulted from the distribution of parasite material. There were a number of planned transfers of NPV between 1939 and 1950, with several in Newfoundland, three in Quebec and three in Ontario (McGugan and Coppel, 1962). The rapid decline of the outbreak after 1942 removed the threat to spruce stands in eastern Canada.

European spruce sawfly is still endemic in eastern Canada but has been held in check by a combination of parasites and NPV since the late 1930s. This is one of the few outstanding examples of successful classical biological control, even if the introduction of the virus was fortuitous. The original introductions of parasites and the NPV appears to have permanently solved the European spruce sawfly problem in eastern North America.

Harrisina metallica (*brillians*) (GV)

A GV of *H. metallica* the western grapeleaf skeletonizer, was discovered in California by Steinhaus and Hughes (1952). Biological controls of this pest were based on introductions of a parasitic fly, a parasitic wasp and the GV, which was inadvertently introduced with them (Clausen, 1961). Further outbreaks, which began in the 1970s, were attributed to termination of a chemical control programme and disappearance of the GV from the insect population. Losses of US $8 million per year created a renewed interest in the GV, which small-scale field trials showed was highly effective (Stern and Federici, 1990). A single application of 5.8 g/ha of virus-infected larval cadavers in 19–95 l water early in the larval period gave 90% or more control with no crop loss for the entire season.

This GV is extremely effective because, unlike other GVs, it infects the midgut epithelium (Smith *et al.*, 1956); many larvae develop diarrhoea 2 to 4 days after infection so that disease spreads rapidly among the gregarious larvae. The GV biology is, therefore, similar to that of NPVs in diprionid sawflies (Federici and Stern, 1990). Safety testing has still to be conducted and an Experimental Use Permit has not yet been obtained. Only about 4 ha are treated annually and about 40 ha have been treated since the mid-1980s.

Helicoverpa zea and *Heliothis virescens* (NPV)

H. zea and *H. virescens* are variously known as cotton bollworms, tobacco budworms and tomato fruitworms. The singly embedded NPV of *Heliothis* spp. *sens. lat.* (SHNPV) was the first virus to be considered for commercialization in North America. The only insects found to be susceptible are seven species of *Heliothis* (*sens. lato*, i.e. including *Helicoverpa*), namely *H. armigera*, *paradoxa*, *peltigera*, *phloxiphaga*, *puctigera*, *virescens* and *zea* (Ignoffo and Couch, 1981). Species of *Heliothis*/*Helicoverpa* are major global pests that attack at least 30 different food and fibre crops and in many instances are resistant to most chemical insecticides (Ignoffo, 1973).

The largest commercial venture in North America with a viral insecticide started with the development of Viron H for control of *Heliothis* spp. by International Minerals and Chemical Corp. (IMC) in the 1960s (Greer, Ignoffo and Anderson, 1971; Ignoffo, 1973; Ignoffo and Couch, 1981). Around this time, Nutrilite Products Inc. had a similar experimental product called Biotrol-VHZ. Limited commercial production of Viron H commenced in 1971. The biological products division of IMC was purchased by Sandoz Inc. and the trade name of Viron H was changed to Elcar TM with a label granted for use on cotton in 1976 (Shieh, 1989). The target pests were *H. zea* and *H. virescens* on cotton, although these ubiquitous species damage other major crops such as corn and soybeans. Sales of Elcar increased until 1980 but dropped substantially in 1981 owing to the introduction of new synthetic pyrethroid insecticides, which gave a rapid kill compared with the viral insecticide. The manufacture of Elcar was discontinued in 1982, although Sandoz Agro Inc. still holds its registration. The Elcar story was hailed as a great success by scientists working in this field and the decision to stop production was a major blow to the development of viral insecticides generally.

Recommended dosages of Elcar ranged from 6×10^{11} to 1.2×10^{12} OB/ha. This amount of virus can be obtained from 25–50 insect larvae. The cost in 1977 was US $7.5 to 10.0 per ha, which was comparable to chemical insecticides (Sheih, 1989). Usually three applications are made at 3- to 7-day intervals starting when the first eggs are laid and continuing as long for as eggs are being deposited. Elcar has been applied both from the ground and from the air in volumes ranging from 18.8 to 188 l/ha. Droplet sizes have been in the 150–300 μm VMD range. A considerable variety of anti-evaporants, stickers, UV protectants and gustatory stimulants have been used under different environmental conditions (Ignoffo and Couch, 1981; and Chapter 7). Between 150 and 200 field trials were conducted with SHNPV in the 1960s and 1970s with about 60% on cotton (see Entwistle and Evans (1985) for an analysis of some of these trials), 30% on corn and the remainder on soybean, sorghum, tobacco and tomato. The total area treated with the virus is estimated at over 1 million ha and in one particular year over 200 000 ha were treated (C. M. Ignoffo, personal communication). Another company, Crop Genetics International, submitted a registration petition to the EPA in 1994 for an SHNPV product. Work continues and, for instance, NPV persistence and control studies have been conducted on grain sorghum (Young and McNew, 1994).

A new strategy has been applied for the use of Elcar on *Heliothis* spp. which produce a first seasonal generation on wild geraniums in the Mississippi delta area. The geraniums are found along field margins, rights of way and railway tracks. The USDA Agricultural Research Service produced virus that was processed by Sandoz Agro Inc. under the Elcar label. In 1990, two 100 square mile (25 900 ha) plots were established in the Mississippi delta; one was an untreated control area and the second plot was aerially sprayed with virus in zones where geraniums proliferate. The dosage was 4×10^{11} OB/ha in 2.4 l. This application gave 30–35% reduction in the F1 generation of both *H. zea* and *H. virescens* and some reduction of the F2 generations, which establishes on crops. This research continues and further aerial spraying is planned (Bell, 1991; M. R. Bell, personal communication). Regulation of *Heliothis* on *Geranium dissectum* was potentially further improved (cage tests) using AfMNPV with a whitening agent: there was 73% less adult emergence than in controls (Bell, 1995).

Scientists at the USDA, Agricultural Research Service in Tifton, Georgia have used honeybees, *Apis mellifera*, to disseminate SHNPV against larvae of both *Heliothis* species on crimson clover (*Trifolium incarnatum*). Bees leaving the hive are surface contaminated with a virus–talc formulation. Clover flowers collected from exposed fields when bioassayed in the laboratory showed high levels of virus activity (H. R. Gross, personal communication).

Lymantria dispar (NPV)

Gypsy moth virus (LdNPV) was considered as a potential biocontrol agent in the USA in the early 1900s (Reiff, 1911). The USDA Forest Service commenced an extensive field testing programme in 1964 (Rollinson, Lewis and Waters, 1965), which culminated in the registration of an NPV insecticide called Gypchek by the EPA in 1978. The use of Gypchek has been reviewed extensively (Lewis, McManus and Scheenberger, 1979a; Lewis *et al.*, 1979b; Lewis, 1981; Podgwaite, 1985). Over 11 000 ha have been treated in the USA since registration. In 1979, Gypchek was applied by state agencies in six states, Pennsylvania, Vermont, Massachusetts, New York, Wisconsin and Michigan, with 4000 ha treated (F. B. Lewis, personal communication). Although a registered product, research to improve formulation and to enhance application methods is still being conducted in the USA. Lack of a commercial product has been the main deterrent to wider use of Gypchek. There is a pressing need for Gypchek in environmentally sensitive habitats where *B.t.* cannot be employed because of important non-target species of Lepidoptera. The current recommendations in the USA are for two applications of 1.5×10^{12} OB/ha in 9.4 l/ha using Novo Nordisk Virus Carrier 038. The cost of producing this dosage is about US $23 per ha (J. D. Podgwaite, personal communication).

Gypsy moth was an insignificant, quarantined insect pest in Canada until 1981, when an outbreak of over 1000 ha was reported in Ontario. Since this initial outbreak, it has spread widely in Ontario and the Canadian Forest Service has developed a viral insecticide named Disparvirus. A Canadian registration petition, based on USA toxicological data, was submitted in 1990 and is currently being evaluated. From 1982 to 1994, 81 plots with a total of 1280 ha were treated experimentally in Ontario with either Gypchek or Disparvirus. Tests of both reduced dosages and emitted spray volumes gave acceptable results and new formulations are being evaluated in an effort to make Disparvirus a viable alternative to *B.t.* in Ontario (Cunningham *et al.*, 1991, 1993; J. C. Cunningham and K. W. Brown, unpublished).

Melanopus sanguinipes (entomopoxvirus)

An entomopoxvirus of *M. sanguinipes*, the migratory grasshopper (Henry and Jutila, 1966; Henry, Nelson and Jutila, 1969), has undergone tier 1 toxicology tests, has been screened against non-target insects and has been granted an Experimental Use Permit (D. A. Streett, personal communication). The virus is formulated in bait; 35 nymphs are required to produce a dose sufficient for 1 ha and there is horizontal transmission by cannibalism. Both aerial and ground spray trials have been conducted in the western USA with 4, 400, 140 and 600 ha treated in 1987, 1988, 1989 and 1990, respectively. The protozoan, *Nosema locustae*, is already registered for grasshopper control in the USA and spores from a single grasshopper can be used to treat 4 ha. It is, therefore, more economical to produce than the entomopoxvirus, but the latter kills faster at high dosages. High application rates of entomopoxvirus may be desirable where rapid reduction of a high population is required, but lower dosages may be of greater value at low densities when horizontal transmission retards population growth (Woods, Streett and Henry, 1992; D. A. Streett, personal communication). At this time, it is not known if commercialization of the EPV will be pursued.

Neodiprion lecontei (NPV)

An intensive research effort to develop redheaded pine sawfly, *N. lecontei*, NPV as a viral insecticide started in Canada in 1976. This diprionid is a pest of red pine and jack pine plantations. All the efficacy trials between 1976 and 1980 were with aerially applied virus (Cunningham, de Groot and Kaup, 1986). Safety tests were also conducted at this time and a product named Lecontvirus was registered under the Pest Control Products Act (Canada) in 1983. Since 1980 all sprays have been applied from the ground only.

The virus is produced as described for *N. sertifer*, below (Cunningham and McPhee, 1986). The recommended dosage for first and second instar larvae is 5×10^9 OB/ha applied in 9.4 l water when sprayed aerially and in 20 l when applied from the ground. This dosage is obtained from about 50 dead larvae and is estimated to cost about Canada $2.5 per ha. Dosage is increased to 1×10^{10} OB/ha when applied to fourth instar larvae. Lecontvirus is the only virus that is used routinely in Canada, and between 1976 and 1994, 686 plantations with a total of 6008 ha were treated in Ontario and Quebec.

Neodiprion sertifer (NPV)

European pine sawfly, *N. sertifer*, can be a serious pest in Christmas tree (*Pinus* spp.) plantations and ornamentals. An NPV from this species from Sweden was first described by Bird (1953). It was used extensively in both Canada and the USA in the 1950s and 1960s but no records are available. The USDA Forest

Service contracted the safety testing of *N. sertifer* NPV and a product named Neochek-S was registered by the EPA in the USA in 1983. Recently, because of the lack of use of Neochek-S, this registration was not renewed by the USDA Forest Service. The Neochek-S registration petition was reworked and submitted in Canada for a product called Sertifervirus; this petition is currently under review.

Presently *N. sertifer* is only a minor pest in North America. The USDA Forest Service recommends 2.5×10^9 OB/ha applied from the ground in 187 l/ha (Podgwaite *et al.*, 1984) (which it is interesting to contrast with the 1.1 l/ha ULV application recommended in the UK) and the Canadian Forest Service recommends 5×10^9 OB/ha in 9.4 l/ha from the air and in 20 l from the ground. Little use of the virus has been made in recent years, with 258 ha treated between 1976 and 1987 in the USA and 152 ha between 1975 and 1993 in Canada. In 1975, 125 ha were aerially sprayed in a provincial park in Ontario to protect Scots, jack and red pines from severe defoliation (Cunningham *et al.*, 1975). Since 1983, small amounts have been used annually to treat ornamental trees in the city of St John's, Newfoundland.

As it replicates in the midgut, *N. sertifer* NPV infection can spread rapidly through colonies of the gregarious larvae while OBs released from colonies readily infect other colonies lower down trees (Cunningham and Entwistle, 1981). Virus should be applied as soon after egg hatch as possible to prevent defoliation and to give sufficient time for virus spread and the development of secondary infection before pupation. Only a single application is required.

Orgyia pseudotsugata (NPV)

Douglas-fir tussock moth, *O. pseudotsugata*, is an intermittent pest that occurs from the southern interior of British Columbia, Canada to parts of Washington, Idaho, Oregon, California, Nevada, Colorado, Arizona and New Mexico, USA. Outbreaks invariably terminate with an NPV epizootic, but not before serious damage and tree mortality occur (Brookes, Stark and Campbell, 1978). An obvious strategy is to apply virus early in the outbreak cycle to terminate the infestation before serious damage occurs.

Singly and multiply enveloped types of NPV occur in *O. pseudotsugata* (Hughes and Addison, 1970). The latter was developed as a viral pesticide by the USDA Forest Service. It was registered as TM BioControl-1 by the EPA in 1976. Product efficacy testing was conducted in British Columbia in 1985 as the host population had collapsed in the USA (Stelzer *et al.*, 1977).

Using mainly USA data and the same virus strain, a product called Virtuss was registered in Canada in 1983. The Canadian Forest Service also secured a Canadian registration for TM BioControl-1 in 1983 to facilitate importation. The latter is produced in *O. pseudotsugata* but Virtuss is produced in larvae of *Orgyia leucostigma*, the whitemarked tussock moth. Between 1971 and 1979, 1508 ha were treated with TM BioControl-1 in the USA and Canada; between 1974 and 1982, 165 ha were treated with Virtuss in British Columbia (Cunningham, 1988).

Sufficient virus-killed larvae to treat 160 000 ha have been stockpiled at the production facility at Corvallis, Oregon, USA (J. Hadfield, personal communication). British Columbia Forest Service purchased enough TM BioControl-1 to treat

8000 ha and have sufficent Virtuss to treat a further 1400 ha. The recommended dosage of both products is a single aerial application of 2.5×10^{11} OB/ha in 9.4 l/ha using an aqueous tank mix containing 25% animal-feed-grade molasses and Orzan LS as a sunscreen agent. The same dosage, but in a higher volume, can also be applied from the ground. The production cost for a 1 ha dose of TM BioControl-1 is US $19.6 (J. Hadfield, personal communication).

A new outbreak of douglas-fir tussock moth started in California in 1990 and 68 000 ha were treated with *B.t.* in 1991 because western spruce budworm, *C. occidentalis*, was also present at outbreak numbers. Populations also reached a control threshold level (Shepherd and Otvos, 1986) around Kamloops, British Columbia. which called for control action in 1991. About 300 ha were treated in 1991 using both TM BioControl-1 and Virtuss, 900 ha in 1992 and 650 ha in 1993.

Penaeus spp.

For convenience, *Penaeus* spp. (cultivated shrimps) are discussed collectively in Chapter 15.

Pieris rapae (GV)

Imported cabbage worm GV was used experimentally as early as 1963 in Nova Scotia (Fox and Jaques, 1966) and in Ontario in the 1970s and 1980s, often in combination with AcMNPV and low dosages of chemical insecticides (Jaques, 1971, 1977, 1988; Jaques and Laing, 1978; Jaques *et al.*, 1989). The recommended dosage is 7.5×10^{12} OB/ha in about 100 l/ha. There are two to six applications a year made on cole crops (cruciferous crops), depending on dates of planting and harvest. The control of *P. rapae* by GV, measured by crop protection equals or exceeds control by chemical insecticides or *B.t.* Widespread natural GV epizootics develop in late summer and, although the virus persists for several years in the soil, its half life on foliage is only about 2 days. Field trials were conducted with the GV in Missouri in 1971 and 1972; dosages of 2×10^{11} OB/ha, or greater, reduced damage to the leaves and heads of cabbage. No toxicological testing has been conducted in North America. In the 1970s, up to 20 ha were treated annually in Canada, but in the 1990s this has declined to only about 1 ha.

P. rapae GV has been employed with great success in the People's Republic of China (Chapter 16).

Plodia interpunctella (GV)

Indian meal moth, *P. interpunctella*, is a pest of stored products, the GV of which was first described by Arnott and Smith (1968). It was reported to be highly pathogenic to neonate larvae (Hunter, 1970) and effective in protecting stored in-shell nuts (Hunter, Collier and Hoffman, 1973, 1977) and stored raisins (Hunter, Collier and Hoffman, 1979; Vail *et al.*, 1991). A freeze-dried preparation was found to be effective for protection of almonds and raisins. The virus has been tested extensively on dried fruit and nuts and, to a lesser extent, on grains. It is used as a prophylactic material and persistence is adequate to provide protection from packaging to consumption of foods. *P. interpunctella* populations can be reduced by 95–100% and damage to products reduced to a non-detectable level. A formulation of this virus has been patented by the USDA Agricultural Research Service and this patent has been licenced to John Evans, Boulder, Co. (P. V. Vail, personal communication). Tier 1 toxicology tests and environmental impact assessments have yet to be conducted. A method of autodisseminating this virus by attracting male moths to virus contaminated pheromone lures may have potential for Indian meal moth control (Vail *et al.*, 1991).

Spodoptera exigua (NPV)

Beet armyworm NPV was used by Yoder Bros in Florida on chrysanthemums in their shade houses. The virus causes epizootics, spreads and persists and has become endemic in Yoder's chrysanthemum fields. The need for its application has been considerably reduced in recent years. Yoder Bros produce the virus for their own use on their own property and have sufficient on hand to treat 40 ha. They joined Espro (Environmentally Safe Products) Inc. in 1989 and Espro in turn was taken over by Crop Genetics International in 1991. *S. exigua* is a primary pest of lettuce in Florida, Arizona and California and of tomatoes and several flower crops in both Florida and California; it also damages sugar beet, asparagus and a variety of cole crops. It is resistant to many chemical insecticides and *B.t.* is not very effective for its control; hence, there is a potential market for a viral insecticide (D. Kolodny-Hirsch, personal communication).

Some field trials have been conducted in Alabama, with dosages of about 5×10^{11} OB/ha. Crop Genetics International conducted trials in 1991 using dosages of 2.5×10^{11} and 1.25×10^{12} OB/ha applied weekly and both gave significant reductions in larvae (D. Kolodny-Hirsch, personal communication). There are several generations of *S. exigua* a year in the southern states, with generations usually overlapping.

An insecticidal viral product called Spod-X R WP was registered by Crop Genetics International in 1993. A registration petition for Spod-X R LC has been submitted to the EPA. Sale of this product in the USA is planned for the late 1990s (D. Kolodny-Hirsch, personal communication).

Trichoplusia ni (NPV)

There was interest in *T. ni*, cabbage looper, NPV (TnNPV) in the 1960s and experimental formulations such as Biotrol VTN (Nutrilite Products Inc.) and Viron T (International Minerals and Chemical Corporation) were developed. In the 1950s, there was a large cottage industry in California where growers collected virus-killed larvae and sprayed them on cotton, potato, brassica and other crops (Falcon, 1982; Yearian and Young, 1982). Sandoz Inc. produced an experimental product, San 405, but, along with Elcar and their other experimental viral insecticides, it was discontinued in 1982.

Trials were conducted in the 1970s in southwestern Ontario with about 20 ha treated annually. *T. ni* NPV causes extensive natural epizootics, e.g. in Ontario and New York State, infecting up to 95% of larvae by the end of the season. Such epizootics are probably initiated by virus persisting from year to year in the soil (Jaques, 1970). Protection of cabbage by the NPV or by AcMNPV equalled or surpassed protection obtained with chemical insecticides or with *B.t.* (Jaques, 1977). No recent field trials have been conducted because *B.t.* is routinely used for control of lepidopterous pests on brassica crops. If a virus is developed for control of *T. ni*, it will probably be AcMNPV or AfNPV, as these viruses have a much broader host range than *T. ni* NPV and are just as effective on *T. ni*.

REFERENCES

Allen, G.W. and Knell, J.D. (1977) The nuclear polyhedrosis virus of *Anticarsia gemmatalis*: ultrastructure, replication and pathogenicity. *Florida Entomologist* **60**, 233–240.

Anon. (1983) *Pesticide Assessment Guidelines. Subdivision – M – Biorational Pesticides.* No. PB83–153965. Environmental Protection Agency. Springfield, VA: NTIS.

Anon. (1993) *Regulatory Proposal: Registration Guidelines for Microbial Pest Control Agents*. Agriculture and Agri-Food Canada. Pro93–4.

Arnott, H.J. and Smith, K.M. (1968) An ultrastructural study of the development of a granulosis virus in the cells of the moth *Plodia interpunctella* (Hübner). *Journal of Ultrastructural Research* **22**, 136–158.

Balch, R.E. and Bird, F.T. (1944) A disease of the European spruce sawfly, *Gilpinia hercyniae* (Htg.), and its place in natural control. *Scientific Agriculture* **25**, 65–80.

Bell, M.R. (1991) Effectiveness of microbial control of *Heliothis* spp. developing on early season wild geraniums: field and field cage tests. *Journal of Economic Entomology* **84**, 851–854.

Bell, M.R. (1995) Effects of an entomopathic nematode and a nuclear polyhedrosis virus on emergence of *Heliothis virescens* (Lepidoptera: Noctuidae). *Journal of Entomological Science* **30**, 243–250.

Betz, F.S. (1986) Registration of baculoviruses as pesticides. In *The Biology of Baculoviruses*, Vol. II, R.R. Granados and B.A. Federici (eds). Boca Raton, FL: CRC Press, pp. 203–222.

Bird, F.T. (1953) The use of a virus disease in the biological control of the European pine sawfly *Neodiprion sertifer* (Geoffr.). *Canadian Entomologist* **85**, 437–455.

Bishop, D.H.L. (1986) UK releases of genetically marked virus. *Nature* (*London*) **323**, 496.

Bohmfalk, G.T. (1982) Progress with the nuclear polyhedrosis virus of *Heliothis zea* by commercialization of Elcar®. In *Proceedings of the 3rd International Colloquium on Invertebrate Pathology and 15th Annual Meeting of the Society for Invertebrate Pathology*, Brighton, UK, pp. 113–115.

Brookes, M.H., Stark, R.W. and Campbell, R.W. (1978) *Technical Bulletin 1585: The Douglas-fir Tussock Moth: a Synthesis*. Washington, DC: USDA Forest Service Science and Education Agency.

Carner, G.R. and Turnipseed, S.G. (1977) Potential of a nuclear polyhedrosis virus for control of the velvetbean caterpillar in soybean. *Journal of Economic Entomology* **70**, 608–610.

Clausen, C.R. (1961) Biological control of western grape leaf skeletonizer (*Harrisina brillians* B. and McD.) in California. *Hilgardia* **31**, 613–638.

Cunningham, J.C. (1985a) Biorationals for control of spruce budworms. In *Proceedings of the CANUSA Spruce Budworms Research Symposium: Recent Advances in Spruce Budworms Research*. C.J. Saunders, R.W. Stark, E.J. Mullins and J. Murphy (eds). pp. 320–349.

Cunningham, J.C. (1985b) Status of viruses as biological agents for spruce budworms. In *General Technical Report NE-100: Proceedings of the Symposium on Microbial Control of Spruce Budworms and Gypsy Moths*. Broomall, PA: USDA Forest Service NE Forest Experimental Station, pp. 61–71.

Cunningham, J.C. (1988) Baculoviruses: their status compared to *Bacillus thuringiensis* as microbial insecticides. *Outlook on Agriculture* **17**, 10–17.

Cunningham, J.C. and Entwistle, P.F. (1981) Control of sawflies by baculovirus. In *Microbial Control of Pests and Plant Diseases 1970–1980*. H.D. Burges (ed.). London: Academic Press, pp. 379–407.

Cunningham, J.C. and McPhee, J.R. (1986) *Technical Note No. 4: Production of Sawfly Viruses in Plantations*. Sault Ste. Marie, Ontario: Canadian Forest Service, Forest Pest Management Institute.

Cunningham, J.C., Kaupp, W.J., McPhee, J.R., Sippell, W.L. and Barnes, C.A. (1975) Aerial application of a nuclear polyhedrosis virus to control European pine sawfly. *Canadian Forest Service, Bi-monthly Research Notes* **31**, 39–40.

Cunningham, J.C., de Groot, P. and Kaupp, W.J. (1986) A review of aerial spray trials with Lecontvirus for control of redheaded pine sawfly, *Neodiprion lecontei* (Hymenoptera: Diprionidae) in Ontario. *Proceedings of the Entomological Society of Ontario* **117**, 65–72.

Cunningham, J.C., Kaupp, W.J. and Howse, G.M. (1991) Development of nuclear polyhedrosis virus for control of gypsy moth (Lepidoptera: Lymantriidae) in Ontario. I. Aerial spray trials in 1988. *Canadian Entomologist* **123**, 601–609.

Cunningham, J.C., Kaupp, W.J., Fleming, R.A., Brown, K.W. and Burns, T. (1993) Development of a nuclear polyhedrosis virus for control of the gypsy moth (Lepidoptera: Lymantriidae) in Ontario. II. Reduction in dosage and emitted volume (1989 and 1990). *Canadian Entomologist* **125**, 489–498.

Entwistle, P.F. and Evans, H.F. (1985) Viral control. In *Comprehensive Insect Physiology, Biochemistry and Pharmacology*, Vol. 12. G.A. Kerkut and L.I. Gilbert (eds). Oxford: Pergamon Press, pp. 347–412.

Falcon, L.A. (1982) The baculoviruses of *Autographa*, *Trichoplusia*, *Spodoptera* and *Cydia*. In *Proceedings of the 3rd International Colloquium on Invertebrate Pathology and 15th Annual Meeting Society for Invertebrate Pathology*, Brighton, UK, pp. 125–128.

Federici, B.A. and Stern, V.M. (1990) Replication and occlusion of a granulosis virus in larval midgut epithelium of the western grapeleaf skeletonizer, *Harrisina brillians*. *Journal of Invertebrate Pathology* **56**, 401–414.

Fox, C.J.S. and Jaques, R.P (1966) Preliminary observations on biological insecticides against imported cabbageworm. *Canadian Journal of Plant Science* **46**, 497–499.

Greer, F., Ignoffo, C.M. and Anderson, R.F. (1971) The first viral insecticide: a case history. *Chemtechnology*, **June**, 324–327.

Hamblin, M., van Beek, N.A.M., Hughes, P.R. and Wood, H.A. (1990) Co-occlusion and persistence of a baculovirus mutant lacking the polyhedrin gene. *Applied Environmental Microbiology* **56**, 3052–3062.

Henry, J.E. and Jutila, J.W. (1966) The isolation of a polyhedrosis virus from a grasshopper. *Journal of Invertebrate Pathology* **8**, 417–418.

Henry, J.E., Nelson, B.P. and Jutila, J.W. (1969) Pathology and development of the grasshopper inclusion body virus in *Melanopus sanguinipes*, *Journal of Virology* **3**, 605–610.

Hostetter, D.L. and Puttler, B. (1991) A new broad host spectrum nuclear polyhedrosis virus from a celery looper, *Anagrapha falcifera* (Kirby) (Lepidoptera: Noctuidae). *Environmental Entomology* **20**, 1480–1488.

Hughes, K.M. and Addison, R.B. (1970) Two nuclear polyhedrosis viruses of the Douglas-fir tussock moth. *Journal of Invertebrate Pathology* **16**, 196–204.

Hunter, D.K. (1970) Pathogenicity of a granulosis virus of the Indian meal moth. *Journal of Invertebrate Pathology* **16**, 339–341.

Hunter, D.K, Collier, S.J. and Hoffmann, D.F. (1973) Effectiveness of a granulosis virus of the Indian meal moth as a protectant for stored inshell nuts: preliminary observations. *Journal of Invertebrate Pathology* **22**, 481.

Hunter, D.K., Collier, S.S. and Hoffmann, D.F. (1977) Granulosis virus of the Indian meal moth as a protectant for stored inshell almonds. *Journal of Invertebrate Pathology* **70**, 493–494.

Hunter, D.K., Collier, S.S. and Hoffmann, D.F. (1979) The effect of a granulosis virus on

Plodia interpunctella (Hübner) (Lepidoptera: Pyralidae) infestations occurring in stored raisins. *Journal of Stored Products Research* **15**, 65–69.
Ignoffo, C.M. (1973) Development of a viral insecticide: concept to commercialization. *Experimental Parasitology* **33**, 380–406.
Ignoffo, C.M. and Couch, T.L. (1981) The nucleopolyhedrosis virus of *Heliothis* spp. as a microbial insecticide. In *Microbial Control of Pests and Plant Diseases 1970–1980*. H.D. Burges (ed.). London: Academic Press, pp. 330–363.
Jaques, R.P. (1970) Natural occurrence of viruses of the cabbage looper in field plots. *Canadian Entomologist* **102**, 36–41.
Jaques, R.P. (1971) Control of the cabbage looper and imported cabbageworm by viruses and bacteria. *Journal of Economic Entomology* **65**, 757–760.
Jaques, R.P. (1975) Persistence, accumulation, and denaturation of nuclear polyhedrosis and granulosis viruses. In *Baculoviruses for Insect Pest Control: Safety Considerations*. M. Summers, R. Engler, L.A. Falcon and P.V. Vail (eds). Washington, DC: American Society for Microbiology, pp. 90–99.
Jaques, R.P. (1977) Field efficacy of viruses infectious to the cabbage looper and imported cabbageworm on late cabbage. *Journal of Economic Entomology* **70**, 111–118.
Jaques, R.P. (1988) Field tests on control of the imported cabbageworm (Lepidoptera: Pieridae) and the cabbage looper (Lepidoptera: Noctuidae) by mixtures of microbial and chemical insecticides. *Canadian Entomologist* **120**, 575–580.
Jaques, R.P. and Laing, D.R. (1978) Efficacy of mixtures of *Bacillus thuringiensis*, viruses, and chlorimeform against cabbage insects. *Canadian Entomologist* **110**, 443–448.
Jaques, R.P., MacLellan, C.R., Sanford, K.J., Proverbs, M.S. and Hagley, E.A.C. (1977) Preliminary orchard tests on control of codling moth larvae by a granulosis virus. *Canadian Entomologist* **109**, 1079–1081.
Jaques, R.P., Laing, J.E., MacLellan, C.R., Proverbs, M.D., Sanford, J.H. and Trottier, R. (1981) Apple orchard tests on the efficacy of the granulosis virus of the codling moth, *Laspeyresia pomonella* (Lep.: Olethreutidae). *Entomophaga* **26**, 111–118.
Jaques, R.P., Laing, J.W., Laing, D.R. and Yu, D.S.K. (1987) Effectiveness and persistence of the granulosis virus of the codling moth *Cydia pomonella* (L.) (Lepidoptera: Olethreutidae) on apple. *Canadian Entomologist* **119**, 1063–1067.
Jaques, R.P., Laing, D.R. and Maw, H.E.L. (1989) Efficacy of mixtures of microbial insecticides and permethrin against the cabbage looper (Lepidoptera: Noctuidae) and the imported cabbageworm (Lepidoptera: Pieridae). *Canadian Entomologist* **121**, 809–820.
Jaques, R.P., Hardman, J.M., Laing, J.E., Smith, R.F. and Bent, E. (1994) Orchard trials in Canada on control of *Cydia pomonella* (Lep.: Tortricidae) by granulosis virus. *Entomophaga* **39**, 281–292.
Lewis, F.B. (1981) Control of gypsy moth by a baculovirus In *Microbial Control of Pests and Plant Diseases 1970–1980*. H.D. Burges (ed.). London: Academic press, pp. 363–377.
Lewis, F.B., McManus, M.L. and Scheenberger, N.F. (1979a) *Research Paper NE-441: Guidelines for the use of Gypchek to Control Gypsy Moth*. Broonall, PA: USDA Forest Service, NE Forest Experimental Station.
Lewis, F.B., Reardon, R.C., Munson, A.S., Hubbard, H.B., Scheenberger, N.F. and White, W.B. (1979b) *Research Paper NE-447: Observations on the use of Gypchek*. Broomall, PA: USDA Forest Service, NE Forest Experiment Station.
Marrone, P.G. and MacIntosh, S.C. (1993) Resistance to *Bacillus thuringiensis* and resistance management. In Bacillus thuringiensis, *an Environmental Biopesticide: Theory and Practice*. P.F. Entwistle, J.S. Cory, M.J. Bailey and S. Higgs (eds). Chichester: Wiley, pp. 221–235.
McGugan, B.M. and Coppel, H.C. (1962) Technical Communication No. 2. Part II – Biological control of forest insects, 1919–1958. In *A Review of the Biological Control Attempts against Insects and Weeds in Canada*. Trinidad: Commonwealth Institute of Biological Control, pp. 90–109.
Moscardi, F. and Ferreira, C. (1985) Biological control of soybean caterpillars. In *Proceedings of the World Soybean Research Conference III*. R. Shibles (ed.). Boulder: Westview Press, pp. 703–711.

Moscardi, F., Allen, G.E. and Greene, G.L. (1981) Control of the velvetbean caterpillar by nuclear polyhedrosis virus and insecticides and impact of treatments on the natural incidence of the entomopathogenic fungus, *Nomuraea rileyi*. *Journal of Economic Entomology* **74**, 480–485.
Nordin, G.L., Brown, G.C. and Sackson, D.W. (1990) Vertical transmission of two baculoviruses infectious to the tobacco budworm, *Heliothis virescens* (F.) (Lepidoptera: Noctuidae), using an auto dissemination technique. *Journal of the Kansas Entomological Society* **63**, 393–398.
Oma, E.A. and Streett, D.A. (1993) Production of a grasshopper entomopoxvirus (Entomopoxvirinae) in *Melanopus sanguinipes* (F.) (Orthoptera: Acrididae). *Canadian Entomologist* **125**, 1131–1133.
Payne, C.C. (1986) Pathogenic viruses as pest control agents. In *Biological Plant and Health Protection*. J.M. Franz (ed.). Stuttgart: FischerVerlag, pp. 183–200.
Podgwaite, J.D. (1985) General Technical Report NE-100: Gypchek: past and future strategies for use. In *Proceedings of Symposium on Microbial Control of Spruce Budworms and Gypsy Moths*. Broomall, PA: USDA Forest Service, N.E. Forest Experiment Station, pp. 91–93.
Podgwaite, J.D., Rush, P., Hall, D., and Walton, G.S. (1984) Efficacy of the *Neodiprion sertifer* (Hymenoptera: Diprionidae) nucleopolyhedrosis virus (Baculovirus) product, Neochek-S. *Journal of Economic Entomology* **77**, 525–528.
Reiff, W. (1911) The wilt disease or flacherie of the gypsy moth. In *Contributions of the Entomology Laboratory*, Harvard, MA: Bussey Institute, Harvard University.
Rollinson, W.D., Lewis, F.B. and Waters, W.E. (1965) The successful use of a nuclear polyhedrosis virus against the gypsy moth. *Journal of Invertebrate Pathology* **7**, 515–517.
Rollinson, W.D., Hubbard, H.B. and Lewis, F.B. (1970) Mass rearing of the European pine sawfly for production of the nuclear polyhedrosis virus. *Journal of Economic Entomology* **63**, 343–344.
Sheih, T.R. (1989) Industrial production of viral pesticides. *Advances in Virus Research* **36**, 315–343.
Shepherd, R.F. and Otvos, I.S. (1986) *Information Report BC-X-270: Pest management of Douglas-fir Tussock Moth: Procedures for Insect Monitoring, Problem Evaluation and Control Action*. Victoria, BC: Canadian Forest Service, Pacific Forest Research Center.
Smith, O.J., Hughes, K.M., Dunn, P.H. and Hall, I.M. (1956) A granulosis virus disease of the western grapeleaf skeletonizer and its transmission. *Canadian Entomologist* **88**, 507–515.
Steinhaus, E.A. and Hughes, K.M. (1952) A granulosis virus of the western grapeleaf skeletonizer. *Journal of Economic Entomology* **45**, 744–745.
Stelzer, M., Neisess, J., Cunningham, J.C. and McPhee, J.R. (1977) Field evaluation of baculovirus stocks against Douglas-fir tussock moth in British Columbia. *Journal of Economic Entomology* **70**, 243–246.
Stern, V.M. and Federici, B.A. (1990) Biological control of the western grape leaf skeletonizer with a granulosis virus. *California Agriculture* **44**, 21–22.
Tanada, Y. (1964) A granulosis virus of the codling moth *Carpocapsa pomonella* (Linnaeus) (Olethreutidae: Lepidoptera). *Journal of Insect Pathology* **6**, 378–380.
Vail, P.V. and Collier, S.S. (1982) Comparative replication, mortality, and inclusion body production of the *Autographa californica* nuclear polyhedrosis virus in *Heliothis* sp. *Annals of the Entomological Society of America* **75**, 376–382.
Vail, P.V. and Jay, D.L. (1973) Pathology of a nuclear polyhedrosis virus of the alfalfa looper in alternate hosts. *Journal of Invertebrate Pathology* **21**, 198–204.
Vail, P.V., Jay, D.L. and Hunter, D.K. (1971) Cross infectivity of a nuclear polyhedrosis virus isolated from the alfalfa looper, *Autographa californica*. *Proceedings of the 4th International Colloquium on Insect Pathology*, College Park, MD, pp. 297–304.
Vail, P.V., Jay, D.L., Stewart, F.D., Martinez, A.J. and Dulmage, H.T. (1978) Comparative susceptibility of *Heliothis virescens* and *H. zea* to the nuclear polyhedrosis virus isolated from *Autographa californica*. *Journal of Economic Entomology* **71**, 293–296.
Vail, P.V., Tebbets, J.S., Cowan, D.C. and Jenner, K.E. (1991) Efficacy and persistence of

a granulosis virus against infestations of *Plodia interpunctella* (Hübner) (Lepidoptera: Pyralidae) on raisins. *Journal of Stored Products Research* **27**, 103–107.

Vail, P.V., Hoffmann, D.F., Streett, D.A., Manning, J.S. and Tebbets, J.S. (1993a) Infectivity of a nuclear polyhedrosis virus isolated from the celery looper, *Anagrapha falcifera* (Kirby), against production and postharvest pests and homologous cell lines. *Environmental Entomology* **22**, 1140–1145.

Vail, P.V., Hoffmann, D.F. and Tebbets, J.S. (1993b) Autodissemination of *Plodia interpunctella* (Hübner) (Lepidoptera: Pyralidae) granulosis virus by healthy adults. *Journal of Stored Products Research* **29**, 71–74.

van Frankenhuyzen, K. (1990) Development and current status of *Bacillus thuringiensis* for control of defoliating insects. *Forestry Chronicle* **66**, 498–507.

Woods, S.A., Streett, D.A. and Henry, J.E. (1992) Temporal patterns of mortality from an entomopoxvirus and strategies for control of the migratory grasshopper (*Melanopus sanguinipes* F.). *Journal of Invertebrate Pathology* **60**, 33–39.

Yearian, W.C. and Young, S.Y. (1982) Control of insect pests of agricultural importance by viral insecticides. In *Microbial and Viral Pesticides*. E. Kurstak (ed.). New York: Marcel Dekker, pp. 387–423.

Young, S.Y. and McNew, R.W. (1994) Persistence and efficacy of four nuclear polyhedrosis viruses for corn earworm (Lepidoptera: Noctuidae) on heading grain sorghum. *Journal of Entomological Science* **29**, 370–380.

21 Central America and the Caribbean

PHILIP F. ENTWISTLE

PERSPECTIVE

Pest control in Central America has been dominated by problems on its principal cash crop, cotton, and by the need to protect major food crops, notably such basic grains as beans, maize and sorghum, which essentially are grown for local use. These two areas of activity often interact. Pests on cotton have rapidly acquired resistance to insecticides because of the intense use of chemicals. As some of the species involved (e.g. *Heliothis* and *Helicoverpa* spp. *Spodoptera albula* (*sunia*) and *Spodoptera exigua*) are also pests on other crops such as some of the basic grains, vegetables and tomatoes, resistance has become a fairly general problem. In addition, there has been much public concern over the high incidence of human toxaemia resulting from pesticide misuse and accident. This unsupportable situation has been well documented (Swezey, Murray and Daxl, 1986) and has provided an agro-social atmosphere favourable to the development and introduction of alternative pest control strategies. The high cost of chemical insecticides and their demand on usually limited foreign exchange in hard currencies have been supporting factors. Indeed, chemical resistance together with pesticide costs drove the Nicaraguan and Guatemalan cotton industry into severe decline.

The area is notable for two distinct types of response to this situation. In Nicaragua, a strong laboratory for biological control (Laboratorio a Control Biologico) within a department of integrated control (Departmento de Control Integrado de Plagas) has been established in Leon within the University (Universidad National Autonoma de Nicaragua, UNAN). At present, the emphasis of work in the laboratory is on the BV control of basic grain pests and also to some extent of dry rice and of cotton. There are also other long-term studies such as the production and use of *Trichogramma* (Hymenoptera: Trichogrammatidae), lepidopterous egg parasites, particuarly for liberation against *Heliothis* (*sens. lat.*) in cotton, an approach that is very much more compatible with the use of viruses than with chemical insecticides. The Laboratorio a Control Biologico covers most aspects of BV development, such as strain search, characterization (e.g. by REN), bioassay, culture of pest Lepidoptera through the development of semi-synthetic diets based as far as possible on inexpensive local ingredients, techniques of virus production, formulation and field trials. By contrast, the programme in Guatemala, where the use of biological control is encouraged by law, has since about 1981 been conducted commercially, albeit so far on a small but increasing scale. One company, Agricola 'El Sol', is involved and operates principally by contracting to both supply and apply BVs, thus ensuring that the somewhat more exacting requirements for success with viruses, compared with chemicals, are observed. In recent years products from Agricola 'El Sol' have been successfully tested in Honduras, Nicaragua and El Salvador. Intrinsically, the future for both the Nicaraguan and the Guatemalan initiatives seems good, especially when they are considered as a part of a wider, vigorous, Latin American movement.

A single short season of work on brassica pest control in Trinidad (conducted by staff of the Institute of Virology and Environmental Microbiology, Oxford, UK) followed recognition that the major components of the local pest complex, *Hellula phidilealis*, *Plutella xylostella* and *Trichoplusia ni*, are all sensitive to a single BV, AcMNPV. Though this study demonstrated the potential of virus control on brassicas it has not been followed up. However, the attitude to biological control, including microbiological control, in Trinidad is now more favourable so that propects for further work have improved.

The range of pests under study in the region is shown in Table 21.1.

VIRUS PRODUCTION

In Guatemala, small-scale commercial production of the NPVs of *A. californica* (VPN-80) and *S. albula* (VPN-82) (both Agricola 'El Sol') is conducted under controlled conditions on a semi-synthetic diet including pinto beans specially grown locally to be insecticide free. In the Laboratorio a Control Biologico in Nicaragua a considerable amount of work has been invested in development of

Table 21.1 NPV control of insect pests in Central America and the Caribbean: investigations that have reached the field trial stage or beyond

	El Salvador	Guatemala	Honduras	Nicaragua	Trinidad and Tobago[a]
Bucculatrix thurberiella		+			
Chrysodeixis includens		+		+	
Diaphania hyalinata		+		+	
Estigmene acraea		+		+	
Hellula phidilealis					+
Plutella xylostella		+	+		+
Spodoptera albula	+	+		+	
Spodoptera exigua	+	+		+	
Spodoptera frugiperda				+	
Trichoplusia ni		+		+	+

[a] Work on brassica lepidoptera was conducted in 1984 only.

techniques appropriate to national circumstances and of a semi-synthetic diet employing inexpensive local ingredients wherever possible. At present, a single UNAN diet supports development of *Helicoverpa zea*, *Chrysodeixis (Pseudoplusia) includens*, *S. albula, S. exigua* and *Spodoptera frugiperda* and some other Lepidoptera of lesser economic importance. Insects reared on this diet are used for production of several NPVs, but especially those of the *Spodoptera* species.

With the aim of locating viruses with a host range and infectivity level adequate for satisfactorily control of lepidopterous pest complexes, rather than just particular species, on individual crops, considerable attention has been paid to the isolation, collection and selection through bioassay results. However, findings from Nicaragua and from cooperative studies in the Institute of Virology and Environmental Microbiology, Oxford, UK so far suggest that while AcMNPV (VPN-80, Agricola 'El Sol' in Guatemala) will be generally useful, separate viruses are still needed for use against *S. albula* and *S. frugiperda*. In Guatemala, an *S. albula* NPV isolate (VPN-82) is employed against *S. albula, S. exigua* and, recently, *Diaphania hyalinata*, a pest of melons.

FORMULATION

In Nicaragua, NPVs are prepared by maceration and filtration of infected larvae and application is with a small quantity of Triton-X as an adhesive. A closely similar simple process was followed in Guatemala where, however, the sticker 'Plyac' is sometimes added to the tank mix while molasses or sugar (1–2%) may be used as feeding stimulants. In both countries, the virus concentrate is taken to the field in cold boxes.

However, more recently, Agricola 'El Sol' has abandoned liquid formulations and has adopted a wettable powder (WP) formulation that does not need refrigeration and can be kept at room temperature. This is modied from Dr Flavio Moscardi's (Brasil) formulation and consists of a kaolin base with a small quantity of Tween 80. The addition of molasses or sugar is still recommended in Guatemala.

Despite the short-term nature of the work in Trinidad, considerable attention was paid to formulation. A combination of AcMNPV with charcoal (2%) or Tinopal (1%) provided the best performance. Some sugars decreased physical persistence on cabbage (Small, 1984).

SAFETY TESTS AND REGISTRATION

No safety tests have been conducted in Central America. However, to comply with Guatemalan registration regulations, bibliographic data on safety and ecological impact from outside the country have been acceptable. In addition to this, registration requires:

1. Physical, chemical and biological description
2. Certified product analysis, or OB count and bioassay certificate
3. Effectiveness test data for the pest(s) concerned

Table 21.2 NPV control of lepidopterous pests in Guatemala (Agricola 'El Sol' practice)

Crop	Virus	Dosage[a] (OB/ha)	No. applications	Approximate total dose/crop season (OB)
Cotton	VPN-82	4.2×10^{10}	8	3.3×10^{11}
Soybean	VPN-82	4.2×10^{10}	2[b]	8.4×10^{10}
Broccoli	VPN-80	4.2×10^{10} to 8.4×10^{10}	3[c]	1.26×10^{11} to 2.52×10^{11}
Cabbage	VPN-80	4.2×10^{10} to 8.4×10^{10}	3–5	1.26×10^{11} to 4.20×10^{11}
Melon	VPN-82	4.2×10^{10}	8	3.3×10^{11}
Water melon	VPN-82	4.2×10^{10}	10	4.2×10^{11}

[a] Application volumes are ground: 5–200 l/ha (Turbair Fox, ULVA + MICRON, motorized knapsacks, pneumatic knapsacks); air: 15.9–26.5 l/ha (Micronair).
[b] Two applications in a 10-day pre-flowering period.
[c] Three applications in a 15-day pre-flowering period.

Table 21.3 Costs of NPV control of lepidopterous pests of various crops in Guatemala, employing Agricola 'El Sol' preparations

Crop	Cost (Q/ha/treatment)[a]	Seasonal total (Q)	Seasonal total (US$)
Cotton	7.0	56.0	20.7[b]
Cabbage	21.0	75.0	27.8
Broccoli	21.0	63.0	23.3
Soybean	7.0	14.0	5.2
Water melon	7.0	70.0	25.9

[a] Q, quetzales; in 1994 US$1 = 5.85Q.
[b] Cost of standard insecticide (Jupiter, an insect growth regulator) is 2.7 times more.

4. Label information
5. Name of professional responsible for registration
6. Name of registered product
7. Payment of registration taxes.

USAGE

So far, commercial usage is restricted to Guatemala and solely through the activities of Agricola 'El Sol'. AcMNPV (VPN-80) is largely used on cabbage (*Brassica oleracea*), broccoli (*B. oleracea* var *botrytis*) and water melon (*Citrulus lanatus*) and only to a lesser extent on cotton and soybean. SaNPV (VPN-82), by contrast, is largely used on the last two crops and melons. Annual levels of application were: 1980–5, 840 ha; 1986, 7000 ha; 1987–9, 2450 ha; 1990, 13 600 ha; 1991, 10 800 ha; 1992, 16 800 ha; 1993, 19 350 ha and 1994, 20 000 ha (estimated). Only about one tenth of this was VPN-80.

A very notable feature of Guatemalan practice is the low dosage of OBs per hectare and the high frequencies of application (Table 21.2), which together result in a total OB usage per season much below that often expended elsewhere in the world. The approximate costs for various crops are shown in Table 21.3.

FIELD EXPERIMENTATION

Throughout most of Part two, pest species are reported on individually. However, in Central America, with the exception of *S. frugiperda*, control experimentation tends strongly to have been concerned with control of lepidopterous species complexes on particular crops and so it seems best to employ crops as subdivisions for discussion below.

Cotton

Field trials using VPN-80 and VPN-82 were conducted at three sites in Guatemala during 1981–2. Each experimental area was divided into 135 m^2 plots on which five treatments were replicated four times using five serial log dosages from 6.4×10^8 to 6.4×10^{12} OB/ha. Spray was applied by pneumatic knapsacks at 280 l/ha and contained 2% (v/v) molasses and the sticker 810 Bayer (35% Nonilfenol-poliglicoleter) at 1 ml/l. Two applications were made per week. Collecting larvae onto diet in individual sterile pots at the end of the second week after the second application, showed no statistically significant response differences between doses from 6.4×10^9 to 6.4×10^{12} OB/ha: VPN-80 gave 53–100% (mean 74.5%) for infection of *T. ni* and 41.0–63.0% (mean 52.0%) for infection of *S. exigua* while VPN-82 gave 41.0–55.6% (mean 47.0%) for infection of *S. exigua* and 43.0–66.0% (mean 54.3%) for infection of *S. albula*.

The fungus *Nomauraea rileyi* is always an important mortality factor in the Guatemalan wet season, kills rising to 100% under some conditions.

Along the south coast of Guatemala where spray deposits are subject to erosion by heavy rainfall, it is recommended that NPV applications be made five to eight times per season (to induce epizootics). Growers usually apply chemical insecticides 15–25 times and it is recommended that NPVs are applied with chemicals where these are directed against bollweevil and white fly. Very good results were observed using the kaolin-based formulation of VPN-82 on 2100 ha cotton in 1993 in El Salvador (Cooperativa Algodonera Salvadorena) with five treatments.

Soybean

In Nicaragua, NPV applications principally against *S. exigua* using ULV have been conducted since 1988. At rates varying from 2.7×10^{11} to 5.4×10^{11} OB/ha, infection varied from 73.8 to 95.0% (mean 79.2%). There was evidence for additional mortality from parasitoids and for spread of virus from treated areas (Mullock*et al.*, 1990). In Guatemala, experiments in 1983 with VPN-80 gave control of *T. ni*, *C. includens* and *Estigmene acrea* when applications were made in the pre-flowering stage. Recently *Anticarsia gemmatalis* has become a pest and is currently being controlled with *B.t.*

Maize

During 1988 and 1989 in Nicaragua, coarse spray applications to the whorl of *S. frugiperda* NPV at 7.5×10^{10} and 1.5×10^{12} OB/ha during *S. frugiperda* early larval instars gave 60–70% control and yields significantly greater than the controls.

Experiments in 1991 showed the NPV at 500 LE/ha to provide adequate control. The level of protection achieved is equal to, or better than, a standard insecticide, chlorpyrifos (Mullock *et al.*, 1990; personal reports, Director, Laboratorio a Control Biologico, UNAN).

Cruciferous crops

Studies in Guatemala have led to control recommendations for a lepidopterous pest complex on cabbage and broccoli. AcMNPV (VPN-80) is principally used. As in other Guatemalan work, fairly frequent applications at rather low dosages have proved optimal. Timing is also shown to be important and, for example, for broccoli it is essential to protect the flowers so that three applications are made in a 15-day pre-flowering period. The short-term study in Trinidad (Small, 1984) also used AcMNPV but here against *H. phidilealis* and *P. xylostella*. Because of the plant tissue-boring habit of the larvae, both these species are potentially difficult control targets. To minimize carriage of pests to the field on transplants, it is essential protection commences in the vulnerable and crowded nursery stage of growth. The best treatment was a high dose, equivalent to 1×10^{14} OB/ha (though, of course, nurseries tend to be quite small so that the real OB 'expenditure' is not excessive), every other day. However, it must be appreciated that the cost of this is shared between a very large number of potential transplants and also that at the time that the experiments were conducted the general pest pressure in the area was very much higher than will be the case once a proper IPM system is finally in place. Once this is instituted, it is possible the dose can be reduced. Preliminary results suggested a dose of around 2×10^{12} OB/ha would be suitable for the crop once planted out (though see below), but further work is required. UV protectants, e.g. Tinopal or carbon, were very beneficial.

In the highlands of Guatemala, the export enterprise ALCOSA, working with small farmers in an IPM programme, is using the WP formulation of VPN-80 against *P. xylostella* and *T. ni* on about 2000 ha. The NPV (at 1×10^{11} OB/ha per application) is alternated with *B.t.*, presumably to minimize the now well-known risk of *B.t.* resistance development in *P. xylostella*. Recent work in Honduras showed the Guatemalan AcMNPV (VPN-80) WP formulation controlled *P. xylostella* larvae as well as Dipel (a commercial *B.t.* product) and better than synthetic insecticides.

Melons

Working in Nicaragua in 1992–3, EAP, ZAMORANO found very good control of *D. hyalinata* with the WP formulation of VPN-82. Trials were also conducted in Guatemala where on the south coast it is used in IPM programmes at six to eight applications per season.

Water melons

Ten applications at 5 day intervals of *S. albula* NPV (VPN-82) at 7×10^9 OB/ha kept infestations of *Spodoptera* species below the economic threshold in Guatemala: the control plots were almost completely destroyed.

Tomatoes

In Honduras in 1993 'Cultivos Palmerola' evaluated VPN-82 together with releases of the egg parasitoid *Telenomus remus* against *S. albula* with excellent results and are planning to use the virus in future. In Guatemala, however, pest problems are dominated by white fly and tomato virus diseases and so BVs have not yet been used commercially, though their potential value has been demonstrated by Agricola 'El Sol'.

REFERENCES

Mullock, B.S., Swezey, S.S., Narvaez, C., Castillo, P. and Rizo, C.M. (1990) Development of baculoviruses as a contribution to biological control of lepidopterous pests of basic grains in Nicaragua. In *Proceedings 5th International Colloquium on Invertebrate Pathology and Microbial Control*, Adelaide, Australia, pp. 179–183.

Small, D.A. (1984) *The Use of a Baculovirus for the Control of Cabbage Pests. Report on a Field Study in Trinidad March 17 to April 18, 1984*. Oxford, UK: Institute of Virology, Natural Environment Research Council.

Swezey, S.L., Murray, D.L. and Daxl, R.G. (1986) Nicaragua's revolution in pesticide policy. *Environment* **28**, 6–36.

22 South America

M. REGINA V. DE OLIVEIRA

PERSPECTIVE

Virus control programmes have been pursued with energy and determination in South America and progress has been impressive. As is well known, the best regional example is the way in which, via technologies accessible to farmers and also to industry, NPV control of the principal pest of soybean, the velvetbean caterpillar, *Anticarsia gemmatalis*, has been achieved. While this particular programme is centred on Brazil it has also been 'exported' to other countries in the region.

Also in Brazil, considerable progress has been made in control of the corn leafworm (fall armyworm), *Spodoptera frugiperda*, on maize and of the cassava hornworm, *Erinnyis ello*, on cassava and rubber. Progress on these is also increasingly being made elsewhere in the region. In Peru, a programme for the GV control of potato tuber moth, *Phthorimaea opercullela*, has resulted in a system designed to be especially applicable to the protection of seed potato in store. Like Brazilian developments, this is being 'exported' to other countries in the region, notably Colombia and Paraguay, and also to other regions of the world. Although in Brazil a GV of the sugar cane borer, *Diatraea saccharalis*, has been shown to be a promising control agent, the project has been abandoned because of formulation problems.

A joint project among 'Cone Sul' countries with Chile is underway to control *Rachiplusia nu*, which is a serious defoliator of soybean, sunflower, flax and lucerne.

The distribution of work at field trial and implementation levels within the region is shown in Table 22.1, which emphasizes the concentration of effort in Brazil.

The carriage of these very low-cost technologies to the farmer through successful cooperation between research and extension workers and the provision, in some cases, to the farmers of free 'seed' virus are key elements of these successful programmes. The speed at which projects have travelled from the laboratory stage to practical field implementation also to an extent reflects lack of preoccupation with the development of oversophisticated production technologies. In addition, compared with some programmes elsewhere in the world, there has been rather

Table 22.1 Virus control of insect pests in South America: investigations that have reached field trial stage, or beyond

Pest species	Virus	Argentina	Brazil	Chile	Colombia	Paraguay	Peru	Uraguay
Acharia spp.	Pv[a]		+		+			
Anticarsia gemmatalis	NPV		+		+	+		
Chrysodeixis includens	NPV					+		
Cydia pomonella	GV	+		+				+
Diatraea saccharalis	GV		+					
Dione spp.	NPV		+					
Eacles spp.	PV		+					
Erynnis ello	GV		+		+			
Heliothis virescens	NPV		+		+			
Phthorimaea opercullela	GV				+	+	+	
Rachiplusia nu	NPV	+	+	+				+
Scrobipalpula absoluta	GV			+				
Spodoptera frugiperda	NPV		+		+			

[a] Parvovirus (densonucleosis virus).

little time spent on formulation studies or on questions of virus interactions with chemical pesticides. The results of such studies can be added at a future date to improve present achievements further.

VIRUS PRODUCTION

Some of the viruses of interest in pest control in South America are very well suited to farmer production by collection of larvae from crops sprayed operationally with virus or sprayed deliberately to provide a source of inoculum for later use on larger areas. *A. gemmatalis* NPV (AgNPV) and *E. ello* GV are good examples. However, because of the cryptic feeding habits of some species, the recovery of infected larvae from crop plants would be very difficult and expensive. Therefore a simple semi-synthetic diet system is much preferred and this is the situation for production of the NPV of *S. frugiperda*. *P. opercullela* is also a rather difficult insect, but as no adequate semi-synthetic diet system has been developed, production has relied on the use of the growing potato plant or its tubers.

FORMULATION

The process of formulation of insect viruses for control has been kept at a very simple level in all examples. For *A. gemmatalis*, the NPV is applied by farmers as a suspension of macerated, crude-filtered, infected larvae with kaolin in water. The final commercial product Multigen (for joint control of *A. gemmatalis* and *D. saccharalis*), a purified preparation, is simply suspended in distilled water. A similar simple field collection and crude water suspension approach is taken with *E. ello* GV production. The activity of both types of preparation is maintained by low temperature storage, usually at around 6–8°C or, if possible, frozen.

Where dust formulations are required, as with *S. frugiperda* NPV for inoculation of measured quantities directly into maize leaf whorls and for surface treatment of potato tubers with *P. opercullela* GV, crude aqueous, filtered suspensions of infected larva are mixed with kaolin or talc, air dried and pulverized. Surfactants may be employed to aid even dispersion of particles.

SAFETY TESTS

During the earlier stages of large-scale farmer implementation of control with viruses of *A. gemmatalis*, *E. ello*, *S. frugiperda* and *P. opercullela*, safety tests have not been prominent. However, in Brazil EMBRAPA (Brazilian Organization for Agricultural Research) has conducted dermal and oral acute toxicity testing of AgNPV and there have also been several investigations indicating the absence of adverse impact of these virus preparations on beneficial predatory and parasitic insects, honey bee and the silkworm, *Bombyx mori*. Where commercial development is concerned, for example with AgNPV, compliance with stringent safety test protocols is a prerequisite to registration.

FIELD EXPERIMENTATION AND IMPLEMENTATION

Acharia (*Sibine*) spp. (parvovirus and CPV)

The larvae of nettle caterpillars (Limacodidae) belonging to the genus *Acharia* are common defoliators of oil palm in South America. A parvovirus infection of larval *Acharia fusca* was detected in Colombia in the early 1970s (Genty, 1972; Genty and Mariau, 1973; Meynadier, Amargier and Genty, 1977). The virus is transmissible to the sympatric *Acharia apicalis*. Though epizootics may terminate *A. fusca* infestations, their slow development makes spraying necessary. It was recommended that 120 ml of viral concentrate (prepared from 20 g dead larvae – roughly equal to 12 final stage larvae – ground in 50 ml distilled water, passed through a 500 μm mesh filter and a further 220 ml water added) should be applied per hectare. A helicopter fitted with Micronair AU 3000 spray units was employed and 100% mortality developed in 1 month. The concentrate may be effectively stored at 4°C for a long period (Genty and Mariau (1973) and summarized in Entwistle (1987)).

Severe economic damage attributed to *Acharia* species has also been observed in northern Brazil and in 1983 2500 ha were treated with a total of 73 500 kg trichlorfon and carbaryl: in each season more insecticide was needed and the pest attack worsened year by year. However, a parvovirus was detected and when applied to palm fronds (stock inoculum of 1 g larva/10 ml water; treatments of 2.5, 5.0 and 10.0 ml/l) induced 88.9 to 92.0% mortality in 9 days. In a field experiment, 25 ml of this inoculum in 4 l water produced 92.4% control in 14 days (Lucchini *et al.*, 1984). A CPV was also noted but seems not to have been tested.

Agraulis vanillae

Agraulis vanillae is discussed with *Dione juno juno*.

Anticarsia gemmatalis (NPV)

Though this is commonly called the velvetbean caterpillar it is more seriously known as a pest of soybean, alfalfa, groundnut and cowpea (Herzog and Todd, 1980) and where soybean and velvetbean (*Stizolobium deeringianum*) are grown in alternate rows soybean is the preferred foodplant (Hinds, 1930; Ellisor, Gayden and Floyd, 1938). In Brazil, where it is more realistically known as 'lagarta da soja', it is the major defoliator of over 12 million ha soybean.

AgNPV was first detected in alfalfa fields in Peru (Steinhaus and Marsh, 1962; Moscardi, 1989) followed by isolations in Campinas, Sao Paulo, Brazil (F. Moscardi, unpublished; Gatti, Silva and Corso, 1977). Using material also collected in southern Brazil, the virions were shown to be multiply enveloped (Carner and Turnipseed, 1977); replication and pathogenesis have been further studied by Allen and Knell (1977). Comparisons of the genome of several AgNPV isolates have been made (Pavan, Zanotto and Ribeiro, 1986; Pinheiro *et al.*, 1990; Castro *et al.*, 1991; Araujo, Castro and Souza, 1994), and a physical map of the genome constructed (Johnson and Maruniac, 1989; Zanotto *et al.*, 1992). Cell line susceptibility to a range of AgNPV isolates has been investigated (Araujo, Santana and Pinheiro, 1992).

The first control trials were conducted in the southern USA and demonstrated the control potential of AgNPV (Carner and Turnipseed, 1977; Moscardi, 1977; Moscardi, Allen and Green, 1981). Beginning in 1979, the emphasis of studies shifted to the National Soybean Research Centre (CNPSo) of the EMBRAPA employing an isolate obtained in Londrina, State of Parana.

Evidence so far obtained suggests the effective host range of AgNPV to be rather narrow. Some Lepidoptera, e.g. *Chlosyne lacinia saundersii*, *Helicoverpa zea*, *Chrysodeixis* (*Pseudoplusia*) *includens*, *Spodoptera* spp. and *Trichoplusia ni*, were infectable only at very high dose; *B. mori* was the least susceptible species. However, *Heliothis virescens* was almost as susceptible as *A. gemmatalis* (Carner, Hudson and Bernett, 1979; Moscardi, 1989; Moscardi *et al.*, 1991). Rather surprisingly, studies on the susceptibility of some soybean and cotton pests in Colombia suggested *Semiothisa abydata*, *Alabama argillacea* and *Anomis formax* were very susceptible (Anon., 1987a).

The predators *Calosoma granulatum*, *Lebia concinna*, *Callida* spp., *Eryophis connexa*, *Nabis* spp. and *Geocoris* sp. as well as the pentatomid *Podisus connexivus* were not susceptible and the virus was not pathogenic to the African strain of the honeybee, *Apis mellifera*, when fed in honey plus sugar at 5×10^7 OB/bee or ULV-sprayed at 1×10^8 OB/ml (S. B. Alves, unpublished data).

Tested in pilot trials in several areas, the virus was shown to be effective in achieving greater than 80.0% control and in providing crop yields equal to those achieved using standard chemical insecticide treatments. The control parameters finally adopted were 50 LE/ha (*c.* 1.0×10^{11} to 1.5×10^{11} OB/ha) to be applied to larvae less than 1.5 cm in length at a maximum population density of 20/metre row. Normally only one application is needed per crop season to maintain *A. gemmatalis* populations at below the economic threshold. Larval mortality usually begins 6 days after spraying and peaks at day 7–8: larvae usually cease to feed after day 4. In laboratory experiments, infected larvae consumed an average of 27 cm^2 of leaf surface compared with 108 cm^2 eaten by healthy larvae (Moscardi, 1989); a higher degree of feeding reduction is thought to occur with younger larvae.

Though the half life on the leaf surface of crude virus with added clay is only about 6–7 days (4 days for purified material), the virus load of plants is increased when sprayed larvae die of infection. In addition, virus persisting in soil contributes to larval infection, probably by rain splashed and wind-blown contaminated soil particles. A 2-year study on a heavy clay soil showed survival of 40.0% original activity after 14 months and 28.0% after 2 years in a no-till system and 13.0% and 8.0%, respectively, in a system with normal tillage (Moscardi, 1989).

Nomuraea rileyi is a frequent cause of fungal epizootics and is important as a natural regulator of *A. gemmatalis* larval populations. In simultaneous infections by *N. rileyi* and AgNPV, there was a clear dominance of virus infections, which was interpreted as a possible antagonism. However, in both laboratory and field experiments, presenting the virus 24 h after *N. rileyi* inoculation tended to cause an anticipated level of fungal mortality but was accompanied by a less than expected level of virus infection. Hence the use of AgNPV over a large area may delay *N. rileyi* epizoosis if this process is not already well underway (Moscardi *et al.*, 1981).

Studies to develop a simple formulation conducive to standardization and to NPV stability commenced in 1984. An initial preparation employing the method of Dulmage, Martinez and Correo (1970), in which OBs and lactose are co-precipitated from water by addition of acetone, showed loss of activity. An economical, effective formulation was achieved by air-drying and milling a kaolin-based impure AgNPV slurry. Activity was maintained for more than 4 months at ambient conditions and for more than a year at 4–6°C and much longer when frozen (F. Moscardi, unpublished data). Further studies on the effectiveness of various formulations of AgNPV have been reported. For instance, infected macerated larvae lyophilized in the presence of skimmed sterilized powdered milk gave both greater effectiveness and greater stability than preparations with other forms of milk or just water (Batista-Filho, Cruz and Oliveira, 1986). A formulation in soybean oil with an emulsifier lost 50% of its activity over 12 months of storage, while the activity of a WP preparation (with talc) did not significantly deteriorate over the same period (Batista-Filho *et al.*, 1991). Further tests demonstrated the superior stability of the WP over a purified aqueous suspension of the virus (Batista-Filho *et al.*, 1994). Laboratory tests also indicated that soybean leaf consumption was less in the presence of a WP than with liquid formulations (Batista-Filho *et al.*, 1992).

In the laboratory, larvae reared on semi-synthetic diet can be transferred in groups of 25 larvae at the fifth instar into 300 ml cardboard cups containing diet sprayed at 4×10^7 OB/ml. Following incubation for 7–8 days at 27–28°C, the dead larvae are collected and stored at –18°C (Moscardi, 1989). However, as this is too expensive for most purposes, production has usually been by collecting virus-killed larvae in the field. In this way it has been possible to achieve 1 kg dead larvae/collector/day; in the 1987–8 season during 30 days 1500 kg were collected, enough to treat 75 000 ha. Since 1986–7 all virus produced at EMBRAPA-CNPSo has been kaolin-formulated prior to release to farmers. This WP contains about 2×10^{10} OB/g (Moscardi, 1989). However, extension workers encourage farmers to collect dead larvae for present use or to store frozen against future needs. A simple quality monitoring programme to ensure that only infective material is used is in place (Moscardi, Bono and Paro, 1988). Because of the very rapid growth and present scale of AgNPV usage, in 1989 a liquid formulation was registered under the trade name Multigen in Brazil by AGROGGEN Biotechnologia Agricola. However, this product did not show the expected efficacy at the farmer level, and its production and commercialization has been discontinued. However, four private companies (Nitral, Nova Era, Tecnivita and Geratec) and the Farmers Cooperative Organization for the State of Parana (OCEPAR) started production and commercialization of AgNPV WP formulations that have proved to be successful (Moscardi and Sosa-Gomez, 1993). Nevertheless, crude preparations are expected to continue to play an important part especially in the south among farmers already familiar with the technology.

The growth of usage in Brazil is shown in Table 22.2. From 1989 up to 1993 1 million ha has been treated each season. The prediction for 4 million by 1993 (Moscardi, 1990) was not achieved because studies with AgNPV demonstrated a possibility of resistance by *A. gemmatalis*. Genomic variations may occur in AgNPV after successive passages through larvae in the laboratory or after field applications. Although the effects of these variations on AgNPV virulence to its host still demand further investigation, initial results indicate that the virus has maintained its high virulence after approximately 10 years of massive applications by farmers. The important question of the potential of *A. gemmatalis* under pressure from prolonged exposure to NPV to develop resistance has been the subject of a laboratory study simultaneously contrasting two populations from Brazil and one from the USA (Moscardi and Sosa-Gomez, 1993). After having been selected at a level of around the LD_{80}, all populations developed significant resistance within three to four generations. However, while within 13–15 generations of exposure the Brazilian populations developed resistance rates of greater than 1000 times the original level, the resistance ratio for the USA population levelled at five times greater after four generations. Therefore, there appears to be a higher potential for the development of resistance in Brazil where the virus occurs naturally than in the USA (Abot *et al.*, 1996).

At the end of the 1980s, the cost of a season's protection by AgNPV was US $2.00 compared with about US $5.00 for chemical protection (Moscardi, 1990).

In Paraguay in 1985, the Cooperativa Colonias Unidas initiated an AgNPV programme, using Brazilian virus, achieving treatment of about 45 000 ha in 1993 with 856 700 g of frozen virus available for future applications (Cardoso and

Table 22.2 Estimated area of soybean treated with NPV of velvetbean caterpillar, *A. gemmatalis*, in Brazil

Season	Treated area (ha)
1982–3	200
1983–4	20 000
1984–5	200 000
1985–6	200 000
1986–7	350 000
1987–8	500 000
1988–9	700 000
1989–0	1 000 000
1990–1	1 000 000
1991–2	1 000 000
1992–3	1 000 000

From Moscardi (1990) and Moscardi and Sosa-Gomes (1993).

Candia, 1994). In 1986 an initiative was developed in Colombia to employ AgNPV to control lepidopterous pests of both soybean and cotton (Anon., 1987a).

Studies related to BVs of other defoliating caterpillars are more recent, and efforts have been focused on the more important Plusiinae (i.e. *C. includens* and *R. nu*). Laboratory studies with an NPV of *C. includens* indicated great potential for control. Plusiine species occur in association with the velvetbean caterpillar and can achieve population densities exceeding the economic threshold in southern Brazil. *R. nu* is very important in Argentina and Uraguay, attacking soybean, sunflower (*Helianthus annus*) and flax (*Linum usitatissimum*). Owing to the importance of these pests, especially *R. nu* in temperate regions, a joint project is at present underway involving research institutions such as EMBRAPA-CNPSo in Brazil, the National Institute of Agricultural Technology, Argentina and the Department of Agriculture, Uraguay aiming at developing NPVs of Plusiinae as microbial pesticides (Moscardi and Sosa-Gomez, 1993).

Cydia pomonella (GV)

Work on the virus control of codling moth on pears and walnuts in Chile (now apparently discontinued) demonstrated suppression equal to a standard diazinon treatment following knapsack spraying of 0.5–0.8 l/tree at 1×10^{11} OB/l (two applications). However, a third application seemed desirable for walnuts. Work has also been conducted on apples (Ripa, 1982). In Argentina, CpGV introduced from the USA and from France in 1987 is being used in apple and pear orchards, work which has been conducted by the Agriculture Institute of Microbiology and Zoology (IMIZA) and private industries. At present, CpGV is the first and only entomopathogen formulation at the 'Secretaria de Agricultura Ganaderia y Pesca de la Nacion' to achieve field experimentation. Laboratory studies indicated good potential to control codling moth (Alvarado and Lecuona, 1994). In Uraguay, both laboratory and field tests were similarly promising (Nunez, 1994).

Diatraea saccharalis (GV)

Sugar cane is one of the economically most important crops in Brazil, being a major source of alcohol to fuel internal combustion engines. *D. saccharalis*, sugar cane borer (SCB), is the most important pest, especially in the south where at an infestation level of 2.5% losses can reach US $100.00/ha. There appear to be no endemic pathogens in Brazil and the GV investigated was obtained in the southern USA (Pavan *et al.*, 1983; Macedo, Botelhio and Pavan, 1989). Research has been conducted at the Molecular Biology Centre, University of Campinas, Sao Paulo State in cooperation with the Sugar and Alcohol Institute (IAA/Planalsucar) (Moscardi, 1989). A selection process by serial passage in host larvae resulted in an LD_{50} reduction from 40 OB/larva to about 1 OB/larva for second instar larvae (Pavan *et al.*, 1983; Macedo *et al.*, 1989).

Increasing efficiency has been reported in successive field trials with the GV, control rising from an initial 15% to 50% (Dinardo *et al.*, 1988; Moscardi, 1989).

Eight noctuid species tested were not susceptible to *D. saccharalis* GV but *Diatraea grandiosella* was (Macedo *et al.*, 1989). Tests showed a considerable loss of sprayed inoculum activity after 16 days but at 37–64 days there was a resurgence of activity consequent on release of virus from larvae killed by the original inoculum.

A combination of *D. saccharalis* GV and the fungus *Beauveria bassiana* was found to be particularly effective in controlling sugar cane borer. *D. saccharalis* GV appeared to act more quickly in the presence of *B. bassiana*, than when applied alone (Lecuona and Alves, 1988). Another aspect of integrated control is the relationship of the GV with the forficulid *Doru luteipes* (see *S. frugiperda*) and with the ant *Solenopsis saevissima*, predators in the sugarcane ecosystem. *D. saccharalis* GV was inactivated in the gut of the ant (pH >10.0) but not in that of the forficulid (pH 5.0–6.0) (S. B. Alves, unpublished data). *D. luteipes* is thus likely to be the more efficient vector.

Dione juno juno and *Agraulis (Dione) vanillae* (NPV)

The nymphalid butterflies of the genera *Agraulis* and *Dione* are very closely related and are common defoliators in the rapidly expanding Brazilian passion fruit (*Passiflora edulis* f. *flavicarpa*) industry. A natural NPV epizootic was noted in Araguari, MG (south-east) and this virus is now said to occur everywhere in Brazil (S. B. Alves, unpublished data). The NPV was very infectious in laboratory tests (Kitajima, Ribeiro and Zanotto, 1986; Kitajima and Moscardi, 1988) and in a field experiment using about 1×10^{11} OB/ha (*c.* 80 LEs) it controlled *D. juno juno* at below the economic threshold. In the northern region *Baculovirus dione* has shown high mortality levels. Larvae stop feeding 4 days after treatment, with 96% mortaility after the 7th day (Ohashi, Figuerio and Farias, 1994c). Some chemical insecticides, carbaryl, endosulfan and trichlorfon, are said to be synergistic with this NPV (Ohashi, Batista and Rodriques, 1994b).

Eacles imperialis magnifica (CPV and parvovirus)

Previously known as a coffee defoliator in Brazil, *Eacles imperialis magnifica*, a large attacine moth, has become a prominent and potentially decimating defoliator of cashew (*Anacardium occidentale*) following expansion of this crop (Chagas *et al.*, 1989). From a natural epizootic observed in 1984, CPV and a putative parvovirus (densonucleosis virus) were identified (Kitajima, 1989). Two laboratory and one field test employed larval macerates and so presumably did not differentiate between the impact of the two viruses. In the laboratory, the inoculum gave high (72.0–80.0%) and rapid (4–5 days) mortality when administered to fourth and fifth instar larvae. The field experiment on fourth instar larvae (at 1 and 2 LE/l water) by the 5th day gave 70.0 and 100.0% mortalities, respectively (Chagas *et al.*, 1990). Much further work is required but the possibility of control with a single virus preparation on coffee and cashew is attractive.

Erinnyis ello (GV)

E. ello is a large migratory sphingid moth that has a wide New World distribution and attacks over 35 species of plant. In South America, it is known especially as a pest of cassava and rubber. A GV has been isolated from larvae on cassava in Colombia in 1980 (Bellotti, Arias and Reyes, 1988; Arias *et al.*, 1989; Bellotti and Arias, 1990) and in southern Brazil (Schmidt, 1985) and on rubber in northern Brazil (Ohashi *et al.*, 1991; Ohashi, Batista and Rodrigues, 1994a).

In Brazil, a programme was started in 1980–1 at the Agricultural Research Organization (EPAGRI), Santa Catarina but the virus is now also being produced by the Agricultural Research Institute at Parana (IAPAR) and is being used on over 2000 ha of cassava. Production is in collaboration with farmers collecting dead larvae. A standardized formulation consists of 20 ml (18 g) of a crude, filtered, macerate which for spraying is diluted in 200 l water/ha or for ULV in 3 l water/ha. Field application for virus production is on third and fourth instar larvae (greater than 5.0 cm in length) (R. Pegoraro and A. Schmidt, personal communication). Virus for control is applied at a top population density of five to seven small larvae per cassava plant.

In Colombia, 50–70 ml larval macerate in 200 l water with 0.2 ml Triton-ACT is sprayed per hectare. The speed of action in Colombia is phenomenal, as over 80.0% mortality is recorded in 48 h. In southern Brazil, a mortality of 90% is usually attained by 4 days after spraying 20 ml of larval macerate: the difference in response of *E. ello* in these two regions may reflect very rapid growth response to temperature and the likelihood that infection development is quite closely tied to this (the larval period is 19.6 days at 20°C and 9.6 days at 30°C (Bellotti and Arias, 1990)). Cassava treatment costs about US $2.00/ha. It seems possible that a powdered preparation of the virus can be less effective (Anon., 1987b).

Considerable virus foliar and soil persistence is claimed. An increase in numbers of natural enemies following GV application is thought to explain the observed spread of infection from sprayed areas.

Laboratory tests indicate control on rubber is likely to be similarly effective (Ohashi *et al.*, 1991). *E. ello* GV has been shown to be compatible with such fungicides as Cupravit, Dithane and Cercobin (Ohashi *et al.*, 1994b).

Heliothis spp. (NPV)

Compared with some other parts of the world, little attention has been paid to virus control of *Heliothis* spp. in South America, even though the technology is available, especially following intensive development in the USA. Aerial application in Colombia on cotton of Viron-H (a commercial NPV preparation from International Minerals Corp., USA) resulted in a mean of 49.0% infection in *H. virescens* larvae and was not regarded as successful (Cujar, 1973). However, in small-scale laboratory and greenhouse tests in Brazil, NPV performed as satisfactorily as *B.t. kurstaki* and the nematode *Steinernema feltiae* (Carrano-Moreira and All, 1995).

Phthorimaea opercullela (GV)

Potato tuber moth, *P. opercullela*, originated in South America but is now a major pest of potato and several other solanaceous crops in all the warmer regions of the world. It attacks both the green parts of the growing potato crop and the tubers in the field and in store. A GV has been isolated in South America (especially in Peru), Australia and India and it is known to be cross-infective to another pest gelachiid moth, *Scrobipalpula absoluta*, which in Chile is the most important tomato pest. Comparison of the genome of eight geographic isolates of *P. opercullela* GV indicated very close identity (Vickers, Cory and Entwistle, 1991) but the possibilities of biological variation (e.g. in infectivity) has yet to be explored.

Production of *P. opercullela* GV is in larvae on potato plants or tubers, as an adequate method of rearing potato tuber moth on semi-synthetic diet has yet to be developed. Two methods are available. Firstly, tubers may be dipped in an aqueous suspension of GV, air dried and infested with young larvae, which are then harvested at about the fourth instar. Secondly, in the Centro International de la Papa (CIP), Lima, Peru, adult moths were liberated into screen houses supplied with potted potato plants. Later GV was sprayed on the now infested plants and in due course infected larvae were collected (K. V. Raman, personal communication). This method has some similarities with that devised by Mattheissen *et al.* (1978) in Australia.

Trials on field crops of potatoes were successfully conducted in Australia (Reed and Springett, 1971). In Peru, control work has centred on protection of seed potatoes in store, which is seen as a real farmer need. A dust formulation is made by air drying a slurry of infected, filtered larvae with talc in shallow trays, followed by milling. Small-scale applications can be effected by shaking dust and tubers together in a plastic bag before storing the tubers in a monolayer in trays (at reduced light intensity to discourage shooting) (K. V. Raman, personal communication). Good results have been obtained and, starting with Paraguay, the technology is progressively being widely tested and utilized (K. V. Raman, personal communication). This virus is also under investigation in Bolivia (Calderon and Andrews, 1994).

Rachiplusia nu (NPV) and *Chrysodeixis (Pseudoplusia) includens*

In South America, the moths *R. nu* and *C. includens*, along with *A. gemmatalis*, are serious defoliators of such important crops as soybean, sunflower, lucerne and flax. *C. includens* is the only plusiine pest in Paraguay and Bolivia (Anon., 1989). It is an important pest of tobacco, cotton, graminae (pastures) and in horticulture (Combe and Perez, 1978). *R. nu* can be found in Chile attacking beans, lucerne, cucumber, potato, pea, rapeseed and pumpkin.

In Brazil, their role varies according to the region and crop. In soybean, *C. includens* is more important than *R. nu* in the states of Parana, Mato Grosso do Sul and Goias, while *R. nu* can be more important than *A. gemmatalis* during January in the state of Rio Grande do Sul (Anon., 1989). *R. nu* is more important on flax.

In Uraguay, *R. nu* is the more important; it is called the sunflower caterpillar. On soybean it is more important than *A. gemmatalis*.

In Argentina, *R. nu* is also more important than *C. includens*, the latter being especially found in the warmer, more northerly areas such as the provinces of Misiones and Corrientes (Anon., 1989).

A natural NPV isolate of *R. nu* was detected in Argentina in 1980 and a project involving Argentina, Brazil, Uraguay and Chile is being directed towards developing a viral insecticide (Alvarado and Lecuona, 1994). In Uraguay, the production of *R. nu* NPV is a priority among the biological control projects as the result of an accord between the 'Programa de Validacion de Alternativas Agricolas' (PROVA) and the group of 'Consumidores de Productos Biologicos'. In 1993, an area of about 100 ha was sprayed with good results (Nunez, 1994). *R. nu* NPV and AcMNPV, alone or in combination, will be used against these plusiine moths.

Scrobipalpula absoluta (GV)

The moth *S. absoluta*, a tomato pinworm, is the most important pest of tomatoes in Chile. The activity of a GV is being studied in young larvae, and plant surface persistence is being inspected. (R. Ripa, personal communication). The possibility of a common identity with *P. opercullela* GV appears not yet to have been examined.

Sibine

Sibine spp. are discussed as *Acharia* spp.

Spodoptera frugiperda (NPV)

Corn is a very important crop in South America; 21.3 million tons are produced annually in Brazil alone. *S. frugiperda* (lagarta-do-cartucho in Brazil, fall armyworm in North America) is the most important pest, the larvae feed concealed in the whorl and damage emergent leaves and flowering stems. It is also a pest of rice, wheat and pasture.

Approaches to virus control of *S. frugiperda* in Brazil led quite rapidly to the NPV treatment of 20 000 ha by 1992 (I. Cruz, personal communication). However, two BVs are known. A GV was first observed in Sete Lagoas, Minas Gerais (MG), south-eastern Brazil (Valicente *et al.*, 1986) and both laboratory and field trials demonstrated its promise as a control agent. An NPV was also isolated in Brazil (Moscardi and Kastelic, 1985) and preliminary investigations and subsequent developmental work has been carried out at the National Research Centre of Corn and Sorghum (EMBRAPA-CNPMS), Sete Lagoas, MG. A comparison of Brazilian isolates of NPV and GV indicated the superiority in infection tests of both NPV and GV from Patos de Minas, with an overall superiority of the NPV. The relative field performances of the NPV and GV have been compared in several field trials with the standard chemical insecticides chlorpirifos and deltamethrin (Cruz, 1992). The response time (LT_{50}) to the viruses varied with larval age, temperature and dose but in general was in the region of 4–6 days (Cruz and Valicente, 1991). Studies showed that NPV placed on the base of the maize leaf resulted in the highest infection rates while deposits on the free end of the leaf were least effective.

Field applications have been made by sprayers drawn by tractor or buffalo. In addition, a simple hand-held device designed to deliver a pre-determined dose of a dried powder formulation into the whorl has been developed (I. Cruz, personal communication). The dried powder consists of kaolin plus dried BV powder and the dose applied is 2.5×10^{11} OB/ha in 300 l water/ha (F. H. Valicente, personal communication). When sprayed at 8.75×10^{11} OB/ha, control is, in general, equal to that from chemicals in terms of yield (*c.* 4200 kg/ha). There is usually a need for two applications per crop, especially during dry seasons or when temperatures are high. Control with non-purified preparations is better than that with purified virus.

Attention has been paid to the interaction of NPV and GV with parasites and predators of *S. frugiperda*. For instance, NPV application at 3×10^3 OB/ml in 3, 5 or 7 mm irrigation water gave about 56% infection and there was an associated parasitism rate of 28% and hence a total mortality of about 84%. When NPV was not applied, parasitism was about 59% (Cruz, 1992). The forficulid *D. luteipes* is a predator of all developmental stages of *S. frugiperda* (and also of some other important pests such as *Heliothis* spp. and *D. saccharalis*) and quite low populations can result in 59–82% host mortality. Its use is found to be compatible with NPV application and it seems very likely that in addition to predatory activity *D. luteipes* will act as a BV dispersal agent. Some interest also centres on the manipulation of *Chelonus insularis* (Hym.: Braconidae), a parasite common in the region of Sete Lagoas, MG. Its relationships to virus infection in *S. frugiperda* have not yet been reported.

In southern Brazil, the fall armyworm is a serious pest in irrigated rice. Field experiments with 30, 60, 90 or 120 g of the *S. frugiperda* NPV formulation plus 2.5 g a.i./ha of a pyrethroid insecticide (Deltamethrin) gave 34, 52, 61 and 91% mortality, respectively (Gatti *et al.*, 1994).

Development of NPV control of *S. frugiperda* in laboratory and field tests is also in progress in Colombia (Jimenez and Bustillo, 1982), Argentina (Alvarado and Lecuona, 1994) and Paraguay. In Paraguay, work is in corn using a Brazilian NPV isolate, a project that commenced in 1990. However, in laboratory tests, an NPV isolate collected at Itapua and Caacupoe in Paraguay has shown greater aggressiveness than the Brazilian isolate (Cardoso and Candia, 1994).

Other insect pests

Ceratoma arcuata *and* Diabrotica speciosa These are chrysomelid beetles of especial importance as vectors of some legume viruses. In Brazil, a CPV-like virus has been found in *C. arcuata* and a non-occluded BV-like virus in *D. speciosa* (Kitajima and Moscardi, 1988) but there has been no further work.

Epinotia aporema *(GV)* There are initial reports of work on this bean shoot moth in Chile (R. Ripa, personal communication) and recently both an NPV and a GV have been detected in Argentina (Alvarado and Lecuona, 1994).

Mythimna (Pseudaletia) sequax *(NPV)* Wheat is a major crop in southern Brazil where *P. sequax* is a severe pest. Preliminary laboratory infectivity tests have been promising (Tonet and Fiuza, 1991).

Scapteriscus acletus *and* S. vicinus *(iridovirus)* *Scapteriscus* is a New World genus of mole cricket. Damage to plants usually occurs close to or below ground level. Iridovirus epizootics have been observed in Sao Paulo State, Brazil. The virus is cross-infective to a termite species (Kitajima and Moscardi, 1988). No further work has been reported.

Forest pests Epizootics of NPV diseases are a frequent mortality factor in *Eupseudosoma abberans*, *E. involuta* and *Sarcina violacens* in Brazil and their viruses are being considered for large-scale production both by government and private industry. A GV of the geometrid moth *Glena bisulca* infesting the amenity and slope-stabilizing tree *Cupressus lusitanica* in Colombia has long been considered a potential control agent (Smirnoff *et al.*, 1977).

REFERENCES

Abot, A.R., Moscardi, F., Fuxa, J.R., Sosa-Gomez, D.R. and Richter, A.R. (1996) Development of resistance by *Anticarsia gemmatalis* from Brazil and the United States to a nuclear polyhedrosis virus under laboratory selection pressure. *Biological Control* **7**, 126–130.

Allen, G.E. and Knell, J.D. (1977) A nuclear polyhedrosis virus of *Anticarsia gemmatalis*: I: ultrastructure, replication and pathogenicity. *Florida Entomologist* **60**, 233–240.

Alvarado, L. and Lecuona, R. (1994) Control biologico en la Republica Argentina. In *Instituto Interamericano de Cooperacion para la Agricultura. Programa Cooperativo para el Desarrollo Tecnologico Agropecuario del Cono Sur-PROCISUR*. L.C. Belarmino, R.M.D.G. Carneiro and J.P. Puignau (eds). Pelotas, RS: IICA-PROCISUR/EMBRAPA-CPACT, pp. 3–51.

Anon. (1987a) Control microbiologico, biologico y quimica de *Anticarsia gemmatalis*. In *Instituto Colombiano Agropecuario. Informe de Labores 1986B–1987A*. Palmira, Colombia: Seccion Entomologia CNI, pp. 83–86.

Anon. (1987b) Biological control halts cassava hornworm. *CIAT Report*, 1987, pp. 34–36.

Anon. (1989) *Reuniao International de Controle Biologico de Plusiinae (Lepidoptera, Noctuidae) 1*. Pelotas, RS. Ata. Pelotas: EMBRAPA-CPATB (not paginated).

Araujo, S., Santana, E.F. and Pinheiro, M.L.S. (1992) Susceptibility of insect cell lines to *Baculovirus anticarsia*. In *Programa e Resumo 6th Encontro Nacional de Virologia*. Novembro, Sao Lourenco, MG, Brazil, p. 84.

Araujo, S., Castro, M.E.B. and Souza, M.L. (1994) Analise comparativa de isolados de *Baculovirus anticarsia*. In *Anais IV Simposio de Controle Biologico*. Brazil: Maio, Gramado, RS, p. 114.

Arias, V.B., Bellotti, A.C., Garcia, F., Heredia, A., Reyes, J.A. and Rodrigues, N.S. (1989) Control de *Erinnyis ello* (L.) (gusano cachon de la yuca) mediante el uso de *Baculoviris erinnyis* en El Patia (Cauca). *Nota Cientifica. Bulletin de Societe Colombiana Entomologia* **62**.

Batista-Filho, A., Cruz, B.P.B., and Oliveira, D.A. (1986) Estudos preliminares relacionados ao emprego da liofilizacao como processo de preservacao de *Baculovirus anticarsia*. *Biologico* **51**, 263–269.

Batista-Filho, A., Alves, S.B., Augusto, N.T. and Cruz, B.P.B. (1991) Estabilidade de formulacaoes de *Baculovirus anticarsia* oleo emulsionavel e po molhavel em condicoes de laboratorio. *Arquivos do Instituto Biologico Sao Paulo* **58**, 17–20.

Batista-Filho, A., Alves, S.B., Augusto, N.T. and Cruz, B.P.B. (1992) Efeito de duas formulacoes de *Baculovirus anticarsia* sobre o consumo foliar de *Anticarsia gemmatalis* Hübner, 1818. *Ecossistema* **17**, 73–78.

Batista-Filho, A., Alves, S.B., Augusto, N.T. and Cruz, B.P.B. (1994) Estabilidade de *Baculovirus anticarsia* formulado como po molhavel. *Ecossistema* **19**, 30–39.

Bellotti, A.C. and Arias, V.B. (1990) Biological control of the cassava hornworm, *Erinnyis ello* (Lepidoptera: Sphingidae) with emphasis on the hornworm virus Programa de Yuca. Colombia: CIAT, Cali.

Bellotti, A.C., Arias, V.B. and Reyes, J.A. (1988) Biological control of the cassava hornworm, *Erinnyis ello* (Lepidoptera: Sphingidae) with emphasis on the hornworm virus. In *Proceedings of the 8th Symposium of the International Society of Tropical Root and Tuber Crops*. R.H. Howeler (ed.). Bangkok, Thailand, pp. 354–362.

Calderon, R. and Andrews, R. (1994) Granulosis virus multiplication (*Baculoviris phthoromaea*) for bio-insecticide formulation. In *Anais IV Simposio de Controle Biologico*, Maio, Gramado, RS Pelotas, RS: EMBRAPA-CPACT, p. 358 (EMBRAPAS-CPACT, Documentos 5 p. 104).

Cardoso, R. and Candia, S.M. (1994) Control biologico de plagas en el Paraguay. In *Instituto Interamericano de Cooperacion para la agricultura. Programa Cooperativo para el Desarrollo Tecnologico Agropecuario del Cono Sur-PROCISUR*. L.C. Belarmino, R.M.D.G Carneiro and J.P. Puigau (eds). Maio, Pelotas, RS: IICA-PROCISUR/ EMBRAPA-CPACT. pp. 55–62.

Carner, G.R. and Turnipseed, S.G. (1977) Potential of a nuclear polyhedrosis virus for control of velvetbean caterpillar in soybean. *Journal of Economic Entomology* **70**, 608–610.

Carner, G.R., Hudson, J.S. and Bernett, O.W. (1979) The infectivity of a nucleopolyhedrosis virus for control of the velvetbean caterpillar for eight other noctuid hosts. *Journal of Invertebrate Pathology* **33**, 211–216.

Carrano-Moreira, A.F. and All, J. (1995) Screening of biopesticides against the cotton bollworm on cotton. *Pesquisa Agropecuaria Brasileira* **30**, 307–312.

Castro, M.E.B., Oliveira, M.D., Irineu, B.P. and Pinheiro, M.L.S. (1991) Analise molecular de isolados geograficos de *Baculovirus anticarsia*. In *Anais XVIIth Reuniao de Genetica de Microorganismos Brasilia*, DF, Brazil, Resumos Marco, p. B2.

Chagas, M.C.M. das, Filgueira, M.A., Costa, O.G. and Medeiros, A.K.P. de (1989) Aspectos de *Eacles imperialis magnifica* Walker, 1856, em cajueiro *Anacardium occidentale* L. *EMPARN Bolletim de Pesquisa* **16**, 10.

Chagas, M.C.M. das, Filgueira, M.A., Costa, O.G. and Medeiros, A.K.P. de (1990) Utilizaças do virus no controle do *Eacles imperiales magnifica* Walker, 1856 (Lepidoptera, Attacidae). *EMPARN Boletim de Pesquisa* **18**, 17.

Combe, I. and Perez, G. (1978) Biologia del 'gusano medidor' *Pseudoplusia includens* (Walk.) (Lep., Noctuidae) en Colombia. *Revista Peruana de Entomologia, Lima* **2**, 61–62.

Cruz, I. (1992) Control biologico (*Spodoptera frugiperda*). In *Relatorio Tecnico Anual do Centro Nacionale de Pesquisa de Milho e Sorgo*. Periodo 1988–1991.

Cruz, I. and Valicente, F.H. (1991) Efeito de diferentes dosagens do virus VPN em lagartas de *Spodoptera frugiperda* de diferentes idades. In *Anais XIII Congresso Brasileiro de Entomologia*, Rio de Janeiro. Resumos, p. 228.

Cujar, A. (1973) Comportimiento del virus de la poleidrose nuclear (VPN) de *Heliothis* spp. en areas comerciales de algodon. *El Algodonero Bolletin* (Bogota) **60**, 16–21.

Dinardo, L.L., Teran, F.O., Pavan, O.H.O. and Pazelle, A.C. (1988) Avaliacao entomologica do virus da granulose para controle de *Diatraea saccharalis*. *Boletim Tecnico Coopersucar*, Piracicaba, SP, **41/48**, 18–22.

Dulmage, H.T., Martinez, A.J. and Correa, J.A. (1970) Recovery of the nuclear polyhedrosis virus of the cabbagelooper, *Trichoplusia ni*, by coprecipitation with lactose. *Journal of Invertebrate Pathology* **16**, 80–83.

Ellisor, L.O., Gayden, H.J. and Floyd, E.H. (1938) Experiments on the control of the velvetbean caterpillar, *Anticarsia gemmatalis* (Hbn.). *Journal of Economic Entomology* **31**, 739–742.

Entwistle, P.F. (1987) Virus diseases of Limacodidae. In *Slug and Nettle Caterpillars*. M.J.W. Cock, H.C.J. Godfray and J.D. Holloway (eds). Ascot, UK: CAB International, pp. 213–221 and references in pp. 245–256.

Gatti, M.M., Silva, D.M. and Corso, I.C. (1977) Polyhedrosis occurrence in caterpillars of *Anticarsia gemmatalis* Hübner, 1818 in south of Brazil. In *IRCS Medical Science: Cell Membrane Biology; Environmental Biology and Medicine; Experimental Animal Microbiology; Parasitology and Infectious Diseases* Vol. 5. Lancaster, UK: IRCS, p. 136.

Gatti, M.M., Botton, M., Carbonari, J.J., Machado, A.E., Belarmino, L.C. and Martins, J.F. da S. (1994) Avaliacao de *Baculovirus spodoptera* no controle de laragrta-da-folha na cultura do arroz irrigado. In *Anais IV Simposio de Controle Biologico*. Miao, Gramado, RS. Pelotas, RS: EMBRAPA-CPACT, p. 358 (EMBRAPA-CPACT Documentos 5, p. 118).

Genty, P. (1972) Morphologie et biologie de *Sibine fusca* Stoll, lepidoptere defoliateur du palmier a huile en Colombie. *Oleagineaux* **27**, 65–71.

Genty, P. and Mariau, D. (1973) Les ravageurs et maladies du palmier a huile et du cocotier. Les Limacodes de genre *Sibine*. *Oleagineaux* **28**, 225–227.

Herzog, D.C. and Todd, J.W. (1980) Sampling velvetbean caterpillar in soybean. In *Sampling Methods in Soybean Entomology*. M. Kogan and D.C. Herzog (eds). New York: Springer, pp. 107–140.

Hinds, W.E. (1930) Occurrence of *Anticarsia gemmatalis* as a soybean pest in Louisiana in 1929. *Journal of Economic Entomology* **23**, 711–714.

Jiminez, J. and Bustillo, A.E. (1982) Preliminary studies towards the microbial control of *Spodoptera frugiperda* in cotton and other crops. MSc thesis, ICA-UN, Bogota.

Johnson, D.W. and Maruniak, J.E. (1989) Physical map of *Anticarsia gemmatalis* nuclear polyhedrosis virus gene region: sequence analysis, gene product and structural comparisons. *Journal of General Virology* **70**, 1877–1883.

Kitajima, E.W. (1989) Classification, identification and characterization of insect viruses. *Memorias Instito Oswaldo Cruz*, Rio de Janeiro **84** (suppl. III), 9–15.

Kitajima, E.W. and Moscardi, F. (1988) The use of viral bioinsecticide in Brazil. In *Proceedings of the 6th Symposium*, Brazil–Japan Science Technology, pp. 131–133.

Kitajima, E.W., Ribeiro, B.M. and Zanotto, P. (1986) Virus de poliedrose nuclear em *Dione juno juno* (Lepidoptera-Nymphalidae). In *Anais XIII Congresso Brasileiro de Microbiologia*. Sao Paulo, 4–6 Janeiro. Resumos.

Lecuona, R. and Alves, S.B. (1988) Efficiency of *Beauveria bassiana* (Bals.) Vuill., *B. brongniartii* (Sacc.) Petch. and granulosis virus on *Diatraea saccharalis* (F., 1794) at different temperatures. *Journal of Applied Entomology, Berlin* **15**, 223–228.

Lucchini, F., Morin, J.P., Rocha Souza, R.L., Lima, E.J and Silva, J.C. (1984) Perspectivos de uso de entomovirus para o combate de *Sibine* sp. (Lepidoptera: Limacodidae) desfolliador de dende no Est Para. In *Res 90 Congresso Brasileiro Entomologica* (Londrina, PR), p. 153.

Macedo, N., Botelhio, P.S.M. and Pavan, O.H.O. (1989) Viral insecticides and an insect

growth regulator in *Diatraea saccharalis* (Fabr) control through aerial spraying. In *Proceedings of the 25th ISSCT Congress*, Sao Paulo, Brazil, pp. 661–667.
Matthiessen, J.N., Christian, R.L., Grace, T.D.C. and Filshie, B.K. (1978) Large-scale field propagation and the purification of the granulosis virus of the potato moth, *Phthorimaea operculleła* (Zeller) (Lepidoptera: Gelachiidae). *Bulletin of Entomological Research* **68**, 385–391.
Meynadier, G., Amargier, A. and Genty, P. (1977) Une virose de type densonucleose chez le lepidoptere *Sibine fusca* Stoll. *Oleagineaux* **32**, 357–361.
Moscardi, F. (1977) Control of *Anticarsia gemmatalis* Hübner on soybean with a baculovirus and selected insecticides and their effect on natural epizootics of the entomogenous fungus *Nomuraea rileyi* (Farlow) Samson. MSc thesis, University of Florida, Gainsville, USA.
Moscardi, F. (1989) Use of viruses for pest control in Brasil; the case of the nuclear polyhedrosis virus of the soybean caterpillar, *Anticarsia gemmatalis. Memorias Instito Oswaldo Cruz, Rio de Janeiro*, **84** (suppl. III, **4**), 51–56.
Moscardi, F. (1990) Development and use of soybean caterpillar baculovirus in Brazil. In *Proceedings and Abstracts 5th International Colloquium on Invertebrate Pathology and Microbial Control*, Adelaide, Australia, pp. 184–187.
Moscardi, F. and Kastelic, J.G. (1985) Ocurrencia de virus de poliedrose nucleare e virus granulose em populacoes de *Spodoptera frugiperda* atacando soja na regiao de Sertaneja, PR. In EMBRAPA-CNPSo Resultados de Pesquisa de Soja 83/84 (EMBRAPA-CNPSo, Documentos 15), p. 128.
Moscardi, F. and Sosa-Gomez, D.R. (1993) A case study in biological control: soybean defoliating caterpillars in Brazil. In *International Crop Science I. Crop Science in America*. D.R. Buxton, R. Shibles, R.A. Forsberg, B.L. Blad, K.H. Asay, G.M. Paulsen and R.F. Wilson (eds). Madison, WI: Crop Science Society of America, pp. 115–119.
Moscardi, F., Allen, G.E. and Greene, G.L. (1981) Control of the velvetbean caterpillar by nuclear polyhedrosis virus and insecticides and impact of treatments on natural incidence of the entomopathic fungus *Nomuraea rileyi. Journal of Economic Entomology* **74**, 480–485.
Moscardi, F., Bono, I.L.S. and Paro, F.E. (1988) Atividades biologicas de lotes de *Baculovirus anticarsia* formulado, segundo processo desenvolvido no CNPSo-EMBRAPA. *Documentos Centro National de Pesquisas de soja, EMBRAPA* **36**, 33–34.
Moscardi, F., Sosa-Gomez, D.R., Soldorio, I.L. and Paro, F.E. (1991) Avaliacao de formulacoes de *Baculovirus anticarsia*, destinadas ao controle da lagarta de soja, *Anticarsia gemmatalis*. In *Anais XIII Congresso Brasileiro de Entomologia*, Rio de Janeiro, Resumos, p. 233.
Nunez, S. (1994) Situacion del control biologoco en el Uraguay. In *Instito Interamericano de Cooperacion para la Agricultura. Programa Cooperativa para el Desarrollo Tecnologico Agropecuario del Cono Sur-PROCISUR*. L.C. Belarmino, R.M.D.G. Carneiro and J.P. Puignau (eds). Pelotas, RS: IICA-PROCISUR/ EMBRAPA-CPACT, pp. 87–108.
Ohashi, O.S., Coutinho, J.C.B., Quieroz, A.S.S. and Rodriguez, R.S. (1991) Controle biologico do mandarova da seringuera *Erinnyis ello* (Lep., Sphingidae) com *Baculovirus erinnyis*. In *Annais XIII Congresso Brasileiro de Entomologia*, Rio de Janeiro, Resumos, p. 230.
Ohashi, O.S., Batista, T.F.C. and Rodrigues, R.C. (1994a) Eficiencia do *Baculovirus erinnyis* no controle de *Erinnyis ello*, o mandarova da seringueira. In *Anais IV Simposio de Controle Biologico*, Maio, Gramado, RS. Pelotas, RS: EMBRAPA-CPACT, p. 358 (EMBRAPA- CPACT Documentos 5, p. 108).
Ohashi, O.S., Batista, T.F.C. and Rodrigues, R.C. (1994b) Compatibilidade do *Baculovirus erinnyis* e fungicidas pulverizados sobre *Hevea brasiliensis*. In *Anais IV Simposio de Controle Biologico*, Miao, Gramado, RS. Pelotas, RS: EMBRAPA-CPACT, p. 358 (EMBRAPA- CPACT. Documentos 5, p. 109).
Ohashi, O.S., Figuerio, C.L.M. and Farias, P.R.S. (1994c) Patogenicidade do *Baculovirus* a lagarta *Dione juno juno* (Cramer, 1779). In *Anais IV Simposio de Controle Biologico*, Miao, Gramado, RS, Pelotas, RS. EMBRAPA-CPACT, p. 358 (EMBRAPA-CPACT Documentos 5, p. 102).

Pavan, O.H.O., Boucias, D.G., Almeida, L.C., Gaspar, J.O., Botelho, P.S.M. and Degaspari, N. (1983) Granulosis virus of *Diatraea saccharalis* (DsGV): Pathogencity, replication and ultrastructure. In *Proceedings of the 18th International Congress ISSCT* Havana, Cuba, Vol. II (tomo 22) pp. 644–659.
Pavan, O.H.O., Zanotto, P.M.A. and Ribeiro, H.C.T (1986) Analise de resticoes do DNA de AgMNPV usado em controle biologico de *D. saccharalis*. In *Anais XIII de Societe Brasileiro Genetica: Reuniao de Genetica de Microrganismos*, p. 80.
Pinheiro, M.L.S., Castro, M.E.B., Sihler, W., Irineu, B.P. and Moscardi, F. (1990) Analysis of DNA and proteins of four geographical isolates of *Anticarsia gemmatalis* multiple nuclear polyhedrosis virus. In *Proceedings and Abstracts of the 5th International Colloquium on Invertebrate Pathology and Microbial Control*, Adelaide, Australia, p. 264.
Reed, E.M. and Springett, B.P. (1971) Large-scale field testing of a granulosis virus for the control of the potato moth (*Phthorimaea operculella* (Zell.) (Lep., Gelachiidae)). *Bulletin of Entomological Research* **61**, 223–233.
Ripa, R. (1982) Field evaluation of a granulosis virus for the control of codling moth. *Abstracts of the 3rd International Colloquium on Invertebrate Pathology*, Brighton, UK, p. 184.
Schmidt, A.T. (1985) Eficiencia das aplicacoes de *Baculovirus erinnyis* no controle do mandarova da mandioca. *EMPASC, Florianopolis Com. Tecnico* **88**, 1–7.
Smirnoff, W.A., Londono, L.L. and Ackermann, H.-W. (1977) A granulosis virus of *Glena bisulca* (Lepidoptera: Geometridae), a serious defoliator of *Cupressus lusitanica* in Colombia. *Journal of Invertebrate Pathology* **30**, 349–390.
Steinhaus, E.A. and Marsh, G.A. (1962) Report of diagnosis of diseased insects, 1951–1961. *Hilgardia* **33**, 349–390.
Tonet, G.L. and Fiuza, L.M. (1991) Acao de baculovirus (VPN) sobre a largarta do trigo *Pseudaletia sequax* Frankleimont, 1951 (Lep., Noctuidae). In *Anais XIII Congresso Brasileiro de Entomologia*, Rio de Janeiro Resumos, p. 229.
Valicente, F.H., Peixoto, M.J., Dinez, V.V. and Kitajima, E.W. (1986) Identificacoes e purificacoes der um virus der granulose em *Spodoptera frugiperda* (Lepidoptera, Noctuidae). In *Anais XVI Congresso Nacional de Milho e Sorgo*, Belo Horizonte, Resumos, p. 47–48.
Vickers, J.M., Cory, J.S. and Entwistle, P.F. (1991) DNA characterization of eight geographic isolates of granulosis virus from the potato tuber moth (*Phthorimaea operculella*) (Lepidoptera, Gelachiidae). *Journal of Invertebrate Pathology* **57**, 334–342.
Zanotto, P.M., Sampaio, M.J.A., Johnson, D.W., Rocha, T.L. and Maruniak, J.E. (1992) The *Anticarsia gemmatalis* nuclear polyhedrosis virus product and structural comparisons. *Journal of General Virology* **73**, 1049–1056.

PART THREE

PRACTICAL TECHNIQUES

23 General laboratory practice

INTRODUCTION

It is extremely important to work in clean conditions and to use aseptic technique whenever possible. This minimizes the possibility of either introducing unwanted infections into stocks of insects or cross-contaminating stocks of virus or insects. A summary of methods for sterilizing and decontaminating a range of materials is given in Table 23.1. All equipment should be cleaned thoroughly and stored in such a way that it does not become contaminated by chemical impurities or microorganisms.

Before disposal, all unwanted stocks of insects, irrespective of developmental stage, should be killed by freezing at –20°C for at least 24 h, followed by incineration or autoclaving at 15 psi for 20 min.

Disposable containers should be autoclaved before being discarded. Non-disposable plastic containers used for rearing insect stocks or those infected with virus should be immersed in disinfectant, e.g. 5% sodium hypochlorite solution or 1–2% Virkon (Antec International), for a minimum of 24 h before washing and recycling. Contaminated glassware and that used for rearing insects should be autoclaved before being cleaned and recycled.

Various detergents are available, including Decon 75 (Decon Laboratory Ltd) and Pyroneg (Diversey Ltd). Metal objects should be cleaned with a non-ionic (neutral pH) detergent, e.g. RBS GL or RIBS VIRO (Chemical Concentrates Ltd) to avoid problems with corrosion. Following detergent treatment where maximum chemical purity is required, objects should be rinsed 10 times in tap water followed by five times in glass-distilled or deionized water.

SAFETY

Safe conduct in the laboratory and the correct handling of equipment, materials and chemicals are of prime importance. The safe handling and disposal of all chemicals should be determined before any procedure is initiated. Information concerning handling procedures can be obtained by reference to the manufacturer, supplier or books such as *Chemical Safety Matters*, prepared jointly by the International Union of Pure and Applied Chemistry and the International Labour

Table 23.1 Guide to techniques employed for sterilizing and decontaminating different materials (see text for details of procedures and disinfectants)

	Dry heat		Wet heat		Chemical		Filtration
Material	Hot air	Incineration	Autoclave	Steam/ boiling	Liquid	Gas	
Artificial diet[a]	–	–	+ (D)	+ (P)	+	–	–
Glassware							
Screw-capped bottles	–	–	+ (P/D)	–	–	–	–
Cell culture vessels	+ (P)	–	+ (P/D)	–	–	–	–
Muslin filters	+ (P)	–	+ (P/D)	–	–	–	–
Pipettes	+ (P)	–	–	–	+ (D)	–	–
Silicone rubber-tipped cell scrapers (rubber policemen)	+ (P)	–	+ (P)	–	+ (D)	–	–
Haemolymph							
Infected with budded virus for inoculating cells	–	–	–	–	–	–	+
Insect cultures							
Eggs	–	–	–	–	+(d)	–	–
Pupae	–	–	–	–	+(d)	–	–
Insect-rearing equipment							
Filter paper	+(P)	–	+(D)	–	–	–	–
Metal mating cages	+(P)	–	+(P/D)	+(P/D)	–	–	–
Plastic mating cages and larval rearing containers	–	–	–	–	+(D)	+(D)	–
Paintbrushes	–	–	–	–	+(P/D)	+(P/D)	–
Scalpels, scissors	+(P)	–	+(P/D)	+(P/D)	+(P/D)	–	
Media							
Bacteriological media	–	–	+	+	–	–	–
Insect cell culture	–	–	+ (rarely)	–	–	–	+
Water	–	–	+	–	–	–	–
Plant material	–	–	+(D)	–	+(P)	–	–
Plasticware							
Micropipette tips	–	–	+(P/D)	–	+ (D)	–	–
Polystyrene e.g. centrifuge tubes	–	–	+(P/D)	–	–	+ (D)	–
Re-usable containers	–	–	–	–	+ (D)	+ (D)	–
Silicone rubber tubing	–	–	+(P/D)	–	+ (D)	–	–
Syringes[b]	–	+(D)	–	–	+ (D)	–	–
Working areas							
Surfaces (including accidental spillage)	–	–	–	–	+(P/D)	–	–
Entire working area	–	–	–	–	–	+ (D)	–
Rotors (centrifuge)	–	–	–	–	+	–	–

P, preparation for sterile use; D, decontamination of equipment and material after use: d surface decontamination.

[a] Gamma radiation can be used to sterilise agar-based diet following its preparation.

[b] Care should be taken to ensure that needles are covered adequately to prevent accidental damage to handlers.

Organization. Data bases such as those prepared and published by SilverPlatter Information Services provide information on chemical safety that can assist compliance with any associated risk assessments that are required. In the UK, the safe handling of chemicals is regulated by COSHH (Control of Substances Hazardous to Health). By law, all experimental procedures that pose a significant risk of exposure to substances hazardous to health must be recorded, together with the correct handling of every chemical substance.

PIPETTES AND PIPETTING DEVICES

Pipettes and pipetting devices are an essential part of any microbiological laboratory and are available in various forms and sizes. Glass (reusable) and plastic (disposable) pipettes are most commonly used for delivering volumes of from 1 to 25 ml. Such pipettes may be either blow-out (graduated to tip) or non-blow-out.

Pasteur pipettes are frequently used for transferring non-quantifiable volumes of up to about 2 ml but can be modified by attachment to a 1 ml syringe (usually plastic) through a small piece of silicone rubber tubing (Figure 23.1). A different modification results in a pipette that can be used to deliver drops of a known volume, most frequently 0.02 ml, i.e. 20 μl (Figure 23.2). A pipettes that delivers 0.02 ml per drop is known as a '50' dropper, i.e. 1 ml comprises 50 drops. The processing and sterilization of pipettes is illustrated in Figure 23.3.

For safety reasons, it is essential that mouth pipetting is forbidden. Various aids, both mechanical and electrical, are available for attachment to pipettes so that liquids can be dispensed safely and without the necessity for mouth pipetting. At their simplest these include rubber bulbs (up to 10 ml capacity) and teats (1–2 ml capacity) that are attached to pipettes and are manipulated manually. Slightly more sophisticated devices for delivering volumes of up to 20–25 ml include attachments such as the pipette fillers marketed by Merck. Some of these have the added advantage of being autoclavable. Electrically operated devices such as the Powerpette Plus (Jencons Scientific Ltd), Pipetboy ACU (Integra Biosciences) or rechargeable pipette controller (Bibby) are also available and are extremely useful for repetitive delivery of a range of volumes.

A range of sophisticated pipetting devices is available for delivering very small volumes (1–1000 μl). Some are electronically controlled while others are mechanically operated. The most well-established of these include the Gilson and Finn pipettes. These are used with sterile disposable tips, which can be bought as such or can be sterilized by autoclaving.

Volumes of sterile liquids from 1–10 ml can be delivered repetitively using a Cornwall syringe or similar device (Figure 23.4) while even larger volumes (up to 250 ml) can be delivered using the apparatus illustrated in Figure 23.5.

All pipettes and pipetting devices should be checked for accuracy on a regular basis. Those delivering small volumes are particularly prone to error, even minor discrepancies will have a marked effect on accuracy and hence on results obtained. The simplest way to check for accuracy is to weigh measured volumes of water (the density of water is 0.99823, 0.99707, 0.99567 and 0.99406 g at 20, 25, 30, and 35°C, respectively).

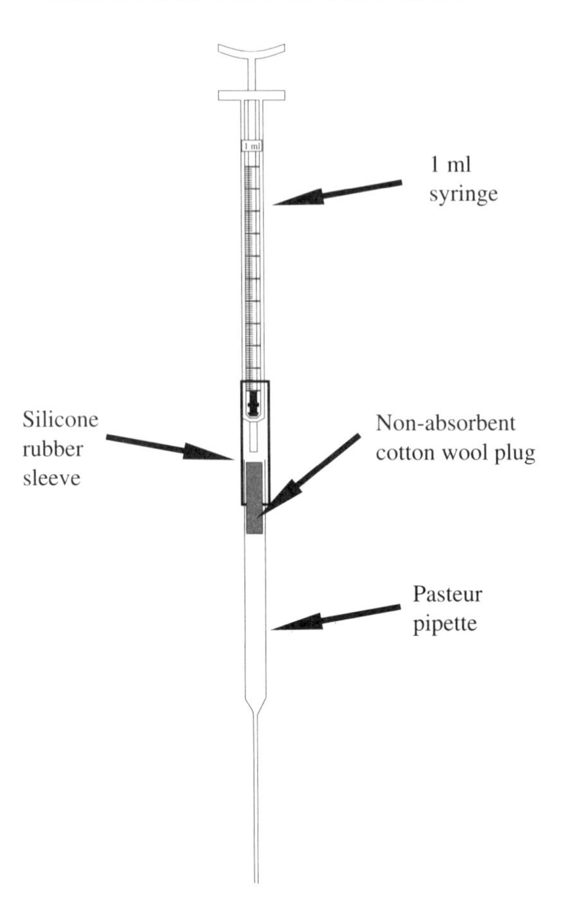

Figure 23.1 Modification of Pasteur pipette to deliver up to 1 ml in volume.

ASEPTIC TECHNIQUE

Aseptic technique involves the practice of precautionary measures designed to prevent the unintentional introduction of microorganisms (contaminants) into any procedure or preparation. Possible sources of contamination include the air, non-sterile or incompletely sterilized equipment and materials, and microorganisms carried by the investigator. It is important not to touch the mouths of bottles, tips of pipettes, edges of caps and any surface that will come into contact with sterile materials and equipment. Measures that should be taken include working near a burner, e.g. a Bunsen burner (see Figure 23.6) and flaming glassware by passing any area likely to have become contaminated through a flame between each operation. Extra care must be taken with plasticware that cannot be flamed.

Bacteriological loops should be heated to red heat along the full length of the wire to the base of the handle. Caps, stoppers and cotton wool plugs should be held in the hand, preferably between fourth finger and the palm during operations. If this is not possible, they should be placed on a sterile surface or in a sterile container.

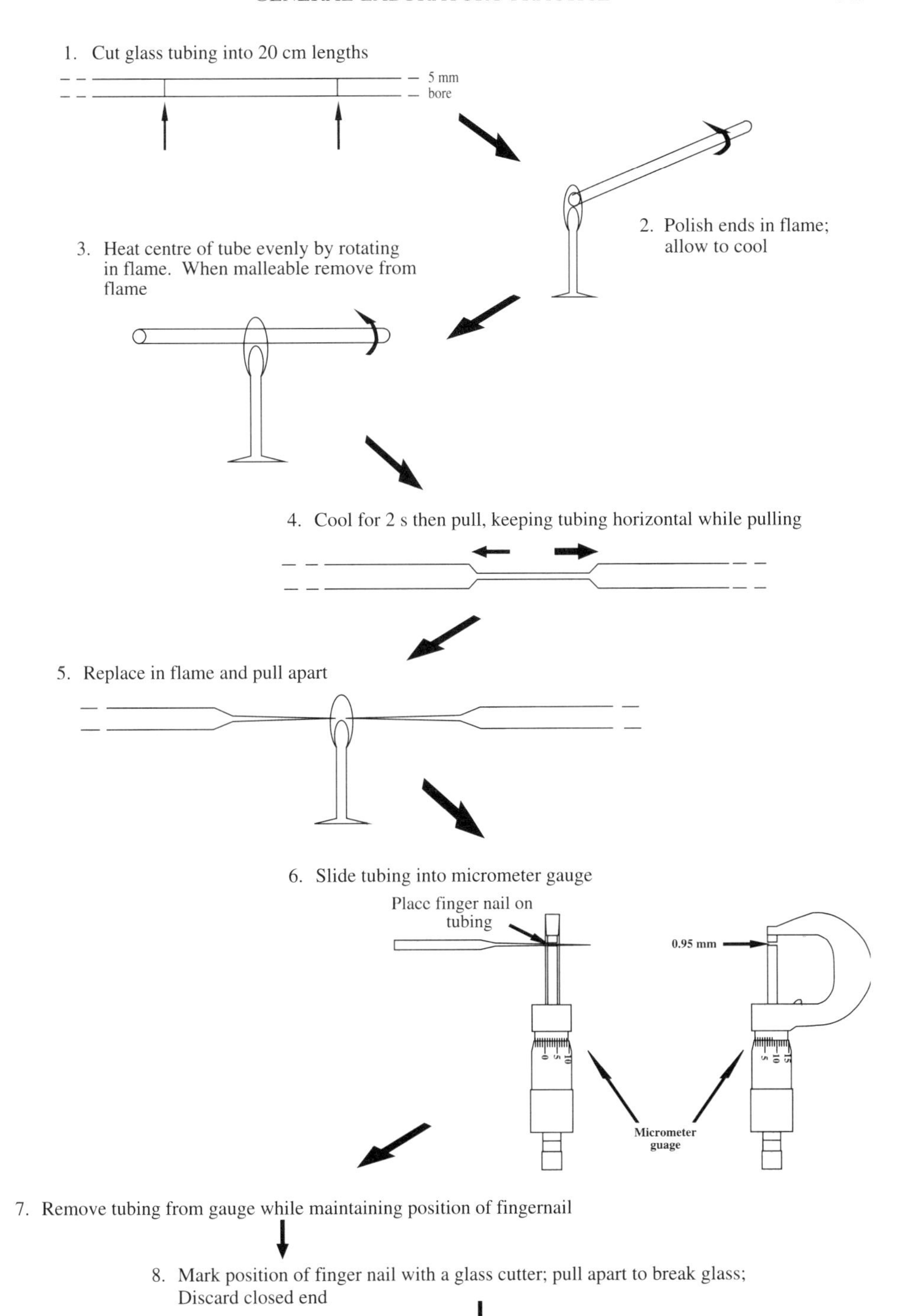

Figure 23.2 Modification of pipette to deliver drops of known volume.

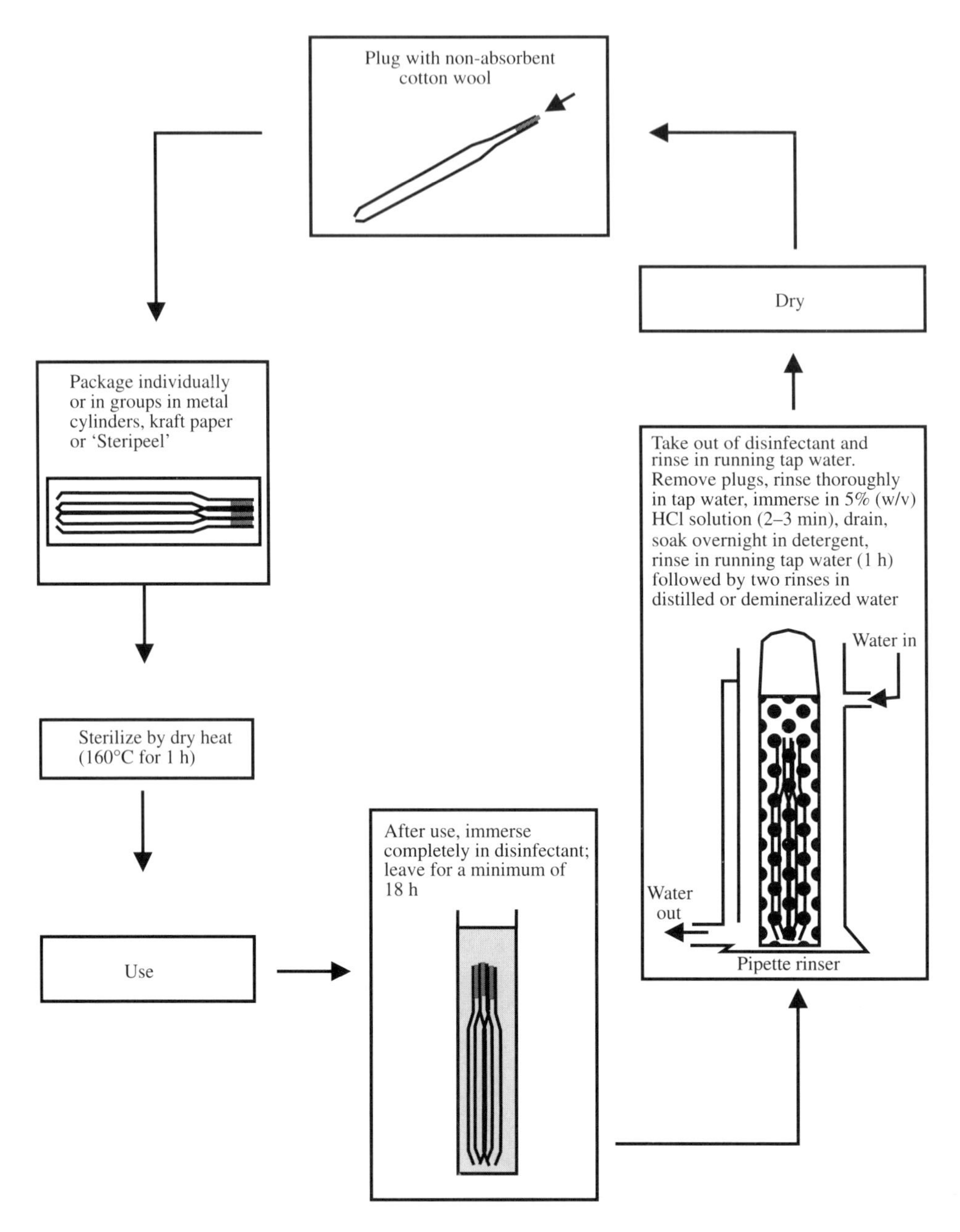

Figure 23.3 Processing and sterilizing glass pipettes.

STERILIZATION TECHNIQUES

There are three main methods of sterilization. These are (a) killing organisms by heat, (b) killing by chemicals and (c) removal by filtration. Heat is employed

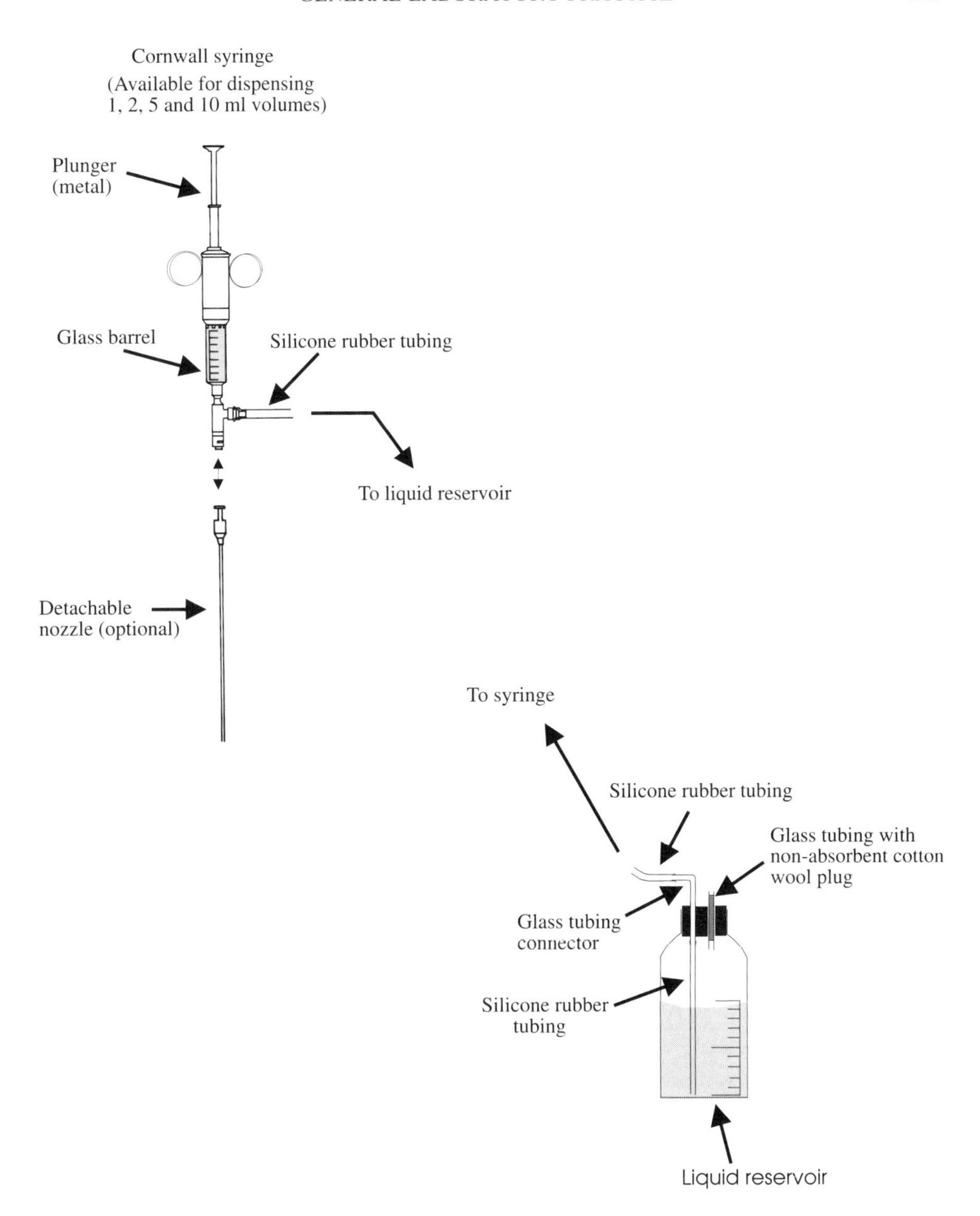

Figure 23.4 Cornwall syringe for repetitive delivery of sterile liquids.

widely as it is generally the easiest and most reliable method. Chemical disinfectants are used mostly for disinfecting the skin, floor, furniture and other surfaces that cannot be heated without damage. Filtration is used for sterilizing liquids that would be harmed by heat e.g. blood serum and antibiotics.

Articles that require sterilizing are packed in such a way that they are protected from subsequent contamination by non-sterile objects including dust carried by

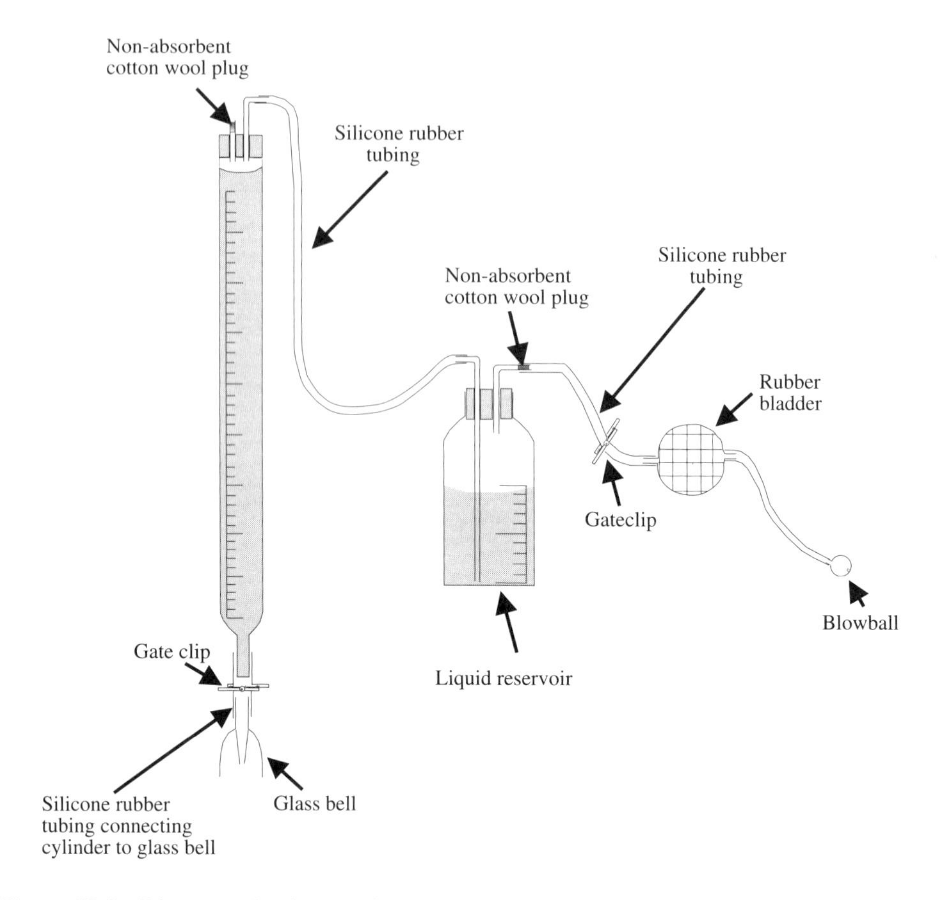

Figure 23.5 Dispenser for large volumes of sterile liquid.

air. This usually takes the form of dust-proof containers or wrapping (e.g. kraft paper or aluminium foil).

Plasticware such as Petri dishes, cell culture vessels and other equipment that is used once only is usually supplied pre-packed and sterilized by gamma irradiation.

It is important to check sterilization equipment regularly. At the very least, maximum thermometers should be placed at strategic positions, e.g. at the top, middle and bottom of each load. More sophisticated equipment is monitored electronically via thermocouples and provides a permanent read-out of each run.

Heat-sensitive tape (autoclave or oven), which changes colour if sterilization is satisfactory, can be fixed to individual items before heating.

Sterilization by heat

Heat is applied in two forms: (a) dry heat in a hot air oven (Table 23.2) or in a flame, and (b) moist heat, in saturated steam in a steamer at 100°C or autoclave (Table 23.2), or in boiling water. Killing of the most resistant spores requires moist heat at 121°C for 15–30 min or dry heat at 160°C for 60 min.

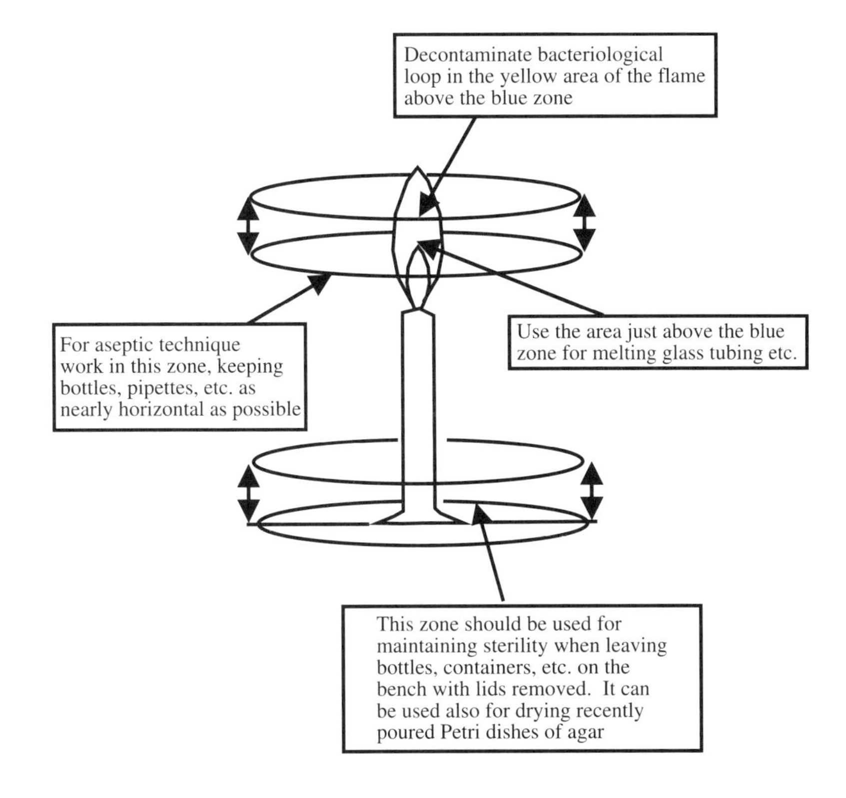

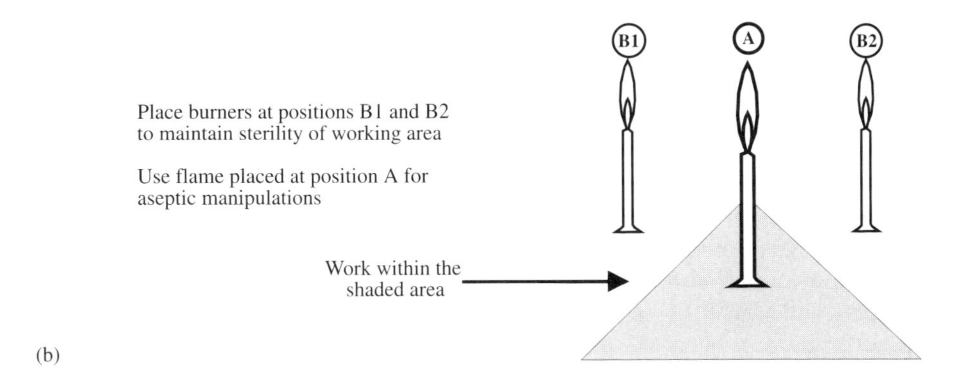

Figure 23.6 Use of a bunsen burner in aseptic techniques: (a) working with a burner; (b) working with cell cultures on a bench.

Dry heat

Hot air oven Dry heat is used mainly for glassware, metal instruments and paper or aluminium-wrapped instruments. It is also used for sterilizing anhydrous fats, oils and powders, which are impermeable to moisture. The time required for an object to reach sterilization temperature will depend on its position within the

Table 23.2 Minimum holding times to sterilize in moist and dry heat

Moist heat		Dry heat	
Temperature (°C)	Sterilizing time[a]	Temperature (°C)	Sterilizing time[a]
100	20 h	120	8 h
110	2.5 h	140	2.5 h
115	50 min	160	1 h
121	15 min	170	40 min
125	6.5 min	180	20 min
130	2.5 min		

[a] The times given do not include the time necessary to attain the temperatures indicated.

oven and its proximity to other items. To sterilize by dry heat at 160°C, for example, a minimum running time of 3 h is advised.

Flame A naked flame is used for sterilizing and decontaminating such articles as bacteriological loops and straight wires. They must be heated until red hot and allowed to cool before being used for transferring bacteria. A more gentle method of sterilization using a flame involves dipping instruments (e.g. glass rods, metal forceps etc.) into absolute ethanol or industrial methylated spirits and setting fire to them by momentarily touching a naked flame. The objects should not be left in the flame as this is liable to crack glass and will spoil the temper of metal. This process, which is called 'flaming', can also be used for glass objects such as spreaders.

Moist heat

Three methods are commonly used. These are (a) boiling in a waterbath at 100°C for 5–10 min, (b) steaming in free steam at atmospheric pressure and 100°C for 90 min or intermittently (3 successive days) for shorter periods (20–45 min), and (c) autoclaving at 121°C for 15–45 min in pure saturated steam at 15 psi above atmospheric pressure.

Steam sterilization is particularly suitable for culture media and aqueous solutions since the steam atmosphere prevents the loss of water by evaporation. It is used less than dry heat for sterilizing glassware and other equipment that is required to be dry when used. For satisfactory sterilization, materials must come into contact with the steam. If they do not, conditions for sterilization by dry heat will apply. To be effective, therefore, caps and tops should be loosened before containers are placed in the steamer or autoclave. Alternatively, for sterilization by autoclave, containers with liquids can be hermetically sealed (e.g. ampoules) or tightly screwed (e.g. bijou bottles). The presence of some air will not interfere with the conditions for moist heat sterilization provided by the presence of the aqueous contents. Furthermore, water loss owing to evaporation during the cooling period is prevented. Except for the smallest bottles, however, the tight application of screw caps increases breakage during autoclaving and slows down the cooling period.

Waterbath Boiling in a waterbath at 100°C for 5–10 min does not ensure sterility but is sufficient to kill all non-sporing microorganisms and most sporing organisms. It has been found satisfactory where absolute sterility is not essential; it can be used for sterilizing metal instruments (e.g. forceps, scalpels and scissors) and types of glass that cannot withstand higher temperatures.

Steaming Although it is convenient, steaming at 100°C is not as surely effective as autoclaving. It is, however, very useful for dissolving and melting agar for use in insect diets.

Autoclaving There are many different types of autoclave, ranging from the domestic pressure cooker, which can be used on the bench, to large, complex, electronically controlled models. Whatever the type, careful loading and routine checking of the equipment are essential. Empty containers and those containing liquids should be processed separately as should materials that are being sterilized and those that are to be decontaminated. All air must be eliminated from the chamber and from the articles being treated. A mixture of air and steam will not reach as high a temperature as unadulterated steam at the same pressure. Furthermore, air not only hinders penetration of steam but tends to separate from it, sinking to the bottom of the chamber and forming a lower cooler layer that will not satisfy conditions for sterilization. The quality of the steam in a simple bench pressure cooker can be tested by attaching a tube to the steam outlet and running it into a beaker of water. The steam condenses in the water while the air bubbles to the top. Once all the air has been discharged the bubbles will cease to form and the steam outlet can be closed.

Media should be autoclaved for the shortest time possible since many ingredients are sensitive to high temperatures. In addition, large and small volumes should not be autoclaved in the same load as they will heat up differently with the smaller volumes reaching the sterilizing temperature first. Bottles should not be filled to more than 75–80% of their volume to prevent loss caused by expansion during heating.

The pressure in the chamber at the end of the run should be reduced slowly (over a period of 15–20 min) to prevent excessive loss (>3–5%) of water by evaporation. Small losses in volume can be compensated for by adding 5% extra distilled or deionized water to media and solutions before sterilization.

When empty bottles and impervious materials are autoclaved, the pressure can be released quickly at the end of the run as this will aid drying. Since articles that are still damp will not become impervious to microbial contamination, they should be dried in an oven before coming into contact with sources of contamination. More advanced types of autoclave will have facilities for drying articles *in situ*, e.g. by removing steam under vacuum.

Chemical sterilizing agents

Dissecting instruments such as scissors, scalpels, etc. can be sterilized by storage in 70–75% ethanol. When required for use they are carefully removed from the alcohol and are placed in a sterile atmosphere such as that provided by a laminar airflow cabinet, where the ethanol is allowed to evaporate. Alternatively they can be washed in sterile water before use.

Chemical disinfectants

The most widely used laboratory disinfectants are clear phenolics and hypochlorites; others are used for special purposes. Clear phenolics such as Hycolin are suitable for killing bacteria (excluding endospores) but will not inactivate viruses. Their action is not affected by organic matter.

Hypochlorites, such as Chloros, are effective against both bacteria (but not *Mycobacterium tuberculosis* and closely related species) and viruses. They are inactivated by organic matter and corrode most metals, including some steels.

Vircon (Antec International) is a balanced, stabilized blend of peroxygen compounds, surfactant and organic compounds in an inorganic buffer system. It has activity against bacteria, mycoplasma and fungi as well as against many vertebrate viruses. It is effective against OBs provided that it is left in contact for a minimum of 1 h. It is non-corrosive, non-tainting and of low toxicity. It can be used as a surface sterilant and may be applied as an aerial spray. It is supplied as a powder that is diluted with water (0.5% solution for routine cleaning/disinfection or 1.0–2.0% for disinfection of contaminated objects). Heavily contaminated materials (e.g. as a result of spillage) can be treated with the powder in the first instance.

Disinfectants should be stored in concentrated form and should not be diluted to working strength until used. Rubber gloves and safety spectacles should be worn when handling concentrated solutions. Gloves should also be worn when swabbing down benches with working strength solutions.

Decontamination by disinfectants

Safety precautions

It must be remembered that these substances are toxic and some (e.g. the vapour generated when hypochlorite and formaldehyde are mixed) are recognized carcinogens They should be treated with great care.

Pipette and microscope slide discard jars

Hypochlorites are most suitable for discard jars. A minimum concentration of 1000 parts per million (ppm) available chlorine should be used. Commercial hypochlorites such as Chloros normally contain 100 000 ppm available chlorine and should be used as a 1% solution. The concentration should be increased to 2.5% (2,500 ppm) for pipettes soiled with small amounts of blood. Greater concentrations should be employed for heavily soiled equipment.

Pipettes and slides should be totally immersed in disinfectant for a minimum of 18 h. The solution should be changed regularly. Chloros, for example, can be purchased with a colour indicator. It is no longer effective when it changes from pink to colourless. Similarly, for hypochlorite solutions that do not contain an indicator, the addition of sufficient potassium permanganate solution to give a pink coloration, will give some indication of available chlorine. Vircon (1%) can be substituted for a hypochlorite solution.

Swabbing benches and safety (laminar airflow) cabinets

An iodophor such as 0.5% Tegodyne or a clear phenolic such 2.5% Hycolin are both suitable for swabbing benches and cabinets. Ethanol (70–75%) may be preferred for swabbing the interior of safety cabinets that are vented to the outside.

Spills and broken cultures

Spills and broken cultures should be treated with a clear phenolic (e.g. Hycolin (minimum concentration 2.5%)) for bacteria or an iodophor (e.g. Tegodyne (0.5%) or Vircon (1%)). The disinfectant must be left in contact with the spillage for a minimum of 20 min (Hycolin or Tegodyne). If using Vircon for disinfecting materials heavily contaminated with NPV or GV, a minimum period of 2 h must be allowed.

Centrifuges and centrifuge rotors

Care must be taken not to damage the centrifuge, rotors, etc. during disinfecting. Before starting any centrifugation work, it is essential to refer to the manufacturer or the manufacturer's manual for information concerning sterilization of rotors and decontamination of the machine in the event of accidental spillage or breakage involving infective material.

In the event of a breakage while the machine is in operation, it must be switched off and the rotor allowed to come to rest. The door must not be opened for at least 1 h after the rotor has stopped in order to allow dispersal of any aerosol that has formed. Once it is safe to open the door, the rotor, buckets, trunnions and any broken glass should be removed and decontaminated, if possible by autoclaving. Gloves and safety spectacles should be worn and the articles for decontamination should be placed directly into an autoclaving vessel for transportation to the autoclave. If it is not possible to autoclave the centrifuge parts, they should be immersed in a suitable non-corrosive disinfectant. Cidex (Johnson and Johnson Medical Ltd), a buffered glutaraldehyde solution, is recommended for this purpose and for swabbing the centrifuge bowl. It must, however, be used strictly in accordance with the manufacturer's instructions.

Rotors and buckets should be wiped each time they are used with Cidex or, if this is unavailable, a solution consisting of 100 ml 40% formaldehyde, 350 ml ethanol, and 50 ml glass-distilled water. This must be done in a fume cupboard. as formalin is toxic. Sufficient time should be allowed for the formalin to evaporate before removal from the fume cupboard. Gloves should be worn.

Disinfection of rooms

Spraying with formalin A room requiring disinfecting should be sealed completely by covering cracks, ventilators etc. with brown paper and adhesive tape.

Wearing an efficient anti-gas mask, spray all surfaces with a 10% formaldehyde solution (1 volume formalin to 3 volumes water) and finally saturate the atmosphere by spraying undiluted formalin at the rate of 1 l/1000 cubic feet. The room is closed by sealing the door and it is left for 24 h. A basin of ammonia solution is then introduced and left for several hours to neutralize the formaldehyde and paraformaldehyde. The ammonia should be used at a rate of 1 l S.G. 880 ammonia mixed with 1 l of water per litre of 40% formaldehyde applied. Remove excess ammonia by ventilating the room.

Disinfection of rooms by formaldehyde vapour The room is sealed as above and heated to 18°C if necessary. Formalin is boiled within the room in an electric boiler with a safety plug that kicks out when the vessel boils dry together with a time switch that cuts off the electricity supply just beforehand. A mixture of 500 ml 40% formaldehyde plus 1 l water is boiled per 1000 cubic feet of air space. The mixture is placed in the boiler, the electricity is switched on and the room is vacated immediately and sealed. After 4–24 h an operator wearing a respirator introduces a cloth soaked in ammonia solution (250 ml/l of formalin used) into the room. This should be left for 2 h before ventilating the room. The same procedure can be used to disinfect laminar air and safety cabinets. It is essential to seal the cabinet completely to prevent the escape of fumes into the laboratory. At the end of the procedure it is preferable, and may be a legal requirement, to vent the cabinet to the outside.

Sterilization by UV irradiation

Cabinets and rooms are sometimes fitted with lamps emitting UV light. This type of irradiation is only effective for surface sterilization and if the object to be sterilized is close to the source of radiation. The efficacy of the lamps decreases with time, necessitating regular changing. One method for determining effective working distance and monitoring the performance of a lamp is to test its effect on the viability of an actively growing bacterial culture. The culture is exposed to UV light at predetermined distances from the lamp and samples are removed at measured time intervals until the culture is no longer viable.

The lamp should be tested when first installed and an inactivation curve constructed by plotting $\log_{10}$ cfu/ml (y axis) against length of time of exposure (x axis). Further tests should be done at regular intervals and the inactivation curves compared with that for the lamp when new. The lamp should be replaced when a marked difference between inactivation curves (slopes) is observed.

Alternatively, a UV-recording meter may be employed (Chapter 27).

For safety reasons UV lamps should not be operated in the presence of personnel who are not adequately protected against the effects of UV irradiation.

TESTING THE EFFICACY OF AN UV LAMP

Equipment and materials

Items marked with an asterisk should be sterile.

1. Overnight broth culture of bacteria, e.g. *Escherichia coli*
2. Nutrient broth*
3. Pipettes for diluting bacteria*
4. Diluent: Ringer's solution*
5. Glass Petri dish*
6. Stop clock
7. Incubator (preferably shaking).

Method

1. Dilute the broth culture by adding nutrient broth and incubate at 37°C until an optical density (OD_{550}) value of 0.4–0.5 is reached.
2. Transfer the culture to a glass Petri dish, remove 0.1 ml and place on ice (Time 0 sample).
3. Place the Petri dish at a pre-determined distance from the source. Remove the lid and expose culture to the source of UV radiation for a measured period of time while rotating the plate to keep the contents mixed.
4. Replace the lid and remove plate from the UV light.
5. Remove 0.1 ml from the culture and place on ice.
6. Expose the culture to further periods of UV radiation and remove a sample at each time point.
7. Prepare serial tenfold dilutions of each sample in Ringer's solution and estimate the number of viable organisms using the method of Miles and Misra (p. 523).
8. Plot $\log_{10}$ colony-forming units (y axis) against time (x axis) to determine rate of inactivation (slope). The steeper the slope the greater the rate of inactivation.

REFERENCE

International Union of Pure and Applied Chemistry, International Programme on Chemical Safety. *Chemical Safety Matters*. Cambridge: Cambridge University Press, (ISBN 0–521–41375–3).

24 Working with the host

ESTABLISHING STOCK CULTURES OF INSECTS

Stock cultures of insects may be built up by requesting insects from an established laboratory culture or by collecting specimens from the field. In either case it is essential to quarantine all insects on arrival at the laboratory and to check for signs of disease. Unhealthy individuals should be removed immediately and the cause of disease identified (Chapter 25). Any pathogens identified that have potential as biological control agents should be documented and stored at –70°C for further investigation. All materials that have come into contact with diseased insects should be decontaminated or destroyed.

For insects that are not endemic, ensure that any licences required for holding or transporting such insects are obtained well in advance of the intended work.

INSECT STOCK CULTURES

Equipment and materials

1. Container(s) for transporting cultures
2. An adequate supply of either natural food or artificial diet for rearing the culture
3. Containers for use on receipt of the culture
4. Incubator facility with suitable light conditions.

Method for insects from an established culture

1. Identify a reliable source of insects and request permission for a culture. The stage of life cycle transported will depend on distance between laboratories, duration of journey, and conditions and means of transportation.

2. Ensure that explicit instructions for rearing the insects are received, together with or in advance of the culture. Prepare an adequate supply of food and request that a sample of diet, or its constituents, be sent for comparison. Construct or obtain appropriate containers for rearing all stages of the life cycle.
3. Advise the sender of any documentation required by, for example, importation or postal authorities.
4. On receipt, transfer insects to the appropriate containers and incubate suitably. Check for signs of disease. Remove any containers with unhealthy individuals, identify disease agents and either destroy or store at –70°C for further investigation as potential control agents. Quarantine until sure that there is no possibility of introducing pathogens into established cultures.

Method for insects collected from the field

1. Identify fields that have not been sprayed recently and search for evidence of infestation.
2. Collect insects
   (a) Adult stages. Use a light trap if necessary. Transfer to the laboratory and keep in a container in suitable conditions. Provide them with nutrient solution, e.g. 10% honey or sucrose solution, if thought advisable, and a site for laying eggs, such as plant material, sterile filter paper or nappy liners. Once laid, surface sterilize eggs if possible (p. 378), incubate suitably and allow to hatch. Transfer newly hatched larvae to a suitable diet (natural or artificial as applicable) and allow to develop to pupation. Rear according to the general instructions that follow but allowing for variation between species.
   (b) Larval stages. Collect larvae with sufficient food to sustain them until reaching the laboratory. Transfer to suitable rearing containers. Ensure an adequate supply of sterilized natural food (see below) or artificial diet to sustain them until they pupate. Rear according to the instructions given below and taking into account specific requirements for type of insect. Check all stages for signs of disease (Chapter 25).

REARING INSECTS

Stock cultures should be kept well away from insects used for propagating virus, preferably in a different building. The use of differently coloured laboratory coats in the two areas is helpful and aids discipline.

It is essential that rearing facilities (Figure 24.1) are kept scrupulously clean. If at all possible, equipment should be disinfected and washed thoroughly following

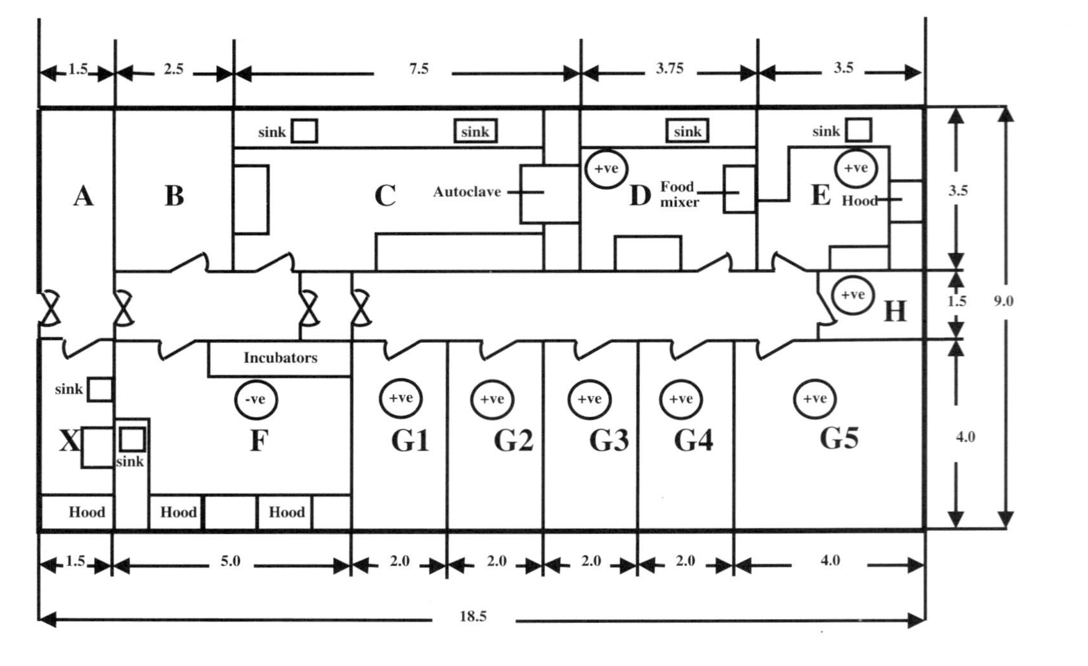

Figure 24.1 Suggested layout for an insectary (measurements in metres). Air pressure should be maintained as: +ve, positive to minimize the risk of air-borne infection of stock insects; –ve, negative to prevent escape of infectious agents into, and contamination of, the environment. A. Lobby for changing clothes when entering and leaving facility. B. Office. C. Central Services room for decontaminating used equipment and materials; processing and sterilizing glassware, equipment, etc.; and preparation of basic diet, which is then fed into the autoclave, processed and removed into room D. E. Handling of adult breeding cages (use of hood reduces exposure of operator to lepidopterous scales) and surface sterilization of eggs and pupae. F. Quarantine room for retention of incoming insect stocks while under examination for diseases. G. Incubation rooms for eggs, larvae and pupae; each room individually controlled for heat, light and humidity. H. Store for plastic rearing-ware. X. Cell culture facility (optional).

use, and before being recycled. As a minimum requirement, it should be washed thoroughly. When not in use, clean equipment should be stored in a dust-free environment. Any sick insects, contaminated materials and equipment must be sterilized (Chapter 23) before being either discarded or recycled. All working surfaces must be wiped with disinfectant before and after use.

A suggested layout for a rearing facility is given in Figure 24.1. It is intended that this can be adapted to suit the particular requirements of the user. If at all possible, provision should be made for a supply of air, preferably filtered, that is (a) under negative pressure in the quarantine area (to prevent dissemination of pathogens) and (b) under positive pressure in other areas (to prevent the introduction of pathogens and contaminants).

Newly hatched larvae are best handled by paintbrush if required individually or, if required in batches, by gently shaking over the food on which they are to be reared. The type of diet and procedure employed for rearing each species is dependent on its habits and preferences. In most instances, the use of an artificial diet is recommended. If not already available, it is worthwhile attempting to develop one. The primary objectives in rearing insects on semi-synthetic diets and the main components of basic diets are illustrated diagrammatically in Figures 24.2 and 24.3, respectively. Suggested components of an ideal simple diet are tabulated in Table 24.1. Contamination of food used for rearing stocks of insects may be controlled by the addition of antibiotics or formaldehyde (Table 24.2). Formaldehyde, however, must not be added to food used in the production of virus.

A general recipe for rearing insects of field crops, which can be adapted to suit particular needs, is given at the end of this section. Insects of stored products can be reared on a diet such as that described for the Indian meal moth *Plodia interpunctella* (see below).

Table 24.1 Insect diets: a survey of lepidopterous diets

Ingredients	Component in 1 l water	
	Quantity (as % of water used)	Frequency used (%)
Agar	2.0–2.5	90
(or Gelcarin HWG)	0.75–1.75	10
Beans	15.0–20.0	30
Beans + wheatgerm		80
Beans	10.0	
Wheatgerm	7.0–10.0	
Yeast (various forms)	2.0–4.0	80
Vitamin mixtures (Bs + C)	0.45	70
Wesson salt mixture	0.8	60
Methyl paraben	0.25	90
Sorbic acid	0.3	70
Antibiotics		45
Aureomycin	0.01	
or streptomycin	0.027	
or tetracycline	0.005	
Optional		
Casein	2.5	70
Sucrose	2.5	60

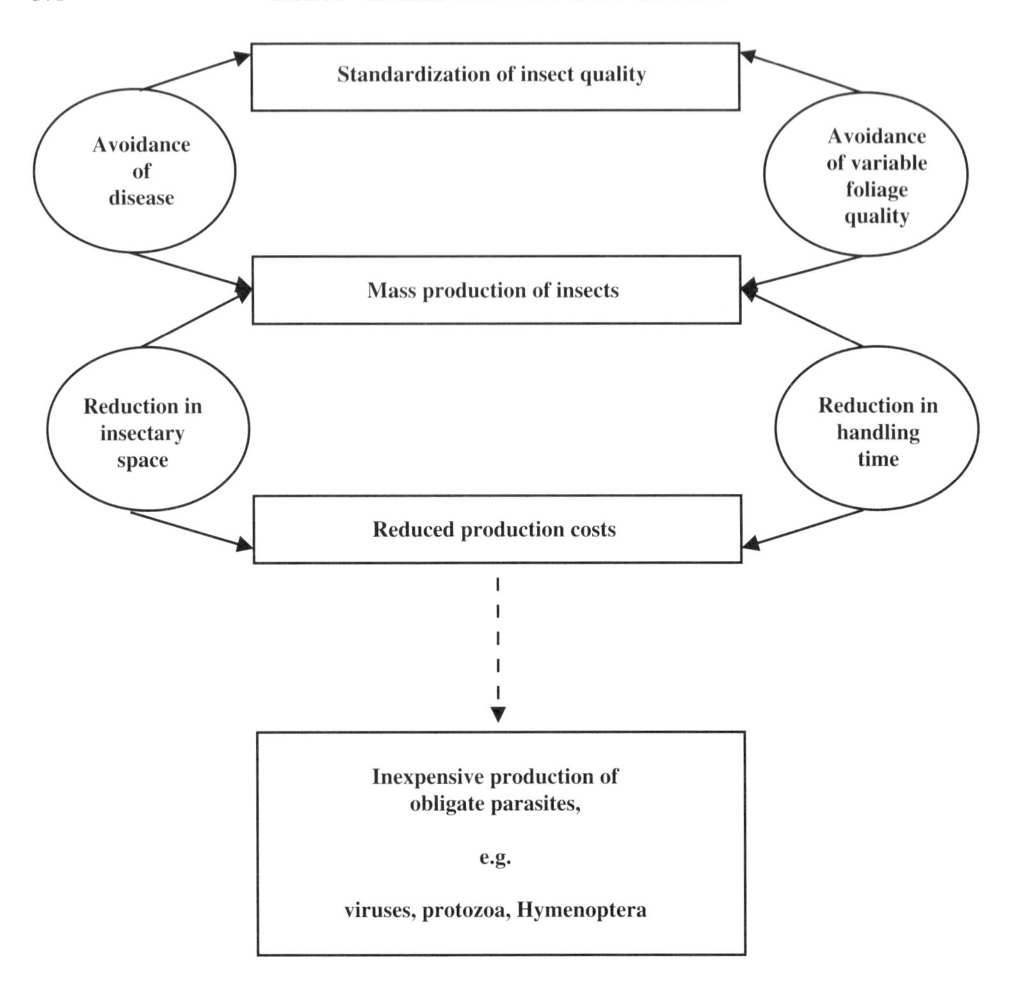

Figure 24.2 The primary objectives in rearing insects on semi-synthetic diet.

STERILIZATION OF EGGS, PUPAE AND FOLIAGE

Materials

Items marked with an asterisk should be sterile.

1. Nappy (diaper) liners (Superdrug Plc., Beddington Lane, Croydon, Surrey) or similar*
2. Triton X (Sigma Chemical Co., PO Box 14508, St Louis, MO 63178, USA)
3. Distilled or deionized water*
4. Buchner flask and funnel

Table 24.2 Insect diets: additives to control microbial load

Additive	Approximate concentration (% (w/v))[a]	Microorganisms suppressed					
		Bacteria	Rickettsiae	Yeasts	Fungi	Protozoa	Viruses
Aureomycin	0.1	+[b]	–	–	–	+[b]	–
Benlate(benomyl or carbendazim)	0.045	–	–	–	+	–	–
Benzoic acid	0.3	–	–	–	+	–	–
Formaldehyde	0.2	+	–	+	+	?	+
Fumidil 'B'	0.004	–	–	–	–	+[b]	–
Methyl paraben (methyl P-hydroxy-benzoate)	0.3	+?	–	+?	+?	–	–
Sorbic acid (or potassium sorbate)	0.25 (0.13)	–	–	+	+	–	–
Streptomycin	0.027	+[b]	–	–	–	–	–
Tetracycline	0.005	+[b]	+[b]	–	–	+[b]	–

[a] Weight/volume of H_2O.
[b] Probably interferes during replication only.

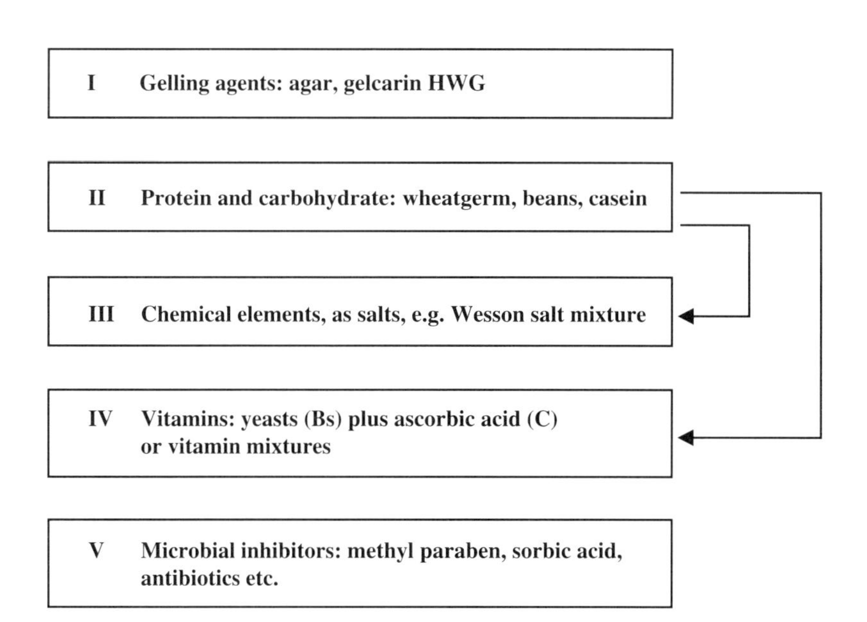

Figure 24.3 Main components of basic diets. The roman numerals refer to the different groups of components, the arrows indicate possible combinations of the different groups during preparation of the diet.

5. Beaker of required volume (depending on volume of eggs, pupae or foliage to be sterilized)
6. Filter paper (Whatman No. 1).*

Method for eggs

1. Place double thickness nappy liner onto a Buchner funnel connected to a Buchner flask.
2. Wash eggs from papers and nets in cages onto the nappy liner with 0.2% sodium hypochlorite solution.
3. Rinse thoroughly with sterile distilled water.
4. Place liner with eggs in a beaker and cover with dilute formalin solution (200 ml formalin in 600 ml distilled water with one drop Triton X); leave for 40 min. (This stage must be done in a fume cupboard since a mixture of sodium hypochlorite and formaldehyde vapours is carcinogenic. If no fume cupboard is available this stage should be omitted.)
5. Place on Buchner funnel and wash thoroughly with distilled water. Leave to dry.

Method for pupae

1. Soak in 1% sodium hypochlorite solution for up to 5 min.
2. Rinse thoroughly with distilled water.
3. Dry on filter paper.

Method for foliage

1. Soak in 0.25% sodium hypochlorite solution for 10–20 min.
2. Rinse well with distilled water.
3. Present to larvae, dry, or standing in water.

Note: If the situation warrants, it is worth attempting to eliminate BV contamination from foliage and the surfaces of egg masses or pupae by agitating in a warm aqueous solution of 0.1 M Na_2CO_3 for 15 min (or as long as is necessary without causing damage to the objects being treated) followed by rinsing in running tap water for 1 min.

As examples, procedures for rearing a field pest, the Egyptian cotton leaf worm, *Spodoptera littoralis*, and a pest of stored products, the Indian meal moth, *P. interpunctella* (Hübner) are given below. A modification of the Hoffman's tobacco hornworm diet is included as an another example of a diet that can be used for rearing field pests.

METHOD FOR REARING *SPODOPTERA LITTORALIS*

This method is based on that of McKinley, Smith and Jones, 1984.

Larval stages

Preparation of fine and coarse diets

The fine (for rearing first to third instars) and coarse diets (late third–final instars) contain the same ingredients but in different proportions. For convenience, it is often better to weigh out several batches of the different ingredients at the same time and to combine them when preparing the complete diet. The components of the two diets are given below together with a method for their preparation.

Dry solids mixture Weigh out and thoroughly mix the following ingredients: wheat germ, 360 g; dried baker's yeast (autoclaved previously at 10 psi for 20 min), 330 g; casein (light white soluble), 60 g; sucrose, 60 g; cholesterol, 1.5 g. Weigh the appropriate amounts for either the coarse (330 g) or fine (200 g) diet and use immediately or transfer to airtight containers, e.g. glass honey pots with screw tops, and store at 4°C until required.

Solution of Vanderzant vitamin mixture Add 70.6 g dry Vanderzant vitamin mixture (supplied by, for example, Sigma) to a medical flat or stoppered bottle containing 200 ml sterile water. Mix and dispense aseptically into sterile bottles in 21 ml (coarse diet) and 15 ml (fine diet) volumes. Store frozen or at 4°C.

Vitamin mixture Weigh out and mix for the coarse diet: ascorbic acid, 15 g; Wesson salt mixture (below), 12 g; choline chloride, 1.2 g; for the fine diet: ascorbic acid, 11.5 g; Wesson salt mixture, 9 g; choline chloride, 0.9 g. Use immediately or store in airtight containers in a cool place, e.g. at 4°C.

Wesson salt mixture $CaCO_3$, 1200 g; K_2HPO_4, 1290 g; $CaHPO_4.2H_2O$, 300 g; $MgSO_4.7H_2O$, 408 g; NaCl, 670.0 g; $FeC_6H_5O_7.6H_2O$, 110 g; KI, 3.2 g; $MnSO_4.4H_2O$, 20 g; $ZnCl_2$, 1 g; $CuSO_4.5H_2O$, 1.2 g.

Agar Weigh out 25 g (coarse diet) or 18 g (fine diet) agar and add to 600 ml (coarse diet) or 500 ml distilled water (fine diet). Autoclave at 15 psi (121°C) for 15 min, boil for 45 min or steam until dissolved. If required immediately, transfer to a waterbath at 70°C and allow to equilibrate, otherwise cool and store at 4°C. For convenience, it may be better to divide the agar for each diet into two equal quantities (i.e. 2×12.5 g or 2×9 g for coarse and fine diets, respectively) and to transfer to a 500 ml medical flat each, together with half the volume of water stated above.

Other ingredients These include: methyl *p*-hydroxybenzoate (methyl paraben), 6 g (coarse diet) or 4.5 g (fine diet); linoleic acid (linseed oil), 12 drops (coarse diet) or 9 drops (fine diet); and distilled water to a total of 1350 ml (coarse diet) or 1110 ml (fine diet).

Equipment

1. A laminar airflow cabinet or a clean room free from draughts and the risk of aerial contamination by microorganisms
2. Trays 24.5 cm × 36 cm × 3.5 cm deep, in which to pour the coarse diet
3. Plastic pots (8 oz) or their equivalent for the fine diet
4. Kitchen spatula
5. Mixing bowl.

An electric food mixer, while being a great asset, is not essential.

Method

1. If not already molten, melt agar by steaming or boiling and then equilibrate to 70°C.
2. Thaw the Vanderzant vitamin mixture.
3. Check that all equipment is clean. Wipe benches with disinfectant, e.g. 0.05% Tegodyne, 1% Vircon, 70% ethanol or industrial methylated spirits.
4. Add 570 ml (coarse diet) or 400 ml (fine diet) distilled water to the dry ingredients plus the methyl paraben. Mix thoroughly and while mixing add the linseed oil.
5. Add a further 160 ml (coarse diet) or 115 ml water (fine diet) to the molten agar to cool it, mix well and pour quickly onto the mixture from step 4 above, with constant stirring. Check that the temperature is below 60°C and then stir in the vitamin and Vanderzant vitamin mixtures. Dispense fine diet into 8 oz pots and pour coarse diet into a tray to a depth of approximately 2–2.5 cm.
6. Allow to set and dry. Stack pots and cover them and the tray to prevent microbial contamination. Store at 4°C.
7. Wipe benches with disinfectant.

Rearing larvae

Equipment and materials

1. Plastic tubs (8 oz) or the equivalent containing fine diet to a depth of approximately 4 mm and with a ventilation slit in the lid

2. Plastic 'polypots' (1 oz), plastic jampots (3 cm deep and 4 cm internal diameter) or equivalent, for rearing larvae individually. Slits should be cut in the lids to provide ventilation
3. Paintbrushes (preferably sable, No. 2) and forceps for transferring larvae
4. Trays of coarse diet for rearing third to final instar larvae
5. Incubator or controlled temperature room (25°C) with controlled lighting (cycles of 10 h light followed by 14 h dark)
6. Boiling waterbath or bench autoclave for sterilizing forceps
7. Disinfectant, e.g. 0.05% Tegodyne, 1% Vircon or formaldehyde vapour (Chapter 23), for sterilizing paintbrushes
8. Small strips (approximately 70 mm × 20 mm) of filter paper to absorb moisture in polypots.

Method

1. Gently shake newly hatched larvae over a pot of fine diet (50–100 per pot) or use a paintbrush. Incubate at the required temperature (25°C) and allow to develop until the third instar is reached (usually 5 days).
2. Cut blocks of food approximately 10 mm × 20 mm × 20 mm and place one piece into each container (Figure 24.4) or, alternatively, use pots into which diet has been poured previously to a depth of approximately 7 mm.
3. Using forceps, transfer one larva to each pot, place a piece of sterile filter paper into the lid and fit on to pot. Transfer healthy larvae only.
4. Incubate as previously and when nearing pupation (17–27 days after hatching), examine daily. Abnormally damp conditions result in large

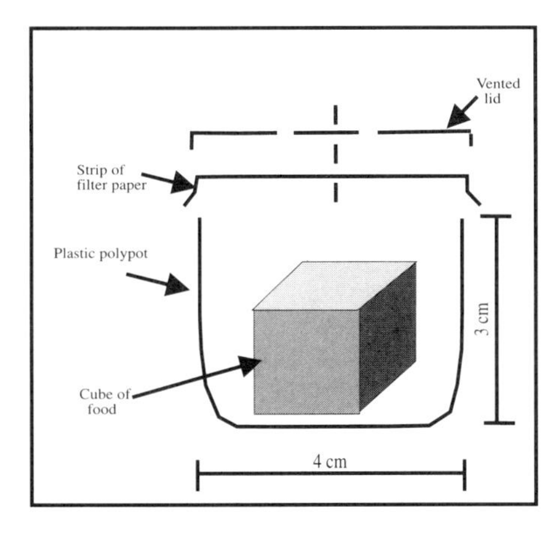

Figure 24.4 Assembly for rearing larvae individually.

numbers of deformed pupae and should be avoided. The diet should not be wet to the touch and should be changed if it is.

Pupae

Equipment and materials

1. Suitable container for pupae such as a plastic sandwich box (9 cm × 15 cm × 9 cm), lined with filter paper
2. Forceps
3. Magnifying glass for sexing pupae (optional)
4. Sodium hypochlorite solution
5. Vermiculite
6. Sterile or boiled water for moistening vermiculite
7. Moth mating cages 30 cm high and 20 cm in diameter (Figure 24.5)
8. Strips of filter paper (28 cm ×1.6 cm).

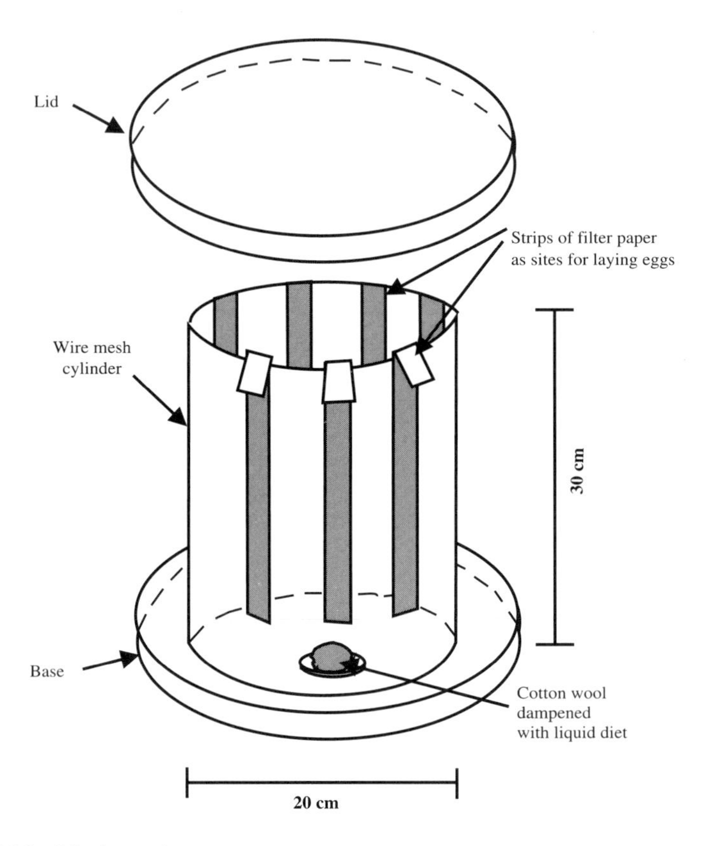

Figure 24.5 Moth mating cage.

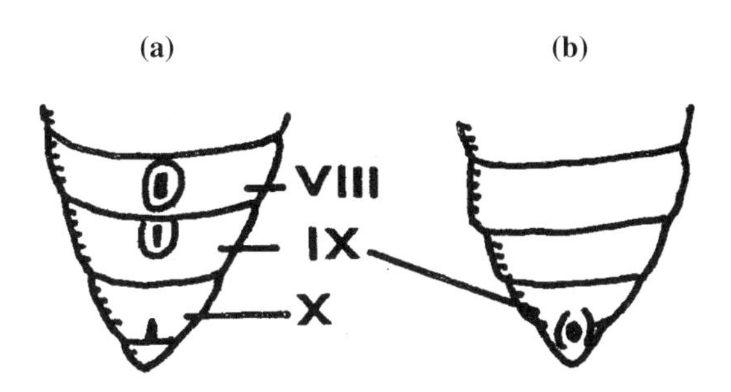

Figure 24.6 Appearance of female (a) and male (b) pupae (ventral side).

Method

1. Once the pupal case has hardened (turned from green to brown) gently remove each pupa from the diet with the aid of forceps and transfer to a sandwich box lined with filter paper. Use one container per week.
2. At the end of each week, wash pupae in 0.5% sodium hypochlorite solution for 5–10 min, dry on filter paper and determine sex (Figure 24.6). Discard dead pupae, those formed later than 27 days after larvae have hatched and those showing physical abnormalities.
3. Pour vermiculite into an 8 oz plastic pot and moisten with sterile or boiled water. Place 20 pairs of healthy pupae on top of the vermiculite and transfer to a mating cage lined with strips of filter paper suspended vertically from the top (Figure 24.5).
4. Examine, daily, for emergence of moths.

Moths

Preparation of liquid diet

Materials and ingredients

1. Yeast, honey, brown sugar, molasses, dried peach halves or apricots
2. Glass beaker 1 l capacity
3. Water or equivalent at 72°C
4. Glass funnel and filter paper.

Method

1. Add 2 teaspoonfuls of yeast, 2 teaspoonfuls honey, 2 teaspoonfuls brown sugar, 1 teaspoonful molasses, 2 dried peach halves or 3 dried apricot halves to 600 ml of distilled water in a glass beaker (1 l volume) (1 teaspoon is 5 ml).

2. Stand overnight.
3. Heat the diet to 72°C and maintain at this temperature for 10 min to inactivate the yeast.
4. Strain the diet through a funnel lined with filter paper into a sterile glass bottle. Cap bottle and keep at 4°C when not in use.

Mating moths and collecting eggs

Equipment and materials

1. Absorbent cotton wool pads on 'polypot lids' moistened with liquid diet for moths
2. Containers suitable for storing eggs
3. Cylinder of CO_2 for anaesthetizing very active moths.

Method

1. Once moths begin to emerge (5–7 days), place pads of absorbent cotton wool moistened with liquid diet in the cage. Replenish pads as necessary.
2. Collect, daily, the strips of filter paper on which eggs have been laid. If moths are very active while eggs are being collected, it may be necessary to anaesthetize them by a short exposure to CO_2 but this is not recommended. Keep each day's batch of eggs separately.
3. Sterilize eggs as described (p. 378) and examine daily for hatching (usually 4 days after being laid).
4. Transfer newly hatched larvae to fine diet as described above. Eggs that are placed on the food before hatching usually fail to develop. This can be overcome by inserting a small piece of plastic, e.g. a piece of a 'polypot' lid, between them and the food.

METHOD FOR REARING *P. INTERPUNCTELLA* (HÜBNER) AND OTHER PESTS OF STORED PRODUCTS

Larval, pupal and moth stages

Preparation of larval diet

Method taken from Hunter and Boraston (1979).

Equipment and materials

1. Sandwich boxes (9 cm × 15 cm × 9 cm or similar) in which to store food
2. Kitchen spatula
3. Mixing bowl
4. Farley's First Timers Farex baby cereal (H. J. Heinz Co. Ltd Hayes, Middlesex, UK). Ingredients consist of rice flour, soya flour, maize flour, calcium carbonate, yeast, vitamin C, niacin, zinc sulphate, vitamin E, iron, vitamin B_6, riboflavin (B_2) vitamin A, folic acid, thiamin (B_1), vitamin B_{12} and vitamin D. Each 100 g Farex supplies: protein, 13.4 g; fat, 5.3 g; carbohydrate, 73 g; sodium, 0.01 g; vitamin A, 450 μg; thiamin; 0.5 mg; riboflavin, 0.6 mg; niacin, 7 mg; vitamin C, 60 mg; vitamin D, 10 μg; vitamin E, 4 mg; calcium, 600 mg; iron 7 mg, zinc, 4 mg; and vitamin B_6, 0.9 mg.
5. Honey
6. Glycerol
7. Dried yeast powder, e.g. Yeastamin (English Grains Ltd, Overseal, Burton on Trent, UK).

An electric mixer is useful but not essential.

Method

1. Weigh dry ingredients (First Timers, 200 g; Yeastamin, 50 g), transfer to the mixing bowl and stir thoroughly.
2. Weigh and mix the glycerol (100 g, 90 ml) and honey (100 g, 70 ml). Pour onto the dry ingredients and mix thoroughly.
3. Dispense in plastic boxes and store overnight at –20°C to kill any insects, eggs, etc. that may have been picked up inadvertently. Store at 4°C.

Rearing all stages

Equipment and materials

1. Clear plastic sandwich boxes (9 cm × 15 cm × 9 cm) for mating and rearing larvae
2. Incubator
3. Semi-synthetic diet
4. Filter paper or paper towelling
5. Forceps

6. Incubator at 26–27°C and with a relative humidity of 40–50%
7. Pooter, small glass vial or bijou for transferring moths to mating box.

Method

Rear all stages at a temperature of 26–27°C and relative humidity of 40–50%.

1. Add diet to incubating eggs or to larvae within 24 h of hatching (3–4 days after laying).
2. Sex larvae at the wandering stage and transfer males (the testes are visible as a dark band on the dorsal side) and females to separate boxes

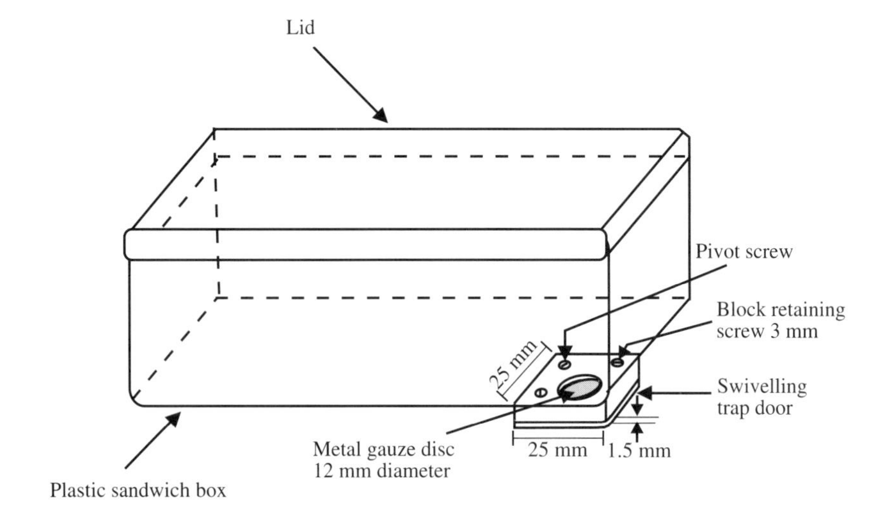

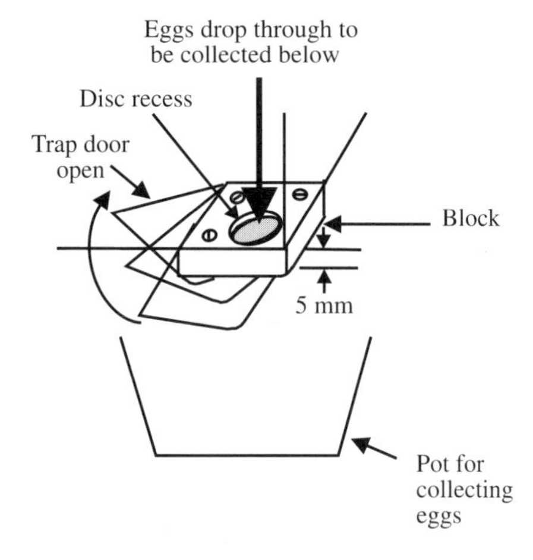

Figure 24.7 Mating cage for moths of *P. interpunctella*.

containing a small quantity of food and filter paper or paper towelling fluted to provide pupation sites.

3. Allow larvae to pupate.
4. When moths emerge (approximately 5 to 8 days after pupation) transfer to clean boxes, 50 pairs per box, and allow to mate.
5. Remove moths once eggs are laid. Alternatively, transfer moths to specially adapted boxes so that eggs can be removed without disturbing the moths (Figure 24.7).

Wesson salt mixture This mixture (p.381) is used at 1.0–1.2% (w/v).

General vitamin mixture The following mixture is used at 0.11–1.2% (w/v): folic acid, 1.25 g; nicotinic acid (niacin), 5 g; aneurine hydrochloride (B_1, thyamine), 1.25 g; riboflavin (B_2), 2.5 g; calcium pantothenate (B_5), 5 g; pyridoxine hydrochloride (B_6), 1.25 g; cyanocobalamine (B_{12}), 0.01 g; ascorbic acid (C), 40 g; and D-biotin (H), 0.10 g.

HOFFMAN'S TOBACCO HORNWORM DIET

The Hoffman diet is given as modified by C. F. Rivers and R. Warner (personal communication).

Ingredients

1. Group 1: agar, 50 g; water, 1500 ml
2. Group 2: casein, 88.0 g; wheatgerm, 187.0 g; Wesson's salts, 25.0 g; cholesterol, 2.5 g; dried brewer's yeast, 38.0 g; sucrose, 78 g; sorbic acid, 4 g; water, 800 ml; 4 M KOH, 12.5 ml
3. Group 3: nicotinic acid (niacin), 5.00 g; calcium pantothenate (B_5), 5.00 g; riboflavin (B_2), 2.50 g; aneurine hydrochloride (B_1, thiamine), 1.25 g; pyridoxine hydrochloride (B_6), 1.25 g; folic acid, 1.25 g; D-biotin (H), 0.10 g; cyanocabalamine (B_{12}), 0.01 g
4. Group 4: Group 3 ingredients (1.0 g) to which are added and well mixed, streptomycin, 2.0 g; aureomycin (veterinary soluble powder (25 g/lb), 18.0 g; ascorbic acid, 40 g. Store at 4°C
5. Formaldehyde (10%), 11 ml (omit from diet when propagating virus)
6. Linseed oil, 5 ml
7. Choline chloride, 2.5 g.

All solid ingredients can be prepared in advance and stored refrigerated until required.

Method

1. Mix agar and cold water (Group 1 ingredients) in a beaker or similar container and boil for a minimum of 5 min. Alternatively, dissolve by autoclaving for 15 min at 15 psi.
2. Finely blend the dry ingredients (Group 2). Add water and 4 M KOH. Mix thoroughly.
3. Place mixture in a covered container, e.g. beaker covered with aluminium foil, and autoclave for 20 min at 15 psi.
4. Cool to 80°C, add formaldehyde (optional, see above), and linseed oil (dropwise) with stirring.
5. Cool to 60–70°C, add Group 4 ingredients, stir well and pour into trays or other suitable containers while molten.
6. Allow to set and cool.
7. Use immediately or cover to exclude contaminants and store at 4°C until required (10 days).

REFERENCES

Hunter, F.R. and Boraston, R.C. (1979) Application of the laurel immunoelectrophoresis technique to the study of serological relationships between granulosis viruses. *Journal of Invertebrate Pathology*, **34**, 248–256.

McKinley, D.J., Smith, S. and Jones, K.A. (1984) *The Laboratory Culture and Biology of* Spodoptera littoralis *Boisduval.* Tropical Development and Research Institute, Overseas Development Administration, London: ODA (ISBN: 0 85135 138 7).

25 Working with the virus

GENERAL

Certain pieces of equipment are essential for working with insects and viruses. These include refrigerators, deep freezes, incubators, waterbaths, benchtop centrifuges and microcentrifuges. Purification of OBs, virus particles and nucleocapsids requires the use of high-speed and ultra-centrifuges. The hazards associated with handling any chemicals used should be checked and safety precautions strictly adhered to throughout. The wearing of disposable gloves is recommended when using stains.

Unless otherwise stated, reagent grade water or double-distilled water should be used.

SCREENING INSECTS FOR VIRUS INFECTION

The procedure for screening insects for virus infection is summarized in Figure 25.1. Practical details are given below.

Isolation of virus

COLLECTION AND TRANSPORTATION OF INFECTED LARVAE AND VIRUS-CONTAMINATED MATERIAL FROM THE FIELD

Materials and equipment

1. Several pairs of sterile forceps for transferring infected larvae and virus-contaminated foliage to collection bottles
2. Containers, e.g. microcentrifuge tubes, plastic vials with screw caps or universal bottles, for collection and storage of diseased larvae

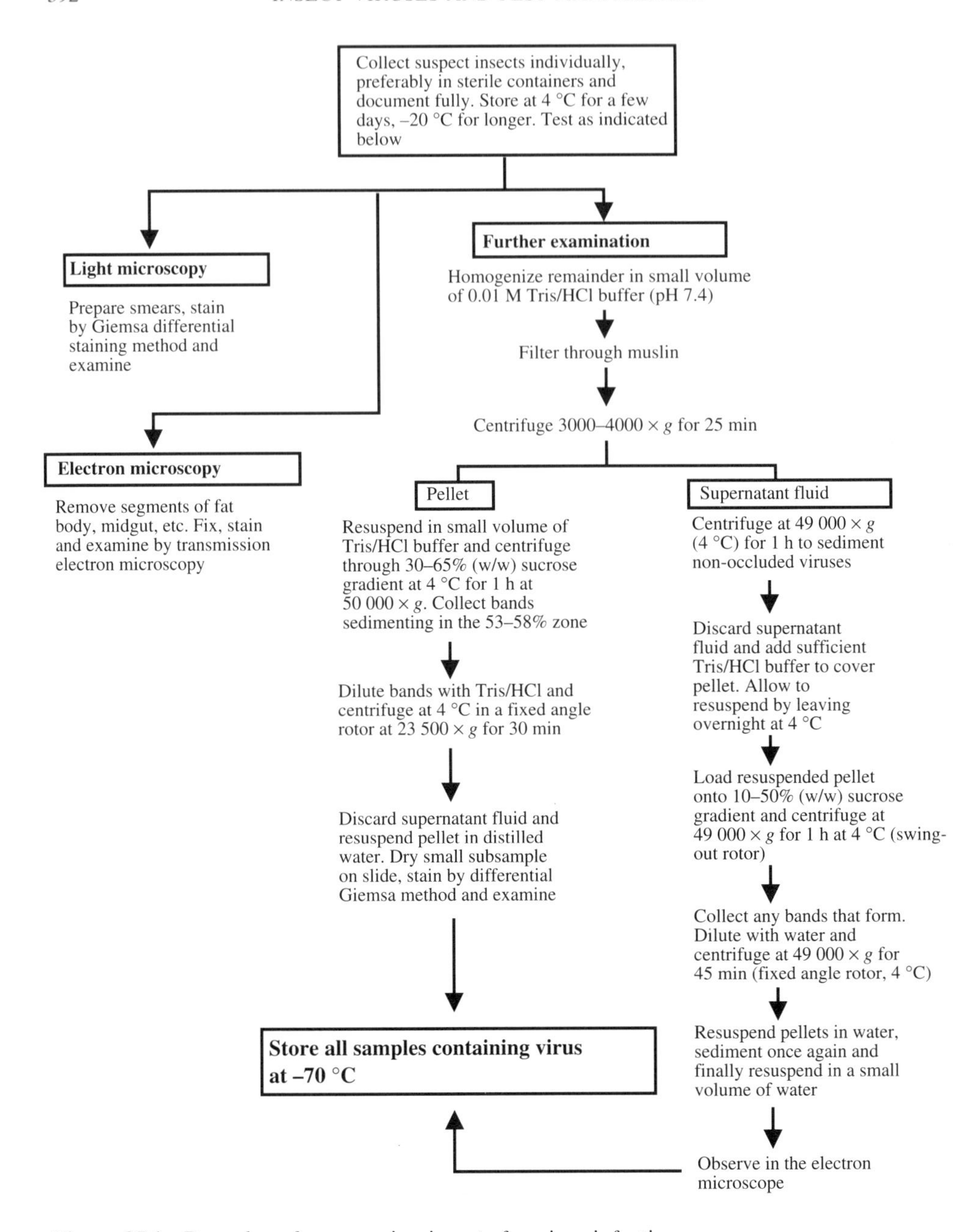

Figure 25.1 Procedure for screening insects for virus infection.

3. Cooler box and freezer blocks for keeping samples cool during collection and transportation back to the laboratory
4. Sterile water, pipettor (e.g. Gilson or similar) and sterile tips for rinsing virus from plant material
5. Autoclave bags, or similar, for transporting used forceps and any virus-contaminated equipment and materials back to the laboratory for decontamination
6. Marker pen, labels, notebook and pen for labelling and documentation of samples.

The type of container chosen for collection of samples will depend on the size of sample to be collected. Leakproof containers are preferable in all circumstances and essential if samples are to be transported by air. Sealing around the lids with waterproof tape, or parafilm, will give extra security.

Procedure

1. Identify plants and areas with diseased larvae.
2. Transfer infected larvae or larval remains to containers using sterile forceps. If larval liquefaction has occurred, it may be necessary to collect virus-contaminated foliage for extraction of virus in the laboratory. Failing this, it may be possible to rinse virus from contaminated material using sterile water and a pipette. (Never leave collected specimens in direct sunlight, always stand in the shade.)
3. To prevent cross-contamination of material, use a fresh pair of forceps for each new location or sample. Once used, place forceps in an autoclave bag or other suitable container for later decontamination, cleaning and re-sterilization.
4. Discard used pipette tips into disinfectant or an autoclavable container for processing in the laboratory (Chapter 23).
5. Transport sample back to the laboratory, in a cool box if necessary.
6. On arrival at the laboratory, any living infected larvae should be killed by freezing, preferably by snap freezing, i.e. immersion in liquid nitrogen or a mixture of dry ice and ethanol. Store all samples at 4°C or –20°C and process as soon as possible (Figure 25.1). Suspensions of virus that are too dilute for direct investigation should be passaged once in the host species before testing. Similarly, virus-infected material, or extracts of it (suspended in water), should be fed to the host species and infected individuals treated according to the protocol in Figure 25.1.

Care should be taken to ensure that all samples are documented correctly.

If a virus-free culture of the species of interest is not already being reared in the laboratory, attempts should be made to establish one from uninfected individuals collected in an area free from infection (Chapter 24).

Storage of virus

For long-term storage, virus is best maintained at –70°C or -20°C. If facilities for achieving such low temperatures are unavailable, storage at 4°C may prove satisfactory. This temperature is certainly acceptable for temporary storage of samples. To maintain maximum infectivity, samples should be frozen as quickly as possible, either by being dispensed in small volumes or by immersing containers with larger volumes in a mixture of ethanol and dry ice. It is essential that the thawing process is also achieved quickly, i.e. by agitation in warm water (*c.* 37°C). The sample should be placed in a plastic bag to prevent it getting wet.

Screening specimens by light microscopy for the presence of virus

SIMPLE GIEMSA STAINING

Materials

1. Microscope slides
2. Staining racks and jars
3. Giemsa fixative (absolute ethanol, 94%; formalin, 5%; acetic acid, 1%)
4. Phosphate buffer 0.02 M
5. Gurr's Improved R66 Giemsa, 10% in 0.02 M phosphate buffer
6. pH meter.

Method

The 0.02 M phosphate buffer is prepared from two solutions:

Solution A: NaH_2PO_4, 29.39 g made up to 1 l with glass-distilled water (0.2 M solution)

Solution B: $Na_2HPO_4.2H_2O$, 31.21 g made up to 1 l with glass-distilled water (0.2 M solution).

1. Add 55 ml of Solution A to 45 ml of Solution B and make up to 1 l with glass-distilled water.
2. Check that the pH is between 6.9–7.0. Adjust to this value if necessary by dropwise addition of either HCl or NaOH (0.1 M or concentrated) depending on the volume required to make the adjustment.

Treatment of specimens

1. Prepare a thin smear of infected insect tissue on a microscope slide making sure to rupture the midgut.

2. Dry the smear in air.
3. Immerse for 1–2 min in Giemsa's fixative.
4. Rinse under running tap water 5–10 s.
5. Stain 25–60 min in 10% Gurr's Improved R66 Giemsa in 0.02 M phosphate buffer.
6. Rinse off stain in running tap water, 5–10 s.
7. If overstained, stand in 0.02 M phosphate buffer until red colour on glass slide disappears.
8. Gently blot with absorbent paper, dry and examine with an oil immersion objective (total magnification of about × 1000).

OBs of NPVs appear as clear, round objects, those of GVs as very small ellipsoidal objects, also clear, while bacteria and other contaminants stain purple.

This and other staining methods can be used to examine the progress of virus infection in cells grown in culture. At seeding of cell culture vessels, place some small pieces of washed, sterile microscope cover glass on the base of the vessel. At desired intervals, these pieces can be removed bearing their intact cell sample. For minimum disturbance during fixation, simply invert the cover glass while still wet over an open vessel containing 3.0% osmic acid for 1–5 min (in a fume cupboard, as osmic acid vapour can injure the eyes) and then proceed with the desired staining method.

DIFFERENTIAL GIEMSA STAINING

The different Giemsa staining method (Wigley, 1980) is used for OBs and Rickettsiae in smears of insect tissue. The method can be used to differentiate between occluded insect viruses as well as to identify other groups of small pathogenic microorganisms. It depends on comparing their staining characteristics with Giemsa stain alone and Giemsa stain in combination with picric acid, or picric acid and Naphthalene Black. A thin smear of insect tissue or homogenate is fixed in Carnoy's fluid, rinsed in alcohol and dried. The cold slide is dipped to two-thirds of its width in hot saturated aqueous picric acid, dried and heated. Then the same slide is dipped to one-quarter of its width in hot acidic Naphthalene Black 12B. After rinsing, the entire smear is stained with Giemsa stain. The staining characteristics of occluded viruses under oil immersion bright field and phase-contrast microscopy are shown in Table 25.1. Bacterial spores do not stain black in Naphthalene Black. Microsporidian spores generally stain a dense black in Naphthalene Black and are relatively uniform in size and shape. Fungal hyphae vary in size and shape. Budding can be seen frequently.

Table 25.1 Appearance of occluded viruses and Rickettsiae in the various zones of pre-treated Giemsa smears (× 1000 magnification, oil immersion objective)

Pathogen	Appearance		
	Zone A: no pre-treatment	Zone B: picric acid	Zone C: picric acid + Naphthalene Black
NPV	Colourless	Yellow to colourless	Black
CPVs	Colourless or light blue/grey	Reddish blue to blue-grey to deep purple	Black
GVs	Colourless	Colourless	Black
Entomopoxviruses	Blue	Blue	Black
Rickettsiae	Red	Red	Grey

Materials

1. Microscope slides
2. Slide racks (metal) and staining jars
3. Absorbent paper, e.g. filter paper, paper tissues
4. Carnoy's fluid (1:3:6 mixture of glacial acetic acid, chloroform and absolute ethanol) or 1:9 mixture of formalin and ethanol
5. Absolute ethanol
6. Aqueous picric acid (saturated at 20–23°C)
7. Gurr's improved R66 Giemsa, 10% in 0.02 M phosphate buffer
8. Hot plate
9. Naphthalene Black 12B 1.5% (w/v) in 35% glacial acetic acid (in glass-distilled water)
10. Phosphate buffer, 0.02 M, pH 6.0 (see p. 394 for method of preparation).

Picric acid is poisonous and should be treated with great care. It is highly explosive if not stored as an aqueous solution.

Method

1. Prepare a thin smear of infected tissue or tissue homogenate (in water), covering the full width of a microscope slide. If using tissue, make sure that the midgut is ruptured. Several smears can be placed on one slide.
2. Place slide in a slide rack and dry smear in air. Then place slide rack on a hot plate and heat to 75–80°C.
3. Transfer rack with slide into a staining jar containing either Carnoy's fluid or a 1:9 mixture of formalin : ethanol. Leave for 2–3 min.

4. Rinse in clean absolute ethanol for 20–30 s then dry in air.
5. Immerse slide sideways for two-thirds of its width in picric acid at 40°C (38–42°C) for 2 min.
6. Rapidly immerse slide in two baths of tap water, the first for 0.5 s and the second for 1–2 s.
7. Bathe the slide for 30 s in 0.02 M phosphate buffer.
8. Drain and air dry.
9. Heat the dry slide to 75°C (70–80°C).
10. Immerse the hot slide to one-quarter of its width for 1 min in 1.5% w/v Naphthalene Black 12B in 35% glacial acetic acid at 40°C (38–42°C), i.e. the same side of the slide previously immersed in picric acid.
11. Immerse the slide in a bath of tap water for 15 s.
12. Bathe in running tap water for 60 s.
13. Bathe for 20 s in absolute ethanol.
14. Bathe for 30 s in a bath of rapidly running water.
15. Stain the whole slide for 45 min in 7% Gurr's Improved R66 Giemsa stain in 0.02 M phosphate buffer at pH 6.9.
16. Rinse off stain in running tap water.
17. Bathe for 15–30 s in tap water.
18. Remove rack and stand on absorbent paper.
19. Quickly remove slides from rack and place them face down on absorbent paper to blot dry.
20. Blot reverse side of slides.
21. Dry and examine by light microscopy (×700 to ×1000 magnification with oil immersion objective).

STAINING TECHNIQUE FOR OCCLUSION BODIES OF GRANULOSIS VIRUSES (KAUPP AND BURKE)

Materials

1. Microscope slides
2. Egg albumin, 50%
3. Bouin's fixative
4. Aqueous iron ammonium sulphate, 2.5% (w/v)

5. Eosin Y (0.4%, w/v) in 45% ethanol
6. Ethanol, 95%
7. Naphthalene Black 12B in 35% glacial acetic acid (in water).

Method

1. Homogenize infected larvae in water.
2. Dilute further with water and mix 500 μl with 500 μl 50% egg albumin.
3. Transfer 5 μl to a microscope slide and spread.
4. Air dry and then place in Bouin's fixative for 24 h.
5. Rinse with water.
6. Bathe in iron ammonium sulphate (2.5%) for 2 h (mordant).
7. Rinse in water for 30 s.
8. Bathe in Eosin Y solution (0.4% in 45% ethanol) for 3 min.
9. Rinse in 95% ethanol for 1 min.
10. Place in Naphthalene Black 12B in 35% glacial acetic acid at 40°C for 5 min.
11. Wash in water, dry in air and examine using a red filter with oil immersion objectives and phase-contrast optics.

GV OBs appear as brilliant black sausage-shaped objects on a flat red background.

THE USE OF NAPHTHALENE BLACK 12B (ACID BLACK, BUFFALO BLACK) FOR STAINING OCCLUSION BODIES

Materials

1. Solution of Naphthalene Black, 1.5 g; glacial acetic acid, 40 ml; glass-distilled water, 60 ml
2. Microscope slides
3. Hot plate.

Method

1. Prepare a thin smear of infected insect tissue making sure to rupture the midgut.
2. Dry the smear in air.

3. Immerse the slide in the Naphthalene Black solution heated to 40–44°C, for 5 min.
4. Wash the slide under running tap water 5–10 s.
5. Blot on absorbent paper, dry and examine under oil immersion at a magnification of × 1000.

Proteinaceous bodies, NPV, CPV, GV and entomopox OBs stain black against a greyish-green to light black background.

RAPID DIAGNOSTIC TEST FOR NPV INFECTIONS IN SAWFLY LARVAE AND FOR CPV INFECTIONS OF LEPIDOPTEROUS LARVAE

During the course of midgut infections, the OBs of NPV and CPV are commonly found in large quantities in the gut fluids.

Taking advantage of defensive regurgitation, a type of behaviour common in, for instance, larval dipronid sawflies and Lepidoptera, it is possible to diagnose NPV and CPV infections before they become fatal and without killing test larvae.

This method is excellent for collecting field samples. Slides can be labelled with a diamond marker, placed in a slide box in the field and can be stored for long periods before staining.

Materials

1. Microscope slides
2. Forceps
3. Materials for staining by the Naphthalene Black method given above.

Method

1. Lift larva with forceps and squeeze gently over a microscope slide so that regurgitated fluids are deposited on the slide. With practise, fluids of 10 or more larvae can be accommodated on the same slide. Those regurgitated by dipronid sawflies contain terpenes and resins from their coniferous host plants that, when dry, secure OBs to the slide sufficiently strongly to withstand the staining process. Those regurgitated by Lepidoptera may need to be fixed after drying, by adding a thin coating of Mayer's albumin or simple egg albumin solution to the sample.
2. Stain by the Naphthalene Black method given above.
3. Observe under oil immersion at a magnification of × 700 to × 1000.

Screening by electron microscopy

Electron microscopy calls for the use of specialized equipment, technical skills and knowledge. These are best provided by a trained electron microscopist. The preparation of samples for observation in the electron microscope is relatively easy, however, and is described below. If the services of a trained electron microscopist are not available, reference should be made to a manual such as that of Bozzola and Russell (1992).

EMBEDDING INVERTEBRATE TISSUES AND CELLS IN CULTURE FOR ELECTRON MICROSCOPY

The method for processing invertebrate tissues and cultured cells is similar. Where there are differences, these are indicated.

Materials and equipment

1. Insect tissue
2. Dissecting tray and dissecting pins (insect tissue only)
3. Bottle(s) or similar of cultured insect cells (cultured cells only)
4. Phosphate buffer, 0.2 M and 0.1 M (pH 7.2)
5. Forceps
6. Fine pointed scissors (insect tissue only)
7. Scalpel
8. Cell scraper, also known as a rubber policeman (cultured cells only)
9. Pasteur pipettes or micropipettes with disposable tips
10. Surgical gloves
11. Glutaraldehyde solution comprising glutaraldehyde (25%), 5 ml; 0.2 M phosphate buffer, 15 ml; glass-distilled water, 20 ml
12. Osmium tetroxide solution comprising osmium tetroxide (2%), 5 ml; 0.2 M phosphate buffer, 5 ml
13. Bench centrifuges to take 10–15 ml and 1 ml volumes
14. Embedding capsules
15. Acetone: 50%, 70%, 90% and 100%
16. Spurr resin comprising a mixture of ERL, 10 ml; DER, 6 ml; NSA, 26 ml; S1, 0.4 ml
17. Fume cupboard
18. Oven at 60°C

19. Conical centrifuge tubes of suitable volume to accommodate cells and medium harvested from culture (cultured cells only).

Method

Preparation of resin Wearing gloves and working in the fume cupboard, mix the required quantities of the components of the Spurr resin in a plastic beaker: 25 ml should be sufficient to embed up to 16 specimens.

Disposal of materials Insect and cell culture materials and containers contaminated with virus (e.g. cell culture supernates) must be decontaminated by autoclaving or similar procedures before recycling or disposal. Any pipettes and syringes used for handling resin should be added to the beakers containing used resin and acetone/resin mixture. Place in an oven at 60°C and allow polymerization to occur before discarding them and their contents.

Insect tissues

1. Dissect insect under 0.2 M phosphate buffer (immediately after killing if possible).
2. Cut tissues for embedding into very small pieces (maximum 3 mm × 3 mm × 3 mm) for optimum penetration of solutions.
3. Process as described below.

Cell cultures

1. Scrape cells off substrate using rubber policeman and transfer suspension into a conical centrifuge tube.
2. Centrifuge at 1500–2000 × *g* for 5 min to sediment cells.
3. Remove supernate.
4. Resuspend cells in up to 1 ml 0.2 M phosphate buffer and transfer to a microcentrifuge tube.
5. Sediment cells by centrifugation.
6. Remove supernate and process as described below.

If cells resuspend during subsequent processing it may be necessary to recentrifuge them between each stage. Large pellets should be resuspended at each stage to ensure penetration of solutions.

Very small quantities of cells can be transferred to embedding capsules after dehydration and while suspended in acetone. They should then be centrifuged to form a discrete pellet before infiltration with resin. Embedding capsules can be centrifuged by placing in microcentrifuge tubes or large conical tubes. It is advisable to tie a piece of thread around each capsule if the latter are used so that they can be retrieved afterwards.

Fixation Between each change of solution the sample should be sedimented by centrifugation, the supernate removed, the next solution added and the pellet resuspended. This ensures that the whole of the specimen is fixed properly. All stages apart from centrifugation (room temperature) should be done at 4°C.

1. Treat with glutaraldehyde solution for 2 h.
2. Wash in two changes (5 min each) of 0.1 M phosphate buffer.
3. Treat with osmium tetroxide solution for 1 h.
4. Wash in two changes (5 min each) of 0.1 M phosphate buffer.

Caution: gloves must be worn and all stages must be done in a fume cupboard. Pipettes used for handling osmium tetroxide must be washed in running tap water for at least 20 min before being discarded.

Dehydration This step is carried out at room temperature. Treat in two changes of 10 min each, with 50%, 70% and 90% acetone followed by three changes of 100% acetone, each of 15 min.

Embedding

1. Transfer samples to capsules and treat for 1 h with Spurr:acetone resin (1:1) at room temperature.
2. Replace with Spurr resin and leave overnight in fume cupboard.
3. Replace with fresh Spurr resin and polymerize by placing in oven at 60°C for 24 h or longer.
4. Remove block from capsule and trim.
5. Mount block in ultramicrotome and cut thick (1–2 μm) sections (for light microscopy, below) followed by thinner (60–80 nm) sections.
6. Stain thin sections with 2% aqueous uranyl acetate, prepared by dissolving overnight in deionized water followed by filtration and centrifugation to remove any particulate matter.

The thin sections are then observed in the electron microscope. As the field of view at the high magnifications of an electron microscope is usually small, it is helpful to be able to observe sections at the lower magnifications provided by the light microscope first. Such observations make it easier to orientate specimens when viewed in the electron microscope. The thick sections are stained for observation by light microscopy (below).

STAINING THICK RESIN SECTIONS FOR OPTICAL INSPECTION

This technique, which is simple to perform, results in stained sections that are of a particularly sharp quality and are excellent for photography using the light microscope.

Method

1. Using a sliver (say 2 mm × 8 mm) of very thin cover glass (allows subsequent use of oil immersion objectives, which tend to have a short working distance), pick up a thick resin section from the water reservoir into which the sections are cut. This is easily done by touching a section gently with the glass slip.
2. Dry in air (a few minutes).
3. Place section downwards onto a drop of 0.5–1.0% aqueous Toluidene Blue in a small Petri dish and replace the lid while staining. Leave for 2–4 min, or up to 30 min if the shorter time gives unsatisfactory results.
4. Rinse in tap water.
5. Rinse in two changes of ethanol.
6. Clear in xylene.
7. Place the glass slip with tissue downwards, onto a small drop of mountant (e.g. DPX or Canada balsam) on a microscope slide.
8. Observe in the light microscope.

EXTRACTION OF OCCLUSION BODIES FROM SOIL

The procedure for extracting OBs from soil is illustrated in Figure 25.2.

PROPAGATION AND PURIFICATION OF BACULOVIRUSES

The procedure for rearing larvae for infection will depend on the species. An artificial diet is preferable and larvae should be transferred from the rearing facility to the virus production area when at a suitable stage of development for infecting.

A plan for the possible layout of a virus production unit is given in Figure 25.3. This is intended for guidance, only. Anyone wishing to establish such a unit should take into account local conditions and facilities before embarking on the work.

In the practical production of BVs, the advantage of the larger PR (productivity ratio) of young larvae must be set against the problems of handling greater numbers of small larvae. An additional consideration is that larval supply is usually

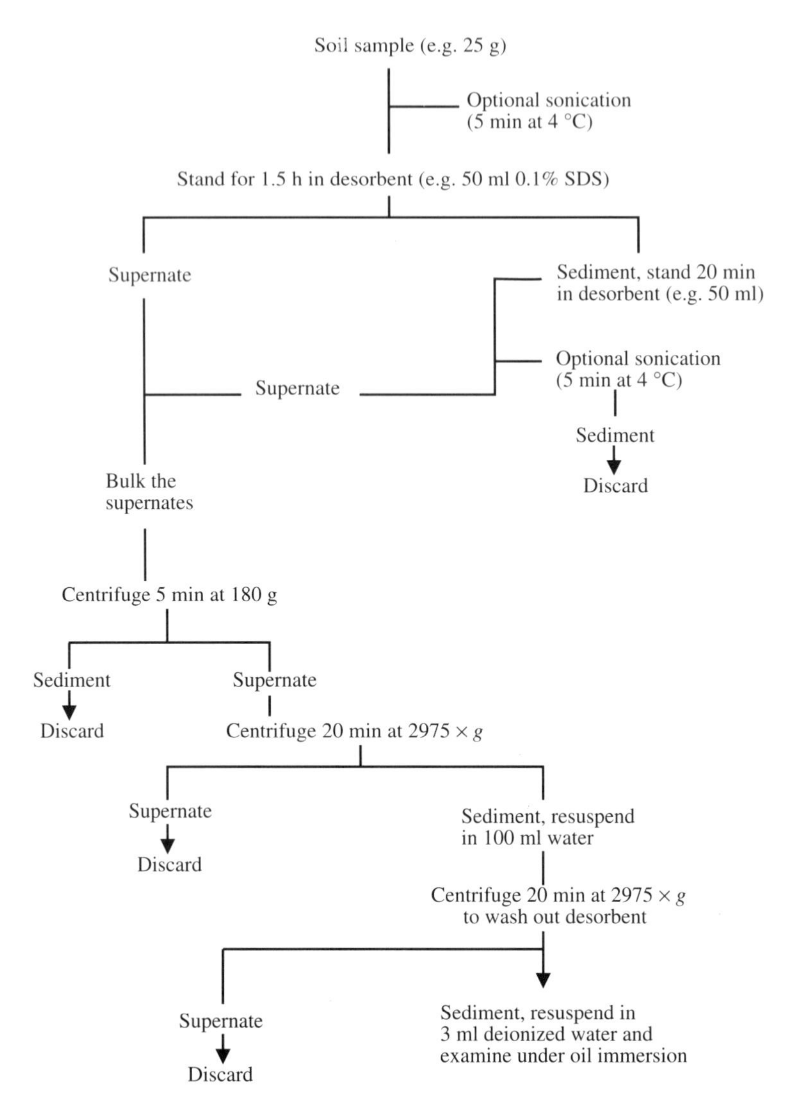

Figure 25.2 Extraction of OBs from soil (adapted from Evans *et al.* (1980) *Journal of Invertebrate Pathology*, reproduced by permission of H.F. Evans *et al.* and Academic Press Inc.).

finite. It is, therefore, common practice to infect larvae at the stage (usually late third instar) that will result in maximum productivity per individual at death. Propagation of virus in larvae is depicted diagrammatically in Figure 25.4.

Infected larvae are commonly harvested shortly before death. Although yields of virus may be less than from dead larvae (Ignoffo and Shapiro, 1978), the latter are generally very fragile and may rupture, resulting in loss of virus. In addition, the purification of OBs (Table 25.2) is easier and the quality of the final product better if the larvae have not putrified or melanized. Where the objective is to make final preparations by freeze-drying and grinding, such early harvesting may

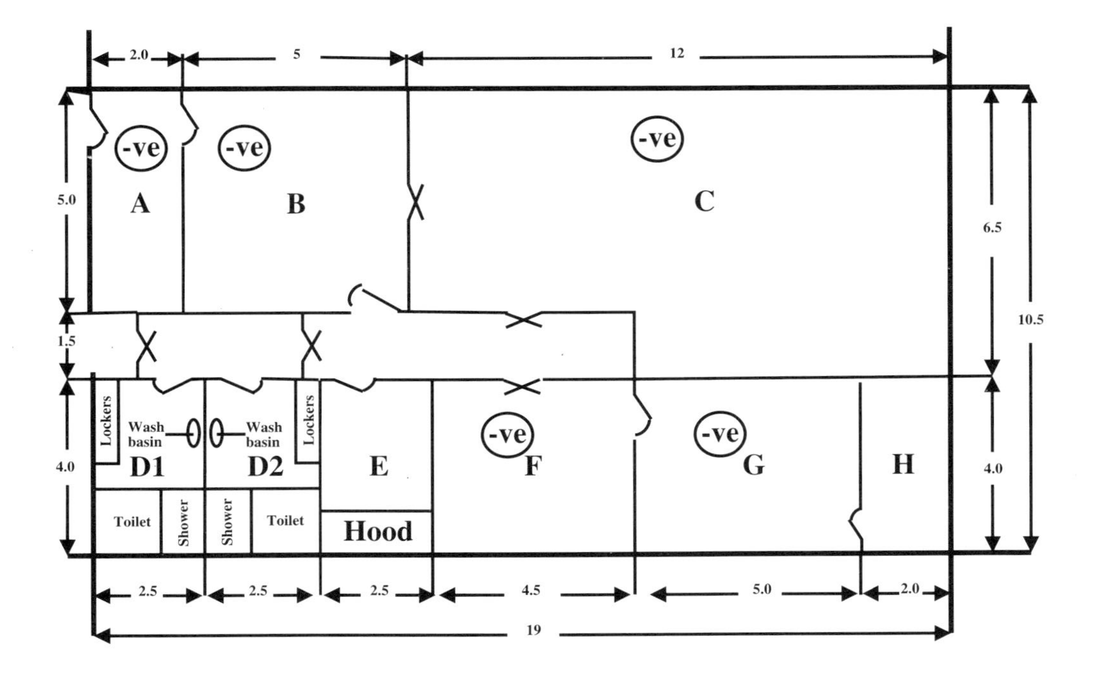

Figure 25.3 A suggested layout for a virus production unit (measurements in metres). Negative air pressure (–ve) is maintained in certain rooms to prevent escape of virus into, and contamination of, the environment. A. Area for holding diet and larvae before infection with virus. Entrance through top left hand corner from separate insectary. B. Virus inoculation of larvae. The inclusion of more than one hood allows for several viruses to be handled simultaneously. (–ve pressure, hoods and room). C. Incubation of inoculated larvae. This area is large enough for subdivision into several separate cubicles. D. Separate facilities for male and female staff (D1 and D2) for changing clothing on entering and leaving work areas, includes showers. E. Virus-stock retention and dilution for purposes of inoculation (using hood). F. Harvesting virus-infected larvae. G. Preparation of virus, or virus-infected larvae, for storage. Equipment should include that for macerating larvae, filtration, centrifugation and freeze drying, as appropriate. H. Cold storage for virus stock (e.g. 'walk-in' deep freezer and cold room). This should include a freeze-file system.

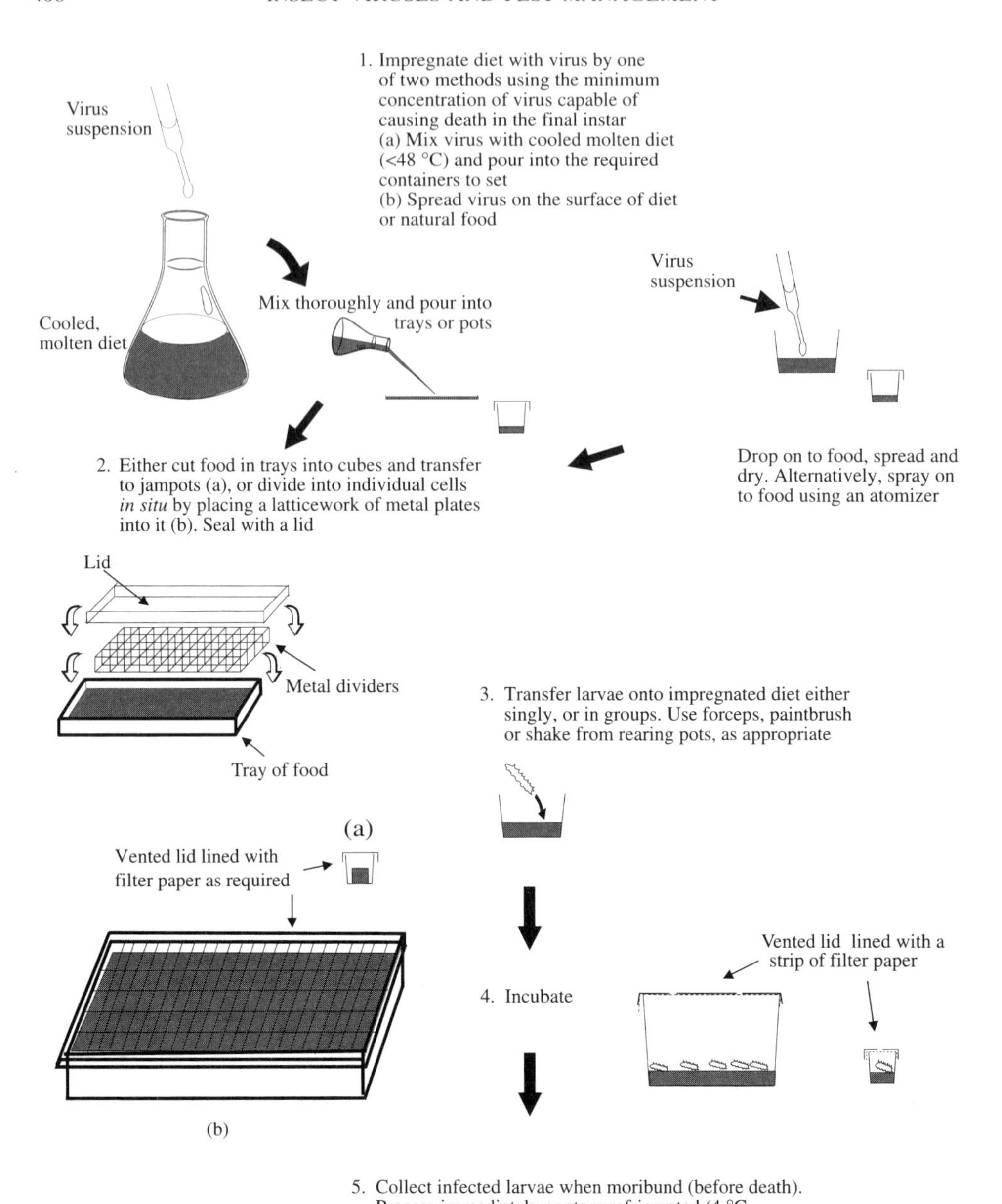

Figure 25.4 Propagation of virus in larvae.

be essential, since a higher than stipulated count of bacteria (pp. 107 and 523) would contravene the regulations of registration authorities.

Preparation of the preliminary homogenate of infected larvae from which the virus is to be extracted and purified can be achieved in various ways depending on the number of infected larvae being handled. For a single larva, a small tissue

Table 25.2 Purification of BV OBs

Protocol	Relative velocities: NPV	Relative velocities: GV
Weigh and macerate larvae in an equivalent volume of SDS[a] (Figure 25.6) ↓		
Filter through muslin to remove gross debris ↓		
Centrifuge to remove large contaminants ↓	$100 \times g$[b], 0.5 min	$400 \times g$, 5 min
Resuspend pellet and repeat centrifugation; combine supernatant fluids and discard final pellet ↓		
Centrifuge to remove lipid, soluble material and small contaminants ↓	$2500 \times g$, 10 min	$10\,000 \times g$, 30 min
Discard supernatant fluid; resuspend pellet in small volume SDS ↓		
Centrifuge on rate zonal gradients 30–80% (v/v) glycerol[c] in SDS or 25–65% (w/w) sucrose (Figure 25.7b) ↓	$1500 \times g$, 8 min (glycerol) $50\,000 \times g$, 45 min (sucrose)	$12\,000 \times g$, 40 min (glycerol) $50\,000\,000 \times g$, 60 min (sucrose)
Collect bands containing virus, dilute at least 1/2 (glycerol) or 1/3 (sucrose) with SDS and centrifuge ↓	$2500 \times g$, 10 min	$10\,000 \times g$, 30 min
Discard supernate and resuspend pellet in SDS: centrifuge on equilibrium gradients 45–60% (w/w) sucrose layers in SDS ↓	$50\,000 \times g$, 60 min	$50\,000 \times g$, 60 min
Collect bands containing virus, dilute with SDS and centrifuge ↓	$2500 \times g$, 10 min	$10\,000 \times g$, 30 min
Discard supernatant fluid; wash OBs twice by resuspending in sterile distilled water and centrifuging; resuspend to required volume in water ↓	$2500 \times g$, 10 min	$10\,000 \times g$, 30 min
Check purity in electron microscope if available. Estimate protein concentration by the BCA protein assay method (see text and Table 25.5); store at –20°C		

[a] 0.1% (w/v) Sodium dodecyl sulphate.
[c] Glycerol gradients are considered preferable for GV in the first stages of purification since the conditions for sucrose are close to reaching equilibrium. They are unlikely, therefore, to separate granules from aggregates of granules, which may contain impurities.
[b] The g or relative centrifugal force (RCF) for any radial position in a rotor can be calculated from the following formula:

$$\mathrm{RCF} = 11.18 \times r \times (\mathrm{rpm}/1000)^2$$

where r is the distance from the centre of rotation in cm and rpm is in revolutions per min.

Table 25.3 Preparation and purification of BV particles

Protocol	Relative velocities
Dissolve OBs at a concentration of 10 mg/ml in 0.05 M Na_2CO_3 at 37°C for 30 min; centrifuge on 10–50% (w/w) sucrose gradient	50 000 × *g*, 45 min
↓	
Collect band (GV and SNPV) or bands (MNPV) corresponding to 1, 2, 3 or more nucleocapsids; centrifuge on 30–60% (w/w) sucrose gradient	50 000 × *g*, 3 h
↓	
Resuspend in desired volume of sterile distilled water	

Table 25.4 Preparation of BV nucleocapsids

Protocol	Relative velocities
Incubate *c.* 1 mg virus particles/ml with 0.05 M Tris-HCl buffer, pH 7.4 containing 1% (w/v) Nonidet P40 (BDH Chemicals Ltd) for 30 min at 30°C; centrifuge on 10–45% (w/w) sucrose gradient in 0.05 M Tris-HCl buffer pH 7.4	40 000 × *g*, 45 min
↓	
Dilute band of nucleocapsids in sterile distilled water and sediment	50 000 × *g*, 60 min
↓	
Resuspend nucleocapsids in distilled water	

grinder or similar is sufficient. Small numbers of larvae can be macerated in universal bottles by shaking in diluent with glass Ballotini beads (2 mm diameter), while larger numbers are most easily homogenized with the aid of a Stomacher (e.g. Colworth, Stomacher 80). Once homogenized, the preparations are treated as indicated in Tables 25.2–25.4. This will necessitate the use of both low (100–2000 × *g*) and high-speed (up to 50 000 × *g*) centrifuges. Schemes for harvesting infected larvae, extracting and purifying OBs are illustrated in Figures 25.5–25.7.

Purified OBs can be stored in suspension for several years at –20°C with little loss of infectivity. Neither virus particles nor nucleocapsids can be stored for long. They should be used as soon as possible after being purified.

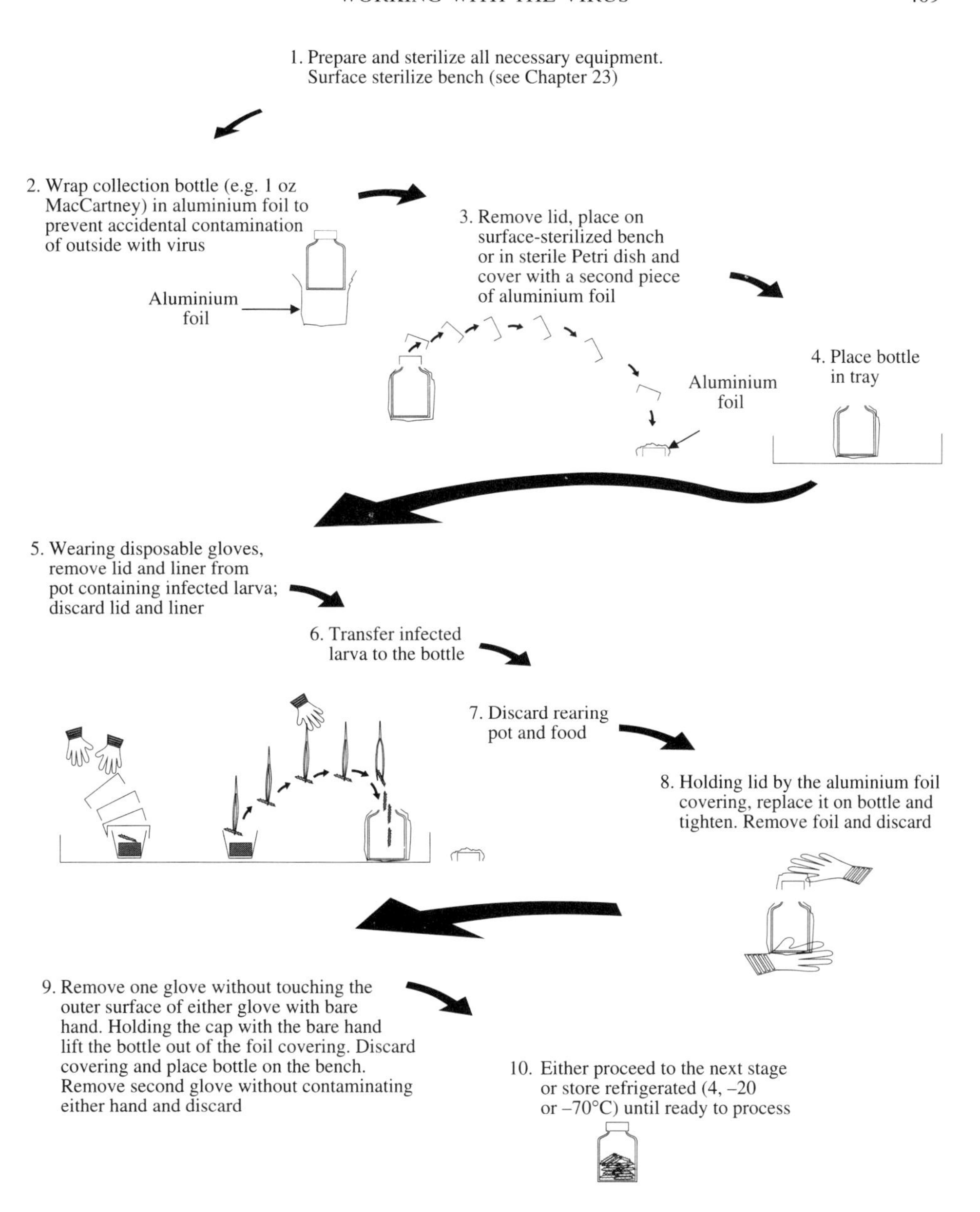

Figure 25.5 Harvesting small batches of infected larvae. Working in a cabinet if at all possible. All items potentially contaminated with virus must be decontaminated before disposal or recycling.

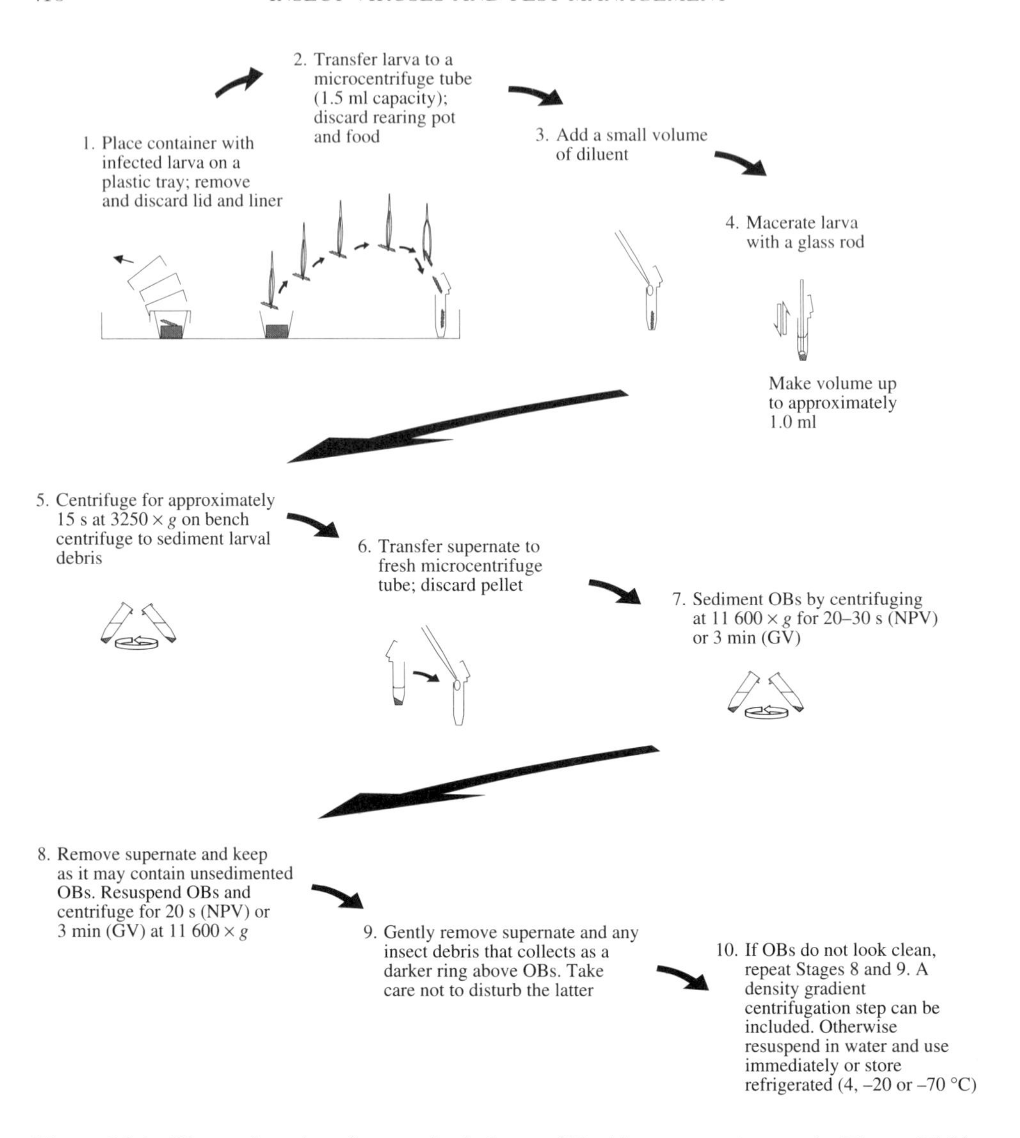

Figure 25.6 Harvesting virus from a single larva. (Working precautions as in Figure 25.5.)

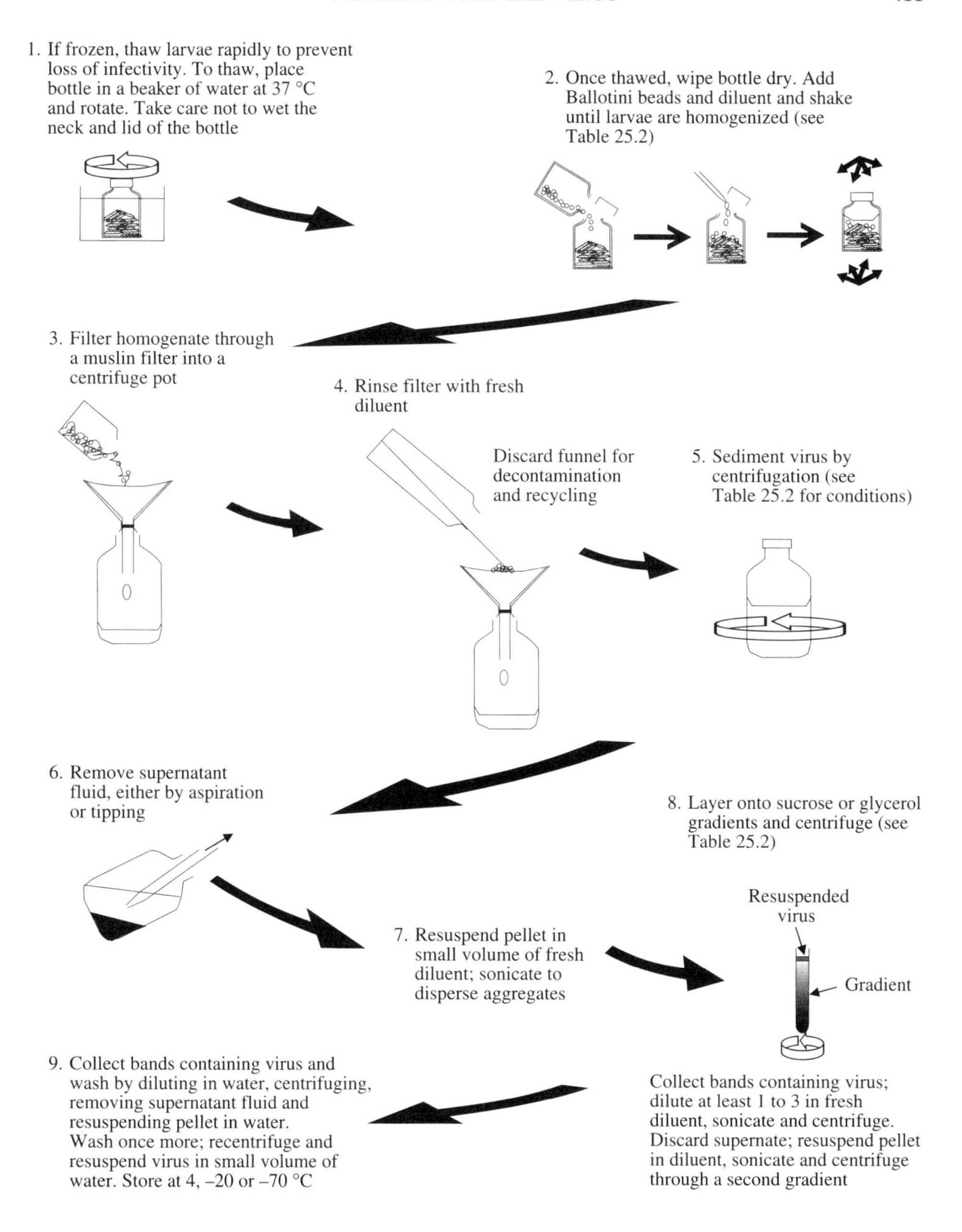

Figure 25.7 (a) Purification of virus from small batches of infected larvae. (b) Preparation of density gradients by the diffusion method and harvesting of samples.

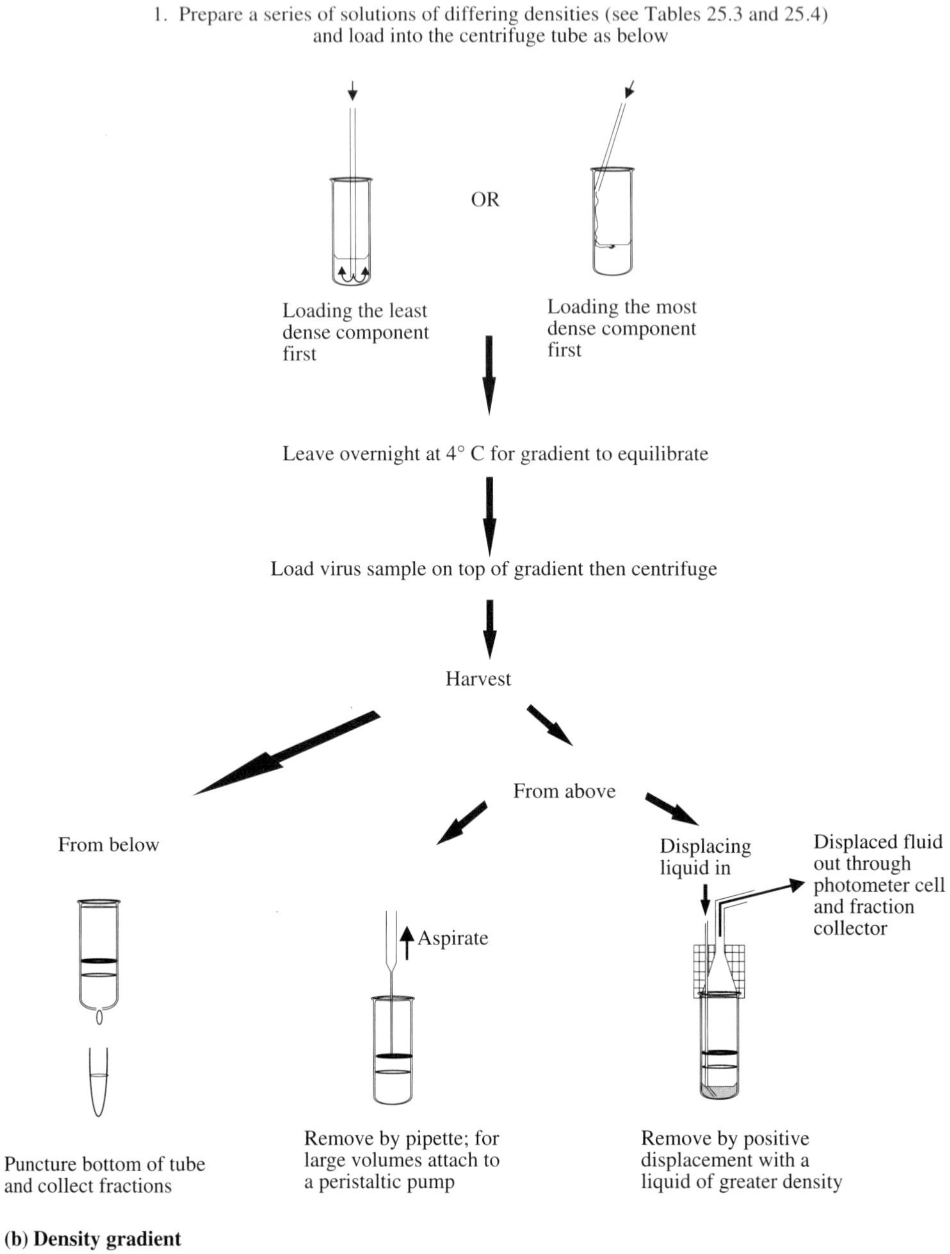

1. Prepare a series of solutions of differing densities (see Tables 25.3 and 25.4) and load into the centrifuge tube as below

OR

Loading the least dense component first

Loading the most dense component first

Leave overnight at 4° C for gradient to equilibrate

Load virus sample on top of gradient then centrifuge

Harvest

From above

From below

Aspirate

Displacing liquid in

Displaced fluid out through photometer cell and fraction collector

Puncture bottom of tube and collect fractions

Remove by pipette; for large volumes attach to a peristaltic pump

Remove by positive displacement with a liquid of greater density

(b) Density gradient

Figure 25.7(b)

Determining the protein concentration of purified virus

BCA PROTEIN ASSAY

The BCA method of protein assay can be used for determining the protein concentration of purified virus. It was developed by Pierce (UK) Ltd and is based on the method of Lowry *et al.* (1951) and is simpler to use. The further advantage claimed is that it is compatible with non-ionic detergents and with a number of other compounds that interfere with the Lowry method. It is supplied by Pierce (UK) Ltd as Reagent A (containing Na_2CO_3, BCA and NaOH) and Reagent B (4% $CuSO_4.5H_2O$), which have a shelf life of at least 6 months at room temperature. The working reagent (50 volumes A + 1 volume B) is stable for 1 week at 4°C.

Materials

1. Standard aqueous solution of bovine plasma albumin (BPA, 1 mg/ml (1 μg/μl))
2. Test sample(s)
3. Glass-distilled or deionized water
4. Chemically clean glass tubes, 10 ml capacity
5. Spectrophotometer or Corning 252 colorimeter
6. Pipettes, pipettors and tips capable of delivering volumes of 4 ml and 20–200 μl, respectively.

Method

1. Add ingredients as indicated in Table 25.5. Mix well and incubate in a water bath at 37°C for 30 min.
2. Cool and measure the absorbance at 540 nm.
3. Plot absorbance (*y* axis) against concentration (*x* axis) for the standard (BSA) sample. Construct the best straight line by eye, or by linear regression. Use this line to determine the concentration of the sample (Figure 25.8).

This method can be scaled down to employ microtitre plates (Microtest III flexible assay plate, 96 U-bottom wells, approximately 0.3 ml volume per well) and a microtitre plate reader (e.g. Titertek Multiscan). In this instance, 10 μl volumes of standard protein and test protein are reacted with 200 μl of working reagent.

Provided that the sample of virus is pure, the concentration of OBs can be determined by reference to the reaction with standard virus for which

Table 25.5 BCA protein assay for estimation of protein concentration: volumes and order of addition of the various components

Reagents and order added	Tubes for standard BPA						Tubes for test sample	
Standard BPA (1 mg/ml) (μl) ↓	0	40	80	120	160	200	0	0
Test sample (suitably diluted) (μl) ↓	0	0	0	0	0	0	200	200
Glass-distilled water (μl) ↓	200	160	120	80	40	0	0	0
BCA Working Reagent (ml)	4	4	4	4	4	4	4	4
Protein concentration (μg)	0	40	80	120	160	200	[a]	[a]

[a] To be determined from the standard curve (Figure 25.8).

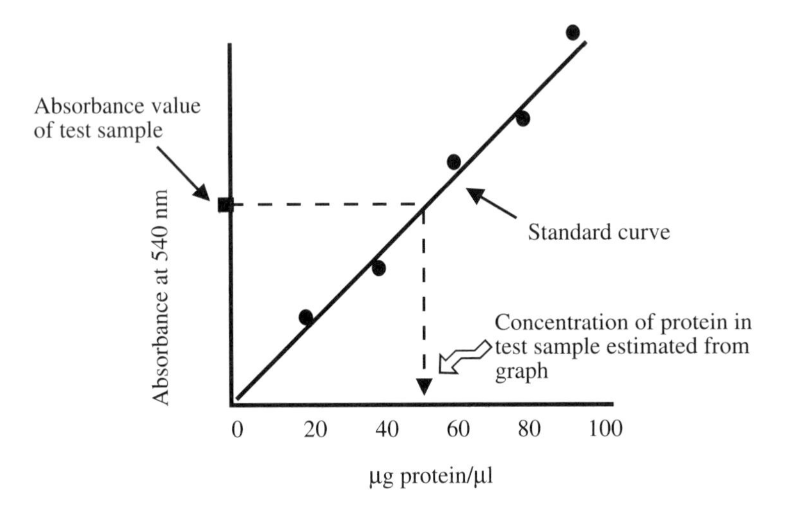

Figure 25.8 Standard curve for protein estimation by the BCA method.

the number of OBs has been determined by light or electron microscopy. For GVs a concentration of 10 mg virus protein per ml is equivalent to approximately 5.0×10^{11} OB/ml.

RECOVERY OF BACULOVIRUSES FROM INFECTED INSECTS

The following method (Dulmage, Correa and Martinez, 1970), which uses acetone to recover BVs from infected insects, has the advantage that expensive centrifuge equipment is not necessary. The final product will, however, have a higher level of impurities than for centrifuge preparations. The infectivity of BVs prepared by lactose/acetone precipitation should always be carefully checked against control samples.

RECOVERY OF *TRICHOPLUSIA NI* NPV BY CO-PRECIPITATION WITH ACETONE

Method

1. Macerate infected larvae (e.g. in Waring blender) in water allowing about 0.5 ml water per larva.
2. Filter through fine cloth to remove insect debris and preserve filtrate.
3. Dissolve 6 g lactose/100 ml filtrate.
4. While slowly stirring, add 4 volumes of acetone.
5. Continue to stir for 30 min.
6. Allow to stand for 10 min.
7. Filter with suction and discard the filtrate.
8. Stir the residue with a small volume of acetone.
9. Filter with suction.
10. Again stir the residue with a small volume of acetone.
11. Filter with suction.
12. Allow the residue to dry overnight.

In the original study, this method yielded 6.3×10^{10} OB/g, and an estimated 81% OB recovery from the larvae.

EXTRACTION OF VIRAL DNA

Viral DNA can be extracted from infected larvae or from cells in cell culture.

EXTRACTION OF VIRUS DNA FROM INFECTED LARVAE

One of the most common reasons for extracting virus DNA from infected larvae is to determine its REN profile. This is an essential aid in identifying and distinguishing between isolates of virus. The method given below is for extraction and purification of virus (Figure 25.6) and virus DNA from a single larva. For other methods of preparation of purified virus see Table 25.2 and Figures 25.6 and 25.7.

Materials and equipment

Items marketed with an asterisk should be sterile.

1. Infected larva
2. Water*
3. Vortex mixer
4. 1 M $NaCO_3$ (prepared daily)
5. Microcentrifuge tubes (1.5 ml, e.g. Scotlab, No. 96.4232.902)*
6. Microcentrifuge bench centrifuge
7. Tissue grinder*
8. Micropipettes calibrated to deliver the required volumes
9. Pipette tips*
10. Dialysis tubing
11. Waterbath at 37°C
12. EDTA, 0.5 M (stored at room temperature or –20°C)
13. Proteinase K, 20 mg/ml (stored at –20°C)
14. Sodium dodecyl sulphate (SDS, sodium lauryl sulphate, stored at room temperature)
15. Dialysis buffer (freshly prepared): 1 M Tris buffer, pH 8.0, 5 ml; 0.5 M EDTA, 1 ml; glass-distilled water, 500 ml
16. Phenol equilibrated to pH >7.8 and stored at 4°C (purchased as such or prepared as given below)
17. Glass beaker (600 ml) for dialysis buffer
18. Dialysis chambers made from the caps of microcentrifuge tubes (Figure 25.9)
19. Phenol:chloroform plus isoamyl alcohol (1:1) purchased as such or prepared by mixing together equal volumes (150 μl of each per sample) of phenol and chloroform : isoamyl alcohol (24:1), centrifuging in a bench centrifuge at 6500 rpm for 1 min and removing the aqueous bubble
20. Chloroform:isoamyl alcohol (24:1) stored at room temperature
21. Pair of forceps*

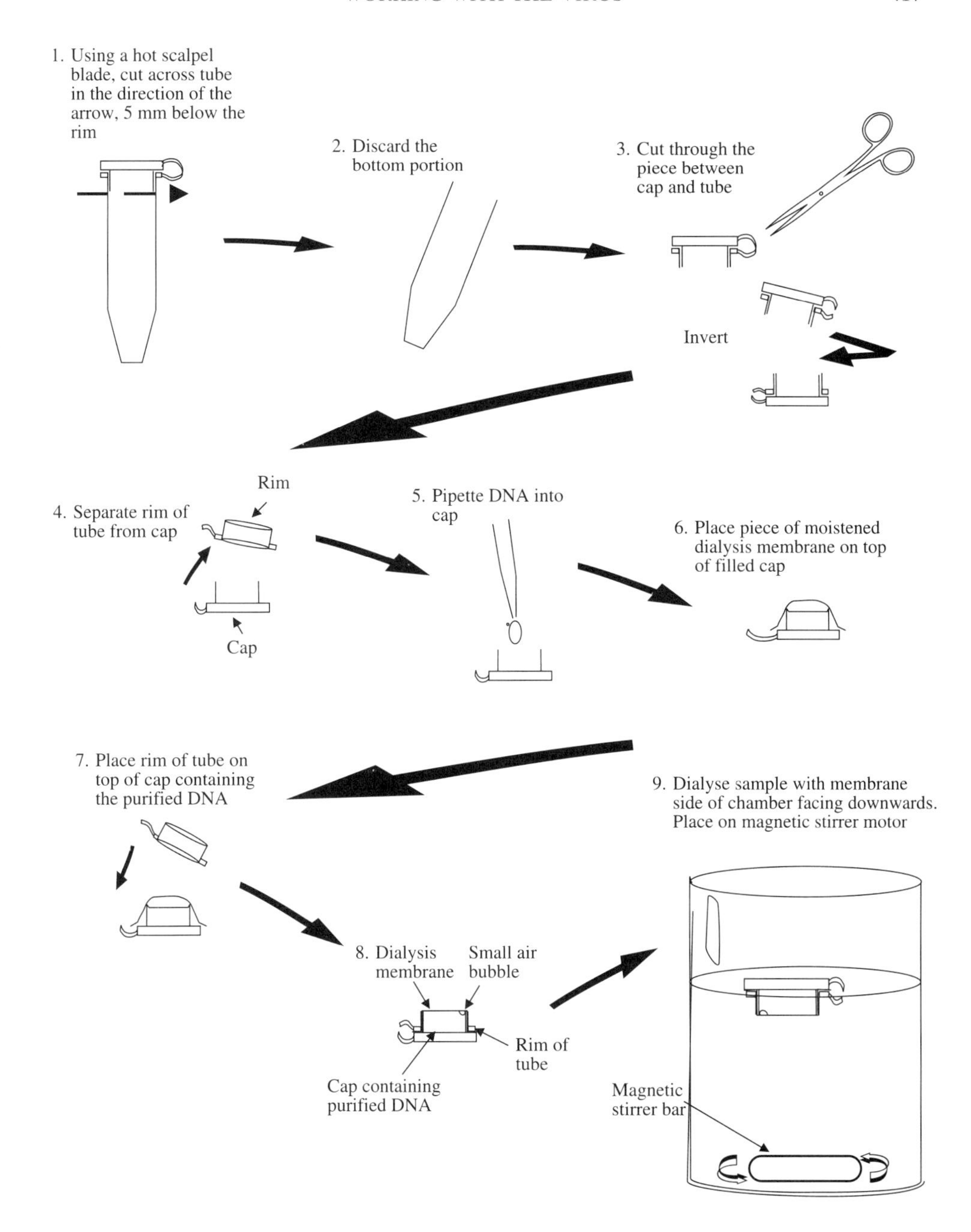

Figure 25.9 Construction of a dialysis chamber from a microcentrifuge tube.

22. Magnetic stirrer bar (4 mm length) and motor
23. Surgical blades.

Method

1. Homogenize a larva in a tissue grinder, or equivalent, with approximately 500 μl 0.1% SDS.
2. Transfer homogenate to a 1.5 ml microcentrifuge tube and sediment larval debris by centrifuging for approximately 15 s at 3250 × *g* (6550 rpm in a microcentrifuge bench centrifuge). Transfer the supernate containing virus to a fresh microcentrifuge tube.
3. Add approximately 300 μl 0.1% SDS to the pellet. Resuspend by vigorous mixing and centrifuge at 3250 × *g* for 15 s. Add the supernate to that obtained in Step 2. Discard the tube and pellet for decontamination and safe disposal.
4. Sediment OBs by centrifuging at 11 600 × *g* for 20–30 s (NPV) or 3 min (GV).
5. Remove supernate and keep, as it may contain some non-sedimented OB.
6. Resuspend the pellet of OBs, which is usually whitish in colour in 750 μl water. Centrifuge again for 20 s (NPV) or 3 min (GV) at 11 600 × *g* (high speed in a microcentrifuge bench centrifuge). Gently remove the supernate and any insect debris that collects as a darker ring above the pellet. Avoid disturbing the latter.
7. Resuspend the pellet in 155 μl water and remove 5 μl for storage at –20°C and subsequent passage. Add 30 μl 0.5 M EDTA and 3 μl proteinase K (digests any nucleases present) to the remainder.
8. Incubate at 37°C for 90 min then add 17 μl 1 M $NaCO_3$ to disrupt the OBs and liberate virus particles. The suspension should become clear. If it does not do so immediately, replace the microcentrifuge tube at 37°C for approximately 20 min. If the suspension still remains cloudy, add a further 10 μl 1 M $NaCO_3$. Incubate a further 20 min, or longer (several hours), as necessary. The time of incubation is not critical.
9. Add 20 μl 10% SDS and incubate for 15 min at 37°C to disrupt the virus particles and release the DNA. During the incubation period, cut pieces of dialysis tubing approximately 25 mm in length and dip in dialysis buffer. Once moistened, slit the sides of each piece of dialysis tubing by holding with forceps and cutting with a scalpel, to make two membranes of equal size (one per sample of DNA). Soak in buffer for 10–15 min.
10. Centrifuge the mixture of virus and SDS at 11 600 × *g* for 20 s (NPV) or 3 min (GV). Collect and transfer the supernate containing the viral

DNA to a fresh microcentrifuge tube, add 2 μl phenol and mix thoroughly by gently tilting backwards and forwards. Centrifuge at 3250 × *g* for 1 min. Transfer the top, aqueous layer containing the DNA to a fresh microcentrifuge tube using a pipette. Take care not to transfer any phenol with it. If the DNA is very concentrated, the phenol will tend to be drawn into the pipette with it. In this instance it is better to leave a small (10–20 μl) volume of aqueous phase above the phenol, thus reducing the chance of contaminating the DNA.

11. Add 200 μl of phenol : chloroform mixture, mix gently and centrifuge at 3250 × *g* for 1 min. Transfer the aqueous phase to a fresh microcentrifuge, again taking care not to contaminate it with any of the chloroform : phenol mixture.
12. Add 200 μl chloroform : isoamyl alcohol mixture, mix gently and centrifuge at 3250 × *g*. Transfer the upper, aqueous phase containing the DNA (100 μl) to a dialysis chamber (Figure 25.9).
13. Using forceps, cover the chamber with a piece of dialysis membrane that has been soaking in dialysis buffer and fasten in position (Figure 25.9). Place, membrane side downwards, in a beaker of dialysis buffer with a magnetic stirrer bar. Place on a magnetic stirrer and stir gently for at least 24 h with three changes of buffer.
14. Remove dialysis chamber from the buffer, blot with paper tissue to soak up excess buffer and prevent contamination of sample. Carefully make a slit in the membrane with a clean surgical blade and transfer the dialysed DNA to a fresh microcentrifuge tube.
15. Store DNA at 4°C rather than frozen.

Note 1: if the concentration of DNA is too low, e.g. because the larvae are small, halve the volumes of reagents used from Step 7 onwards.

Note 2: if the quality of the DNA is not satisfactory, insert a further purification step between Steps 6 and 7. Load a suspension of virus from Step 6 onto a sucrose or glycerol gradient, centrifuge (Table 25.2), harvest the relevant bands, sediment, wash and proceed to Step 7.

EXTRACTION FROM CELL CULTURE

Materials and equipment

Items marked with an * should be sterile.

1. Culture of cells infected with virus
2. Refrigerated bench centrifuge

3. Centrifuge tubes (12 ml, Falcon 2059 or similar)*
4. Phosphate-buffered saline (PBS, Dulbecco A): NaCl, 8.0 g; KCl, 0.2 g ; Na_2HPO_4, 1.15 g; KH_2PO_4, 0.2 g; water, up to 1 l, or Oxoid BR14a Dulbecco 'A' tablets
5. RNA extraction buffer (NaCl, 0.14 M; $MgCl_2$, 1.5 mM; TrisCl (pH 8.6), 10 mM, Nonidet P40 (NP40), 0.5%)
6. Tris-EDTA buffer (TE): Tris-HCl (pH 8.0); EDTA (pH 8.0), 1.0 mM; in water
7. Items 5–23 as listed as *Materials and equipment* for Extraction of virus DNA from infected larvae (above).

Method

1. Dislodge cells from the surface of the culture vessel (e.g. 50 cm^2) with a rubber policeman and decant into a 12 ml centrifuge tube.
2. Centrifuge at $3250 \times g$ for 5 min (4°C).
3. Carefully pour off supernate and discard.
4. Resuspend pellet in 0.5 ml ice-cold PBS.
5. Transfer suspension to a microcentrifuge tube and centrifuge at $3500 \times g$ for 30 s (microcentrifuge bench centrifuge).
6. Remove and discard supernate.
7. Resuspend pellet in 200 μl RNA extraction buffer.
8. Vortex for 15 s and centrifuge at $6500 \times g$ for 90 s.
9. Remove and discard supernate.
10. Resuspend pellet in 148 μl TE buffer.
11. Add 40 μl EDTA, 0.5 M; 10 μl 10% SDS; 2 μl proteinase K (20 mg/ml).
12. Incubate at 37°C for 2 h.
13. Extract using phenol; phenol : chloroform and chloroform : isoamyl alcohol and prepare for dialysis as described in Steps 10–13 of the *Method* for Extraction of virus DNA from a single larva (above).
14. Dialyse for 12 h in TE with three changes of buffer.
15. Collect sample of DNA as described above (in the *Method* for Extraction of virus DNA from a single larva, Step 14).

PREPARATION OF PHENOL SOLUTION EQUILIBRATED TO PH >7.8

A method for preparing equilibrated phenol is given for completeness. From the safety point of view, however, it is recommended that it is purchased in the pre-equilibrated form.

Materials and equipment

1. Tris-base (1 M Tris buffer pH 8.0)
2. Glass-distilled water (water)
3. Phenol
4. Concentrated HCl
5. Glass beakers, 1 l and 200 ml
6. Pipettes, 25 ml
7. Gloves and face mask
8. Volumetric flask, 1 l
9. Dark-coloured glass bottle and stopper for storing phenol
10. 8-Hydroxyquinoline.

Method

1. Prepare a 1 l solution of 1 M Tris buffer (pH 8.0) as follows: dissolve 121.1 g Tris-base in 800 ml water. Adjust the pH to 8.0 by adding approximately 42 ml of concentrated HCl. Make the volume up to 1 l with water.
2. Wearing gloves and face mask, weigh 50 g of phenol into a beaker.
3. Dissolve phenol in 50 ml 1 M Tris (pH 8.0), cover beaker and place on a magnetic stirrer. Stir for approximately 20 min.
4. Remove beaker from stirrer, allow the two phases to separate, remove upper aqueous layer by aspiration and discard safely.
5. Transfer phenol solution (lower layer) to a storage bottle and add 50 ml 1 M Tris (pH 8.0). Mix the phases and allow them to settle once more.
6. Remove and discard the upper phase. Add 50 ml of 1 M Tris (pH 8.0) to the lower (phenol) layer, mix and allow layers to separate. Remove and discard the upper layer as before. Check that the pH is greater than 7.8 with pH indicator paper. If it is, proceed to Step 7. If not, repeat Step 6 once more.
7. Repeat Step 6 twice more but with 50 ml of 0.1 M Tris (pH 8.0) instead of 1 M Tris (pH 8.0). Finally add 20 ml 0.1 M Tris (pH 8.0) and 50 mg

8-hydroxyquinoline (final concentration 0.1%). Shake to dissolve and leave to settle.

8. Store at 4°C in a light-proof bottle; this solution can be used for 1–2 months. For use over longer periods of time store at –20°C.

Precautions Wear gloves, protective clothing and safety spectacles when handling phenol and chloroform. All manipulations should be done in a fume cupboard. Waste phenol and chloroform should be collected separately from other chemicals and should be disposed of safely. Phenol is inactivated by light and should be stored in a dark coloured bottle.

RESTRICTION ENZYME DIGESTION OF DNA

Once a sample of DNA has been extracted it can be treated with the chosen restriction enzymes. The method for digestion with the restriction enzyme *Eco*RI is given here and can be adapted for other enzymes.

DIGESTION WITH *ECO*RI

Equipment and materials

1. Sample of DNA (0.1 μg/ μl or 500 ng)
2. The required restriction endonuclease and reaction buffer supplied with the enzyme, e.g. *Eco*RI and Rct 3
3. Bromophenol Blue loading dye (50% glycerol; 0.1 M EDTA; 2% Bromophenol Blue, stored at room temperature)
4. Waterbath at 37°C
5. Agarose
6. TAE (Tris-acetate/EDTA) electrophoresis buffer (50 × working strength, containing Tris base, 242 g; glacial acetic acid, 57.1 ml; 0.5 M EDTA (pH 8.0))
7. Water
8. Electrophoresis equipment
9. Autoclave tape
10. Powerpack
11. Ethidium bromide, 0.5% aqueous solution
12. Transilluminator

13. Safety mask to protect against UV irradiation (from item 12 above)
14. Boiling water bath or equivalent
15. Polaroid camera, film and stand
16. The appropriate automatic pipettors and pipette tips (sterile)
17. Microcentrifuge tube (sterile)
18. Latex examination gloves or equivalent (Sherwood Surgical Supply, Peacehaven, East Sussex, BN10 8JQ, UK).

Method

1. Add 0.3 μl *Eco*RI (10 units/μl, stored at –20°C) and 0.7 μl Rct3 buffer (stored at room temperature) to 6 μl virus DNA in a microcentrifuge tube. Incubate at 37°C for 1.5–2 h.
2. Stop reaction by adding 0.9 μl Bromophenol Blue (final concentration approximately 13%).
3. Seal the ends of an electrophoresis tray with autoclave tape to form a receptacle for molten agar. Place comb in the tray.
4. Add 0.28 g agarose and 800 μl 50× TAE electrophoresis buffer to 39.2 ml water (to give a 0.7% gel) and heat in a conical flask in a water-bath until the agarose has dissolved. Cool to approximately 55°C and pour into the electrophoresis tray prepared above.
5. Leave to set for approximately 20 min. While setting, fill an electrophoresis tank with 1× TAE buffer (4 ml 50× TAE and 392 ml water). The same buffer can be used three to four times before being discarded.
6. Once the agar has set, remove autoclave tape and comb and place gel on tray in tank ensuring that buffer covers the wells.
7. Load gel by placing the tip of a pipette containing the sample of DNA (total volume 7.9 μl) into a well and expelling gently. Take care not to trap air bubbles in the well and not to contaminate adjacent wells.
8. Switch on apparatus and electrophorese sample at 60 V until the Bromophenol Blue dye is about three-quarters of the way down the gel. Switch off.
9. Using a plastic spatula, gently slide gel from tray and immerse in a bath of 0.5% ethidium bromide. Leave to stain for a minimum of 30 min. Maximum staining is achieved after 1 h. The gel will not overstain if left longer but the DNA will begin to diffuse out of the gel after 18–24 h.
10. Wearing protective gloves, slide gel onto transilluminator, put on protective mask, turn on source of illumination and view gel for bands of digested DNA. If satisfactory, and if required, take a photograph.

Follow the manufacturer's recommendations for correct exposure and development times.

11. Allow gel to dry before disposing of it by incineration.

Precautions

1. Restriction enzymes are unstable and must be kept at –20°C. Take out of the refrigerator momentarily when removing a sample.
2. Ethidium bromide is a powerful mutagen and is moderately toxic. It must be treated with great caution. Gloves must be worn when working with this substance. Solutions of ethidium bromide must be decontaminated before being discarded. Two methods for doing this are given below.
3. Ethidium bromide decomposes at above 262°C. Standard incineration conditions are usually sufficient to render it harmless.
4. Slurries of Amberlite XAD-16 or activated charcoal can be used to decontaminate surfaces that become contaminated with ethidium bromide.

DECONTAMINATION OF CONCENTRATED (>0.5 MG/ML) SOLUTIONS OF ETHIDIUM BROMIDE

Method

1. Add sufficient water to reduce the concentration of ethidium bromide to <0.5 mg/ml.
2. Add 0.2 volume of fresh 5% hypophosphorous acid and 0.12 volume of fresh 0.5 M sodium nitrite. Mix carefully. Check that the pH of the solution is less than 3.0.
3. Leave at room temperature for 24 h.
4. Add a large excess of 1 M sodium bicarbonate.
5. Discard solution.

DECONTAMINATION OF DILUTE (<0.5 MG/ML) SOLUTIONS OF ETHIDIUM BROMIDE

This method can be used for decontaminating electrophoresis buffer, for example, containing 0.5 μg/ml ethidium bromide.

Method

1. Add 100 mg powdered activated charcoal for each 100 ml of solution.
2. Leave at room temperature for 1 h with intermittent shaking.
3. Filter solution through a Whatman No. 1 filter.
4. Discard the filtrate.
5. Seal the filter and activated charcoal in a plastic bag. Dispose of safely.

DETECTION AND IDENTIFICATION OF VIRUS DNA BY THE POLYMERASE CHAIN REACTION[†]

Techniques for the identification of virus by REN analysis are well established and can be used to identify species and strains of NPV or GV. However, the relatively large amount of viral DNA required for this technique often means that low level infections and the presence of latent virus remain undetected. In contrast, detection of a particular viral gene sequence by the polymerase chain reaction (PCR) requires only a small quantity of target DNA. A single copy of a gene can, in theory, provide sufficient starting material even when it is part of a highly complex sample.

The PCR technique exploits the ability of DNA polymerase to synthesize many complementary strands of DNA from a very small amount of DNA template. The DNA sequence to be amplified is identified and two short oligonucleotide sequences (primers) are constructed, each being complementary to one or other of the 3′ ends of the template sequence (Figure 25.10). By subjecting the reaction mix to cycles of heating and cooling at selected temperatures, for the appropriate periods of time, amplification of the template DNA can take place (Figure 25.10). The original dsDNA is first denatured at a high temperature so that the two strands separate. Following this, the temperature is reduced allowing the two primers to anneal to complementary strands and synthesis of a new copy of the DNA to take place across the region flanked by the primers, beginning at the 3′ end of each primer. Each strand has now been copied to beyond the primer site on the opposite strand and the amount of target DNA to which the primers can

[†] Contributed by Margaret Brown.

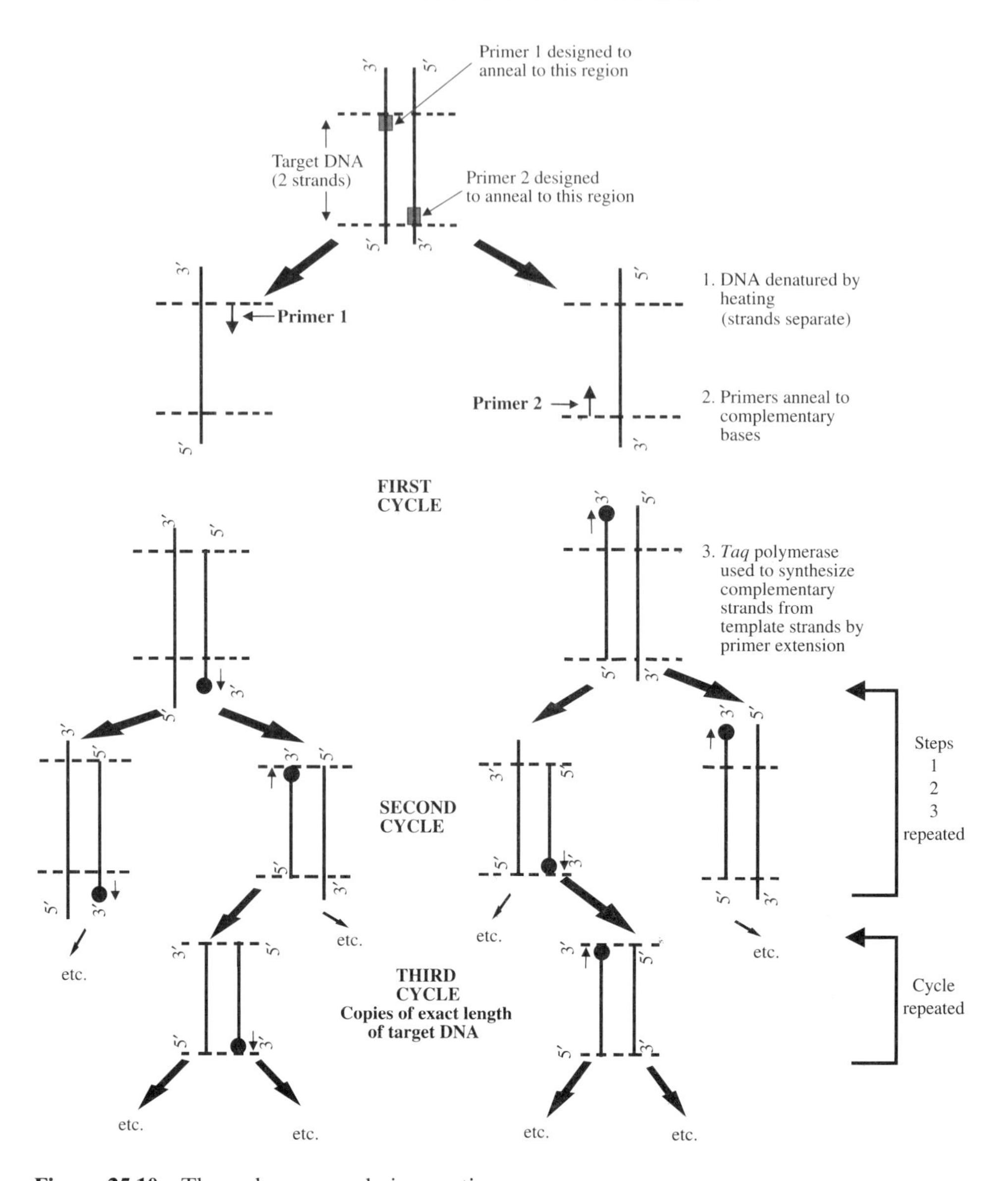

Figure 25.10 The polymerase chain reaction.

anneal has been doubled. By cycling through the separation, annealing and synthesis (or extension) temperatures n times, it is theoretically possible to produce 2^n copies of the DNA region between the primers.

At the end of the reaction the resulting PCR (or amplification) product can be electrophoresed and visualized on an agarose gel. The successful amplification of a DNA fragment will be indicated by a discrete band of the same size as the target length, i.e. the sequence flanked by the primers.

PCR materials and methods

In addition to the DNA sample and oligonucleotide primers, other reaction components are required. Deoxynucleotide triphosphates (dNTPs) provide the raw material for synthesis and are usually provided in concentrations of 50–200 μM each of dATP, dCTP, dGTP and dTTP. About 2 units of *Taq* DNA polymerase per 100 μl reaction is generally recommended, but a range of concentrations should be assessed. *Taq* polymerase is usually supplied with the appropriate buffer containing KCl, Tris-HCl, $MgCl_2$ and Triton X-100. However, the final $MgCl_2$ concentration may need to be altered, often by increasing it. Although 0.5–2.5 mM $MgCl_2$ is usual, EDTA and other inhibitors in the DNA template may affect the optimum concentration. This is particularly so for one-step extracts from larvae or crude lysates.

The design of primers to detect NPV or GV is dependent on knowing the sequence of at least part of the virus genome. At the present time there are few published sequences for regions other than the polyhedrin and granulin genes and most primer designs are based on these. By comparing known sequences of the polyhedrin gene for a range of BVs it should be possible to design primers specific to a particular virus or to a highly conserved region within the polyhedrin gene to use for the non-specific identification of a range of viruses.

There are a number of basic rules for the design of primers and it is often necessary to try more than one pair. The length of the primer for each end of the chosen template section is usually from 18–28 nucleotides with a 50–60% G + C content. Certain features should be avoided, i.e. complementarity and runs of three or more G or C nucleotides at the 3′ ends, palindromic sequences within the primers and significant secondary structures in the template DNA. Further information on primer design is readily available in publications dealing specifically with PCR (Rohrman, 1986; Innis *et al.*, 1990; Hughes, Possee and King, 1993; Noguchi, Kobayashi and Shimada, 1994). Once the primer sequences have been chosen the nucleotide sequences can be synthesized by any of a number of specialist companies (e.g. Cruachem Ltd, Glasgow, UK; R & D Systems Europe Ltd, Abingdon, UK).

It is important to optimize the conditions for each new application of the PCR technique since the specificity and the amount of amplification product can be dramatically affected by the concentration of the components, the annealing temperature and the time allowed for each stage of thermal cycling. The way in which different parameters affect the outcome should be fully understood before commencing.

A PCR PROTOCOL USED FOR DETECTION OF NPV

Pre-sterilized pipette filter tips and microcentrifuge tubes should be used.

Method

1. Make up a reaction mix of all the reagents except the DNA (Table 25.6). The quantities used should be sufficient for the number of samples

Table 25.6 Reagents for the PCR protocol

Reagent	Sample (μl)
Sterile water	33.6
10 × PCR buffer	5.0
$MgCl_2$, 25 mM	3.0
dNTPs, 2 mM	2.5
Primer 1, 20 μM	0.5
Primer 2, 20 μM	0.5
Taq polymerase (10 units/ml)	0.4
DNA	5.0

being reacted, plus two: one for a control and one to allow for pipetting errors.

2. Pipette 45 μl of reaction mix into clean, sterilized 0.5 ml microcentrifuge tubes: one for each sample, plus a control.
3. Add 5 μl of virus DNA to each tube, except the control.
4. Add a drop of mineral oil to each tube to prevent evaporation during the thermal cycling.
5. Centrifuge briefly.
6. Place the tubes into a thermal cycling machine programmed as follows:
   (a) Initial denaturing 94°C 2 min
   (b) Denaturing 94°C 1 min
   (c) Annealing * 60°C 1 min
   (d) Extension * 72°C 2 min
   (e) To ensure dsDNA 72°C 6 min

   Repeat Steps (b)–(d) for 40 cycles. (*See notes at end of method on these steps.)
7. Remove to 4°C until required.
8. To view the PCR products, make a 1.5% gel with agarose and 0.5× TBE buffer pH 8.5 (Tris base, 5.4 g; boric acid, 2.75 g; 0.5 M EDTA (pH 8.0), 2 ml; made up to 1 l with water). To do this, pour the molten agarose dissolved in TBE buffer into a prepared mould (see manufacturer's instructions), placing a comb at one end to form wells.
9. Once the agarose has set, remove end plates or tape and place in an electrophoresis tank containing 0.5× TBE buffer.
10. Pipette a 3 μl drop of 5× loading buffer, one for each sample, onto a piece of parafilm. Add 12 μl of each PCR product, or the control, to the drops, including as little oil as possible.
11. Load samples from parafilm into the wells. Include a molecular weight marker, made up according to the manufacturer's instructions, to act

as a standard for measuring the size of the amplification product. Electrophorese at approximately 90 V for about 1.5 h. (The timing will need to be optimized.)

12. Remove from the electrophoresis tank and stain for 30 min with ethidium bromide (0.1 μg/ml) in 1 × TBE buffer. View and photograph on a UV transilluminator.

Annealing (Step 6c) The annealing temperature should be determined by experimentation. It will depend on the primer composition, length and concentration. For example, a higher temperature reduces the possibilities of mismatch between primer and template, i.e. improved specificity.

Extension (Step 6d) The time required can be estimated at 35–100 nucleotides/s, depending on template structure, salt concentration, etc. Longer times may be beneficial in the early stages if DNA concentration is very low. If the cycling machine does not contain a control tube it is important to remember that there will be a time lag before the temperature in the reaction tubes reaches the temperature of the heating block.

Number of cycles (Step 6b–d) Too many cycles increases the amount of non-specific products caused by mismatches. No more than 45 cycles should be required.

DNA concentrations Reactions should be carried out initially for a range of concentrations of DNA. The optimum amount of DNA should be contained within 2–5 μl of solution to reduce the effect of pipetting errors that may occur when working with very small volumes.

VIRAL PURIFICATION BY CONTROLLED GLASS CHROMATOGRAPHY

PURIFICATION OF *ORYCTES RHINOCEROS* VIRUS

Although not a BV, OrV has been used very successfully as a component in the integrated management of its host species. Also, it has been demonstrated to be cross-transmissible to a number of other dynastid pests (Lomer, 1986; Zelazny *et al.*, 1988) notably *Orcytes monoceros*, another pest of palm, and a species of *Papuana*, a genus that comprises some of the most important root crop pests in the Western Pacific and elsewhere. A method for its purification is, therefore, included. This method, which utilizes controlled pore glass chromatography (CPG) (Porath and Flodin, 1959; Steere, 1963; Haller, 1965,

1967; Barton, 1977), dispenses with the use of centrifugation techniques and utilizes inexpensive, reusable materials. The virus product is stable for 4 weeks at tropical room temperature and, at 4°C, possibly indefinitely.

The technique, as described here, is the outcome of work on the preparation of stabilized extracts of virus (Zelazny, Alfiler and Crawford, 1985, 1987; Crawford, 1988) and was successfully employed by W. R. Carruthers, P. F. Entwistle, J. S. Cory and C. Prior in 1989 (unpublished data) for the introduction of virus to a major infestation of *O. rhinoceros* in Oman.

The methodology described here was developed for the extraction and stabilization of OrV from *O. rhinoceros* midgut but is thought to be equally suitable for recovery of virus from the larval stages, most of the tissues of which are affected.

Propagation of virus

Materials

Items marked with an asterisk should be sterile.

1. Virus inoculum
2. Micropipette (10 μl) and tips*
3. Beetles
4. Plastic beaker*
5. Water*
6. Banana.

Preparation of inoculum

1. Aseptically add 0.5 g sucrose (commercial white cane sugar is adequate) to every 10 ml virus suspension (prepared as described below).
2. Using a 10 ml syringe fitted with a pre-sterile membrane filter, pore size 45 μm, dispense 0.5 ml volumes into sterile plastic vials and store at 4°C pending use.

Note: where an electron microscope is available, the concentration of virus particles can be counted (Chapter 26) and, if required, diluted in 0.05 M Tris/HCl buffer, pH 7.4. A dose of 1×10^4 virions administered in 10 μl of 10% sucrose solution is lethal for adults (Lomer, 1986).

Inoculation of beetles

1. Working with one beetle at a time, hold it on its back and drop a 10 μl dose of virus onto its mouthparts. Allow it to drink the inoculum before

placing it in a beaker containing 3–5 ml water and a small piece of banana.

2. Incubate in the dark at 25°C.
3. Change water every other day and store used water (containing excreted virus) at –20°C.
4. Transfer beetles to clean beakers and add fresh banana as required.
5. Freeze dead beetles at –20°C until virus is harvested.

Harvesting virus-infected guts

Materials

Items marked with an asterisk should be sterile.

1. Dissecting instruments*
2. Dissecting tray
3. Ethanol (75%)
4. Phosphate-buffered saline (PBS, Dulbecco A; p. 420)
5. Petri dish*
6. Infected adult(s)
7. Ice bucket with ice.

Method

1. Surface sterilize the dissecting tray by swabbing with ethanol.
2. Swab beetle (previously killed by freezing) with alcohol and place in dissecting tray.
3. Sever the elytra and wings of the carapace with scissors and remove the carapace just in advance of the scutellum to the apex of the abdomen.
4. Remove the gut from the anterior to the posterior suture after parting the fat body and, in the case of females, the ovariole tissue.
5. Place in PBS held over ice. Infected guts appear swollen, opaque and white. Those of healthy adults are semi-translucent, pale brown and linear in appearance. Guts may be frozen in PBS pending virus extraction. If desired, a small portion of putatively infected gut can be tested by crossover gel electrophoresis (see below) to confirm presence of virus. A sample of purified virus should be included for comparison.

Cross-over gel electrophoresis

Equipment and materials

1. Power pack and electrophoresis tank with two bridges
2. Cellogel strips
3. Barbitone buffer (10.3 g/l barbitone and 1.34 g/l barbitone sodium, pH 8.6)
4. Virus-specific serum in PBS
5. Physiological saline (0.85% w/v)
6. Naphthalene Black (Acid Black 2), 0.5% (w/v) in methanol : acetic acid : water (45 : 10 : 45)
7. Methanol:acetic acid:water (47.5 : 10 : 45).

Preparation of cellogel

1. Prepare barbitone buffer by dissolving 10.3 g barbitone and 1.34 g barbitone sodium in 1 l of water. Check pH and adjust if necessary.
2. Allow cellogel strips (two per tank) to sink into the buffer and soak for a minimum of 10 min.
3. Fill electrophoresis tank with buffer, blot cellogel strips gently and place, porous side up, over the bridge and with their ends dipping in the buffer.

Preparation and electrophoresis of gut samples

1. Thaw and snip off a small portion of gut and add 25 μl PBS.
2. Freeze and thaw twice (in dry ice and ethanol if available) and vortex.
3. Place 5 μl of this antigen and 5 μl of a one-half or one-quarter dilution (or other appropriate dilution) of antiserum 1.0–2.5 cm apart on a strip of cellogel with the antiserum at the anodic end. Include a sample of purified virus particles as a control.
4. Electrophorese for 30 min at 200 V.
5. Remove gels from equipment and wash for 20 min in two changes of physiological saline.
6. Stain for 2 min in Naphthalene Black solution.
7. Destain in methanol : acetic acid solution until bands are apparent.
8. Store in 5% acetic acid.

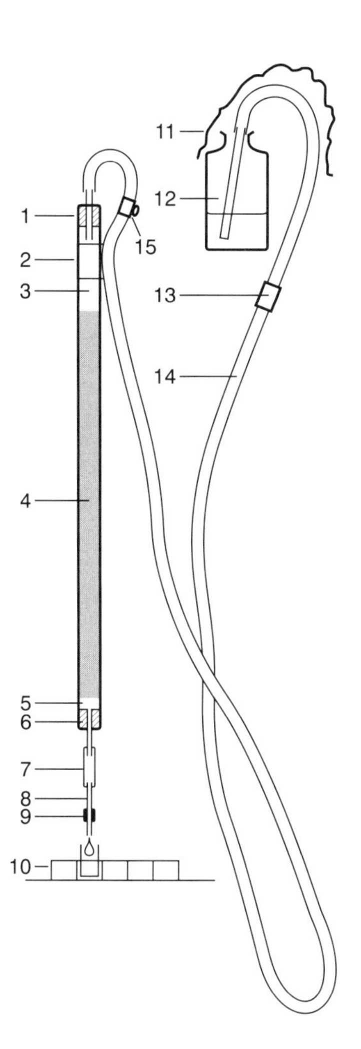

Figure 25.11 CPG bead column prepared for separation of OrV from suspensions of macerated infected adult midgut tissue. 1, Bung; 2, tubing taped to column; 3, 10 cm buffer; 4, CPG beads; 5, glass wool; 6, bung; 7, thicker plastic tube; 8, fine plastic tube; 9, plastic clamp to close column; 10, rack with sterilized fraction collection bottles; 11, loose foil cover; 12, buffer reservoir; 13, plastic clamp; 14, plastic tube; 15, screw clamp to control flow.

Setting up controlled pore glass column

Materials and equipment

1. CPG column with components as shown in Figure 25.11. The glass tubing is 1 m long and has an internal diameter of 15 mm
2. Controlled pore glass, mesh size 120–200 (Sigma)

3. 2 M HCl
4. Phosphate buffer (0.05 M, pH 7.0)
5. Phosphate buffer (0.05 M, pH 7.0) containing 1.0% w/v PEG 20 m (BDH)
6. Vacuum pump (tap or motor driven)
7. Buchner funnel.

Phosphate buffer 0.05 M This is prepared using solution A (305 ml) and solution B (194 ml) (in Method on p. 394) made up to 1 l with water. Add 250 ml of this buffer to 1 l of water to make 0.05 M buffer. Check the pH and adjust if necessary.

Method

1. Wash 150 ml of CPG three or four times with 200 ml phosphate buffer (0.05 M, pH 7.0). Finally resuspend with 400 ml phosphate buffer containing 1.0% (w/v) PEG 20 m in a Buchner flask.
2. Attach flask to a suction pump and de-gas with shaking until bubbling ceases.
3. Wash CPG three times with 200 ml phosphate buffer.
4. Plug the bottom of the column (item 4, Figure 25.11) with a bung (item 6) perforated to receive a fine glass tube (4 mm external diameter). The tube should not stand proud of the bung's upper surface.
5. Attach plastic tubing and clamp (items 7, 8 and 9) to the glass tube.
6. Place glass wool (item 5) above the bung and press down until a depth of 1 cm is achieved.
7. Prepare a second bung (item 1) to fit the top of the column.
8. Pour phosphate buffer in the column until it begins to drip through the glass wool. The column will now be about one-third full. Clamp it at an angle of about 60° to the horizontal and pour in the CPG slurry. Add more as the CPG settles.
9. Clamp the tube upright and add more CPG. The final height of the CPG in the column should be about 90 cm.
10. Fill the safety loop with phosphate buffer by suction and attach to the tube at the top of the column.
11. Pack the column by placing de-gassed phosphate buffer in the reservoir (item 12) and allow to run through the column for 2 h at about 1 ml/min.

12. Check the effectiveness of the column as follows. Clamp the tubing (item 14) at positions 13 and 15, replace the phosphate buffer in the reservoir with de-gassed Tris/HCl (0.05 M, pH 7.5), drain the remaining phosphate buffer from the column until it is just dry, disconnect the tube at the top of the column, carefully add 200 μl Dextran Blue (1% (w/v)) to the surface of the CPG and chase through with a few millilitres of Tris/HCl buffer. Then add more Tris/HCl buffer to the top of the column and reconnect to the reservoir.
13. Open the clamps, pass buffer through the column, collect and measure the eluate. The result is satisfactory if the Dextran Blue exits the column following the addition of 70–85 ml eluant.
14. Flush the column with 150 ml Tris/HCl buffer.

Preparation of virus suspension for purification

Method

1. Prepare virus extraction buffer: 50 mM Tris/HCl buffer, pH 8.5, 100 mM NaCl, 1 mM EDTA, 10 mM Na_2SO_3. To make 250 ml: dissolve 1.514 g Tris in 200 ml deionized water, adjust the pH to 8.5 with HCl, add 1.46 g NaCl, 0.093 g EDTA and 0.315 g Na_2SO_3. Dissolve and make up to 250 ml with water.
2. Grind up five or so heavily infected guts in 4 ml extraction buffer
3. Centrifuge at $5000 \times g$ for 10 min and keep the supernate. Discard the pellet.

Purification of virus

Method

1. Disconnect the tube from the top of the CPG column and gently drain excess buffer.
2. Load the supernate containing virus onto the CPG surface with a syringe or pipette and allow to enter the column.
3. Add more buffer above the CPG, reattach the tube and chase through with Tris/HCl buffer at a rate of approximately 1 ml/min.
4. Collect 50 ml of eluate followed by fractions of 7.5 ml each. Those containing concentrated virus are cloudy and opalescent. Check fractions for absorbance at 260 nm. Confirm the presence of virus, by electron microscopy if possible, or alternatively by bioassay. A typical elution profile is given in Figure 25.12.

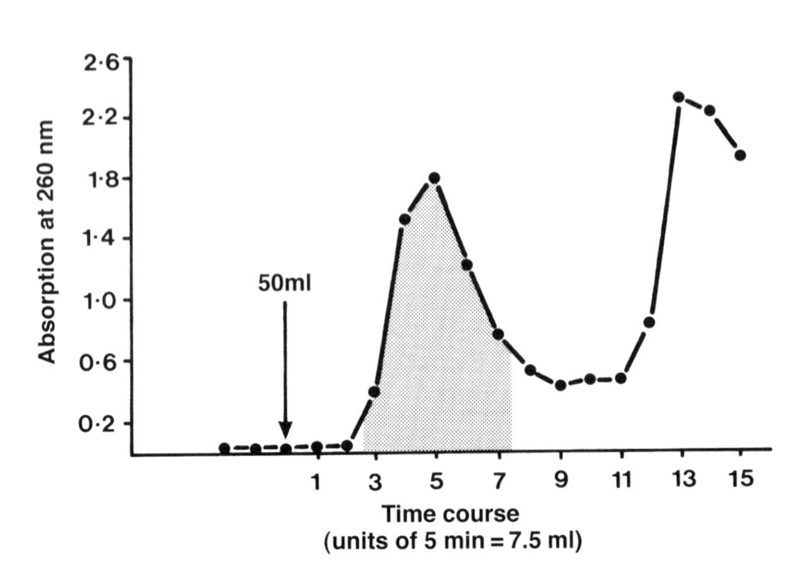

Figure 25.12 An elution profile for OrV from a controlled pore size glass bead column using two infected adult midguts in a 2 ml sample. The shaded area indicates the usable virus-containing fraction of eluant.

Storing the CPG column

Remove the top 0.5 cm CPG and replace with fresh, degassed CPG. Wash the column through with 300 ml of 2 M HCl, seal and store. Before use, flush through with phosphate buffer as described above.

ANALYSIS OF VIRAL STRUCTURAL PROTEINS

The use of REN analysis of virus DNA has largely replaced PAGE of virus polypeptides as a method of distinguishing between strains of virus. The latter technique still has application, however, particularly when combined with immunoblotting in the technique known as Western blotting. For this purpose, identical samples are electrophoresed through duplicate gels. One gel is then stained, for example with Coomassie Brilliant Blue, so that the location of the individual bands of protein can be identified. The (unstained) bands on the second gel are transferred to a nitrocellulose membrane where they are reacted with virus-specific antiserum followed by an anti-species antiserum. Any specific reactions between antibody and antigen are visualized by treating with ECL (enhanced chemiluminescence) reagents (Amersham Life Sciences) and exposing to X-ray film. PAGE electrophoresis and Western blothing are described below.

Electrophoresis in SDS-polyacrylamide gels

Electrophoresis in SDS-polyacrylamide (PAGE) gels is the standard technique for analysing virus structural proteins. The discontinuous Tris-HCl buffer system of Laemmli (1970) is most frequently used with gel concentrations of 10–12% (w/v) acrylamide. Electrophoresis of whole OBs produces only a single band (molecular weight *c.* 3×10^4) since, unless the gel is heavily loaded, only the matrix protein will be detected. If the OBs are first dissolved in carbonate buffer (0.05 M), endogenous alkaline proteases digest the polyhedral protein, resulting in smaller molecular weight bands corresponding to breakdown products. Between 15 and 25 polypeptide bands of molecular weight 12×10^3–100×10^3 are normally detected by electrophoresis of purified virus particles. Although there is much variation in the structural proteins of different viruses, a major band of low-molecular-weight basic protein is common to many of them and is associated with the nucleocapsid.

PREPARATION OF ANTIBODY

The rabbit is most frequently used for the preparation of antibodies and can be injected with whole OBs, although they are usually solubilized before injection. If guinea pigs are used, the OBs must be solubilized before injection.

To prepare antibodies against individual polypeptides, most commonly polyhedrin, purified virus is electrophoresed through PAGE gels and the required bands are eluted individually into PBS (Dulbecco A).

Regulations concerning the care of animals vary from country to country. It is essential that animals are properly housed and maintained. They should be in the care of a qualified animal technician, who must have a thorough knowledge of animal husbandry and who should be competent in the collection of blood. In the UK, work with animals is strictly controlled and licensed through the Home Office (Spring Gardens House, Princes Street, Swindon SN1 2JA). In the USA, information can be obtained from the American Association for Laboratory Animal Science (National Office: 70 Timber Creek Drive, Cordova TN 38018 (world wide web address: http://www.aalas.org/contact.htm)).

Preparation of antibodies against occlusion bodies

Equipment and materials

Items marked with an asterisk should be sterile.

1. Sample of purified whole OBs suspended in PBS (Dulbecco A)* at a concentration of 2 mg protein/ml (for solubilized OBs, treat beforehand with 0.05 M Na_2CO_3 for 30 min at 37°C)
2. Freund's complete adjuvant*

3. Freund's incomplete adjuvant*
4. Pipette suitable for delivering antigen and syringe for measuring adjuvant*
5. Rabbit(s)
6. Waring blender* or bijou* and vortex mixer for preparing emulsions of virus and adjuvant

Method for inoculation of animals

1. For the primary inoculation, prepare an emulsion of purified OBs (antigen) in Freund's complete adjuvant. To do this, add dropwise, an equal volume of Freund's complete adjuvant to the purified preparation of OBs, either whole or solubilized. Between drops, mix in a Waring blender or by vortexing until a white emulsion is formed. Check the quality of the emulsion by floating a small drop on water. When stable, the drop will not disperse.
2. Inject 4×0.1 ml volumes of the emulsion subcutaneously, 1×0.1 ml into each shoulder and 1×0.1 ml into each hind leg.
3. Boost 14, 28, and 42 days later with an emulsion of the same antigen in Freund's incomplete adjuvant injected into the same sites as above.
4. To maintain a continuous supply of antibodies boost thereafter at 28 day intervals.

Preparation of antibodies to individual virus polypeptides

Equipment and materials

1. Sample of purified OBs
2. Pipette (sterile) suitable for delivering antigen*
3. Rabbit(s)
4. Materials and equipment for electrophoresing OBs (see below).

Method

1. Electrophorese duplicate samples of virus according to the method given below (Electrophoresis and Western blotting of virus polypeptides) but using a larger electrophoresis tank (e.g. Biorad, Protean II Slab Cell) and gels (16 cm × 20 cm) in order to increase the amount of sample used.
2. Cut a vertical strip from the gel and stain. Using the stained gel to locate the position of the bands, cut out those of interest and soak (5 min) individually in PBS.

3. Centrifuge in a bench centrifuge at a speed and period of time sufficient to sediment the gel.
4. Remove supernate, replace with fresh PBS and soak for a further 5 min.
5. Repeat centrifugation step.
6. Emulsify gel containing antigen in 1.0–1.5 ml PBS.
7. Inject 2×0.5 ml intramuscularly and 2×0.25 ml subcutaneously (minimum total protein concentration of 100 μg).
8. Boost with a similar preparation of antigen after 4 weeks.
9. Bleed every 2–4 weeks (do not withdraw more that 10 ml blood/kg body weight in any 4-week period).

Steps 2 and 6 A synthetic meltable gel (OligoPrep) can be used. The manufacturer should be consulted for conditions of use (National Diagnostics (UK) Ltd, Fleet Business Park, Hessle HU13 9LX, UK).

Collection of blood

Method

1. Collect blood in clean, sterile glass or plastic vials or bottles.
2. Allow to clot overnight at 4°C.
3. Insert the tip of a Pasteur pipette between the clot and the side of the bottle and circle it. This allows the serum to separate from the clot.
4. Draw off the serum, which should be a thin, straw coloured liquid and transfer to a fresh container. If any red blood cells are present sediment them by centrifugation.
5. Collect the supernate (serum) and store frozen.

Regulation for collection of blood According to the Animals (Scientific Procedures) Act 1986 Inspectorate, Antibody Production – Advice on Protocols for Minimal Severity, no more than 15% total blood volume should be collected in any 1 month and this should be taken from superficial blood vessels.

Isolation of immunoglobulins from serum: ammonium sulphate precipitation

Materials

1. Physiological saline (0.85%)
2. Saturated ammonium sulphate (dissolved at 56°C in glass-distilled water

in a 1 : 1 ratio (w/v), cooled to room temperature and the pH value adjusted to 7.2 with ammonia solution or sulphuric acid)

3. Saturated ammonium sulphate solutions, 45% (v/v) and 40% (v/v), pH 7.2
4. PBS (Dulbecco A)
5. PBS containing 0.02% (w/v) NaN_3.

Method

1. Dilute serum with an equal volume of physiological saline (0.85% (w/v)) and stir for 5 min at 4°C.
2. Slowly add an equal volume of saturated ammonium sulphate and continue stirring for 30 min at 4°C.
3. Sediment precipitate by centrifugation at $1000 \times g$ for 15 min at 4°C.
4. Wash twice by suspending in 45% (v/v) saturated ammonium sulphate and centrifuging at $1000 \times g$ for 15 min at 4°C.
5. After the second wash, redissolve sediment (i.e. the immunoglobulins) in PBS and centrifuge as above to remove any insoluble material.
6. Collect supernate and reprecipitate immunoglobulins by slow addition, with stirring, of saturated ammonium sulphate to give a final concentration of 40% (v/v) saturated ammonium sulphate.
7. Stir for a further 15 min.
8. Centrifuge as above to sediment precipitate.
9. Wash twice in 40% (v/v) saturated ammonium sulphate.
10. Resuspend precipitated immunoglobulins in PBS to the original volume of serum.
11. Remove contaminating ammonium sulphate by dialysing against three changes of 1500 ml PBS containing 0.02% (w/v) NaN_3 over a 24 h period.
12. Store at –20°C.

ELECTROPHORESIS OF VIRUS POLYPEPTIDES

Equipment

1. Electrophoresis equipment such as that manufactured by Biorad under the name of Mini Protean II, (5.5 cm × 6 cm gels) together with a power-pack (0–100 V, 0–200 mA).

2. Pipettes, pipettors and tips for delivering up to 5 ml, and 5–200 μl
3. Square Petri dish (102 mm square).

Materials

1. Purified preparation of OBs, virus particles or nucleocapsids
2. Set of molecular weight markers (available commercially from, for example, Sigma)
3. Acrylamide: Bis 30:0.8 available commercially as a solution and stored at 4°C
4. 1.5 M Tris-HCl, pH 8.8
5. 0.5 M Tris-HCl, pH 6.8
6. Aqueous solution of 10% SDS
7. Sample buffer (prepared as a 4 × concentrated solution): 0.5 M Tris-HCl (pH 6.8) 1 ml; glycerol, 1.6 ml; 10% (w/v) SDS, 1.6 ml; 2β-mercapto-ethanol, 0.4 ml; 0.05% (w/v) Bromophenol Blue, 0.2 ml; glass-distilled water, 4.0 ml. Store at room temperature or lower. Dithiothreitol, 0.496 g, can be substituted for the 2β-mercaptoethanol
8. Running buffer (5 × working strength): Tris base, 45 g; glycine, 216 g; SDS, 15 g; glass-distilled water up to 3 l
9. Glycerol
10. TEMED (*N*,*N*,*N*′,*N*′-tetramethylethylenediamine)
11. Ammonium persulphate (APS, 10%), freshly prepared, daily
12. Separating gel (Table 25.7)
13. Stacking gel (sufficient for two mini-gels): 0.5 M Tris-HCl buffer pH 6.8, 0.83 ml; 10% SDS, 33 μl; Nanopure H_2O, 1.87 ml; 30% acrylamide, 0.55 ml; 10% APS, 33 μl; TEMED (added just before use), 1.66 μl
14. Coomassie Blue R250 (0.25%) in 40% methanol
15. Glacial acetic acid (10%) in 40% methanol
16. Pipettes as appropriate.

Method

1. Clean the glass plates thoroughly with detergent, rinse thoroughly in tap water followed by deionized water and finally glass-distilled water. Dry and wipe with alcohol. Ensure that no dust or fibres are left on the plates.
2. Assemble the plates and spacers (two sets) according to the manufacturer's instructions and place on the casting stand. Check for leaks by

Table 25.7 Separating gel (sufficient for two mini-gels)

Ingredients (add in the following order)	Gel percentage		
	10	12.5	15
1.5 MTris-HCl buffer, pH 8.8 (ml)	2.5	2.5	2.5
SDS 10% (ml)	0.1	0.1	0.1
Glycerol 30% (ml)	3.13	3.125	–
Glycerol 100% (ml)	–	–	0.93
Water (ml)	–	–	1.366
Acrylamide 30% (ml)	3.33	4.15	5.0
APS 10% (freshly made) (ml)	0.1	0.1	0.1
De-gas under vacuum for 15 min			
TEMED (μl	5	5	5

APS, ammonium persulphate.

adding glass-distilled water. If the assembly leaks, take apart and reassemble. When leak-free pour off water.

3. Prepare the acrylamide solution for the separating gel (Table 25.7) and add slowly with a pipette to the space between the assembled plates. Take care to avoid the formation of bubbles. Stop approximately 20 mm from the top of the shorter glass plate to allow room for insertion of the comb.
4. Overlay the separating gel with 200 μl glass-distilled water and allow the gel to set.
5. Pour off the water, insert the comb between the plates and carefully add the stacking gel mixture, again avoiding the formation of air bubbles. Allow to set.
6. Remove the comb and wash the wells so formed with glass-distilled water.
7. Assemble the gel sandwich plates either side of the central core containing the electrodes with the larger glass plate of each pair to the outside.
8. Place this assembly in the tank and fill the central reservoir formed by the plates with 1 × concentration running buffer.
9. Check that the central reservoir is not leaking. If it is, take equipment apart and reassemble.
10. Add 20 μl of 6 × concentrated sample buffer to 100 μl test sample in an microcentrifuge tube and boil for 5 min.
11. Pulse centrifuge for 10–20 s in bench centrifuge to ensure that any liquid that has condensed on the lid is returned to the bottom of the tube.
12. Load 10 μl volumes of sample into each well (duplicates samples, one for each gel) using a fine pipette tip or Hamilton syringe.

13. Pour working strength running buffer into the bottom of the large tank; try to ensure that air bubbles are not trapped under the lower edge of the plates. The buffer should cover the bottom 20 mm of the glass plates.
14. Attach the lid, ensuring that positive and negative connections match up.
15. Connect to power pack and set the voltage at 200 V to run for approximately 45–60 min until the dye front reaches the bottom of the gel.
16. Turn off the power. Dismantle the plates and gently peel off one gel into a square Petri dish containing 0.25% Coomassie Brilliant Blue R250 in 40% methanol. Stain with gentle agitation for 30 min. Remove excess stain by soaking until the background is clear (1–3 h). Transfer the individual polypeptides in the second gel to a nitrocellulose membrane as described below.

WESTERN BLOTTING

Equipment

1. Transblotter, e.g. Biorad S Minitransblotter
2. Rocking table
3. Square Petri dishes (102 mm square).

Materials

1. Hybond-C nitrocellulose membrane (Amersham Life Science, 0.45 μm, 20 cm × 3 m roll) or similar
2. Virus-specific antiserum (e.g. rabbit)
3. Anti-species antibodies (e.g. goat anti-rabbit) labelled with horseradish peroxidase
4. Blocking buffer I (5% skimmed milk powder in TTBS): skimmed milk powder, 10 g; 1 M NaCl, 30 ml; 0.5 M Tris (pH 7.5), 20 ml; Tween, 40 μl; glass-distilled water, 150 ml
5. TBS (150 mM NaCl; 50 mM Tris, pH 7.5): 1 M NaCl, 30 ml; 0.5 M Tris (pH 7.5), 20 ml; glass-distilled water, 150 ml
6. Transfer buffer: Tris base, 6.0 g; glycine, 28.8 g; glass-distilled water, 2000 ml
7. TTBS (150 mM NaCl; 50 mM Tris pH 7.5; 0.02% Tween): 1 M NaCl,

30 ml; 0.5 M Tris (pH 7.5), 20 ml; Tween, 40 μl; glass-distilled water, 150 ml

8. ECL Western blotting detection reagents (Amersham Life Sciences)
9. Autoradiography film (e.g. Fuji medical X-ray film)
10. Film cassette.

Method

1. Cut nitrocellulose membrane to fit size of gel (use gloves and forceps to handle the membrane).
2. Place in transblotter assembly according to the manufacturer's instructions, ensuring there are no air bubbles between the gel and the membrane. Insert in tank filled with transfer buffer.
3. Blot for 1 h at 159 mA in transfer buffer.
4. Remove membrane from assembly and place in square Petri dish.
5. Add 40 ml blocking buffer and soak for 1 h at 25°C or overnight at 4°C followed by 30 min at 25°C.
6. Remove blocking buffer and add 20 ml of blocking buffer containing 20 μl first antibody (antibody specific to test protein) and warmed to 25°C.
7. Incubate 1.0–1.5 h with rocking (on rocking table).
8. Remove blocking buffer containing antibody and wash filter briefly in TTBS (warmed to 25°C) followed by soaking in three changes of 30 ml TTBS, the first for 15 min and the others for 5 min.
9. Remove TTBS and add 20 ml blocking buffer containing 2 μl second antibody warmed to 25°C.
10. Incubate at 25°C for 1 h with rocking.
11. Remove blocking buffer and wash membrane briefly in TTBS.
12. Soak in 30 ml TTBS for 15 min and then in two changes of TTBS, each of 15 min duration, all with rocking.
13. Prepare detection reagents according to the manufacturer's instructions (the final volume required is 125 μl/cm^2 membrane).
14. Drain excess buffer from the washed membrane and place in a fresh Petri dish.
15. Add the detection reagent directly to the membrane and incubate for precisely 1 min at room temperature.
16. Drain off excess detection reagent, blot dry with paper tissue and wrap membrane in SaranWrap or place in a plastic bag. Gently smooth out air pockets.

17. Working quickly, place the membrane in film cassette, protein side upwards.
18. Switch off lights and carefully place a sheet of autoradiography film on top of the blots. Mark the film in some way to ensure correct orientation with the blot after developing, e.g. by cutting a small piece off one corner. Close the cassette and expose for 15 s.
19. Remove the film and immediately replace with a fresh piece of unexposed film and close the film cassette. Develop and examine exposed film and, from the appearance of the blots, determine the length of time required to expose the second piece of film. This may vary from 1 min to 1 h depending on the amount of target protein on the membrane.

CLONING VIRUS

In order to obtain a population of virus that is genetically homogenous it is necessary to clone a single genotype.

Two cloning strategies are possible. One, the limiting dilution method, relies on infecting the whole organism or unit of test cells in culture (e.g. bottle, test tube or well) with a dose of virus estimated to contain one infective unit. The other strategy depends on the serial passage of progeny from a single focus of infection such as plaques in cell culture. Provided that the dose–response curve (concentration of virus versus number of plaques formed) is linear, each plaque results from infection with a single virus particle.

Cell cultures are probably easier to handle than larvae and, although liable to accidental microbial contamination owing to faulty equipment or technique, are less likely to be carrying occult or latent virus. Virus passaged through cells in culture may, however, acquire elements of foreign DNA such as transposable elements at a faster rate than virus passaged through larvae. In addition, cytopathic effect in cell culture does not necessarily correlate with pathogenicity in the whole organism. A method based on that of Smith and Crook (1988) for cloning virus through larvae is given below. One for cloning virus in cell culture is given in Chapter 27.

CLONING VIRUS THROUGH LARVAE

Equipment and materials

Items marked with an asterisk should be sterile.

1. Susceptible larvae
2. Virus suspension

3. Diluent, e.g. glass-distilled water*
4. Eppendorf tubes or microtitre plates in which to dilute the virus*
5. Pipettes with which to dilute and administer virus*
6. Containers, e.g. 1 oz plastic polypots, with the appropriate diet on which to rear the test larvae
7. Containers, e.g. Eppendorf tubes or similar, in which to collect infected larvae*
8. Bench microcentrifuge
9. Narrow bore glass rods adapted to macerate larvae*.

Method

1. Estimate the concentration of OBs in a purified suspension of the virus to be cloned (Chapter 26).
2. Titrate virus, choosing an instar that will result in maximum virus production per individual. Use the result obtained to determine the concentration of virus that will cause disease in 10% of larvae inoculated.
3. Inoculate a minimum of 50 larvae with the dose determined in Step 2.
4. Collect larvae showing signs of infection, individually. Extract and purify virus from each larva, according to the protocol shown in Figure 25.6. Extract the DNA from a small portion of each sample of purified virus and prepare an REN profile (p. 436). Store the remainder of each sample at –20°C or –70°C.
5. Select the most promising samples, based on their REN profiles, and estimate the concentration of OBs in each. Dilute each to the same concentration as calculated from the results obtained in Step 2. Repeat Steps 3 and 4. Store remainder of each sample at –20°C or –70°C.
6. Repeat Step 5 until no submolar bands can be detected in the REN profiles of at least three consecutive passages of the viruses being tested. At this point it can be assumed that they are pure.

SIMULATION OF SOLAR UV FOR LABORATORY WORK

Many workers have conducted laboratory studies on the effect of UV irradiation on insect viruses and on the possibilities of virus protection by addition of a range of substances to spray formulations. In the earlier studies, there was a strong tendency for the use of 'germicidal' medium-pressure mercury arc lamps as a convenient and easily obtained source of UV, albeit with other wavelengths usually represented. Because such lamps have a quartz or Vycor envelope, peak irradiational emission is in the UV-C region at 253.7 nm, a wavelength not represented in solar irradiation reaching the earth's surface. As has been mentioned elsewhere in this book, the damaging effect of UV varies with wavelength, the shorter wavelengths in general being more injurious than the longer ones. It is, therefore, fundamentally incorrect to attempt to derive the biological effect of sunlight from studies on exposure to radiation from such lamps. Similarly, the value of those protectant materials that function by UV absorption cannot necessarily be adequately assessed by exposure to wavelengths shorter than those in incident sunlight. Substances that are not absolute screens to all radiation absorb particular UV wavebands. Hence a potential additive that provides a degree of protection from 253 nm will not necessarily do so from UV-B at, say, 305 nm.

It is, therefore, essential that the simulated radiation source closely mimics the solar spectrum, both in terms of waveband and the distribution of intensity across this waveband. Alternatively, for more closely analytical work, sources producing more defined emissions are required. Here a monochromator probably provides the ultimate analytical tool. The possible values of studying the response of a virus to various discrete segments of the solar UV spectrum should not be overlooked. Firstly, this permits definition of the most injurious wavelengths and so allows a more focused approach to the search for protectant materials. Secondly, it allows the possibility of detection of radiational activation effects. If and where such exist it might be considered important that UV-protectant materials which do not block activational wavelengths be selected. Finally, the possibility of inactivational synergy between wavebands should be considered as this also would have relevance to protection of viruses from solar radiation.

Heat may be an additional product of the irradiation source and hence the problem of temperature stabilization during experimental exposures may have to be resolved. There are various possible approaches. The virus may be exposed in thin layers on a refrigerated surface. A quartz screen, permeable to UV but providing some heat protection, may be interposed between virus and radiation source (Mubuta, 1985) or a temperature controlled chamber with air circulation might be considered.

Finally, the possibilities of hazard to the operator should be carefully considered. Exposure of skin and eyes to shortwave irradiation is irreversibly injurious but it is a feature of this problem that often the more serious effects (dermal cancers and cataract) may not become apparent until many years have elapsed. It is essential, therefore, that the 'experimental area' be completely screened off by UV-opaque material. UV irradiation of certain atmospheric pollutants can result in ozone production. Ozone is a health hazard so where there is a possibility of its generation adequate ventilation should be ensured.

Fluorescent lamps

A fluorescent lamp is a very low pressure mercury vapour lamp the inside of which is coated with phosphor. Passage of an electric current produces radiation at 253.7 nm, which on collision with the phosphor coating produces visible fluorescence. UV-C is filtered out and the broad-spectrum radiation emitted externally is mainly in the visible region but usually has notable bands in UV-B and UV-A. Narrower band radiation is available from fluorescent 'black light' lamps working on the same principle but with an integral glass envelope that filters out most of the visible components. Escaping radiation peaks approximately at 360 nm but with a minor line at about 315 nm in UV-B. It should be noted that the emission spectrum of fluorescent lamps, visible or black-light producers, varies with the manufacturer and model. Details of spectral output will usually be provided by the manufacturer. Witt and Hink (1979) used a thin plastic film to cut out a small amount of radiation in the region of 190–300 nm from such a lamp. An advantage for experimentation using fluorescent lamps is that they produce little heat.

Total solar simulation

For a fairly close simulation of the solar spectrum, high-pressure xenon lamps may be used. (The emission of xenon sources may change with use so that periodic calibrations should be conducted.) An experimental disadvantage of such sources is the generation of heat. However, most infrared and visible wavelengths can be removed by use of a dichroic mirror. Any UV-C present can be removed with a blocking filter so that targets can, if required, be exposed mainly to a combination of UV-A and UV-B. An example of equipment for solar simulation which also provides options for infrared and visible light elimination is the Oriel Solar Simulator (Oriel Corporation, Stratford, CT, USA), which has been used in several viral and fungal studies (Pozgay *et al.*, 1987; Carruthers, 1988; Moore *et al.*, 1993).

Simulation of UV only

Using a screened 150 watt xenon lamp, UV emitters have recently been developed that will emit air-cooled total solar UV or UV-A only. Various facilities permit the simultaneous delivery, to small adjacent target areas, of several radiational doses and the precise timing control of deliveries (Solar Light Co. Inc., Philadelphia, PA, USA). The UV-B portion of the emitted spectrum may require some screening to eliminate a small wavelength fraction <290 nm.

Monochromators

Exposure of virus samples to a range of monochromatic wavebands permits a proper analysis of spectral damage response. However, this approach has seldom been taken with insect viruses. A xenon arc lamp is used as the radiational source with gratings to regulate waveband emission (e.g. a Bausch and Lomb grating monochromator). Employing such equipment both David (1969) and Griego, Martignoni and Claycomb (1985) exposed an insect BV to monochromatic light, from 250–330 and 290–320 nm, respectively, in 10 nm-wide bands. Exposed virus samples were bioassayed and the degree of inactivation expressed as the percentage of original activity remaining (OAR). Because of the sigmoid nature of the host mortality–virus dose response curve, such units cannot be directly translated into degrees of virus inactivation *per se* (Chapter 4).

MEASURING UV RADIATION: SOLAR AND SIMULATED

There are three main approaches to the measurement of solar UV radiation. *Direct* measurements are made using a radiometer to which a range of interference filters or defraction gratings can be fitted. These measure the incoming radiation spectrum in the wavebands determined by the individual filters. An *indirect* measure can be obtained by measuring the total ozone (O_3) column using a Dobson meter and estimating the incoming radiation using a simple atmospheric radiative transfer model. Since ozone principally absorbs UV-B, the most active biological spectral region, it is of especial relevance. *Biological* assessments are obtained by use of a radiometer or photochemically active substance sensitive to that spectrum of UV which is biologically active. An essential aspect of this approach is that the devices are selected to respond in terms of a biologically active intensity profile. Examples are the Robertson–Berger meter (R–B meter) and polysulphone film.

Direct measurement

The basis of instruments providing direct measurement of UV is a radiation detector surface, a photodiode. A range of interference filters each permitting passage of selected wavebands are interposed, usually sequentially, to measure incoming radiation at any one time and place. For a complete assessment, radiometer measurements are taken by pointing the instrument at the sun to record direct irradiation (E_a) and, also, to obtain a measure of diffuse irradiation (E_d), at each of four cardinal sky directions. Such instruments are also suitable for monitoring UV in laboratory work. Because of the possibility of photo-deterioration of filters over time, periodic manufacturers, checks are usually recommended. Easily portable versions of such instruments are available (e.g. from Macam Photometrics Ltd, Livingston, Scotland).

Similar instruments offering continuous UV monitoring are also available (e.g. Solar Light Co., Inc, Philadelphia, USA). Only a single filter is usually available: according to need this may be either fairly broad band or it may be chosen to pass a narrower section of UV-B selectively. In the first case, real intensity differences across the incoming UV spectrum are not equally represented and in the second case true intensity is measured only over the restricted waveband passed by the filter. Direct metering has been extensively employed in studies related to the impact of UV-B on insect NPVs and UV protectants (Killick and Warden, 1991). Killick (1987) illustrated such a local DNA action spectrum for global radiation at 50° solar elevation in the north of Scotland around the summer solstice indicating a biologically injurious maximum at about 307 nm (Figure 25.13). A quite similar value is obtained by substituting the occurrence of erythema (skin reddening) for DNA data.

Indirect measurement

Direct measurement of column ozone is achieved by extrapolation from intensity measurements using appropriate interference filters, as in the Dobson meter. Here

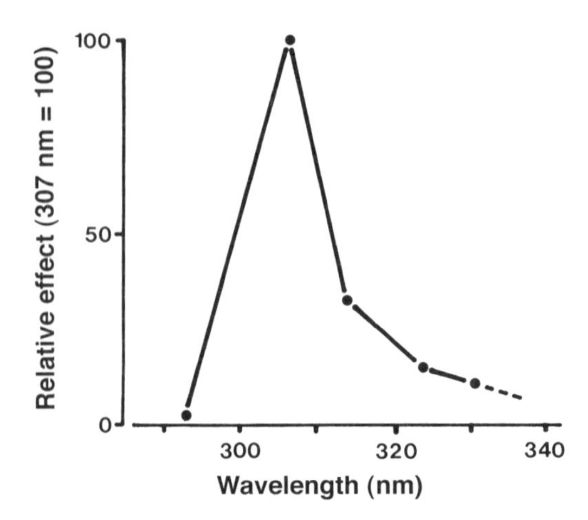

Figure 25.13 Product of the amount of solar UV energy (global radiation at 50° solar elevation (Bener, 1960, 1964)) and its relative damaging power to DNA (Harm, 1980). The plot indicates a biologically injurious maximum at 307 nm. (After Killick (1987) *Forestry Commission Bulletin* **67**; reproduced by permission of the Forestry Commission, H. J. Killick, 1987.)

ozone quantity is expressed in terms of Dobson units (1 Dobson unit = 0.01 mm ozone). There is often a disparity between results obtained from Dobson meters and other forms of atmospheric UV metering, e.g. using R–B meters. It has been noted, for instance, that over cities values obtained from R–B meters can depart, especially in winter, from Dobson-generated data. This is thought to indicate seasonal changes in levels of aerial pollutants absorbing incoming radiation. Therefore, while ozone measurement may be appropriate in the absence of pollution, direct metering is a more generally realistic means of obtaining a measure of actual UV at the earth's surface.

Biological measurement

The response, or action, spectrum is how the substance/organism responds to radiation as a function of wavelength. It is independent of the shape of the incoming radiation, which will be likely to vary latitudinally, seasonally, with solar elevation throughout the day, and in relation to cloud, atmospheric dust and certain pollutants. Detectors are relevent to the extent to which their output correlates with a photobiological effect. Selecting or adjusting recording devices for appropriate biological *weighting* of spectra is of great importance. Calkins (1982) succinctly defined this problem: 'The essence of a biological dosimetry system is a transformation function for physical units such that equal exposures expressed in the chosen dosimetric units produce equal biological responses regardless of the spectral composition of the incident radiation. Two major limitations become immediately evident; 1) different solar UV wavelength bands produce qualitatively different effects; 2) species vary in their relative sensitivity to the various components of solar UV.'

At present biological UV dosimetry is especially geared to human health questions. Solar irradiation of human skin (especially the relatively unpigmented skin of Caucasian types) induces sunburn (commencing with reddening (erythema)), photo-ageing, minor skin cancers such as basal cell carcinoma, and, much more importantly, melanomas (Ambach and Blumthaler, 1993). The key recording device concerned, the R–B meter, has been developed (and is still being modified) to have an output as closely biologically relevant to these problems as possible. While measurements made with the R–B meter probably have wide biological significance, it is important for workers in areas other than human skin problems to determine the action spectra that are relevant to the particular questions with which they are concerned. They should then assess the degree to which the R–B meter would be a satisfactory monitor.

In investigating the biological impact of solar radiation, it is important to consider the possibilities of synergy between different spectral components. It is also important to be aware of the possibility of photoactivation (e.g. Ramoska, Stairs and Hink, 1975) and photoreactivation (e.g. Harm, 1980) by the longer wavelengths, e.g. by UV-A and even visible light. So far it appears there have been no developments of metering that take into account such spectral effects and interactions.

The Robertson–Berger meter

The R–B meter was initially developed by Berger (1976). Light enters through a dome-shaped quartz cover and passes through a broad solar visible filter before hitting a magnesium tungstate phosphor surface that responds to radiation in the UV-B region. A second filter is then used to absorb any stray UV before the green light that has been generated reaches a photodiode. Here a current proportional to the incident light is produced. Once a certain voltage threshold has been reached, the detector sums the current in terms of counts, which it prints for half-hourly intervals before resetting to zero (Grainger, Basher and McKenzie, 1993). The spectral response closely corresponds to the action spectrum for UV-B-induced human erythema.

The R–B meter now plays a major role in monitoring medically meaningful solar UV-B reaching the earth's surface. A global R–B meter network has come into existence and, for the original ten sites, records go back continuously to 1973 (Berger, 1982). This network represents the only long-term attempt to monitor solar radiational injurious effects so that, especially in view of ozone changes detected since the late 1970s, considerable interest has been expressed in the extent to which it realistically represents incident UV-B insolation levels. In attempting to equate R–B response to incident UV-B, various instrumental imperfections, such as the cosine response, have had to be taken into account. Progress in this area is reported by Grainger *et al.* (1993).

R–B meters are available from a number of sources, the original one being the Solar Light Co., Inc., Philadelphia, PA, USA.

Polysulphone films

Polysulphone is sensitive to solar UV because its high aromatic content results in considerable absorption. This response is temperature independent and can be measured by the increase in optical density at 330 nm. Polysulphone film 'badges' can be used as personnel UV dosimeters as their response is similar to the erythemal action spectrum. They can be laid horizontally to capture all the UV, both direct and diffuse, from all parts of the sky and exposed for as long as desired, up to a certain saturation limit (Davis, Deane and Diffey, 1976; Davis and Sims, 1983). Polysulphone, or Davis, films can be placed on or within the canopy of plants to provide time summations of biologically active UV-B. Killick *et al.* (1988) developed and described a simple automatic device capable of effecting hourly film changes over a period of 7 days. This machine is suitable, for instance, for remote sites.

Measurement of microorganism mortality

Various workers have suggested that standard microbial cultures exposed under standard conditions could, by bioassay, provide a biologically realistic measure of UV impact. Calkins (1982) discussed the relative sensitivity of various microorganisms and lower organisms (e.g. larvae of the mosquito *Aedes aegypti*) to equal exposures as measured by the R–B meter and found a maximum two-fold interspecific difference. However, such labour intensive methods are not in general use.

REFERENCES

Ambach, W. and Blumthaler, M. (1993) Biological effectiveness of solar UV radiation in humans. *Experientia* **49**, 747–753.

Barton, R.J. (1977). An examination of permeation chromatography on columns of controlled pore glass for routine purification of plant viruses. *Journal of General Virology* **35**, 740–743.

Bener, P. (1960) ACFRL Contract AF61 (052)-54. *Technical Summary Report* No 1, Dec., 1960.

Bener, P. (1964) Diurnal and annual course of the spectral intensity of total and diffuse ultraviolet radiation for cloudless sky in Davos. *Strehlentherapie* **123**, 306–316.

Berger, D.S. (1976) The sunburning ultraviolet meter: design and performance. *Photochemistry and Photobiology* **24**, 587–593.

Berger, D.S. (1982) The sunburn UV network and its applicability for biological predictions. In *NATO Conference Series, Series IV: Marine Sciences. The Role of Solar Ultra-violet Radiation in Marine Ecosystems*. J. Calkins (ed.). New York: Plenum Press, pp. 181–192.

Bozzola, J.J. and Russell, L.D. (1992). *Electron microscopy. Principles and Techniques for Biologists*. Boston, MA: Jones and Bartlett, pp. 542.

Calkins, J. (1982) Measuring devices and dosage units. In *NATO Conference Series, Series IV: Marine Sciences. The Role of Solar Ultra-violet Radiation in Marine Ecosystems*. J. Calkins (ed.). New York: Plenum Press, pp. 169–179.

Carruthers, R.I., Ziding, F., Ramos, M.E. and Soper, R.S. (1988) The effect of solar radiation on the survival of *Entomophaga grylli* (Entomophthorales: Entomophthoraceae) conidia. *Journal of Invertebrate Pathology* **52**, 154–162.

Crawford, A.M. (1988). Detection of baculovirus infection in rhinoceros beetle (*Oryctes rhinoceros)* and the purification and identification of virus strains. In *Integrated Coconut Pest Control Project Annual Report 1988*. Manado, North Sulavesi, Indonesia: UND/FAO Coconut Research Institute, pp. 120–141.

David, W.A.L. (1969) The effect of ultraviolet radiation of known wavelengths on a granulosis virus of *Pieris brassicae*. *Journal of Economic Entomology*, **14**, 336–342.

Davis, A. and Sims, D. (1983) *Weathering of Polymers*. London: Applied Science.

Davis, A., Deane, G.H.W. and Diffey, B.L. (1976) Possible dosimeter for ultraviolet radiation. *Nature, London* **261**, 169.

Dulmage, H.T., Correa, J.A. and Martinez, A.J. (1970) Coprecipitation with lactose as a means of recovering the spore-crystal complex of *Bacillus thuringiensis*. *Journal of Invertebrate Pathology* **15**, 15–20.

Evans, H.F., Bishop, J.M. and Page, E.A. (1980) Methods for the quantitative assessment of nuclear polyhedrosis virus in soil. *Journal of Invertebrate Pathology* **35**, 1–8.

Grainger, R.G., Basher, R.E. and McKenzie, R.L. (1993) UV-B Robertson–Berger meter characterization and field calibration. *Applied Optics* **32**, 343–349.

Griego, V.M., Martignoni, M.E. and Claycomb, A.E. (1985) Inactivation of nuclear polyhedrosis virus (Baculovirus subgroup A) by monochromatic UV radiation. *Applied and Environmental Microbiology* **49**, 709–710.

Haller, W. (1965) Chromatography on glass controlled pore size. *Nature, London* **206**, 693–696.

Haller, W. (1967) Virus isolation with glass of controlled pore size: MS-2 bacteriophage and Kilham virus. *Virology* **33**, 740–743.

Harm, W. (1980) *Biological Effects of UV Radiation*. Cambridge: Cambridge University Press.

Hughes, D.S., Possee, R.D. and King, L.A. (1993). Activation and detection of a latent baculovirus resembling *Mamestra brassicae* nuclear polyhedrosis virus in *M. brassicae* insects. *Virology* **194**, 608–615.

Ignoffo, C.M. and Shapiro, M. (1978) Cheracterisation of baculovirus preparations from living and dead larvae. *Journal of Economic Entomology* **71**, 186–188.

Innis, M.A., Gelfand, D.H. Sninsky, J.J. and White, T.J. (eds) (1990) *PCR Protocols: A Guide to Methods and Applications*. London: Academic Press.

Kaupp, W.J. and Burke, F.R. *Technical Note, No 1, Biological Control Methods*, Sault Ste. Marie, Canada: Canadian Forestry Service (ISSN 0826–0532).

Killick, H.J. (1987) Ultraviolet light and *Panolis* nuclear polyhedrosis virus: a non-problem? In *Forestry Commission Bulletin 67: Population Biology and Control of the Pine Beauty Moth*, S.R. Leather, J.T. Stoakley and H.F. Evans (eds). Forestry Commission, pp. 69–75.

Killick, H.J. and Warden, S.J. (1991) Ultraviolet penetration of pine trees and insect virus survival. *Entomophaga* **36**, 87–94.

Killick, H.J., Primarolo, A.A., Mackenzie, R.M. and Forkner, A.C. (1988) An inexpensive automatic recorder for biologically important solar UV. *Photochemistry and Photobiology* **48**, 555–557.

Laemmli, U.K. (1970) Cleavage of structural proteins during the assembly of the head of bacteriophage T4. *Nature, London* **227**, 680–685.

Lomer, C.J. (1986). *Baculovirus oryctes*: biochemical studies and field use. PhD Thesis, University of London.

Lowry, O.H., Rosenbrough, N.J., Farr, A.L. and Randall, R.J. (1951) Protein measurement with folin phenol reagent. *Journal of Biological Chemistry* **193**, 265–275.

Moore, D., Bridge, P.D., Higgins, P.M., Bateman, R.P. and Prior, C. (1993) Ultraviolet radiation damage to *Metarhizium flavoviride* conidia and the protection given by vegetable and mineral oils and chemical sunscreens. *Annals of Applied Biology* **122**, 605–616.

Mubuta, D. (1985) Ultra violet protectants of *Panolis flammea* nuclear polyhedrosis virus. MSc thesis, University of Newcastle-upon-Tyne.

Noguchi, Y., Kobayashi, M. and Shimada, T. (1994) An application of the polymerase chain reaction for practical diagnosis of the nuclear polyhedrosis in large-scale culture of *Bombyx mori*. *Journal of Sericultural Science of Japan* **63**, 399–406.

Porath, J. and Flodin, P. (1959). Gel filtration: a method for desalting and group separation. *Nature, London* **183**, 1657–1659.

Pozgay, M., Fast, P., Kaplan, H. and Carey, P.R. (1987) The effect of sunlight on the protein crystal from *Bacillus thuringiensis* var *kurstaki* HD1 and NRD12: a Raman spectroscopic study. *Journal of Invertebrate Pathology* **50**, 246–253.

Ramoska, W.A., Stairs, G.S. and Hink, F.W. (1975) Ultraviolet light activation of insect nuclear polyhedrosis virus. *Nature, London* **253**, 628–629.

Rohrman, G. F. (1986) Polyhedrin structure. *Journal of General Virology* **67**, 1499–1513

Smith, I.R.L. and Crook, N.E. (1988) *In vivo* isolation of baculovirus types. *Virology* **166**, 240–244.

Steere, R.L. (1963). Tobacco mosaic virus: purifying and sorting associated particles according to length. *Science* **140**, 1089–1090.

Wigley, P.J. (1980) Practical: counting microorganisms. In *Microbial Control of Insect Pests*. J. Kalmakoff and J.F. Longworth (eds). Wellington, New Zealand: DSIR, pp. 29–35.

Witt, D.J. and Hink, W.F. (1979) Selection of *Autographa californica* nuclear polyhedrosis virus for resistance to inactivation by near ultraviolet, far ultraviolet and thermal radiation. *Journal of Invertebrate Pathology* **33**, 222–232.

Zelazny, B., Alfiler, A. and Mohamed, N.A. (1985) Glass permeation chromatography for purification of the baculovirus of *Oryctes rhinoceros* (Coleoptera: Scarabaeidae). *Journal of Economic Entomology* **78**, 992–994.

Zelazny, B., Alfiler, A. and Crawford, A.M. (1987) Preparation of a baculovirus inoculum for use by coconut farmers to control rhinoceros beetle (*Oryctes rhinoceros*). *FAO Plant Protection Bulletin*, **35** 36–42.

Zelazny, B., Autat, M.L., Singh, R. and Malone, L.A. (1988) *Papuana uninodis*, a new host for the baculovirus of *Oryctes*. *Journal of Invertebrate Pathology* **51**, 157–160.

26 Enumeration of virus

Methods for bioassay can be divided conveniently into those where lethal dosage must be tightly controlled in order to produce LD_{50} values and others where a comparative measure is required and only a lethal concentration is calculated (LC_{50}). A third category concentrates on time to response as measured by death (LT_{50} or ST_{50}). This can generally be accommodated within the other assay methods although it is possible that more detailed study will be required to determine LT_{50} for rapidly acting pathogens. The use of the 50% response is a statistical convenience providing the greatest information on the population susceptibility of a particular insect species to challenge by virus.

In the great majority of bioassays, a population of test organisms is challenged by a range of dosages ideally spanning the LD_{50} or LC_{50} and avoiding 0 or 100% responses. Indeed, a response of 0 or 100% provides almost no statistical value since it is not known whether the organism has received a dose just sufficient to kill it or whether there is an excess of virus: overkill. Ranging bioassays to determine the approximate value of the LD_{50} or LC_{50} are, therefore, necessary before the full-scale assays are set up.

Data obtained by LC_{50} assays probably cannot be used for accurate comparison of mortality responses in different host species as these are likely to have differing feeding rates and so will ingest different numbers of OBs at any one concentration presented. LD_{50} tests, however, will provide an absolute dose–mortality response that is indifferent of the rate of host feeding. Such a test provides a basis for comparisons both within and between species.

DETERMINATION OF VIRUS CONCENTRATION

In all cases, it is essential to determine as accurately as possible the concentration of virus OBs in the preparation to be tested. A number of methods, detailed below, are available for this.

Owing to their size (1–2 μm or larger in diameter), enumeration of numbers of OBs in suspensions of purified NPVs is relatively easy to achieve by light microscopy with a standard haemocytometer such as an improved Neubauer (0.1 mm in depth). Counting numbers of OBs (granules or capsules) of GV is less easy because of their smaller size (less than 500 nm in length). It requires considerable experience but can be done with a specially manufactured haemocytometer of smaller depth (0.01 mm). For purified preparations of GV, there is good correlation between numbers of granules and protein concentration. For routine work, it is often easier to determine the protein concentration of a suspension and to derive an estimate of the number of granules by reference to a standard for which the concentrations of granules and protein are known. For example, for purified *Plodia interpunctella* GV, a protein concentration of 10 mg/ml is equivalent to 5.03×10^{11} granules/ml. A method for determining the protein concentration of virus is given in Chapter 25.

An alternative method for determining concentrations of OBs by light microscopy involves the preparation and staining of dried films of virus. Enumeration of numbers of OBs can also be done by electron microscopy. This requires additional expertise and equipment and is more costly and time-consuming than light microscopy.

LIGHT MICROSCOPY TO DETERMINE VIRUS IN SUSPENSION

Materials

1. Suspension of virus suitably diluted (containing approximately 2×10^7 to 5×10^7 OB/ml)
2. Haemocytometer, e.g. Improved Neubauer (Gallenkamp)
3. Pasteur pipette or equivalent for transferring the suspension of virus to the counting chamber
4. Dark ground light microscope with $\times 40$ objective and $\times 10$ eyepieces ($\times 400$ total magnification) or $\times 90/100$ oil immersion objective and $\times 10$ eyepieces ($\times 900$ or $\times 1000$ total magnification) (NB some microscopes also incorporate a magnification factor in the barrel)
5. Disinfectant (e.g. 1% Vircon, minimum immersion time 30 min) or sterilizing agent (e.g. 70% ethanol) in beaker for decontaminating the haemocytometer
6. Tally counter.

Method

1. Ensure that the haemocytometer and coverslip are clean by rinsing with 70% ethanol or distilled water and wiping with a paper tissue, paper towelling or similar.
2. Place coverslip on top of slide so that it is covering the chamber into which the virus suspension is going to be placed (Figure 26.1).
3. Press coverslip down firmly onto either side of the chamber until

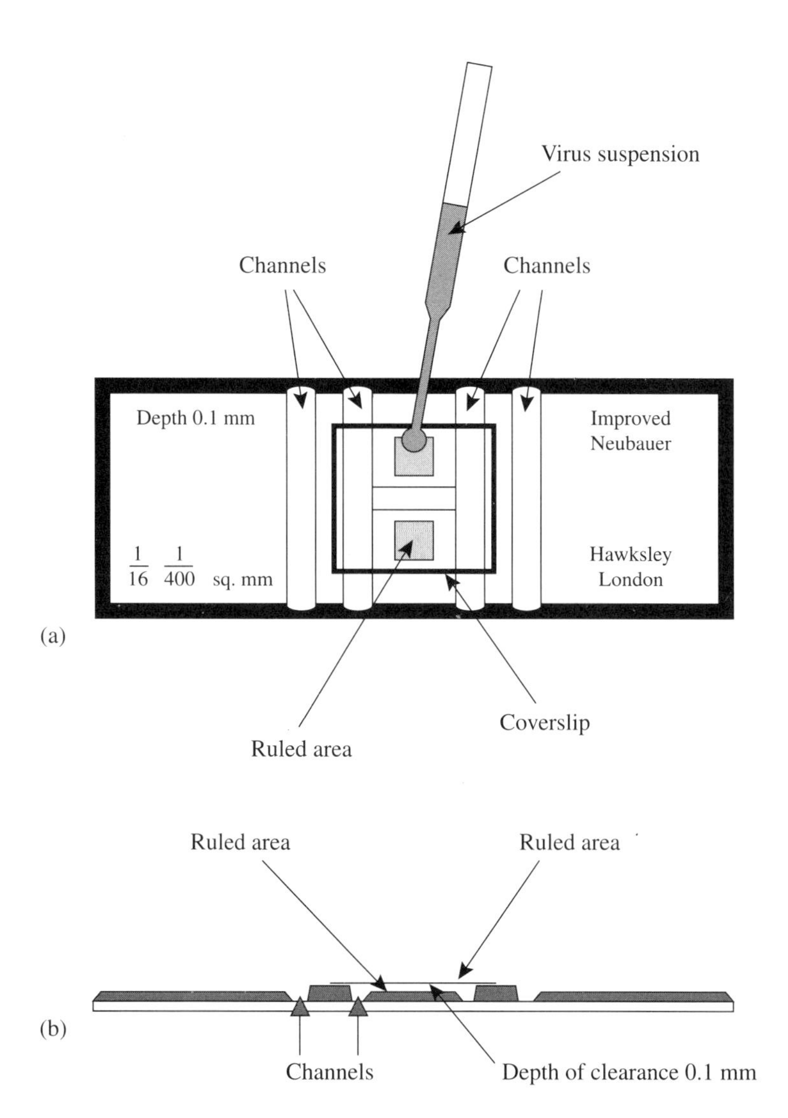

Figure 26.1 Improved Neubauer haemocytometer: (a) double-sided counting chamber; (b) side view of chamber; (c) Neubauer ruling within chamber, channels depth of clearance 0.1 mm; (d) centre portion of Neubauer ruling.

Newton's rings are visible; this ensures that the chamber is of the correct depth.

4. Disperse any aggregates of virus that may have formed by sonicating at maximum setting in an ultrasonic water bath (e.g. Kelly Pulsatron, model KS 400) for 30 s to 1 min (p. 595).
5. Withdraw a sample of suspension with a Pasteur pipette or similar; place

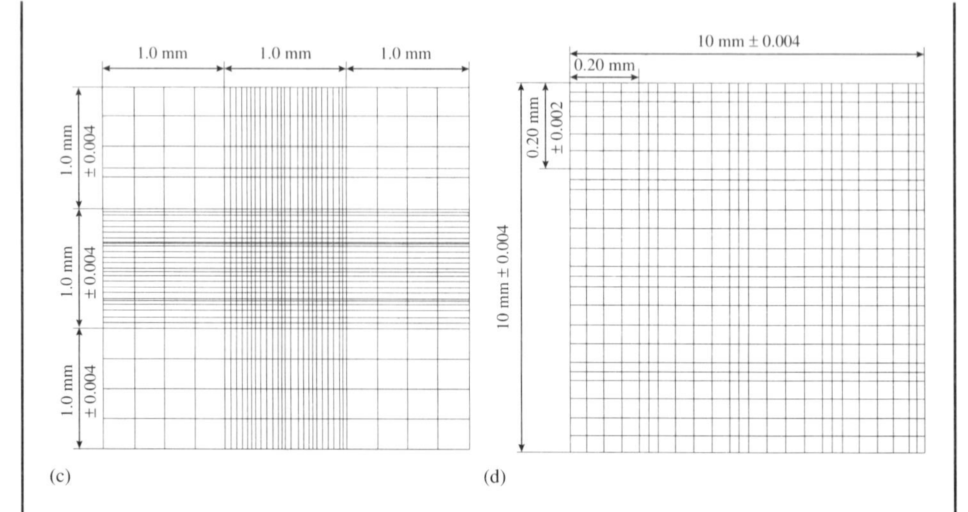

Figure 26.1(c) and (d)

the tip of the pipette at the junction of the coverslip and the chamber (Figure 26.1).

6. Expel sufficient suspension to fill the chamber completely; excess suspension will drain into the channels at the side.
7. Leave the OBs for 10 min to settle and reduce the degree of Brownian motion of the particles.
8. Count the number of OBs in the four large corner squares of the chamber, each of which comprises 16 smaller squares (Figure 26.1c,d). For each small square, count the OBs within the square, including those touching the top and left-hand sides but not those touching the bottom or right-hand sides.
9. Calculate the mean number of OBs in each large square and from this the concentration of OBs per millilitre.

Each preparation of virus should be counted at least three times.

Calculation

Area of each small square = 0.0025 mm^2
Volume in each small square = 0.00025 μl (for chamber with a depth of 0.1 mm)

$$\text{Number of OBs per ml} = (D \times X) \div (N \times V)$$

where D is the dilution of the suspension; X is the number of OBs counted; N is the number of small squares counted; V is the volume, in millilitres, above a small square (for an Improved Neubauer haemocytometer this is 0.00025 μl).

If the virus suspension has been mixed properly and is homogeneous, the distribution of OBs within the squares will be random and will follow a Poisson series:

$$Ne^{-m}\left(1,\ m,\ \frac{m^2}{2!},\frac{m^3}{3!},\frac{m^4}{4!},\ \ldots\ \text{etc.}\right)$$

where N is the number of samples examined; m is the mean number of items per sample (estimated by $\bar{x}$); e = 2.71828 and the symbol ! signifies 'factorial'; thus $4! = 4 \times 3 \times 2 \times 1 = 24$.

For a Poisson distribution, the variance of the counts is equal to the mean and the standard deviation is equal to the square root of the mean count per small square, i.e. the standard deviation (SD) is

$$\sqrt{X} \div N = \bar{x}$$

and the Standard Error of the mean of N counts is:

$$\sqrt{\bar{x}} \div N$$

If the suspension is not homogeneous, the values obtained for numbers of particles in each small square will not agree with those expected for a Poisson distribution. To check this, calculate the mean number of particles in each square ($\bar{x}$) and the variance s^2 as $[(\Sigma x^2 - (\Sigma x)^2/N]/(N - 1)$. If the values for $\bar{x}$ and s^2 thus calculated are approximately the same, it can be assumed that the particles are randomly distributed. If s^2 is markedly greater than $\bar{x}$ the particles are aggregated in patches of relatively high density. In this instance, the suspension should be rehomogenized and further counts made.

LIGHT MICROSCOPY TO DETERMINE VIRUS IN DRIED FILMS

This method (Wigley, 1980) is adapted from those used for counting micro-organisms in milk and can be used with either Giemsa's stain (using 14 mm templates) or Buffalo Black stain (using 15 mm templates).

Materials

1. A glass microscope slide cleaned by rinsing in absolute ethanol
2. Metal template of the same dimensions as the microscope slide, in which a line of 4 holes each 14 mm in diameter has been drilled (Giemsa method)

3. Suspension of virus
4. 0.1% (w/v) aqueous solution of gelatin
5. Pipetting device to deliver 5–20 μl volumes
6. Blunt seeker
7. Carnoy's fixative (glacial acetic acid, 100 ml; chloroform, 300 ml; absolute ethanol, 600 ml)
8. 7% (v/v) solution of Gurr's improved Giemsa stain (BDH) in 0.02 M phosphate buffer (pH 6.9)
9. Staining baths
10. Blotting paper or absorbent paper towelling
11. Light microscope with oil-immersion objective (× 1000 magnification)
12. Eyepiece grid consisting of 100 squares (Gallenkamp) the dimensions of which have been calibrated (for the magnification selected) using a stage micrometer (Gallenkamp).

Method

Preparation of 0.02 M phosphate buffer pH 6.9 Add 55 ml of an aqueous 0.2 M solution of NaH_2PO_4 (29.39 g/l) to 45 ml of an aqueous 0.2 M solution of $Na_2HPO_4.2H_2O$ (31.21 g/l) and make up to 1 l with glass-distilled water.

Preparation of stained films of NPV using Giemsa stain

1. Place the microscope slide on the metal template.
2. Sonicate the virus suspension for 30 s to 1 min to disperse clumps and add to an equal volume of 0.1% (w/v) gelatin at room temperature.
3. Dispense a known volume (5–20 μl) of the virus suspension in gelatin onto the centre of each area of the slide defined by the holes in the template.
4. Using the blunt seeker and a spiraling motion, spread each drop evenly over the 14 mm circular area allocated to it.
5. Leave the slide to dry at room temperature.
6. When dry, transfer slide to a bath of Carnoy's fixative and leave for 2 min.
7. Immerse slide in absolute ethanol for 30 s and then dry at room temperature.
8. Place in 7% Gurr's improved R66 Giemsa stain for 45 min.
9. Rinse in two changes of tap water.

10. Drain by holding the edge of the slide on a piece of filter paper or absorbent paper towelling.
11. Leave to dry at room temperature.
12. Observe under the light microscope using an oil-immersion objective at a magnification of ×1000.

OBs appear as clear, round objects while bacteria and other contaminants stain purple.

Preparation of stain films of NPV by the Buffalo Black method

1. As above but use Meyer's albumen as a sticker rather than gelatin and use 15 mm templates for spreading a precise 5 μl of the mixed suspension onto the standard circle.
2. Heat-fix the smears on a hotplate at 75°C for 2 min.
3. Stain in Buffalo Black (Naphthalene Black 12B) solution (Chapter 25) at 45°C for 2–5 min, rinse in running tap water and dry.

Counting occlusion bodies

For each film, choose a radius positioned at right angles to those on adjacent films (Figure 26.2). Make counts at 10 predetermined positions along each radius for 14 mm circles and 11 positions for 15 mm circles. For each position, count and record the number of OBs contained within the area defined by the eyepiece grid, including only those within the grid or touching the top and left edges.

Calculation

Since the thickness of the film increases towards its centre, the OBs will not be distributed randomly nor will the proportion of the circle sector sampled be the same at each position. It is necessary, therefore, to weight mathematically each count according to its position along the radius. The weighting values are shown in Table 26.1.

A standard results sheet for each type of circle can be constructed similar

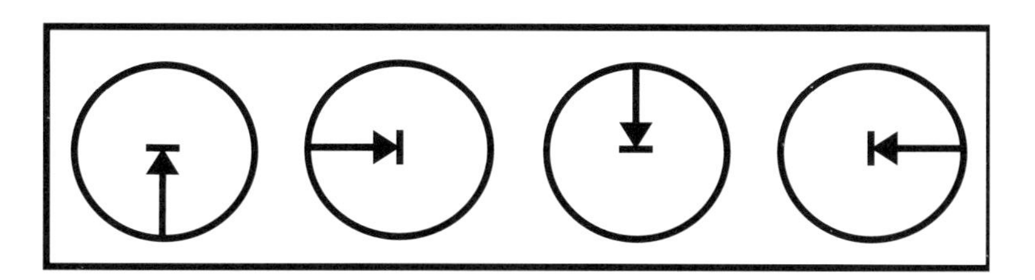

Figure 26.2 Sampling pattern for counting dry films of virus.

Table 26.1 Weighting values for counts of OBs estimated by the dry counting procedure

Count number	Count position (mm from edge of film)	Weight factor for circle diameter (mm) of	
		14	15
1	0.25	0.138	0.129
2	0.75	0.128	0.12
3	1.25	0.117	0.111
4	1.75	0.107	0.102
5	2.25	0.097	0.093
6	2.75	0.087	0.084
7	3.25	0.076	0.076
8	3.75	0.096	0.067
9	4.75	0.092	0.058
10	5.75	0.062	0.106
11	6.75	–	0.054

to that in Table 26.2 for 15 mm circles. The procedure is to calculate the mean value for each count position and to multiply each mean count by its weighting factor, then to total the weighted mean for the entire circle.

Estimate the virus concentration in OB/ml using the following formula:

$$\text{Concentration} = \frac{F}{M} \times C \times \frac{V}{1000} \times D$$

where F is the area of each stained film (mm^2); M is the area of eyepiece grid (mm^2); C is the overall weighted mean count; V is the volume (in µl) dispensed for each film; D is the dilution factor (including the twofold dilution in gelatin or albumin).

The standard error is estimated as:

$$\text{SE} = \sqrt{\sum w^2}\left(\frac{x}{n}\right)$$

where w is a weighting factor at each position; x is the mean count at each position; and n is the number of films counted, i.e. four.

Table 26.2 Standard results' form for the dry counting procedure

Radius/Sector	1	2	3	4	5	6	7	8	9	10	11
1											
2											
3											
4											
Sector mean											
Wt factor	0.129	0.120	0.111	0.102	0.093	0.084	0.076	0.067	0.058	0.106	0.054
Sector mean × wt factor											
					Total of (Sector mean × wt factor)						

TRANSMISSION ELECTRON MICROSCOPY TO DETERMINE THE CONCENTRATION OF OCCLUSION BODIES IN SUSPENSION

Materials

1. Suspension of purified virus (containing 2×10^6 OB/ml or more) in water or 2% ammonium acetate
2. Suspension of polystyrene beads of approximately the same diameter as that of the virus and of known concentration (similar to that of the virus)
3. Nebulizer (Figure 26.3)
4. Safety cabinet that can be sterilized after use to prevent any cross-contamination of viruses
5. Carbon-coated formvar grids
6. Fine tipped forceps (Figure 26.3)
7. Glass Petri dish lined with filter paper (e.g. Whatman No. 1) for transporting virus-coated grids
8. Pipettes, e.g. Pasteur with adapter (Figure 23.1, p. 362), or similar, for measuring volumes of virus and polystyrene beads
9. Pipette for adding stain (microlitre volumes).

Method

1. Sonicate (p. 460) bead and virus suspensions to disperse any aggregates that may have formed.
2. Mix equal volumes of virus and beads.
3. Transfer to nebulizer.
4. Assemble equipment as shown in Figure 26.3.
5. Spray virus onto the appropriate number of grids to give a statistically satisfactory count of particles.
6. Leave to dry in air.
7. Add a small drop (2–5 μl) of 2% phosphotungstic acid (pH 7.2). Remove this from the edge of the droplet with a wedge-shaped piece of filter paper and dry in air for 10 min.
8. Examine in the electron microscope at an accelerating voltage of 60 or 100 kV. Choose droplets in which polystyrene beads and virus particles are clearly distinguishable and countable. For each droplet calculate the ratio of beads to virus particles and determine the concentration of virus particles by proportionality.

Calculation

Number of polystyrene beads = 20
Number of OBs = 40
Ratio of OBs to polystyrene beads = 40/20 = 2
Concentration of polystyrene beads = 4×10^6 particles/ml
Concentration of OBs = $2 \times 4 \times 10^6$ particles/ml = 8×10^6 particles/ml

Fine forceps used for handling grids

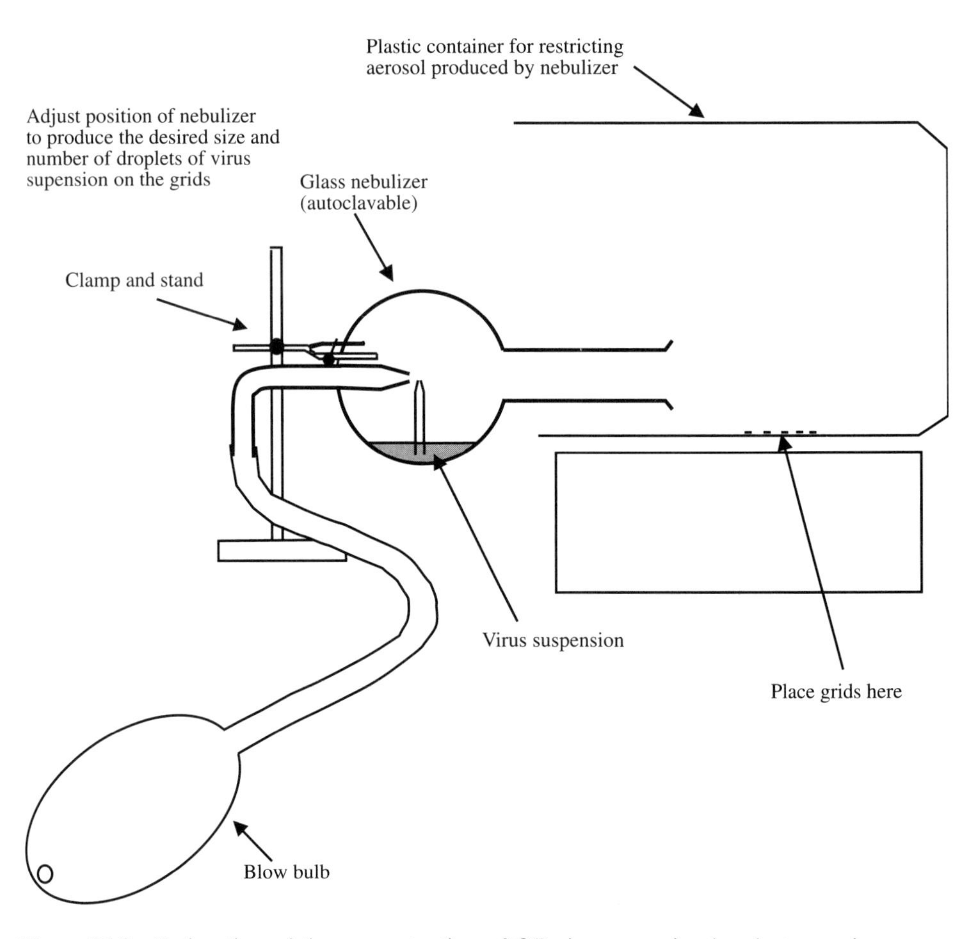

Figure 26.3 Estimation of the concentration of OBs in suspension by electron microscopy.

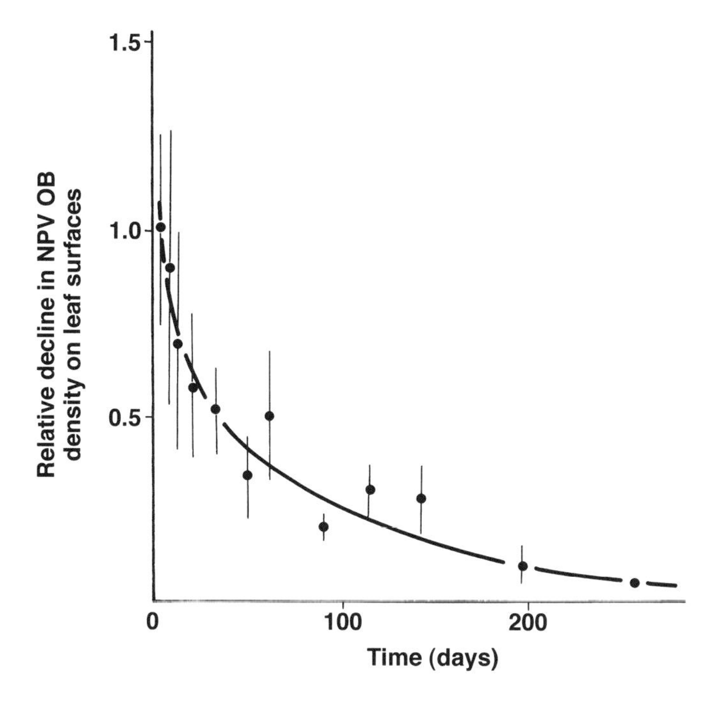

Figure 26.4 Relative decline in leaf surface density of OBs of NPV on pine (*Pinus* sp.). At day '0' foliage was dipped in an aqueous suspension of 1×10^7 OB/ml while *in situ* on the trees. Bars indicate standard deviations of samples from the mean. For adequate definition of decay curves of this form, it is necessary to have a high frequency of sampling in the early stages because of the initial loss of OBs. (P. F. Entwistle and H. F. Evans, unpublished data.)

ESTIMATION OF NUMBERS OF OCCLUSION BODIES ON PLANT SURFACES

It is often useful to be able to measure the density of OBs of insect viruses on plant surfaces, for instance in order to examine the physical rate of decay of deposits (Figure 26.4). This can be done using scanning electron microscopy or by taking impression films. Such estimates do not distinguish between infective and non-infective virus. Infectivity can only be determined by bioassay (see below).

IMPRESSION FILM TECHNIQUE USING THE LIGHT MICROSCOPE

Particles on plant surfaces may be (partially) removed by pressing surfaces onto adhesive tape or a microscope slide treated with an adhesive. Plant surfaces vary in robustness. For example, the surfaces of coniferous needles are very tough and the needles can be removed after pressing onto quite strong adhesives (e.g. Sellotape) without leaving the epidermis behind. More fragile leaves, e.g. cotton, will usually require a gentler adhesive, as is provided by some sticky tapes designed for paper repair. Pressing the same piece of plant tissue onto a successive series of adhesive surfaces will progressively remove more and more of the adherent particles, but the point will soon be reached at which statistically reliable results can no longer be obtained. The method described below is essentially that used by Elleman, Entwistle and Hoyle (1980) for double-sided Sellotape.

Method

1. Clean a slide in alcohol or acetone to remove grease and dry in air in a non-dusty atmosphere.
2. To enable easy location of plant impression, draw a grid on the upper side of the slide using a black waterproof marker pen (Figure 26.5a). For coniferous needles, for instance, a comb-like marking pattern is valuable, the needles being pressed between the 'teeth'. Several (10 or more) impressions can be accommodated on one slide marked in this way.
3. Cut a suitably sized piece of double-sided sticky tape (e.g. Sellotape or Scotch tape) and remove paper from one surface only.
4. Press sticky side very firmly onto the slide, covering the grid drawn on it; slides are now ready for use and may be stored until required.
5. Immediately before use, remove paper from upper surface of tape.
6. Press plant tissue firmly onto the sticky surface using hand pressure or weights as a press. Whichever of these is used, it will be found useful to cover the top side of the leaf with a hard flat surface, e.g. another slide. Remove the tissue carefully so that the epidermis is not removed.
7. Stain in Buffalo Black (Naphthalene Black B12 Chapter 25) for 5 min. Handle very gently to avoid the sticky tape falling from the slide. To deter this, it is best to stain the slide horizontally. Where leaf surfaces leave large deposits of wax on the tape, e.g. *Brassica oleracea*, shorter staining times should be used.
8. Rinse gently in water, tap water is suitable. Dry in air.
9. Examine under oil immersion (at greater than $\times 90$ or $\times 100$ objective) with an eyepiece graticule. Elleman *et al.* (1980) employed a magnification of $90 \times 1.25 \times 10 = \times 1125$, with the result that the actual area of plant surface covered by the eyepiece graticule was 1.44×10^{-4} cm^2. This value will possibly vary with different types of graticule.

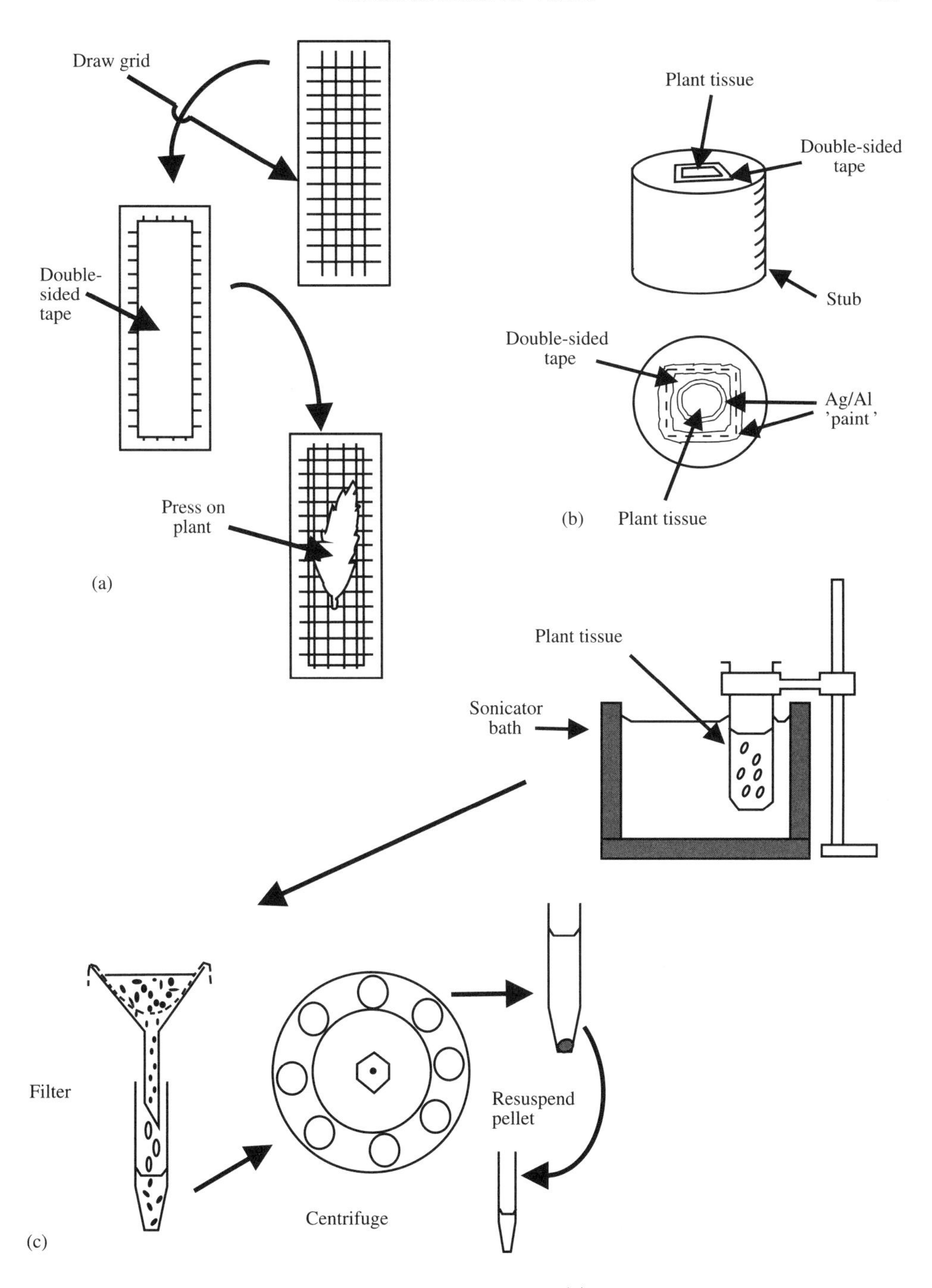

Figure 26.5 Estimation of OBs from plant surface (a) by the leaf surface impression method; (b) by scanning electron microscopy; and (c) by physicochemical recover.

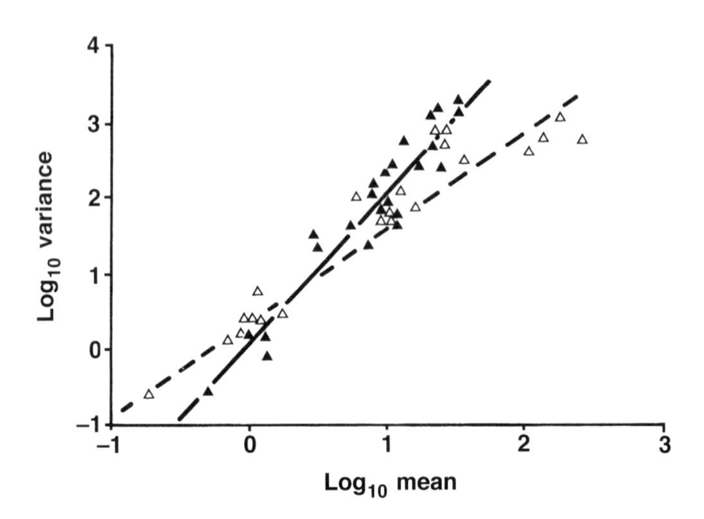

Figure 26.6 Relationship between the $\log_{10}$ mean values and $\log_{10}$ variances of OB counts per graticule for samples taken with double-sided Sellotape from cotton leaves (△) (each sample, 20 counts) and *Pinus contorta* (▲) (each sample, 8 counts made on one needle). (Elleman *et al.* (1980) *Journal of Invertebrate Pathology*; reproduced by permission of C.J. Elleman *et al.*, Academic Press, Inc.)

Accuracy

Under the above conditions, the lower limit of detection will be *c.* 3×10^4 NPV OB/cm^2. Counting becomes very inaccurate at densities greater than about 100 OBs/graticule area, which occurs at 1×10^6 OB/cm^2. This is because of the increasing likelihood of OBs occurring in groups, with resultant difficulty in counting them accurately. Care should be taken in the interpretation of quantitative data. OB distribution is more likely to be aggregated than random. This is a common consequence of aggregation associated with preparations of OBs, or caused by the structure of the plant surface. On pine needles, for example, OBs tend strongly to aggregate on longitudinal tracks, while on broadleaved plants aggregation occurs in the depressed areas over the veins.

The relationship between $\log_{10}$ mean values and $\log_{10}$ variances of counts per graticule are shown in Figure 26.6. For guidance on mathematical treatment see Elleman *et al.* (1980) and Elleman (1983).

THE USE OF A SCANNING ELECTRON MICROSCOPE (SEM)

Method

1. Always include an 'untreated' control; especially in regions where there is aerial pollution, OB-like particulate pollutants may be present. Particulate contamination can be so great that plant material from polluted areas may be unsuitable for some experimental requirements.
2. Select pieces of plant tissue as near flat as is possible to permit ease of examination at high power where there is little depth of focus. Handle with forceps being careful not to touch the surface being examined.
3. Fasten small pieces to SEM specimen holders of stubs using double-sided sticky tape (e.g. Sellotape or Scotch tape) (Figure 26.5b). Dry thoroughly in a desiccator over silica gel or in a critical point drier.
4. 'Paint' junctions of stub/sticky tape and sticky tape/tissue with a colloidal suspension of metallic silver (Ag) or aluminium (Al). This will ensure that a charge does not accummulate when the assembly is exposed to the electron beam. Place in a desiccator while organic solvents evaporate from the Ag/Al 'paint'.
5. Place in sputter coater and deposit a covering of gold (a few nanometres thick) to ensure total conductivity of specimen when in the microscope. Failure to ensure good conductivity will result in a charge build-up and a hazy picture. Difficulty in sputter coating may be experienced if the specimen is inadequately dehydrated or the organic solvent in the paint has not evaporated totally.
6. For OB counting, it is convenient to draw a grid on a clear plastic sheet and to place this over the VDU screen, very like an eyepiece graticule in an optical microscope.
7. Count numbers of OBs at low magnification, e.g. $\times$ 1000 to $\times$ 5000 is suitable for NPV. It is essential to count several fields of replicate samples in order to be able to analyse the data statistically. The advice of a statistician is advisable concerning selection of samples and number of replicates required.

RECOVERY OF OBS FROM PLANT SURFACES

The ability to recover OBs from plant surfaces varies with the nature of the plant surface itself and on whether the OBs have been purified or are in their original state combined with haemolymph and other host materials. The following protocol (Carruthers, Cory and Enwistle, 1988) is recommended.

Method

1. Calibrate the plant sample. The parameters selected will suit the needs of individual projects but for both leaf and stem tissue, they might include (a) the surface area of tissue sample; (b) the age of the plant tissue; and (c) the position on the plant (orientation, height, etc.).
2. Prepare plant tissue for sonication by cutting into pieces of convenient size, e.g. 2 cm lengths of stem.
3. Place pieces of plant tissue stem in a strong tube *c.* 5.0 cm diameter and immerse in 0.1% SDS. Clamp in a sonicator bath in a position that will be held constant for this and all subsequent samples (degree of sonication varies with position in bath and angle held). Make sure that the level of the liquid in the tube is below that of the water in the bath (Figure 26.5c). Maintain at a temperature of ≤15°C by adding ice if necessary.
4. Sonicate for 5 min.
5. Filter suspension through two layers of muslin; rinse plant tissue through the filter with a small volume of 0.1% SDS.
6. Centrifuge at 3500 × *g* for 45 min at 5°C.
7. Resuspend pellet in a small, measured volume of water and, if necessary, store at 4°C.
8. Count the number of OBs using the method of Wigley (1980) (see above).

To assess infectivity of recovered virus, bioassay against a standard virus of known concentration. To determine the quantitative efficiency of the above process, apply a known quantity of OBs to 'clean' plant tissue, allow to dry and then process as described.

Method comparison

A comparison of the relative merits of methods used for estimating OB densities on plant surfaces is illustrated in Figure 26.7.

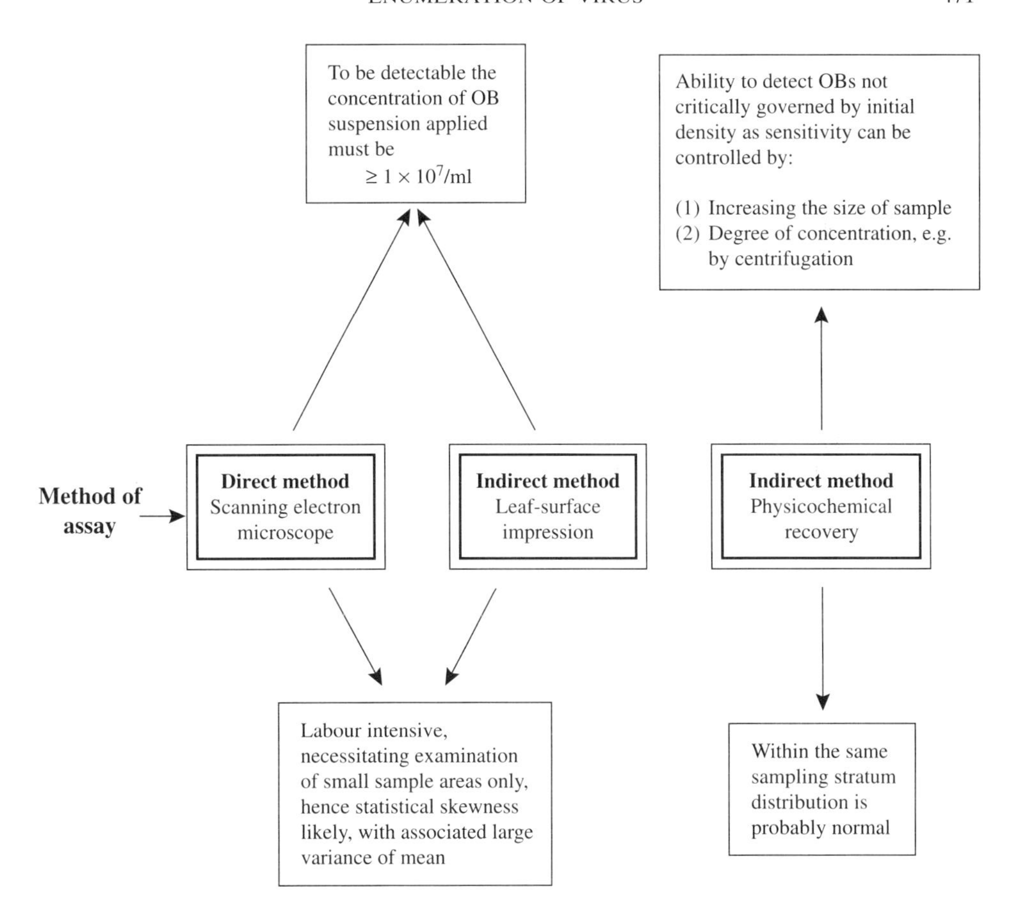

Figure 26.7 Estimation of number of BV OBs on plant surfaces. A comparison of the relative merits of direct observation and two indirect, indexing methods.

DOSAGE–MORTALITY ASSAYS

Dose determination

Determination of the precise dose ingested by a test organism is an essential prerequisite for accurate LD_{50} studies. Central to this is the ability to suspend the inoculum evenly through the carrier fluid and to deliver precise volumes to the feeding substrate. Test organisms, especially the larval stages of insects that will be the subject of most bioassays with BVs, develop very rapidly and a given population will have a range of variability reflecting the degree of synchrony of egg hatch and rate of larval development (feeding rate, possible effects of crowding if larvae are mass-reared, duration of the moult, etc.). Body size, measured by weight, is a convenient measure of variability. The effort put into weighing individual larvae has to be balanced against the needs of the assay (for example,

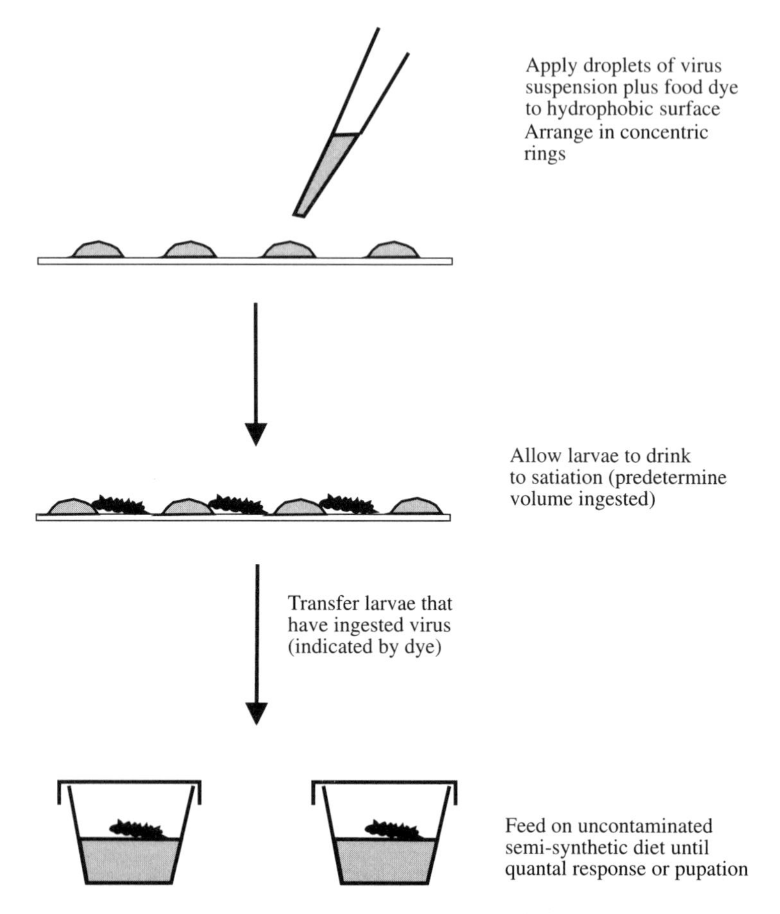

Figure 26.8 Droplet feeding of individual larvae. Droplets can also be delivered using a multiple dispenser.

is it necessary to minimize variability rather than increase numbers to give a more accurate *average* population response?) and of the danger of damage from increased handling. Individual handling and delivery of dosage is, therefore, the main constraint in setting up LD_{50} studies, but there is no other way of determining the precise dosages ingested by individual larvae. Some saving in resources can be made for particular methods of assay and these are pointed out in the discussions below.

Droplet feeding

One of the most effective ways of reducing variability in both the dosage delivered and in the amount of time taken by the test larvae to consume that dosage is to feed the inoculum to test larvae directly in droplets of virus suspension (Figure 26.8).

MORTALITY DETERMINATION BY DROPLET FEEDING

This method, developed by Hughes and co-workers (Hughes and Wood, 1981; Hughes, Wood and van Beek, 1986), is outlined in Figure 26.8. The sequence of work is outlined below for neonate larvae of *Spodoptera littoralis*.

Materials

Items marked with an asterisk should be sterile.

1. Newly hatched larvae (the eggs should not be surface sterilized as larvae emerging from treated eggs tend not to drink)
2. Plastic-coated bench protector (e.g. Benchkote) or Parafilm
3. Virus suspension diluted suitably in an aqueous solution of 1% Brilliant Blue R (Sigma Chemical Co.)
4. 7 cm plastic funnel
5. 30 ml plastic polypots or similar, five per dilution of virus
6. One or more paint brushes, preferably No. 2, sable
7. Vircon (1%) or Tegodyne (0.05%) solution for decontaminating paint brushes (avoid immersing metal parts as the solutions are corrosive)
8. 30 ml plastic polypots containing diet to a depth of 1 cm
9. Pipettes or pipetting device for preparing and dispensing dilutions of virus
10. Containers appropriate for preparing dilutions of virus, e.g. plastic microtitre plate or microfuge tubes*
11. Strips or squares of sterile filter paper to insert in lids of polypots
12. Glass-distilled water*.

Method

1. Prepare a suitable series of dilutions (two- or threefold) of virus in 1% aqueous solution of Brilliant Blue R (or a similar food dye).
2. Surface sterilize bench, for example with 0.05% Tegodyne or 70% ethanol.
3. Place Parafilm or bench protector (plastic side uppermost to form a hydrophobic surface) on bench and surface sterilize.
4. For each dilution of virus, pipette $5 \times 5\ \mu l$ drops on to the bench protector in a circle approximately 2 cm in diameter; as a control prepare one circle of drops of stain alone.

5. Place a piece of filter paper on which larvae are hatching into the mouth of the funnel and hold over the centre of the circle of drops. Tap gently until 100–150 larvae have been dispensed.
6. Invert a 30 ml polypot over each circle of drops. Label with the appropriate dilution and leave undisturbed for 20–30 min. The guts of larvae that ingest the dye, and with it the virus, will appear blue. If an LD_{50} value is required, the mean volume ingested per larva should be determined (Hughes and Wood, 1981).
7. Starting with the control sample (lacking virus), use a paintbrush to transfer larvae with blue dye visible in their guts to pots containing diet. Place filter paper liner into each lid and add to pot. Repeat for the larvae fed the suspensions of virus, working from the most dilute to the most concentrated. If the results of the assay are to be read within 7 days, larvae can be dispensed in groups of 10 per pot. If they are to be followed until death, pupation or emergence of moths, each should be put in a separate pot to avoid cannibalism.
8. Incubate pots at 26 ± 1°C (12 h:12 h, light:dark) and observe for signs of infection, i.e. disintegration of larval bodies and release of milky fluid containing polyhedra and lysed cells.
9. Record the numbers treated and infected for each dilution of virus and for the control sample.
10. Calculate the ED_{50}, ID_{50} or LD_{50} values.

Semi-synthetic diet feeding

The ability to rear insects on semi-synthetic diet has great value in developing bioassay procedures. The two main approaches to use are surface contamination, where virus suspensions are applied only to the diet surface (Figure 26.9) or diet incorporation, where the virus is incorporated fully into the diet while it is still liquid (Figure 26.10). Once the virus is applied by either method there are two pathways that can be adopted, leading to LC_{50} or LD_{50} determination depending on whether a known quantity of virus is consumed. These procedures are illustrated here by reference to *S. littoralis* and *P. interpunctella* for LD_{50} determinations. Each insect species to be tested will need to be assessed for feeding rate, size, etc. and the method adapted accordingly. The procedures below are labour intensive and are necessary only if detailed LD_{50} determination is to be carried out. LC_{50} determinations are easier to set up and require substantially less handling, although greater information can be gained if tight control over larval weight and larval stage can be retained.

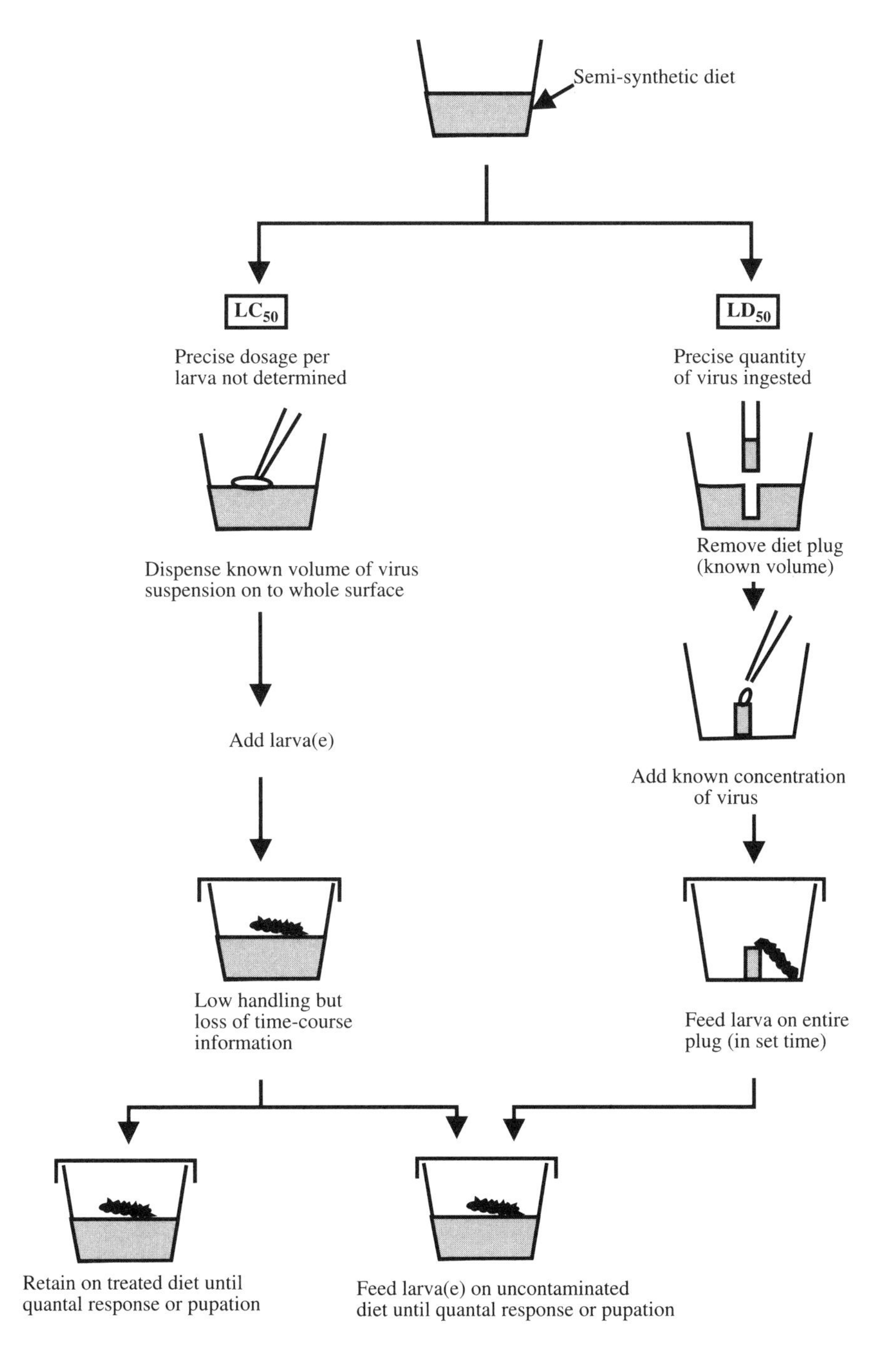

Figure 26.9 Feeding larvae with a semi-synthetic diet: surface treatment of the diet with virus.

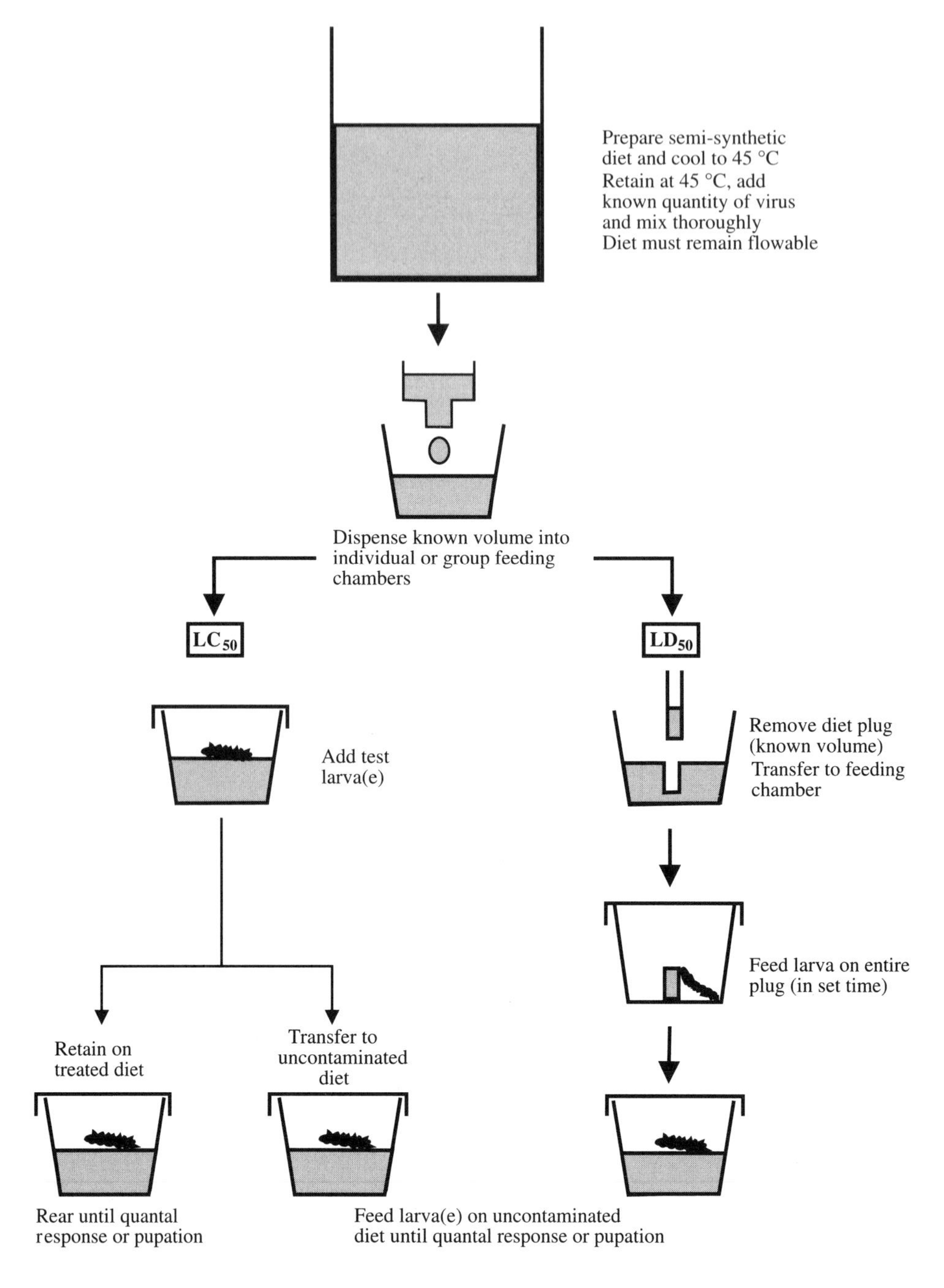

Figure 26.10 Feeding larvae with a semi-synthetic diet: incorporation of virus into the diet.

MORTALITY DETERMINATION IN FOURTH INSTAR AND OLDER *S. LITTORALIS* LARVAE

Materials

Scalpel and forceps should be sterilized between operations by flaming with alcohol (Chapter 23).

1. Larvae at the required stage of development
2. Plastic polypots, or equivalent, one empty and one with food, for each larva to be tested
3. Block of semi-synthetic diet (Chapter 25)
4. Cork borer 5 mm in diameter and scalpel
5. Pipetting device for dispensing 1 μl volumes
6. Forceps for handling larvae
7. Series of dilutions of virus suspension
8. Diluent (sterile), e.g. glass-distilled water, for testing control larvae
9. Plastic tubs (8 oz) containing fine diet.

Method

1. Rear larvae from the same batch of eggs (i.e. laid at the same time) in groups of 100 in 250 ml plastic tubs at 26 ± 1°C.
2. When the correct age or stage of development is reached, choose larvae of uniform size and weigh individually.
3. Once weighed, transfer larvae to individual polypots. Label each pot with the weight of the larva it contains. Allow at least 20 larvae for each dilution of virus to be tested.
4. Arrange pots in order of weight of larvae.
5. Select the five heaviest and five lightest larvae and measure the sizes of their head capsules. Use these measurements to determine the instar of the larval population as a whole (McKinley, Smith and Jones, 1984).
6. Starting with the pot containing the lightest larva and working towards the heaviest, label one pot for each dilution of virus to be tested. Repeat this procedure until all pots have been labelled (minimum of 20 replicates per dilution). This ensures that weights are spread evenly within dose batches.
7. Sterilize a 5 mm diameter cork borer and scalpel by flaming. Use the cork borer to cut plugs from the block of semi-synthetic diet. Slice into plugs 3 mm deep with the scalpel.

8. Arrange plugs of food on the lids of 250 ml plastic tubs, bench protector or similar.
9. Starting with diluent alone (control, lacking virus) and working up to the most concentrated suspension of virus, dispense 1 μl volumes on to the required number of replicate plugs.
10. Leave to dry at room temperature (approximately 20–25°C).
11. Using sterile forceps, transfer one plug to each polypot containing a larva, ensuring that each plug is allocated to the correct larva.
12. Incubate at 26 ± 1°C for 24 h.
13. Discard larvae that have not consumed the complete diet plug and add cubes of fresh food (without virus and approximately 2 cm × 2 cm × 1 cm in size) to the remaining larvae. Alternatively transfer larvae to pots containing semi-synthetic diet.
14. Reincubate and check daily for death caused by virus infection, or for pupation and emergence of adults.
15. Record proportion of larvae dying from virus infection and calculate LD_{50} values.

MORTALITY DETERMINATION FOR GRANULOSIS VIRUS INFECTION IN THE INDIAN MEAL MOTH *P. INTERPUNCTELLA*

Materials

1. Late third to early fourth instar larvae
2. Glass capillary tubing (1.5 mm internal diameter)
3. Plastic polypots or equivalent
4. Semi-synthetic diet for larvae
5. One pair of forceps
6. One paper clip
7. Cotton wool
8. Paint brush, preferably No. 2, sable
9. Suspension of virus to be assayed
10. Sterile diluent, e.g. glass-distilled water
11. Pipettes and suitable containers for preparing dilutions of virus
12. Micropipette for delivering 1 μl volumes

13. Disinfectant suitable for decontaminating paint brush (e.g. 0.05% Tegodyne) and pipetting equipment.

Method

1. Prepare food as described for rearing larvae (Chapter 25).
2. Cut capillary tubing into approximately 17 mm lengths using a glass cutter and wearing safety spectacles. Allow one capillary for each larva.
3. Plug the end of each capillary with diet to a uniform depth of about 2 mm and push the diet about 1 mm along the tube using a straightened paper-clip. Wipe any diet off the outside of the tube. Store at 4°C until required.
4. Transfer capillary tubes to a sandwich box lid and arrange so that the filled ends are touching the edge and the open ends are facing a source of light (Figure 26.11).
5. Decide which dilutions of virus will be suitable for inoculating larvae and label polypots accordingly.
6. Allow one polypot per larva and 10 larvae per dilution. Include 10 tubes for uninoculated control larvae.
7. Lay the polypot lids in rows of 10 on the bench and place sufficient food to sustain a larva for at least 10 days on each.
8. Using a paintbrush, transfer larvae from the rearing box and place individually near the open ends of the capillary tubes.
9. Once a capillary tube is filled, remove to another lid and replace with a fresh tube.
10. Prepare a thread from non-absorbent cotton wool and cut into pieces approximately 1 mm long.
11. Trap the larvae within the tubes by plugging the open end of each with a piece of cotton wool thread (Figure 26.11).
12. Prepare serial twofold or threefold dilutions of virus in the appropriate diluent, e.g. sterile distilled water.
13. For each dilution, place 1 μl of the appropriate dilution of virus onto the food in each of 10 capillary tubes.
14. Place one capillary tube onto each polypot lid, add a polypot and transfer, lid downwards, to an incubating basket.
15. Incubate at 26°C for a minimum of 10 days or until larvae have either died or have developed into moths.
16. If incubating for 10 days, note the proportion of larvae infected with virus at each dilution and either estimate the ID_{50} by the method of Spearman and Kärber (below) or use probit analysis. If incubating until death or emergence of moths, estimate or calculate the LD_{50}.

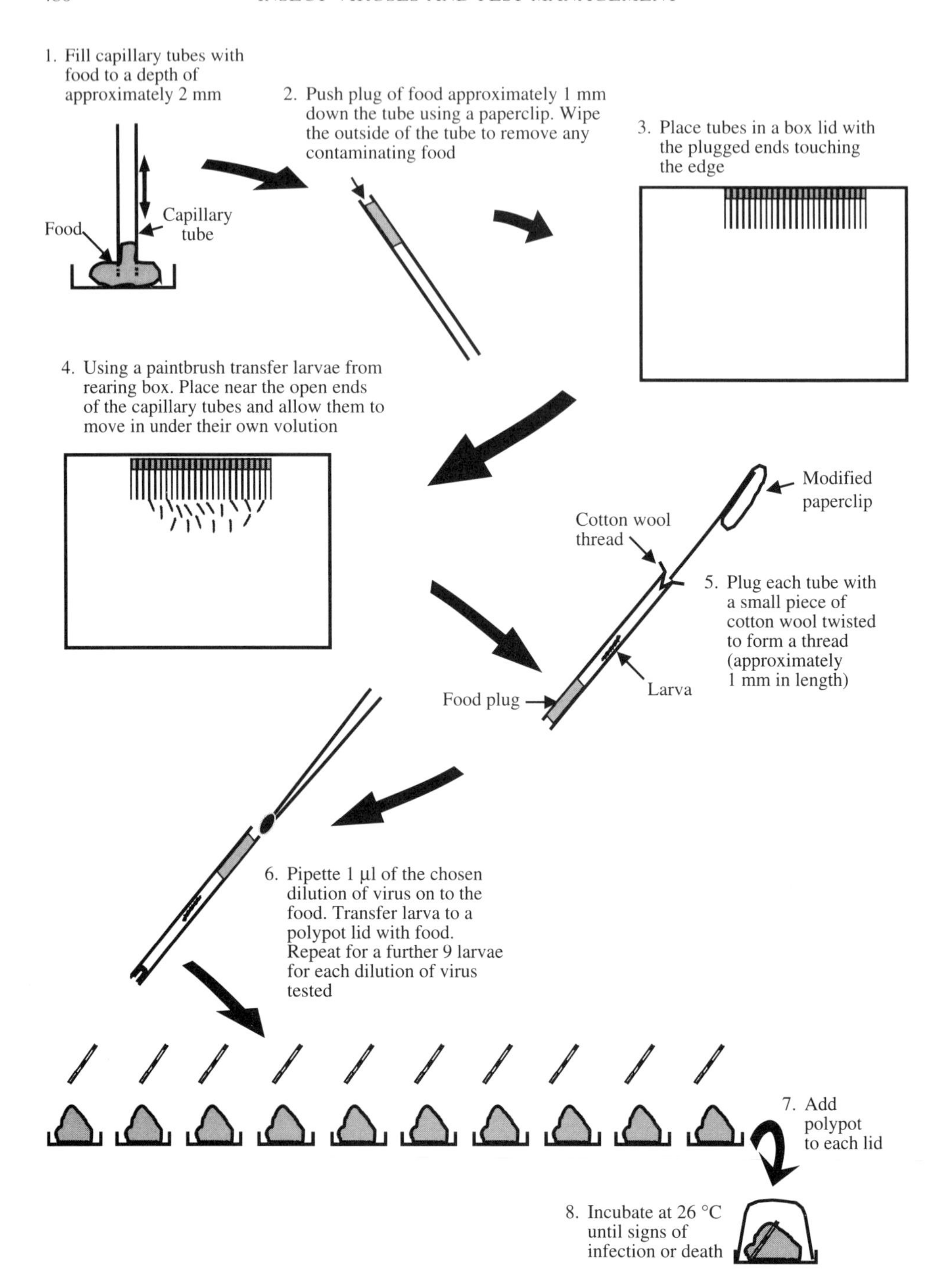

Figure 26.11 Assay of GV in *P. interpunctella*.

Leaf feeding

Assays using leaf as the feeding substrate are required when it is not possible to feed on semi-synthetic diet or where it is necessary to mimic, as closely as possible, the natural uptake of virus that would occur in larvae in a field situation.

Both LD_{50} and LC_{50} methods can be used and are outlined in Figure 26.12. For both, either coniferous or broadleaved foliage can be used and, in the case of LC_{50} studies, the method is the same. For LD_{50} studies, it is necessary to dispense precise dosages to quantities of foliage that will be consumed entirely in 12 h or less and the methods used will depend on the leaf substrate. The two main procedures are outlined in Figure 26.12. The retention of leaf disks on an agar medium helps to maintain turgor long enough for the entire disk to have been consumed. In all other respects the assays are identical to those described for semi-synthetic diet, unless it is necessary to continue rearing on leaf for the whole duration of the experiment.

METHODS OF ANALYSIS

Analysis of dose–response data derived from bioassay is based on assessing the relationship between the quantal response and the dosage applied. As mentioned in Chapter 3, the one-hit theory of virus acquisition and infection suggests that response will not be determined by increased concentration of inoculum but by the probability of a single virus particle commencing the infection process. Such responses are best described by an exponential model, which has been discussed fully by Ridout, Fenlon and Hughes (1993). However, the inherent variability in carrying out bioassays tends to disguise the one-hit response and, therefore, the observed effects are very similar to those where an accumulated dose is ingested. As a result, methods such as probit analysis that have a long pedigree in analysis of dose–mortality data are still effective and can be used routinely. The methods below are selected to illustrate the procedures that can be applied in quantal assays. In general, it is preferable to use computer packages to carry out probit analysis and to produce the associated statistics.

Simple estimation methods

Spearman–Kärber method

An effective and reasonably accurate method of estimating the LD_{50} is that known as Spearman–Kärber. This is simple in operation and also allows estimates of error from the observed data. The LD_{50} is determined by associating the proportion of positive responses at each dose with an average $\log_{10}$ dose. The method is not strictly applicable if the range of doses does not include responses from 0% to 100% but, as explained above, these values cannot be determined precisely and, therefore, it is acceptable to assume, by approximation from doses giving near to 0% or 100% mortality, what the next lower or next higher values were.

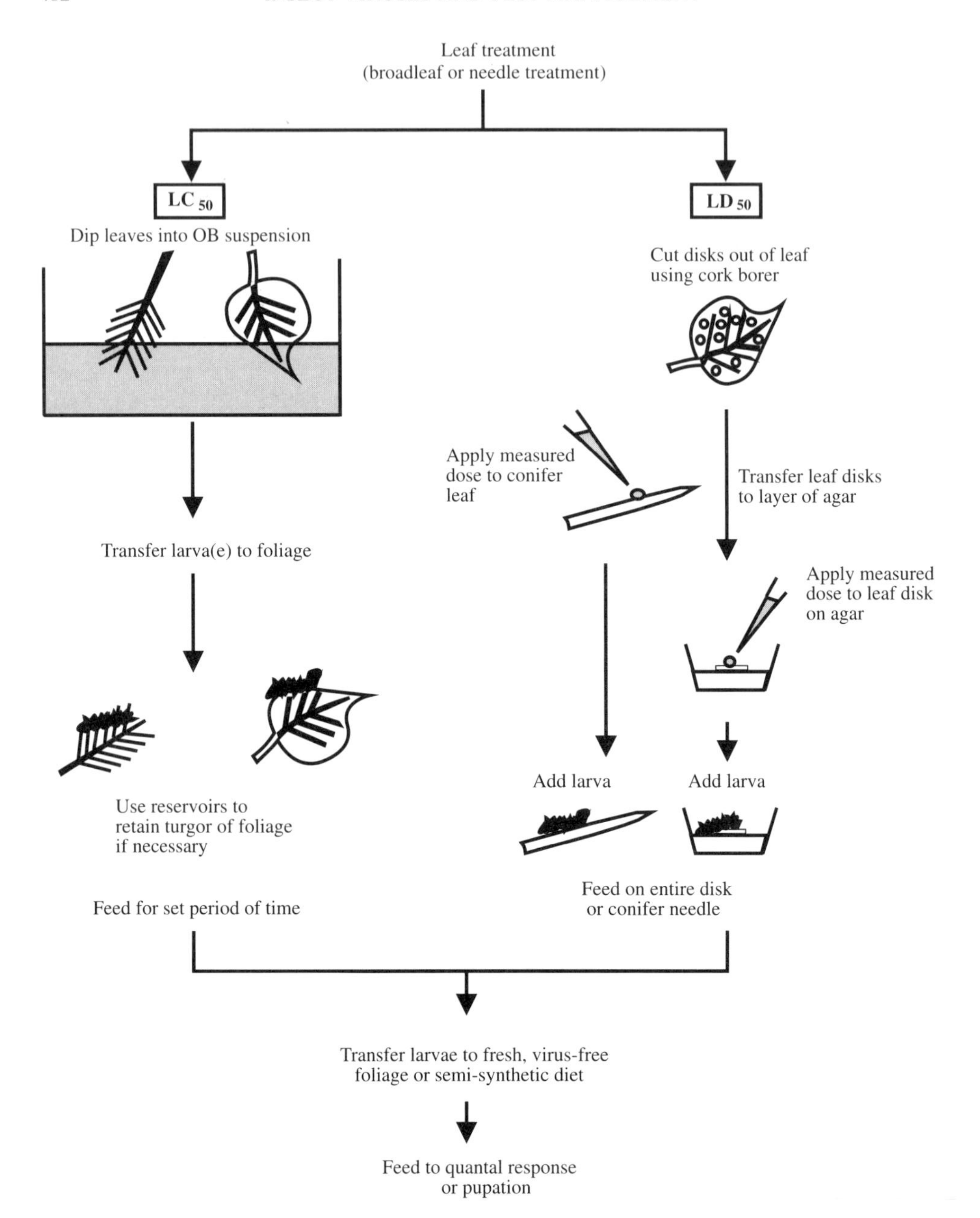

Figure 26.12 Feeding larvae with virus by direct treatment of foliage.

The LD_{50} is calculated from the formula:

$$\text{Log}_{10}\text{LD}_{50} = X_{p=1} - d(\sum p - 0.5)$$

where $X_{p=1}$ is the highest $\log_{10}$ dilution giving 100% quantal response; d is the $\log_{10}$ dilution factor; p is the proportion of positive responses at a given dosage; and Σp is the sum of p for $X_{p=1}$ and all higher dilutions.

The standard error of the LD_{50} can be calculated from the formula:

$$\text{SE}_{\log10}\text{LD}_{50} = d\sqrt{\frac{\sum p(1-p)}{(n-1)}}$$

where n is the number of units (larvae) at each dilution.

The 95% confidence limits (CL) for the $\log_{10}LD_{50}$ are obtained from:

$$\log_{10}(95\%) = \log_{10}\text{LD}_{50} \pm 1.96\ \text{SE}_{\log10}\text{LD}_{50}$$

The method is illustrated by reference to data from a *P. interpunctella* GV assay (Table 26.3). Although tedious, the result can be derived by hand calculator, although it is simple to develop a spreadsheet solution that is quicker and more accurate. The calculation using the data in Table 26.3 becomes

$$\begin{aligned}\text{Log}_{10}\text{LD}_{50} &= -1.6 - (0.3)(4.592 - 0.5)\\ &= -1.6 - 0.3(4.092)\\ &= -1.6 - 1.2276\\ &= -2.8276\end{aligned}$$

$$\text{SE}_{\log10}\text{LD}_{50} = 0.3\sqrt{\frac{1.0603}{85}}$$

$$= 0.0335$$

$$\begin{aligned}\text{Log}_{10}(95\%\ \text{CL}) &= -2.8276 \pm 1.96(0.0335)\\ &= -2.794 \text{ and } -2.8611\end{aligned}$$

Table 26.3 Assay of *P. interpunctella* GV

Log dilution (X)	Number in group (n)	Number positive	Proportion positive (p)	$(1 - p)$	$p\ (1 - p)$
1.6	86	86	1.000	0.000	0
1.9	86	83	0.965	0.035	0.0338
2.2	86	80	0.930	0.069	0.0642
2.5	86	54	0.628	0.372	0.2336
2.8	86	34	0.395	0.605	0.2390
3.1	86	31	0.360	0.640	0.2304
3.4	86	18	0.209	0.791	0.1653
3.7	86	9	0.105	0.895	0.0940
4.0	86	0	0.000	1.000	0

$X_{p=1} = -1.6$
$\Sigma p = 4.592$
$\Sigma p\ (1 - p) = 1.0603$
$d = 0.3$

Probit analysis

It is not appropriate to provide full details on probit analysis since it is a technique that is available in most laboratories that possess computer facilities. Alternatively, it is possible to calculate probit responses and their fiducial limits using a hand calculator, as detailed in Finney (1971). In essence, the method converts the sigmoid dose–response curve for percentage mortality to a linear response based on probits (log normal linearization). This allows an iterative process to be carried out to determine the best fit of the regression of probit against log dose. This can then be used to calculate LD_{50} and the slope of the regression and to determine the fiducial limits and degree of heterogeneity of the response. Hughes and Wood (1987) have reviewed bioassay methods and provide a comprehensive discussion of various methods for calculation of dosage response curves.

REFERENCES

Carruthers, W.R., Cory, J.S. and Entwistle, P.F. (1988) Recovery of pine beauty moth (*Panolis flammea*) nuclear ployhedrosis virus from pine foliage. *Journal of Invertebrate Pathology*, **52**, 27–32.

Elleman, C.J. (1983) The interrelationships between baculovirus of *Spodoptera littoralis* and the leaf surface of *Gossypium hirsutum* with comparative observations in *Brassica oleracea*. DPhil thesis, University of Oxford.

Elleman, C.J., Entwistle, P.F. and Hoyle, S.R. (1980) Application of the impression film technique to inclusion bodies of nuclear polyhedrosis viruses on plant surfaces. *Journal of Invertebrate Pathology,* **36**, 129–132.

Finney, D.J. (1971). *Probit Analysis*. London: Cambridge University Press.

Hughes, P.R. and Wood, H.A. (1981) A synchronous peroral technique for the bioassay of insect viruses. *Journal of Invertebrate Pathology* **37**, 154–159.

Hughes, P.R. and Wood, H.A. (1987) *In vivo* and *in vitro* bioassay methods for baculoviruses. In *The Biology of Baculoviruses*: Vol. II. *Practical Application for Insect Control*, R.R. Granados and B.A. Federici (eds). Boca Raton, FL: CRC Press, pp. 1–30.

Hughes, P.R.; Wood, H.A. and van Beek, N.A.M. (1986) A modified droplet feeding method for rapid assay of *Bacillus thuringiensis* and baculoviruses in noctuid larvae. *Journal of Invertebrate Pathology* **48**, 187–192.

McKinley, D.J., Smith, S. and Jones, K.A. (1984) The laboratory culture and biology of *Spodoptera littoralis* Boisduval. London: Tropical Development and Research Institute, Overseas Development Administration.

Ridout, M.S., Fenlon, J.S. and Hughes, P.R. (1993) A generalized one-hit model for bioassays of insect viruses. *Biometrics* **49**, 1136–1141.

Wigley, P.J. (1980) Practical: counting micro-organisms. In *Bulletin 228: Microbial Control of Insect Pests*, J. Kalmakoff and J.F. Longworth (eds). Wellington, New Zealand: New Zealand Department of Science and Industrial Research, p. 29.

27 Cell culture

JUN MITSUHASHI

INTRODUCTION

Although large quantities of virus can usually be propagated in insects, there are situations where propagation of virus in cell culture is advantageous or essential. Certain host species may, for example, be difficult to rear in the laboratory because of their stringent growth requirements, the length of time between generations or reactivation of occult virus. Molecular biological studies and manipulations and studies of budded virus are more easily performed in cell culture.

Once established, insect cell lines are relatively easy to culture and passage. With experience, healthy cells are clearly distinguishable from unhealthy ones, although the appearance (size, shape and contents) of cells in culture will vary depending on origin.

Although it is easier to work in a laminar airflow cabinet, it is quite possible to passage established cell cultures on the open bench provided that adequate precautions are taken and stringent aseptic technique is used (Figure 23.6, p. 367). It may be necessary to add antibiotics to the medium when working under unfavourable conditions but this should be discouraged as the cells may be affected adversely by such additions. Also, the presence of antibiotics in the medium may mask low levels of microbial contamination. For establishing primary cell or organ cultures, the use of antibiotics is mandatory and the use of a cabinet almost essential. The sterility of the air within the cabinet or working area should be tested regularly (see below). If contamination is discovered, steps should be taken to trace and eradicate its source.

Cells can be grown as monolayers or in suspension, in glass (reusable) or specially treated plasticware (disposable). For small quantities, it is more usual to prepare stationary culture but for large volumes of cells, suspension culture is preferred. It is essential that aseptic technique is used throughout.

Additional information on insect cell culture and its application to the study of viruses can be obtained by reference to King and Possee (1992), O'Reilly, Miller and Lucklow (1992), Shuler *et al.* (1995) and Vlak *et al.* (1996).

CHECKING FOR AERIAL CONTAMINATION OF WORKING SPACE

Media and materials

1. Petri dishes containing media suitable for growing bacteria (e.g. nutrient agar) and fungi (e.g. malt agar).

Method

1. Swab bench or cabinet with 70% ethanol or equivalent.
2. Place Petri dishes at strategic positions over the working area (within a cabinet or on the bench).
3. Remove lids for a range of time periods from a few minutes to several hours, both while the laboratory or cabinet is in use and while empty.
4. Replace lids and incubate the dishes at 25–27°C.
5. Examine plates for growth of colonies.
6. Take steps to remove any source of contamination. If working on the bench, check the air movement within the room and re-site the working area if necessary. If working in a cabinet, check the airflow and replace any filters that show signs of damage. As an interim measure test the antibiotic sensitivity of any contaminant(s) identified and add the relevant antibiotic to the culture medium until the source of contamination has been removed. Care should be taken to ensure that any antibiotic used is not toxic to the cells being treated.

TESTING FOR ANTIBIOTIC SENSITIVITY

Media and materials

1. Overnight broth culture of the contaminant
2. The required number of sterile Petri dishes
3. Molten agar medium, e.g. nutrient agar, in 20 ml volumes in 25 ml MacCartney bottles maintained at 45°C in a water bath
4. A pair of forceps
5. Ethanol or industrial methylated spirits for flaming forceps
6. Filter paper discs prepared with a paper punch and sterilized by dry heat (p. 367)
7. Solutions of the antibiotics under test.

Note: Items 6 and 7 can be replaced by antibiotic sensitivity discs purchased commercially.

Method

1. Inoculate the molten agar medium with two loopfuls of an overnight broth culture of the contaminant.
2. Mix thoroughly and quickly pour into a Petri dish.
3. Rotate Petri dish to ensure an even spread of medium and distribution of organisms.
4. Allow agar to set.
5. Sterilize forceps by flaming, i.e. dip in alcohol, ignite by touching to a flame momentarily and allow to burn. Once flamed, pick up a paper disc, dip into the appropriate solution of antibiotic and then place on the seeded agar plate. Repeat for all antibiotics being tested. At least three discs can be placed on each Petri dish.
6. Incubate plates overnight at 25–30°C or until the contaminant has grown.
7. Examine for clear areas (lack of bacterial growth) around the discs. The larger the zone of inhibition the greater the sensitivity of the contaminant to the antibiotic being tested. The lack of a zone of inhibition indicates that the contaminant is resistant to the action of the antibiotic being tested.

CELL CULTURE MEDIA AND SOLUTIONS

Many media and solutions are available commercially, either as powders or liquids from such sources as Gibco BRL Life Technologies and Sigma Chemical Co. Media prepared from powders must be reconstituted with high-grade water and, since most of the ingredients are heat labile, must be sterilized by membrane filtration. It is advisable to use a pre-filter to remove coarse particles that are likely to clog the pores of a sterilizing filter (0.22 μm pore size). The pore size of the most commonly used pre-filter is 0.45 μm. This can be preceded by filters with pore sizes of 1.00 and 0.60 μm. Serum is usually purchased commercially and has, therefore, been checked for sterility. It is most important that this includes testing for the absence of mycoplasmas. These microorganisms, which are able to pass through 0.2 μm filters, can markedly affect the quality of cell cultures as well as their ability to support virus replication.

The general principles involved in the preparation of culture media are given below by reference to the preparation of Mitsuhashi and Maramorosch's medium (1964). Although containing very few ingredients, it is possible to adapt certain cells e.g. those of *Spodoptera litura*, to grow in it and to support virus multiplication. Aseptic transfer of media and cell suspensions can be effected by the use of such devices as are illustrated in Figures 23.1, 23.4 and 23.5, pp. 362, 365, 366).

MITSUHASHI AND MARAMOROSCH'S MEDIUM

Ingredients

$NaH_2PO_4.H_2O$, 0.2 g; $MgCl_2.6H_2O$, 0.1 g; KCl, 0.2 g; $CaCl_2.2H_2O$, 0.2 g; NaCl, 7.0 g; $NaHCO_3$, 0.12 g; glucose, 4.0 g; lactalbumin hydrolysate. 6.5 g; TC-yeastolate, 5.0 g; fetal bovine serum, 200 ml; penicillin, 1×10^5 units; dihydrostreptomycin sulphate, 0.1 g; glass-distilled water, up to 1 l.

Preparation

1. Dissolve chemicals in the order given in approximately 700 ml glass-distilled water.
2. Add 200 ml fetal bovine serum.
3. Make up to 1 l with glass-distilled water.
4. Adjust the pH to a value of 6.5 with KOH.
5. Filter through a membrane filter of pore size 0.45 μm (pre-filter) followed by one of 0.22 μm pore size (sterilizing filter).
6. Dispense in suitable volumes, e.g. 100 ml. Write the date on each bottle used and identify it by number.
7. Transfer a small sample (e.g. 3 ml) from each bottle to a similarly numbered sterile bottle and check for sterility by incubating it for a few days at 28°C or 37°C.
8. In the meantime, cover caps with aluminium foil or similar to exclude dust and store medium at 4°C. Discard any that appears to have been contaminated once the results of the sterility test are known.
9. Store at 4°C.

Note: It is possible to obtain sterile solutions of antibiotics. These can be stored at –20°C and need not be added until the medium is ready for use.

Carlson's solution

This is used as a physiological saline (Carlson, 1946) and contains: NaCl, 0.7 g; $CaCl_2.2H_2O$, 0.02 g; KCl, 0.02 g; $MgCl_2.2H_2O$, 0.01 g; $NaH_2PO_4.H_2O$, 0.02 g; $NaHCO_3$, 0.012 g; glucose, 0.8 g; distilled water, 100 ml.

Rinaldini's salt solution

Rinaldini's salt solution is used for trypsin solution (Rinaldini, 1959) and contains NaCl, 0.8 g; $NaH_2PO_4.H_2O$, 0.005 g; $NaHCO_3$, 0.1 g; KCl, 0.02 g; glucose, 0.1 g and sodium citrate, 0.0076 g dissolved in 100 ml glass-distilled water. Store at 4°C.

Preparation of 0.1% trypsin solution

1. Add 0.1 g trypsin to 100 ml Rinaldini's solution.
2. Dissolve by stirring overnight at 4°C.
3. Sterilise by filtration through a membrane filter, pore size 0.22 μm.
4. Dispense in 5 ml volumes and store at –20°C.

PREPARATION OF CONDITIONED MEDIUM

The equipment required will depend on the volume of conditioned medium to be prepared. Requirements for volumes of up to 50 ml are given.

Equipment and materials

Items marked with asterisk should be sterile.

1. Cell cultures that are almost confluent
2. Pipettes for transferring medium*
3. Aspirator (optional) for collecting medium
4. Syringe, 5, 10, 20 or 50 ml capacity*
5. Membrane filters with pore sizes of 0.40 μm and 0.22 μm*
6. Bench centrifuge (optional)
7. Centrifuge tubes sufficient to accommodate the volume of medium being harvested (optional)*
8. Beakers 10, 20 or 50 ml volume, as required, with aluminium foil lids*
9. Containers for storing filtered, conditioned medium, e.g. sterile bijoux, ½ or 1 oz MacCartney bottles*.

Method

1. Examine cultures and choose those that are healthy.
2. Transfer the spent medium into a sterile container by pouring, aspiration or pipette and allow any cells that may be present to settle under gravity. Alternatively, dispense into centrifuge tubes and centrifuge at 500–1000 × *g* for 10 min.
3. Transfer supernatant fluid to sterile beaker.
4. Load syringe with medium and attach to 0.45 μm pore size filter.

5. Discharge medium through the filter into a second beaker.
6. Remove filter from syringe without touching any part of the syringe that is going to come into contact with the medium.
7. Load the syringe with filtered medium and attach to a 0.22 μm pore size filter.
8. Discharge conditioned medium into a sterile container.
9. Use immediately or store at 4°C.

Note: Mycoplasma will not be removed by filtration. It is, therefore, very important not to touch any part of the equipment that will come into contact with the medium.

COUNTING NUMBERS OF VIABLE CELLS

USE OF HAEMOCYTOMETER

Several types of counting chamber are available, of which the Improved Neubauer haemocytometer is one of the most common; its use is described on p. 456.

Equipment

Items marked with asterisk should be sterile.

1. Improved Neubauer haemocytometer with thick coverslip
2. Pipettes and diluent for diluting cells*
3. Pasteur pipette or similar for transferring cells to haemocytometer*
4. Light microscope.

Method

1. Load a Pasteur pipette, or similar, with suitably diluted cell suspension.
2. Gently discharge some of the suspension into the chamber ensuring that it is filled. Excess suspension will flow into the channels and should not be removed.
3. Allow cells to settle.

4. Count the number of cells in the 4 large squares (1.0 mm square × 0.1 mm depth). Include cells that touch two of the outside borders, e.g. the upper and left-hand side. A minimum of 200 cells should be counted.
5. Determine the number of cells per large square. This is the number of cells in 10^{-4} ml.

The number of cells/ml = mean number of cells per large square × 10^4 × dilution factor.

DETERMINATION OF THE CONCENTRATION OF VIABLE CELLS BY USE OF FLUORESCEIN DIACETATE

Equipment and materials

1. Haemocytometer (e.g. Improved Neubauer)
2. Fluorescence microscope
3. Fluorescein diacetate, 5 mg/ml in acetone, stored in a tightly capped container at –20°C
4. PBS (Dulbecco A)
5. Cells diluted suitably in growth medium (approximate concentration 10^6 cells/ml)
6. Pipettes for transferring and diluting cells.

Method

1. Dilute the fluorescein diacetate solution 1/50 in PBS at room temperature.
2. Add 100 μl of the diluted fluorescein diacetate solution to 900 μl cell suspension and leave at room temperature for 15 min. The fluorescein diacetate will be hydrolysed within viable cells, which will fluoresce green. Dead cells will not.
3. Load the cell suspension into the haemocytometer and count a minimum of 200 cells.

Number of viable cells/ml = mean number of cells in large square × 10^4 × dilution factor. Percentage viable cells = number of viable cells ÷ (number of viable cells + number of dead cells) × 100%.

DETERMINATION OF CONCENTRATION OF VIABLE CELLS BY USE OF TRYPAN BLUE EXCLUSION

Viable cells appear colourless with Trypan Blue while non-viable cells are blue. The cells must be counted within 3 min of being stained, however, since viable cells will begin to take up the dye after this time. Serum should be omitted from the diluent as Trypan Blue has an affinity for proteins and its presence will affect the accuracy of the count.

Equipment and materials

1. Equipment as for the fluoresent method above with the exception that an ordinary light microscope is employed
2. Cell suspension at a concentration of approximately 2×10^6 to 5×10^6 cells/ml
3. Trypan Blue, 0.2% (w/v) in water
4. Concentrated saline (5 ×): 4.25% NaCl (w/v).

Method

1. On day of use, mix 4 parts of 0.2% Trypan Blue with 1 part 5× saline.
2. Dilute cells 1/2 in the diluted Trypan Blue solution.
3. Load cells into the haemocytometer and count.

DETERMINATION OF CONCENTRATION OF VIABLE CELLS BY NIGROSIN EXCLUSION

Equipment and materials

1. Equipment and cell suspension as for the Trypan Blue exclusion method
2. Nigrosin, 1% (w/v) in water (stock solution, filtered through Whatman filter paper No. 1)
3. Saline: 0.85% (w/v) NaCl.

Method

1. Dilute the 1% Nigrosin stock solution 1/10 in medium containing 2–5% fetal bovine serum just before use.
2. Dilute the cell suspension between 1/2 and 1/10 with 0.1% Nigrosin solution.
3. Leave for 5–10 min, load into haemocytometer and determine numbers of living (colourless) and dead (brown-black) cells.

PRIMARY CELL CULTURE OF LEPIDOPTERAN CELLS

Cell lines have been established from various tissues, for example embryos, haemocytes, fat bodies, ovaries, imaginal discs, guts, embryos and pupal ovaries. Successful development of a cell line from a particular species requires trial and error. Adjustments to temperature, composition of medium and experimental procedure may be necessary. The methods given below are intended to form a base from which to start.

Sterilization of insect material

Some insects can be reared under aseptic conditions and can be used without sterilization. Those that are reared in non-sterile conditions or are collected from the field should be surface-sterilized by immersion in a solution such as 70% ethanol, 0.1% mercuric chloride, 5% formaldehyde or 2% sodium hypochlorite. Eggs and pupae usually have smooth surfaces and can be sterilized intact. For hairy insects, it is advisable to remove the hairs first, by shaving or burning. Insects that are inclined to regurgitate or excrete body contents after surface sterilization should be ligated behind the head and at the tip of the abdomen beforehand. Insects with waxy cuticles that prevent wetting with the sterilizing solution should be dipped in ethanol before treatment. Wetting agents such as 1% sodium chloride may be used in 0.1% mercuric chloride.

Establishment of cell cultures

CELL CULTURES FROM EMBRYOS

Equipment and materials

Items marked with asterisk should be sterile.

1. Eggs (1–3 days old) laid on polythene sheet. A minimum of 200 eggs of smaller species of Lepidoptera is required for establishing a 25 cm^2 culture. Smaller numbers may not establish well and can result in cells that are less susceptible to virus. Allow extra eggs for testing for viability and freedom from microbial contamination (see *Method*, Step 6 below)
2. Paintbrush*
3. Glass jar with lid
4. Filter paper disc
5. Metal sieve (150 mesh), e.g. a small copper cylinder with a copper mesh bottom, in a Petri dish*
6. Pasteur pipettes*
7. Petri dish, glass tube or beaker*

8. Fine forceps (two pairs) or mounted needles*
9. Glass rod with ground end*
10. Dissecting microscope
11. Flask*, 25 cm^2.

Note: Some butterflies, e.g. *Pieris rapae*, will only lay their eggs on leaves and must be tricked into laying eggs on any other type of support (D. Winstanley, personal communication). For *P. rapae* this can be achieved by sandwiching a leaf between a sheet of perforated plastic of the type used for wrapping bread.

Media and solutions

1. Formaldehyde, 5%
2. Sodium hypochlorite solution, 2%
3. Sterile water
4. Ethanol, 70%
5. Penicillin, streptomycin and amphotericin B solutions to give final concentrations of 50 U/ml, 50 μg/ml and 2.5 μg/ml, respectively.

Method

Strict aseptic technique must be used for all stages from Step 6 onwards.

1. Cut polythene sheet, with eggs attached, into conveniently sized pieces and place into the glass jar.
2. Pipette 5% formaldehyde onto a filter paper disc in the ratio of x μl formaldehyde (e.g. 5 μl) to x^2 ml (e.g. 25 ml) jar volume.
3. Place filter paper disc inside the lid of the jar and seal the jar.
4. Leave for 6 h to fumigate the eggs.
5. Remove eggs from the plastic sheet with a paintbrush and immerse in 2% sodium hypochlorite solution in a Petri dish, beaker or tube for 15 min and then wash with water.
6. Transfer eggs to the metal seive in a Petri dish and wash with 70% ethanol for 5–10 min. Then wash with two changes of sterile water or cell culture medium. Eggs should still be viable at this stage. To confirm viability and sterility, remove a proportion of them and test by incubating at the optimum temperature for the species. Observe for hatching. Check for degree of bacterial contamination by placing a few eggs on a nutrient agar plate and incubating for 3 days at 37°C. Low

levels of bacterial contamination may still be present but should not cause a problem if antibiotics are used as indicated below.

7. Using a dissecting microscope, tear chorion in culture medium with two pairs of fine forceps, or mounted needles, to release the embryos.

8. Wash the released cells through the sieve into the Petri dish. A glass rod with a ground end can be used to squash the eggs against the gauze, thus releasing more cells. These, also, should be washed into the Petri dish.

9. Transfer the culture medium containing the cells into a 25 cm^2 flask. Add fetal bovine serum to a final concentration of 15% and antibiotics (volume of complete medium, 5 ml).

10. Incubate the culture at 27°C. After 2 weeks replace half the medium. Thereafter replace half the medium every 10 days to 2 weeks depending on the growth of the cells. Antibiotics and 15% serum should be used for the first two changes of medium. After this, the concentration of serum may be reduced to 10% and antibiotics may be omitted.

11. Maintain the primary culture in the same flask at 27°C until the first passage becomes possible. Passage cells when they become completely confluent, according to the method given below. At this stage, the cells are delicate and it is best to use a rubber policeman as modified in Figure 27.4c (below) if mechanical detachment of cells is required. Otherwise, chemical detachment by the use of a proteinase such as trypsin, pancreatin or dispase is generally applied for passaging cells (see p. 504). During the first few months, there may be several changes in the appearance of the cells, with different types predominating at different stages. Some cell types are not able to survive well in culture and will die out. Eventually a small number of cell types will predominate and grow well.

Note 1. The purpose of treating eggs with formaldehyde followed by sodium hypochlorite solution is to inactivate any contaminating virus that may be present on their surfaces.

Note 2. Incubation temperature may be a crucial factor in maintaining susceptibility to virus in some cell lines. Cultures derived from *Cydia pomonella* embryos, for example, remained highly susceptible to *C. pomonella* GV when maintained at 21°C. At this temperature, they retain a stable, differentiated form and show colonial growth. When maintained at 27°C for several months, however, they change morphologically and gradually lose their susceptibility to the virus (Winstanley and Crook, 1993).

CELL LINES DERIVED FROM PUPAL OVARIES

Equipment and materials

Items marked with asterisk should be sterile.

1. Middle stage pupae
2. Dissecting pins*
3. Mounted needles and fine forceps*
4. Absorbent cotton wool*
5. Dissecting tray (sterilized by filling with 70% ethanol, leaving for 5 min and rinsing with sterile glass-distilled water)
6. One pair of ophthalmologist's scissors*
7. Two Maximov slides or small glass Petri dishes*
8. Cell culture flask(s) (25 cm^2)*.

Media and solutions

1. The appropriate cell culture medium, e.g. TC100 containing 15% fetal bovine serum and antibiotics (penicillin, streptomycin amphotericin B solutions to give final concentrations of 50 U/ml, 50 μg/ml and 2.5 μg/ml, respectively)
2. Carlson's salt solution.

Method

1. Select female pupae. Middle-stage pupae are recommended (Figure 27.1). It is best not to use mature ovaries since they are difficult to remove intact because of the fragility of the ovarial sheath and because the ovaries are firmly fixed to the tracheoles.
2. Submerge pupae in 70% ethanol for 5 min.
3. Decant ethanol and wipe pupae dry with sterile cotton wool (absorbent).
4. Fix each pupa onto a dissecting tray with sterile dissecting pins.
5. Pour Carlson's salt solution into the tray until the pupa is fully immersed.
6. Make an incision in the pupal case and cut as shown in Figure 27.2. Take care not to damage the gut.
7. Pull the cuticle upwards and downwards so as to expose the inside of the abdomen and pin to the tray in several places. When immature, the ovaries are located dorsally in masses of fat bodies.

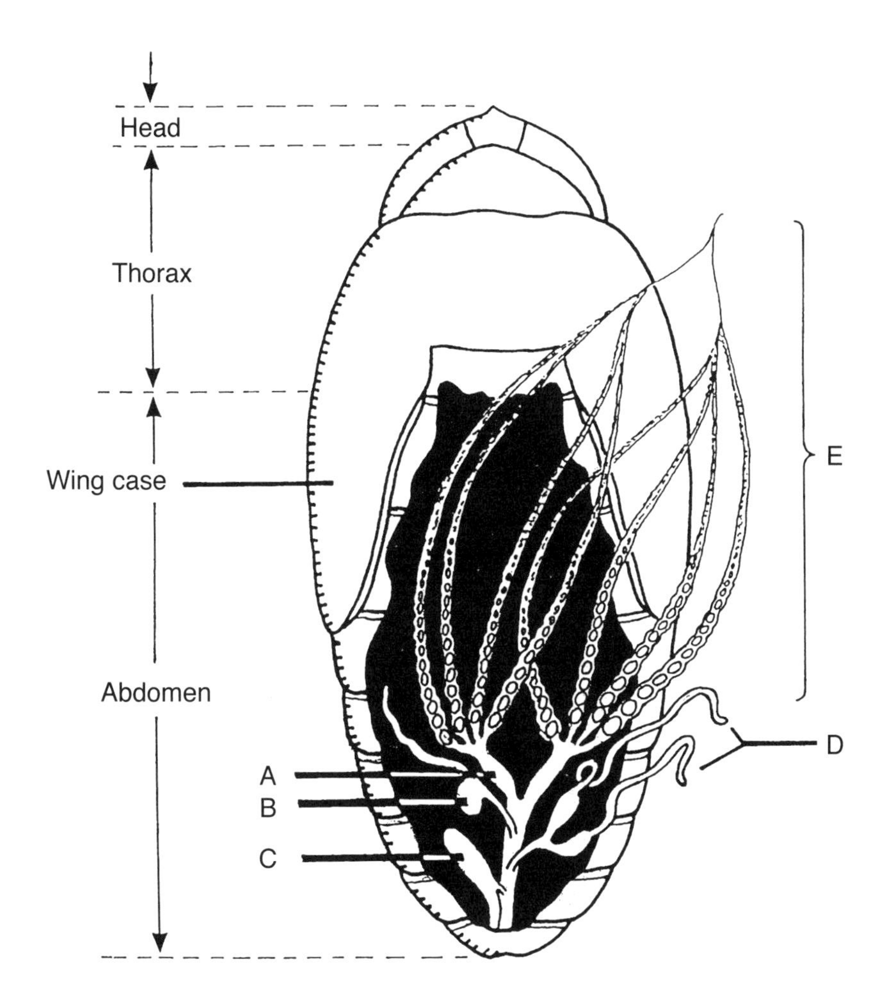

Figure 27.1 Dorsal view of a generalized female lepidopterous pupa dissected to show the reproductive system. For clarity, fat body and tracheal system have been omitted. A, an oviduct; B, receptaculum seminis (spermatheca); C, bursa copulatrix; D, accessory glands; E, ovarioles. Four ovarioles arise from each of the two oviducts, the latter opening posteriorly into vagina. Anteriorly, the ovarioles of each side are connected together and these themselves, finally, are united.

8. Remove the ovaries using forceps and needles. First find the common oviduct and cut it at its proximal end. Next catch the proximal end of the common oviduct and pull gently, bringing the ovaries with it. Care must be taken not to damage the gut in any way during this procedure although it is safe to cut the trachea and tracheoles.
9. Transfer the ovaries to Carlson's salt solution on a Maximov slide or similar.

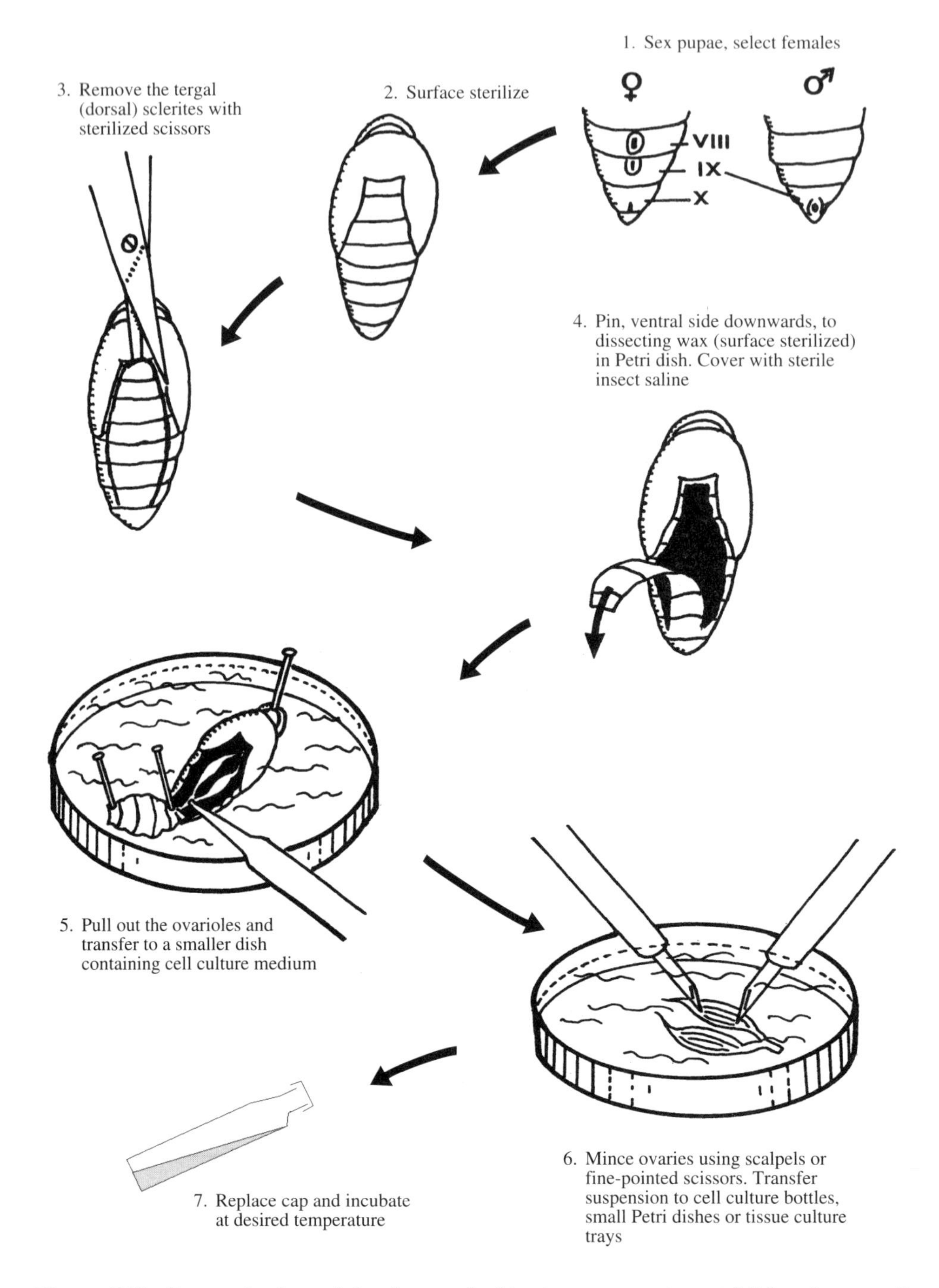

Figure 27.2 Removal of ovarioles from a lepidopterous pupa to establish primary cell cultures.

10. Separate the fat bodies, trachea or Malpighian tubes attached to the ovaries with forceps and needles and transfer the ovaries by needle to fresh Carlson's salt solution on a second Maximov slide.
11. Cut the ovaries into pieces about 1 mm long.
12. Remove the Carlson's salt solution by Pasteur pipette.
13. Add culture medium.
14. Distribute the suspension of ovarial fragments into culture flasks ($25\ cm^2$, 5 ml per flask) and seal.
15. Incubate at 27°C.
16. Replace half of the medium in each culture with an equal volume of fresh medium at appropriate intervals (from 1 week to 1 month).
17. Subculture cells when they have multiplied sufficiently.

CELL PASSAGE

The term cell passage refers to the procedure by which a culture of cells that has become confluent or has reached its maximum growth potential is subcultured. The spent medium is removed and the cells are diluted and resuspended in fresh medium (Figure 27.3). Colloquially, the process of passaging cells is referred to as cell 'splitting'. If the cells from one bottle are divided between two new bottles they are said to have been split one to two, and if divided between three bottles, one to three, and so on.

In some circumstances cells may grow better if resuspended in a mixture of fresh and used (so-called conditioned) medium (see above for preparation). The proportion of fresh to conditioned medium can be varied depending on experience, but a ratio of 1 : 1 is frequently used.

The method given below refers to cells growing in $25\ cm^2$ bottles. Table 27.1 gives suggested volumes for different types of flask and tissue culture plates. The

Table 27.1 Volume of medium required for growing cells in different types of vessel

Type of vessel	Volume of medium
Flasks	
$25\ cm^2$	4–5 ml (minimum 3 ml)
$50\ cm^2$	10 ml
$75\ cm^2$	20 ml
Plates	
6 well	up to 3 ml per well
24 well	1.0–1.5 ml per well
96 well	100–150 μl per well
Petri dishes	
100 mm	10 ml

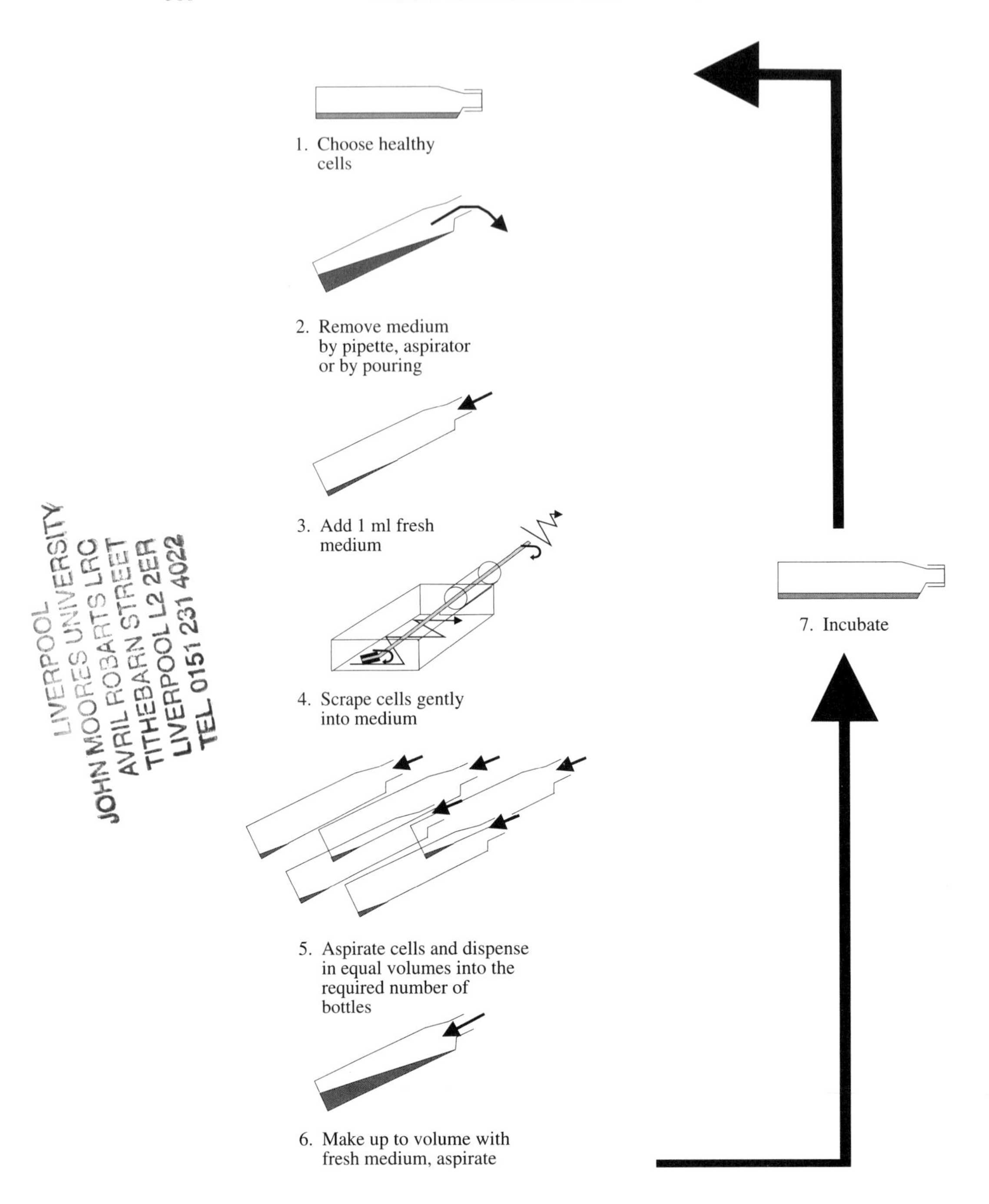

Figure 27.3 Cell passage: the subculture of a fully grown culture.

term aspiration, in the context used here, refers to the repeated sucking up and expulsion of cells and medium from a pipette until an homogeneous suspension is achieved.

ROUTINE PASSAGING OF CELLS GROWN AS MONOLAYERS

Equipment

Items marked with asterisk should be sterile.

1. Laminar airflow cabinet if available
2. Inverted microscope (×10 eyepieces, ×20 objective, camera optional but very useful)
3. Bunsen burner or equivalent (mains or portable gas, portable spirit burner): optional for working in a cabinet, essential for working on the bench
4. Aspirator, optional but useful when removing large volumes of spent culture medium from cells
5. Culture flasks (25 cm^2; glass or plastic)*
6. Rubber policeman, one per bottle passaged (Figure 27.4)*
7. Pipettes and pipetting devices, various volumes and types as required.

Materials and media

Items marked with asterisk should be sterile.

1. Ethanol (70%) or equivalent for sterilizing surfaces of equipment
2. Separate pots of disinfectant for decontaminating pipettes (1% sodium hypochlorite or equivalent) and rubber policeman (0.5% Tegodyne or 1% Vircon). Soak for a minimum of 18 h before recycling
3. Petri dishes containing a range of agar media for checking for aerial contamination (required periodically)*
4. Confluent monolayers of cells
5. The appropriate medium for the cells being propagated*
6. The required number of cell (tissue) culture flasks into which to passage the cells*. This can include the bottles in which the cells are growing.

Method

1. If using a laminar airflow cabinet, turn on power 1 h before use. Surface sterilize the outsides of pipette containers, bottles of medium, etc. with disinfectant (e.g. 70% ethanol or 0.5% Tegodyne) and place in cabinet.

2. Examine cells with an inverted microscope and choose healthy confluent monolayers for passaging.
3. Record date and new passage number on the upper face of all bottles into which the cells will be passaged.
4. Pour off spent medium or remove with the aid of a pipette or aspirator.
5. Add approximately 0.8 ml fresh medium.
6. Detach the cells from the side of the bottle by gently shaking or scraping them into the medium with a rubber policeman. Discard rubber policeman into disinfectant.
7. Aspirate cells, e.g. with a Pasteur pipette and adapter, and dispense equal volumes into the required number of appropriately labelled bottles.
8. Add fresh medium to each bottle to a total volume of 4 ml (or 5 ml if this is preferred by the cells).
9. Swirl gently, or reaspirate, to mix cells and medium.
10. Place bottles in incubator at 25–27°C with labelled sides uppermost and leave cells to grow.

Note: Mechanical detachment as in Step 6 may be damaging to certain cells. If so, treatment with an enzyme is advised. Trypsin is detrimental to many lepidopteran cells but is not necessarily toxic to cells of insects belonging to other orders. Pancreatin is less toxic to lepidopteran cells while dispase is satisfactory for most insect cells (see p. 504).

Alternative method

This is not recommended for cells that are to be used for titrating virus.

Stages 1–3 as above

4. Gently scrape cells into medium in which the cells have been growing.
5. Aspirate cells with a 5 ml pipette and dispense into fresh bottles in the desired volumes, e.g. 1 ml cell suspension per bottle.
6. Add the required volume of fresh growth medium (e.g. 3 ml added to 1 ml of cells will result in a one to four split) and aspirate.
7. Incubate at the required temperature.

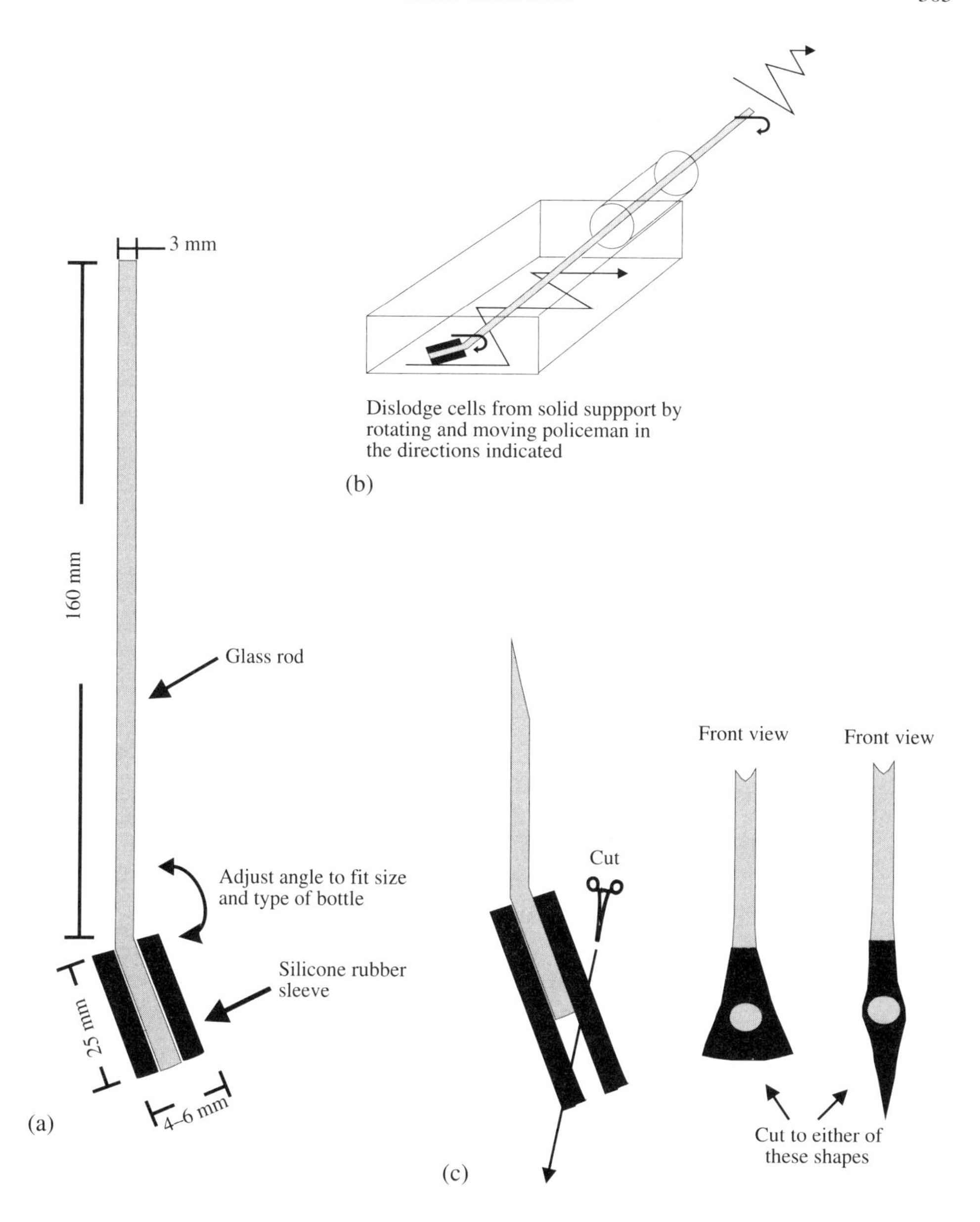

Figure 27.4 (a) A silicone rubber policeman (cell scraper). The dimensions given are those suitable for a 25 cm^2 cell culture bottle. (b) Use of the policeman to dislodge cells. (c) Method to make a more pliable scraper. This type is particularly suitable for cells that are damaged easily.

PROTEINASE TREATMENT OF CELLS

Equipment

This equipment list is additional to that given for routine passaging of cells.

1. Bench centrifuge
2. Sterile centrifuge tubes or bottles.

Media and materials

1. A 1 : 1 mixture of 0.1% trypsin and 0.02% EDTA both in Rinaldini's salt solution, or 0.03% pancreatin (Scientific Protein Laboratories Inc., Waunakee, WI, USA) and 0.02% EDTA in Rinaldini's salt solution, or a 0.5% aqueous solution of dispase (Godo Shusei Co. Ltd, Ginza 6–2–10, Chuo-ku, Tokyo, Japan)
2. Growth medium.

Method

1. Remove spent medium.
2. Add sufficient of the desired proteinase solution to cover the cells.
3. Leave until cells attached to substrate become round (approximately 10–30 min).
4. Stop reaction by adding growth medium containing serum.
5. Aspirate gently to resuspend cells evenly.
6. Transfer to centrifuge tubes and centrifuge at $150 \times g$ for 5 min to sediment cells. The speed and time can be varied depending on conditions and cells. The slower the speed and the faster the operation, the better.
7. Remove and discard supernatant fluid.
8. Resuspend cells in fresh growth medium and dispense in culture vessels.
9. Incubate at required temperature.

CLONING CELLS

Early passage cultures, especially from embryos, are likely to contain a variety of cell types. More uniform cell lines can be obtained by cloning the cells. Several different cloned cell lines may be obtained from early passage cells derived from a single primary culture and these may vary in their sensitivity to virus infection, productivity for budded virus or occluded virus and in their cytopathology following infection. Similarly, established cell lines that have not been cloned may

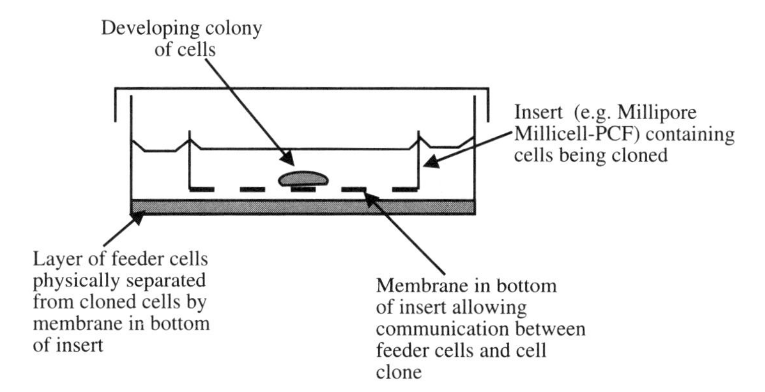

Figure 27.5 Technique for cloning fastidious cells with the aid of feeder cells.

contain a mixture of cell types with varying abilities to support virus replication, and it may be considered important to clone them. Cloning insect cells from a primary culture is extremely difficult, if not impossible.

Some cell lines are much easier to clone than others. It is particularly difficult, and may be impossible, to clone cells that depend on others for certain growth factors. The simplest method for cloning non-fastidious cells depends on preparing limiting dilutions of cells in growth medium, dispensing in small tissue culture wells at an estimated concentration of one cell per well, incubating at a suitable temperature and observing for growth. Fastidious cells can sometimes be cloned by growing them in association with, but physically separated from, a feeder layer of cells (Figure 27.5). Cells should be cloned a minimum of three times for each cell line.

CLONING CELLS BY LIMITING DILUTION

Equipment, media and materials

Items marked with asterisk should be sterile.

1. Culture of healthy cells, e.g. as a monolayer in a 25 cm^2 tissue culture flask
2. Rubber policeman*
3. Pipettes for delivering 5 ml, 1 ml and 100–200 μl volumes. The last can most easily done with a Pasteur pipette and adapter (Figure 23.1, p. 362)*
4. Suitable culture medium, including conditioned medium if thought necessary*
5. Penicillin, streptomycin and amphotericin B solutions added to the medium to a final concentration of 50 U/ml, 50 μg/ml and 2.5 μg/ml, respectively

6. Diluent e.g. PBS (Dulbecco A) or Rinaldini's salt solution, for suspending cells for counting
7. Improved Neubauer haemocytometer for determining concentration of cells
8. Bijoux*
9. Vital stain (optional)
10. Phase-contrast microscope
11. Tissue culture plate (96 well)*
12. Incubator at optimum temperature for the cells that are to be cloned (normally 25–27°C)
13. Laminar airflow cabinet
14. Inverted microscope
15. Aspirator for changing medium (optional)*
16. Parafilm
17. Clingfilm.

Method

1. Choose a culture of healthy cells.
2. Remove spent medium.
3. Add 1 ml of fresh growth medium.
4. Gently suspend cells in the medium using a rubber policeman.
5. Transfer to a bijou bottle.
6. Prepare a 1/10 dilution of the cells by aseptically transferring 100 μl (0.1 ml) of suspension to 900 μl diluent (with or without vital stain) and estimate the concentration of viable cells by one of the methods described above.
7. Calculate the dilution necessary to give a suspension of cells that contains 1 cell in 100 μl of medium.
8. Prepare a series of dilutions (see below) of cells in growth medium estimated to give 1, 5 and 10 cells per well.
9. Dispense 100 μl volumes of each concentration into the required number of rows of a 96 well tissue culture plate.
10. Make volume up to 200 μl with fresh growth medium, with a mixture of fresh growth medium and conditioned medium, or with conditioned medium only.
11. Seal the lid to the base of the plate with parafilm and cover with

clingfilm to prevent evaporation of medium. Alternatively, place in a tin with a beaker of water.

12. Incubate at the required temperature, e.g. 25–27°C.
13. Observe daily and note those wells that contain one cell only.
14. Change medium once a week and allow cells to grow until a monolayer is formed.
15. Passage cells into fresh wells in the ratio 1 : 2.
16. When sufficient cells have been grown, passage into four 6-well plates in 2–3 ml volumes per well.
17. Finally passage into larger flasks and reclone.

Note 1: Cells from individual wells can be passaged into plastic bijou bottles in 2 ml of medium instead of 24-well plates (Step 16).

Note 2: It may be possible to re-clone cells at Step 16 above rather than at Step 17.

Note 3: Each viable cell will give rise to a single colony, which, with time, will expand until the entire area of the bottom of the well is covered. If there is more than one cell in a well, the colonies that develop will eventually merge. It may be possible before this stage is reached, to pick cells from individual colonies by pipette with the aid of a dissecting microscope and to passage them.

Note 4: Cells can be cloned in tissue culture dishes (60 mm diameter) as an alternative to 96 well plates. If using this technique, seed the dishes at a low seeding density and allow cells to form colonies. Collars can be used to isolate individual colonies, which are carefully picked off, transferred to bijoux, shaken to distribute the cells and incubated until monolayers are formed.

PRESERVATION OF CELLS

When establishing new cell lines, it is essential to preserve cells at early passages. It is also wise to store adequate stocks of any cells with which one is working.

CELL PRESERVATION

Equipment

Items marked with asterisk should be sterile.

1. Liquid nitrogen refrigerator

2. Pipettes for dispensing cells*
3. Rubber policeman, one per bottle of cells (Figure 27.4)*
4. Plastic ampoules or heat-sealed plastic drinking straws* (1.5 ml capacity).

Media and materials

1. Cultures of healthy cells, e.g. in 25 cm^2 flasks
2. Growth medium (used for culturing cells)
3. Freezing medium: the same as that used for culturing cells, but containing 10% glycerol (or 5% dimethyl sulphoxide).

Method

1. Remove spent medium by pipette or aspirator.
2. Add a small volume of growth medium, e.g. 1 ml, gently scrape the cells into it and aspirate.
3. Centrifuge the cell suspension at $100 \times g$ for 5 min and discard the supernatant fluid.
4. Suspend the cells in the freezing medium. A cell density of approximately 10^6 cells/ml is preferable.
5. Dispense immediately into the ampoules in 1 ml volumes. (Sterile, heat-sealed plastic drinking straws can be used instead of plastic ampoules. The contents of one straw is sufficient to seed one 25 cm^2 tissue culture flask.)
6. Freeze at a rate of 1°C/min until the temperature reaches –50°C. Alternatively, place cells in a box of expanded polystyrene and leave overnight at –70°C.
7. Place in the gas phase of a liquid nitrogen refrigerator for 1 h then place in the liquid phase.
8. Store until required.

If a liquid nitrogen refrigerator is not available, cells may be stored at –70°C, but preferably at a temperature of between –80 and –100°C.

RESUSCITATION OF CELLS

Method

1. Remove an ampoule from the refrigerator and immerse in hot (90°C) water. Leave until almost totally thawed.

2. Wipe ampoule with 70% ethanol.
3. Transfer cell suspension to a centrifuge tube.
4. Sediment the cells ($100 \times g$ for 5 min or as suitable for the particular type of cell) and remove supernatant fluid.
5. Resuspend cells in fresh medium.
6. Transfer to an appropriate bottle and incubate.
7. Passage when confluent.

PROPAGATION OF VIRUS IN CELL CULTURE

PREPARATION OF INOCULUM FROM INFECTED LARVAE

Equipment and materials

Items marked with an asterisk should be sterile.

1. Larvae infected with the required virus
2. Syringes (1 ml or larger as relevant) and membrane filters (0.45 μm and 0.22 μm pore size*
3. Ethanol (70%)
4. Filter paper in a Petri dish*
5. Phenyl thiocarbamide (1-phenyl-2-urea)
6. Medium, e.g. TC100 with antibiotics (see above) but without serum*
7. Syringe needle or fine scalpel blade*
8. Bijou bottle*
9. Ampoules for storing infected haemolymph*.

Method

1. Surface sterilize larvae by dipping briefly in 70% ethanol and drying by placing on sterile filter paper.
2. Either amputate a proleg with scalpel blade or prick with syringe needle.
3. Gently squeeze out haemolymph and drain into a bijou bottle or Petri dish containing a small crystal of phenyl thiocarbamide to prevent melanization.

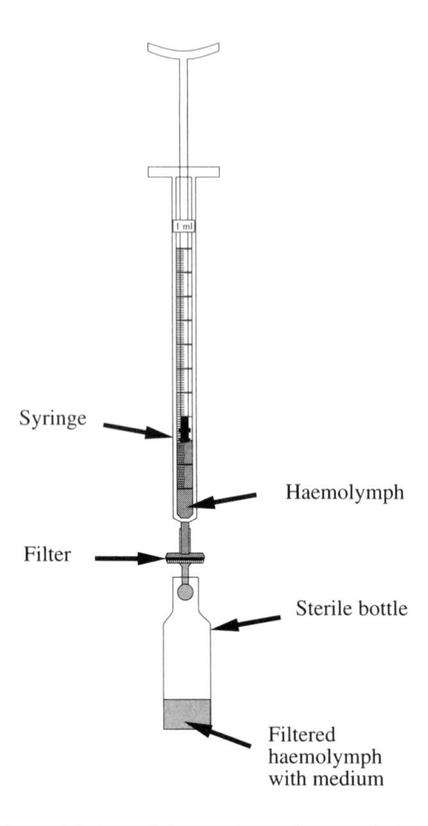

Figure 27.6 Filtration of infected haemolympth containing budded virus.

4. Dilute with medium and filter through a membrane filter attached to a syringe (Figure 27.6).
5. Dispense in small volumes and store at –70°C.

The degree of dilution in Step 4 may need to be determined. For *S. littoralis* MNPV, a 1 : 1 ratio of haemolymph to medium has been found to be satisfactory (unpublished) while Winstanley and Crook (1993) found a ratio of 1 : 20 to be most satisfactory for *C. pomonella* GV.

PROPAGATION OF INOCULUM

Equipment

Items marked with an asterisk should be sterile.

1. Tissue culture flasks of the required volume (e.g. 25 cm^2)*

2. Pipettes for delivering cells, adding and removing medium (e.g. Pasteur pipette with adapter, 5, 10 or 25 ml capacity with mechanical or electrical pipetting device) and for inoculating cells with virus (e.g. Gilson pipettor or equivalent for delivering 100, 200 μl volumes, with sterile tips)*
3. Microcentrifuge tubes, or similar, for preparing dilutions of virus*
4. Waterbath at 40°C
5. Inverted microscope
6. Phase-contrast microscope with ×10 eyepieces and objective.

Media and materials

1. Inoculum of extracellular virus obtained from virus propagated either in insects (see above) or in cell culture (cell culture harvest)
2. Confluent monolayers of healthy cells
3. Sterile medium for propagating cells, e.g. TC100 with 10% fetal bovine serum (complete medium)*
4. Sterile medium lacking serum as diluent.

Method

1. Passage cells (see above) and incubate overnight (27°C) or until the monolayer is semi-confluent.
2. Remove complete medium and add inoculum. This can be either infected cell culture harvest or infected haemolymph and can be used either undiluted or diluted further with medium lacking serum, (total volume 1.0 ml for a 25 cm^2 culture flask).
3. Incubate at 27°C for 4 h with gentle rocking.
4. Remove inoculum and replace with complete medium.
5. Incubate at 27°C until characteristic cytopathic effects are observed (Figure 27.7).
6. Harvest medium, allow cells to settle, collect supernate containing extracellular (budded) virus, dispense in small volumes and store at –70°C. Virus will remain infective for short periods of time (up to 1 year) if stored at 4°C.

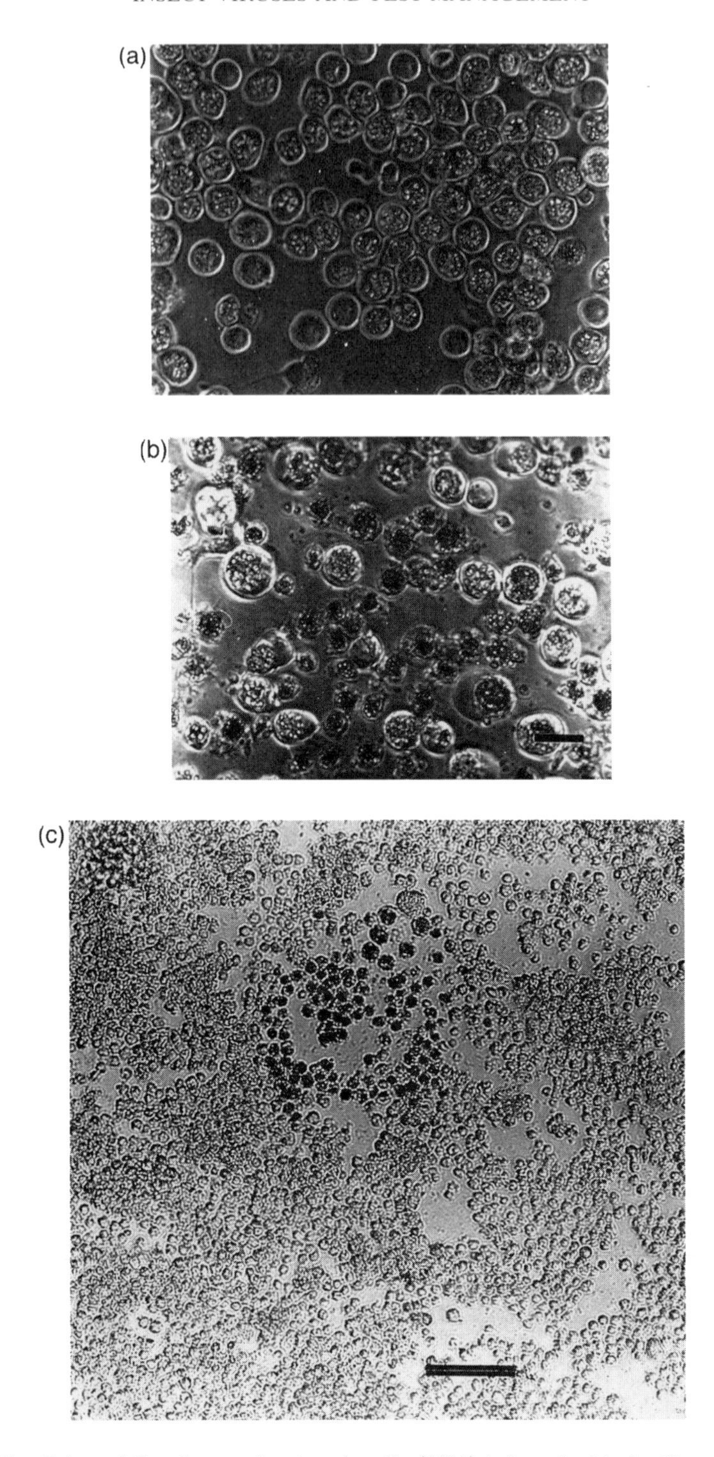

Figure 27.7 Cultured *Spodoptera frugiperda* cells (Sf21) infected with the Ogasawara strain of *S. litura* NPV and observed at 48 h (a) and 120 h (b) after inoculation. OBs are clearly visible as large refractile bodies within infected cells. Size bar = 25μm. (c) Plaque morphology of AcMNPV in Sf21 cells. Size bar = 100μm.

PLAQUE ASSAY OF EXTRACELLULAR (BUDDED) VIRUS

Equipment

Items marked with an asterisk should be sterile.

1. Tissue culture plate(s), 6 × 8 well*
2. Equipment for counting number of viable cells (see above)
3. Pipettes for delivering cells, adding and removing medium (e.g. Pasteur pipette with adapter, 5, 10 or 25 ml capacity with mechanical or electrical pipetting device) and for inoculating cells with virus (e.g. Gilson pipettor or equivalent for delivering 100, 200 μl volumes, with sterile tips)*
4. Microcentrifuge tubes, or similar, for preparing dilutions of virus*
5. Waterbath at 40°C
6. Inverted microscope
7. Humidity incubator or chamber (e.g. a tin containing a beaker of water) to prevent evaporation of medium from cell culture wells. Alternatively, the plates can be sealed with parafilm, clingfilm, autoclave tape or similar
8. Phase-contrast microscope with ×10 eyepieces and objective.

Media and materials

Items marked with asterisk should be sterile.

1. Inoculum of extracellular virus obtained from virus propagated either in insects or in cell culture (see above)
2. Confluent monolayers of healthy cells
3. Medium for propagating cells, e.g. TC100 with 10% fetal bovine serum (complete medium)*
4. Serum-free medium (e.g. TC100 lacking serum) for diluting virus*
5. Sea plaque agarose (5% in glass-distilled water)*.

Method

1. Prepare a suspension of cells as if for passaging (see above).
2. Determine the concentration of viable cells (see above).
3. Dilute cells in growth medium to give a concentration of 6×10^5 viable cells/ml.
4. Dispense 0.5 ml cell suspension into each well of the required number of 24-well tissue culture plates.

5. Incubate overnight at 27°C.
6. Check that monolayers are semi-confluent and appear healthy.
7. Prepare a suitable range of dilutions of virus (serial 10-fold for preliminary assays otherwise serial two- or threefold) in serum-free medium.
8. Gently remove medium from cells.
9. Add 100 μl of the appropriate dilution of virus to each well and rock gently to spread evenly over the cell sheet. Leave for 15 min.
10. Add 200 μl serum-free medium to each well to prevent cells from drying out.
11. Incubate at 27°C for 3 h.
12. While the cells are being incubated, warm complete medium to 40°C. Melt agarose and equilibrate to 40°C. Finally mix medium and agarose in the ratio 4:1. Maintain at 40°C until required.
13. Gently remove the inoculum from each well and add 0.5 ml sea plaque agarose.
14. Leave plates undisturbed for 20–30 min so that agar sets.
15. Seal each plate with tape and wrap in clingfilm or place in a humid atmosphere to prevent dehydration.
16. Incubate at 27°C.
17. Feed cells after 2 days by adding 100 μl complete medium to each well.
18. Reincubate until plaques are well defined, usually 6–7 days after inoculation (Figure 27.7)
19. Before counting plaques, remove any medium remaining on the surface of the agarose and discard in disinfectant, invert plates and drain on paper towelling or filter paper in a plastic tray for 30 min. Decontaminate paper towelling and tray by immersion in 5% Chloros.
20. Observe cells with a phase-contrast microscope and count plaques. Calculate titre in plaque-forming units per millilitre.

For better visibility, plaques can be stained with Neutral Red.

STAINING PLAQUES WITH NEUTRAL RED

Method

1. Dilute stock solution of 0.5% (w/v) Neutral Red solution 1/20 with PBS.
2. Add approximately 0.5 ml diluted stain to each well.

3. Incubate at 28°C for 2–4 h.
4. Remove stain by aspirator or pipette and discard into disinfectant.
5. Store plate in the dark at room temperature from 2 h to overnight.
6. Count plaques.

INFECTIVITY OF BUDDED VIRUS

Infectivity can be assessed as median tissue culture infective dose ($TCID_{50}$).

Equipment and materials

1. Tissue culture plate (96 well)
2. Virus inoculum (cell culture harvest or infected haemolymph)
3. Medium, both complete and without serum
4. Equipment for passaging cells, preparing a range of dilutions of virus and inoculating cells.

Method

1. Passage cells and seed each well (96 well tissue culture tray) with 2×10^4 cells.
2. Incubate at 27°C until cells attach.
3. Prepare a range of dilutions of virus (preferably serial two- or threefold but not greater than serial 10-fold).
4. Remove complete medium from wells and replace with 100 μl of virus inoculum (eight replicates per dilution).
5. Incubate at 27°C with gentle rocking for 4 h.
6. Remove inoculum and replace with 100 μl complete medium.
7. Incubate at 27°C. Replace medium after 6 days.
8. When fully developed, score each well for the presence of a cytopathic effect. For ease of observation, the cells can be stained first with Neutral Red as described above. Record the number of wells positive for each dilution and calculate the $TCID_{50}$ by probit analysis (p. 484).

CLONING EXTRACELLULAR (BUDDED) VIRUS

Equipment, media and materials

Equipment, media and materials are the same as those used for the *Plaque assay of extracellular virus* (above) with the exception that 6-well cell culture plates or glass or plastic tissue culture dishes (e.g. 35 mm $\times$ 10 mm) are used.

Method

1. Prepare a suspension of cells as if for passaging (see above).
2. Determine the concentration of viable cells.
3. Dilute cells in growth medium (TC100) to give a concentration of 1×10^6 to 1.3×10^6 viable cells/ml and dispense 1.5 ml volumes (1×10^6 to 2×10^6 cells) into the required number of wells or Petri dishes.
4. Incubate overnight at 27°C.
5. Check that the cells look healthy.
6. Dilute virus suspension in TC100 lacking serum to give a concentration that will yield plaques that are well separated from each other.
7. Remove medium from each well and add 100 μl of virus suspension to each monolayer of cells.
8. Gently rock three to five times to ensure contact between virus and cells.
9. Add 1 ml of TC100 lacking serum to each well.
10. Incubate cells at 27°C for 1 h.
11. While the cells are incubating, prepare a 5% suspension of agarose (sea plaque) in glass-distilled water, or equivalent; dissolve and sterilize by autoclaving at 121°C for 15 min.
12. Warm some TC100 to 40°C and mix with molten agarose in the ratio 4 : 1. Maintain at 40°C in a waterbath.
13. At the end of the incubation period (1 h), remove the virus inoculum from the cells and add 1.5 ml of the agarose overlay to each well.
14. Allow to set (15–20 min) then seal plate with Parafilm to prevent loss of moisture and incubate at 27°C in a humid atmosphere until the plaques are well formed. Feed as necessary by adding medium onto the top of the agar.
15. View plaques with the aid of a light microscope ($\times$ 10 eyepieces and objective) and mark, on the underside of each well, the position of well-isolated plaques containing OBs. Use a marker pen for this purpose.
16. Using a Gilson P-200 micropipettor (or equivalent), aspirate each

suitable plaque together with the agarose above it and suspend in 1 ml complete TC100.

17. Repeat the above procedure using the virus from Step 16 as inoculum. Continue the same process until one type of virus only, as determined by its REN profile, is present. A minimum of three rounds, but preferably four, is recommended. Store suspensions at 4°C.

REFERENCES

Carlson, J.G. (1946) Protoplasmic viscosity changes in different regions of the grasshopper neuroblast during mitoses. *Biological Bulletin* **90**, 109–121.

King, L.A. and Possee, R.D. (1992) *The Baculovirus Expression System. A laboratory Guide*. London: Chapman and Hall.

Mitsuhashi, J. and Maramorosch, K. (1964) Leafhopper tissue culture: embryonic, nymphal, and imaginal tissues from aseptic insects. *Contributions of the Boyce Thompson Institute* **22**, 435–460.

O'Reilly, D.R., Miller, L.K. and Luckow, V.A. (1992) *Baculovirus Expression Vectors*. New York: W.H. Freeman.

Rinaldini, L.M. (1959) An improved method for the isolation and quantitative cultivation of embryonic cells. *Experimental Cell Research* **16**, 477–505.

Shuler M.L., Granados R.R., Hammer D.A. and Hood H.A. (eds) (1995) Insect cell culture methods and their use in virus research. In *Baculovirus Expression Systems and Biopesticides*. New York: Wiley, p. 259.

Vlak, J.M., de Gooijer, C.D., Tramper, J. and Miltenburger, H.G. (eds) (1996) *Insect Cell Cultures: Fundamental and Applied Aspects* (Reprinted from *Cytotechnology*), Dordrecht, the Netherlands: Kluwer Academic, pp. 325.

Winstanley, D. and Crook, N.E. (1993) Replication of *Cydia pomonella* granulosis virus in cell cultures. *Journal of General Virology* **74**, 1599–1609.

28 Mass production, product formulation and quality control

At the present time, the only realistic way for most laboratories to produce virus is by propagation *in vivo*. This section, therefore, deals with aspects of the mass production, formulation and quality control of virus propagated in, and extracted from, the insect host.

MASS PRODUCTION

The system adopted for mass production of a particular BV will depend on local conditions. Consideration should be given to the availability and degree of expertise of a possible workforce, the amount of laboratory space at one's disposal, facilities for servicing the laboratory including energy and water supplies, availability and degree of sophistication of equipment, and facilities for storing, testing and processing the virus during the various stages from inoculation of host to release of final product for application in the field. Factors that will aid any decision that has to be made are covered more fully in Chapters 23–25. Reference can also be made to Chapter 6.

FORMULATION

From information given in Chapter 7, it should be possible to devise a wide variety of formulations. The following are, however, examples of methods of preparation of salient types of formulation.

Preparation of virus OBs for inclusion in formulations is often achieved by high-speed shearing of thawed cadavers in water followed by filtration. The ratio of insect tissue to water and blend time can affect the OB yield; for example, an optimal combination for *Lymantria dispar* NPV was 1 g larvae to 10 ml water blended for 15 s, which gave a 95–98% yield of OBs (Shapiro, Bell and Owens, 1980). By using a series of size-graded filters, a large amount of the host particulate material may be removed. This process may require the use of filtration under pressure, or with suction, to avoid premature filter clogging. The incorporation of pre-filters may also confer advantage. The resultant material may be submitted to centrifugation for further purification and concentration. It may then, if desired, be air dried, freeze dried or spray dried (as in some commercial preparations) as required.

Detailed descriptions of virus separation and recovery, including a method not involving the use of centrifuges (p. 415; Dulmage, Martinez and Correa, 1970) are given in Chapter 25.

Wettable powders

Examples of simple wettable powders are provided by *L. dispar* NPV as Gypchek (Shapiro *et al.*, 1980; Lewis, 1981) and *Orgyia pseudotsugata* NPV as TM Biocontrol-1.

FORMATION OF A WETTABLE POWDER

Method

1. Homogenize infected larvae in water (ratio 1 g : 10 ml) in a blender.
2. Pass the slurry through a series of sieves (cheesecloth or nylon). Retain the filtrate and the filter matt.
3. Rehomogenize the matt and filter again.
4. Combine these two filtrates and either (a) lyophilize, or (b) centrifuge at $7000 \times g$ for 25 min and air-dry pellet(s).
5. Mill to fine powder.
6. Package under vacuum.
7. Store at approximately 4°C.

Dusts

The example selected is *Helicoverpa zea* NPV using a procedure which permits incorporation of an adjuvant (Montoya, Ignoffo and McGarr, 1966).

FORMATION OF *H. ZEA* NPV DUST

Method

1. Combine adjuvant and OB suspension (here at about 3×10^{10} OB/ml) in the proportion of 1.6 : 1.0 and add 1 part attapulgite clay.
2. Mix to a slurry.

3. Lyophilize the slurry.
4. Triturate and pass through a 200 mesh sieve.
5. Homogenize in a blender until of even particle size.

Granules

The example employs *H. zea* NPV as Elcar (Hostetter and Pinnell, 1983). This particular formulation includes an adjuvant, Coax, which could be omitted, though it is logical with a granular formulation to include all formulants at the beginning. Filler makes up the bulk and a specific choice of filler may be determined by availability or price, e.g. the authors found wheat bran eight times more expensive than rice mill feed. The preparation of starch-based granules is described under *Microencapsulation techniques*, below.

FORMATION OF *H. ZEA* NPV GRANULES

Method

1. Make a paste from: filler, 100 g; adjuvant (e.g. Coax), 8 g; gelatin, 2.0% (or Decagin 1.0% (see Table 7.4, p. 130); NPV (here as Elcar WP at 4×10^9 OB/g), 2 mg; water, approximately 135 ml.
2. Spread out paste (e.g. on aluminium foil) and allow to dry.
3. Grind and pass through 40 mesh sieve (screen).

Note. The fact that this formulation was less effective on cotton than a Coax/Elcar spray is irrelevant to the process of granule formation itself. The question of retention of virus infectivity can only be tested in an appropriately devised bioassay.

Flowable suspensions

There have been no fully published accounts of this form of preparation. However, the literature indicates that a flowable may be approached through the use of gelling agents such as xanthan gum, gum arabic, and Kelzan (Table 7.4, p. 130) (Couch and Ignoffo, 1981; Smith *et al.*, 1982). As with all liquid formulations, microbial stabilization is necessary, for which methods are given in Table 7.1 (p. 122). Difficulty with microbial contamination was encountered when *H. zea* NPV was formulated as a flowable powder (Bull, 1978). To prevent OB aggregation a dispersant should be used (Table 7.4, p. 130).

Micro-encapsulation techniques

Two techniques have been selected for quotation as micro-encapsulation techniques as they appear to be simple and effective. It is recommended that others be further explored and developed.

MICRO-ENCAPSULATION USING STYRENE MALEIC ANHYDRIDE HALF ESTER DISSOLVED IN ETHYL ACETATE

The method described is derived from Bull *et al.* (1978).

Method

1. Suspend OBs, together with UV protectants, etc., in 10% (w/v) ethyl acetate solution of styrene maleic anhydride half ester (e.g. SMA-2625A, ARCO Chemical Co. USA) using a high-shear stirrer.
2. Deliver the mixture to a disc rotating at high speed. Small droplets are flung off the disc margin, the ethyl acetate evaporates and the resultant dry capsules are collected following impingement on a solid shield.
3. Sieve capsules, seal in containers and store frozen.

MICRO-ENCAPSULATION USING STARCH AND WATER

The starch and water method is taken from Dunkle and Shasha (1988, 1989) as slightly modified by Ignoffo, Shasha and Shapiro (1991).

Method

1. To 10 ml deionized water add 1 g refined corn oil (into which UV protectants can previously be mixed) and 20 ml of an aqueous suspension of BV OBs.
2. Vigorously mix with 10 g pre-gelatinized starch (e.g. Miragel 463, A.E. Staley Mfg., Decatur, USA)
3. Hold at 5°C for 3 days to promote retrogradation (in this process, the linear chains of amylose and the highly branched chains of amylopectin, which jointly compose starch, associate, become insoluble and so precipitate).
4. Dry at room temperature.
5. Blend with 5 g pearl corn starch (e.g. CPC International, USA).

6. Pulverize in a mill (a Wiley mill was used (A.K. Thomas, Philadelphia, USA)) with dry ice.
7. The authors used only particles (micro-capsules) retained on a 106 μm mesh screen but suggested that increasing the concentration of corn oil might reduce particle size.

Other micro-encapsulation processes

Provided that BV infectivity is not adversely affected, probably most micro-encapsulation processes listed in Table 7.6 (p. 141) may be considered where a persistent preparation is required and especially where the capsule size need not necessarily be very small, e.g. NPV of *Spodoptera frugiperda*, fall armyworm, specifically for placement in corn leaf whorls.

QUALITY CONTROL

The aim, finally, of quality control is to satisfy the registration requirements of the relevant authorities (Chapter 30). In order to achieve this, it is essential to maintain a good standard of quality throughout the production process.

Best laboratory practice must be used and frequent checks made to ensure that the inoculum is pure, its infectivity is up to standard, the microbial load resulting from the microbial flora of the host insect is at a minimum and that cross-contamination with other viruses and microorganisms does not occur.

Laboratory personnel must be well trained in the techniques and safety procedures employed and also in the correct reporting procedures to be followed. Any adverse reactions that occur while working with the virus or product should be noted and serum samples should be taken from personnel before any work is initiated and at intervals thereafter. This should be tested for the presence of virus-specific antibodies (p. 556).

In order to generate the necessary data to fulfil the registration requirements, reference should be made to the techniques described in Chapter 24–26. These do not cover testing for allergenicity, toxicity, carcinogenicity and teratogenicity for which the aid of a specialist should be sought.

Temperature stability can be tested in the laboratory by incubating a sample in a waterbath at the required temperature and removing and bioassaying samples at defined time intervals. A graph of LD_{50} (y axis) versus time (x axis) should be plotted and the slope calculated to determine the inactivation gradient. Similarly, inactivation curves should be constructed using data for virus applied in the field. If temperatures greater than the temperature of incubation are being tested, virus should be placed on ice immediately after sampling to prevent further inactivation.

Simulation of solar UV for laboratory use is described on p. 446.

BACTERIOLOGICAL EXAMINATION†

The purpose of a bacteriological examination is to monitor the various stages of the production process for the presence of human, veterinary and insect pathogens that may reduce efficiency or render the product unacceptable. It involves the identification of species and the determination of viable numbers of organisms in each sample tested. Both selective and differential solid media are used, with and without pre-enrichment and enrichment.

It is often cheaper to prepare liquid media, agar plates and slopes from dehydrated powder rather than to use kits or ready-prepared items. If facilities for the preparation of sterile media are unavailable, however, the convenience and time saved in using ready-prepared items will often outweigh their higher unit costs.

Random samples should be taken that are representative of each stage of the production process. If procedures such as, for example, centrifugation, drying and formulation are involved, then samples from each should be tested separately. Feeding stimulants, adjuvants such as fishmeal, or UV protectants such as agricultural by-products added into the final product may be significant sources of contamination.

It is preferable to examine samples from pooled homogenates of insects rather than from homogenates of selected individuals, since microbial contamination of a small proportion of the population may be overlooked in the latter instance although it might be sufficient to render the final product unacceptable.

The most convenient and economical method for determining numbers of viable organisms is that of Miles and Misra (1938). By careful selection of different media it can be used to estimate numbers of viable organisms belonging to specific groups as well as estimating the total number of viable organisms in the sample being tested.

MILES AND MISRA METHOD FOR DETERMINING NUMBERS OF VIABLE BACTERIA

Media and materials

Items marked with an asterisk should be sterile.

1. Plastic or glass Petri dishes*
2. Pipettes* and appropriate filling device e.g. rubber teat or pipette pump
3. Ingredients for preparing agar plates appropriate for detecting the desired species of bacterium
4. Containers for diluting the sample to be investigated e.g. capped or plugged test tubes (150 mm × 16 mm)*
5. Diluent, e.g. glass-distilled water*

†Contributed by David Grzywacz

6. Containers with sufficient disinfectant (e.g. 5% sodium hypochlorite solution) to submerge used pipettes
7. Containers for disposing of used Petri dishes
8. Incubators at the required temperatures
9. Test tube racks
10. Bunsen burner or similar for aseptic technique and flaming equipment
11. Vortex mixer
12. Plastic-backed bench surface protector, e.g. Benchkote (supplied by BDH), absorbent cloth or paper towelling on which to work if the bench cannot be decontaminated directly
13. Disinfectant (e.g. 0.5–1.5% Hycolin or 0. 5% Tegodyne) for surface sterilizing the bench
14. Ice bucket for keeping samples cool if ambient temperature is high (over 30°C).

Note. Various types of pipetting device are available. The procedure described below utilizes 10 ml and 1 ml glass pipettes graduated to tip, and '50' dropper glass pipettes (p. 361).

Preparation of agar plates

1. Prepare and sterilize medium according to the manufacturer's instructions.
2. Aseptically dispense equal (approximately 20 ml) volumes of molten agar medium into the required number of Petri plates.
3. Allow to set and then dry by incubating overnight at 37°C or by placing inverted in an incubator at 50°C for 30 min, as shown in Figure 28.1.
4. Store unused plates at 4°C.

Preparation of serial tenfold dilutions

If working on an open bench swab with disinfectant. If preferred, cover the working area with cloth or paper towelling soaked in disinfectant, or absorbent bench coating, to prevent splashing and reduce the resultant formation of aerosols.

1. Use 10 test tubes for each sample to be tested. Mark each with the batch or identification number of the sample together with the appropriate dilution (one tube per serial tenfold dilution from 10^{-1} to 10^{-9}).
2. Place tubes in a rack in descending order of dilution.
3. Using a 10 ml pipette, transfer 9 ml of sterile glass-distilled water to all tubes.

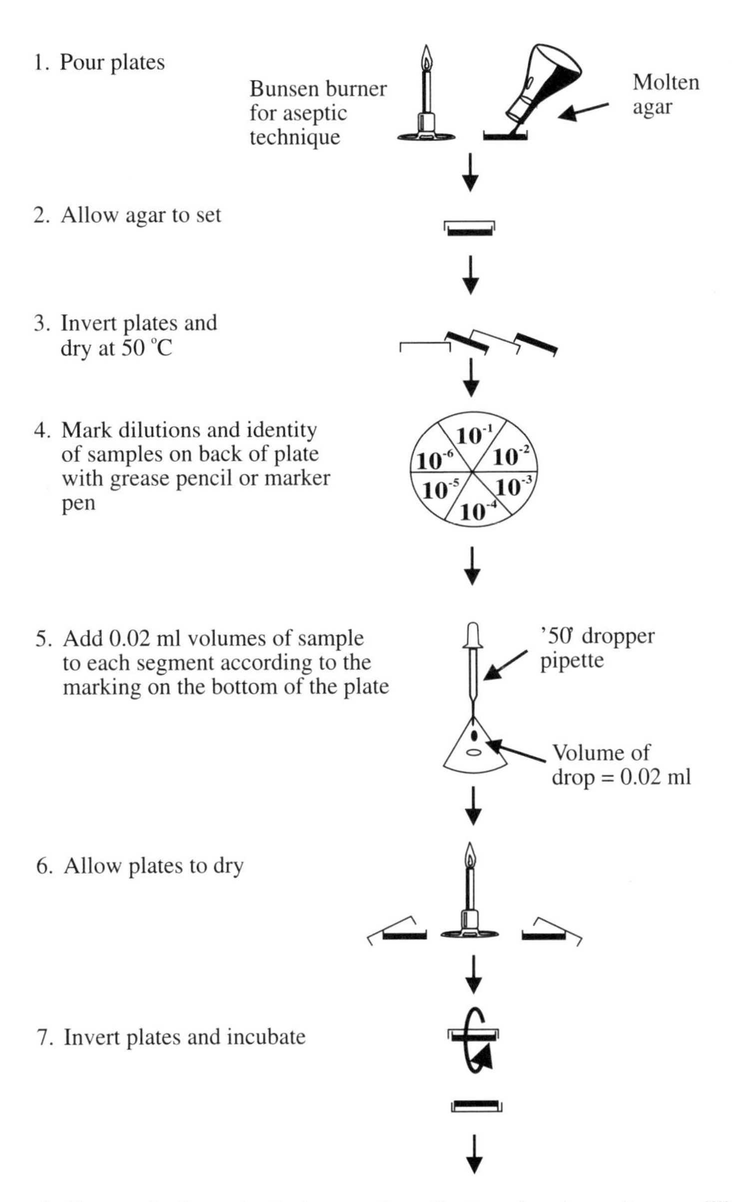

Figure 28.1 Procedure for the Miles and Misra technique.

4. Weigh 1.0 g sample and add to the tube labelled 10^{-1}.
5. Mix thoroughly by placing on a vortex mixer for 30 s. then transfer 1.0 ml of the suspension into the second tube (labelled 10^{-2}). Avoid touching

the tube or the water with the pipette while doing this. Discard the pipette into disinfectant. To avoid making an aerosol, remove the teat only after the tip of the pipette is below the level of the liquid. Once removed, make sure that the pipette is completely immersed.

6. Take a fresh pipette and mix the suspension thoroughly by pipetting up and down for a standard number of times (at least five and preferably ten; if preferred, the suspension can be mixed on a vortex mixer for a standard period of time, e.g. 2 s). Using the same pipette, transfer 1.0 ml of this suspension to the tube labelled 10^{-3}. Discard the pipette into disinfectant.
7. Continue the dilution series until the last tube is reached.

Suspensions prepared as above can be used both for screening for pathogens and for estimating the numbers of viable organisms. All dilutions should be plated out within 30 min of being prepared. If left longer, multiplication of some species may occur while others may die.

Plating out samples for viable counts

The media used will depend on the purpose of the tests to be performed and on the types of organism to be isolated (Figure 28.1).

1. Mark the underside of each plate into six equal segments by grease pencil or marker pen and label each segment with a different dilution (Figure 28.1). Also label each plate with the date and sample number. Prepare six replicate plates for each dilution to be examined. Initially it will be necessary to plate out all dilutions, but with experience dilutions containing too many or too few bacteria can be omitted.
2. Choose the largest dilution (smallest concentration) of suspension to be plated and mix thoroughly on a vortex mixer or by aspirating with a '50' dropper pipette. Draw a small volume into the pipette and transfer a single drop onto the appropriate segment of each plate. Deliver the drops at a rate of one per second while holding the pipette vertically 2.5 cm above the surface of the agar.
3. Using the same pipette, repeat the process for the other dilutions working from the most dilute to the most concentrated.
4. Leave plates on the bench until the drops have dried, then invert so that the lids are underneath and the bottoms of the plates are uppermost (Figure 28.1).
5. Incubate overnight at 35°C.
6. Count by eye or with the aid of a colony counter the number of colonies formed by each type of bacterium. For better visibility, it is helpful to place the plates on a light box.

7. Make counts in the drop areas showing the largest number of colonies without confluence (up to 20 or more). It will not be possible to count as many colonies for fast-growing species, e.g. *Bacillus cereus*, as it will for slow-growing species such as faecal enterococci.
8. After counting, replace plates in incubator and leave for a further 24 h.
9. Examine for the appearance of small, slower growing colonies and count if evident.

Calculation of results

The results of a representative total viable count are given in Table 28.1. As can be seen, overcrowding has occurred at the higher concentrations (smallest dilutions) resulting in an underestimate of the numbers of viable bacteria present. At the lowest concentrations (largest dilutions), sampling errors give rise to unreliable results. In the example given, the counts at the 10^{-4} dilution should be used. The mean of the six counts gives the viable count in 0.02 ml of the dilution. The titre in colony-forming units per millilitre is, therefore:

$$17.8 \times (1.0 \div 0.02) \times \text{the dilution factor} = 17.8 \times 50 \times 10^4 \text{ cfu/ml}$$
$$= 8.9 \times 10^6 \text{ cfu/ml}$$

Table 28.1 Results of viable count using the method of Miles and Misra (1938)

Dilution	Viable counts at varying dilutions				
	10^{-2}	10^{-3}	10^{-4}	10^{-5}	10^{-6}
Total number of colonies (six replicates)	600	300	107	13	4
Mean number of colonies per drop (0.02 ml)	100	50	17.8	2.2	0.67

Screening for specific bacterial types

It must be remembered that colonial morphology can be very variable, depending on conditions of incubation, degree of overcrowding and the precise formulation of the media. If possible, published atlases of bacterial morphology (Olds, 1986; Varnam and Evans, 1992) and the data sheets or manuals supplied by the manufacturers of media should be consulted (Difco; Oxoid).

PREPARATION OF BACTERIAL SMEARS FOR STAINING

Materials

1. Wire loop for transferring bacteria
2. Glass microscope slides for making smears of bacteria
3. Staining rack
4. The appropriate stains, e.g. Gram's and counterstain, for visualizing bacteria
5. Light microscope with ×100 oil immersion objective
6. Immersion oil
7. Blotting or filter paper
8. Diamond marker.

Method

1. Clean a microscope slide and scratch an identification mark on the right-hand side with a diamond marker or similar.
2. For microorganisms from a solid medium, put a loopful of tap water or sterile water on the slide at the point where the smear is to be made. Omit this step for liquid cultures. With experience, up to five smears can be made on a single slide.
3. Flame the loop, touch onto an uninoculated part of the medium to ensure it is cool and then onto the chosen colony. Withdraw a *small* quantity of colony on the loop.
4. Emulsify the bacteria on the loop in the drop of water on the slide and spread sufficiently to give a thin smear. Flame the loop and replace on the bench.
5. Dry the smear by holding the slide, with forceps, above the Bunsen flame. (The slide should not be too hot to touch). Once dry, fix the material by passing the slide slowly, three times, through the Bunsen flame.
6. Cool and place on a staining rack.

GRAM STAIN

The Gram stain method depends on the ability of some bacteria to retain the Crystal Violet–iodine complex when treated with alcohol. Such bacteria are designated Gram positive. Bacteria that are decolorized by alcohol after being stained with Crystal Violet and treated with iodine are designated Gram negative. These are visualized by counter staining with 0.5% aqueous solution of saffranine.

Materials

1. Crystal Violet solution. Solution A: Crystal Violet, 2 g; absolute ethanol, 100 ml. Solution B: ammonium oxalate, 1 g; distilled water 100 ml. Add 25 ml of Solution A to 100 ml of Solution B
2. Gram's iodine solution: iodine, 1 g; potassium iodide, 2 g; distilled water, 300 ml
3. Saffranine, 0.5 g; distilled water, 100 ml.

Method

1. Prepare a smear of bacteria as described above.
2. Stain with Crystal Violet for 1 min.
3. Pour off stain and wash away precipitate of stain with Gram's iodine. Cover with Gram's iodine for 1 min.
4. Wash off iodine with 95% ethanol until the washings are a very pale violet.
5. Wash immediately in running tap water.
6. Counterstain with 0.5% saffranine for 30 s.
7. Wash in running tap water, blot and dry.
8. Examine under the oil immersion objective.

OXIDASE TEST

Control organisms Positive, *Pseudomonas aeruginosa* ATCC$ 27853; negative, *Staphylococcus aureus* ATCC$ 25923.

Method

1. Rub a loopful of pure culture from a colony grown on a non-selective medium onto a piece of filter paper. If possible smear examples of both positive and negative reference strains onto the same piece of paper.
2. Soak the paper with 1% (w/v) aqueous solution of *N*,*N*-tetramethyl-*p*-phenylene-diamine dihydrochloride that has been freshly made and is no darker than pale blue in colour.

Oxidase positive cultures will turn a deep colour almost immediately.

Precaution: care should be taken to avoid all contact of *N*,*N*-tetramethyl-*p*-phenylene diamine dihydrochloride with skin and clothing, as the reagent is extremely toxic. If preferred, oxidase identification sticks (Oxoid BR 64), which are easier to handle, can be used instead.

CATALASE TEST

Control organisms Negative, *Escherichia coli* ATCC$25922.

Method

1. Take a loopful of colony grown on a non-selective agar medium and smear onto the bottom of a Petri dish.
2. Place a drop of 3% aqueous H_2O_2 solution 1 cm from the smear. Close the Petri dish and tilt so that the drop runs onto the smear.

The immediate production of bubbles indicates a positive reaction.

COAGULASE TEST

The traditional coagulase test has been superseded by proprietary agglutination tests for the quick identification of coagulase-positive strains of *Staphylococcus aureus*. These are manufactured and marketed by both Oxoid (Unipath) 'Staphylase test' and BioMerieux (API) 'Staphyslide test'.

Some common contaminants of BV preparations have a typical colonial appearance on nutrient agar.

Faecal enterococci These are small, round, convex colonies 1–2 mm in diameter after incubation at 37°C for 18 h. Most strains produce white opaque colonies but some strains produce yellow tinted colonies.

Bacillus cereus Large white flat spreading colonies form with an irregular outline and a characteristic 'ground glass' appearance; they are 3–5 mm in diameter after 18 h incubation at 37°C.

Bacillus sphaericus The round colonies are 2–3 mm in diameter after 24 h at 37°C and have a flattened convex shiny surface.

Acinetobacter sp. Colonies have a diameter of 0.1–0.3 mm after 48–72 h incubation at 37°C. They may not be apparent after 18–24 h.

Isolation and identification of coliforms, *Salmonella* and *Shigella* spp.

The coliforms, salmonellae and shigellae are Gram-negative, rod-shaped (1.0 μm in length), catalase-negative, oxidase-negative, facultative anaerobes belonging to the family Enterobacteriaceae. This family includes both pathogens and common, usually harmless, inhabitants of the human and animal gut. Serial tenfold dilutions of test sample in sterile glass-distilled water should be plated onto general media with low selectivity, such as MacConkey's or Eosin Methylene Blue agar. They should also be plated onto one or two of a selection of moderately selective/differential media, such as deoxycholate citrate agar (DCA), *Shigella–Salmonella* agar (SS), hektoin enteric agar (HE) or xylose lysine deoxycholate agar (XLD), and onto media such as bismuth sulphate and Brilliant Green agar that are highly selective for *Salmonella* spp.

It is advisable to include an enrichment step for duplicate test samples to aid recovery of organisms that are either present in small numbers or have been injured by certain stages in the production process, e.g. by freezing and thawing.

PROCEDURE FOR ENRICHMENT

Method

1. Pre-enrich by adding 1 ml of sample to 9 ml lactose broth.
2. Incubate for 24 h at 35–37°C.
3. Selectively enrich by adding 1 ml of culture from above to 9 ml each of tetrathionate broth and selenite broth.
4. Incubate for 24 h at 35–37°C.
5. Streak samples from the enrichment broths onto a range of selective and differential media.
6. Examine colonies.
7. Test further as necessary.

Suspect organisms should be stained by the method of Gram. Gram-negative rod-shaped organisms should be subcultured onto nutrient agar and incubated for 35–37°C for 14–18 h before staining and testing for motility using the hanging drop technique (see below). Oxidase and catalase tests should also be performed.

HANGING DROP TECHNIQUE FOR INVESTIGATING BACTERIAL MOTILITY

Materials

1. Glass slide with central concavity (cavity slide)
2. Broth or plate culture of the organism being investigated
3. Vaseline
4. Coverslip.

Method

1. Dip a match into some vaseline and mark a circle or square (depending on the shape of the coverslip) around the concavity.
2. With a wire loop place a drop of broth culture in the centre of the coverslip or, if using a plate culture, emulsify a small amount in nutrient broth or saline and place on the coverslip.
3. Invert the slide over the coverslip, making sure the vaseline adheres, then invert quickly so that the coverslip is uppermost.
4. Place on the stage of a light microscope under low illumination and focus on the edge of the drop using low power.
5. Change to high power (× 40 objective) and refocus. Obtain the best illumination by lowering or raising the condenser and secure sharp definition by reducing the diaphragm aperture.
6. Observe for motility.

Appearance of colonies on general, selective and differential media

MacConkey agar (Oxoid No. 3, CM115)

Control organisms Positive, *Escherichia coli* ATCC$ 25922, *Shigella sonnei* ATCC$ 25931; negative, *Enterococcus faecalis* ATCC$ 29212.

Appearance of colonies after incubation at 35 °C for 18–24 h
Salmonella/Shigella spp.: colourless colonies; coliforms, pink or red colonies 2–3 mm in diameter
Streptococci: deep red colonies 1 mm in diameter that may develop white edges or turn completely white over 2–3 days
Pseudomonas spp.: green-brown fluorescent colonies
Klebsiella spp.: high-domed, pink mucoid colonies growing to 4–6 mm in diameter after 48 h.

DCA (Hynes) (Oxoid CM227)

Control organisms Positive, *Salmonella typhimurium* ATCC$ 14028, *Shigella sonnei* ATCC$ 25931; negative, *Enterococcus faecalis* ATCC$ 29212.

Appearance of colonies after incubation at 35 °C
Shigella sonnei: smooth convex colonies 1 mm in diameter at 18 h that increase to 2 mm at 36 h; initially colourless but turning pale pink by 36–48 h owing to late fermentation of lactose
Shigella flexneri: similar to those of *S. sonnei* but with a narrow, planar periphery around a central dome
Shigella paratyphi B: colony diameter increases from 1 mm at 18 h to 2–4 mm by 48 h; at this point, colonies are slightly opaque, dome-shaped and with a central black dot
Salmonella typhi: pale pink colonies 0.25 mm in diameter at 18 h; they increase in diameter to 2 mm at 48 h while becoming colourless, slightly opaque and flat with a conical centre
Non-pathogenic species, e.g. *Proteus* and *Pseudomonas*: may grow on DCA producing colonies similar to those above
Escherichia coli: most strains inhibited by this medium, a few produce umbilicated colonies 1–2 mm in diameter that may be surrounded by a zone of precipitate
Aerogenes: domed and mucoid colonies.

SS agar modified (Oxoid CM533)

Control organisms Positive, *Salmonella enteritidis* ATCC$ 13076, *Shigella sonnei* ATCC$ 25931; negative, *Enterococcus faecalis* ATCC$ 29212.

Appearance of colonies after incubation at 35 °C
Shigella spp.: transparent colonies
Salmonella spp.: transparent colonies usually with black centres
Proteus and *Citrobacter* spp.: colonies similar to those of salmonellae may form but the centres tend to be grey-black.

Hektoen enteric agar (Oxoid CM419)

Control organisms Positive, *Salmonella typhimurium* ATCC$ 14028, *Shigella flexneri,* ATCC$ 12022; negative, *Escherichia coli* ATCC$ 25922, *Enterococcus faecalis* ATCC$ 29212.

Appearance after incubation at 37°C
Salmonella sp.: blue-green colonies with or without black centres
Shigella sp.: produce green moist raised colonies. incubation for 48 h improves differentiation between *Salmonella* and *Shigella* spp.
Commensal enteric organisms: salmon pink colonies
Pseudomonas sp.: green or brownish, flat, irregular colonies

Coliforms: salmon-pink to orange colonies surrounded by a zone of bile precipitation.

XLD agar (Oxoid CM469)

Control organisms Positive, *Salmonella typhimurium* ATCC$ 14028; negative, *Escherichia coli* ATCC$ 25922.

Appearance of colonies after incubation at 37°C
Salmonella, *Arizona*, and *Edwardsiella* spp.: red colonies with black centres after 24 h
Shigella, some strains of *Salmonella, Proteus* and *Providencia*: red colonies
Escherichia, *Enterobacter*, *Klebsiella* and *Citrobacter* spp.: yellow opaque colonies.

Brilliant green agar (Oxoid CM263)

Control organisms Positive, *Salmonella enteritidis* ATCC$ 13076, *Salmonella typhimurium* ATCC$ 14028; negative, *Escherichia coli* ATCC$ 25922, *Proteus vulgaris* ATCC$ 13315.

Appearance of colonies after incubation at 35°C Most species (including *Shigella*) other than *Salmonella* are inhibited by this very selective medium.

Salmonella spp.: red colonies surrounded by a bright red zone
Pseudomonas spp.: some strains may grow and form small crenated colonies
Proteus spp.: almost completely inhibited but those that do grow are red and do not swarm
Other lactose or sucrose fermenters: any that grow form yellow or green colonies.

Bismuth sulphite agar (Oxoid CM201)

Control organisms Positive *Salmonella enteritidis* ATCC$ 13076, *Salmonella typhimurium* ATCC$ 14028; negative, *Escherichia coli* ATCC$ 25922, *Citrobacter freundii* ATCC$ 8090.

Appearance of colonies after incubation at 37°C The appearance of colonies other than those of *Salmonella* sp. varies after 18 h: they may be black, green or clear and mucoid. After 48 h they become uniformly black with widespread staining of the medium and have a pronounced metallic sheen. Most other species of bacterium are inhibited but occasionally may produce dull green or brown colonies without a metallic sheen or staining of the medium.

Salmonella typhi: characteristic 'rabbit eye' colonies formed at 18 h, with a black zone and a metallic sheen around each colony. After 48 h colonies are uniformly black and 2–3 mm in diameter.

Further tests for Enterobacteriaceae

If facilities are limited or the number of suspect isolates is small, commercial kits may be used to identify isolates (for example API Enterobacteriaceae kits API 20E, Rapid 20 E), or to screen for *Salmonella/Shigella* (API Z). Alternatively, Gram-negative, catalase- and oxidase-positive bacteria should be subcultured onto nutrient agar and tested as follows.

UREASE TEST

Method

1. Inoculate slants of Christensen's agar.
2 Include an uninculated control tube for comparison.
3. Incubate at 37°C and examine at 2, 4, 6 and 8 h. A distinct pink colour indicates a urease-positive strain. Urease-negative cultures should be tested further for the presence of *Salmonella, Shigella* and *Arizona* spp.

TRIPLE SUGAR IRON (TSI) TEST (OXOID CM277)

Method

1. Stab and streak slants and butts of TSI agar.
2. Incubate at 35°C and examine at 18 and 48 h.

Table 28.2 provides the reaction types that enable the isolate to be identified. Any isolate that does not affect TSI cannot ferment lactose, dextrose or sucrose and is not, therefore, a member of the Enterobacteriaceae. It probably belongs to one of the non-fermenter groups such as the pseudomonads.

LYSINE IRON AGAR (OXOID CM381)

Method

1. Stab and streak one each of a lysine agar butt and slant.
2. Incubate at 35°C overnight in aerobic conditions (ensure that the test tube cap, if used, is loose).

Bacteria that decarboxylate lysine produce an alkaline (purple) colour throughout the medium while those that do not give an acid reaction (yellow colour). Reactions of some members of the Enterobacteriaceae are given in Table 28.3.

Control organisms Lysine decarboxylation, *Enterobacter aerogenes* ATCC$ 13048; deamination, *Proteus vulgaris* ATCC$ 13315; negative control, *Enterobacter cloacae* ATCC$ 23355.

Table 28.2 Triple sugar iron reactions of the Enterobacteriaceae

Organism	Slant	Butt	H_2S production
Enterobacter aerogenes	A	A + G	–
Enterobacter cloacae	A	A + G	–
Escherichia coli	A	A + G	–
Proteus vulgaris	A	A + G	+
Morganella morgianii	NC or Alk	A or A + G	–
Shigella dysenteriae	NC or Alk	A	–
Shigella sonnei	NC or Alk	A	–
Salmonella typhi	NC or Alk	A	–
Salmonella paratyphi	NC or Alk	A + G	–
Samonella enteritidis	NC or Alk	A + G	+
Salmonella typhimurium	NC or Alk	A + G	+

Alk (alkaline), red colour; A (acid), yellow colour; NC (neutral), no colour; A+G, production of acid and gas; + , production of H_2S (blackening of medium).

Table 28.3 Reaction of some members of the Enterobacteriaceae on Lysine Iron Agar

Species	Slant	Butt	H_2S production
Arizona	Alk	A	+
Citrobacter	Alk	A	+ (+/–)
Escherichia	Alk	A or neutral	–
Klebsiella	Alk	Alk	–
Providence	Red	A	–
Proteus	Red	A	–
Salmonella	Alk	Alk	+
Shigella	Alk	A	–

Alk (alkaline), purple colour; A (acid), yellow colour; red, red slant over an acid butt results from deamination of lysine; +, production of H_2S (blackening of medium).

Isolation and identification of *Staphylococcus aureus*

Staphylococcus aureus is an aerobic, Gram-positive, catalase-positive, coagulase-positive species that forms cocci which are 1.0 μm in diameter; both pathogenic and non-pathogenic strains occur. The pathogenic strains can cause boils, wound infections and food poisoning. The bacterium is commonly found in the nasal passages, mouth and hands, where it is non-symptomatic, Its presence in preparations of BVs is, therefore, indicative of poor hygiene.

Staphylococcus aureus can be detected by plating dilutions directly onto selective media such as Manitol salt, Baird–Parker or Vogel–Johnson agar. Enrichment procedures should be employed for samples that could contain *Staphylococcus aureus* damaged by, for example, freezing.

NON-SELECTIVE PRE-ENRICHMENT AND ENRICHMENT

Method

1. Add 1 ml of test sample to 2 ml double-strength brain heart infusion broth (brain heart infusion (BFI) or Oxoid nutrient broth No. 2, CM67).
2. Incubate for 2 h at 35–37°C.
3. Add 2 ml broth containing 20% NaCl.
4. Incubate for 24 h at 35–37°C.
5. Plate onto selective medium (e.g. Baird–Parker or Vogel–Johnson medium).
6. Examine colonies.
7. Test further as necessary (see below).

SELECTIVE PRE-ENRICHMENT AND ENRICHMENT

Method

1. Add 1 ml of sample to 9 ml of rich broth containing 10% NaCl.
2. Incubate for 48 h at 35°C.
3. Plate onto selective media.
4. Incubate for 24–48 h at 35–37°C.
5. Examine colonies.
6. Test further as necessary (see below).

Appearance of *Staphylococcus aureus* colonies on selective media

Manitol salt agar (Oxoid CM85)

Control organisms Positive, *Staphylococcus aureus* ATCC$ 25923, *Staphylococcus epidermis* ATCC$ 12228; negative, *Escherichia coli* ATCC$ 25922.

Appearance of the colonies Manitol salt agar medium relies for its selective action on the ability of *Staphylococcus aureus* to grow in the presence of high salt concentrations. Cultures should be incubated for 36 h at 35°C or 3 days at 32°C. Pathogenic strains are differentiated from non-pathogenic strains by the ability of the former to ferment manitol so that the colonies are surrounded by a yellow zone (presumptive coagulase positive) in the medium compared with a red or purple zone for coagulase-negative strains. Pick off suspect colonies and subculture in a medium not containing an excess of salt (e.g. Oxoid Nutrient Broth No. 2, CM67) to avoid interference with coagulase and other diagnostic tests. Some salt-tolerant *Bacillus* spp. may also grow on this medium, but the organisms can be distinguished from those of *Staphylococcus aureus* by being rod-shaped.

Baird-Parker agar (Oxoid Baird-Parker agar base CM 275 with Egg Yolk-Tellurite Emulsion SR54)

Control organisms Positive, *Staphylococcus aureus* ATCC$ 25923; negative, *Bacillus subtilis* ATCC$ 6633, *Staphylococcus epidermidis* ATCC$ 12228.

Appearance of the colonies *Staphylococcus aureus* forms round flat shiny grey-black convex colonies 1.0–1.5 mm in diameter (18 h) and up to 3 mm after 48 h incubation at 35–37°C. Each colony is surrounded by a narrow white entire margin surrounded by a clear zone 2–5 mm in width, owing to proteolytic action. *Staphylococcus saprophyticus* may also grow on this medium but its egg yolk reaction is different. Other micrococci may grow but are often brown in colour and colonies show no yolk clearing. Some *Bacillus* spp. may develop with clearing of the medium but the colonies are brown.

Vogel-Johnson agar (Oxoid CM641)

Control organisms Positive, *Staphylococcus aureus* ATCC$ 25923*; negative, Escherichia coli* ATCC$ 25922.

Appearance of the colonies *Staphylococcus aureus* produces shiny black colonies surrounded by a yellow coloured zone in the medium after 24 h incubation at 37°C.

Biochemical kits

All isolates comprising Gram-negative, coagulase-negative cocci should be identified using biochemical kits such as API Staph and the identification confirmed by performing agglutination tests specific for *Staphylococcus aureus* such as the ‘Slidex Staph-kit’ and ‘Staphyslide’ test marketed by bioMérieux or the ‘Staphylase’ test marketed by Unipath (formerly Oxoid). These utilize sensitized red blood and/or latex particles.

Bacillus spp.

Since it can be difficult to detect *Bacillus cereus* in material that contains large numbers of other species at greater concentrations, it is advisable to use an agar medium that is specific for this organism. This will not suppress the growth of all other species likely to be encountered but will reduce numbers of colonies sufficiently for the distinct colony form of *Bacillus cereus* to be visualized. Positive confirmation can be obtained by staining (see below) or on the basis of a biochemical profile using a commercial kit (BioMérieux, API50 CHB).

Bacillus cereus *selective agar (Oxoid CM617 with SR99)*

Bacillus cereus produces crenated colonies, 5 mm in diameter, which are a distinctive turquoise to peacock blue in colour and surrounded by a good egg yolk precipitate of the same colour 24 h after incubation at 35°C.

STAINING FOR *BACILLUS CEREUS* SPORES

Materials

1. Aqueous Malachite Green 5%
2. Sudan Black 0.3% (w/v) in 70% ethanol
3. Aqueous saffranin 0.5% (w/v).

Method

1. Prepare a smear from a 24–48 h colony and fix by flaming.
2. Place slide over boiling water and flood with Malachite Green for 2 min.
3. Wash in running tap water and blot dry.
4. Stain with Sudan Black for 15 min.
5. Wash slide in xylene for 5 s and blot dry.
6. Counterstain with Saffranin for 20 s.
7. Wash in running water, dry and examine under an oil immersion objective.

B. cereus cells are seen as large bacilli (4–5 μm long by 1.0–1.5 μm wide) with square ends and rounded corners. The spores stain pale green to mid green, are central or paracentral in position and do not swell the bacillus. The cytoplasm stains red and contains black-stained lipid inclusions.

Detection, isolation and enumeration of yeasts and moulds

Yeasts and molds can be characterized using malt extract agar and malt extract broth. The latter is recommended for the cultivation of moulds and yeasts especially during tests for sterility.

Control organisms Positive, *Aspergillus niger* ATCC$ 9642, *Candida albicans* ATCC $10231; negative, *Bacillus cereus* (at pH 3.5), ATCC$ 10876.

MALT EXTRACT AGAR (OXOID CM59)

Method

1. Inoculate each specimen in duplicate.
2. Incubate one set at 22–25°C and the other at 35°C, for 5–30 days. Ensure adequate moisture to prevent the agar drying out but do not seal the plates.
3. Examine plates every 2–4 days and describe colony morphology.
4. If wished, subculture onto appropriate media for further identification.

Bacterial growth can be suppressed by adding 10% lactic acid (e.g. Oxoid SR21) to the molten malt extract agar just before it is poured.

REFERENCES

Bull, D.L. (1978) Formulations of microbial insecticides: microencapsulation and adjuvants. *Miscellaneous Publications of the Entomological Society of America* **10**, 11–20.

Bull, D.L., Ridgway, R.L., House, V.S. and Pryor, N.W. (1976) Improved formulations of the *Heliothis* nuclear polyhedrosis virus. *Journal of Economic Entomology* **69**, 731–736.

Couch, T.L. and Ignoffo, C.M. (1981) Formulation of insect pathogens. In *Microbial Control of Pests and Plant Diseases 1970–1980*. H.D. Burges (ed.). London: Academic Press, pp. 621–634.

Dulmage, H.T., Martinez, A.J. and Correa, J.A. (1970) Recovery of the nuclear polyhedrosis virus of the cabbage looper, *Trichoplusia ni*, by coprecipitation with lactose. *Journal of Invertebrate Pathology* **16**, 80–83.

Dunkle, R.L. and Shasha, B.S. (1988) Starch-encapsulated *Bacillus thuringiensis*: a potential new method for increasing environmental stability of entomopathogens. *Environmental Entomology* **17**, 120–126.

Dunkle, R.L. and Shasha, B.S. (1989) Response of starch-encapsulated *Bacillus thuringiensis* containing ultraviolet screens to sunlight. *Environmental Entomology* **18**, 1035–1041.

Hostetter, D.L. and Pinnell, R.E. (1983) Laboratory evaluation of plant-derived granules for bollworm control with virus. *Journal of the Georgia Entomological Society* **18**, 155–159.

Ignoffo, C.M., Shasha, B.S. and Shapiro, M. (1991) Sunlight ultraviolet protection of the *Heliothis* nuclear polyhedrosis virus through starch-encapsulation technology. *Journal of Invertebrate Pathology* **57**, 134–136.

Lewis, F.B. (1981) Control of the gypsy moth by a baculovirus. In *Microbial Control of Pests and Plant Diseases 1970–1980*. H.D. Burges (ed.). London: Academic Press, pp. 363–377.

Miles, A.A. and Misra, S.S. (1938) The estimation of the bacteriocidal power of the blood. *Journal of Hygiene* **38,** 732.

Montoya, E.L., Ignoffo, C.M. and McGarr, R.L. (1966) A feeding stimulant to increase effectiveness of, and a field test with, a nuclear-polyhedrosis virus of *Heliothis*. *Journal of Invertebrate Pathology* **8**, 320–324.

Olds, R.J. (1986) *A Colour Atlas of Microbiology*, London: Wolfe Publishing.

Shapiro, M., Bell, R.A. and Owens, C.D. (1980) *In vivo* mass production of gypsy moth nucleopolyhedrosis virons. In *The Gypsy Moth: Research Toward Integrated Pest Management, Technical Bulletin 1584*. C.C. Doane and M.L. McManus (eds). Washington, DC: USDA Forest Service, pp. 633–655.

Smith, D.B., Hostetter, D.L., Pinnell, R.E. and Ignoffo, C.M. (1982) Laboratory studies of viral adjuvants: formulation development. *Journal of Economic Entomology* **75**, 16–20.

Varnham, A.H. and Evans, M.G. (1991) *Foodborne Pathogens*. London: Wolfe Publishing.

29 Spray application

SPRAY DROPLET TRACING: DYES AND FLUORESCENT PARTICLES

In studies on droplet formation and deposition, it is often desirable to use a tracer or marker substance to aid droplet recognition and sizing. For instance, an important use is the measurement of total spray deposits (independently of counting droplet density and content) by washing leaves and estimating recovered tracer substance colourimetrically (e.g. Richards, 1984).

In theory, a very wide range of tracer substances is available but in practice for soluble entities the choice is limited by the question of compatibility with the virus (or other test microorganism) (e.g. Morris, 1977). For insoluble particulate tracers, it is limited by their physical performance in the spray fluid employed. Table 29.1

Table 29.1 Spray tracers or markers: the following substances have mostly been reported as used in studies on the fate of spray droplets containing microorganisms. The insoluble fluorescent particles are almost certainly benign to viruses. Soluble tracers may be more hazardous but probably not at the concentrations indicated here

Name	Concentrations reported (%)	Comments
Dyes		
Uvitex 2B/120	1.0	Water soluble; may be sunlight degradable
Erio Acid Red XB	0.1	A possible sunlight screen; at <0.1% can be toxic to *B.t.*
Nigrosine	0.5	Water- and alcohol-soluble types available
Rhodamine	0.1	Harmful. Dilute solutions strongly fluorescent
Fluorescent particles		
Brilliant Sulfo-flavine	<0.5	Safe for NPV and *B.t.*
Lumogen: Light Yellow	1.0	Difficult to mix with water
Saturn Yellow	0.6	–
Zinc sulphide	0.2–1.0	Not yet tested with microorganisms?

lists a range of substances most of which have been used in spray studies with insect pathogens.

For some purposes, the use of soluble tracers is preferable and for others insoluble particulates will be more appropriate. Soluble tracers are more likely to be selected for recognizing the position of droplets on artificial targets such as Millipore filter material, paper or glass trapping surfaces and for estimating the volumetric deposit rate (VDR) on the target crop or selected regions thereof. When a tracer used for the latter purpose is also fluorescent, e.g. Uvitex 2B/120 (Richards, 1984), there is the added benefit of direct droplet recognition and counting especially on difficult surfaces (e.g. very dark foliage or bark) using UV irradiation on plant samples brought back to the laboratory.

If a tracer is to provide a realistic picture of droplet behaviour, it is essential that it does not affect any factor, for example viscosity and surface tension, which may influence the size spectrum of the droplets produced under any given set of conditions. This problem may not be restricted to the use of soluble tracers. For instance, satisfactory suspension of particulate tracers may be aided by use of a surfactant and this may influence the size of the droplets and the efficiency of their collection by naturally reflective plant surfaces (Wirth, Storp and Jacobson, 1991) and even the retention within the droplets of the particles themselves (Smith and Bouse, 1981).

Counts of particles of insoluble tracers, which are usually selected for their fluorescent qualities in aiding droplet visualization, can be employed to size droplets. However, the possibilities of loss of particles before droplet capture and of a particle distribution not directly proportional to droplet volume (p. 162) must be considered.

Soluble tracers

Important properties of soluble tracers are that they should not posses human or animal toxicity, they should not affect the viruses or other microorganisms being studied and they should not be appreciably subject to environmental decay over the desired test period. For instance, Richards (1984) mentioned the possibility that Uvitex 2B/120 is sensitive to sunlight, though he himself did not encounter this effect. Not all dyes are suitable for colorimetric analysis. Erio Acid Red XB has low sensitivity at 0.1% (Morris, 1977).

Particulate tracers

All the particulate, insoluble tracers so far employed have fluorescent qualities. Such tracers are available in a very wide range of colours, which has the attraction that under exactly the same set of experimental field or forest conditions spray clouds produced by different methods can be colour coded. However, carbon in its various forms (Table 7.5, p. 134) will act as a visual marker, especially useful for the identification of droplets on colour-contrasting test surfaces. In working with particulate tracers three main questions should be considered: particle size, concentration and surface qualities.

Particle size

For realism particle size should usually be in the range of the microorganism under test. A desired particle size can easily be obtained by milling in the laboratory.

Particle concentration

Particle concentration should be calculated so that even the smallest droplets likely to impact on the chosen targets will contain some identifiable tracer (usually a single fluorescent particle (fp) can be identified using a UV fluorescence microscope, even at a low magnification). Because some particles, notably fps, may remain in the sprayer mechanism it may not be possible to calculate directly the tank concentration necessary to give adequate droplet labelling. Instead the task may have to be approached empirically (e.g. Payne, 1983).

Particle surface qualities

Many fps are hydrophobic. As a consequence, satisfactory mixing with water alone may be impossible and a surfactant may be required. Payne (1983) recommended first forming a paste with a little of the diluent and then diluting this to the working concentration. Even so, he experienced adhesion to the sides of the beaker with Lumogen and subsequent rather rapid settling. Incorporation will be easier with an oily carrier fluid. Fps will tend to aggregate and to lie on the surface of watery spray droplets and so may be difficult to count, but there will be fewer problems with oil. Fps cannot be satisfactorily employed in the micro-thread technique, at least where the thread is gossamer, as they tend to aggregate on the thread. Fortunately, NPV OBs appear not to do this.

The use of fps in droplet work was reviewed by Aston (1989), who also discussed estimation by image analysis systems.

Estimation of fluorescent markers

Quantification of the soluble Uvitex 2B/120 is described in detail by Richards (1984). The spray deposit containing Uvitex can be efficiently recovered, for instance by shaking leaf discs in 10 ml distilled water in 25 ml screw-cap bottles for 3 h (e.g. at speed No. 7 on a Stuart flask shaker). Richards used a fluorescence spectrophotometer (Model 1000, Perkin-Elmer) with a Schott UG-1 excitation filter to produce excitation light at 300–400 nm wavelength (peak 355 nm). Following zeroing with distilled water, a range of standard Uvitex aqueous dilutions, 0.01–1.0 mg/l, were prepared and readings taken at 435 nm. Leaf washings were measured in the same way after dilution by an appropriate factor (100–1000 in the example given below). Emission readings and the concentrations of Uvitex in the standard dilutions were analysed by linear regression after log transformation. Estimates of Uvitex concentration in undiluted washings from sprayed leaves were calculated from the formulae of regression lines. The 95% fiducial limits were attached to these estimates by inverse estimation (Sprent, 1969) according to the formula:

$$\bar{\Theta} = \frac{x\beta^2 \pm \left[ts \sqrt{\frac{(n+1)\beta^2}{n} + \frac{\beta^2 x^2}{n^2} - \frac{(n+1)t^2 s^2}{n\Sigma x^2}} \right]}{\left[\beta^2 - \frac{t^2 s^2}{x^2} \right]}$$

where $\bar{\Theta}$ is the 95% fiducial interval for the calculated x value; β is the slope of the regression line; n is the number of standard x values contributing to the regression line; $t = t^{0.95}$ d.f. (degrees of freedom); s^2 = variance of residues; and x^2 = sum of squares of standard x values.

From such data, the volumes of spray retained on a given leaf area can be obtained, taking into account the leaf area washed and the relative concentration of Uvitex and virus in the spray formulation. Richards employed this technique to compare spray arriving at the target with the quantity of codling moth GV, which itself was estimated by bioassay (Richards, 1984).

Droplets with fps can be caught on slides, dried in the dark and examined under UV light using a fluorescence microscope through a range of magnifications, e.g. $\times 25$, $\times 100$ and $\times 400$. In the image analysis system, it was found that use of a particle end-count mode mitigated the problem posed by overlapping fps (Aston, 1989).

COLLECTING SPRAY DROPLETS

The possibility of loss of solid particles from spray droplets or of a particle distribution that is not directly proportional to droplet volume is referred to in Chapter 8. The existence of these phenomena makes it necessary to devise techniques that will permit collection of spray droplets in such a way that their size can be measured and their BV OB content can be accurately counted.

Information on droplet volume may be obtained in the following ways.

1. A prior knowledge of droplet number and size spectra emitted by the spray machinery being used can be obtained, for instance from droplet laser sizing techniques, etc. The disadvantage here is the counts of particles in a sample of collected droplets can only be generally related to spray cloud size characteristics. The exception is when droplets are caught very soon after formation, for instance under controlled laboratory conditions (Aston, 1989) before significant evaporation, so that 'spot size' on a target collecting surface can be realistically related to initial droplet volume. Droplets involved in this type of assessment will usually be caught on planar surfaces (e.g. metal foil for SEM assessments of particle content, Millipore filter – which can be rendered translucent – or glass surfaces, these last two for assessment by optical microscopy). Also see approach 2, below.
2. The size of droplets of a non-volatile or partly volatile fluid can be estimated from measuring contact angle and deposited 'lens' diameter (Chapter 32).
3. The final diameter (after evaporation of any volatile component has ceased) of droplets of a non-volatile or partly volatile composition can be measured directly (e.g. on a micro-thread).

Three methods of droplet capture permitting assessment of their solid particle content are available whilst a fourth is suggested.

Capture of droplets on Millipore filter surfaces

Morris (1973, 1977) identified the position of spray droplets on a Millipore filter by incorporation of a dye (e.g. Rhodamine B or Erio Red XB, both at 1%) in the spray fluid. The BV OBs were then stained *in situ* on the filters.

MILLIPORE FILTERS TO CAPTURE DROPLETS

Materials

1. Stock stain solution: basic Fuschin, 1 g in 10 ml tertiary butanol mixed with 5 g phenol dissolved in 90 ml deionized water
2. Working stain solution: 30% stock solution in water filtered through a 0.45 μm Millipore filter
3. Counterstain: tertiary butanol, 50 parts; distilled water, 45 parts; acetic acid, 5 parts; Amido Black 90B, 1.5 g filtered through a 0.45 μm Millipore filter
4. A decolorizer: filtered 3% concentrated HCl in ethanol.

Method

1. Expose Millipore filters to spray and dry in air.
2. Place air-dried filters in a Petri dish on filter paper flooded with 4 ml working strength stain. Leave for 30 min.
3. Dip several times in 200 ml distilled water.
4. Dip several times in three successive batches of decolorizer and rinse in fresh distilled water.
5. Dip for 8 s in counterstain, rinse in water and air dry.
6. Place pieces of filter of suitable size on a microscope slide in a drop of clarifying fluid (e.g. Protex Mounting Medium (Scientific Products, Evanton, IL: USA) is recommended, but try conventional immersion oil).
7. Observe in the light microscope.

Polyhedra of NPV (and of CPV) stain a light to deep pink. Fungal spores, microsporidial spores and *B.t.* spores stain similarly, but *B.t.* crystals stain a deep purple.

Morris (1977) successfully used this technique in the forest to estimate deposition of *B.t.*, NPV and entomopoxvirus. Millipore filter targets were placed on the ground but could equally well have been placed in trees and could have been cut into shapes closely similar to tree needles (in this case of *Picea glauca*) to determine realistically the distribution of droplet deposits on the tree.

Because staining processes have not yet been devised for the two following methods, their value is dependent upon being able to distinguish between OBs and any contaminant particles reliably. In our experience, this is not generally a problem with, at least, semi-purified NPV preparations.

Capture of droplets on glass surfaces

Droplets may be captured in the laboratory or the field on clean glass surfaces, the position of the droplets being identified by incorporation of a soluble dye (e.g. 0.5% Nigrosin) in the spray fluid. Where an emulsifiable or pure spray oil is the carrier fluid, the position of droplets may be determined from the residual oil droplet, but this is possibly less easy than when a dye is employed. Ideally, the geometry of the target glass surface will be selected to be very similar to that of the target plant surface so that the capture efficiencies are the same. The possibility that the capture efficiency of the plant and the glass target surfaces may differ should be considered. We have found that glass microslides are very suitable as a model for pine needles. Microslides are bilaterally flattened capillary tubes with good optical properties, about 1.25 mm wide and available in a variety of lengths (Camlab, Cambridge, UK). Other dimensions can be used in broadleaved crops.

GLASS MICROSLIDES TO COLLECT DROPLETS

The method is illustrated in Figure 29.1 (and see Killick, 1990).

Method

1. Degrease targets in ethanol.
2. Dry targets on fluff-free, absorbent paper.
3. Apply two strips of double-sided sticky tape (e.g. Scotchtape) across a microscope slide, remove paper from the upper surface of one strip and attach several microslide targets.
4. Place ensemble *in situ* in spray targeted zone.
5. Recover after spraying and remove paper from second strip of double-sided sticky tape and place microslides so that they are attached by both ends.
6. Place in microscope slide box pending transfer to the laboratory.
7. Examine by light microscopy.

Note. Use clean fine forceps throughout to avoid contamination or smudging of targets.

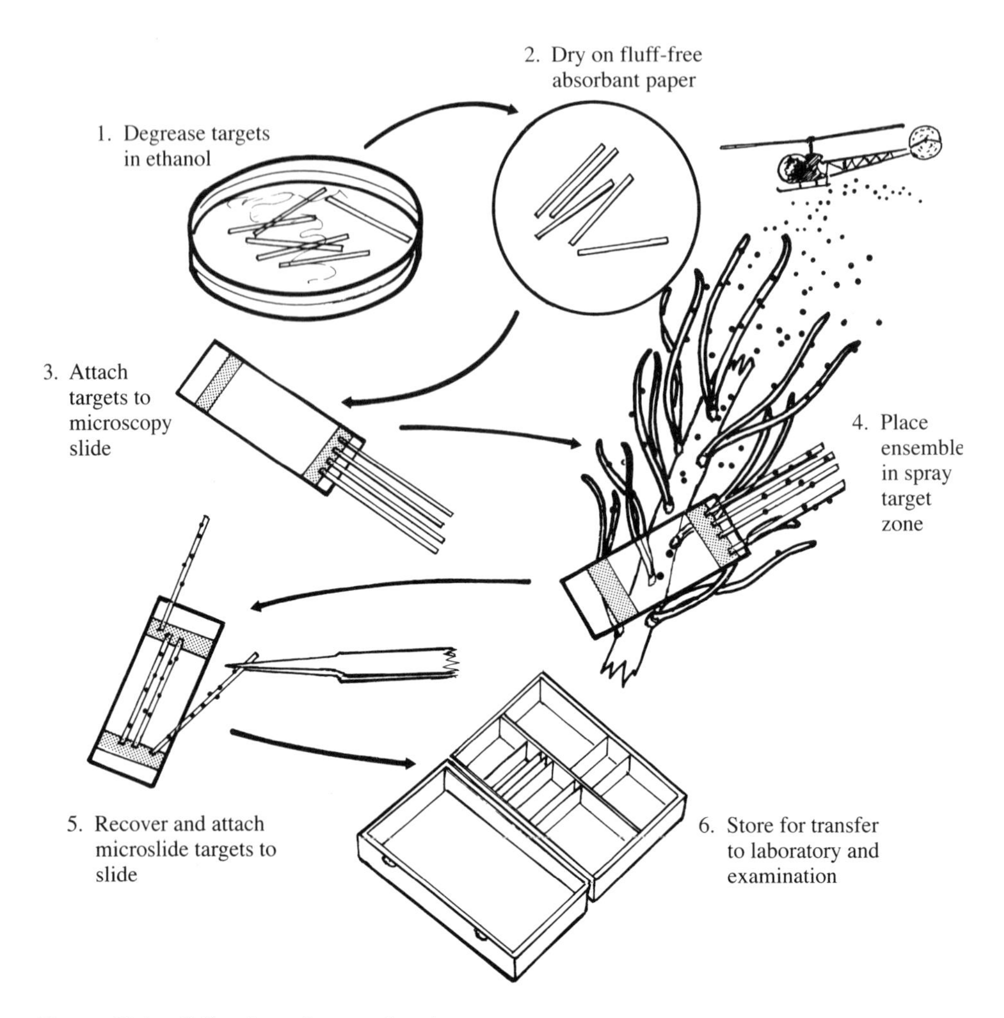

Figure 29.1 Collection of spray droplets.

Incorporation of a dye (e.g. 0.5% Nigrosin; 0.1% Rhodamine B) in the spray fluid permits easy recognition of the position of the spray droplet on the target but does not hinder the counting of OBs. OBs can be identified and counted directly on these slides using phase contrast microscopy. Also of interest is the position of the OBs within the droplets relative to, say, particulate UV protectants: the two may not be coincident. In experimental studies on NPV loss from droplets, Payne (1983) used a similar technique, trapping droplets on 1.5 mm diameter glass cylinders up to 60 m downwind. He measured the contact angle, and the lens diameter of impacted droplets to calculate droplet diameter just before landing.

For broadleaved crops, alcohol-washed and air-dried microscope cover glasses of various sizes and shapes may be attached *in situ*, for example to upper and lower and upwind and downwind leaves, or whatever positions are required to give a picture of the pattern of deposition on a crop. Droplet recognition and particle content counting would be conducted as above.

Capture of droplets on micro-threads

Capture of droplets on micro-threads is appropriate only where spray fluids are of low volatility or incorporate an anti-evaporant such as spray oil. The technique involves the capture of droplets on threads of very fine diameter on which droplets remain essentially spherical so that their diameter can be accurately measured. The technique has a long history (Dessens, 1949; May and Druett, 1968; Payne, 1983).

A MICRO-THREAD METHOD TO CAPTURE DROPLETS

The technique is extended here by measuring the diameters of individual droplets *in situ* on the threads, then transferring them to microscope slides for counting NPV OBs using phase-contrast microscopy (Figure 29.2). The application of this technique depends on incorporation of an anti-evaporant spray oil, or similar substance, to prevent evaporation of the droplet below a certain minimum size, at least during the time required to measure the droplet diameter.

Method

1. Generate a droplet spray cloud.
2. Using a small spider (very fine gossamer) spin a few strands of gossamer onto the cut-out template or 'harp'. Set up harp in the spray cloud path. This may be close to, or distant from, the sprayer.
3. Measure the diameter of the droplets using a compound microscope with a linear scale in one eyepiece: the original droplet volume (i.e. before evaporation) can be calculated from the known proportion of non-volatile(s) in the spray fluid tank.
4. Pick up gossamer strands bearing droplets that have been measured onto a microscope slide. With practice, individual droplets may be picked up on a sliver of cover glass held by fine forceps and transferred, droplet-uppermost, to a microscope slide for counting. A drop of immersion oil previously placed on the slide will hold the cover glass firmly in position.
5. Count numbers of OBs in the individual droplets at a magnification of $\times 600$ or so. A squared graticule in the eyepiece will be helpful in this.

Note. A thread of gossamer is *c.* 1 μm in diameter and can support droplets down to 5 μm without distorting them into an ellipse (Payne, 1983).

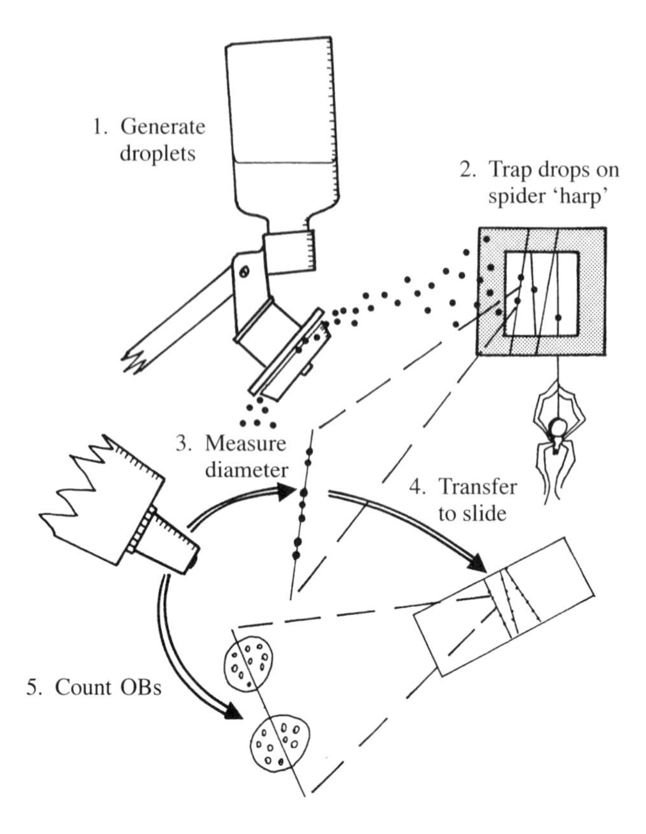

Figure 29.2 Measurement of droplet diameter and estimation of NPV OB numbers using a micro-thread technique (partly based on May and Druett, 1968).

Validity of the micro-thread technique

The ability of the micro-threads to sample representatively the spectrum of droplet sizes in a spray cloud has been questioned. Results obtained by Payne (1983) suggested that droplets in the size range 10–100 μm diameter do not remain on threads or are not deposited in winds greater than about 3 m/s. He also stated that impaction efficiency is probably determined by the tendency of droplets to bounce off a thread. This is especially likely where molasses is present in the formulation since this forms a solid or semi-solid shell on the droplet surface that is likely to result in droplets bouncing off micro-threads and also off leaves. Surface tension, viscosity, speed of approach and the ratio of thread diameter to droplet diameter are all variable factors that may affect droplet capture. However, though the utility of the technique has been questioned in the field, we have found it valuable in the laboratory under controllable conditions where droplets of a wide size range were successfully trapped and processed. Hence, coupled with knowledge obtained by other techniques of the droplet size composition of the spray cloud, this method may provide the information on particle content in relation to a range of droplet sizes necessary to compute the OB population actually present in a spray cloud.

Capture of spray cloud material by a suction device

It is suggested that the BV content of a spray cloud may be estimated using a suction device such as a cyclone air scrubber or sampler. The volume of the spray droplet material collected would be estimated from spectrophotometric measurements of a soluble dye (e.g. Uvitex, Rhodamine etc., see above and Table 29.1). Counts of OBs would be made by the methods described elsewhere in this book (Chapter 26), probably following prior concentration by centrifugation and dilution of the sedimented OBs in a volume of fluid suitable for counting purposes. A comparison would then be made between real and expected OB presence.

Such a technique has wide possibilities. For instance, a collecting device mounted not too far from a spray head on an aircraft could provide information of the spray cloud as directly emitted, while one placed over the crop could provide a comparison with the spray cloud actually arriving. Consideration should also be given to the use of such a system for detecting off-target drift, perhaps at considerable distances. Initial studies of the latter type need employ only a tracer dye (again for spectrophotometric measurement). Definitive tests where the virus is incorporated in the formulation could follow.

REFERENCES

Aston, R.P. (1989) The use of *Bacillus thuringiensis* (Berliner) for the control of *Heliothis armigera* (Hübner) (Lepidoptera: Noctuidae) on cotton. PhD thesis, Department of Bioaeronautics, College of Aeronautics, Cranfield Institute of Technology, Bedford, UK.

Dessens, H. (1949) The use of spider's threads in the study of condensation nuclei. *Quarterly Journal of the Meteorological Society* **75**, 23.

Killick, H.J. (1990) Influence of droplet size, solar ultraviolet light and protectants, and other factors on the efficacy of baculovirus sprays against *Panolis flammea* (Schiff.) (Lepidoptera: Noctuidae). *Crop Protection* **9**, 21–28.

May, K.R. and Druett, H.A. (1968) A microthread technique for studying the viability of microbes. *Journal of General Microbiology* **51**, 353–366.

Morris, O.N. (1973) A method of visualizing and assessing deposits of aerially sprayed insect microbes. *Journal of Invertebrate Patholology* **22**, 115–121.

Morris, O.N. (1977) Relationship between microbial numbers and droplet size in aerial spray applications. *Canadian Entomology* **109**, 1319–1323.

Payne, N.M. (1983) A quantification of turbulent dispersal and deposition of coarse aerosol droplets over a wheat field. PhD thesis, Ecological Physics Research Group, Cranfield Institute of Technology, Bedford, UK.

Richards, M.G. (1984) The use of a granulosis virus for control of codling moth, *Cydia pomonella*; application methods and field persistence. PhD thesis, University of London.

Smith, D.B. and Bouse, L.F. (1981) Machinery and factors that affect the application of pathogens. In: *Microbial Control of Pests and Plant Diseases 1970–1980*. H.D. Burges (ed.). London: Academic Press, pp. 635–653.

Sprent, P. (1969) *Models in Regression and Related Topics*. London: Methuen.

Wirth, W., Storp, S. and Jacobsen, W. (1991) Mechanisms controlling leaf retention of agricultural spray solutions. *Pesticide Science* **33**: 411–420.

30 Registration requirements

The registration of BV preparations for pest control purposes is essentially similar in all countries where these agents have been developed. The processes have evolved from existing requirements for chemical pesticides but, in the light of experience, have been modified to take account of the biological nature of the active ingredients involved. In particular, the ability of the active ingredient to reproduce in both target and, potentially, non-target hosts and hence the potential for persistent and widespread effects have to be taken into account. The safety of microbial insecticides has been reviewed comprehensively by Laird, Lacey and Davidson (1990). Although there have been developments, notably in gradually introducing fast-track procedures for the testing of toxicology, teratology, etc. (based on experiences in use of the tier 1 approach adopted in the USA), the review still provides detailed and pertinent information on many of the aspects necessary to include in registration packages.

Registration requirements for countries within the European Union (EU) provide an useful model for procedures and processes that are almost universally included in legislation internationally. For example, in the UK, the Pesticide Safety Directorate (PSD) and the Health and Safety Executive (HSE) administer registration of all pesticides, including microbial agents. This falls within various UK areas of legislation that, taken together, fulfil UK requirements under EU law, specifically the EC *Plant Protection Products Directive* (91/414/EEC), which harmonises approval across member states. This directive is still evolving with respect to microbial control agents, and current draft documents are being considered and discussed by member states.

REGISTRATION REQUIREMENTS FOR MICROBIAL AGENTS, INCLUDING BACULOVIRUSES

Full details on the registration criteria for biological agents used as pesticides in the UK can be found in Part Three/A3/Annex II Part B, Part Three/A3/Annex III Part B and Part Three/A3/Appendix 4 of *The Registration Handbook*, published by PSD and HSE. These documents provide detailed guidance on the steps required to fulfil registration requirements for both the UK and the rest of the

EU. A summary of the main steps are included here, taken mainly from Part Three/A3/Appendix 4. It may not be necessary to provide information under all headings, and applicants are encouraged to discuss the scale of data requirement with the authorities before submitting applications. Published and, with full documentary support, unpublished information should be used to support the applications. The following is reproduced, with only minor amendment, from the UK guidance notes.

1. Identity of and information on the formulated proprietary product

1.1. This section deals with the product as sold, since it is the product which is approved rather than the active ingredient. This reflects the need to account for any formulation that may, in itself, require testing as an active ingredient.

- 1.1.1. Proprietary name of product.
- 1.1.2. Type of formulation.
- 1.1.3. Composition of formulation, nature and quantity of diluent, purpose and identity of all non-active ingredients such as ultra-violet protectors, water retaining agents, etc.
- 1.1.4. Stability of formulated product, effects of temperature change, method of packaging and storage, retention of biological activity in storage.

2. Identity of the active agent

2.1. It is necessary to provide information on the taxonomy of the agent for precise identification, for establishing biological purity for approval purposes and ultimately for quality control of the commercial product. Further details on these requirements are provided in Annexes II and III of Part three of the Handbook.

- 2.1.1. Systematic name and strain for bacteria, protozoa and fungi, whether a mutant strain; for viruses the taxonomic designation of the agent, serotype, strain or mutant.
- 2.1.2. Common name or alternative and superseded names.
- 2.1.3. The appropriate test procedures and criteria used for identification, such as morphology, biochemistry and/or serology. In this case, increasing use is being made of DNA technology to provide definitive identification.
- 2.1.4. Composition of the unformulated material, microbiological purity, nature and identity of any impurities and content of extraneous organisms.

3. Biological properties of the active agent

3.1. It is important to know which species are attacked and the degree of specificity for the target pest(s), where the agent is naturally occurring and in what circumstances and, where appropriate, geographical distribution.

Information on the likely biological effects arising from use is required in order to assess possible long-term changes in the ecology of the crop and in the environment in general.

3.1.1. Target host species of pest. Pathogenicity or antagonism to pest, infective dose, transmissibility and information on mode of action; history of agent and its use.

3.1.2. Natural occurrence and geographical distribution.

3.1.3. Host specificity range and effects on species other than the target pest including species most closely related to the target species to obtain the taxonomic boundary of susceptibility – to include infectivity, pathogenicity and transmissibility.

3.1.4. Infectivity and physical stability in use by specific application method. Effects of temperature, exposure to air, radiation, etc. Persistence under the likely environmental conditions of use.

3.1.5. If the agent is closely related to a crop pathogen or to a pathogen of a vertebrate species, laboratory evidence of genetic stability, ie laboratory evidence of mutation rate using appropriate tests should be produced. The latter is unlikely to apply to a baculovirus agent, although it is recommended that consultation with the Registration Department should take place at an early stage to ensure that all requirements are met.

4. Manufacture and formulation

4.1. Manufacturers should outline how the agent is produced in bulk and describe in detail the quality controls that are applied and the methods employed including:

4.1.1. Descriptions of the techniques used to ensure a uniform product and of assay methods for its standardisation. If the active ingredient is a mutant, the steps involved in its production and isolation should be described, together with all known differences between the mutant and the parent wild strains. This will enable decisions to be made about whether any extra safety tests will be required because it is a mutant.

4.1.2. Methods for establishing the identity and purity of seed stock from which batches are produced.

4.1.3. Methods to show microbiological purity of the final product and for showing that contaminants have been controlled to an acceptable level.

4.1.4. Methods to show freedom from any human and mammalian pathogen.

4.1.5. Methods for determining storage stability and shelf-life and stability (if appropriate) after reconstitution of the active agent.

5. Application

5.1. This section calls for detailed information on how the agent is to be used, the field of use, identity of target pest species. Circumstances of use are relevant to registration, for instance a method which may be appropriate for controlling a pest of forest trees might not be acceptable for use by home gardeners.

5.1.1. Agriculture, horticulture and forestry.
  5.1.1.1. Species of pest controlled.
  5.1.1.2. Crops protected.
  5.1.1.3. Application rate.
  5.1.1.4. Number and timing of applications.
  5.1.1.5. Method of application (eg high volume, ultra low volume or aerosol, release of infected host, etc.).
  5.1.1.6. Information on compatibility of the formulated product where the notifier intends to recommend on the label use with other products. Chemical pesticides which have been shown to be positive mutagens in bacterial or viral systems should not be mixed with biological agents.
  5.1.1.7. Phytopathogenicity.
  5.1.1.8. Phytotoxicity.

5.1.2. Food storage.
  5.1.2.1. Species . . . etc. as appropriate.

5.1.3. Animal husbandry.
5.1.4. Public health pest control.
5.1.5. Household use.
5.1.6. Home garden use.

6. Experimental data on efficacy

6.1. In order to make a meaningful risk/benefit assessment of the proposed use of a biological agent, registration departments will need a certain amount of data on the efficacy of the agent. Basic information generated from laboratory and field trials during the development of the agent will normally be sufficient. Information on possible advantages of the product would be helpful.

6.1.1. Laboratory experiments.
6.1.2. Field experiments under practical conditions.
6.1.3. Any information on resistance.

7. Experimental data on infectivity, allergenicity and toxicity including carcinogenicity and teratogenicity

7.1. In addition to establishing that the active agent is not a known pathogen of man or other mammals and that the preparation does not contain such pathogens as contaminants or mutants as determined by acceptable tests, information is required on the effects of the agent on laboratory mammals.

7.2. This information will differ in several respects from that required for chemical pesticides and must cover the infectivity of the living agent, multiplication *in vivo*, and its allergenic potential. Some assessment of acute or subacute toxicity will normally be required while more prolonged studies including carcinogenicity and teratogenicity may be appropriate for those agents that are either known or may be shown to produce toxic substances.

7.3. All biological control materials may contain proteins which are potentially allergenic; appropriate safety measures will therefore be mandatory to ensure the safety of those engaged in their production and application. Limited tests for allergenicity are proposed but these will be modified as more appropriate procedures capable of discriminating between the allergenic potential of different materials become available.

7.4. Each application will be considered on its merits and expert advice sought both on the requirements of testing and the methods to be employed. It is anticipated that the requirements for viruses will differ in some respects from those for bacteria/fungi/protozoa.

7.4.1. Acute toxicity and infectivity. For viruses this will include oral, percutaneous, inhalation and intraperitoneal single doses, skin (or eye) irritation and skin sensitisation, cell culture studies using purified infective virus and primary cell cultures of mammalian, avian and fish cells.

7.4.2. Subacute toxicity. For viruses this includes tests over 90 days as in 7.4.1 plus tests for infectivity carried out by bio-assay or on a suitable cell culture at least seven days after the last administration to the test animals.

7.4.3. Special procedures for teratology and carcinogenicity, etc. (detailed under Part A of Annexes II and III).

8. Effects on humans

8.1. Information is required for the purpose of assessing possible effects on the health of workers handling the biological agent and will depend initially on the firm's experience in producing, formulating and applying the agents.

8.1.1. Information on workers manufacturing and formulating the biological agent, with particular attention to allergic response, as judged clinically and by questionnaire (at least annually), results of regular clinical skin tests where appropriate, and serological examination of workers, initiated where possible before exposure (to set baseline).

8.1.2. Where appropriate, information on farm and forestry workers and including examination for any allergic responses.

9. Residues

9.1. It may be necessary to identify and measure residues remaining on an edible crop at harvest as a result of the use of the biological agent if consideration of information provided in paragraphs 7 and 8 suggests a hazard to consumers. Although no known human pathogens are involved, there may be objections to or reasons for determining the presence of biologically active or inactive material on food. If the biological agent remains active for a significant period the question of further multiplications must be considered, as must contamination of non-target crops, water and the environment generally.

9.1.1. Identification of viable and non-viable (eg toxins) residues in treated crops, the viable residue by culture or bioassay and non-viable by appropriate techniques.
9.1.2. Likelihood of multiplication in or on crops or food, and its effect on food quality.
9.1.3. Extent of indirect contamination of adjacent non-target crops, wild plants, soil and water.

10. Information on environmental and wildlife hazards

10.1 A pest control agent may harm beneficial fauna in and beyond a treated crop if it causes disease that is likely to spread in beneficial organisms. Tests are required on relevant beneficial invertebrates and vertebrates. Firms are required to provide as complete an account as possible on what is already known of the biological 'side effects' on the environment of the use or natural occurrence of the biological agent. Infectivity of the agent to non-pest invertebrates closely related to the pest species should be dealt with under host specificity.
10.1.1. Effects on relevant invertebrates, including infectivity.
10.1.1.1. Important parasites and predators of the target species.
10.1.1.2. Honey bees: acute toxicity and infectivity.
10.1.1.3. Earthworms.
10.1.1.4. Other non-target organisms believed to be at risk.
10.1.2. Effects on fish including: effects on one indigenous fish species; for viruses this is necessary only if primary fish cell cultures are infected.
10.1.3. Effects on birds. Acute feeding studies on two species of birds, one of which will accept insects in its diet, such as hen, partridge or quail. For viruses this is necessary if primary avian cell cultures are infected.
10.1.4. Effect on livestock animals including immunological response.
10.1.5. A review of *existing* general knowledge describing expected effects on the environment and its fauna. This should include comments on the expected spread, and on persistence in soil and water, also possible fate in food chains. Any likely undesirable pollution and any likely species at risk should be mentioned.

11. Labelling

11.1 This section deals with the information to be made available to the user by means of the label on the nature of the product, the crops for which it is intended, how to use it effectively, handling and safety precautions, layout and manufacturer's details. It should contain all the types of information on a good chemical pesticide label.
11.2. The standard phrases used for labelling chemical pesticides and advice on labelling are provided in the *Registration Handbook*, Appendix 8.

REGISTRATION REQUIREMENTS IN THE USA

Developments in the use of microbial agents have been greatest in the USA so that, consequently, the regulations for registration of these pesticides have evolved more rapidly there than they have elsewhere in the world. The regulatory framework includes responsibility by both the United States Department of Agriculture (USDA) and the US Environmental Protection Agency (EPA). Two acts, namely the Federal Insecticide, Fungicide and Rodenticide Act (FIFRA) and the Toxic Substances Control Act (TSCA), cover the registration of pesticides, although microbial agents are dealt with mainly under FIFRA. The interactions between the two agencies and the two Acts, particularly in relation to genetically modified microbial insecticides, are discussed by Persley (1992).

The US system differs from that used in Europe by placing greater reliance on a tier testing procedure that uses a range of stringent tests of infectivity, pathogenicity, teratogenicity, toxicity, etc. If they prove negative at one tier it removes the need for further testing. In addition, there is increasing reliance placed on the generic safety of BVs so that tests, as in the UK, must be based on the interaction of the microbial agent with any other active ingredients in the formulation. Betz, Forsyth and Stewart (1989) provide a useful summary of US requirements, including the main tests required in the US Tier testing system. This is summarized in Table 30.1. The level of stringency differs depending on whether full registration or an experimental use permit (EUP) is required. There is also a presumption that the active ingredient, particularly for BVs, provided it has been identified with certainty, may not require such stringent testing as a formulated product. In the light of experience, the USA is gradually developing a fast-track approach so that some microbial agents that have been well characterized and have a record of safe field use may require lower stringency, leading to a relaxation of a requirement for an EUP. This level of flexibility is not yet available in the EU. The OECD is initiating work on a consensus document on BVs (Richards and Kearns, 1977) as part of its programme on the harmonization of data requirements for registration of microbial pesticides.

REGULATIONS CONCERNING THE RELEASE OF GENETICALLY MODIFIED ORGANISMS

Deliberate release of genetically modified bioinsecticides in Britain is covered under Statutory Instruments No. 3280 Environmental Protection: *The Genetically Modified Organisms (Deliberate Release) Regulations, 1992 (The 1992 Regulations)* and *The Genetically Modified Organisms (Deliberate Release) Regulations 1995 (The 1995 Regulations)*. These are British interpretations of EU Council Directives 90/220/EEC (*The Deliberate Release into the Environment of Genetically Modified Organisms*; OJ No. L117, 08.05.1990, p. 15) and 94/730/EC (*First Simplified Procedure (crop plants) Decision*, OJ No. L292, 12.11.94, p. 31), which, in the UK, are implemented by Part VI of the Environmental Protection Act 1990.

Guidance notes on experimental releases of genetically modified organisms (GMOs) were published in 1993 as *DOE/ACRE Guidance Note 1* (DOE/ACRE,

Table 30.1 Data required for registration of microbial pesticides in the USA (based on Betz *et al.* (1989)

Data	Full Registration	Experimental use permit
Product analysis		
Identity and method of manufacture	✓	✓
Assess presence of other ingredients	✓	✓
Analysis of samples	☑	☑
Certification of limits	✓	✓
Analytical methods	✓	✓
Physical and chemical properties	✓	✓
Submission of samples	☑	☑
Residue analysis	☑	☑
Tier I: toxicology tests		
Acute oral	✓	✓
Acute dermal	✓	–
Acute pulmonary	✓	✓
Acute intravenous	✓	✓
Primary eye irritation	✓	✓
Hypersensitivity	✓	–
Tissue culture (for viruses only)	✓	✓
If any of the above fail then Tier II or III tests are required	☑	☑
Tier I: non-target organisms and fate in the environment		
Avian oral	✓	✓
Avian inhalation	☑	–
Wild mammal	☑	–
Freshwater fish	✓	✓
Freshwater aquatic invertebrate	✓	✓
Estuarine and marine animals	☑	–
Non-target plants	✓	✓
Non-target insects	✓	✓
Honey bees	✓	✓
If any of the above tests indicate potential problems then Tiers II to IV may be required		
Tier II: Environmental fate	☑	☑
Tier III: Ecological effects	☑	☑
Tier IV: Simulated and/or actual terrestrial or aquatic field studies	☑	☑

Tests:
✓, required; –, not required; ☑, requirement depends on results of other tests.

1993) and recently, reflecting the changes in regulations on release of genetically modified plants, in *DOE/ACRE Guidance Note 7* (DOE/ACRE, 1995b). Further guidance on experimental releases of genetically modified microorganisms (excluding viruses and similar agents) was published in 1994 as *DOE/ACRE Guidance Note 5* (DOE/ACRE, 1994b) and, in 1995, for genetically modified BVs (DOE/ACRE, 1995a). In the light of experience, changes to the methods of working outlined in the *Deliberate Release Regulations* have been included in practice, in particular the development of fast track procedures for releases of

GMOs where experience has shown that risks are small or easily contained (DOE/ACRE, 1994a).

The full UK regulations require applicants to complete a 90 point check list spanning the process of genetic modification, through release and finally to post-release monitoring. Data provided are evaluated by the Advisory Committee on Releases to the Environment (ACRE), which is a committee appointed by the Secretary of State for the Environment to advise him on the safety of the release. ACRE includes scientific experts and representatives of various interests (for example, industrial, public and environmental interests). Other Government departments, such as the Ministry of Agriculture and the Forestry Commission act as assessors to cover their respective interests. The DOE co-ordinates the processing of release applications, provides the ACRE secretariat and acts as a single point of contact for those releases that may require permission under more than one Act. For example, release of a genetically modified bioinsecticide may also require clearance under the Control of Pesticides Regulations. A decision on release is provided within 90 days of receipt, unless there are substantial queries to the applicants during which time the clock is stopped. Applicants are required to issue an advertisement in a newspaper announcing that application has been made for release within 10 days of the date of application (except Commercial in Confidence information). It is also a requirement to place specific information on applications for consent to release on the Public Register and to notify relevant organizations of a proposed release. Additionally, a Summary Notice Information Format (SNIF), containing a summary of the release proposal, must be sent to the European Commission and other Member States.

With regard to international harmonization, the Organization for Economic Cooperation and Development (OECD) has been very active in developing standards following publication in 1986 of the Blue Book on recombinant DNA safety considerations (OECD, 1986). Their main proposal was the use of a 'case by case' approach to evaluate the risks posed by release of genetically modified organisms.

The greatest number of releases, especially of genetically modified plants, have taken place in the USA (>1400 approvals at approximately 5000 sites in the 8 years to March 1995 compared with approximately 250 approvals in the EU).

The original USA regulatory framework was developed in 1986 and set out the respective responsibilities of the two principal agencies involved in evaluating use of unmodified and genetically modified bioinsecticides, namely the USDA and the EPA. The legislation dealt with microbial products of biotechnology under the FIFRA and the TSCA within a 'Coordinated Framework' developed by a Cabinet Council Working Group (Kingsbury, 1988). Microbial pesticides tend to be dealt with under FIFRA, especially those involving small-scale field testing of genetically modified, non-indigenous and pathogenic microbial pesticides. Others, such as microorganisms intended for general commercial and environmental applications (e.g. pollutant degradation etc.) are subject to TSCA. The necessity to determine which act and which agency to contact has led to some confusion among scientists and commercial companies applying for field releases, although the co-ordinated framework has provided a mechanism to direct applicants to the correct set of rules and the appropriate agency. Persley (1992) discussed this problem, pointing out that both state and government agencies were attempting to balance

Table 30.2 Regulatory responsibility for tests of microbial agents in the USA

Type of microbial product	FIFRA		TSCA	
	<10 acres	>10 acres	<10 acres	>10 acres
Genetically engineered microorganisms				
Formed by deliberate combinations of genetic material from dissimilar source organisms (intergeneric combinations)	✓	✓	✓	✓
Formed by genetic engineering other than intergeneric combinations				
Pathogenic source organisms[a]	✓	✓	✓	✓
Non-pathogenic source organisms	×	✓	×	×
Non-engineered microorganisms				
Non-indigenous pathogens[a]	✓	✓	×	✓
Non-indigenous non-pathogens	×	✓	×	×
Indigenous pathogens[a]		✓	×	✓
Indigenous non-pathogens		✓	×	×

✓ indicates that the microorganism will be subject to EPA review before small-scale (<10 acres) or large-scale (>10 acres) release. Under TSCA, submitters would only notify the Agency once, unless during the original review EPA specify that further reporting is needed.
× indicates that the microorganism will be subject to abbreviated review before release. Under FIFRA, this is effective immediately. Under TSCA, the abbreviated notification will be implemented through rulemaking.
[a] Pathogens used solely for non-pesticidal agricultural purposes will not be subject to EPA notification requirements. They will be subject only to USDA review.

their responsibilities for human health and the environment with their desire to encourage innovative science and, ultimately, improved crop protection.

Table 30.2 summarizes the main US responsibilities by department. A major difference between the USA and UK regulations is the well-defined cut-off in certain requirements depending on the area to be tested. A test area of 10 acres (approximately 4 ha) would be regarded as very large under current UK regulations for release of genetically modified bioinsecticides.

Releases must comply with a set of requirements that, in the majority of respects, are also included in the EU and UK regulations. Further refinement, in the light of experience and in development of fast-track procedures, has been produced in the form of *EPA 40 CFR Part 172, Microbial Pesticides; Experimental Use Permits and Notifications; Final Rule*, published in September 1994, covering EPA responsibilities in assessing use of microbial pesticides. This is little changed from the original ruling in June 1986, the main modifications being rearrangement and expansion of some of the original requirements and deletion of a specific reference to a detailed description of the proposed testing programme, which was already adequately covered elsewhere. Provision of information on the parental organism and the relative competitiveness of the genetically modified organism have received greater emphasis. In the light of experience, microbial pesticides resulting from deletions or rearrangements within a single genome that are brought about by the introduction of genetic material that has been deliberately modified do not require a notification.

Useful address

Pesticides Safety Directorate, Mallard House, Kings Pool, 3 Peasholme Green, York YO1 2PX, UK.

REFERENCES

Betz, F.S., Forsyth, S.F. and Stewart, W.E. (1989) Registration requirements and safety considerations for microbial pest control agents in North America. In *Safety of Microbial Insecticides*, M. Laird, L.A. Lacey and E.W. Davidson (eds). Boca Raton, FL: CRC Press, pp. 3–10.

DOE/ACRE (1993) The Regulation and control of the deliberate release of genetically modified organisms. *DOE/ACRE Guidance Note* **1**, 1–80.

DOE/ACRE (1994a) Fast track procedures for certain GMO releases. *DOE/ACRE Guidance Note* **2**, 1.

DOE/ACRE (1994b) Guidance for experimental releases of genetically modified micro-organisms (excluding viruses and similar agents). *DOE/ACRE Guidance Note* **5**, 1–21.

DOE/ACRE (1995a) Guidance for experimental releases of genetically modified baculoviruses. *DOE/ACRE Guidance Note* **6**, 1–21.

DOE/ACRE (1995b) Guidance to the *Genetically Modified Organisms (Deliberate Release) Regulations 1995. DOE/ACRE Guidance Note* **7**, 1–47.

Kingsbury, D.T. (1988) Regulation of biotechnology in the United States: one and a half years of using the 'coordinated framework'. *Trends in Ecology and Evolution* **3**, 39–42.

Laird, M., Lacey, L.A. and Davidson, E.W. (eds) (1990) *Safety of Microbial Insecticides*. Boca Raton, FL: CRC Press.

OECD (1986) *Recombinant DNA Safety Considerations: Report*. Paris: OECD.

Persley, G. (1992) Regulation of biotechnology. *INFORM (International News on Fats, Oils and Related Materials)* **3**, 242–265.

Richards, J. and Kearns, P. (1997) Work on micro-organisms within the OECD's environmental health and safety programme. In *BCPC Symposium Proceedings No. 68: Microbial Insecticides: Novelty or Necessity?* Chaired by H.F. Evans. Farnham, UK: British Crop Protection Council.

PART FOUR

ENVIRONMENTAL FACTORS INFLUENCING VIRAL SURVIVAL

31 Solar radiation, with emphasis on the ultraviolet

The nature of the solar electromagnetic spectrum is shown in Figure 31.1. Most of the sun's emissions are blocked by various atmospheric factors and either do not penetrate to the earth's surface or do so at greatly diminished intensities. However, a comparatively narrow window permits penetration of a waveband region encompassing some infrared, visible light and some UV. This chapter is concerned principally with providing information on the nature of UV, its interactions with the earth's atmosphere and its biological activity, especially as this relates to that portion of the UV spectrum penetrating to the earth's surface. Units and terms are defined at the end of the chapter.

FACTORS DETERMINING ULTRAVIOLET PENETRATION OF THE ATMOSPHERE

The principal atmospheric component responsible for absorption of solar UV radiation is ozone. Ozone absorbs UV at 220–300 nm and there is a penetration cut-off point at about 290 nm. In combination with atmospheric oxygen, absorption is extended to 170–340 nm. This is the region in which ground irradiance could be influenced by changes in ozone quantity. Some slight availability of UV below 290 nm exists and has been considered in relation to photo-modification of polymers. Though only about 1 ppm of sunlight energy consists of <290 nm, the solar flux density at this wavelength is 10^{10} photon/cm^2 per s or 10^{16} photons/cm^2 per month. This will produce 100 ppm damage sites/month in polymers if radiation is absorbed within about 4 μm of the surface (Barker, 1968). The biological and ecological significance of such short UV wavelengths has yet to be investigated.

The general pattern of distribution of atmospheric ozone is indicated in Figure 31.2. The total amount of ozone present in the atmosphere is very small. It also varies with latitude and with the season. As a standard, a value of 0.35 cm ozone, totalled over the atmospheric column, at normal temperature and pressure has been recommended (Elterman, 1968) as the annual mean at mid-latitudes. In the UK, the approximate minimum is 0.2 cm, but in equatorial latitudes the minimum falls to 0.15 cm while in the Antarctic 'ozone hole' it may be less. Longer

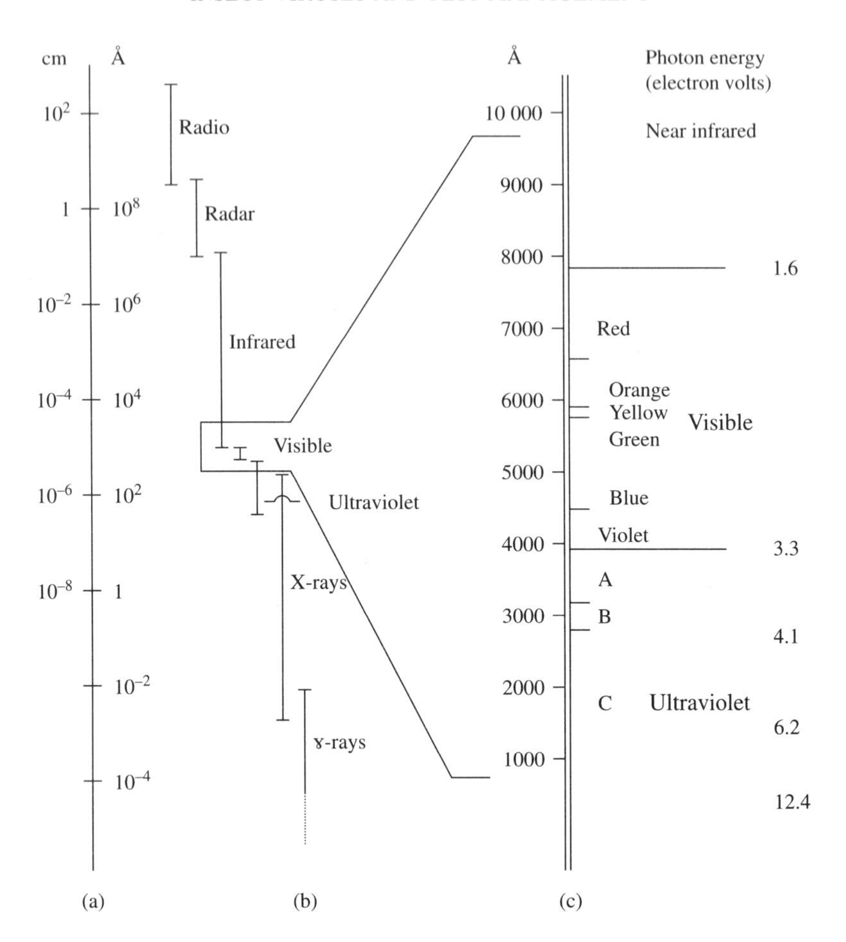

Figure 31.1 The solar electromagnetic spectrum. (a) Total spectrum; (b) definitions of the various regions; (c) classification for the near infrared, through the visible to the ultraviolet. Units are given in centimetres, Ångstroms and photon energy (electron volts).

wavelengths are attenuated more gradually than is UV, through a series of absorption bands in the infrared, caused by water vapour, CO_2, ozone etc. Ozone has some slight absorption at 440–740 nm (Chappin's bands).

Figure 31.3 shows the seasonal and geographic distribution of ozone concentrations in the northern hemisphere over the Americas. The typical annual pattern of UV intensity shows a peak at the summer solstice, about June 21st, which is the date of highest solar elevation in the northern hemisphere. However, because of seasonal asymmetry in ozone concentration, the amount of UV insolation is also seasonally asymmetrically distributed (Figure 31.4).

Most UV falls on the earth's surface between 09.00 and 15.00 h local solar time and this is the period of day when maximum biological damage is inflicted. This reflects the *air mass* through which radiation must penetrate to reach the earth's surface. Air mass is defined as the ratio of the slant path length to the vertical path. Clearly, the air mass value varies with solar elevation both on a diurnal and

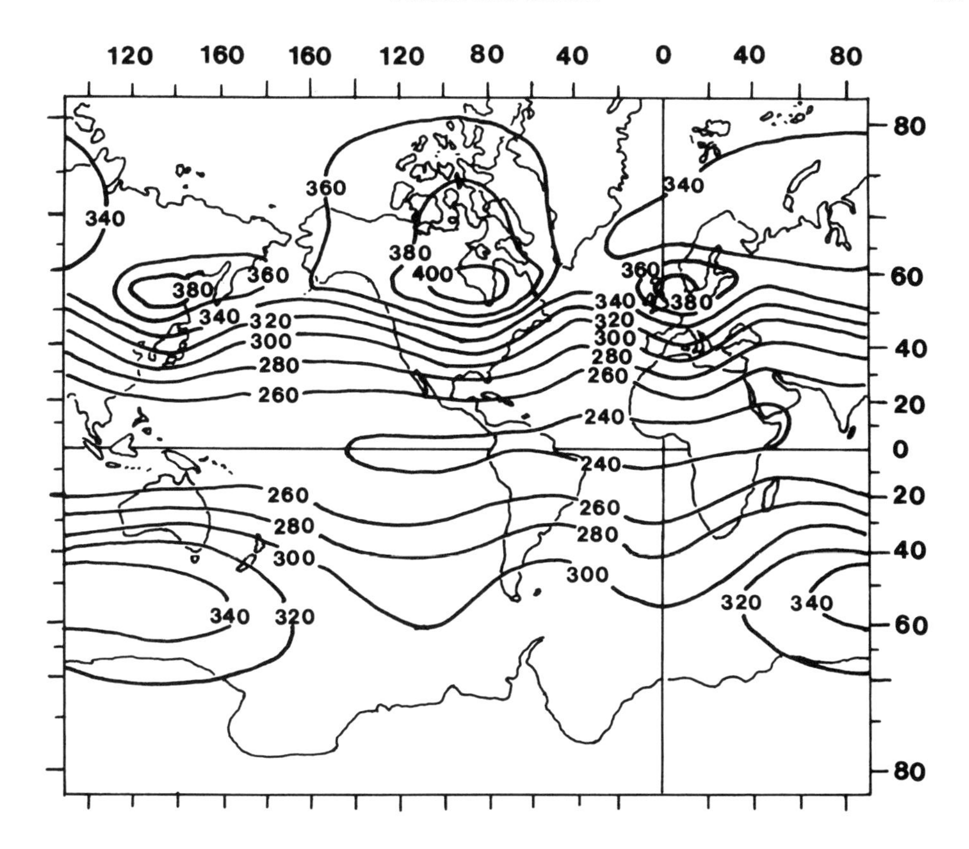

Figure 31.2 Global distribution of total ozone averaged over the period 1957–67. (Source: London *et al.* (1976), adapted from Cutchins (1982) with permission from P. Cutchins and the Plenum Press.)

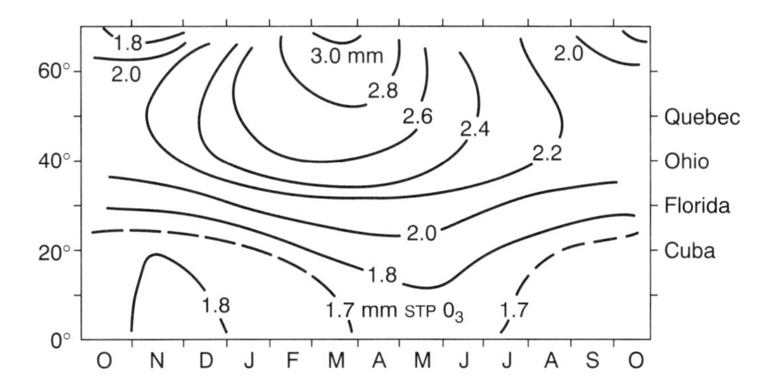

Figure 31.3 Seasonal and latitudinal variations in ozone (at STP) in the northern American hemisphere. (Based on data from Haurwitz, London and Valouin in Nawrocki and Papa (1961) as illustrated in Barker (1968). Reproduced by permission of the American Society of Photobiology, R. E. Barker.)

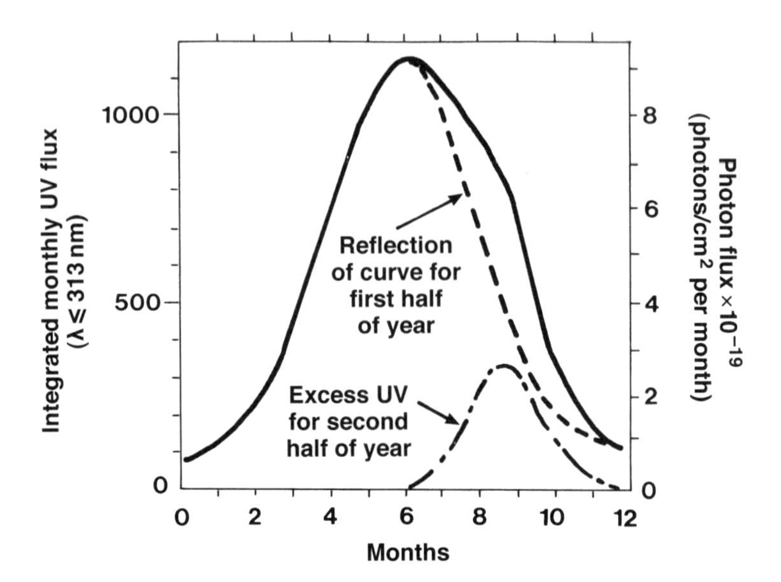

Figure 31.4 The 3-year average of total monthly UV fluxes below 313 nm onto a horizontal surface at Washington DC. The total 'area' is 423 J/cm^2-year. The 'excess' UV has peaked in August and accounted for 49 J/cm^2-year of the total. (After Barker (1968). Reproduced by permission of the American Society of Photobiology, R. E. Barker.)

a seasonal pattern and also varies with latitude. For instance, at 40° N in the winter at noon, sec $\theta = 2$, which is equivalent to having twice the actual ozone concentration as occurs at a latitude where the sun is at the zenith at noon. (However, latitudinal variations in ozone concentrations should be taken into account in considering this principle.) These factors have also an impact on the wavelengths reaching the earth's surface. At 52° N, London experiences practically no radiation less than 300 nm.

Tables are available providing values for UV radiation reaching the earth's surface as a function of wavelength, latitude and season (Johnson, Tsan and Green, 1975). These are on the basis of a clear sky and seasonally and latitudinally averaged ozone quantities. However, UV insolation values may be modified by *cloudiness*. A correction factor of $1 - 0.056C$, where C is the number of tenths of sky covered by cloud, has been deduced by Buttner (1938) and Bener (1964). Further discussion of cloudiness is provided by Johnson *et al.* (1975). The level of UV flux also varies with altitude: for a wavelength of 290 nm at 2 km there is an increase in UV flux to 1.5 times that at sea level (Figure 31.5).

THE BEHAVIOUR OF ULTRAVIOLET IN THE ATMOSPHERE AND AT THE EARTH'S SURFACE

In the environmental monitoring of UV *irradiance* (E, in units of W/m^{-2}) it is important to record not only that coming from the sun direction (E_a) but also

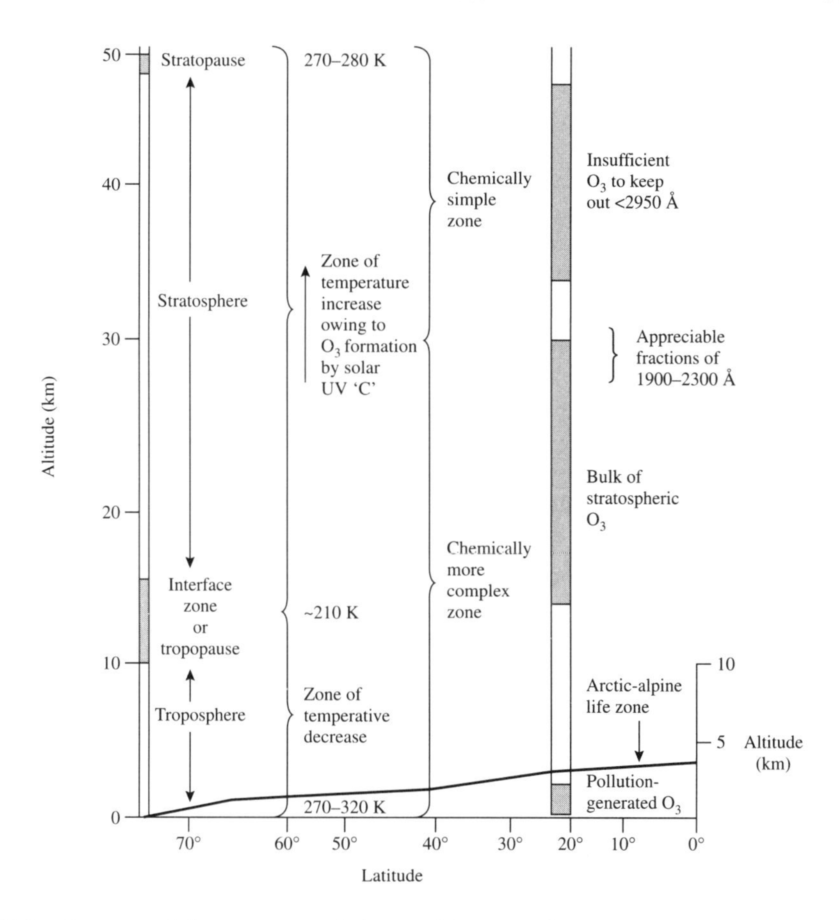

Figure 31.5 Ozone and the structure of the earth's atmosphere

that present diffusely from other sky directions (E_d). The combination of these provides a measure of total UV and is known as global radiation (E_g).

Diffuse sky radiation is partly polarized and illuminates surfaces in all directions. It is relatively rich in blue as well as in UV and has peak values between 330 and 410 nm. Scattering is brought about through various interactions. Rayleigh, or molecular, scattering occurs by reaction to nitrogen, oxygen, etc. and is proportional to the fourth power of the frequency. Mie, or large particle, scattering is caused by the presence of dust, aerosols, water droplets and other particles comparable to the wavelength of the radiation.

Surface reflectivity, or **albedo**, is defined as the proportion of reflected to received radiation and it can strongly influence the flux of **sky light**. For instance, a reflectivity of 0.25 increases the intensity of scattered sky light by about 35% relative to the intensity of sky light above a black surface. Therefore, surface type (e.g. vegetation, soil, sand, etc.) is important.

OZONE, ITS FORMATION AND ROLE

Oxygen-containing compounds, especially molecular oxygen and ozone and, to a lesser extent, water and CO_2, account for almost all absorption of solar UV radiation. The UV absorption of molecular oxygen begins weakly near 242 nm but becomes very strong near 190 nm. The process of absorption results in molecular dissociation into two atoms of O, which can combine with molecular oxygen to form ozone. When newly formed, the ozone atom is extremely unstable unless its excess energy is removed by collision with another molecule (M), usually the atmospherically abundant molecular nitrogen. This reaction is proportional to the concentration of nitrogen, which decreases with increasing altitude. (As the atmospheric pressure decreases by approximately ten times for every 15 km altitude, pressure at the top of the troposphere is 10^{-1} atmosphere and at the top of the stratosphere about 10^{-3} atmosphere.)

The reaction is as follows:

$$O + O_2 + N_2 \text{ (or M)} \rightarrow O_3 + N_2 \text{ (or M)} \quad (31.1)$$

and hence nitrogen or M is quantitively unchanged. Ozone absorbs in the infrared, visible and the UV spectrum but does so especially strongly in the region 350–260 nm. During this process

$$O_3 \rightarrow O + O_2 \quad (31.2)$$

but the atomic oxygen rapidly recombines with molecular oxygen so that the total quantity of ozone is little affected. During reaction 31.1, solar UV is converted to heat and this accounts for most of the stratospheric warming. As shown in Figure 31.5, the troposphere cools with altitude from 270–320 K at the earth's surface to about 210 K at 10–12 km altitude, a zone which is known as the **tropopause**. Reaction 31.1 causes a reversal of this process in the stratosphere, the temperature of which rises to 270–280 K at 50 km. This altitude, where temperature increase ceases, is known as the **stratopause**. As Rowland (1982) wrote 'The very existence and location of the stratosphere is [therefore] closely tied to the distribution with altitude of ozone in the atmosphere'. The exact height of the tropopause and stratopause varies within season and latitude.

ATMOSPHERIC POLLUTION

The role of those processes of atmospheric pollution that influence the amount of ozone present in the atmosphere, and hence the intensity and to some extent spectral composition of solar UV irradiance at the earth's surface, must be of direct interest to the present day biologist. This is especially so since even when there is a total cessation in the release, for instance, of anthropogenic chlorine-containing gases, pollution-based imbalances could be felt for at least the next century.

Two main radical types are involved in ozone destruction: nitrogen oxides (No_x) and chlorine oxide (ClO_x).

The NO_x-catalysed free radical chain

When nitrogen/oxygen mixtures pass over hot surfaces, as in internal combustion engines, there is partial conversion to NO leading to NO_x introduction into the atmosphere. The origination of NO and NO_x from transport at ground level tends to lead to short-lived pollution since the troposphere is regularly cleansed by rain. However, when introduced into the tropopause (10–12.5 km altitude) by subsonic jet aircraft or into the lower stratosphere by supersonic aircraft designed to cruise at about 15 km altitude, these molecules are more persistent.

NO interacts with ozone:

$$NO + O_3 \rightarrow NO_2 + O_2 \tag{31.3}$$

NO_2 is a free radical, i.e. it has an odd number of electrons (in this case, 23); free radicals are chemically very reactive and their presence can give rise to chain reactions, e.g.

$$NO_2 + O \rightarrow NO + O_2 \tag{31.4}$$

This duo of reactions (31.3 and 31.4) is by definition catalytic because there is no change in the concentration of NO or NO_2; however, it results in the removal of one molecule of ozone and an atom of oxygen with the formation of two molecules of molecular oxygen.

The ClO_x-catalysed free radical chain

The production worldwide of 'freeons', CCl_3F, CCl_2F_2 and CH_3CCl_3, as chemically non-reactive gases with which to power aerosol containers and for use in refrigerator and some fire extinguisher systems increased greatly during 1955–80, at a rate equal to a doubling of atmospheric releases.

Of these gases, CH_3CCl_3 reacts with HO radicals in the troposphere and so has an average lifetime of (only) about 7 years. Nevertheless, the chlorine released by such reactions and the escape of the less active CCl_3F and CCl_2F_2 into the stratosphere contributes to an increasingly abnormal level of ClO and HCl in the upper regions. The innate process of removal of chlorine from the stratosphere depends on diffusion of the water-soluble molecules such as HCl (and in the case of NO_x of $HONO_3$) into the troposphere, from which they can be rapidly deposited by rainfall. For the correction of imbalances accrued only to the present, this process is expected to extend for at least a century.

The ClO_x-catalysed free radical removal of ozone proceeds as follows:

$$Cl(17\ \text{electrons}) + O_3 \rightarrow ClO(23\ \text{electrons}) + O_2 \tag{31.5}$$

$$ClO + O \rightarrow Cl + O_2 \tag{31.6}$$

(Other chain reactions involve HO_x and carbon-containing radicals also contribute to this process.)

The rapidity of the ClO_x process of ozone depletion in the stratosphere (e.g. at 40 km) is such that every chlorine atom can catalyse the conversion of thousands of ozone molecules into molecular oxygen every day.

Despite the relatively simple chemistry of the upper stratosphere, where conditions are inimical to survival of polyatomic molecules, the whole process is more complex than described above. For instance, free radical chains based on Cl and NO can interact with each other; also when a chain termination occurs it is usually temporary because photo-dissociation of the even-electron products can restart events.

Recently, the pesticide methyl bromide has been implicated as a factor in ozone destruction. Possibly, therefore, all halogens pose a threat.

A very valuable review of the current global status of UV radiation and the changing ozone situation is provided by McKenzie (1995).

SUBDIVISION OF ULTRAVIOLET WAVEBANDS

Below about 190 nm, both air and water vapour absorb UV and the region 100–190 nm is often known as extreme UV. It is also sometimes called vacuum UV, as experiments in this region need usually to be conducted in vacuum. Incoming solar UV radiation is cut off at about 290 nm by atmospheric ozone, providing a convenient physically based point at which to separate UV: between 190 and 290 nm is known as far UV and that above 290 (approximately 290–380 nm) as near UV. Other systems of subdivision of the total UV band of radiation will often be encountered in the literature but seem not to be as precisely and consistently defined. Two systems that are in common use in photobiology are compared in Figure 31.6 with the near/far/extreme system. They are near/middle/far and the A/B/C systems and the figure illustrates at least some of the variation in waveband widths commonly found associated with these in the literature. It is evident that there is some commonality in the two systems but that, in view of the variation, the use of semantic terms alone can never be a sufficient description and that actual wavelength numbers should always be provided.

For photobiology, it would probably be sensible to define UV-B as 290–320 nm, i.e. that waveband span reaching the earth's surface to which the bulk of biological activity can be attributed. It follows from such a definition that UV-C is <290 nm while UV-A is >320 nm: the upper limit of UV-A is commonly taken as 380 nm.

Additional systems are in use; for instance, the region 190–380 nm is known to physicists as near-UV and 4–190 nm as far-UV. Photobiologists sometimes refer to radiation in the region 280–315 nm as middle-UV, especially as this is the principal component responsible for erythema in human. Finally, 300–400 nm has been defined as longwave UV. These terms are to be deprecated.

THE BIOLOGICAL IMPACT OF ULTRAVIOLET

DNA

The most important adverse effect of solar UV radiation is the induction of molecular changes in DNA. Two adjacent thymine bases on a DNA strand may

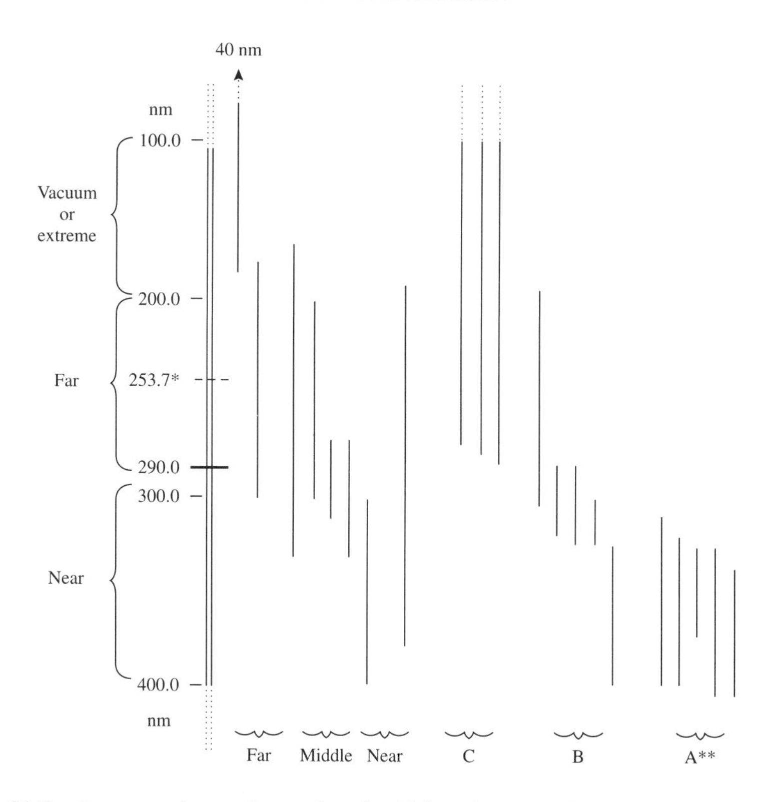

Figure 31.6 A comparison of waveband widths often employed in two systems of UV classification that are commonly used in photobiology. Note the variable numeric values that are attributed to the various semantic groups. *253.7 nm is the principal wavelength generated by 'bactericidal lamps'. **This wavelength does not reach the earth's surface. 290 nm is approximately the shortest UV wavelength reaching the earth's surface.

be fused by the formation of a cyclobutane ring between the bases. This creates a block to synthesis of normal DNA and a high rate of mutation. Less commonly, strand breakages may occur (Harm, 1980). Even near-UV, where absorption by DNA is small or even unmeasurable, causes lethal effects either directly or through sensitization by other UV-activated molecules. A DNA UV absorption curve is shown in Figure 31.7. However, the possibility of a degree of activation of NPV exposed to 320 nm should be noted (Ramoska, Stairs and Hinks, 1975).

RNA

The biological consequences of UV-induced alterations are much less than similar alterations in DNA. This is partly because while DNA is unique (in an individual cell) there are many copies of messenger RNA, transfer RNA or ribosomal RNA. However, some plant and animal viruses (examples for insects include picornaviruses, nodaviruses and CPVs) are exceptions to this because RNA is their sole genetic material.

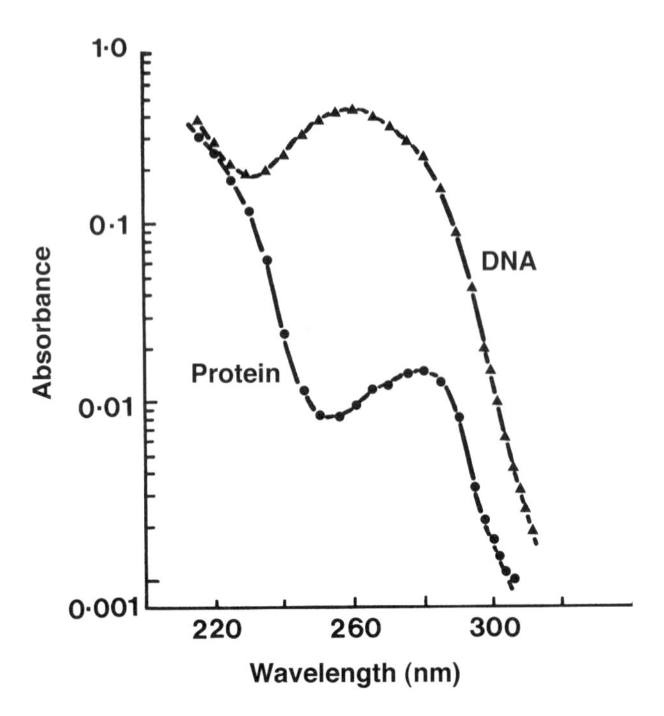

Figure 31.7 Absorption spectra of DNA (calf thymus DNA) and a protein (bovine serum albumin), both at concentrations of 19.3 μg/ml (after Harm (1980) *Biological Effects of Ultraviolet Radiation*; reproduced by permission of W. Harm and Cambridge University Press).

Proteins

Absorption of UV by proteins in the region 240–300 nm is only about one tenth that of nucleic acid solutions of equal concentration. This is because only a few of the amino acid residues of proteins absorb measurably in this region. Tryptophan and tyrosine (NPV OB protein is especially rich in the latter) are among the most absorptive amino acids of UV wavelengths reaching the earth's surface. A typical absorption curve is shown in Figure 31.7. Because proteins, as with RNA, may be present in many copies, photochemical alterations in only a fraction of them may still permit biological functions to occur.

However, energy-absorbing molecules may interact and 'activate' other molecules so that the impact of UV may be indirect.

Photo-repair of DNA UV-induced damage

Strong and enzymatically complex DNA repair systems in both eukaryotes and prokaryotes permit life to continue normally in the presence of solar UV irradiation. Where this does not occur, as in the genetic deficiency disease xeroderma pigmentosa in humans, the accumulation of mutations leads to premature death. However, as far as insect viruses are concerned, photo-repair is almost certainly something that occurs, if it occurs at all, only during the virus replication cycle in the host and not, for instance, when the isolated virus is exposed on plant surfaces.

Interactions and restrictions

There are reports on some RNA viruses of vertebrates in which exposure to UV can cause cross-linkages of RNA to viral proteins. This occurs in the picornavirus polio type 1.

Isolated solubilized RNA of tobacco mosaic virus (TMV) suffers the usual type of damage when exposed to UV radiation. However, this does not take place when suspensions of intact virus particles are exposed under similar conditions. It has been suggested that the rigidity of the TMV protein coats prevent molecular 'distortion' of the RNA.

The biological and climatic impacts of changes in ozone prevalence are far from resolved. It has been calculated that a 10% reduction in ozone increases non-melanoma skin cancer by 26% and cataract by 6–8% (Ambach and Blumthaler, 1993) but the more general question of impact on ecosytems has yet to be evaluated, even supposing such a complex task to be possible. An ecosystem investigation by Bothwell, Sherbot and Pollock (1994) indicated how irradiation of a predator and its prey, both sensitive to UV but at different levels, can actually lead to increase in the prey population! In addition to the so-called greenhouse gases and their involvement in global warming, depletion of ozone in the upper stratosphere could lead to warming at lower altitudes. The consequences of this for global climate are currently the subject of world scientific interest and concern (Rowland, 1982).

DEFINITIONS OF UNITS AND TERMS

A publication (Anon., 1978) of the Radiation Commission of the International Association of Meteorology and Atmospheric Physics will be found useful in the understanding and definition of some of the terms involved.

Wavelength The units of UV measurement are essentially those used for the electromagnetic spectrum as a whole. There has been, however, a tendency to express wavelength (λ) in the visible region, and to a less extent in the UV region, in terms of ångstroms, though nanometres are increasingly used and are used here (1 Å = 0.1 nm). Where wavelength tends to be very long, for example in radio frequencies, centimetres or even metres are used; where it is very short, for example in gamma irradiation, picometres may be employed (1 pm = 10^{-19} m).

Ångstrom unit (AU, Å or A) Approximately equal to the diameter of a hydrogen atom; 0.10 nm or 1×10^{-8} cm.

Nanometre (nm) Equals 1×10^{-7} cm or 1/1000 μm.

Micrometre (μm) Equals 1×10^{-4} cm or 1/1000 mm.

Radiant flux density This is often simply referred to, in biological literature, as 'flux'; it is indicated by the symbols *M* or *E*. It is the radiant flux of any origin crossing an area element. Units are W/m^{-2}. However, megajoules (mJ) or photons have also been used.

Photon The *quantum* of electromagnetic radiation is referred to as a **photon** or the **energy of a photon** (see *Radiant energy*, below).

Langley The term **pyron** has theoretically replaced Langley, though the latter is still encountered *in lit*. The Langley is defined as a unit of solar radiation density: 15°C g/cal/cm per min or 697.8 W/m^{-2} per s. The mean radiation density of the sun at the earth's surface is about 2 Langley/min (Jerrard and McNeill, 1986).

ABSORPTION OF ULTRAVIOLET AND THE EXTINCTION COEFFICIENT

Electromagnetic radiation that is monochromatic (i.e. of a single wavelength) is absorbed exponentially:

$$\frac{U}{U_o} = e^{-p} \tag{31.7}$$

where U_o is the radiant energy entering a front surface of an absorbing 'barrier' and U is the unabsorbed energy incident upon the back surface; p is a parameter known as absorbancy or optical density and is given by

$$p = kcx \tag{31.8}$$

where k is a constant, the absorption coefficient, c is the concentration of a dissolved solute and x is the thickness of the absorbing layer.

This introduces the use of the Bouguer–Lambert law:

$$I_f = I_i \exp(-\alpha Cl) \tag{31.9}$$

This explains the emergent intensity (I_f) in terms of the incident intensity (I_i), the absorption coefficient (α), the concentration (C) and the thickness (l). This is essentially a restatement of equations 31.7 and 31.8, above.

Equation 31.9 may be given in logarithmic (base 10) terms:

$$I_f = I_i\, 10^{-\epsilon Cl} \tag{31.10}$$

or

$$I_g(I_f/I_i) = -\epsilon Cl \tag{31.11}$$

Here, ϵ is ($\alpha/2.303$) and is called the extinction coefficient (Cl is the optical density or absorbance of a sample).

However, absorption may extend over a range of frequencies and the extinction coefficient at a single frequency will not give a full picture of the intensity of absorption. Here, the **integrated absorption coefficient** (A), the sum of coefficients for all the frequencies in the appropriate band, is preferable.

$$A = \int \alpha(\nu)d\nu \tag{31.12}$$

Radiant energy: photons and wavelengths

Radiant energy (E) is defined as:

$$E = h\nu \qquad (31.13)$$

where h is a universal constant (Plank's constant), ν is the frequency of vibration (vibrations/s) of an oscillator and E is the energy of the photon. The frequency of vibration, ν, can be given by:

$$\nu = \frac{U}{\lambda} \qquad (31.14)$$

where λ is wavelength and U is the speed of light, or the velocity of the propagation of the wave. Substituting this for ν in equation 31.13 it follows that the energy of the photon is inversely proportional to the wavelength of radiation: $E = 6.624 \times 10^{-27}\ U/\lambda$.

A significant message in this for photobiology is that though the representation of shorter wavelengths (say, 290–300 nm) in UV-B is small, their energy value is high and any increase in this spectral area is biologically very important.

From the Bouguer–Lambert law the monochromatic start path transmission is

$$\lambda T = e^{-r_{\lambda} m} \qquad (31.15)$$

where T is the 'monochromatic extinction optical thickness' at a wavelength λ along a path normal to the earth's surface, rm is the air mass (ratio start path to vertical path). For solar zenith angles, z, of less than 72°, the air mass is adequately described by $m = \sec z$, but for larger angles corrections must be made for atmospheric refraction (Bemporad, 1907).

The sum of Rayleigh and Mie scattering together with ozone absorption gives the monochromatic extinction optical thickness: it includes selective absorption by water vapour and CO_2 and can be calculated by processes given by Leckner (1978).

REFERENCES

Anon. (1978) *Terminology and Units of Radiation Quantities and Measurements*. Boulder, CO: International Association of Meteorology and Atmospheric Physics (IAMAP), Radiation Commission.

Ambach, W. and Blumthaler, M. (1993) Biological effectiveness of solar UV radiation in humans. *Experientia* **49**, 747–753.

Barker, R.E. (1968) The availability of solar radiation below 290 nm and its importance in photomodification of polymers. *Photochemistry and Photobiology* **7**, 275–295.

Bemporad, A. (1907) Calculation of refraction correction for zenith angles >72°. *Meterological Zoology* **24**, 309.

Bener, P. (1964) Diurnal and annual course of the spectral intensity of total and diffuse ultraviolet radiation for cloudless sky in Davos. *Strahlentherapie* **123**, 306–316.

Bothwell, M.L., Sherbot, D.M.J. and Pollock, C.M. (1994) Ecosystem response to solar ultraviolet-B radiation: influence of trophic-level interactions. *Science* **265**, 97–100.

Buttner, K. (1938) *Physik, Broklimat*. Leipzig.

Cutchins, P. (1982) A formula for comparing annual damaging ultra-violet (DUV) radiation doses at tropical and mid-latitude sites. In *NATO Conference Series, Series IV:*

Marine Sciences: The Role of Solar Ultra-violet Radiation in Marine Ecosystems J. Calkins (ed.). New York: Plenum Press, pp. 213–228.

Elterman, L. (1968) *Environmental Research Paper*. Bedford, MA: Office of Aerospace Research, US Air Force, AFCRL-68–0135, No. 285.

Harm, W. (1980) *Biological Effects of Ultra-violet Radiation*. Cambridge: Cambridge University Press.

Jerrard, H.G. and McNeill, D.B. (1986) *A Dictionary of Scientific Units*, 5th edn. London: Chapman & Hall.

Johnson, F.S., Tsan, M. and Green, A.E.S. (1975) Average latitudinal variations in ultra-violet radiation at the earth's surface. *Photochemistry and Photobiology* **21**, 179–188.

Leckner, B. (1978) The spectral distribution of solar radiation at the Earth's surface – elements of a model. *Solar Energy* **20**, 143–150.

McKenzie, R.L. (lead author) (1995) Surface ultraviolet radiation. In *UNEP/WMO Scientific Assessment of Ozone Depletion: 1994*, Ch. 9, pp. 9.1–9.15.

Nawrocki, P.J. and Papa, R. (1961) *Atmospheric Processes*. Bedford, MA: Air Force Cambridge Research Laboratory, AFCRL-595, Office of Aerospace Research, pp. 1-23–2-14.

Ramoska, W.A., Stairs, G.S. and Hink, F.W. (1975) Ultraviolet light activation of insect nuclear polyhedrosis virus. *Nature, London* **253**, 628–629.

Rowland, F.S. (1982) Possible anthropogenic influences on stratosphere ozone. In *NATO Conference Series, Series IV: Marine Sciences, The Role of Solar Ultra-violet Radiation in Marine Ecosystems*. J. Calkins (ed.). New York: Plenum Press, pp. 29–48.

32 Plant surfaces

INTRODUCTION

The relationship between the BV OB and the plant surface has never been addressed in any comprehensive way. This is surprising because the plant surface is the most common environmental site onto which inoculum is released and from which it is later acquired during feeding by a potential host insect. It is also the site from which inoculum, possibly now affected by its sojourn, eventually reaches that long-term environmental reservoir, the soil. The physical and chemical complexity of the plant surface has become apparent, especially as a result of investigations since the early 1950s.

For us, an area of major interest is the degree of retentivity of OBs by plant surfaces. Plant-related factors that may affect this are hydrophobicity, micro-roughness, the nature and concentration of the leachate, pH and surface charge. Once on the leaf surface, the biological survival of infectivity becomes of overlapping interest. Here some of the above factors (especially leachates and pH) may be of significance but others are also involved, notably micro-climate and angle of declination of the leaf, which will affect interception of solar radiation.

Another area of strong interest lies in the receptivity of leaf surfaces to incoming spray droplets. The deposition potential of spray droplets is not only a function of the wettability of the surface but also of leaf shape (sometimes referred to as target geometry). Wettability itself has complex origins. It is determined partly by the innate waxiness, and the type of wax, by the micro-structure of the waxes and by other factors not yet fully evaluated, e.g. possibly leachate types and levels. When spray droplets themselves are rejected (by reflection or by 'pearling', followed often by rapid run-off, see p. 125) or simply sit with very high contact angle (see below), what is the fate of OBs within them? An understanding of these areas of interaction is very important both to epizootiological and to control studies.

BIBLIOGRAPHIC NOTE

Basic questions of the microstructure of the leaf surface were first comprehensively dealt with in a classic work by Martin and Juniper (1970). Other detailed texts on this subject are Cutler, Alvin and Price (1982) and Jeffree (1986). Harr *et al.* (1991) is valuable as an account of certain aspects, especially pH, waxes and hydrophobicity (contact angles) of the leaf surfaces of a wide range of species of major weeds. Rodriguez, Healey and Mehta (1984) discuss the biology and chemistry of plant trichomes in their many modifications. Notable briefer reviews of plant surface anatomy are Holloway (1970b), Hallam and Juniper (1971), Esau (1977) and Wilkinson (1979).

A detailed treatment of plants and micro-climate is that of Jones (1992) while Burrage (1971, 1976) and Wilmer (1986) provide briefer but very valuable reviews, especially as these are written in the context of phylloplane microbial communities and, in the last instance, also of the insect.

Book-length accounts of the microbiology of plant surfaces are Preece and Dickinson (1971), Dickinson and Preece (1976), Blakeman (1981), Fokkema and van den Heuvel (1986) and Andrews and Hirano (1991). These accounts vary in content and so even the earlier ones are valuable. Of course, since they concern organisms resident on plant surfaces or pathogenic to plants via surface penetration, BVs are not specifically covered.

Recent work on the adhesion of microorganisms to plant surfaces seems to be scattered. However, Ellwood, Melling and Rutter (1979) provide a very wide coverage, while an in-depth description and discussion of the complex of longer range and shorter range forces involved in particulate acquisition and attachment are provided by Small (1985), who also describes experimental investigations related to the attachment of OBs of BVs (Small, 1986; Small, Moore and Entwistle, 1986; Small and Moore, 1987).

The chemistry of plant surfaces is complex. It is dealt with by Tukey (1971), whose review is updated by Godfrey (1976); both should be consulted.

Techniques for the visualization of plant surfaces and their microorganisms by scanning electron microscopy are discussed by Royle (1976) while Harr *et al.* (1991) also provide a plant surface SEM preparation protocol.

Finally, the relationships of insects with plant surfaces was the subject of a meeting, a full account of which was published by Juniper and Southwood (1986).

LEAF SURFACE MACRO- AND MICRO-STRUCTURE

A generalized picture of leaf structure is shown in Figure 32.1. At a macro level, the presence of trichomes of various types may favourably influence retention of OBs as they do with bacteria, fungi and yeasts (Allen *et al.*, 1991). The stomatal cavity, more correctly termed the substomatal cavity, can provide a particularly protective locale for OBs provided that the stomata are open at their arrival (e.g. in a spray droplet), the aperture is large enough to permit entry (which may be generally easier for GVs than for NPVs) and the surface tension of carrier fluids is low enough to allow penetration. Stomata in most plant species are on the underside (adaxial) leaf surface. (They are on both surfaces in the potato and some other plants.) Appropriate presentation of spray droplets for stomatal entry of OBs will, therefore, mainly arise from leaves at a steep angle of declination, when sprays are applied by certain small droplet techniques or when crops are

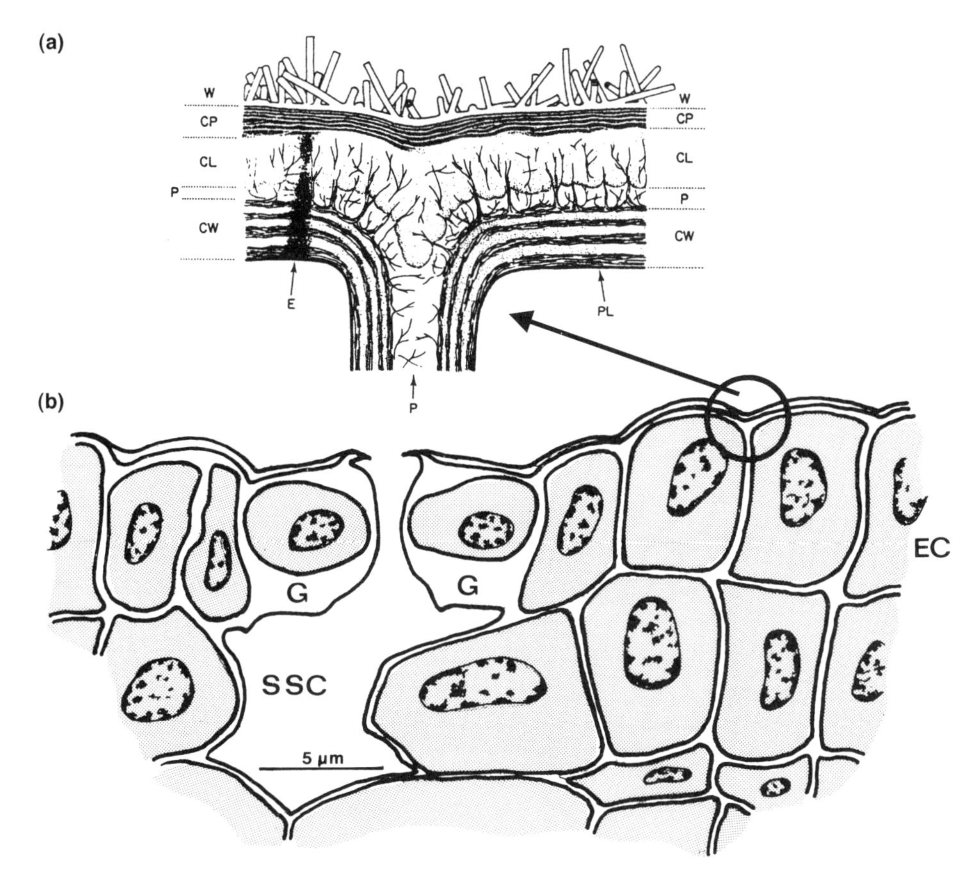

Figure 32.1 Generalized structure of (a) the leaf epidermis and (b) the cuticle. EC, epidermal cells; G, stomatal guard cells; SSC, substomatal cavity; W, epicuticular wax; CP, cuticle; CL cuticular layer or reticulate region traversed by cellulose microfibrils; P, pectinaceous layer and middle lamella; CW, epidermal cell wall; PL, plasmalemma; E, ectodesma. (Epidermal section adapted from Juniper (1991); plant cuticular micro-structure from Jeffree (1986). (Reproduced from C. E. Jeffree, in *Insects and the Plant Surface*. B. E. Juniper and T. R. E. Southwood (eds), by kind permission of Cambridge University Press. (Originally published by Edward Arnold.)

drift-sprayed on an easterly wind in the evening when the leaves are orientated at right angles to the declining sun (Chapter 8).

At the micro-structural level, we are mainly concerned with the epicuticular layer, but lower layers such as the cuticular layer, and the intermediate zone containing chemical elements of both, may be exposed. As Holloway (1969) showed, the low wettability aspect of leaves is by no means solely associated with the nature of the epicuticle. When organic solvents are used to remove epicuticular wax, the leaf surface may still be very hydrophobic, sometimes even more so.

The surface of an individual leaf may be innately variable. For instance, some areas seem devoid of wax while the structure of the wax may vary from area to area. Generally wax with a smooth surface is more easily wetted than a rough wax surface (Figure 32.2).

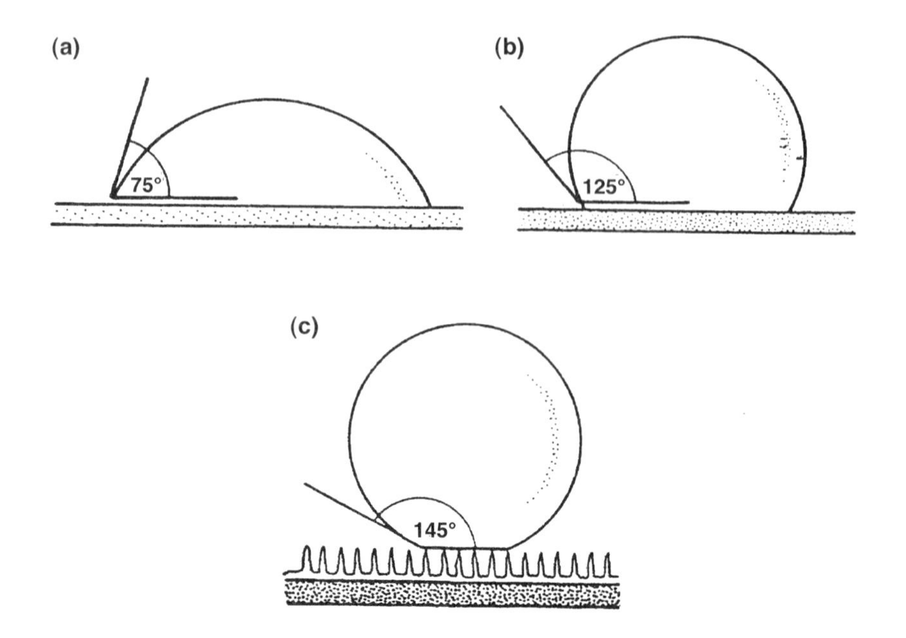

Figure 32.2 Effect of contact angle size on a water drop on a leaf surface. (a) Wax-free plant cuticle: small contact angle, water drop covers a large surface area. (b) Smooth wax surface: large contact angle, droplet covers a smaller area. (c) Crystalline wax surface of ornamented cuticle: largest contact angle, droplet is in contact with the tips of the projections only. (From Juniper, 1991.)

Varying plant growth conditions will often result in modification of surface structure. Increased UV-B radiation can result in the growth of a thicker wax layer (Steinmuller and Tevini, 1985). Epicuticular wax erodes under the influence of wind-borne dust particles and so wettability may change. There may be differences in the wettability of upper and lower surfaces of leaves.

Contact angles are used as a measure of wettability (see below) and are listed for a range of plant species by Holloway (1969, 1970a), partially repeated in Matthews (1979) and by Harr *et al.* (1991). The epicuticle is composed of a complex of waxes. Alkanes are less wettable (higher droplet contact angle) than the more polar hydrophilic components such as fatty acids. Epicuticular micro-roughness greatly increases hydrophobicity (Jeffree, 1986). It can be seen that a range of factors is responsible for wettability/hydrophobicity. Holloway (1970a) gives a detailed account of such factors. The possibility that leachates on the leaf surface may influence wettability seems not yet to have been investigated.

CHEMICAL CONDITIONS

The chemical nature of the leaf surface is derived from three main sources. Probably most plant species have leaf surfaces that are permeable to water; as a result organic and inorganic leachates are a common feature of the surface. The leachates may be present in considerable amounts and may constitute an appreciable component in plant–environment nutrient recycling. For example, in apple orchards 25–30 kg/ha potassium, 9 kg sodium and 10 kg calcium in addition to sugars, amino acids and phenolics may be lost from leaves every year (Tukey, 1971). Leachates have an important influence on the microorganisms of the leaf surface. They may either facilitate or, at times, discourage their growth and can also act defensively against certain pathogens (Godfrey, 1976; Elleman and Entwistle, 1985a).

An apparently less frequent source of leaf surface chemicals is the salt secretory gland. Glands of this type are recognized on cotton (*Gossypium* spp.) (where they are known to be multicellular) and *Nolana mollis*, an Atacama Desert shrub, and on some other plants (e.g. Berry, 1970; Sakai, 1974; Shimony, Fahn and Reinhold, 1973). In *N. mollis*, sodium is the principal exudate. Deliquescence occurs at atmospheric humidities less than saturated and it has been shown that the captured moisture is absorbed by the leaves (Mooney *et al.*, 1980). It is, therefore, a xerophytic physiological mechanism. Cotton by origin is a plant of arid zones. The salt gland product is complex but it is very rich in magnesium and calcium and is also deliquescent (Elleman and Entwistle, 1985b). As cotton has no obvious xerophytic modifications, it is possible that secretion of salts is an osmoregulatory function. Whether the moisture is taken up by the leaf is unknown. These exudates impart a highly alkaline condition to cotton leaf surfaces (Harr *et al.*, 1980) and as this is a common phenomenon throughout the Malvaceae (Harr, Guggenheim and Boller, 1984); possibly the presence of salt glands is equally widespread. Most plants, however, have a rather pH-neutral surface (Harr *et al.*, 1991).

The third contribution to the leaf surface chemical complex is derived from the atmosphere. Soluble substances may be deposited on leaf surfaces in rain and mist. Atmospheric pollution by sulphur dioxide and other chemical groups may lead to a lowering of leaf surface pH and, in the presence of atmospherically low-altitude pollutant ozone, to leaf damage and to adverse effects on leaf surface microorganisms (Magan and McLeod, 1991). In addition, solid particles may be captured by raindrops and deposited on leaf surfaces. This takes place even when the plant surface is hydrophobic, especially if the particles themselves are also hydrophobic (Davies, 1961). Carter (1965) has noted that hydrophobic spores on the surface of water droplets 'migrate' to the hydrophobic leaf surface and adhere.

MICRO-METEOROLOGICAL CONDITIONS

Solar UV radiation

Solar UV radiation is discussed on p. 40 and in Chapter 31.

Temperature

A substantial portion of total solar energy is in the near infrared (700–1500 nm) but as leaves absorb weakly in this region the potential for heat increase is greatly reduced (Sutcliffe, 1977). Also leaves have low thermal capacity and respond very rapidly to changes in solar radiation levels. Increase in wind speed, by increasing the diffusion coefficient, decreases leaf surface temperature. Evaporative cooling also exerts control and is naturally most effective for plants with a good water status. Peripherally, leaves may be at their coolest (Burrage, 1971, 1976; Wilmer, 1986). In relation to local air temperatures, leaf surface temperatures tend to be at their highest during the day but lower during the night, though when air temperatures are high (>33°C) leaf temperature may be lower than ambient.

When the leaf is considered in the context of the whole crop canopy, it is usually found that there is considerable vertical variation in leaf surface temperature: this tends to be higher during the day at intermediate crop heights and least at the top of the canopy, reflecting the lower density of leaves (and so of surfaces to intercept solar radiation) and greater air disturbance there. A night, there is something of a reversal of this profile (Burrage, 1976). Leaf surface structure may influence temperature through scattering of solar radiation, e.g. from white particulate epicuticular wax and by the white tomentum of trichomes often seen in desert plants.

Humidity

At the leaf surface, humidity depends on the rate of water vapour transfer through the boundary air layer. This is controlled by stomatal aperture and interactions involving the thickness of the boundary air layer (Burrage, 1969, 1971). While little seems to be known about the levels of humidity close to the leaf surface, it appears that under conditions of dry air it is usually higher than that of the air itself (Burrage, 1976). Wilmer (1982, 1986) gives data on relative humidity profiles for upper and lower leaf surfaces in the day and the night. It is, of course, influenced by wind speed, especially wind close to the leaf surface.

Relative humidity is least during the day and highest at night, when, if it approaches saturation, a very small decline in temperature (*c.* 1°C) will result in dew formation. Because the centre of the leaf is coolest, dew formation is often greatest here, though transfer of water vapour can result in greater formation of dew at the leaf periphery.

Wind speed and leaf shape

Wind speed is reduced by the drag of physical structures both on a macro- and a micro-scale. Thus there may be brisk air movement at the top of a crop canopy but this may decrease to near zero at ground level. Very great speed reductions also occur as the surface of the individual leaf is approached. Wind speed near the leaf surface is controlled by leaf size and surface topography. The presence of trichomes can have a slowing effect. Wooley (1965) removed the trichomes from soybean leaves and found a 65% increase in wind velocity at 0.5 mm from the surface.

Wind speeds are higher near the leaf margin, where both temperature and humidity profiles are consequently less stable. Wind speeds, with leaf shape and size, influence the extent of deposition of particles from the atmosphere. In general, finer structures acquire larger numbers of particles. Information on target geometry and particle capture is to be found in Gregory (1951, 1971, 1973), Gregory and Stedman (1953), Carter (1965) and Allen *et al.* (1991): the last should be consulted for a recent survey of literature, especially as far as leaf surface features are concerned.

THE COMPLEX ISSUES ASSOCIATED WITH PLANT SURFACES FOR BACULOVIRUSES

Many aspects of the plant surface environment as they affect BVs have been treated elsewhere in this book e.g. temperature (p. 43), UV radiation (pp. 40 and 565) chemical questions (p. 49) and wettability (pp. 124 and 582).

In working with BVs at this important environmental site, it is useful to be aware of the possible unrealism of single-factor studies. For example, while moisture levels *per se* may not induce biological decay, wetting and drying cycles may (Elleman, 1983). Moisture also exacerbates the attritional influence of UV-B (Ignoffo and Garcia, 1992).

It should also be remembered that on both spatial and temporal scales conditions within a crop are not uniform: as referred to above, there is not only variation of temperature, wind speed and humidity on a vertical scale within whole crops but such factors also vary at the small scale of the individual leaf.

MEASURING LEAF SURFACE CONDITIONS

pH

The test plant is held in a saturated atmosphere. A drop of distilled water is placed onto the underside of a flat-bottomed electrode which is lowered onto the test leaf surface. pH should be recorded at brief intervals during 45–60 min until a clear equilibrium is reached. Temperature and pH should be continuously monitored. Harr *et al.* (1991) used a METROHM EA 156X electrode and a Metrohm Model 691 pH meter, which was attached to an IBM Ps/2 model 50Z computer. Oertli, Harr and Guggenheim (1977) should also be consulted.

Temperature

Temperature has usually been measured by thermocouples threaded into the leaf (Waggoner and Shaw, 1952) but infrared thermometers (Lourance, Pruitt and Burraga, 1965) and thermal imaging cameras (Wigley and Clarke, 1974) have also been successfully employed. Burrage (1971) used copper–constantin thermocouples, 0.01 mm wide, clipped to upper and lower leaf surfaces. Three millimetres or more of the sensor tip was pressed close to the surface, air temperature being measured by a similar thermocouple.

Epicuticular waxes

The collection and chromatographic analysis of epicuticular waxes, with identification of polar and non-polar components is described by Jeffree (1986) and by

Harr *et al.* (1991). At least some epicuticular waxes can be recrystallized in forms morphologically similar to those present on the leaf (Jeffree, Baker and Holloway, 1976; Jeffree, 1986). By recrystallization onto chemically neutral surfaces, this may prove a useful analytical technique in BV acquisition and retention studies.

Wettability

The wettability of a surface (Figure 32.2) is measured by the angle of contact (θ) of a water droplet (1 μl) of standard surface tension (72 mN/m). One minute after deposition, θ is measured using a dissecting microscope fitted with a special 90° reflecting mirror and a goniometric 10× eyepiece. It is advisable to repeat each measurement a number of times (Holloway, 1969; Harr *et al.*, 1991).

Chemical conditions

Probably most of the leachates on the plant surface are water soluble and so can be easily washed off for chemical analysis. To an extent, the elemental composition of the leaf surface can be investigated by X-ray micro-analysis on an electron microscope fitted with an EDAX attachment. Specimens are mounted on a carbon–nylon grid and carbon coated. The lowest detectable atomic number is 23, i.e. sodium. The technique permits analysis of very small areas of tissue, for example a single salt gland (Elleman and Entwistle, 1982).

REFERENCES

Allen, E.A., Hoch, H.C., Steadman, J.R. and Stavely, R.J. (1991) Influence of leaf surface features on spore deposition and the epiphytic growth of phytopathogenic fungi. In *Microbial Ecology of Leaves*. J.H. Andrews and S.S. Hirano (eds). New York: Springer-Verlag, pp. 87–110.

Andrews, J.H. and Hirano, S.S. (eds) (1991) *Microbial Ecology of Leaves*. New York: Springer-Verlag.

Berry, W.L. (1970). Characteristics of salts secreted by *Tamarix aphylla. American Journal Botany* **57**; 1226–1230.

Blakeman, J.P. (ed.) (1981) *Microbial Ecology of the Phyllophane*. New York: Academic Press.

Burrage, S.W. (1969). Dew and the growth of the uredospore germ tube of *Puccinia graminis* on the wheat leaf. *Annals of Applied Biology* **64**: 495–501.

Burrage, S.W. (1971) The micro-climate at the leaf surface. In *Ecology of Leaf Surface Micro-organisms*. T.F. Preece and C.H. Dickinson (eds). London: Academic Press, pp. 91–101.

Burrage, S.W. (1976) Aerial microclimate around plant surfaces. In *Microbiology of Aerial Plant Surfaces*. C.H. Dickinson and T.F. Preece (eds). London: Academic Press, pp. 173–184.

Carter, M.V. (1965) Ascospore deposition in *Euphyta cirmeniacae. Australian Journal of Agricultural Research* **16**, 825–836.

Cutler, D.L., Alvin, K.L. and Price, C.E. (eds) (1982) *Linnaean Society Symposium Series* No. 10: *The Plant Cuticle*. London: Academic Press.

Davies, R.R. (1961) Wettability and the capture, carriage and deposition of particles by raindrops. *Nature, London* **191**, 616.

Dickinson, C.H. and Preece, T.F (eds) (1976) *Microbiology of Aerial Plant Surfaces.* London: Academic Press.

Elleman, C.J. (1983). The inter-relationships between baculovirus of *Spodoptera littoralis* and the leaf surface of *Gossypium hirsutum*, with comparative observations on *Brasssica oleracea.* DPhil thesis, University of Oxford.
Elleman, C.J. and Entwistle, P.F. (1982) A study of glands on cotton responsible for the high pH and cation concentration of the leaf surface. *Annals of Applied Biology* **100**, 553–558.
Elleman, C.J. and Entwistle, P.F. (1985a) Inactivation of a nuclear polyhedrosis virus on cotton by the substances produced by the cotton leaf surface glands. *Annals of Applied Biology* **106**, 83–92.
Elleman, C.J. and Entwistle, P.F. (1985b) The effect of magnesium ions on the solubility of polyhedral inclusion bodies and its possible role in the inactivation of the nuclear polyhedrosis virus of *Spodoptera littoralis* by the cotton leaf gland exudate. *Annals of Applied Biology* **106**, 93–100.
Ellwood, D.C., Melling, J. and Rutter, P. (eds) (1979) *Adhesion of Microorganisms to Surfaces.* New York: Society of General Microbiology and Academic Press.
Esau, K. (1977). *Anatomy of Seed Plants.* New York: Wiley.
Fokkema, N.J. and van den Heuvel. J. (eds) (1986) *Microbiology of the Phyllosphere.* Cambridge: Cambridge University Press.
Godfrey, B.E.S. (1976) Leachates from aerial parts of plants and their relation to plant surface microbial populations. In *Microbiology of Aerial Plant Surfaces.* C.H. Dickinson and T.F. Preece (eds). London: Academic Press, pp. 433–439.
Gregory, P.H. (1951) Deposition of air-borne *Lycopodium* spores on cylinders. *Annals of Applied Biology* **38**, 357–376.
Gregory, P.H. (1971) The leaf as a spore trap. In *Ecology of Leaf Surface Microorganisms.* T.F. Preece and C.H. Dickinson (eds). London: Academic Press, pp. 239–243.
Gregory, P.H. (1973) *The Microbiology of the Atmosphere.* New York: Wiley.
Gregory, P.H. and Stedman, O.J. (1953) Deposition of air-borne *Lycopodium* spores on plant surfaces. *Annals of Applied Biology*, **40**, 651–674.
Hallam, N.D. and Juniper, B.E. (1971). The anatomy of the leaf surface. In *Ecology of Leaf Surface Microorganisms.* T.F. Preece and C.H. Dickinson (eds). London: Academic Press, pp. 3–38.
Harr, J., Guggenheim, R., Boller, T. and Oerrtli, J.J. (1980) High pH values on the leaf suface of commercial cotton varieties. *Cotton et Fibres Tropicales* **35**: 379–384.
Harr, J., Guggenheim, R. and Boller, T. (1984) High pH values and secretions of ions on leaf surfaces: a characteristic of the phylloplane of Malvacae. *Experientia* **40**, 935–937.
Harr, J., Guggenheim, R., Schulke, G. and Falk, R.H. (1991) *The Leaf Surface of Major Weeds.* Switzerland: Sandoz Agro Ltd.
Holloway, P.J. (1969) The effects of superficial wax on leaf wettability. *Annals of Applied Biology* **63**, 145–153.
Holloway, P.J. (1970a) Surface factors affecting the wetting of leaves. *Pesticide Science* **1**, 156–163.
Holloway, P.J. (1970b) The chemical and physical characteristics of leaf surfaces. In *Ecology of Leaf Surface Organisms* T.F. Preece and C.H. Dickinson (eds). London: Academic Press, pp. 39–53.
Ignoffo, C.M. and Garcia, C. (1992) Combinations of environmental factors and simulated sunlight affecting the inclusion bodies of the *Heliothis* (Lepidoptera: Noctuidae) nucleopolyhedrosis virus. *Environmental Entomology* **21**, 210–213.
Jeffree, C.E. (1986) The cuticle, epicuticular waxes and trichomes of plants, with reference to their structure, function and evolution. In *Insects and the Plant Surface.* B.E. Juniper and T.R.E. Southwood (eds). London: Edward Arnold, pp. 23–64.
Jeffree, C.E., Baker, E.A. and Holloway, P.J. (1976) Origins of the fine structure of plant epicuticular waxs. In *Microbiology of Aerial Plant Surfaces.* C.H. Dickinson and T.F. Preece (eds). London: Academic Press, pp. 119–158.
Jones, H.G. (1992) *Plants and Microclimate.* Cambridge: Cambridge University Press.
Juniper, B.E. (1991) The leaf from the inside and the outside: a microbe's perspective. In

Microbial Ecology of Leaves. J.H. Andrews and S.S. Hirano (eds). Berlin: Springer-Verlag, pp. 21–42.

Juniper, B.E. and Southwood, T.R.E. (1986) *Insects and the Plant Surface*. London: Edward Arnold.

Lourence, F., Pruitt, W.O. and Burrage, S.W. (1965). *Final Report 1965 USAE PGN*, Ch. xii. DA-AMC-282–043–65–612. Davies, CA: University of California.

Magan, N. and McLeod, A.R. (1991). Effect of atmospheric pollutants on phyllosphere microbial communities. In *Microbial Ecology of Leaves*. J.H. Andrews and S.S. Hirano (eds). Berlin: Springer-Verlag, pp. 379–400.

Martin, J.T. and Juniper, B.E. (1970) *The Cuticles of Plants* London: Edward Arnold.

Matthews, G.A. (1979) *Pesticide Application Methods*. London: Longman.

Mooney, H.A., Gulman, S.L., Ehleringer, J. and Rundel, P.W. (1980) Atmospheric water uptake by an Atacama desert shrub. *Science* **209**, 693–694.

Oertli, J.J., Harr, J. and Guggenheim, R. (1977) The pH value as an indicator for the leaf surface microenvironment. *Zeitschift für Pflanzenkrakheit und Pflanzenschutz* **84**, 729–737.

Preece, T.F. and Dickinson, C.H. (1971) *Ecology of Leaf Surface Microorganisms*. London: Academic Press.

Rodriguez, E., Healey, P.L. and Mehta, I. (Eds) (1984) *Biology and Chemistry of Plant Trichomes*. New York: Plenum Press.

Royle, D.J. (1976) Scanning electron microscopy of plant surface microorganisms. In *Microbiology of Aerial Plant Surfaces*. C.H. Dickinson and T.F. Preece (eds). London: Academic Press, pp. 569–605.

Sakai, W.S. (1974) Scanning electron microscopy and energy X-ray analysis of chalk secreting leaf glands of *Plumbago capensis*. *American Journal of Botany* **61**, 94–99.

Shimony, C., Fahn, A. and Reinhold, L. (1973). Ultrastructure and ion gradients in the salt glands of *Avicennia marina*. *New Phytology* **72**, 27–36.

Small, D.A. (1985) Aspects of the attachment of a nuclear polyhedrosis virus from the cabbage looper (*Trichoplusia ni)* to the leaf surface of cabbage (*Brassica oleracea*). DPhil thesis, University of Oxford.

Small, D.A. (1986) Identification of subpopulations within baculovirus polyhedra: possible relevance to attachment to leaf surfaces. *Microbios Letters* **32**, 91–96.

Small, D.A. and Moore, N.F. (1987) Measurement of surface charge of baculovirus polyhedra. *Applied Environmental Microbiology* **53**, 598–602.

Small, D.A., Moore, N.F. and Entwistle, P.F. (1986) Hydrophobic interactions in attachment of a baculovirus to hydrophobic surfaces. *Applied Environmental Microbiology*, **52**, 220–223.

Steinmuller, D. and Tevini, M. (1985) Action of ultraviolet radiation (UV-B) upon cuticular waxes in some crop plants. *Planta*, **164**, 557–564.

Sutcliffe, J. (1977) *Institute of Biology, Studies in Biology No. 86, Plants and Temperature*. London: Edward Arnold.

Tukey, H.B. (1971) Leaching of substances from plants. In *Ecology of Leaf Surface Microorganisms*. T.F. Preece and C.H. Dickinson (eds). London: Academic Press,. pp 67–80.

Waggoner, P.E. and Shaw, R.H. (1952) Temperature of potato and tomato leaves. *Plant Physiology, Lancaster* **27**, 710–723.

Wigley, G. and Clark, J.A. (1974) Heat transport co-efficients for constant energy flux models of broad leaves. *Boundary-Layer Meteorology* **7**, 139–150.

Wilkinson, H.A. (1979) The plant surface (mainly leaf). In *Anatomy of Dicotyledons*. C.R. Metcalf and L. Chalk (eds). New York; Oxford University Press, pp. 97–165.

Wilmer, P.G. (1982) Microclimate and the environmental physiology of insects. *Advances in Insect Physiology*, **16**, 1–57.

Wilmer, P.G. (1986) Microclimatic effects on insects at the plant surface. In *Insects and the Plant Surface*. B.E. Juniper and T.R.E. Southwood (eds). London: Edward Arnold, pp. 65–80.

Wooley, J.T. (1965) Water relations of soybean leaf hairs. *Agronomy Journal* **57**, 569–571.

Glossary

ABBREVIATIONS USED FOR NPVS

AcMNPV	*Autographa californica* MNPV
BmNPV	*Bombyx mori* NPV
CfMNPV	*Choristoneura fumiferana* MNPV
HzSNPV	*Helicoverpa zea* SNPV
LdMNPV	*Lymantria dispar* MNPV
MbMPNV	*Mamestra brassicae* MNPV
SeMNPV	*Spodoptera exigua* MNPV
SfMNPV	*Spodoptera frugiperda* MNPV
TnMNPV	*Trichoplusia ni* MNPV
TnSNPV	*Trichoplusia ni* SNPV

OTHER TERMS

Actipron (Ulvapron)	light mineral oil containing surfactant(s) to allow emulsification with water. Manufactured by BP (British Petroleum Co. Ltd)
additive effects	two or more factors (e.g. pathogens, toxicants) that in concert produce an effect which is precisely the sum of their individual activities
a.i.	active ingredient
aliquot	here used in the sense of one of a number of equal parts
antagonism	two or more factors (e.g. pathogens, toxicants) acting together to produce an effect less than the sum of their individual effects
antibodies	in the vertebrates, modified blood globulins produced in response to alien material (the antigen) invading the body
anti-evaporant	here in the sense of a spray additive slowing or preventing evaporation of carrier fluid from a spray droplet
antigens	alien material (especially proteins) entering the blood (especially of a vertebrate) and stimulating the production of antibodies
apoptosis	a normal physiological form of cell death (also known as programmed cell death) required for embryonic development and the maintenance of adult tissue. As a defence mechanism, it provides a means of destroying virus-infected cells, thus limiting the spread of virus within the host
APS	ammonium persulphate, initiates polymerization of acrylamide; TEMED which is added with it, acts as an accelerator
atomizer	simple device for producing small spray droplets by passage of air

	across a narrow orifice through which spray fluid is being delivered
auto-dissemination	strictly, spatial dissemination of pathogen inoculum by the (homologous) host animal but also applied to situations where non-homologous hosts disseminate the pathogen
Ballotini beads	small glass beads obtainable in a range of sizes used, for example, for the disintegration of cells and release of virus OBs. For this purpose, beads (0.5 mm diameter), diluent (e.g. water), and infected larvae are placed in a water-tight container and are shaken together manually until an homogenate is produced
biological assay, bioassay	measurement of the response produced in an organism by an agency (here pathogenic); in essence, bioassays are quantitative and so depend on observing the response in a group of similar test organisms
B.t.	*Bacillus thuringiensis*
BV	a member of the Baculoviridae family of insect pathogenic viruses (NPV or GV); the abbreviation can also be used (but not here) for budded virus from cells
carrier fluid	here the main fluid component in a spray delivering an insect pathogen or toxicant
CDA	see controlled droplet application
cling film	a very thin impermeable plastic film used for wrapping
cole crops	a North American term for brassica crops
controlled droplet application (CDA)	implying a spray with control over droplet size and density, the idea being to generate droplets that are as near mono-sized as possible; usually employed with ultra-low-volume applications, both being obtainable with, for instance, spinning disc equipment
CPV	cytoplasmic polyhedrosis virus
cyclone air scrubber	device for collecting very small air-borne particles, which are sucked into a chamber out of which they are washed by a very fine spray of a chosen collection fluid; this then drains down the walls of the vessel into a collecting tube
cytopolyhedrosis virus	a synonym for cytoplasmic polyhedrosis virus that is now rarely used
delayed density-dependent factors	factors for which the proportionate impact (e.g. level of infection) is dependent on the density of the host population but which are manifest only after the host population has begun to decline as a consequence of the interaction
density-dependent factors	factors for which the proportionate effect on a population is dependent upon the density of that particular population
desorbant	a substance, the addition of which causes release of some factor, for example virus, from the substrate to which it is adsorbed
diprionid, diprionid sawfly	Hymenoptera, Symphyta of the family Diprionidae, all the members of which feed on conifers
dry ice	solid, frozen CO_2; at atmospheric pressure it sublimes at –78.5°C
EC_{50}	*see* median effective concentration
ED_{50}	*see* median effective dose
ELC	*see* expected lethal concentration
ELD	*see* expected lethal dose
ELISA	enzyme-linked immunosorbant assay
Elkay capsule	a small plastic capsule of around 1–2 ml volume with a snap lid
elution	dissociation of an adsorbed particle
entomopathogen	a pathogenic microorganism attacking insects
enzootic	usually used to suggest long-term presence of disease in an animal population but also to suggest disease at a low level of incidence
EPA	Environmental Protection Agency in the USA
epicentre	the point in space at which a disease outbreak commences and from which it spreads

epizootic	term describing an outbreak of disease in an animal population; also frequently employed to imply a high incidence of disease (*cf.* epiphytotic, disease in plant populations, and epidemic, disease in human populations)
EPV	entomopoxvirus
expected lethal concentration (ELC)	the concentration of pathogen (BV) that will kill a given proportion of the individuals (insects) tested
expected lethal dose (ELD)	the particular dose of pathogen (BV) ingested over a known period of time that will lead to the death of a given proportion of individuals (insects) tested
Finn pipette	a micro- and semi-micro-pipette, with disposable plastic tips, adjustable to deliver particularly a range of very small volumes. Other makes of this type of pipette are also available, e.g. Gilson etc.
fixative	chemical substance used to stabilize proteins etc. before staining and in other stages of preparation for microscopic examination
fp	fluorescent particle, used to identify spray droplets for quantification
freeze dry	a process by which (biological) material is first frozen and from which water is then volatilized under greatly reduced air pressure
g	the force of gravity; *see* relative centrifugal force
genome	the genetic material, especially the genes, contained in an organism/microorganism
gradients	as applied to centrifugation, these are used for separating particles according either to size and shape (rate-zonal density-gradient sedimentation) or to buoyant density (isopycnic density-gradient sedimentation). Sucrose and glycerol gradients are commonly employed for the former and caesium chloride for the latter
granulosis	especially an infection by a granulosis virus (GV), which is typified by the presence of very large numbers of granules each of which is a capsule containing, usually, a single virion
graticule	here an insertion, for a microscope eyepiece, on which a linear scale or a square subdivided into smaller squares (for counting particulate material) has been scribed; the latter is often called a 'squared' eyepiece
gustatory stimulant	a substance that stimulates feeding
GV	granulosis virus or granulovirus (genus *Baculovirus*)
haemocytometer	a counting chamber in which, employing an optical microscope, the density of blood cells in a sample can be estimated; also used extensively to count viral OBs and other particulate microorganisms
heterologous host/virus	firstly, any permissive host species from which the virus was not initially described; secondly, a virus infecting a host other than that in which it was originally detected
HI_{50}	*see* median infectivity temperature
histopathology	a study of abnormal changes especially at the cell and tissue level
homologous host/virus	firstly, the original host species in which the virus was found; secondly, a virus infecting the host species in which it was originally found
horizontal passage	transmission of a pathogen between individuals of the same host generation
hydathode	a gland occurring on the tips or edges of leaves of many plants; secretes water, especially at high atmospheric relative humidity
IC_{50}	*see* median infective concentration
ID_{50}	*see* median infective dose
immunity	usually regarded as an acquired state in which an organism is able

	to resist the disease-causing capacity of a pathogen or other potentially injurious element
impression film	here the use of sticky tape to take an impression from a plant surface, at the same time removing on the tape, for purposes of visualization and counting, virus OBs
inapparent infection	one giving no overt indication of its presence
inclusion body (IB)	often used synonymously with occlusion body (q.v.)
incubation period	strictly the period of time elapsing between acquisition (e.g. by ingestion) of a pathogen and the appearance of the signs of infection; in insects often (loosely) used to imply the period between pathogen acquisition and host death
infection	invasion of and multiplication in the body of the host by a microorganism which may or may not cause overt disease
infectivity	a capacity to cause an infection, *see* infection
instar	a developmental stage of an insect, but most often used here in relation to the five or six larval stages (instars)
integrated pest management (IPM)	pest control usually by a collection of environmentally acceptable methods that minimize, but do not necessarily entirely remove, the need to use chemical pesticides
in vitro	literally, in glass; commonly used here in relation to cell culture as opposed to whole insect culture
in vivo	literally, in life; commonly used here in relation to culture in insects as opposed to in cell or tissue culture
IPM	*see* integrated pest management
isopycnic	sedimentation of particles by centrifugation in a gradient the maximum density of which exceeds that of the particles being sedimented, *cf.* rate-zonal sedimentation, where the maximum density of the gradient is less than the buoyant density of the particles being sedimented
larval equivalent (LE)	the quantity of (virus) inoculum present in one insect (here usually a larva) at death or harvesting and, unless otherwise stated, usually implying a larva dying at 'maximum' size
latent infection	an infection in which the pathogen is not multiplying or is multiplying at a rate that will not produce overt disease or pathology until the implied balance between the host and pathogen is disturbed
LC_{50}	*see* median lethal concentration
LD_{50}	*see* median lethal dose
LE	*see* larval equivalent
LP_{50}	*see* median larval period
LT_{50}	*see* median lethal time
lyophilize	*see* freeze dry
macerate	break up by grinding, by use of spinning blades, or by stomacher for the purpose of freeing (viruses) from cells
median effective concentration (EC_{50})	the concentration producing a response in half the individuals in a test sample
median effective dose (ED_{50})	the particular dose of virus ingested over a known period of time that produces a quantal response in half the individuals in a test sample; an indirect measure of the mean tolerance (Steinhaus and Martignoni, 1962)
median infective concentration (IC_{50})	the concentration of virus causing an infection in half of the individuals tested over a known period of time
median infective dose (ID_{50})	for BV, the particular dose of virus ingested over a known period of time that causes an infection in half of the individuals tested

median infectivity temperature (HI_{50})	the temperature at which 50% of the infectivity of a viral preparation is lost
median larval period (LP_{50})	the time taken for 50% of the larval population to complete development
median lethal concentration (LC_{50})	the concentration (of pathogen, e.g. BV) required to kill half of a batch of a test (insect) sample
median lethal dose, (LD_{50})	for BV, the particular dose of virus ingested, over a known period of time, that will lead to the death of half of the test (insect) sample
median lethal time (LT_{50})	the time elapsing between acquisition of a particular dose of a pathogen and death of half of the test (insect) sample; calculated by reference to probit analysis (used synonomously, but erroneously, with ST_{50})
median survival time (ST_{50})	the time elapsing between acquisition of a particular dose (of pathogen) and survival of half of the test (insect) sample; calculated by reference to life expectancy tables
methyl paraben	methyl *p*-hydroxybenzoate
Micronair sprayer	spraying equipment, usually aircraft-mounted, generating spray droplets by impingement of a jet of spray fluid onto a spinning metal mesh cage. Sprayers are designated AU4000 to AU8000 and are specialized for different delivery systems
Micron sprayers	*see* spinning disc equipment
microtitre plate	also known as a multi-well plate and comprising eight rows of 12 wells; designed to contain as small a volume as practicable and used for assessing, for example, the protein concentration of a virus sample (120 μl volume of reactants per well) or concentration of virus by ELISA (50 μl per well)
microvilli	finger-like processes forming the brush border of the gut lumen-side of midgut cells
midgut	the central, absorptive region of insect gut; it is of mesodermal origin and does not have a chitinous lining, but *see* peritrophic membrane
Milli Q water	reagent-quality water produced by use of the Milli Q system, one of many commercially available systems for purifying water
MNPV	multinucleocapsid (multiply enveloped) NPV
nappy liner	non-absorbent, but water-permeable, small mesh cloth
negative staining	technique in elecron microscopy
Newton's rings	concentric coloured rings caused by interference between light waves reflected from the upper surface of the film of air between the coverslip and flat glass plate (e.g. Neubauer haemocytometer slide) and light waves reflected from the lower surface of the film of air
number median diameter (NMD)	the diameter that divides the droplets into two equal parts by number regardless of their volumes
nomogram	chart or diagram of scaled lines used to help in calculations, for example of relationships between viruses; it comprises three scales in which lines joining values on two determine the third
NOV	non-occluded virus
NPV	*see* nuclear polyhedrosis virus
NRI	Natural Resources Institute, University of Greenwich, Chatham, UK
nuclear polyhedrosis virus	virus belonging to the genus nucleopolyhedrovirus of the family Baculoviridae
nucleopolyhedrosis virus	syn. of the above
OAF	*see* open air factor
OAR	*see* original activity remaining
occlusion body (OB)	generally preferred term for the proteinaceous body (polyhedron, capsule) in which the virions of some virus groups are embedded; the virus particles are thus 'occluded'

oil immersion (objective)	a high magnification (×90–100) microscope objective lens that functions only when a film of immersion oil unites it to the material to be examined on the microscope slide
open air factor (OAF)	an ill-defined complex of factors inimical to the free survival of (pathogenic) microorganisms in the external environment
original activity remaining (OAR)	the disease-causing capacity of, in this case, a pathogen that remains after exposure to an injurious or attritional influence; usually expressed as a percentage of the activity prior to any such treatment
paratrophic membrane	*see* peritrophic membrane
Pasteur pipette	an open-ended glass tube (internal diameter *c.* 5 mm) with one end drawn out to an internal diameter of *c.* 1 mm; the wide end is fitted with a rubber bulb. It is used for non-quantitative transfer of liquids
pathogenicity	the potential to produce disease; applied to groups or species of microorganisms, while 'virulence' is used in the sense of degree of pathogenicity within the group or species. Some authorities regard pathogenicity as the genetically determined ability to produce disease, and 'virulence' as disease-producing ability that is not genetically determined (Steinhaus and Martignoni, 1962)
pathology	'the study of the cause, nature, processes and effects of disease'; 'if biology is defined as that branch of science which deals with the origin, structure, functions and life history of organisms, then pathology might be defined as "biology of the abnormal"' (Steinhaus and Martignoni, 1962)
PBS	phosphate buffered saline
PCR	*see* polymerase chain reaction
PDV	polyhedral derived virus
PEG	polyethylene glycol
peritrophic membrane	a very thin continuously secreted membrane the function of which is thought to be to protect midgut cells from mechanical abrasion but which almost certainly also provides some protection from pathogenic microorganisms: syn., paratrophic membrane
per os	by way of the mouth; a pathogen may be administered *per os*
Petri dish	a round, shallow, flat-bottomed glass or plastic dish (often 10 cm in diameter) with a vertical side, together with a similar slightly larger loose fitting lid
phenology	here the timing of natural biological events, especially in relation to climate or season
phylloplane	in the region of the leaf surface
PIB	*see* polyhedral inclusion body
poikilotherm	of an animal whose body temperature is largely determined by the ambient temperature
polyhedral inclusion body (PIB)	used to describe the OBs of nuclear polyhedrosis viruses; *see* occlusion body
polyhedron,	a faceted crystal-like proteinaceous body usually enclosing a number of (often many) virus particles; especially characteristic of NPVs and CPVs; however, sometimes polyhedra form without enclosing viruses
polyhedrosis	a virus disease characterized by the formation of polyhedra (see above)
polymerase chain reaction (PCR)	a method by which a specific nucleic acid sequence can be exponentially amplified *in vitro* resulting in a large quantity of DNA sequence; the technique allows the detection and retrieval of a DNA sequence that constitutes a minor component of the original DNA mixture

polypot	small plastic vessel, volume about 30 ml, with snap lid
PR	*see* productivity ratio
probit	a mathematical transformation of the percentage response of an organism when challenged by a given dosage of pathogen. Converts the sigmoid curve of a population response to challenge by virus to a linear response
productivity ratio (PR)	the ratio between dosage administered to cause lethal infection and the numbers of infectious units produced on death. Usually expressed as ratio of numbers of OBs for ease of measurement
quantal response	all-or-nothing response of an organism to a stimulus (in this case, usually challenge by a virus). Quantal response is pre-determined by the researcher but is usually death of the test organism
rate-zonal sedimentation	*see* isopycnic
relative centrifugal force (RCF)	for any radial position in a rotor RCF = $11.18 \times r \times (\text{rpm}/1000)^2$ where r is the distance of particles from the centre of rotation measured in cm
REN (RE)	*see* restriction endonuclease analysis
resistance	the capacity of an organism to resist, here, potentially infective microorganisms; the actual site(s) in the host at which resistance is manifest is variable
restriction endonuclease analysis (REN)	analysis of DNA by treatment with specific restriction endonucleases and separation by gel electrophoresis. Individual genes and sequences are usually identified by use of specific probes. Also known as RE analysis
rotor	container in which tubes or other vessels containing the material to be sedimented are mounted for spinning in a centrifuge
rotor, fixed angle	standard rotor in which the tubes are held at a fixed angle (from 20° to greater than 40°) during centrifugation
rpm	revolutions per minute
SDS	sodium dodecyl sulphate (sodium lauryl sulphate)
sediment	that which settles at the bottom of a liquid; to deposit as sediment
SEM	scanning electron microscopy
serology	the study of blood serum, especially in relation to antibodies and the immune response
SNPV	single nucleocapsid (singly enveloped) NPV
sonication	the use of high-frequency sound waves to disperse aggregations of small particles and to disintegrate tissues and cells in liquid suspension. Two types of equipment are available: water bath and probe. The former is recommended for use with infectious material since it can be maintained within a closed vessel, e.g. a universal bottle, during operation and does not give rise to aerosols
spinning disc equipment	spraying equipment in which spray fluid is delivered to the centre of a very fast spinning disc with fine toothed margin; produces small droplets of regular size off the edge of the disc
ST_{50}	*see* median survival time
stomacher	a device designed to separate bacteria from samples, for example, food, fabrics, swabs. Here used to harvest virus from infected insects. The sample and diluent are temporarily sealed in a sterile, disposable plastic bag which is squeezed between a platen and moving paddles
stress	a condition of (often subtle) bodily changes induced by inimical extrinsic or intrinsic factors: here sometimes thought to mediate the onset of pathogen activity and hence disease in insects
sucrose gradients	*see* gradients
supernate, supernatant	the material remaining above the pellet after centrifugation of a suspension of macromolecules

Svedberg unit	the unit in which the sedimentation coefficient (s) of a particle (e.g. a macromolecule) is commonly quoted. $$s = (dx/dt)/\omega^2 r$$ where dx/dt is the the measured sedimentation rate (in ultracentrifuge), ω is the angular velocity (radians/s) and r is the distance (cm) between a point within the sample and the axis of rotation. The units of s are reciprocal seconds, the basic unit is termed one Svedburg unit (S) (10^{-13} s).
syndrome	a group of indications (symptoms) characteristic of a particular disease or abnormal condition; however, the causation of a syndrome may be variable
synergy	two or more factors (pathogens, toxicants) working together and producing a result greater than the sum of their actions separately
transmission	*see* horizontal passage; vertical passage
transovarian (transovarial) transmission	form of vertical passage of a pathogen within the egg
trans-ovum transmission	form of vertical passage of a pathogen on the egg surface
trans-stadial transmission	transmission (of pathogens) from one life stage of the host to the next, throughout all or a part of the host's life cycle
trichome	a monocellular or multicellular plant surface structure, often hair-like, possibly branched, possibly secretory; *see also* hydathode
triturate	grind or divide to a fine powder or pulp
Ulva-fan sprayer	portable spinning disc sprayer with directed air supplied by fan; battery or petrol motor powered (micron sprayers)
Ulva-sprayer	hand-held spinning disc sprayer with battery operation; highly portable (micron sprayers)
ULV	*see* ultra low volume
ultra low volume (ULV)	a type of spraying involving very low volumes of fluid (<5 l/ha but as high as 50 l/ha for certain crop types) in which crop coverage is ensured by generation of very small spray droplets in large numbers; usually combined with controlled droplet application, CDA
Ulvapron	*see* Actipron
VDR	volume deposit rate
virion	virus particle
volume median diameter (VMD)	for a given number of droplets the diameter at which half the volume is in droplets smaller that the VMD and half the volume is in droplets larger than the VMD
USDA	United States Department of Agriculture
vertical passage	passage of a pathogen from one generation to the next
viraemia	the presence of virus in the blood (in the haemolymph in insects)
virulence	the disease-producing power of a microorganism, i.e. its ability to invade and injure the tissues of the host; the relative capacity of a microorganism to overcome the body defences of the host; the degree of pathogenicity within a group or species: strains may be described as avirulent, virulent and highly virulent within a group or species of microorganisms that are said to be pathogenic; according to some authors, disease-producing ability that is not genetically determined (Steinhaus and Martignoni, 1962); *see also* infectivity
wipfelkrankheit	tree-top disease, indicating the ascent of insects dying of (virus) infections to the tops of trees or other plants (from the original German description)
WP	wettable powder

X-15 and X-30	forms of spinning disc sprayers with, respectively, 15 and 30 'stacked' (on a common spindle) discs; *see* spinning disc equipment
zonal rotor	used for centrifuging large volumes of sample, both rate-zonal and isopycnic separations; the gradient is made in the bowl of the rotor rather than in individual tubes

REFERENCE

Steinhaus, E.A. and Martignoni, M.E. (1962) *An Abridged Glossary of Terms Used in Insect Pathology. Mimeographed Series, No. 6*. Berkeley, CA: Department of Insect Pathology, University of California.

FURTHER READING

Anon. (1990) *A Concise Dictionary of Biology*. Oxford: Oxford University Press.
Jerrard, H.G. and McNeill, D.B. (1986) *A Dictionary of Scientific Units: Including Dimensionless Numbers and Scales*. London: Chapman & Hall.
Singleton, P. and Sainsbury, D. (1987) *Dictionary of Microbiology and Molecular Biology*, 2nd edn, Chichester: Wiley.

Further reading

Adams, J.R. and Bonami, J.R. (1991) *Atlas of Invertebrate Viruses*. CRC Press, Boca Raton, FL.
Aruga, H. and Tanada, Y. (eds) (1971) *The Cytoplasmic Polyhedrosis Virus of the Silkworm*. University of Tokyo Press, Tokyo.
Bailey, L. (1981) *Honeybee Pathology*. Academic Press, London.
Bailey, L. and Ball, B. (1991) *Honeybee Pathology*, 2nd edn, Academic Press, New York.
Beckage, N.E., Thompson, S.N. and Federici, B.A. (1993) *Parasites and Pathogens of Insects*, Vols 1 and 2. Academic Press, San Diego.
Burges, H. (ed.) (1981) *Microbial Control of Pests and Plant Diseases*. Academic Press, London.
Burges, H.D. (ed.) (1998) *Simulation of Microbial Biopesticides, Beneficial Microorganisms, Nemotodes and Seed Treatments*. Chapman & Hall, London.
Burges, H. and Hussey, N.W. (eds) (1971) *Microbial Control of Insects and Mites*. Academic Press, London.
Cantwell, G.E. (ed.) (1974) *Insect Diseases*, Vol. 1 and 2. Marcel Dekker, New York.
Davidson, E.W. (1981) *Pathogenesis of Invertebrate Microbial Diseases*. Allanheld and Osmun, Totowa, NJ.
Deseo Kovacs, K.V. and Rovesti, L. (1992) *Lotta Microbiologica Contro i Fitofagi. Teoria e Practica*. Edagricole-Edizioni Agricole, Bologna.
Doerfler, W. and Bohm, P. (1986) *Current Topics in Microbiology and Immunology 131: The Molecular Biology of Baculoviruses*. Springer-Verlag, Berlin.
Franz, J.M. (ed.) (1986) *Biological Plant and Health Protection*. Fischer Verlag, Stuttgart.
Fuxa, J.R. and Tanada, Y. (eds) (1987) *Epizootiology of Insect Diseases*. New York: Wiley.
Gibbs, A.J. (ed.) (1973) *Viruses and Invertebrates*. North Holland, Amsterdam.
Gooser, M.F.A., Daugulis, A.J. and Faulkner, P. (1993) *Insect Cell Culture Engineering*. Marcel Dekker, New York.
Granados, R.R. and Federici, B.A. (eds) (1986) *The Biology of Baculoviruses*, Vol. 1 and 2. CRC Press, Boca Raton, FL.
Hedin, *et al.* (1994) *ACS Symposium Series 551: Natural and Engineered Pest Management Agents*. American Chemical Society, Washington, DC.
Kalmakoff, J. and Longworth, J.F. (1980) *Microbial Control of Insect Pests*. (*Research Bulletin 228*) New Zealand Department of Science and Industry, Wellington, New Zealand.
Kawase, S. (1976) *Viruses and Insects*. Journal of Japanese Chemistry Monograph Series 12 (in Japanese.)
Kerkut, G.A. and Gilbert, L.I. (eds) (1985) *Comprehensive Insect Physiology, Biochemistry and Pharmacology*, Vol. 12, *Insect Control*. Pergamon Press, Oxford.
King, L.A. and Possee, R.D. (1992) *The Baculovirus Expression System: a Laboratory Guide*. Chapman & Hall, London.

Kurstak, E. (ed.) (1982) *Microbial and Viral Pesticides*. Marcel Dekker, New York.
Kurstak, E. (ed.) (1991) *Viruses of Invertebrates*. Marcel Dekker, New York.
Laird, M., Lacey, L.A. and Davidson, E.J. (1989) *Safety of Microbial Insecticides*. CRC Press, Boca Raton, FL.
Maramorosch, K. (1968) *Current Topics in Microbiology and Immunology*, Vol. 42: *Insect Viruses*. Springer-Verlag, Berlin.
Maramorosch, K. (1977) *The Atlas of Insect and Plant Viruses*. Academic Press, New York.
Maramorosch, K. (ed.) (1987) *Biotechnology of Invertebrate Pathology and Cell Culture*. Academic Press, New York.
Maramorosch, K. and Sherman, K.E. (eds) (1985) *Viral Insecticides for Biological Control*. Academic Press, Orlando, FL.
Martignoni, M.E. and Steinhaus, E.A. (1961) *Laboratory Exercises in Insect Microbiology and Insect Pathology*. Burgess Publishing, Minneapolis, MN.
Mitsuhashi, J. (1989) *Invertebrate Cell System Applications*. CRC Press, Boca Raton, FL.
O'Reilly, D.R., Miller, L.K. and Luckow, V.E. (1992) *Baculovirus Expression Vectors, a Laboratory Manual*, Freeman, New York.
Poinar, G.O. and Thomas, G.M. (1978) *Diagnostic Manual for the Identification of Insect Pathogens*. Plenum Press, New York.
Poinar, G.O. and Thomas, G.M. (1984) *Laboratory Guide to Insect Pathogens and Parasites*, 2nd edn. Plenum Press, New York.
Sambrook, J., Fritsch, E.F. and Maniatis, T. (1989) *Molecular Cloning: A Laboratory Manual*, 2nd edn. Cold Spring Harbor Laboratory Press, Cold Spring Harbor, MA.
Singh, P. and Moore, R.F. (eds) (1985) *Handbook of Insect Rearing*, Vol. 1. Elsevier, Amsterdam.
Smith, K.M. (1967) *Insect Virology*. Academic Press, New York.
Smith, K.M. (1976) *Virus–Insect Relationships*. Longman, London.
Sparks, A.K. (1985) *Synopsis of Invertebrate Pathology Exclusive of Insects*. Elsevier Science, Amsterdam.
Steinhaus, E.A. (1946) *Insect Microbiology*. Comstock, Ithaca, New York.
Steinhaus, E.A. (1949) *Principles of Insect Pathology*. McGraw Hill, New York.
Steinhaus, E.A. (ed.) (1963) *Insect Pathology. An Advanced Treatise*, Vols 1 and 2. Academic Press, New York.
Summers, M., Engler, R., Falcon, L.A. and Vail, P. (1975) *Baculoviruses for Insect Pest Control: Safety Considerations*. American Society for Microbiology, Washington DC.
Tanada, Y. and Kaya, H.K. (1993) *Insect Pathology*. Academic Press, New York.
Weiser, J. (1969) *An Atlas of Insect Diseases*. Czechoslovakian Academy of Sciences, Praha.
Weiser, J. (1977) *An Atlas of Insect Diseases*, 2nd edn. Junk, Den Haage. Also reprinted by the Irish University Press, Shannon.

General virology texts with insect virus information

Hull, R., Brown, F. and Payne, C.C. (1989) *Virology: Directory and Dictionary of Animal, Bacterial and Plant Viruses*. Macmillan, London.
Webster, R.G. and Granoff, A. (1994) *Encyclopedia of Virology*. Academic Press, London.

Insect morphology and histology

Eaton, J.L. (1988) *Lepidopteran Anatomy*. Wiley-Interscience, New York.
Freeman, W.F. and Bracegirdle, B. (1971) *An Atlas of Invertebrate Structure*. Heinemann, London.
Neville, A.C. (ed.) (1970) *Insect Ultrastructure*. Blackwell Scientific, Oxford.
Smith, D.S. (1968) *Insect Cells: their Structure and Function*. Oliver and Boyd, Edinburgh.

Index

Note: page numbers in *italics* refer to figures and tables